MAY -- 2019

REFERENCE
DO NOT CIRCULATE

Principles of Climatology

Principles of Climatology

Editor
Richard Renneboog

SALEM PRESS

A Division of EBSCO Information Services, Inc.
Ipswich, Massachusetts

GREY HOUSE PUBLISHING

Cover Image: Hurricane between Florida and Cuba. By Harvepino (via iStock).

Copyright ©2018, by Salem Press, A Division of EBSCO Information Services, Inc., and Grey House Publishing, Inc.

All rights reserved. No part of this work may be used or reproduced in any manner whatsoever or transmitted in any form or by any means, electronic or mechanical, including photocopy, recording, or any information storage and retrieval system, without written permission from the copyright owner. For permissions requests, contact proprietarypublishing@ebsco.com.

∞ The paper used in these volumes conforms to the American National Standard for Permanence of Paper for Printed Library Materials, Z39.48 1992 (R2009).

Publisher's Cataloging-In-Publication Data
(Prepared by The Donohue Group, Inc.)

Names: Renneboog, Richard, editor.
Title: Principles of climatology / editor, Richard Renneboog.
Description: [First edition]. | Ipswich, Massachusetts : Salem Press, a division of EBSCO Information Services, Inc. ; [Amenia, New York] : Grey House Publishing, [2018] | Series: Principles of | Includes bibliographical references and index.
Identifiers: ISBN 9781682179437 (hardcover)
Subjects: LCSH: Climatology.
Classification: LCC QC861.3 .P75 2018 | DDC 551.6—dc23

FRANKLIN TOWNSHIP PUBLIC LIBRARY
485 DEMOTT LANE
SOMERSET, NJ 08873
732-873-8700

First Printing
Printed in the United States of America

CONTENTS

Publisher's Note..........................ix
Editor's Introduction......................xi
Contributors............................xvii

Abrupt climate change......................1
Acid rain and acid deposition................3
Aerosol Pollutants.........................8
Air pollution.............................12
Albedo feedback..........................17
Anthropogenic air pollution and
    pollutants..........................20
Anthropogenic climate change...............23
Aral sea desiccation......................27
Arctic and global warming..................30
Arctic ocean.............................35
Atlantic heat conveyor....................40
Atlantic Multidecadal Oscillation (AMO)......42
Atlantic Ocean...........................45
Atmosphere..............................50
Atmosphere and climate change..............54
Atmosphere: Evolution.....................58
Atmosphere: Global circulation.............63
Atmosphere: Structure and thermodynamics....68
Atmospheric and oceanic oscillations........76
Atmospheric boundary layer.................81
Atmospheric chemistry and climate
    change..............................82
Atmospheric circulation...................86
Atmospheric dynamics......................89
Atmospheric inversions....................92
Atmospheric opacity......................94
Atmospheric oxygen.......................97
Atmospheric physics......................99
Atmospheric sciences....................105
Atmospheric structure and
    evolution..........................111
Average weather in climate studies.........116

Barometric pressure and barometry.........118

Carbon and carbon dioxide equivalent......123
Carbon dioxide in Earth's atmosphere......125
Carbon equivalent.......................127
Carbon-oxygen cycle.....................128
Carbon sink............................133
Catastrophist-cornucopian debate..........135

Chlorofluorocarbons.....................136
Chlorofluorocarbons and related
    compounds..........................138
Climate and the climate system...........141
Climate change and global warming........143
Climate change theories..................145
Climate feedback........................149
Climate models and modeling..............151
Climate prediction and projection.........153
Climate reconstruction...................156
Climate refugees........................158
Climate sensitivity......................160
Climate variability.....................162
Climate zones...........................165
Climatology............................167
Clouds and global climate................172
Continental climate.....................174
Coriolis Effect.........................177
Cryosphere.............................178
Cyclones and anticyclones................180

Dating methods for climate change.........185
Deep ocean currents.....................187
Deglaciation...........................191
Drought................................193

Earth-Sun relations.....................200
Ecosystems' influence on climate..........203
El Niño-Southern Oscillation and
    global warming.....................207
Enhanced greenhouse effect...............210
Extreme weather events...................212

Fabry quantifies ozone in the upper
    atmosphere.........................215
Floods.................................219
Fluid dynamics.........................225
Fog and regulation of the global
    climate............................228
Forests and the mitigation of
    anthropogenic climate change........229
Fronts and climate change................232

General circulation models (GCM).........234
Glaciations............................236
Glaciers...............................238

v

Contents

Global climate	242
Global Climate Coalition (GCC)	245
Global economy and climate change	247
Global monitoring	250
Global positioning system (GPS)	253
Global surface temperature	256
Global Warming in Antarctica	258
Great Lakes	262
Greenhouse gases and global warming	265
Ground ice and climate change	269
Gulf Stream	271
Gyres and climate change	275
Hadley circulation and climate change	278
Heat vs. temperature	280
Humidity and greenhouse gases	282
Hurricanes and typhoons	284
Hydrogen power	289
Ice ages and glaciations	292
Ice cores	296
Ice shelf collapses	298
Icebergs and Antarctica	300
Ice-out studies	304
Indian Ocean	305
Indicators of Climate Change and Global Warming	310
Interglacial period	312
Kilimanjaro's ice cap	315
La Niña	318
Lakes	319
Little Ice Age (LIA)	325
Long-term weather patterns	327
Maritime climate	331
Mean sea level	333
Medieval Warm Period (MWP)	333
Mediterranean climate	335
Megacities' effects on local climates	337
Meridional overturning circulation (MOC)	339
Meteorology and climate policy	341
Methane's global warming potential	343
Modes of climate variability	346
Monsoons	348
North Atlantic Oscillation (NAO)	353
North Sea	355

Observational data of the atmosphere and oceans	360
Ocean dynamics	363
Ocean life and rising water temperatures	365
Ocean structure and atmosphere coupling	369
Ozone depletion and ozone holes	374
Pacific Ocean	380
Paleoclimates and paleoclimate change	385
Physics of weather	388
Plankton and the mitigation of the greenhouse effect	393
Pleistocene climate	395
Polar bears and climate change	398
Polar climate	401
Polar stratospheric clouds (PSCs)	403
Polar vortex	405
Precipitation	406
Probability and statistics	411
Rain forests and the atmosphere	417
Recent climate change research	420
Remote sensing of the atmosphere and oceans	424
Satellite meteorology	430
Sea ice and the global climate	436
Sea levels	438
Sea sediments and climate change	443
Sea surface temperatures (SST)	445
Seasonal changes	447
Seawater composition	448
Severe storms	452
Solar cycle	459
Storm surges	461
Surface ocean currents	462
Thermohaline circulation (THC) and the Thermocline	468
Tornadoes	471
Tree rings and climatological information	475
Tropical climate	477
Tropical weather	479
Troposphere	483
Urban heat island (UHI)	486
Urban heat island (UHI) and Megacities	488

Volcanoes and climate 491

Weather forecasting: Numerical weather
  prediction 498
Weather modification 503
Weather vs. climate 506
Wetlands and sea-level rise 510
Wind 512
World ocean circulation experiment 516

Timeline of Periods and Events in
  Climate History 523
Glossary 527
General Bibliography and Additional
  Reading 551
Index 585

Publisher's Note

Salem Press is pleased to add *Principles of Climatology* as the eleventh title in the *Principles of* series that includes *Chemistry, Physics, Astronomy, Computer Science, Physical Science, Biology, Scientific Research, Sustainability, Biotechnology,* and *Programming & Coding.* This new resource introduces students and researchers to the fundamentals of climatology using easy-to-understand language for a solid background and a deeper understanding and appreciation of this important and evolving subject. All of the entries are arranged in an A to Z order, making it easy to find the topic of interest.

Entries related to basic principles and concepts include the following:

- Fields of Study to illustrate the connections between the topic and the various branches of science related to climatology;
- A Summary that provides brief, concrete summary of the topic and how the entry is organized;
- Principal Terms, to introduce terminology used in the entry;
- Text that gives an explanation of the background and significance of the topic to climatology by describing events such as monsoons, fronts, typhoons, and droughts;
- Illustrations that clarify difficult concepts via models, diagrams, and charts of such key topics as Atmospheric inversions; Barometry; Carbon-oxygen cycle; Thermohaline circulation; Volcanoes and the evolution of Earth's climate; and
- Further reading lists that relate to the entry.

This reference work begins with a comprehensive introduction to climatology, written by volume editor Richard Renneboog.

The book includes helpful appendixes as another valuable resource, including the following:

- Timeline of Periods and Events in Climate History;
- Glossary;
- General Bibliography and Additional Reading; and
- Subject Index.

Salem Press and Grey House Publishing extend their appreciation to all involved in the development and production of this work. The entries have been written by experts in the field. Their names and affiliations follow the Editor's Introduction.

Principles of Climatology, as well as all Salem Press reference books, is available in print and as an e-book. Please visit www.salempress.com for more information.

Editor's Introduction

Climatology based in science is older than is perhaps commonly realized by most people, with its origins in the last quarter of the 1800s. Smith's publication of 1872 begins an association of the air of the atmosphere with its behavior and the different materials that it may contain, specifically of air and rain in the context of chemical interactions. In a more general sense, climatology describes the environment that is most immediately and organically experienced in everyday life. Weather, a subset of climatology, refers to climate effects experienced on a local scale. Primarily, the science of climatology focuses on the physical behavior and properties of a planetary atmosphere over time. As such the science of climatology is applicable to all planets having any type of gaseous atmosphere. In all cases, climate properties and behaviors are determined by the interaction of the atmospheric gases with energy from different sources. On Earth, the primary source of energy is the Sun, while Earth itself provides a secondary energy source and the rest of the Universe provides small, but nonetheless important, energetic interactions. In their many combinations, these energy sources interact with the planetary atmosphere to drive its motions both locally and globally.

It is common to perceive the atmosphere as a consistent mass of air punctuated by various forms of precipitation, particulate matter, and pollution. This is an entirely understandable perception, since this is precisely how people experience the atmosphere in their daily lives. The actual atmosphere, however, is very different from this perception, having several defined layers throughout its depth, each with certain well-defined characteristics. As altitude increases from ground level, the composition of the atmosphere undergoes continual change and decreases in density until it effectively becomes non-existent in the "vacuum of space." Paradoxically, even in space there is an atmosphere of sorts, since molecules of the gaseous components of the denser atmosphere around a planet exist there as well, but are so widely separated that their concentration or abundance is effectively zero.

On all planets having a gaseous atmosphere, gravity acts to retain the atmosphere as a relatively narrow layer surrounding the planet. While the density of the atmosphere gradually dissipates without boundaries into space as an open system, the planetary atmosphere is effectively a closed system due to the constraints of gravity. The behavior of the gaseous matter in such a system is well described by the ideal gas law

$$PV = nRT$$

relating pressure, volume, and temperature to the quantity of matter in the gaseous state. An examination of this relationship will demonstrate that the volume of any gas increases or decreases when the temperature increases or decreases. Similarly, the pressure exerted by any gas increases or decreases when the temperature increases or decreases. However, the pressure and volume of any gas follow an inverse relationship with each other, in that decreasing the volume of a particular quantity of a gas increases the pressure that the gas experiences, while increasing the volume available to a particular quantity of a gas reduces the pressure that it exerts. All of these effects are involved in the mechanics of any planetary atmosphere, giving rise to high and low pressure cells, the movement of winds, and so on.

Solar energy entering Earth's atmosphere as sunlight is absorbed by the ground and radiates out from there at wavelengths that raise the temperature of the atmosphere above ground level. Some of this re-radiated thermal energy is captured by the carbon dioxide, water, and methane molecules that make up part of the atmosphere, and is returned toward the surface to further add to the thermal energy of the atmosphere. This is the so-called "greenhouse effect," and without this process in operation in Earth's atmosphere over the past billions of years Earth would be just another frozen ball of water ice. This process has run rampant on the planet Venus, and has long since turned it into a planet where the surface temperatures are higher than the melting point of lead and some other metals, and rocks are red-hot. The atmosphere of Venus is many times denser than that of Earth, and is composed primarily of carbon dioxide and sulfuric acid. The atmosphere of Mars, on the other hand, is so thin that Mars has no practical atmosphere. The planets Jupiter and Saturn are termed "gas giant" planets, and appear to be just large balls of atmosphere held together by gravity. Yet in all of these different scenarios, the gas law relationship holds true, and is essential for understanding the mechanics of any planetary atmosphere. Studying the atmospheres of other planets, inasmuch as this is possible with present-day technology, is crucial to understanding how the atmosphere of Earth functions as well.

The average temperatures over the surface of Earth vary with latitude. They are highest in the equatorial region where sunlight strikes Earth directly and lowest in the polar regions where sunlight does not reach the surface for several months each year. In general, this distribution is fairly constant, yet because Earth is a very dynamic planet there are significant variations in this distribution. Colder, denser air displaces warmer, less dense air and the interaction is felt locally as wind. Colder air, because it is denser, represents a region of higher pressure within the atmosphere. Relatively warmer air, being less dense, therefore represents a region of lower pressure. It is intuitive to think of the air in a high pressure cell as flowing outward radially from the center in a uniform manner, and of the air flowing inward radially toward the center of a low pressure cell. This is, to a first approximation, essentially what happens. Other influencing factors affect the movement of air in the atmosphere, particularly in regard to pressure cells, rendering air movement much more complex within the atmosphere.

Earth spins on a central axis that is tilted with respect to the plane of the planet's orbit about the Sun. In addition, the uneven distribution of the continental massifs, the gravitational interaction of the Earth-Moon system and other such factors perturb Earth's axial rotation in such a way that the planet precesses about the rotational axis as it spins. In other words, the planet "wobbles" around its central axis rather than spinning true. The important relationship for climatology and weather is that the atmosphere is a continuous fluid layer within which the planet rotates. The atmosphere is physically affected by friction with the surface, but as altitude above the surface increases the influence of friction with the surface decreases quickly. Movement of the atmosphere is thus relative to the movement of the planetary surface and is not completely synchronized with it. This relative motion gives is characterized by the atmospheric Coriolis Effect.

The Coriolis Effect is observed when rotational forces are applied to linear motion. This results in a curvilinear direction in the motion of objects. The classical demonstration of the Coriolis Effect is to toss a ball across a rotating platform to someone on the other side. The ball is not attached or fixed to the platform in any way once it has been tossed, and so must follow a perfectly linear trajectory. But because the platform is rotating beneath the ball in flight, the ball is seen to follow a curved path relative to the platform. The same effect is observed in the flow of air to and from pressure cells as Earth rotates within the atmosphere. Air movement from high pressure cells in the Northern Hemisphere is seen to be clockwise, or anticyclonic, while in low pressure cells it is seen to be counterclockwise, or cyclonic. Typically, the more northerly portion of the cell draws colder, drier air down from farther north, while the more southerly portion of the cell draws warmer, damper air up from farther south. In the Southern Hemisphere, the opposite is true, as high and low pressure cells rotate in the opposite sense of their northern hemisphere counterparts. The combination of these movements results in a much more complex "dance" of air currents and moisture effects that often makes local weather prediction more a matter of "educated guessing" than an exact science.

Prediction of local weather, and hence the overall state of the planetary climate, has been greatly facilitated by the development of advanced technology to augment the most basic and fundamental methods of measurement. From simple weather stations marking such factors as wind speed and direction (i.e.: wind velocity), humidity, cloud observations, temperature, precipitation and length of the daylight period, to Doppler radar imaging of cloud formations and movements, to satellite-based observations of atmospheric and ground-based phenomena with GPS precision, the corresponding data is utilized in mathematical modeling for the short- and long-term prediction of local weather and broader weather patterns. Several computational methodologies are used to model weather patterns, but the millions of data points required can only be adequately processed by sophisticated supercomputers. Nevertheless, even the most robust and dependable models currently available are not able to accurately predict weather patterns more than a week into the future. By that time, the inherent complexity of computing millions of data points, the chaotic nature of the atmosphere itself and the random factors that influence it combine to degrade the reliability of any particular prediction.

Random and unpredictable influences on the atmosphere of Earth come from a variety of sources, from within Earth itself and from the expanses of outer space. Volcanic eruptions can inject large quantities of gases and dust into the atmosphere, to heights of several kilometers, where they become distributed by the prevailing winds. Once in the atmosphere, those gases and the dust change the existing balance and composition of the atmosphere, however slightly, and the magnitude of resulting atmospheric phenomena such as acid rain and the "greenhouse effect," as well as Earth's albedo. A planet's albedo is a measure of the amount of solar energy that is reflected back from it into space by the surface and atmosphere of the planet. Earth's neighboring planet Venus has a very high albedo due to its thick, dense, and highly reflective cloud cover. Earth's albedo is due to the combination of clouds, water, snow and ice cover on the planet. Changes to the planet's albedo can have the effect of raising or lowering the average temperature of the planetary atmosphere over time. The Sun also provides unpredictable perturbations of Earth's atmosphere as the result of solar flares, surface eruptions and mass ejections of immense quantities of charged particles. The matter and radiation from these phenomena impinge upon Earth's atmosphere, adding energy to whatever face of the planet intercepts them. One of the effects of this interaction is visible as the aurora borealis and aurora australis, the Northern and Southern Lights. These phenomena are due to the emission of visible light from molecules of nitrogen ad oxygen in the upper atmosphere that have been electronically excited by absorbing energy from impacts with charged particles from the Sun and other stars in the depths of space.

It is tempting to think of these effects as minor phenomena, and in comparison to the amount of energy Earth acquires from the Sun, they are. Yet they are vitally important for the existence of life on Earth. The oxygen in the upper atmosphere surrounding the planet undergoes a molecular change by its interaction with charged cosmic particles and electromagnetic radiation, as molecules of oxygen (O_2) are converted to molecules of ozone (O_3). The "ozone layer" that this produces performs the essential function of absorbing the ultraviolet radiation that continually bathes the planet. Without that protective layer of ozone in Earth's upper atmosphere, the ultraviolet radiation would quickly render Earth a sterile wasteland where neither plant nor animal can live.

The ozone layer has protected Earthly life for many millions of years, and it is with this layer that the modern science of climatology begins on Earth. Climatology is the study of the long-term average behavior of the atmospheric environment of Earth. Because Earth's atmosphere is a single fluid layer, every part of it is influenced by what happens in any other part. The behavior of liquid water provides a good analogy. If so much as a drop of liquid water is removed from the ocean, that action does not leave a hole where that water was removed. Instead, every remaining molecule of water in the ocean adjusts its position to fill in that space. The fluid atmosphere behaves in exactly the same way. While local effects—the weather—are important, the movement of a body of air such as a pressure cell in one part of the atmosphere necessarily results in the movement of all other parts of the atmosphere to compensate. Within that fluid layer, a number of strata and global phenomena are identifiable, characterized by their temperature and movement. The most familiar of these are the "jet streams," high velocity air currents that encircle Earth and effectively mark boundaries between climatic regions. In the Northern Hemisphere, the jet stream separates cold Arctic air from the more temperate air mass to the south. Seasonal fluctuations of this boundary allow cold air to move farther south in the winter, often bitterly cold air flowing eastward from Siberia. The movement of is normally from west to east, allowing cold air to sweep down across the Canadian prairies and the central plains of the United States. Farther east, this same effect draws moist air from the Caribbean region and drops is as heavy snowfalls or rain on the northeast coastal regions of North America. In 2017, this "polar vortex" moved west from Siberia rather than eastward, and brought unusually bitter cold and heavy snowfalls to Europe while North America experienced a period of unseasonably mild winter weather. The shift even covered parts of the Sahara Desert with a very rare

xiii

coating of snow. In early 2018, winds from the Sahara Desert returned the favor by sending dust from a large sandstorm to color the snow-covered ski hills about Sochi, in southern Russia.

This planet Earth is a very dynamic and complex physical system, in a constant state of change. Over the billions of years that it has existed, it has achieved a "steady state" with regard to atmospheric composition and phenomena. The atmospheric composition is a constant combination of nitrogen and oxygen in a ratio of almost exactly four to one. That ratio is not quite perfect, however, as approximately one percent of Earth's atmosphere is composed of several other gases, including water vapor, carbon dioxide, argon, helium, neon, ozone, some nitrogen oxides and methane as the major constituents. Many other gases, vapors from innumerable other compounds, can also be identified in air, albeit in very small quantities, within that one percent of the atmosphere that is not nitrogen and oxygen. Methane is a characteristic gas produced by carbon-based life forms on Earth, and presumably on other planets as well. Similarly, carbon dioxide is the primary oxidation product from carbon-based materials, especially as the end product of metabolism. That these gases should exist in Earth's atmosphere is an inevitable consequence of the existence of carbon-based life existing on this planet. Indeed, these gases, along with water vapor, are singularly responsible for the steady-state climate of Earth in which water is able to exist in an equilibrium of its gaseous, liquid, and solid states. The mechanism that enables this is the so-called "greenhouse effect," by which a certain amount of heat energy is captured and radiated back toward the surface instead of being lost to space. Without this effect operating in the atmosphere, Earth would have an average temperature well below the freezing point of water, and a very different climate.

The long age of Earth's atmosphere has been mentioned several times, and it should be understood from this that the climate of a planet refers to the long-term cumulative effects of the multitude of weather effects and affective factors that occur continually over that time span. Through the study of the physical traces left behind in the rocks, ice and even the plants of the planet's past, it is apparent that the climate of Earth has undergone drastic changes many times in the past. "Ice ages" lasting thousands of years, in which the world climate as a whole became just a few degrees cooler than average, have come and gone and undoubtedly will come again. Similarly, periods of time in which the climate was just a few degrees warmer than average have also come and gone, and at the present time it seems inevitable that such a period is bound to reappear in the foreseeable future as Earth's climate changes. The record of these different times is found in the changes to the flora and fauna that inhabit this planet, in air bubbles trapped in ice formed hundreds of thousands, even millions, of years ago. Past environmental conditions are reflected in the growth patterns of fossilized trees and in their mineral content, as well as in living trees. Fluctuations in sea level are recorded in the structural features of the continental shelves of the oceans, and in coastal mountain ranges, indicating sea level falls and rises of two to three hundred meters when ocean water became sequestered in glacial icecaps as the planetary climate cooled and was subsequently released as it warmed once more.

FUTURE PROSPECTS

The investigation of climatology is very much a forensic science as clues from the past are collected and interpreted as evidence to provide an understanding of this dynamic system in which we live. The impact of human activities as a driver of climate change is a particular focus of climatology and has become especially important in regard to human activity since the 18th century and the Industrial Revolution that it ushered in. The central feature of climatology at the present time is characterized by study of the presence of carbon dioxide, methane, and water vapor in the atmosphere. Historically, according to evidence from the past found in nature, the relative quantities of these gases has remained essentially constant, but significant changes of those concentrations are associated with equally significant changes in Earth's climate. Increased carbon dioxide levels overall are associated with warming of the planetary atmosphere due to enhancement of the greenhouse effect. Methane, though in lesser quantities than carbon dioxide, is much more effective as a greenhouse gas. Reduction of carbon dioxide levels and increases of water vapor levels are associated with increased planetary albedo and cooling of the overall climate. The time since the beginning of the Industrial Revolution is of special interest in this regard, because of the massive quantities of additional carbon dioxide that have been freely injected into the atmosphere from the combustion of fossil fuels, particularly coal and petroleum products. These source materials represent huge quantities of carbon that had been sequestered within the planet and so were not available to the atmosphere. Each atom of the carbon in these materials becomes oxidized to a molecule of carbon dioxide during combustion. Estimates to date place the amount of additional carbon dioxide sent into Earth's atmosphere over the past 200+ years as being perhaps more than two billion tonnes. This in itself may seem like proof that human activity is affecting the climate, and there is supporting evidence of this in the form of an observed general increase in the global average temperature of the atmosphere. Yet the problem is that climatologists do not yet have the knowledge they need to predict precisely and accurately how the climate of Earth responds to any particular environmental input. The quantities are so small in comparison to the whole atmosphere and any changes extend over such long periods of time that minor short-term fluctuations of the climate from any number of inputs effectively masks the important long-term effects on the climate. For policy makers in government, this lack of concrete, science-based knowledge is anathema, leaving some to make entirely wrong policy decisions and the rest to wonder what the correct policy decisions should be or if correct policy decisions are even possible.

Climatology is therefore a forensic science of the highest importance for maintaining the health and livability of Earth. Future generations of climate scientists have a rich, though complicated, field in which to practice, in the present, in the near future and, one would hope, into the far distant future as well.

FURTHER READING

Barry, Roger G., and Richard J. Chorley. *Atmosphere, Weather and Climate.* 9th ed. Routledge, 2010.

Barry, Roger G., and Peter D. Blanken. *Microclimate and Local Climate.* Cambridge University Press, 2016.

Berners-Lee, Mike. *How Bad Are Bananas? The Carbon Footprint of Everything.* Greystone, 2011.

Bodri, Louise, and Vladimir Cermak. *Borehole Climatology. A new method on how to reconstruct climate.* Elsevier, 2007.

Dessler, Andrew. *Introduction to Modern Climate Change* 2nd ed. Cambridge University Press, 2016.

Fagan, Brian. *Floods, Famines and Emperors. El Nino and the Fate of Civilizations.* Basic Books, 1999.

Flannery, Tim. *We Are the Weather Makers* Somerville, MA: Candlewick Press, 2010. Print.

Hidore, John J. *Climatology: An Atmospheric Science.* Prentice Hall, 2010.

Mackwell, Stephen J., et al., eds. *Comparative Climatology of Terrestrial Planets.* University of Arizona Press, 2013.

Mooney, Chris. *Storm World. Hurricanes, Politics, and the Battle over Global Warming.* Harcourt, 2007.

Nuccitelli, Dana. *Climatology versus Pseudoscience: Exposing the Failed Predictions of Global Warming Skeptics.* Praeger, 2015.

Oke, T.R., G. Christen Mills, and J.A. Voogt. *Urban Climates.* Cambridge University Press, 2017.

Romm, Joseph *Climate Change. What Everyone Needs to Know.* Oxford University Press, 2016.

Smith, Robert Angus. *Air and Rain. The Beginnings of a Chemical Climatology.* Green & Company, 1872.

Contributors

Tomi Akanle
Michael P. Auerbach
Dennis G. Baker
Victor R. Baker
Anita Baker-Blocker
Melissa A. Barton
Raymond D. Benge, Jr.
Rachel Leah Blumenthal
Jeffrey C. Brunskill
Ewa M. Burchard
Byron D. Cannon
Roger V. Carlson
Rebecca S. Carrasco
James A. Carroll
Jongnam Choi
Thomas Coffield
John H. Corbet
Ralph D. Cross
Anna M. Cruse
Alan C. Czarnetzki
Thomas L. Delworth
Joseph Dewey
Dave Dooling
Steven I. Dutch
Justin Ervin
Dell R. Foutz
C. R. de Freitas
Yongli Gao
Joyce Gawell
Nancy M. Gordon

Daniel G. Graetzer
Kenneth P. Green
Gina Hagler
Joyce M. Hardin
M. Jasper L. Harris
Isaac Held
Thomas E. Hemmerly
Jane F. Hill
Robert M. Hordon
Micah L. Issitt
Bruce E. Johansen
David Kasserman
Amber C. Kerr
David T. King, Jr.
Gary G. Lash
Raymond P. LeBeau, Jr.
M. Lee
Denyse Lemaire
Josué Njock Libii
Chungu Lu
Yiqi Luo
Sergei Arlenovich Markov
W. J. Maunder
Shari Parsons Miller
Joseph M. Moran
Otto H. Muller
M. Marian Mustoe
John E. Mylroie
To N. Nguyen
Zaitao Pan

Robert J. Paradowski
John R. Phillips
George R. Plitnik
Victoria Price
Gail Rampke
P. S. Ramsey
C. Nicholas Raphael
Mariana L. Rhoades
Charles W. Rogers
Julia A. Rosenthal
Kathryn Rowberg
Somnath Baidya Roy
David Royster
R. D. Russell

Virginia L. Salmon
Elizabeth D. Schafer
Rose Secrest
Martha A. Sherwood
R. Baird Shuman
Billy R. Smith, Jr.
Carolyn P. Snyder
Dion Stewart
Toby Stewart
Alexander R. Stine
Paul C. Sutton
Rena Christina Tabata
John M. Theilmann
Lisa A. Wroble

ABRUPT CLIMATE CHANGE

SUMMARY

In 1840, Louis Agassiz published his theory that the Earth had passed through an ice age. As a result of Agassiz's work, the corollary idea that the globe's climate could change dramatically for extended periods entered scientific thinking. Scientists assumed, however, that such change occurred very slowly and smoothly over many millennia. When, in 1922, meteorologist C. E. P. Brooks first proposed that climate can change swiftly, he was largely ignored. During the early 1990s, however, a steady accumulation of data from four main sources strongly supported Brooks's hypothesis. These data suggested that in past epochs the atmosphere went from warm to cool or from cool to warm within decades, perhaps even within a few years.

Proxy Evidence for Abrupt Change

The theory of abrupt change rests on proxy data from ice cores taken from the ice sheets covering Greenland and Antarctica, as well as from tree rings, sediments in oceans and lakes, and coral. In each of these proxies, layers of material are laid down annually and vary in thickness in accordance with annual atmospheric conditions. In tree rings, for instance, wet years foster greater growth in trees, which is reflected in wider rings than those produced during dry years. In addition to such evidence, gases in bubbles trapped in ice reveal the relative abundance of elements at the time they were trapped, which in turn provides clues to atmospheric temperatures at that time.

Taken together, proxy evidence demonstrates not only sudden climate change in past epochs but also frequent change. The most recent of four ice ages lasted from 120,000 to 14,500 years ago. Even during that frigid period, there were twenty-five periods of abrupt warming, called Dansgaard-Oeschger events, and six extended plunges in temperature, called Heinrich events; in all of them, change took place within decades.

The most studied example of abrupt change is the period known as the Younger Dryas, which began about 12,800 years ago. As the Northern Hemisphere was warming from the ice age, it suddenly relapsed into ice-age temperatures and stayed cold until 11,500 years ago, when temperatures over Greenland rose by 10° Celsius within a decade.

Mechanisms for Abrupt Change

The United States National Research Council defines abrupt climate change as occurring when the climate system is forced to cross some threshold, triggering a transition to a new state at a rate determined by the climate system itself and faster than the cause.

Mechanisms for such change are poorly understood. It appears that some physical process forces an aspect of the climate system to pass a tipping point—for instance, in the albedo, average cloud cover, or salinity of ocean water. After the tipping point, positive feedback in the system accelerates the warming or cooling trend.

In the case of the Younger Dryas, scientists know that the water of the North Atlantic suddenly became less salty, which slowed or altered the course of the thermohaline cycle. The warm waters of the Gulf Stream no longer flowed north of Iceland and back down along the European coast, causing the continent to relapse to ice-age temperatures. The freshening probably resulted from a sudden outflow of water from a freshwater inland sea, Lake Agassiz, in north-central North America. The physical event that led to this forcing is a point of controversy. Scientists have proposed the breaking of an ice dam

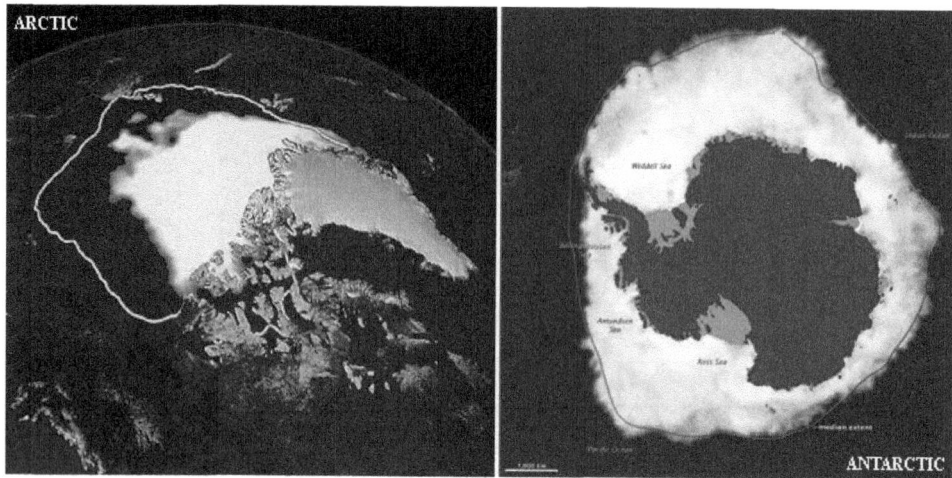

September 2012 witnessed two opposite records concerning sea ice. Two weeks after the Arctic Ocean's ice cap experienced an all-time summertime low for the satellite era (left), Antarctic sea ice reached a record winter maximum extent (right).

after gradual warming or possibly a meteor impact. Forcings for other abrupt changes in past climates include alterations in the salinity of the tropical Atlantic Ocean, evaporation and cloud cover in the South Pacific Ocean, melting of methane clathrates (frozen methane in the ocean beds), and the periodic warming of the South Pacific known as the El Niño-Southern Oscillation (ENSO).

ANTHROPOGENIC GLOBAL WARMING

Scientists worry that increasing levels of greenhouse gases (GHGs) in the atmosphere, much of them released by the burning of fossil fuels, have trapped radiant energy from the Sun in the atmosphere and increased average global temperatures in both the atmosphere and the oceans. This greenhouse effect could lead to abrupt climate change in several ways.

The vast ice sheets in the Arctic and Antarctica have the highest regional albedo on Earth, but they are shrinking rapidly, especially in the Arctic. There, the ice rests primarily on water, which is darker than ice and absorbs more heat. As the ice disappears, there is more exposed ocean surface to absorb solar energy, and the warmed water in turn helps melt the ice faster, creating a positive feedback loop. This melted ice will not affect ocean levels or salinity, but if ice sheets melt off the land of Antarctica or Greenland, ocean levels could rise by dozens of meters within a century, lowering ocean salinity enough to stall the thermohaline cycle, which could cool Europe rapidly and drastically even while the rest of the world warmed. Should ocean water heat up too much, clathrates could melt and send billions of metric tons of methane into the atmosphere, accelerating global warming further. The augmented thermal energy in the atmosphere is likely to redistribute wind and rainfall patterns, plunging some regions into drought while making others wetter; catastrophic storms, such as hurricanes and tornados, could become more frequent and severe.

CONTEXT

Some scientists argue than the Earth is entering a new geological age, the Anthropocene, because humanity itself now takes part in shaping Earth's overall surface conditions, climate in particular. Particulate pollution (especially soot), waste heat, release of GHGs, water consumption, and alteration of soil and plant cover affect not only the land, water bodies, and atmosphere but also modern civilization. If human effects on the environment trigger abrupt climate change, the onset of icy conditions in the Northern Hemisphere, droughts, superstorms, or rising sea level—all of which are possible according to computer models of climate change—would require radical, swift, and comprehensive measures to adapt or relocate much of Earth's human population. Not only would that be an expensive undertaking, but it would also mark a shift in the course of human history as profound as the Industrial Revolution.

KEY CONCEPTS

- *albedo:* the fraction of radiation reflected by a surface
- *feedback:* a process in which any change accelerates further changes of the same type (positive feedback) or counteracts itself (negative feedback)
- *greenhouse gases (GHGs):* atmospheric gases, such as carbon dioxide, water vapor, and methane, that trap heat radiation from Earth's surface by absorbing it and reemitting it
- *proxy:* remnant physical evidence from which past climatic conditions can be inferred
- *thermohaline cycle:* the "great conveyor belt" of ocean currents powered by density gradients created by heat and relative salt content
- *tipping point:* the point at which the transition from one state in a system to another becomes inevitable

—Roger Smith

BIBLIOGRAPHY

"Abrupt Climate Change." *Woods Hole Oceanographic Institution,* 2017, http://www.whoi.edu/main/topic/abrupt-climate-change. Accessed 31 Jan. 2017.

Cox, John D. *Climate Crash: Abrupt Climate Change and What It Means for Our Future.* Washington, D.C.: Joseph Henry Press, 2005.

Flannery, Tim. *We Are the Weather Makers: The Story of Global Warming.* Rev. ed. London: Penguin, 2007.

Lynas, Mark. *Six Degrees: Our Future on a Hotter Planet.* Washington, D.C.: National Geographic, 2008.

Pearce, Fred. *With Speed and Violence: Why Scientists Fear Tipping Points in Climate Change.* Boston: Beacon Press, 2007.

ACID RAIN AND ACID DEPOSITION

FIELDS OF STUDY

Atmospheric Chemistry; Biochemistry; Environmental Chemistry; Geochemistry; Inorganic Chemistry; Physical Chemistry; Phorochemistry; Thermodynamics; Chemical Engineering; Petroleum Refining; Environmental Sciences; Environmental Studies; Waste Management; Physical Sciences; Meteorology; Process Modeling; Fluid Dynamics; Chemical Kinetics; Hydroclimatology; Atmospheric Science; Oceanography; Hydrology; Physical Geography; Ecosystem Management; Ecology; Spectroscopy

SUMMARY

Acid rain is rain that is more acidic than would be natural, as a result of reactions with pollutive, acid-forming gases, such as sulfur dioxide and nitric oxides. Lakes, forests, soils, and human structures in the eastern part of the United States and southeastern Canada have been damaged by acid rain and deposition of sulfuric and nitric acid aerosol on terrestrial objects.

PRINCIPAL TERMS

- **acid deposition:** the depositing of acidic materials on the ground surface through the action of precipitation
- **acid rain:** rain composed of water having a lower-than-normal pH due to having dissolved and reacted with airborne contaminants to produce acidic materials
- **alkaline:** having a pH greater than 7 due to a lower concentration of hydrogen (H^+) ions than are in neutral water
- **bicarbonate:** a negatively charged ion, as HCO_3^-, that effectively neutralizes excess hydrogen ions in natural waters, reducing acidity
- **cap and trade legislation:** legislation that places limits on the emission of acid-producing materials, such as sulfur dioxide, while allowing emitters of excess amounts to purchase and utilize the unused allowances of those whose emissions are below the legislated limit
- **limestone:** a rock containing calcium carbonate that reacts readily with acid rain and tends to

- **neutralize** it, being chemically eroded in the process
- **neutralization:** the adjustment of the concentration of hydrogen ions in solution in order to achieve neutral pH
- **nitric acid:** an acid formed in rain from nitric oxide gases in the air
- **nitric oxide gases:** gases formed by a combination of nitrogen and oxygen, particularly nitrogen dioxide and nitric oxide
- **pH:** a measure of the hydrogen ion concentration, which determines the acidity of a solution; the lower the pH, the greater the concentration of hydrogen ions and the more acidic the solution
- **sulfur dioxide:** a gas whose molecules consist of one sulfur atom and two oxygen atoms, formed by the combustion of sulfur in the presence of oxygen
- **sulfuric acid:** an acid formed as the primary component of acid rain by reaction of sulfur dioxide gas with liquid water in the atmosphere

DEFINITION AND CAUSE

Acid rain is rain that is more acidic than it would be normally, usually because it has reacted with acid-forming pollutive gases. The acidity of rain is measured in pH units, which are the negative logarithmic values of the concentrations of hydrogen ions in solution. Pure water, which is neutral, has a pH of 7, reflecting the natural concentration of hydrogen ions in pure water of 1×10^{-7} moles per liter. Any solution with a pH greater than 7 is basic, or alkaline, and any solution with a pH less than 7 is acidic. The lower the pH, the more hydrogen ions there are and the more acidic the solution is.

The natural acidity of rain is determined by its reaction with carbon dioxide gas in the atmosphere, a reaction that produces carbonic acid. Carbonic acid partly dissociates to produce hydrogen ions and bicarbonate ions. As a result of this reaction, pure rain water is moderately acidic, with a pH of 5.7. Any rain with a pH less than 5.7 is called "acid rain" and has reacted with acid-forming atmospheric gases other than carbon dioxide. Reaction of water with sulfur dioxide, for example, produces sulfuric acid in rain, and reaction of water with nitrogen dioxide produces nitric acid in rain. In some cases, acid rain has been observed to have a pH value as low as 2.4, which is as acidic as vinegar.

In addition to acid rain, there is "dry deposition," which occurs without rain and deposits acidic nitrate and sulfate particles and sulfuric and nitric acid aerosols from the atmosphere. The acidic particles are trapped by vegetation or settle out, and the gases are taken up by vegetation. "Acid deposition" usually refers to dry deposition of acids.

Acid rain was first recognized in Scandinavia in the early 1950s. It was discovered that acid rain (with a pH from 4 to 5) came from winter air masses that were carrying pollution into Scandinavia from industrial areas in Central and Western Europe. Rain became more acidic over the next twenty years, and the area of Europe receiving acidic rains increased. By the mid-1970s, most of northwestern Europe was receiving acid rain with a pH of less than 4.6. As a result of the discovery of acid rain in Europe, scientists began measuring the acidity of North American rain. Initially, around 1960, acid rain was concentrated in a bull's-eye-shaped area over New York, Pennsylvania, and New England. By 1980, however, most of the United States east of the Mississippi River and southeastern Canada was receiving acid rain (pH less than 5.0), and the central bull's-eye was receiving very acidic rain, having a pH less than 4.2. The greatest increase in the acidity of rain was in the southeastern United States.

The primary cause of acidity in U.S. and European rains is sulfuric acid, which comes from pollutive sulfur dioxide gas produced by the burning of sulfur-containing fossil fuels, particularly coal, but also oil and gas. In the United States, much of the sulfur dioxide gas is produced in the industrial area of the Midwest. However, sulfur dioxide gas and the resulting sulfuric acid can be transported for a distance of 800 kilometers to the northeast by the prevailing winds in the atmosphere before precipitating as acid rain in the northeastern United States and southeastern Canada. To reduce the acidity of rain in the East, then, the emissions of sulfur dioxide gas in the Midwest would have to be reduced. Another source of sulfur dioxide gas is smelters that process ores, such as that in Sudbury, Ontario, located north of Lake Huron in Canada, which is one of the largest sources of sulfur emissions in the world. This smelter's high exhaust stack spreads sulfuric acid aerosols over an extensive area hundreds of kilometers downwind. The original intent of building high exhaust stacks was to reduce local air pollution, but the net effect has been to spread the pollution over much larger areas. Acid rains are even found in Alaska, where sulfuric acid particles have been transported from the contiguous United States.

Nitric acid is a secondary cause of acid rain (contributing about 30 percent of the acidity), but it is one that is increasing. Nitric acid comes from the nitrogen oxide gases, nitrogen dioxide, and nitric oxide, which are produced by the burning of fossil fuels. In contrast to sulfur dioxide, 40 percent of the pollutive nitrogen oxide gas comes from vehicles and most of the remainder from power and heating. The production of nitrogen oxides therefore tends to be concentrated in urban areas. Nitric acid is an important component of acid rain in Los Angeles, for example, because air pollution from vehicle exhaust tends to become trapped in this area.

Some acid rain results from natural causes. Reduced sulfur gases, such as hydrogen sulfide and dimethyl sulfide, are produced by organic matter decay and converted to sulfur dioxide and sulfuric acid in the atmosphere. This process results in naturally acid rain. Volcanoes are another natural source of sulfur dioxide gas. Nevertheless, about 75 percent of the sulfur dioxide gas produced in the United States comes from the burning of fossil fuels. Naturally acid rain (with a pH less than 5.5) is uncommon, falling chiefly in remote areas such as the Amazon basin and some oceanic areas.

There are natural factors that work to reduce or neutralize the acidity of rain in certain areas. Windblown dust, particularly that containing limestone particles, tends to make rains in arid areas of the western United States less acidic by reacting with the acid to produce a rainwater solution with a pH of 6 or more. In addition, the presence of ammonia gas produced in agricultural areas by animal waste, fertilizers, and the decomposition of organic matter will reduce or counteract the acidity of rain on a local scale.

EFFECTS OF ACID RAIN AND ACID DEPOSITION
The detrimental effects of dry deposition and acid rain include the corrosion and chemical erosion of structures and buildings made of susceptible materials, changes in soil characteristics, increases in the acidity of lakes, and other biological effects, particularly in high-altitude forests. The corrosive effects of airborne acids are particularly obvious on limestone, a rock composed of calcium carbonate, which reacts easily with acid rain. In many New England cemeteries, tombstones made of marble, a form of limestone, have been badly corroded, although older tombstones made of slate, which is less affected by acid rain, are intact. Limestone components of buildings and other structures are similarly corroded.

The effect of acid rain on soils depends on their composition. Alkaline soils that contain limestone have the ability to neutralize acid rain. Even in soils that do not contain limestone, several processes operate to neutralize acid rain, though such processes invariably alter the chemical nature of the soil. Cation exchange occurs, whereby hydrogen ions from the rainwater are exchanged for metal ions, such as calcium or magnesium, on the surface of clays and other minerals. This exchange removes excess hydrogen ions from soil solutions, rendering them less acidic. Another neutralization process involves the release of soil aluminum into solution and the accompanying uptake of hydrogen ions. This process occurs by dissolution of aluminum bound to clays and organic compounds. Frozen soils and sandy soils containing mostly quartz, which does not react with acid rain, have little ability to neutralize acid rain.

Lakes in certain areas have become acidic (with a pH less than 5) from the deposition of acid rain. Lakes in granitic terrain are most affected by acid rain because the surrounding bedrock has little or no ability to react with or neutralize the excess hydrogen ions in the water. Areas with many acid lakes include the Adirondack Mountains in New York, the Pocono Mountains in Pennsylvania, the Upper Peninsula of Michigan, Ontario, Nova Scotia, and Scandinavia. Generally, the deposition of highly acidic rain having a pH less than 4.6 over a long period of time is required. The effect is enhanced in bodies of water that are maintained by watershed drainage rather than by freshwater springs.

Lake waters that have a tendency to become acidified initially have little ability to neutralize acid rain because they are low in carbonate and bicarbonate ions, which come predominantly from limestone. Such lakes are described as being poorly buffered. (Buffering is the resistance to changes in pH upon the addition of acid or base.) The soil in the drainage area surrounding acid lakes does not neutralize acid rain adequately before it reaches the lake because of a lack of limestone and clay minerals or because

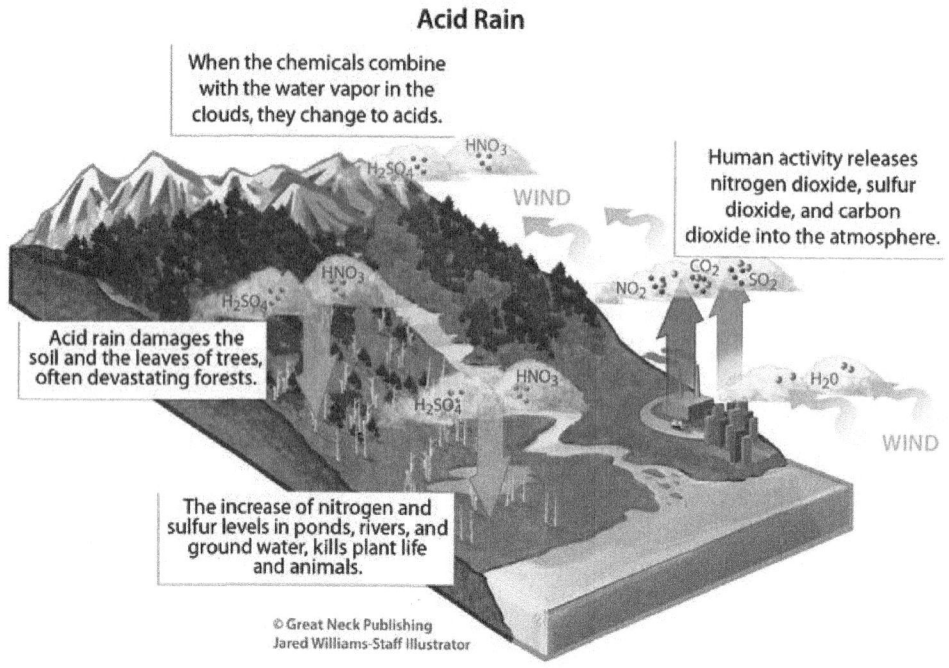

This illustration shows the effects of acid rain on the environment.

the soil cover is thin or lacking altogether. In addition, some lakes, although not usually acidic, may have periods of elevated acidity due to the runoff of snowmelt, which collects acid precipitation stored in the snow. This runoff gives a sudden large pulse of highly acidic water to the lake. In certain areas, such as Florida, acid lakes result partly due to causes other than acid rain, such as the presence of organic acids produced by the decay of vegetation in poorly drained areas and nitric acid formed from nitrate-based fertilizer runoff. The gradual acidification of lakes results in the death of fish populations because of reproductive failure, as well as other changes in the organisms living in the lake. A reduction in the number of species occurs at all levels of the food chain. In some cases, snowmelt acidity has been identified as the cause of a massive, instantaneous fish kill in lakes.

Rivers are also known to become acidic. Eastern U.S. rivers show high concentrations of sulfate and a low pH in cases where the soil cannot neutralize the acid rain it receives. Certain acid rivers are caused by acidic drainage from mine dumps rather than by acid rain. Acid rivers rich in organic matter are found in the eastern United States coastal plain and in the Amazon basin. These rivers have naturally high concentrations of dissolved organic acids.

Acid rain and acid deposition are implicated in the decline and death of certain forests, particularly evergreen forests at high elevations. These forests receive very acidic precipitation from the accumulation of clouds at the mountaintops. It is thought that acid rain does not actually kill the forests but rather provides a stress that causes them to become less resistant and die from other causes. The actual stress provided by acid rain is still being studied. Possible stresses include loss of nutrients from soil and leaves through leaching, destruction of beneficial soil microorganisms, and increased susceptibility to frost damage during winter.

Efforts have been made to reduce the acidity of rain, particularly by controlling sulfur emissions. Power plants have been required to reduce the sulfur content of coal that they burn, thus lowering the amount of sulfur dioxide that is produced. Sulfate concentrations in rain in the northeastern United States have been reduced by this method. Nitrogen oxide emissions from cars have also been reduced through stricter emissions controls and improved engineering design for more efficient combustion. In some cases, acid lakes have been treated with limestone to temporarily neutralize their acidity, but the only permanent solution is a reduction in the acidity of the rain that they receive.

Study of Acid Rain and Acid Deposition

The acidity of rain can be measured directly by an electronic device called a pH meter. A pair of electrodes is inserted into a solution, and the electrical potential, or voltage, is measured between them. This voltage is directly related to the concentration of hydrogen ions in the solution—that is, to the acidity of the solution. To monitor and measure the acidity of rainwater, networks have been constructed to collect rain samples over large geographical areas. The acidity of rainwater over the course of the entire year must be measured because pH varies between rainfalls, both seasonally and according to whether the air masses that produce the rain have passed over significant sources of pollution. The pH of rainwater and other forms of precipitation is also measured over a period of years.

In addition to the concentration of hydrogen ions, the concentration of other ions, such as sulfate from sulfuric acid and nitrate from nitric acid, is measured in the rainwater samples. Such measurements give evidence of the source of the acidity—that is, which proportion is attributable to sulfuric acid and which to nitric acid. The pH levels of samples collected over a large geographical area are plotted on a map, and contours are drawn through equal values of the pH. Such maps show which areas are receiving the most acid rain. The amount of sulfate and nitrate being deposited by rain is also plotted separately. Meteorologists also use information about a storm system's path as it moves across the country. Such atmospheric systems transport pollutive gases from one area to another. Combining deposition patterns on maps with information about the path followed by a storm shows where the gas residues in rainwater may be coming from and suggests sources of the acidity.

Computers have been used to predict where acid rain will fall and how acidic it will be, given the sources and amounts of sulfur and nitrogen emissions, particularly from power plants and smelters, and the weather patterns. Predictions of this type require a detailed knowledge of the atmospheric chemistry by which sulfur dioxide is converted to sulfuric acid and the oxides of nitrogen are converted to nitric acid. This type of modeling is necessary to predict how much reduction in the acidity of rain in a distant area will result from a given reduction in a power plant sulfur source, for example.

The effects of acidity on soils have been the subject of study for many years. Laboratory experiments can demonstrate how soil clays and other minerals react to acid rain, including which chemical species are taken up and which are released. In addition, soil solutions and minerals are collected and analyzed from actual field areas affected by acid rain. Ideally, such an analysis should be carried out over a period of time to determine whether any changes in the soil solution chemistry are occurring. From a knowledge of the soil chemistry, it is possible to predict how long a soil can receive acid rain before it loses particular nutrients or the ability to neutralize the excess acidity.

Measuring the acidity and chemical composition of lakes in various areas over long periods and sampling their fish populations and other biota enables scientists to see increases in lake acidity and to correlate the increases with changes in the populations of affected species. In some areas, lakes have been artificially acidified so that the changes in their chemistry and biological populations can be observed. Apparently, acidic lake water inhibits reproduction in fish and other creatures, in addition to destroying the organisms that they use for food. Computer models of acid rain falling on susceptible drainage areas of lakes are made in order to predict how the drainage area reacts to acid rain and how much reduction in acid rain would be necessary to lower the acidity of the lake to the point where it would support fish. In badly affected areas such as the Adirondacks, it may be necessary to reduce the acidity by half.

To study forest decline, surveys of present forest conditions are compared with historical records for the same areas. For example, in high-elevation areas in New England and the Adirondacks, more than half of the red spruce died between 1965 and 1990. Tree rings, which record annual growth, show reduced growth in certain forests. It is known that acid rain causes changes in the soil, such as the release of aluminum, which is toxic to root tissues and so prevents the uptake of essential nutrients. In addition, acid rain causes the loss of certain nutrients from the soil, such as sodium, calcium, and magnesium.

Another effect of acid rain is the reduction of the numbers of microorganisms in the soil. Yet, because acid rain from nitric acid contains nitrogen, a plant nutrient, it may fertilize the soil if there is a deficiency of soil nitrogen. One problem in studying forests receiving acid rain is determining which of the many changes occurring are contributing most to forest damage. It is often difficult to distinguish between the stresses of acid rain and other stresses, such as those caused by drought, cold, and insects. Field studies in this area may involve artificial acidification of forest environments in order to determine which mechanisms are important.

—*Elizabeth K. Berner*

Further Reading

Ahrens, C. Donald. *Essentials of Meteorology: An Invitation to the Atmosphere*. Brooks/Cole Cengage Learning, 2012.

Alaby, Michael *Fog, Smog, and Poisoned Rain*. Infobase Publishing, 2014. An entry-level book that describes the characteristics of fog formations and their effects.

Berner, Elizabeth K., and Robert A. Berner. *The Global Water Cycle: Geochemistry and Environment*. Prentice-Hall, 1986.

Brimblecombe, Peter, et al., eds. *Acid Rain: Deposition to Recovery*. Springer, 2010. Compiles articles from *Water Air, & Soil Pollution: Focus*, Volume 7 (2007).

Denny, Mark *Making Sense of Weather and Climate. The Science Behind the Forecasts*. Columbia University Press, 2017. An informative yet entertaining book that looks at many readily-visible aspects of climate effects such as fog from a question and answer perspective.

Hill, Marquita K., *Understanding Environmental Pollution*. Cambridge University Press, 2010.

Jenkins, Jerry C., et al. *Acid Rain in the Adirondacks: An Environmental History*. Cornell University Press, 2007.

Sommerville, Richard C. J. *The Forgiving Air: Understanding Environmental Change*. 2nd ed. Boston: American Meteorological Society, 2008.

AEROSOL POLLUTANTS

Fields of Study

Physical Chemistry; Geochemistry

Summary

The effects of aerosol pollutants have been debated for a long time. Release of volcanic dust from the 1783 eruption of a volcanic fissure in Iceland seemed to have been related to an unusually cool summer in France the same year. In 1883, volcanic dust from the explosion of Krakatoa in the East Indies dimmed the sunlight for months, as had the 1815 eruption of Tambora. Some scientists perceived a pattern of temporary cooling from such events. Others asked if pollutants should be expected to warm, rather than cool, the atmosphere.

Principal Terms

- **cloud condensation nuclei:** atmospheric particles such as dust that can form the centers of water droplets, increasing cloud cover
- **dust Veil Index:** a numerical index that quantifies the impact of a volcanic eruption's release of dust and aerosols
- **global dimming:** the effect produced when clouds reflect the Sun's rays back to space
- **stratosphere:** part of the atmosphere just above the troposphere that can hold large amounts of aerosols produced by volcanic eruptions for many months
- **troposphere:** location in the lower atmosphere where the majority of aerosols form a thin haze before being washed out of the air by rain

Characteristics of Aerosols

Aerosols are minute airborne solid or liquid particles suspended in the atmosphere, typically measuring between 0.01 and 10 microns and either natural or anthropogenic in origin. Normally, most aerosols rise to form a thin haze in the troposphere; rain washes these out within about a week's time. Some aerosols, however, are found in the higher stratosphere, where

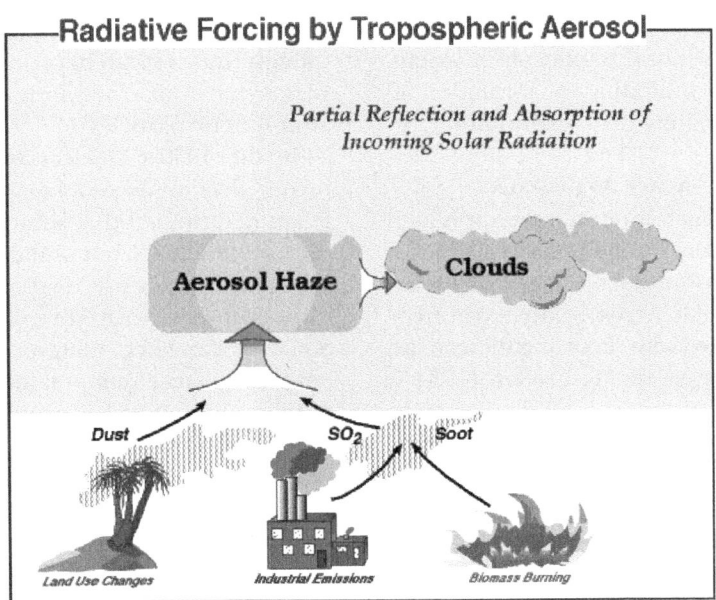

Aerosols can both reflect and absorb solar radiation, making their role in the greenhouse effect particularly complex.

it does not rain. They can remain in this atmospheric layer for months. Aerosols may influence climate either directly, through scattering and absorbing radiation, or indirectly, by acting as cloud condensation nuclei or by modifying the optical properties and lifetimes of clouds.

Natural aerosol sources include salt particles from sea spray; clay particles from the weathering of rocks; volcanic emissions of sulfur dioxide, which oxidizes to form sulfuric acid molecules and sulfate salts; and desert dust. Anthropogenic (human-produced) aerosol sources include industrial pollutants such as sulfates, from combustion of oil and coal; smoke from large-scale burning of biomass, such as occurs in slash-and-burn clearing of tropical forests; and pollution from naval vessels' smokestacks and the transportation sector in general.

On windy days, bubbles created by breaking waves eject salt into the air when they burst, forming aerosols. Salt aerosols scatter sunlight, lessening the amount of energy that reaches Earth's surface which acts to cool the climate. The interaction of sea salt with clouds also causes cooling. The resulting overall 'whitening' of the Earth further reduces the amount of sunlight that can reach the ground. Oceans cover over 70 percent of Earth's surface, and sea salt is a major source of aerosols in areas far distant from land.

Wind also helps form aerosols over land. Particles carried by the wind abrade the surfaces of rock and other landforms. This action wears down rocks and other surfaces progressively over time, converting them eventually into fine dust particles of stone and other materials. When these particles are incorporated into the air, they too form aerosols.

Following major volcanic eruptions, sulfur dioxide gas vented during the eruptions is converted into sulfuric acid droplets. These droplets form an aerosol layer in the stratosphere. Winds in the stratosphere scatter the aerosols over the entire globe, where they may remain for about two years. Because they reflect sunlight, these aerosols also reduce the amount of energy that reaches the troposphere and the Earth's surface, exerting a cooling influence.

Another significant natural aerosol is desert dust. "Veils" of dust stream off deserts in Asia and Africa, and have also been observed on the American continents. These particles typically fall out of the atmosphere after a short flight, but intense dust storms often blow them to altitudes of 4,500 meters or higher, allowing them to reach great distances from their source. Since the dust is made up of minerals, the particles both absorb and scatter sunlight. Absorption warms the layer of the atmosphere where they are located, possibly inhibiting the formation of storm clouds and contributing to desertification.

Early Speculation About Aerosols

Long before there was much interest in aerosols as a factor in climate change or any equipment capable of adequately analyzing aerosol data, a few individuals speculated about a possible aerosol-climate connection. The first man credited with reporting these ideas was Jacques Antoine Morgue de Montredon (1734 – 1818), a French naturalist who in 1783 documented the eight-month-long Laki eruption in southern Iceland. The eruption caused the grass to die. Three-quarters of the region's livestock and one-quarter of its people starved to death. For months, a haze hovered over western Europe. When Benjamin Franklin was visiting in France in 1783, he experienced an unseasonably cold summer and speculated that the Laki volcanic "fog" had noticeably dimmed the sunlight.

A century later, in 1883, the eruption of the Indonesian volcano Krakatoa (Krakatau) sent up a veil of volcanic dust that reduced sunlight globally for months. Scientists were unable to determine what effect the eruption might have had on the average global temperature, but scientists thereafter acknowledged volcanoes as a possible natural influence on Earth's climate.

A few scientists who examined temperatures after major volcanic eruptions between 1880 and 1910 perceived a pattern of temporary cooling. Only later would older records reveal that the 1815 eruption of Tambora in Indonesia had affected the climate more severely than had the eruption of Krakatoa. Speculation led some to ask if volcanic eruptions had precipitated ice ages or had cooled the Earth to the extent that dinosaurs became extinct.

Early Twentieth Century Aerosol Research

Throughout the first half of the twentieth century, it was known that volcanic aerosols could affect climate. As a result, some scientists suspected that other kinds of particles could have similar climatic effects. Physics theory seemed to support the notion that these particles should scatter radiation from the Sun back into space, thereby cooling Earth's overall climate. These ideas remained largely speculative, though some researchers began to focus on the possibility that human activity might be a major source of atmospheric particles.

In the 1950's, nuclear bomb tests provided improved data on aerosol behavior in the stratosphere. It was determined that stratospheric dust would persist for some years, but would stay in one hemisphere. Research in the early 1960's indicated that large volcanic eruptions lowered average annual temperatures. Some researchers, however, deemed those results enigmatic, since temperatures had also decreased during a period of few eruptions. Meteorologists acknowledged that other small, airborne particles could influence climate, but throughout the first half of the century, speculation fell short of conclusion.

Gradually, scientists shifted their focus to anthropogenic atmospheric particles. Measurements by ships between 1913 and 1929 noted that sea air showed an extended decrease in conductivity, apparently caused by smoke and gases from the stacks of ships and possibly from industry on land. Even in 1953, however, scientists were uncertain about the significance of the pollution.

During the 1950's, some scientists asked whether aerosols might affect climate by helping form clouds. Since particulate matter is essential for providing a nuclear surface for water droplets to condense around, the notion of seeding clouds with silver iodide smoke to make rain became widespread. By this time, aerosol science was just coming into its own as an independent field of study, having been given impetus by the concern that disease-carrying aerosols and poisonous gas could be employed with lethal effect. Public concern over urban smog also fueled studies by aerosol experts. By and large, however, scientists avoided the study of cloud formation. Field testing often produced contradictory results and was extremely expensive, and many researchers believed that the effects of aerosols on clouds were too complex to comprehend.

Aerosol Research in the Later Twentieth Century

By the early 1960's, the scientific community was beginning to pay more attention to the possibility that human activities influence clouds. It was predicted that jet airplane contrails would spread, thin, and become indistinguishable from cirrus clouds, and this was borne out by observation. The apparent ability of aircraft to create cirrus clouds sugested the possibility that they might be causing climate changes along major air routes. Others questioned the possibility of anthropogenic activity as the source of pollution settling on polar ice caps. At the time, the theory was not given much credence.

Around 1970, the British meteorologist Hubert Horace Lamb (1913-1997) devised the Dust Veil Index, establishing a connection between atmospheric dust and lower temperatures. While scientific studies at this time did not yet find strong evidence for an increase in global turbidity, they did document regional hazes that spread in a radius of up to one thousand kilometers or more from industrial centers. The scientific debate shifted from the existence of anthropogenic dust to the effects of that dust. It remained a subject of controversy whether and under what circumstances dust would cool or heat the climate, especially after a spacecraft on Mars in 1971 found that a large dust storm had caused substantial warming of the Martian atmosphere.

Deadly droughts in Africa and South Asia in 1973 caused public concern about climate change, but it was not confirmed that sulfate pollution had contributed to the Sahel drought until the end of the century. Scientific publications in the mid- to late 1970's discussed warming or cooling effects without reaching accord, although a majority felt that greenhouse warming would dominate. At this time, only a few researchers noted that aerosol pollution might cancel out some greenhouse warming and thus temporarily mask its effects. Others denied that industrial pollution could mitigate the enhanced greenhouse effect caused by carbon dioxide (CO_2) emissions.

The 1980's brought the realization that additional factors contributed to climate and climate change. For example, climate scientists generally treated aerosols as a globally uniform background, largely of natural origin, when in fact different aerosol properties were obtained in different regions based on relative humidity. Many questions remained.

By 1990, it was acknowledged that from one-fourth to one-half of all tropospheric aerosol particles were anthropogenic. These included industrial soot and sulfates, smoke from forest-clearing fires, and dust from overgrazed or semiarid land turned to agriculture. Impressive advances in laboratory instrumentation made possible much more sophisticated satellite observations, greatly increasing the resolution of climate models. A key paper establishing the net effect of aerosols on Earth's heat balance, published in the early 1990's, concluded that radiation scattering due to anthropogenic sulfate emissions was counterbalancing CO_2-related greenhouse warming in the Northern Hemisphere.

It became apparent that earlier climate projections might be erroneous, because they had not factored in sulfate aerosol increases. Climatologists redoubled their efforts to produce accurate models and projections of Earth's climate. In 1995, for the first time, new results that took into account aerosol influence yielded a consistent and plausible picture of twentieth century climate. According to this picture, industrial pollution had temporarily depressed Northern Hemisphere temperatures around the mid-century. A 2008 study found that black carbon aerosols had exerted a much greater warming effect than had been earlier estimated, because the combined effects of black carbon with sulfate aerosols had not been taken into account. It seemed clear that reducing sooty emissions would both delay global warming and benefit public health.

Context

A number of aerosol specialists have questioned whether they have underestimated the cooling effect of aerosols. If they had, they would have underestimated those aerosols' restraint of greenhouse warming, significantly underestimating the extent of global warming in the absence of anthropogenic aerosol pollution. Much uncertainty remains, and each new study introduces new complexities. It seems clear that reducing sooty emissions would both delay global warming and benefit public health, yet nagging questions remain: Since aerosols and clouds, unlike gases, are not distributed evenly throughout the atmosphere, uniform samples cannot be obtained. Further, the properties of clouds and aerosols are incompletely understood, and scientists are only beginning to understand some of the interactions that take place between aerosols, clouds, and climate. Thus, these interactions have not yet been fully incorporated into climate models.

—*Victoria Price*

Further Reading

Le Quéré, Corinne and Saltzman, Eris S., eds. *Surface Ocean-Lower Atmosphere Processes* John Wiley & Sons, 2013. This 187th volume of the Geophysical Monograph series provides a clear and documented description of processes contributing to aerosol formation at the surface of Earth's oceans.

Seinfeld, John H. and Pandis, Spyros N. *Atmospheric Chemistry and Physics: From Air Pollution to Climate Change* 3rd ed. John Wiley & Sons, 2016. Presents a comprehensive historical perspective and overview of the chemistry and physics of the atmosphere, with multiple chapters covering aerosols and their effects.

Tanvir, Islam, Hu, Yongxiang, Kokhanovsky, Alexander and Wang, Jun, eds. *Remote Sensing of Aerosols, Clouds and Precipitation* Elsevier, 2017. Describes the methods, theoretical foundations and applications of satellite-based monitoring of aerosols, clouds and precipitation events occurring in Earth's atmosphere.

Wang, Yuan. *Aerosol-Cloud Interactions from Urban, Regional and Global Scales* Springer Theses/Springer, 2015. Written for a highly technical, Ph.D-level audience, this is an advanced study examining the aerosol-cloud-precipitation interaction in a quantitative manner. A significant contribution is made to coordinating regional and global climate prediction models, based on radiative forcing by atmospheric aerosols.

AIR POLLUTION

Fields of Study

Atmospheric Science; Atmospheric Chemistry; Photochemistry; Environmental Chemistry; Thermodynamics; Chemical Engineering; Petroleum Refining; Engineering; Environmental Sciences; Environmental Studies; Waste Management; Meteorology; Climate Modeling; Process Modeling; Fluid Dynamics; Hydroclimatology; Hydrometeorology; Bioclimatology; Physical Geography; Ecosystem Management; Ecology; Toxicology; Pathology; Spectroscopy

Summary

Air pollution is generated from both natural and anthropogenic sources. Natural sources include gases, particulate matter from volcanoes and decomposing organic matter, pollen from plants, and windblown dust. Anthropogenic sources include industrial and automobile emissions and airborne particles associated with human-induced abrasion.

Principal Terms

- **acid rain:** precipitation having elevated levels of acidity relative to pure water
- **atmosphere:** the layer of mixed gases that surrounds Earth
- **carbon dioxide:** CO_2, one of many minor gases that are natural components of the atmosphere; the product of the complete oxidation of carbon
- **greenhouse effect:** the environmental process that results when heat energy is absorbed and retained in the atmosphere by various gases and is not radiated out into space
- **inversion:** an unusual atmospheric condition in which temperature increases with altitude
- **off-gassing:** the spontaneous emission of entrained or entrapped gases from within natural and artificial sources
- **oxides of nitrogen:** several gases that are formed when molecular nitrogen is heated with air during combustion, primarily NO and NO_2
- **oxides of sulfur:** gases formed when fuels containing sulfur are burned, primarily SO_2
- **ozone:** a highly reactive compound composed of three atoms of oxygen, as O_3
- **photochemical oxidants:** pollutants formed in air by primary pollutants undergoing a complex series of reactions driven by light energy
- **photochemical reaction:** a type of chemical reaction that can occur in polluted air driven by the interaction of sunlight with various pollutant gases

EARTH'S ATMOSPHERE

Air pollution results from the unusual addition of gases, solids, and liquids to the atmosphere. The concentration of pollutants depends on prevailing atmospheric conditions as well as emission rates. Once pollutants are put into the atmosphere, it is impossible to control them to any significant degree. Thus emissions at the local level contribute to regional and global air pollution problems, such as smog and photochemical oxidants, acid precipitation, the depletion of the ozone layer, and global warming associated with the intensification of the greenhouse effect. Although there are many air pollutants, the major ones are usually associated with burning, particularly the burning of fossil fuels and oil-carbon dioxide based products. They are generally unburned hydrocarbons, oxides of sulfur and nitrogen, carbon monoxide, various photochemical oxidants and reactive compounds, and particulate matter from many different sources.

The atmosphere is a mixture of gases, aerosols, and particulate matter surrounding Earth. The concentration of some of the gases in clean air is fairly constant both spatially and temporally. Consequently, these gases are referred to as stable or permanent gases. Nitrogen and oxygen, the two most abundant permanent gases, account for 78 percent and 21 percent of the total atmosphere by volume, respectively. Gases that experience noticeable temporal and spatial variations are termed variable gases. The two most abundant of these are water vapor and carbon dioxide. The average concentration of carbon dioxide is about 0.034 percent. It varies seasonally in response to the growth cycle of plants, daily in response to plant photosynthesis, and spatially in response to the burning of fossil fuels. Water vapor is also highly variable. Some variable gases have natural origins and tend to have relatively high concentrations in urban areas. They are methane, carbon monoxide, sulfur dioxide, nitrogen dioxide, ozone, ammonia, and hydrogen sulfide.

The atmosphere is stratified according to its vertical temperature gradient. From the ground surface up, the major layers are the troposphere, the stratosphere, the mesosphere, and the thermosphere. The troposphere contains the bulk of atmospheric gases and, under normal conditions, is characterized by a fairly uniform temperature decline from the surface upward. The uppermost limit of the troposphere is called the tropopause, a transition zone between the troposphere and stratosphere where temperatures stabilize with increasing altitude. The troposphere extends up to about 10 kilometers. The next layer encountered is the stratosphere, which extends from about 12.5 kilometers up to about 45 kilometers above the surface. In its lower layer, the temperature gradient is somewhat stable. At an elevation of about 30 kilometers, however, the temperature starts to increase. Located within the stratosphere about 24 to 32 kilometers above the Earth's surface is a zone with a relatively high concentration of ozone, a triatomic form of oxygen. This zone is called the ozone layer. It is important because the ozone absorbs most of the incoming ultraviolet rays emitted by the sun, preventing them from reaching the surface where they would have harmful effects on plant and animal life.

The two uppermost layers, the mesosphere and the thermosphere, have a distinctive temperature gradient. In the mesosphere, temperatures decline steadily with altitude, a condition that continues until its transition zone with the thermosphere, called the mesopause, is reached. At latitudes of 50 degrees north and higher, mesospheric clouds known as noctilucent clouds are sometimes seen during summer. These clouds may be anthropogenic in origin. The last layer, the thermosphere, slowly gives way to outer space and has no defined upper limit.

Smog blankets the Chinese countryside north of Beijing on a late November day. (Sean O'Keefe)

ATMOSPHERIC INVERSIONS AND SMOG
Vertical and horizontal mixing of air is necessary to dilute pollutants in the atmosphere. Under normal conditions, temperatures decline with altitude in the troposphere. This decline in temperature with altitude is referred to as the thermal or environmental lapse rate. The warmer air near the surface rises, mixes with the air above it, and is dispersed upward by winds. This dilution process is important in reducing the concentration of pollution near the surface. Conversely, the vertical mixing of air is inhibited when the temperature profile in the troposphere inverts, developing one type of atmospheric inversion: a temperature or thermal inversion. When an inversion exists, a layer of warmer air becomes sandwiched between two layers of cooler air above and below, and the warmer, less dense air does not rise as it would normally. Pollutants can then accumulate below the warmer air as vertical mixing is prevented.

Conditions for temperature inversions develop when the Earth readily radiates heat energy from its surface on clear nights or when air subsides and warms adiabatically from compression. On cool, clear nights, the Earth readily radiates heat energy to space, cooling the surface. Air near the surface is, in turn, cooled by conduction, while the air above it is still relatively warm. This condition is referred to as a radiation inversion. Radiation inversions are common during autumn and are usually short-lived, as the rising sun in the morning heats the air near the surface, causing the inversion to dissipate as the day advances. Less frequent but more persistent subsidence-type inversions can occur when cooler air subsides in high-pressure systems or in valleys, as cooler, denser air descends along adjacent mountain slopes. Subsidence inversion episodes may last for days, allowing pollutants to concentrate to excessive levels, causing eye irritation, respiratory distress, reduced visibility, corrosion of materials, and soiling of clothes.

The atmosphere has inherent self-cleansing mechanisms. Pollutants are removed from the atmosphere through fallout due to gravitational settling, through rainout in condensation and precipitation processes, through washout as waterdrops and snowflakes accumulate pollutants as they fall to Earth, and through chemical conversion. Solar radiation, winds, and atmospheric moisture are important meteorological factors in these removal processes.

Chemical reactions between two or more substances in the atmosphere produce secondary pollutants, which are those created from other pollutants that have been released directly into the atmosphere from identifiable sources. Smog is a product of such reactions. Stability in the atmosphere that accompanies inversions provides favorable conditions for smog to develop. Smog is produced by chemical and photochemical reactions involving primarily sulfur oxides, hydrocarbons, and oxides of nitrogen. Smog that is characterized by sulfur oxides is called sulfurous smog and is associated with the burning of fuels having relatively high sulfur content. This type of smog is common in undeveloped countries. Photochemical smog develops when oxides of nitrogen and various hydrocarbons undergo photochemical reactions to produce ozone and other chemical oxidizers. Sunlight promotes the reactions, and automobile exhaust is a primary source of nitrogen oxides. This is the type of smog typically encountered in large cities and urban centers such as Los Angeles and Mexico City. In November 2014, the United States Environmental Protection Agency announced tightened restrictions on smog levels, lowering the acceptable level from 75 parts per billion to 65 to 70 parts per billion. According to the agency, a standard of 70 parts per billion would greatly improve the quality of life in high-pollution areas, preventing 1,400 asthma-related emergency room visits and 750 premature deaths.

ACID PRECIPITATION
While smog is a relatively localized phenomenon closely associated with urban areas, acid precipitation is a more widespread phenomenon. Its effects are observed in national parks, agricultural regions, forested areas, and lakes and other bodies of water as well as in urban centers. Acid rain develops when oxides of sulfur or nitrogen combine with water vapor in the atmosphere to form sulfuric and nitric acids that fall back to Earth in precipitation. Once released into the atmosphere, these oxides and the compounds formed from them can travel great distances before returning to Earth in precipitation or as dry particulates, as much as 1,000 to 2,000 kilometers over three to five days. This long-range transport allows time for chemical reactions to convert pollutant gases into components of acid precipitation. Evidence suggests that the pH values in precipitation

have been dropping, becoming more acidic, for some years. Wet precipitation is not the only way pollutants find their way to the surface. Diffusion and settling enable acidic gases and particles to find their way to the ground even under dry conditions. It is now widely accepted that both wet and dry deposition can be traced to human activity.

Much evidence has been gathered documenting the damaging effects of acid precipitation. These effects include damage to wildlife in lakes and rivers, reduction of forest productivity, damage to agricultural crops, and deterioration of human-made materials. Acid precipitation, also suspected of promoting the release of heavy metals from soils and pipelines into drinking water supplies, has different effects on different ecological systems and is most damaging to aquatic ecosystems. Acidity in precipitation at a given time depends not only on the type and quantity of pollutants being produced but also on the prevailing and immediate atmospheric conditions. Stagnant air, resulting from upper-level inversions, tends to cause higher levels of acidity. Furthermore, prevailing and local atmospheric systems are associated with the spread of acid precipitation over broader areas. Higher exhaust stacks, while minimizing levels of air-borne pollutants locally, simply disperse the pollutants over larger areas, thus increasing their residence time in the atmosphere.

OZONE DEPLETION AND GLOBAL WARMING
It is now realized that the impact of air pollution is more far-reaching than the troposphere. Evidence indicates that pollutants making their way up to the stratosphere are causing the ozone layer to break down or dissipate. Even though ozone constitutes a very small portion of the atmosphere, only about one part per million, it absorbs almost all of the ultraviolet rays from the sun, preventing them from reaching Earth's surface. Research findings from satellite-based monitoring systems have shown that there has been a breakdown of the ozone shield over the Antarctic, where a hole has been identified in the ozone layer. More recent research and satellite-based observation has shown that the ozone layer also appears to be thinning over the Arctic.

While early laboratory studies showed that oxides of nitrogen could attack ozone, attention later focused on chlorofluorocarbons (CFCs) as being responsible for the decline in ozone. These compounds were widely used as refrigerants in common household appliances and air conditioning systems, propellants in aerosol sprays, agents for producing foam, and cleansers for electronic products because they have low boiling points, facile compressibility, and very low chemical reactivity. Behaving much like inert gases, they do not degrade readily in the troposphere but eventually make their way into the stratosphere. Laboratory studies have shown that when the CFC molecules come in contact with ozone and ultraviolet light, they enter into a complex series of gas-phase reactions by which they are converted into more reactive gases, such as chlorine. Since these gases tend to linger in the atmosphere for many years, it is believed that even though several nations and the United States have banned the use of CFCs was discontinued, the ozone layer would continue to disintegrate for several years afterward. To further complicate the matter of CFCs and the recovery of the ozone layer, monitoring stations have identified a significant and steady increase in atmospheric levels of certain CFC compounds over the ten-year period from 2008 to 2018. Given the global ban on CFC production initiated by the Montreal Protocol, this indicates that someone somewhere has begun producing and continues to produce the material in bulk quantities in defiance of the global ban.

Further evidence suggests that CFCs not only destroy the ozone but also trap heat energy radiated from the ground and contribute to heating the atmosphere. The trapping of sensible heat energy in the atmosphere by gases is called the greenhouse effect. One of the most important gases contributing to the greenhouse effect, however, is not a chlorofluorocarbon but carbon dioxide. Carbon dioxide moves in a continuous cycle throughout the environment. It provides a link between the organic and inorganic components of the environment. Reacting with water and solar energy through photosynthesis in plants, it forms glucose that is subsequently passed through the food chain as a source of energy required by essentially all animal species.

Carbon dioxide is given off by plants and animals to the atmosphere during respiration. When plant and animal remains decay, carbon dioxide is passed back to the atmosphere and hydrosphere through the most natural processes of decomposition. When fossil fuels are burned, however, those natural processes are short-circuited, and large amounts of carbon dioxide are released directly into the atmosphere.

About 0.04 percent of the total atmosphere is carbon dioxide. Molecules of carbon dioxide in the atmosphere absorb and retain infrared radiation as heat, in much the same way that the glass walls of a greenhouse reflect heat back into the structure rather than allowing it to escape. While it is transparent to shortwave radiation from the sun, carbon dioxide absorbs strongly in the sensible heat or longwave radiation band. It is hypothesized that an increase in atmospheric carbon dioxide causes a decrease in outgoing longwave radiation and thus an increase in the atmospheric temperature.

The consequences of rising global temperatures will greatly alter the Earth's surface. As the atmosphere warms, polar ice and glaciers will begin to melt and the rising oceans could flood many of the world's coastal regions, devastating low-lying countries. As shorelines rise, saltwater intrusion will contaminate the drinking-water supplies of many cities worldwide. Agricultural regions of the middle-latitude countries will migrate farther northward, increasing the length of the growing seasons in Canada and Russia.

STUDY OF AIR POLLUTION

Methods of studying air pollution include controlled laboratory experiments, simulations in fluid-modeling facilities, computer simulations, and mathematical modeling. Controlled laboratory experiments are conducted in laboratories where gases are mixed to determine how they react. Laboratory experiments usually do not provide the definitive answer to what is actually occurring in the ambient environment because many variables cannot be replicated. These studies suggest what should be further studied and monitored in the natural environment. Some laboratory studies have simulated atmospheric conditions in a controlled environment, such as a biosphere where the impact of pollution on plants can be determined by introducing the pollutants at various levels. Simulation studies may also include gathering data from fluid-modeling facilities, where the environment is replicated using miniature models and atmospheric conditions are controlled. These studies often contribute to an understanding of the dispersion and deposition of air pollutants.

Monitoring the atmosphere is an essential component of air pollution studies involving computer simulations and mathematical modeling. These types of studies rely largely on data sources or values from the ambient environment and are constrained by difficulties of measuring ambient levels. Sometimes, vessels containing samples of air are collected and returned to the laboratory for analysis, but continuous monitoring devices that are placed in the ambient environment are more common.

Many of the monitoring devices involve a colorimetric or photometric technique. Air to be analyzed is isolated and subjected to conditions in which carefully selected gas phase reactions can produce specific compounds from the pollutants that are present. The reaction product is then analyzed by photometric techniques, in which the concentration of light-absorbing substances is indicated by the light intensity that reaches the photometer. Particulate matter can be measured by fairly simple collectors that may use adhesive coated paper or filtration, and measuring the increase in weight resulting from the trapped particles. Another method involves passing a known volume of air through filter paper and measuring the intensity of light passing through it. The intensity of light indicates the scattering and absorptive properties of aerosols; it is expressed as a coefficient. Instruments may be located at the surface, mounted on airplanes, or allowed to ascend in balloons. Acidity in precipitation is determined by standard measures of acidity using a pH indicator.

In 2016, researchers found that exposure to air pollution increased the risk of developing metabolic dysfunction and childhood obesity. Specifically, when very young children breathe polluted air, they are much more prone to develop diet-induced weight gain and are more susceptible to insulin resistance as adults. For the study, pregnant laboratory rats and their newborns were placed in chambers containing either the highly polluted air of Beijing, China, or air that had been filtered with most pollutants removed. After almost three weeks of exposure to polluted air, the pregnant rats had heavier lungs and livers and significantly higher cholesterol and triglyceride levels. Additionally, their insulin resistance level, which medical professionals know is a precursor to Type 2 diabetes, was significantly high than their counterparts who were breathing the filtered air.

SIGNIFICANCE

Many industry officials continue to downplay the threat of pollution to the atmosphere and dispute the extent of damage caused by pollution. Yet as greater amounts of pollutants are released into the atmosphere, it becomes increasingly difficult to control their levels and reverse any resulting damage. The depletion of the ozone layer and increasing levels of smog and acid rain are all very real and threatening manifestations of air pollution. Efforts to protect the atmosphere must be made on a worldwide basis, as gases cannot be confined to political boundaries. Air pollution may be lowered by reducing emissions or by extracting pollutants from the atmosphere through natural means. Thus any plans to reduce air pollution should center on one or both of these approaches.

—*Jasper L. Harris*

FURTHER READING

Bandy, A R., and Priestley Conference (7, 1994, Lewisburg, Pa.). "The Chemistry of the Atmosphere: Oxidants and Oxidation in the Earth's Atmosphere ; [Proceedings of the 7th BOC Priestley Conference Organized by the Royal Society of Chemistry..., Held at Bucknell University, Lewisburg, Pennsylvania, USA on 24–27 June 1994." Royal Society of Chemistry, 1998.

Bolius, David. *Paleoclimate Reconstructions Based on Ice Cores: Results from the Andes and the Alps.* SVH-Verlag, 2010.

Brimblecombe, Peter, et al., eds. *Acid Rain: Deposition to Recovery.* Dordrecht: Springer, 2010. Print.

"Exposure to Air Pollution Increases the Risk of Obesity." *Duke Today.* Office of News and Communications, 19 Feb. 2016. Web. 6 June 2016.

Gomez, Alan. "EPA Proposes New Limits to Ozone Air Pollution." *USA Today.* Gannett, 26 Nov. 2014. Web. 4 Oct. 2014.

Haerens, Margaret. *Global Viewpoints: Air Pollution* Farmington Hills: Greenhaven, 2011.

Larssen, Thorjørn, et al. "Acid Rain in China." *Environmental Science & Technology* 40 (2006): 418–25.

McGranahan, Gordon, and Frank Murray, eds. *Air Pollution and Health in Rapidly Developing Countries* Earthscan, 2003.

Nadadur, Srikanth S., and John W. Hollingsworth. *Air Pollution and Health Effects.* London: Humana Press, 2015. Print.

Ostmann, Robert. *Acid Rain: A Plague upon the Waters.* Dillon, 1982.

Pendergrass, Angeline G., and Dennis L. Hartmann. "Changes in the Distribution of Rain Frequency and Intensity in Response to Global Warming." *Journal of Climate* 27.22 (2014): 8372–83. Print.

Radovanović, Milan M., Vladan Ducić, and Saumitra Mukherjee. "Climate Changes Instead Of Global Warming." *Thermal Science* 18.3 (2014): 1055–61. Print.

Sportisse, Bruno. *Fundamentals in Air Pollution: From Processes to Modelling.* New York: Springer, 2009. Print.

Stolarski, Richard S. "The Antarctic Ozone Hole." *Scientific American* 258 (January, 1988): 30–36. Print.

Tiwary, Abhishek, and Jeremy Colls. *Air Pollution: Measurement, Modelling and Mitigation.* New York: Routledge, 2010. Print.

Vallero, Daniel. *Fundamentals of Air Pollution.* 4th ed. Burlington: Academic Press, 2008. Print.

ALBEDO FEEDBACK

FIELDS OF STUDY

Observational Astronomy; Climate Modeling; Oceanography; Atmospheric Chemistry

SUMMARY

Albedo (Latin for "whiteness") is a measure of the amount of incident radiation, such as sunlight, that a surface reflects. Arctic ice, for example, has a high albedo, meaning that most of the light hitting it is reflected back into space, so little of the solar energy striking ice is absorbed into the Earth's surface in the form of heat. When the ice melts into liquid water, however, surfaces darken and absorb more light, warming the planetary surface. As the ocean surface warms, more of the remaining ice melts, further

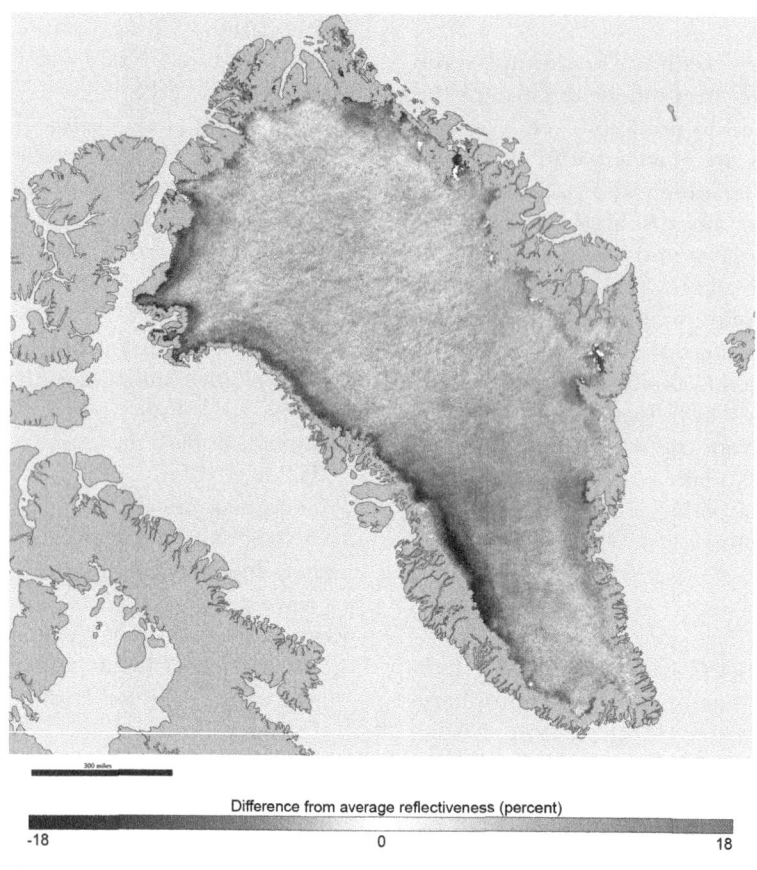

The map shows the difference between the amount of sunlight Greenland reflected in the summer of 2011 versus the average percent it reflected between 2000 to 2006.

increasing the amount of solar heat being absorbed. This cycle in which lower albedo creates conditions that cause a further decrease in albedo is an example of a positive feedback loop.

PRINCIPAL TERMS

- **albedo:** a value between zero and one representing the fraction of light reflected from a body such as a planet.
- **greenhouse effect:** an atmospheric phenomenon in which certain molecular components of the atmosphere capture infrared radiation from a planet's surface and prevent it from radiating back out into space.
- **greenhouse gas:** an atmospheric gas whose molecules absorb infrared radiation from a planet's surface and then emit a significant portion of it back into the atmosphere, most commonly carbon dioxide, methane, water and ozone.

THE ALBEDO AND FACTORS AFFECTING IT

Albedo, measured as the fraction of solar radiation reflected by a surface or object, is often expressed as a percentage. Snow-covered surfaces have a high albedo, the surface albedo of soils ranges from high to low, and vegetation-covered surfaces and oceans have a low albedo. Earth's planetary albedo varies mainly through varying cloudiness, snow, ice, leaf area, and land cover. The warming of the Arctic influences weather throughout North America. Outbreaks of Arctic cold can and have become weaker as ice coverage erodes.

Earth's albedo usually changes in the cryosphere (ice-covered regions), which has an albedo much greater (at around 80 percent) than the average planetary albedo (around 30 percent). In a warming climate, the cryosphere shrinks, reducing the Earth's overall albedo, as more solar radiation is absorbed elsewhere to warm Earth still further. Cloud cover patterns may also change, resulting in further albedo feedback. Changes in albedo have an important role in changing temperatures in any given location and, thus, the speed with which ice or permafrost melts.

SIGNIFICANCE FOR CLIMATE CHANGE

The rate of Arctic warming around the beginning of the twenty-first century has been eight times the average rate during the twentieth century. Changes in albedo are among the factors contributing to this increase. The number and extent of boreal forest fires have also grown, increasing the amount of soot in the atmosphere and decreasing Earth's albedo. As high latitudes warm and the coverage of sea ice declines, thawing Arctic soils also may release significant amounts of carbon dioxide (CO_2) and methane that have been trapped in permafrost for millenia.

Arctic warming has shortened the region's snow-covered season by roughly 2.5 days per decade, increasing the amount of time during which sunlight is absorbed. Gradual darkening of Arctic surfaces thus produces significant changes in the total amount of solar energy that the area absorbs. Scientists have estimated this increase in surface energy absorption at 3 watts per square meter per decade. This means that, in areas such as the Arctic where albedo has changed markedly, the effect of this change on climate has been roughly equal to the effect of doubling atmospheric CO_2 levels. Moreover, the continuation of contemporary trends in shrub and tree expansion would amplify atmospheric heating by two to seven times.

Changes in albedo over a broad area, such as the Arctic, can produce a significant effect, allowing Earth to be "whipsawed" between climate states. This feedback has been called the "albedo flip" by James E. Hansen, director of the Goddard Institute for Space Studies (GISS) of the National Aeronautics and Space Administration (NASA). The flip provides a powerful trigger mechanism that can accelerate rapid melting of ice. According to Hansen, greenhouse gas (GHG) emissions place Earth perilously close to dramatic climate change that could run out of control, with great danger for humans and other creatures. Changes in albedo have been greatest in Earth's polar regions, especially in the Arctic, where snow and ice are being replaced during summer (a season of long sunlight) by darker ocean or bare ground.

Changes in albedo also play a role in increasing Arctic emissions of methane, tropospheric ozone (O_3), and nitrous oxide (N_2O). All of these are GHGs. Tropospheric ozone is the third most influential anthropogenic GHG, after CO_2 and methane.

Black carbon (soot) also has a high global warming potential and deserves greater attention, according to Hansen. Soot's albedo causes massive absorption of sunlight and heat and compounds warming effects, especially in the Arctic. Increases in soot due in part to combustion of GHGs can play a role in accelerating climate change to tipping points, at which feedbacks take control and propel increasing levels of GHGs past a point where human control, or "mitigation," is possible.

—*Bruce E. Johansen*

FURTHER READING

Alley, Richard B. *The Two-Mile Time Machine. Ice Cores, Abrupt Climate Change and Our Future* Princeton University Press, 2014. Explains the manner in which past climate conditions are recorded, trapped in ice formed over long periods of time, and the analytical methods used to extract that information.

Appenzeller, Tim. "The Big Thaw." *National Geographic*, June, 2007, 56-71. Surveys ice melt in the Arctic, Antarctic, and mountain glaciers. Explains albedo feedback, describing why the process has been accelerating.

Kaper, Hans and Engler, Hans *Mathematics and Climate* Society for Industrial and Applied Mathematics, 2013. Discusses the mathematics of climate modeling and the predictive nature of climate statistics.

Pearce, Fred *With Speed and Violence. Why Scientists Fear Tipping Points in Climate Change* Beacon Press, 2007. Explains the dynamic relationships between glaciers, polar ice caps and climate change, and the probable effects arising from disruption of the dynamic balance between ocean water and fresh water.

Pierrehumbert, Raymond T. *Principles of Planetary Climate* Cambridge University Press, 2010. Discusses many aspects of Earth's climate from first principles to contemporary models of climate change.

ANTHROPOGENIC AIR POLLUTION AND POLLUTANTS

FIELDS OF STUDY

Environmental Chemistry; Photochemistry; Environmental Sciences; Waste Management; Engineering; Meteorology; Atmospheric Science; Atmospheric Chemistry; Toxicology

SUMMARY

Air pollution has been a problem since humans began burning carbon-based fuels while living in large cities. The first known air-pollution ordinance was passed in London, England, in 1273 in an attempt to alleviate the soot-blackened skies caused by excessive combustion of wood in the heavily populated city. From the mid-eighteenth through the mid-twentieth centuries, the increasingly heavy use of coal for heat, electricity, and transportation resulted in filthy cities and an escalating crisis of respiratory diseases. It was not until the latter half of the twentieth century that governments began attacking the problem by enacting legislation to control noxious emissions at their source.

PRINCIPAL TERMS

- **aerosols:** minute particles or droplets of liquid suspended in Earth's atmosphere
- **anthropogenic:** deriving from human sources or activities
- **chlorofluorocarbons (CFCs):** chemical compounds with a carbon backbone and one or more chlorine and fluorine atoms
- **greenhouse effect:** global warming caused by gases such as carbon dioxide that trap infrared radiation from Earth's surface, raising atmospheric temperatures
- **ozone:** a highly reactive molecule consisting of three oxygen atoms
- **parts per million:** number of molecules of a chemical found in one million molecules of the atmosphere

CLEAN AIR

Before discussing anthropogenic air pollution, one must first define "clean air." Earth's atmosphere is approximately 78 percent nitrogen (N_2), 21 percent oxygen (O_2), and 1 percent argon. These concentrations may be reduced slightly by water vapor, carbon dioxide (CO_2) and methane (CH_4) which occur naturally in the environment and can make up between 1 percent and 3 percent of the atmosphere. In addition, there are many trace elements present in the atmosphere in concentrations so small that they are measured in parts per million. Among the trace elements near Earth's surface are 0.52 part per million of oxides of nitrogen and 0.02 part per million of ozone, both of which occur both naturally and anthropogenically. This combination of N_2, O_2, argon, water, ozone, and oxides of nitrogen constitutes clean air. Any change in these concentrations or introduction of other compounds into the atmosphere constitutes air pollution, which occurs in one of two forms: gases and particulate matter.

Gaseous Air Pollutants

The primary gaseous pollutants are carbon, sulfur and nitrogen oxides and ozone. Carbon oxides occur whenever a carbon-containing fuel is burned; in general, a carbon-based fuel unites with oxygen to yield CO_2 and water vapor. If the combustion is incomplete as a result of insufficient oxygen, carbon monoxide will also be produced. Although CO_2 is a relatively benign compound, the vast amount of fossil fuels (coal, petroleum distillates and natural gas) burned since the Industrial Revolution began have raised the atmospheric concentration of CO_2 from about 280 parts per million (ppm) to more than 400 ppm. In May, 2017, it was reported to have exceeded 410 ppm. While transparent to visible light coming from the Sun, CO_2 molecules reflect infrared radiation emitted from Earth's surface after the visible light has been absorbed and radiated at heat-producing infrared wavelengths, thus raising Earth's temperature in proportion to the amount of CO_2 in the atmosphere. As CO_2 concentrations continue to increase, this 'greenhouse effect' will continue to increase Earth's average temperature, causing droughts, more frequent severe storms of greater intensity, the shifting of climate zones, and rising sea levels.

Carbon monoxide (CO) is a toxic compound that can cause death by suffocation even when present in relatively small amounts. CO is two hundred times more reactive with hemoglobin than is oxygen; thus CO replaces oxygen in the bloodstream, depriving cells of their necessary oxygen. Deprived of sufficient blood oxygen, an organism will die in about ten minutes.

Since almost all coal contains sulfur, burning coal causes sulfur to react with oxygen to create sulfur dioxide (SO_2), which reacts with water vapor in the atmosphere to produce H_2SO_4, sulfuric acid. This pollutant reaches Earth's surface as a component of rain (acid rain), and it pollutes rivers, lakes, and other bodies of water.

Nitrogen oxides are synthesized whenever air is rapidly heated under pressure and then cooled quickly, as occurs in automobile cylinders and thermoelectric power plants. The two main compounds of this pollution are nitric oxide (NO) and nitrogen dioxide (NO_2); both are toxic, but NO_2 is worse (in equivalent concentrations, it is more harmful than CO). Nitrogen dioxide affects the respiratory system and can lead to emphysema, while nitric oxide often combines with water and oxygen to form nitric acid (HNO_3), another component of acid rain.

NO_2 can also combine with oxygen to form NO and ozone (O_3), a very reactive and dangerous form of oxygen. Combustion-caused ozone is undesirable near Earth's surface, but the compound occurs naturally in the upper atmosphere (about 19 kilometers above the surface) when energetic ultraviolet (UV) light from the Sun interacts with oxygen. Although the ozone composing it constitutes less than 1 part per million of Earth's atmosphere, the ozone layer plays an extremely important role. It prevents most of the Sun's UV light from reaching Earth's surface, a highly desirable effect since it is UV radiation that causes sunburn and skin cancer in small doses, and if the ozone layer were to disappear the full influx of ultraviolet radiation would quickly sterilize the entire planet.

Chlorofluorocarbons

When first synthesized in the 1930's, chlorofluorocarbons (CFCs) were hailed as ideal refrigerants (known collectively as Freon), because they are nontoxic, noncorrosive, nonflammable, and inexpensive to produce. Later, pressurized CFCs were used as the propellant in aerosol cans and as the working fluid for air conditioners. In 1974, the chemists Mario Molina and F. Sherwood Rowland proposed that the huge quantities of CFCs released into the atmosphere from aerosol sprays (500,000 metric tons in 1974 alone) and discarded refrigerant units were slowly migrating to the stratosphere. There, the CFCs were decomposed by the highly energetic UV radiation from the Sun, releasing large quantities of ozone-destroying chlorine.

Any decrease of the ozone layer could increase the incidence of skin cancer, damage crops, and decimate the base of the marine food chain. The reduction of ozone was most pronounced over Antarctica, where an "ozone hole," first detected in the early 1970's, was increasing in size annually. Pressured by environmentalists and consumer boycotts, in 1978 the U.S. government lead several other nations in imposing a ban on aerosol cans and refrigeration units utilizing CFC propellants, forcing the chemical industry to support the ban and to develop alternatives. By 1987, the depletion of the ozone layer had become so problematic that most CFC-using nations met in Montreal, Canada, to produce an international treaty calling for immediate reductions in all CFC use, with a complete phase-out by 2000. This Montreal Protocol, by 2001, had limited the damage to the ozone layer to about 10 percent of what it would have been had the agreement not been ratified.

Smog

The word "smog" is a melding of "smoke" and "fog." When a local atmosphere becomes stagnant—for example, during a temperature inversion—pollution levels in the smog can become severe enough to merit being called "killer fogs." At least three times during the twentieth century, such killer fogs have caused a

The Great Wall of China climbs over rugged mountain terrain at Badaling, north of Beijing. Even on a relatively clear day, smog still hovers over the countryside. (Sean O'Keefe)

statistically significant increase in the death rate, particularly among the old and those with respiratory problems. The first documented killer fog occurred in 1948 at Donora, Pennsylvania, when a four-day temperature inversion stagnated a fog that became progressively more contaminated with the smoky effluents of local steel mills. The second documented case occurred in 1952 in London, England, when fog, trapped by another four-day temperature inversion, mixed with the smoke pouring from thousands of chimneys where coal was being burned. Many elderly people and people with respiratory ailments succumbed to these deadly events. Finally, during Thanksgiving, 1966, New York City experienced an increased death rate due to a choking smog.

A second, completely different type of smog is photochemical smog, a noxious soup of reactive chemicals created when sunlight catalyzes reactions of hydrocarbons and nitrogen oxides. This catalysis first occurred in Los Angeles in the late 1940's, when automotive traffic increased drastically, emitting thousands of metric tons of exhaust daily. As mentioned above, car engines, in addition to emitting carbon oxides, emit nitrogen oxides, ozone, and some residual unburned hydrocarbons from the fuel. When light acts on these chemicals, it produces photochemical reactions that create aldehydes (compounds, such as formaldehyde, that are well known for their obnoxious odors) and other dangerous compounds that can induce respiratory ailments, irritate eyes, damage leafy plants, reduce visibility, and crack rubber. Although photochemical smog was first observed in Los Angeles because of the abundant sunlight and heavy automotive traffic, it has since become prevalent in many other large cities.

Particulates

Particulate matter consists of soot, fly ash, or any other small particles or aerosols suspended in the air that can be breathed into the lungs or ingested with food. It is generated by combustion, dry grinding processes, spraying, and wind erosion. Particulate concentrations in the body can, over time, lead to cancer of the stomach, bladder, esophagus, or prostate.

The human respiratory system has evolved a mechanism to filter out and prevent certain sizes of particulates from reaching the lungs. The first line of defense is the nose and nasal passageway, whose mucus membranes and hairs will catch and remove particles larger than 10 microns (one one-hundredth of a millimeter). After passing through the nasal passages, air travels through the trachea, which branches into the right and left bronchi. Each bronchus is divided and subdivided about twenty times, terminating in the small bronchioles located inside the lungs. These end in 300 million tiny air sacs called alveoli, where oxygen is passed to the bloodstream and CO_2 removed for exhalation.

Particles ranging in size from 2 to 10 microns usually settle on the walls of the trachea, bronchi, and bronchioles, before reaching the alveoli. They are eventually expelled by ciliary action, a cough, or a sneeze. Particles smaller than 0.3 micron are likely to remain suspended in inhaled air and then removed from the lungs with exhaled air, similarly failing to enter the bloodstream. Humans thus have evolved a protective mechanism that shields them from particles of all sizes smaller than 0.3 micron and larger than 2 microns. No defense mechanism evolved for this intermediate size range, because during the long course of human evolution there were very few particles of this size in the environment. In recent centuries, however, many particles in this range—including coal dust, cigarette smoke, and pesticide dusts—have been added to the environment. Since no natural defense exists to eliminate these hazards from the human body, they coat the alveoli, causing such illnesses as black lung, lung cancer, and emphysema.

Context

The issue of whether or not global warming is caused by humans is still being debated, but strong measures were taken in the latter half of the twentieth century to control the noxious gases and particulate emissions known as air pollutants. When it was discovered that the ozone layer was being depleted by CFCs, the Montreal Protocol was ratified by most industrial nations. Both of these historic precedents indicate that strong, effective action and international cooperation is possible when a perceived threat to humanity and the environment is grave enough. Since the preponderance of scientific evidence seems to suggest that global warming is due to humanity's excessive use of fossil fuels, perhaps it would be prudent to err on the side of caution and begin to curtail the disproportionate dependence on nonrenewable resources.

—*George R. Plitnik*

FURTHER READING

Abbasi, S.A. And Abbasi, Tasneem *Ozone Hole Past, Present, Future* Springer, 2017. Describes the observational studies from ground and satellite that identified and continue to monitor the "ozone hole", with discussion of the efforts made to reverse and stabilize the ozone layer.

Jacobson, Mark Z. *Air Pollution and Global Warming. History, Science and Solutions* 2nd ed. Cambridge University Press, 2012. Discusses the relationship between the atmosphere and the ozone layer in the broader context of climate change.

Matyssek, R., Clarke, N., Cudlin, P., Mikkelsen, T.N., Tuovinen, J-P., Wieser, G. and Paoletti, E, eds. *Climate Change, Air Pollution and Global Challenges* Elsevier, 2013. Examines the relationship between atmospheric pollution and climate change as a global problem.

Metcalfe, Sarah and Derwent, Dick *Atmospheric Pollution and Environmental Change* Routledge, 2014. Discusses the manner in which air pollution effects and is affected by environmental change.

Tan, Zhongchao *Air Pollution and Greenhouse Gases* Springer, 2014. This book describes the sources and role of greenhouse gases in the atmosphere.

Vallero, Daniel *Fundamentals of Air Pollution* 5th ed. Elsevier, 2014. This is a basic introduction to the fundamental study of air pollution and its effects.

Yu, Ming-Ho, Tsunoda, Humio and Tsunoda, Masashi *Environmental Toxicology. Biological and Health Effects of Pollutants* CRC Press, 2011. This volume provides a considerable source of reference data and discussion regarding the effects of various environmental and atmospheric pollutants.

ANTHROPOGENIC CLIMATE CHANGE

FIELDS OF STUDY

Environmental Science; Chemical Engineering; Physical Chemistry; Photochemistry; Waste Management; Petroleum Refining; Chemical Kinetics; Climate Modeling; Atmospheric Science; Earth System Modeling; Oceanography; Computer Science

SUMMARY

Human impact on Earth's climate extends many centuries into the past. However, at the beginning of the Industrial Revolution, in the mid-eighteenth century, the pace of human effects on the natural environment increased. The invention of power sources such as the steam engine, the internal combustion engine, and systems for delivering electric power sped anthropogenic environmental alterations in three ways. First, these technologies made it easier to modify the landscape; second, they created a vast demand for energy to fuel them; and finally, emissions from steam and internal combustion engines altered the composition of the atmosphere.

PRINCIPAL TERMS

- **aerosols:** tiny particles suspended in Earth's atmosphere
- **albedo:** a value between zero and one representing the fraction of incident light reflected from a body such as Earth.
- **anthropogenic:** deriving from human sources or activities.
- **fossil fuels:** energy sources such as coal, oil, and natural gas that were formed by the chemical alteration of plant and animal matter under geologic pressure over long periods of time.
- **global dimming:** a reduction in the amount of sunlight reaching the surface of the Earth.
- **greenhouse gases (GHGs):** atmospheric gases that trap heat within a planetary system rather than allowing it to escape into space; principally carbon dioxide, methane, water and ozone.
- **urban heat island:** a spot on Earth's surface that is significantly warmer than the surrounding area as a result of human alterations to the landscape, primarily cities.

HUMAN CAPACITY TO AFFECT CLIMATE CHANGE

By far the largest source of energy available to Earth is the sun. With the single exception of nuclear energy, solar energy is the ultimate source of all energy on Earth's surface, including wind, hydroelectric,

Global Anthropogenic GHG Emissions, 2004

Gas	Percent of Global Emissions
Carbon dioxide (burning fossil fuels)	57
Carbon dioxide (deforestation and biomass decay)	17
Methane	14
Nitrous oxide	8
Carbon dioxide (other sources)	3
F-gases	1
Total	100

Data from Intergovernmental Panel on Climate Change.

and biomass energy. After solar energy, the most important energy source in the earth is geothermal energy, maintained and generated by gravity and the decay of radioactive elements within the mantle. Tidal energy is also generated from interactions between Earth, Moon, and Sun. The amount of solar energy available on Earth is about twenty-seven hundred times the amount of energy provided by geothermal heat and twenty-nine thousand times the amount of tidal energy.

Humans can affect climate by modifying Earth's ability to absorb or reflect heat from the sun and by modifying the ability of the atmosphere to retain heat. Compared to those effects, the direct production of heat by human activities is insignificant. The annual supply of solar energy is more than eight thousand times the total amount of human energy use. Human energy use in one year is equal to roughly one hour of global sunlight. By comparison, the total energy in the world's nuclear arsenals is only about 0.0001% of the annual input from the sun, or less than an hour's worth of sunlight. A large-scale nuclear war would inject smoke and dust into the atmosphere that would have a far greater effect on climate than would the heat given off by the nuclear explosions themselves.

During the 1991 Persian Gulf War, the smoke emitted from burning oil wells in Kuwait had a strong cooling effect on areas under the smoke layer, but the heat from the oil fires had negligible effects on local or regional weather.

How Human Changes to Earth's Surface Affect Climate

When humans modify the landscape, or surface, of Earth, they trigger climate change in a number of ways. Replacing forest with cleared land for use as farmland or pasture, or simply by felling timber for wood or fuel, increases the albedo of Earth's surface: More sunlight is reflected back into space and less is absorbed and retained by the Earth. Old forests, especially conifer forests, are generally darker and less reflective, whereas younger growth or cleared land is much lighter. Generally, making the surface lighter reflects more solar energy back into space and contributes to climatic cooling, whereas a darker surface contributes to warming.

Many human changes to the landscape affect local climate. One of the most familiar examples is the urban-heat-island effect. Cities impede the flow of air, trapping heat. They also contain large areas of materials such as asphalt and concrete that absorb heat. As a result, cities tend to be significantly warmer than nearby countryside. Other effects associated with urban heat islands include increased rainfall and stalling of intense rainstorms.

Among the most extreme climatic effects associated with human changes to Earth's surface are those due to the drying of the Aral Sea in central Asia. Since 1960, diversion of water for agriculture has reduced the area of the Aral Sea (actually a vast lake) by more than 80 percent. In 2014, the Aral Sea's eastern basin of freshwater was completely dry for the first time in 600 years. The original size of the Aral Sea was about 68,000 square kilometers, making it the fourth largest lake in the world, large enough to have significant moderating effects on the regional climate. These effects have almost entirely disappeared with the drying of the lake, and the climate has become more continental, with hotter summers, colder winters, and much less rainfall. Similarly, increased drought has been linked to the vast Three Gorges Dam hydroelectric project in China. The Three Gorges Dam has artificially and significantly altered the natural terrain, resulting in elongation of the vent channel, and disturbing circulation of water vapor. This has altered the temperature balance of the area and lead to drought.

Indirect Human Modifications of the Atmosphere

Human changes to the landscape also affect the atmosphere. Human activities that create dust and smoke, for example, can reflect sunlight back into space and block sunlight from reaching Earth's surface. Such activities thus increase Earth's albedo and act to cool the planet. But they can also prevent radiation emitted from the surface from going back into space and thus trap heat, acting to warm the planet. For example, studies of condensation trails, or contrails, of aircraft have shown that they reflect sunlight back to space but also prevent heat from the surface from escaping, so that their overall effect is to warm the Earth. Tiny particles, or aerosols, can also serve as nuclei for the condensation of water droplets and affect fog, cloud cover, or precipitation.

The degree to which human dust and smoke alter visibility is remarkable. In preindustrial times, it was normal in most places for visibility to exceed 100 kilometers. The Great Smoky Mountains were so named precisely because the persistent haze in the valleys, due to natural emissions by the forests, was unusual. Persistent haze unrelated to local weather was so unusual in preindustrial times that it was recorded by chroniclers and has been used by geologists to pinpoint the dates of large volcanic eruptions in remote areas of the world. Visibility in heavily populated contemporary industrial regions is often only a few kilometers, and even remote national parks in the western United States are threatened with diminished visibility. Some studies have suggested that global dimming, the reduction of sunlight reaching the surface of the Earth by dust, haze, and smoke, may have masked the effects of global warming.

Human changes to the landscape often release greenhouse gases (GHGs). These changes generally result in the destruction of biomass, either by burning or by decay, thereby adding carbon dioxide (CO_2), one of the six major GHGs, to the atmosphere. Drainage of wetlands for agriculture can also result in the decomposition of organic material and also adds CO_2 to the atmosphere. These are, to a large extent, considered to be carbon neutral, since all of the carbon contained in vegetable matter is from carbon dioxide that was captured from the atmosphere in the process of photosynthesis. On a long-term average basis, they do not represent any net change to the carbon dioxide content of the atmosphere. However, in the short-term such introductions of carbon dioxide as would come from burning a large quantity of biomass are able to produce significant spikes in the atmospheric carbon dioxide content.

Human activities are responsible for the release of other GHGs, particularly methane, which is a much more effective GHG than carbon dioxide. Modification of the land can release methane trapped in the soil. Agriculture increases the amount of methane in the atmosphere in several ways. Livestock produce large amounts of methane in their digestive tracts, so increased numbers of cattle lead to increased methane emissions, although it is debatable whether large herds of domestic cattle have any more significant effect than the massive herds of wild animals that preceded them. Clearing of forest lands for agriculture reduces the ability of soils to absorb and oxidize methane. Certain types of agriculture, notably rice production, create oxygen-poor conditions for the decay of organic materials and thus emit methane. Finally, burial of waste rather than incineration results in methane emission. Large quantities of methane are sequestered on the floors of large bodies of water, particularly as methane hydrate, and is occasionally released suddenly as a result of human activity. Alarms have also been raised with regard to the rapid release of the large quantities of methane that have been trapped in permafrost for many centuries. Thawing of the permafrost in Siberia has resulted in the explosive release of methane gas that has accumulated underground, powerful enough to create large craters many meters deep.

According to a controversial theory by William Ruddiman, if it were not for human activities, the Earth would already have passed the peak of the present interglacial period and would be on the way to the start of the next glacial advance. Ruddiman argues that, while clearing forests for agriculture increased the CO_2 content of the atmosphere, increased methane production, especially that due to rice cultivation, is the more important climatic change agent.

Direct Human Changes to the Atmosphere

Beginning with the Industrial Revolution, human activities began modifying the composition of the atmosphere directly on a large scale. The burning first of wood and then of coal and other fossil fuels released increasing amounts of smoke and gases directly into the atmosphere. Among the most important emissions were CO_2, nitrogen oxides, sulfur dioxide, and ozone-depleting chemicals.

The most important anthropogenic GHG contributing to the enhanced greenhouse effect is CO_2. The enhanced greenhouse effect supplements Earth's already significant natural greenhouse effect, in the context of which water vapor is the single most important GHG in Earth's atmosphere.

Nitrogen oxides are the result of high-temperature combustion during which atmospheric nitrogen and oxygen combine. Ordinary fires are typically not hot enough to drive reactions between nitrogen and oxygen, but at high temperatures the two gases can react. Lightning is a natural source of nitrogen oxides, but human activities also create significant quantities that are subsequently injected into the atmosphere, especially from internal combustion engines. Nitrogen oxides react with water to produce nitric acid and with hydrocarbons to generate 'smog' as well as ozone in the lower atmosphere. Ozone high in the atmosphere protects Earth's surface from ultraviolet light, but at ground level ozone is a powerfully oxidizing pollutant that contributes to respiratory problems.

Sulfur dioxide is emitted naturally by volcanoes but also is produced by smelting of sulfide ores or burning fossil fuels that contain sulfur. All fossil fuels contain varying amounts of sulfur due to their origin as once-living species and the presence of sulfur atoms in certain amino acids and proteins. Sulfur dioxide is a GHG, but its most important environmental effect is that it combines with water vapor to create sulfuric acid. Tiny droplets or aerosols of sulfuric acid can aggravate respiratory problems, contribute to atmospheric haze, and make rain and snow more acidic, contributing to acid precipitation. The high albedo and run-away greenhouse effect that characterize the atmosphere of the planet Venus are both due to the dense sulfuric acid clouds that blanket that planet.

Ozone-depleting chemicals include a large number of synthetic chemicals, mostly organic chemicals containing chlorine or bromine. Both of these elements are highly effective at destroying ozone through free radical reactions driven by sunlight. Ozone-depleting chemicals such as the now banned chlorofluorocarbon compounds are sufficiently stable to enable them to reach the high altitudes of the ozone layer. International controls on ozone-depleting chemicals, such as the Montreal Protocol and the Kyoto Protocol, have slowed the depletion of stratospheric ozone, but the quantities of such chemicals in the atmosphere continue to have a significant, though diminishing, effect. Although ozone-depleting chemicals also act as GHGs, in most respects global warming and ozone depletion are separate problems.

Context

Even though the amount of CO_2 and other GHGs in the atmosphere continues to increase and strong evidence exists that the global climate has become warmer in the recent past, the actual web of cause and effect relating to climate change is extremely complex. Many unanswered questions remain. It is currently projected that rising sea levels due to water from melting ice cover as the climate warms will render many atolls, low-lying islands and coastal areas uninhabitable within the next 15 years. The consequences of making wrong policy decisions about climate change are therefore very serious. If human activities are forcing climate change, then failure to act effectively will lead to catastrophic environmental, social, and economic changes. If human activities contribute only insignificantly to climate change, then attempting to halt climate change through government policies could have catastrophic economic effects without ameliorating climate change. The enormous stakes and the sheer complexity of climate are the reasons that the debate about global warming is so fierce.

—*Steven I. Dutch*

FURTHER READING

Arnold, Dennis G., ed. *The Ethics of Global Climate Change* Cambridge University Press, 2011. Discusses the ethics of wealthy nations contributing disproportionately to global climate change that affects the poorest nations and the majority of the global population.

Baer, Hans A. *Global Capitalism and Climate Change. The Need for an Alternative World System* Alta Mira Press, 2012. Examines the manner in which the current global economic system directs the course of global climate change.

Darling, Seth B. and Sisterson, Douglas L. *How to Change Minds About Our Changing Climate* The Experiment LLC, 2014. Addresses the issue of climate change denial in the face of the demonstrable effects of climate change.

Dessler, Andrew *Introduction to Modern Climate Change* 2nd ed., Cambridge University Press, 2016. Provides a good overview of the science and sociology of climate change.

Neelin, J. David *Climate Change and Climate Modeling* Cambridge University Press, 2011. Discusses how climate data is used to identify climate change and to produce predictive models of future climate states.

Washington, Haydn and Cook, John *Climate Change Denial. Heads in the Sand* Earthscan, 2011. Addresses the dangers of policy that denies the reality of climate change.

Zedillo, Ernesto, ed. *Global Warming. Looking Beyond Kyoto* Brookings Institute Press, 2008. Discusses the measures agreed to by nations that signed the Kyoto Accord on Climate Change, and measures that might be taken to address the issue beyond those of the Kyoto Accord.

ARAL SEA DESICCATION

FIELDS OF STUDY

Hydrology; Environmental Sciences; Environmental Studies

SUMMARY

The dramatic drying of Central Asia's Aral Sea is sometimes called one of the greatest ecological disasters of the twentieth century. Over the past ten thousand years, the area and volume of this internal lake have greatly fluctuated as a result of both natural and anthropogenic forces. Anthropogenic forces in particular greatly reduced the sea at the end of the twentieth century, from 67,500 square kilometers in 1960 to 17,382 square kilometers in 2006. It continues to reduce in size today as a result of changing climate, with the eastern arm having recently gone dry for the first time in the past 600 years.

PRINCIPAL TERMS

- **anthropogenic:** caused or produced by humans
- **hazardous:** poisonous, corrosive, flammable, explosive, radioactive, or otherwise dangerous to human health
- **herbicides:** substances or preparations for killing plants, such as weeds
- **pesticides:** chemical preparations that kill pests, including unwanted animals, fungi, and plants
- **Pleistocene epoch:** first half of Quaternary period, beginning about two million years ago and ending about ten thousand years ago

- **Pliocene epoch:** Tertiary period, beginning about ten million years ago and ending about two million years ago, known for its cool climate, mountain building, and increased mammal populations

THE ARAL SEA IN CONTEXT

The Aral Sea is located in an area of cold temperatures and deserts: the Karakumy is to the south, the Kyzylkum Desert is to the southeast. The sea's 1.8-million-square-kilometer drainage basin encompasses six central Asian countries: Iran, Turkmenistan, Kazakhstan, Afghanistan, Tajikistan, and Uzbekistan (including the Karakalpak Autonomous Republic). Kazakhstan and Uzbekistan are physically adjacent to the Aral Sea. Although nine streams flow within the drainage basin, the Syr Dar'ya and the Amu Dar'ya are the major rivers. *Dar'ya* translates from the Turkic languages of central Asia as "river."

In 1918, the Soviet Union decided to develop the area around the Aral Sea to grow cotton. The decision was economic: Cotton, or "white gold," provided revenue for the government. Herbicides, pesticides, and fertilizers were heavily used to bolster crop production, and the Amu and Syr Dar'ya were diverted for irrigation.

The Soviet irrigation programs were inefficient, with open waterways and irrigation basins subject to evaporation. Open-air channels were dug through sandy deserts with the thirteen-hundred-kilometer Karakam Canal diverting between 20 and 30 percent of the Amu Dar'ya's flow west to Turkmenistan.

Orphaned ship in former Aral Sea, near Aral, Kazakhstan.

Between 1987 and 1989, the Aral Sea divided into the small Aral Sea in the north, fed by the Syr Dar'ya, and the large Aral Sea in the south, supplied by the Amu Dar'ya. Using the years 1960 to January, 2006, as a baseline, the water level of the little Aral fell by 13 meters, the large Aral by 23 meters.

DISPLACED FISHING INDUSTRY

Historically, the Aral Sea fishing industry employed several thousand workers and provided, according to commercial fishing reports, one-sixth of the Soviet fish supply. The lowered lake level reduced the industry and increased the distance between the lake and fishing ports. Decreased water flow to the river deltas and wetlands diminished fish spawning and feeding, so that of the thirty-two fish species formerly existing in the lake, only six survived. Those remaining survived by inhabiting small water areas of river deltas that play a large role in regenerating lake fish supplies. Commercial fisheries did not exist after the mid-1980s.

When lake water is reduced by evaporation and freshwater input is negligible, salts within the water are concentrated and approach the salinity of a typical ocean, 35 grams of salt per liter. Since the Aral Sea was becoming more saline in the 1970s, a saltwater fish, the Black Sea flounder (*Platichthys flesus lulscus*) was introduced into the sea. The intent was to enable the lake's fishing industry to survive, but by 2003 the flounder no longer existed in the Aral Sea, whose salinity had reached greater than 70 grams per liter.

HAZARDOUS LAKEBED DEPOSITS

As the Aral Sea shrinks, calcium sulfate, calcium carbonate, sodium chloride, sodium sulfate, and magnesium chloride are deposited on the exposed seafloor. In addition to these salts, pesticide residues of organochlorines, dichlorodiphenyl-tricholoethanes (DDT), hexachloro-cyclohexane compounds (HCH, Lindane), and toxaphene remain. Other toxic materials present are the result of biological weapons testing and failed industrial sites.

The region immediately surrounding the Aral Sea—Uzbekistan, Kazakhstan, and parts of Turkmenistan—are affected by hazardous dust and salt storms. Most major storms occur in a one-hundred-kilometer margin along the north-northeastern coastal zone. Some 60 percent of these storms trend southwesterly, for 500 kilometers, depositing salts, agrochemical dusts, and aerosols on the delta of the Amu Dar'ya. This southern river delta region is densely populated, so toxic storms affect human and animal health and economic stability. Another 25 percent of the storms trend west, moving over and beyond the Ust-Urt plateau, an area of livestock pastures.

Toxic dust storms harm the human food supply and physical health of domestic and wild animals. The human risk associated with airborne salt and dust is high, and greater than average incidences of respiratory illnesses, eye problems, throat and esophageal cancer, skin lesions and rashes, and liver and kidney damage are reported in the Aral Sea region.

ARAL SEA GEOLOGY

The Aral Basin has experienced geologic cycles of diversion and desiccation. During the Pliocene epoch, the ocean withdrew from Eastern Europe and Turkestan, leaving remnant basins such as the Aral, Caspian, and Black seas. Late Pliocene continental crust movements created a more permanent depression in the area of the Aral Sea, which was filled with water, some of which came from the ancestral Syr Dar'ya.

The effects of the Pleistocene epoch are recorded by terrestrial sedimentary deposits. Eolian processes operated in the Aral depression during the early and middle Pleistocene. During the late Pleistocene, fluvial processes filled the depression by inflows from the ancestral Amu Dar'ya; then, the basin was filled for a second time with waters from the Syr Dar'ya.

Both rivers affect lake level changes, but when the 2,525-kilometer course of the Amu Dar'ya migrates away from the basin, lake level drops. Diversions of the Amu Dar'ya are natural, resulting from filling fluvial channels during heavy rains or floods. Some river diversions are the result of human actions, such as improper or failing irrigation systems or intentional destruction of river dams and levees during political upheaval or war.

ARAL SEA RESTORATION

The Aral Sea cannot be reestablished to its pre-1960 status, because to do so would mean curtailing irrigation, which uses 92 percent of all Aral water withdrawals. Curtailment of irrigation would mean crop failure and economic and social collapse in the Aral Basin. After the fall of the Soviet Union in 1991, Kyrgyzstan, Uzbekistan, Turkmenistan, Kazakhstan, and Tajikistan joined together to address the Aral Sea crisis. Two major agencies were formed by these new regional states, with the International Fund for the Aral Sea (IFAS) taking the lead role in 1997. Also, the United Nations, the European Union, and many other international aid agencies operated to improve the region.

In order to regulate flow in the little Aral, Kazakhstan and the World Bank constructed an eighty-five-million-dollar, thirteen-kilometer earthen ditch connected to a concrete dam with gates and spillways. Completed in November, 2005, this system brought early success: The water level increased to 42 meters from 40 meters, and, by summer of 2006, the lake area increased by 18 percent. Salinity decreased by one-half to almost 10 grams per liter in 2006, but future levels will vary, by area, from 3 to 14 grams per liter. Decreased salinity increases fish population, aiding the fishing industry and the Kazakhstan economy. A former fish-processing plant has reopened in Aralsk, Kazakhstan, to process lake carp, Aral bream, Aral roach, Pike perch, and flounder—the top five species caught in autumn, 2007.

CONTEXT

The Aral Sea Basin has become synonymous with irreversible environmental disaster. The entire region demonstrates the potential for humans to act as geomorphic change agents. The area's inhabitants diverted rivers for irrigation; replaced desert vegetation with such crops as melons, cotton, and rice; and altered natural water chemistry to salinities greater than that of ocean water. Anthropogenic environmental degradation has also affected human health in the Aral Basin.

Populations around the sea are in a state of upheaval, dislocation, and poverty as a result of the collapse of the fishing industry and a lack of government support from the former Soviet Union since the early 1990s. Essential medications and adequate hospital facilities are not available when economic conditions are stagnant. Health problems begin in the youngest populations: High infant mortality, low birth weight, growth retardation, and delayed puberty are present in the basin.

Poor quality and insufficient quantities of drinking water in the basin have increased typhoid, hepatitis

A, and diarrhea in all age groups. High levels of mineralized water within the basin may contribute to kidney and liver diseases. Also increasing are acute respiratory diseases—killing almost one-half of all children. Dust storms sourced from former seabeds deliver salt and toxic chemical sediments to humans and areas of human habitation.

Possibly the greatest health risk is from pesticides, which contaminate the water and food supplies and infiltrate during dust storms. Pesticides may be applied to crops, especially cotton, several times during a growing season. Lindane (HCH) has also been used biannually to rid sheep's skin and fleece of vermin.

It is difficult to determine the relative importance of each human health issue, especially when population groups may suffer multiple medical problems and live in impoverished conditions. What is clear is that residents of the Aral Sea are experiencing a pronounced health crisis, not unlike the environmental one that surrounds them.

—*Mariana L. Rhoades*

FURTHER READING

Chen, Dene-Hern. "Once Written Off for Dead, the Aral Sea Is Now Full of Life." *National Geographic*, National Geographic Society, 16 Mar. 2018, news.nationalgeographic.com/2018/03/north-aral-sea-restoration-fish-kazakhstan/?beta=true.

Micklin, P. "The Aral Sea Disaster." *Annual Review of Earth and Planetary Science* 35 (2007): 47–72. Excellent collection of data on the water balance, salinities, and hydrology of the Aral Basin; human and ecological consequences of the disaster; improvement efforts by global aid agencies; and engineering mitigation.

_____. "Desiccation of the Aral Sea: A Water Management Disaster in the Soviet Union." *Science* 241, no. 4844 (September 2, 1988): 1170–1176. Micklin's early work on causes of sea recession and resulting environmental problems; details local water quality improvements and future schemes for preservation of the Aral.

Micklin, P., and N. V. Aladin. "Reclaiming the Aral Sea." *Scientific American* 298, no. 4 (April, 2008): 64–71. Outlines the collapse of the Aral Sea due to wasteful irrigation of the desert; the geography of residual lakes; the 2005 dam construction and future of the Amu Dar'ya; and application of lessons learned from the Aral disaster to other regions with similar risks.

Nihoul, J. C. J., P. O. Zavialov, and P. Micklin, eds. *Dying and Dead Seas: Climatic Versus Anthropic Causes.* Dordrecht, the Netherlands: Kluwer Academic, 2004. Synthesis of sixteen lectures given at the May, 2003, North Atlantic Treaty Organization Advanced Research Workshop in Belgium on dead or dying internal seas and lakes; assesses the past and present roles of natural and anthropogenic causes.

Whish-Wilson, P. "The Aral Sea Environmental Health Crises." *Journal of Rural and Remote Environmental Health* 1, no. 2 (2002): 29–34. Provides general background on the disaster; in-depth health status assessment of the region; list of causes and outcomes of pollution and its effects on humans and the environment; and details of the health community's response.

ARCTIC AND GLOBAL WARMING

FIELDS OF STUDY

Environmental Sciences; Environmental Studies; Meteorology; Climate Modeling; Hydroclimatology; Hydrometeorology; Atmospheric Science; Bioclimatology; Climate Classification; Climate Zones; Oceanography; Physical Geography; Hydrology; Earth System Modeling; Ecology

SUMMARY

According to the Arctic Climate Impact Assessment Scientific Report, produced by 250 scientists under the auspices of the Arctic Council, Arctic sea ice was half as thick in 2003 as it had been thirty years earlier. The melting of ice in the Arctic accelerated through 2007, advancing the projected date of an ice-free summer to perhaps 2020. During September 2007, the Arctic ice cap shrank to what was, at the time, its smallest extent since records have been kept, 4.14 million square kilometers, versus the previous record low of 5.32 million square kilometers in 2005. The shrinkage from 2005 to 2007 represented a loss of more than 20 percent of the Arctic's ice cover, or an area the size of Texas and California combined. These trends continued into the 2010s.

PRINCIPAL TERMS

- **albedo:** proportion of incident radiation reflected by Earth's surface
- **drunken forest:** forest that leans at an odd angle as a result of melting permafrost
- **feedback loops:** climatic influences that compound or retard each other, accelerating or decelerating the rate of global warming
- **permafrost thawing:** defrosting of previously permanently frozen ground, usually in or near the Arctic
- **tipping point:** the point at which feedback loops take control and propel climate change

Shrinking Ice Cover

In August 2012, the sea ice extent had decreased even further than in 2007 to 4.10 million square kilometers, making this record the lowest in satellite history. The U.S. National Oceanic and Atmospheric Administration's (NOAA) 2015 Arctic Report Card found that the maximum extent of sea ice set a new record low in 2015 since record keeping began in 1979. The average air temperature on land in the Arctic that year also set a new record high since 1900. The escalation of global warming in the Arctic paralleled the global pattern (both 2014 and 2015 set average global high temperature records), but the sensitivity of the Arctic environment makes it particularly at risk to major environmental disruption due to climate change.

Following the summer of 2016, scientists were surprised to find that the Arctic ice extent minimum had once again shrunk to a record low of 4.14 million square kilometers, effectively tying the 2007 record for the second-lowest extent since the beginning of the use of satellite analysis. Though the ice had begun to melt at a faster pace subsequent to a record low yearly maximum extent in March, satellites had recorded a more unusual decrease in melting during the summer likely due to atmospheric pressure and cloud cover. Since ice melting usually increases during the months of June and July, scientists were not expecting the sharp receding of the ice back to the low 2007 extent in September. The atypical melting pattern confirmed scientists' conclusions that the loss of older, thicker ice has left the ice more susceptible to changing weather conditions (including storms, such as the two cyclones that passed over the Arctic Ocean in August) later in the season than in the past. A new record low for Arctic sea ice coverage was set once more in 2012, and has remained almost as low since that time, with the 2017 sea ice coverage being only slightly higher than in 2012. The 2017 sea ice coverage, however, is more notable due to the near absence of ice coverage in the Bering Sea and Chukchi Sea.

Retreat of Arctic Ocean Ice

Scientists were shaken by the sudden retreat of the Arctic ice during the summer of 2007, which was much greater than their models had projected. Some said that a tipping point had been reached and that the Arctic could experience ice-free summers within a decade or two. At the annual American Geophysical Union meeting in San Francisco during December 2007, scientists reported that temperatures in waters near Alaska and Russia were as much as 5°C above average.

Scientists at the University of Washington said the sun's heat made the greatest contribution to the record melting of the Arctic ice cap at the end of summer in 2007. Sunlight added twice as much heat to the water as was typical before 2000. Relatively warm water entering the Arctic Ocean from the Atlantic and the Pacific Oceans was also a factor, according to Michael Steele, an oceanographer at the University of Washington. Energy from the warmer water delayed the expansion of ice in the winter as it warmed the air. Similarly, warmer air flowing northwards into the regions of the Bering Sea and Chukchi Sea are believed to have curtailed the formation of sea ice in those regions.

In addition to the retreating extent of sea ice in the Arctic, a greater proportion of the ice that does remain in the region is thin and freshly formed, unlike older, thicker ice that is more likely to survive summer melting. The proportion of older, "durable" ice dropped drastically between 1987 and 2007, according to studies by Ignatius G. Rigor of the University of Washington. In addition to a decrease in the extent of Arctic Ocean ice, by 2007 large areas of the ice that remained were only about one meter thick, half of what they had been in 2001, according to measurements taken by an international team of scientists aboard the research ship *Polarstern*. In 2015 NOAA reported that the amount of first-year new ice in February and March was double that of thirty years prior, reflecting the extreme loss of older ice. The 2015 Arctic Report Card also found that the sea ice sheet reached its maximum extent fifteen days before average and was the smallest on record dating back to 1979.

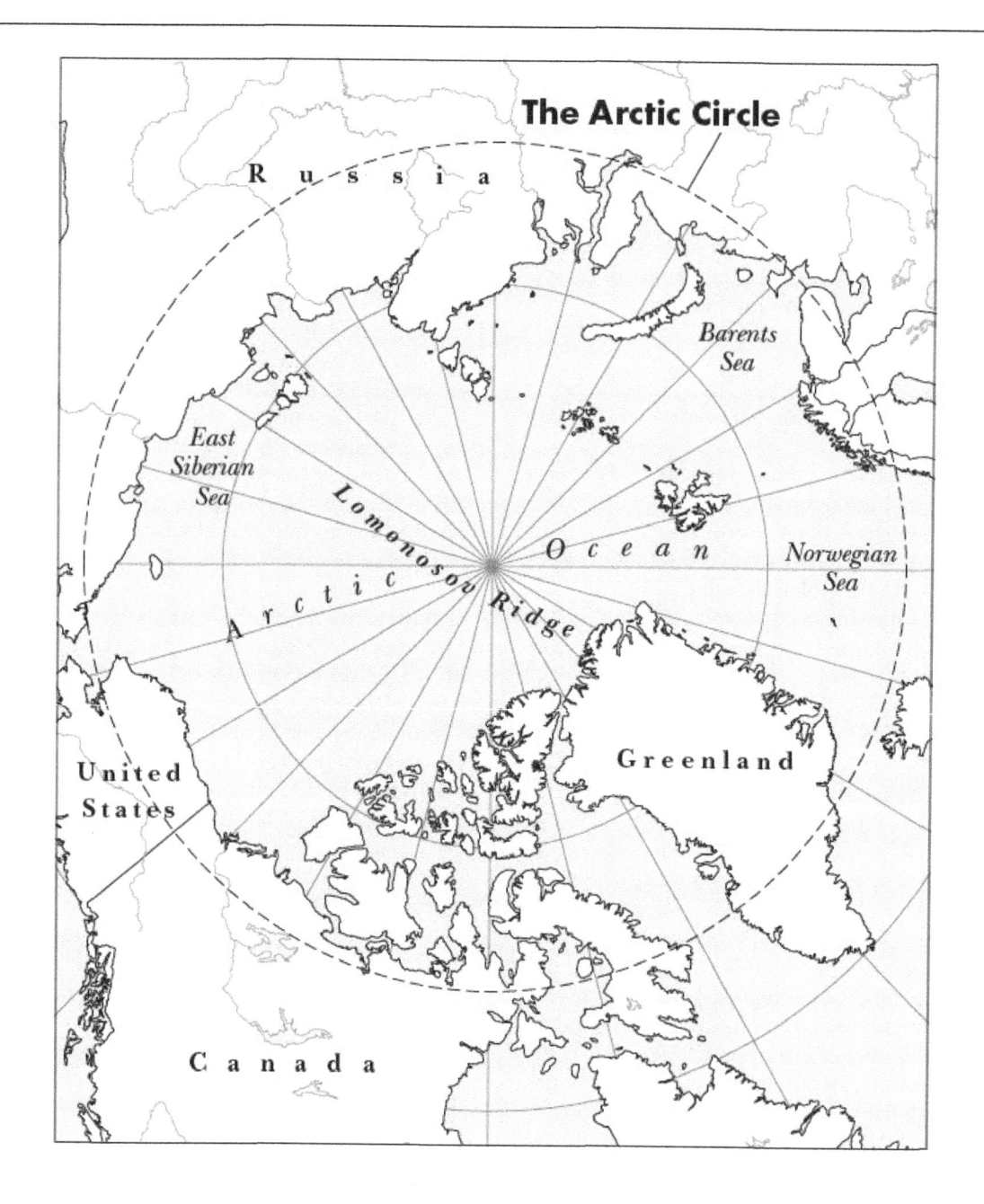

WALRUS DEATHS

With ice receding hundreds of kilometers offshore during the late summer of 2007, walruses gathered by the thousands on the shores of Alaska and Siberia. According to Joel Garlich-Miller, a walrus expert with the U.S. Fish and Wildlife Service, walruses began to gather onshore late in July, a month earlier than usual. A month later, their numbers had reached record levels from Barrow to Cape Lisburne, about 480 kilometers southwest, on the Chukchi Sea. Walruses dive from the ice to feed on clams, snails, and other bottom-dwelling creatures. As a result of increased melting, however, the Arctic ice had receded too far from shore to allow walruses to engage in their usual feeding patterns.

A walrus can dive 180 meters, but water under receding ice shelves is now more than a thousand meters deep by late summer. The walruses have been forced to swim much farther to find food, using energy that would otherwise have been available for the development of healthy calves. This would naturally result in increased calf mortality. In addition, more calves are being orphaned. Russian research observers have reported many more walruses than usual on shore, tens of thousands in some areas along the Siberian coast. These creatures would have stayed on the sea ice in earlier times.

Walruses are prone to stampedes when they gather in large groups. The appearance of a polar bear or a human hunter, or even noise from a low-flying small airplane, can send thousands of panicked walruses rushing to the water. Those that are caught up in such a stampede have a high probability of being severely injured or even killed. Thousands of Pacific walruses were killed on the Russian side of the Bering Strait during the late summer of 2007, when more than forty thousand hauled out on land at Point Shmidt as ice retreated northward. Scientists reported in 2015 that the continued loss of walrus habitat is further complicated by secondary effects of global warming, such as greater development and shipping in the Arctic (including for gas and oil projects), acidification of the ocean, and the spread of disease and contaminants. In October, 2017, the Pacific walrus was not designated as a threatened species, placing the animals at increased risk due to climate-driven changes in the Arctic environment.

FORECASTS
In the past, low-ice years often were followed by recovery the next year, when cold winters or cool summers kept ice from melting further. That kind of balancing cycle stopped after 2002. The year 2004 was the third in a row with extreme ice losses, indicating acceleration of the melting trend. Arctic ice has been declining by about 8 percent per decade as part of this trend.

By the end of 2005, scientific projections for the Arctic were becoming more severe. A climate modeling study published in the *Journal of Climate* indicated that if humanity does not address global warming in the near future, irreversible damage may take place. The paper's lead author, Govindasamy Bala of the U.S. Energy Department's Lawrence Livermore National Laboratory, predicted that it might take twenty or thirty years before the scope of anthropogenic climate change becomes evident, but that after that the damage would be obvious.

The *Journal of Climate* study projected that the global concentration of atmospheric carbon dioxide (CO_2) would be double that of preindustrial levels in 2070, triple in 2120, and quadruple in 2160, based on the prediction of slightly less than present-day rates of increase. It was the first study to assume consumption of all known reserves of fossil fuels. This model anticipates that the Arctic will see the planet's most intense relative warming, with average annual temperatures in many parts of Arctic regions rising by more than 14°C by about 2100.

ACCELERATION OF ICE-CAP MELTING
Comparisons more than a dozen models of Arctic ice-melt found that nearly all of them underestimated the actual speed of ice-melt, in many cases by

A global view of Arctic sea ice in 2000.

large amounts. These findings have two important implications, according to a summary published in *Science*. First, the effect of increasing greenhouse gases may be greater than has been believed. Second, future loss of Arctic Ocean ice may be more rapid and extensive than predicted. Within a decade, projections of the first ice-free summer in the Arctic had moved from the end of the twenty-first century to about the year 2020. Even projections made by the Intergovernmental Panel on Climate Change (IPCC) in its 2007 assessments (forecasting an ice-less Arctic in summer between 2050 and 2100) were out of date weeks after they were made public, according to reports by scientists at the National Center for Atmospheric Research and the University of Colorado's National Snow and Ice Data Center. The study, "Arctic Sea Ice Decline: Faster than Forecast?" was published during early May 2007 in the online edition of *Geophysical Research Letters*. An article published in *Journal of Marine Science*, in August, 2017, stated that current climate models predict that the Arctic will see its first ice-free season in 100,000 years in 2030. However, there is a 10-year uncertainty in the prediction, so it could occur anytime between the years 2020 and 2040.

Beginning in the mid-1990s, scientists observed pulses of relatively warm water from the North Atlantic entering the Arctic Ocean, further speeding ice-melt. Mark Serreze, senior research scientist at the National Snow and Ice Data Center, said that such warm-water pulses represented yet another potential kick to the system that could accelerate rapid sea-ice decline and send the Arctic into a new state. As Arctic ice retreats, ocean water transports more heat to the Arctic, and the open water absorbs more sunlight, further accelerating the rate of warming and leading to the loss of more ice. The entire Arctic system is thus beset by accelerating positive feedback loops that intensify climate change. For example, permafrost has been melting, injecting additional CO_2 and methane into the atmosphere. In Alaska, trees that have become destabilized in melting permafrost lean at angles, creating so-called "drunken forests."

The speed of ice breakup can sometimes be astonishing. For example, the thirty-meter-thick Ayles shelf of floating ice, a shelf roughly sixty-five square kilometers in area, had extended into the Arctic Ocean from the north coast of Ellesmere Island in the Canadian Arctic for roughly three thousand years. The Ayles shelf was detached during the summer of 2005 by wind and waves in warming water. The break-up was observed by Laurie Weir of the Canadian Ice Service in satellite images of Ellesmere Island during August 2005. The images showed a broad crack opening and the ice shelf collapsing and flowing out to sea with a speed that could be observed hour by hour. In 1906, Arctic explorer Robert Edwin Peary surveyed 26,160 square kilometers of ice shelves. Nine-tenths of these have broken up over the last century, according to Luke Copland, director of the Laboratory for Cryospheric Research at the University of Ottawa.

Context

The extent of Arctic Ocean ice has been declining year-round for more than four decades, according to analysis of satellite data by the U.S. National Ice Center. Data sets of show shrinkage in the Arctic Ocean summer ice cover of more than 8 percent per decade, according to Pablo Clemente-Colón, the ice center's chief scientist. While the extent of winter sea ice has also been decreasing, summer shrinkage has also become more pronounced.

A tipping point toward an ice-free Arctic in summer probably already has been passed, according to James E. Hansen, former director of the National Aeronautics and Space Administration's (NASA) Goddard Institute for Space Studies. The complete loss may occur rapidly, on the time scale of a decade, once ice loss has reached such a degree that the albedo feedback becomes a dominant process, according to Hansen. The albedo feedback refers to the fact that loss of some sea ice increases the amount of solar energy absorbed by the Arctic because the liquid ocean is darker than the ice, absorbing more of the Sun's heat and thus increasing ice melt. Hansen and his team could not determine the exact level of added CO_2 necessary to cause complete ice loss, but they declared that once such a state were reached, it would be difficult to return to a climate with summer sea ice, because of the long lifetime of atmospheric CO_2.

— *Bruce E. Johansen*

Further Reading

Alley, Richard B. *The Two-Mile Time Machine. Ice Cores, Abrupt Climate Change and Our Future*. Princeton University Press, 2014. Describes the analytical methods and interpretations of physical clues of past climates trapped in ice that is thousands of years old.

"Arctic Marine Conservation is Not Prepared for the Coming Melt" *ICES Journal of Marine Science* August 2017. https://www.researchgate.net/publication/306300142_How_predictable_is_the_timing_of_a_summer_ice-free_Arctic_PREDICTING_A_SUMMER_ICE-FREE_ARCTIC. Accessed 2 May 2018.

"Arctic Sea Ice Annual Minimum Ties Second Lowest on Record." *NASA*, 15 Sept. 2016, www.nasa.gov/feature/goddard/2016/arctic-sea-ice-annual-minimum-ties-second-lowest-on-record. Accessed 18 Oct. 2016.

Mann, Michael E. *The Hockey Stick and the Climate Wars. Dispatches from the Front Lines.* Columbia University Press, 2014. This very readable book lays out the science behind global climate change and identifies the conflict between climate change proponents and deniers.

Margesin, Rosa, ed. *Permafrost Soils.* Springer-Verlag, 2009. A collection of learned articles describing many aspects of Arctic permafrost soils and the effects of global warming on those soils and their content.

"NASA, NOAA Analyses Reveal Record-Shattering Global Warm Temperatures in 2015." *Goddard Institute for Space Studies.* NASA, 20 Jan. 2016, www.nasa.gov/press-release/nasa-noaa-analyses-reveal-record-shattering-global-warm-temperatures-in-2015. Accessed 18 Oct. 2016.

Serreze, Mark C. *Brave New Arctic. The Untold Story of the Melting North.* Princeton University Press, 2018. Describes the Arctic environment and the climate-related changes to that environment that have occurred since 1980.

Wadhams, Peter *A Farewell to Ice. A Report from the Arctic.* Oxford University Press, 2017. Discusses the role of Arctic and Antarctic ice shields, and how feedback systems function. Special attention is given to the potential for the release of quantities of methane from polar sea floors and Arctic permafrost as Earth's average temperature increases.

ARCTIC OCEAN

FIELDS OF STUDY

Oceanography; Physical Geography; Ecosystem Management; Earth System Modeling; Ecology; Heat Transfer; Thermodynamics; Engineering; Environmental Sciences; Environmental Studies; Waste Management; Meteorology; Climate Modeling; Fluid Dynamics; Hydroclimatology; Hydrometeorology; Atmospheric Science; Bioclimatology; Climate Classification; Climate Zones; Hydrology; Petroleum Refining

SUMMARY

The Arctic Ocean, the fourth largest of the world's oceans with an area of 12,257,000 square kilometers, lies completely within the Arctic Circle. Large segments of it remain frozen throughout the year. It has an average depth of 1,000 meters but in some parts is nearly 5,500 meters deep.

PRINCIPAL TERMS

- **brash:** splinters that become detached from ice floes and float in the Arctic Ocean
- **ice floes:** large formations of ice, usually 2.5 to 3.5 meters thick, that float in the waters of the Arctic Ocean
- **igloo:** a temporary Inuit structure made from blocks of dense snow
- **Inuit:** the indigenous peoples of the northern polar regions, whose name means "the People"; often referred to incorrectly as Eskimos, a word from a more southerly native language
- **phytoplankton:** tiny floating sea plants that are most plentiful in the presence of sunshine and rich nutrients
- **pingo:** a large, stable ice intrusion of the Arctic tundra terrain, appearing as a large, dome-shaped, earth-covered mound with cracks visible at the top, the core being solid ice
- **salinity:** the salt content of such substances as water and food
- **umiak:** a large boat, constructed with animal skins on a wood and bone frame, that the Inuit traditionally used when hunting marine mammals such as seals and whales

Map of the Arctic Ocean. (Courtesy of the CIA World Fact Book)

LOCATION OF THE ARCTIC OCEAN

Lying wholly within the Arctic Circle, the Arctic Ocean was once viewed by geographers as a part of the Atlantic Ocean. It is now viewed as a discrete body of water with definite boundaries. As such, it is the fourth-largest ocean on Earth, smaller than the Pacific, the Atlantic, and the Indian Oceans, which are, respectively, the largest, second-largest, and third-largest oceans in the world. Waters from all oceans intermingle as tides and currents carry them along and disperse them. The Arctic Ocean is unique in that it is almost landlocked. Essentially circular, it extends from the North Pole south to about 80 degrees north latitude or, if one includes its smaller fringe seas, to about 70 degrees north latitude. The main landmasses that it touches are Canada, Alaska, Siberia, Greenland, Iceland, and Norway.

The Arctic Ocean spills about 60 percent of its water into the Atlantic Ocean between Greenland and Spitsbergen, a group of islands that belong to Norway. It is largely the surface waters that are exchanged because a high range of submerged mountains known as the Faeroe-Icelandic Ridge, which in some places breaks the surface and creates islands, blocks the exchange of the deepest water. Of all the major oceans, the Arctic, because of its unique pattern of temperatures, currents, and ice conditions, probably has the most independent existence. There is virtually no flow of water from the Arctic Ocean into the Pacific because land barriers and Earth's rotation prevent such a flow. Some 35 percent of the water that comes to the Arctic Ocean flows in from the Pacific. Most of the other water that enters it comes from the Norwegian Sea.

Crucial to the ecology of the Arctic Ocean is the Greenland Sea. Several large rivers flow from Canada and Siberia into the sea, bringing enormous quantities of water into its somewhat constricted basin. In this part of the world, evaporation is not great, so if the water that comes into the ocean were not expelled into the Greenland Sea, serious problems would ensue.

The water from the Greenland Sea creates a cold current, termed the East Greenland Current, that flows south along Greenland's east coast. The much weaker Labrador Current flows through Smith Sound and Baffin Bay. Yet another weak current flows from the Bering Strait. Water that does not flow out through the Greenland Sea is deflected by Greenland's northern shore. It forms a current that, off the northwest portion of the Arctic Archipelago, runs southwest and west, and then turns again seeking an outlet, which creates a unique circular current in the Arctic Ocean.

This current explains why the part of the Arctic Ocean that is bounded by Siberia has less ice than the same ocean has in Greenland and in parts of the Arctic Archipelago, notably Ellesmere Island. Large ice floes tend to drift southward and westward, many of them melting before they have a chance to drift into the congested shipping lanes of the Atlantic Ocean. Icebergs that reach the Atlantic Ocean are usually brought there by the Labrador Current from western Greenland's fjords. Although some of the ice melts as it moves southward, the polar ice cap that covers part of the Arctic Ocean has not melted in recorded history.

It was once thought that the Arctic Ocean had a considerable effect upon the climate and ecology of all the other oceans. Researchers, however, have questioned this supposition, concluding generally that the Arctic Ocean is more affected by conditions in the world's other oceans than they are affected by conditions in the Arctic Ocean.

Derivation of the Name "Arctic"

In early times, the Arctic Ocean lay in what the ancients called "terra incognita," meaning "unknown land." The areas that it touched were thought to be incapable of sustaining life, although it is now recognized that the Arctic area is teeming with life of many varieties, from complex vertebrates such as humans, seals, and polar bears to simple microorganisms such as phytoplankton, which flourish in both the Arctic and the Antarctic.

The ancient Greeks named the Arctic after the astronomical constellation known as "The Great Bear." The Greek word for bear is *arktos,* hence "Arctic." The ancients observed that the great bear constellation appeared to revolve around the North Pole. Convinced that there must be another pole at Earth's other extremity, the Greeks coined the term "*antarktos,*" or "Opposite Bear," which in time became the term "Antarctic."

In early times, the Arctic Ocean was often referred to as the Frozen Ocean because so much of it was permanently covered with ice extending from the polar ice cap. With winter temperatures typically as low as -33°C and below, it was generally thought that the ocean and the area surrounding it precluded human habitation, a fallacy that has long since been disproved.

People of the Polar Regions

Despite early conjecture that suggested that the polar regions were uninhabited, it is now clear that they have sustained human life for thousands of years. Archaeologists have recently uncovered the remains of Arctic settlements that date to an age of at least 13,000 years and as much as 30,000 years. The Inuit have long been permanent residents of areas around the Arctic Ocean. These people are traditionally hunters who hunt the native animals, particularly seals, for food.

The Inuit are expert at fishing through the ice and take a great deal of their food from the Arctic Ocean, particularly in winter when they tend to dwell along the ocean's shore. They also hunt whales on a limited basis, which they once harpooned from "umiaks," large open boats made of animal skins on a wood-and-bone frame. Whales provided the Inuit with oil to fuel their lamps and with blubber, the large, fatty layers that preserve the whale's buoyancy and body warmth in the frigid Arctic waters. Eating the blubber helped sustain the Inuit during the long winters. In the present day, Inuit cultural traditions live on, but the traditional weapons and boats have largely given way to their modern counterparts, and snowmobiles now run where dogsleds and snowshoes were once the norm.

During the winter, the Inuit often build igloos, domed shelters made from blocks of densely packed snow, that are surprisingly comfortable inside. Caught out in the open with a storm approaching or simply traveling away from home for an extended period, a skilled person could construct an igloo for shelter in a matter of minutes. Sometimes a hole can be cut in the ice of the igloo's floor in order to catch fish protected from the driving winter winds.

In summer, most of the Inuit went inland to the tundra to hunt for game and to fish in the lakes. They gathered the wild berries and roots that grow there. They preserved the meat of some of the fish and game they caught by drying or smoking it so that they would have food during the harsh winter.

In the present day, the vast majority of Inuit live in simple frame houses in scattered villages throughout the north, and the art of building igloos is becoming a rare skill. Concerted action by tribal elders now seeks to preserve the traditional knowledge and pass it down to younger generations as a right of their heritage.

The modern people of the polar regions have been touched by technology, usually living in villages that have schools, stores, churches, and medical centers. They live in houses made from imported materials. Such houses usually have an air lock, an area between the door that opens to the outdoors and the second door that opens into the house so that the cold Arctic air does not penetrate the building's interior warmth.

Only sketchy and inaccurate information was available about the Arctic and its people until the nineteenth century, when exploration of the region began. Earlier European sailors had gotten to the region in their search for a short route to Asia. Most, however, found the area so forbidding that they turned back. In the early twentieth century, a now-classic documentary film titled *Nanook of the North* was shot on location and though much of the action in the movie was staged rather than actual, the film was the first real exposure to Inuit culture experienced by the rest of the world.

EXTENT OF THE ARCTIC OCEAN

The Arctic Ocean occupies an area of more than 12 million square kilometers, about one-seventh the size of the Atlantic Ocean. Many scientists believe that global warming will gradually increase the size of the Arctic Ocean and will cause flooding in the land areas that touch it.

Several seas lie on the fringes of the Arctic Ocean, including the Barents, Beaufort, East Siberian, Greenland, Kara, and Laptev Seas, as well as Baffin Bay. These fringe seas extend 10 degrees farther south than the Arctic Ocean proper, reaching as far as 70 degrees north latitude. Although some geographers consider the polar regions to be those that cannot sustain the growth of trees, the polar circles that most geographers accept are generally calculated as being 66.33 degrees north and south of the equator, so that even the fringe seas are wholly within the Arctic Circle. As global warming progresses, it should be noted that the region that cannot support tree growth is shrinking accordingly, and boreal forest encroaches farther and farther into the northlands.

The whole of the Arctic Ocean is roughly the size of Antarctica, whose ocean is very different from its northern counterpart in that it surrounds a continent, whereas the Arctic Ocean is surrounded by landmasses. That more life thrives in the Arctic than in Antarctica is attributable to the fact that the landmasses around the Arctic Ocean are warmed by the ocean's currents, with much of the water being quite shallow because of the continental shelves. Nevertheless, much of the Arctic Ocean is frozen all year long, with its water temperature hovering between $-1.1°C$ and $+1°C$.

The northern and southern polar regions are generally defined as being those points on Earth that experience at least one day each year when the sun does not set. This phenomenon is true at both the North and South Poles in June and December of each year, respectively.

APPEARANCE OF THE ARCTIC OCEAN

Looking down upon the Arctic landscape from above, one is impressed by how much water and ice constitute the landscape. The land below seems solidly frozen, rimmed with ice from the sea, much of which is covered with ice. The view of the frozen sea from the air is dramatic beyond imagining. At times it is not punctuated for hundreds of kilometers by any stretches of water. As one moves out from the land, the ocean waters, if they are visible at all, appear dark, but they gradually moderate into a lighter green, their surfaces studded by huge chunks of ice called brash that have splintered off from larger ice floes.

If one could look below the surface to the ocean floor, it would present the appearance of a warped, rifted surface with great irregularities. In some places, it sinks to depths of 5,500 meters. Freshwater ice formations, similar to the pingos that mark the Arctic tundra, are present in some of the shallower water, each moored soundly to the bottom on which they rest.

It may seem odd that freshwater islands exist in a saltwater ocean. This peculiarity is explained by the fact that ice resulting from the freezing of saltwater becomes less and less saline in its constitution as it ages. Ice loses one-half of its salinity in its first year of being frozen. Eventually, it reaches the point where it has virtually no salinity, so that such ice can be returned to its liquid state and drunk with no ill effects.

Beneath its surface, the Arctic Ocean consists of two basins. They are believed to have developed separately more than 100 million years ago when the tectonic plates of Earth's surface drifted apart. The Eurasian Basin resulted when the sea floor spread in a line along the Nansen Cordillera, a range of submerged mountains that constitutes the northernmost part of the Mid-Atlantic Range. In time, this movement pushed a narrow portion of Asia away from the mainland. Known as the Lomonosov Ridge, this sliver lies directly beneath the North Pole. At roughly the same time, the land that is now Alaska moved away from North America and left a basin, known as the Canada Basin, on the other side of the Lomonosov Ridge. This basin is also sometimes referred to as the Amerasian Basin.

Continental Shelves

The Arctic Ocean contains the widest continental shelf in the world, between 490 and 1,780 kilometers in width around the Eurasian basin, stretching north from Siberia toward the North Pole. A similar shelf, in the Amerasian Basin, which is from 97 to 200 kilometers wide, extends north from North America. Its exposed portions form the Arctic Archipelago, which consists of Wrangel Island, the Franz Josef Archipelago, the New Siberian and Lyakhov Islands, Severnaya Zemlya, Novaya Zemlya, and Spitsbergen.

The continental shelf that extends north from North America eventually drops off into a deep, oval basin reaching south from the North Pole to the Bering Sea. This and parts of the Greenland Sea east of Greenland are the deepest parts of the Arctic Ocean. It was once thought that no form of life could exist at such depths, but it has now been established that in some of the deepest parts of the oceans, hot water, sometimes as hot as 367 degrees Celsius, flows out of deeply submerged structures called hydrothermal vents that are comparable in some ways to volcanoes. Life exists in many forms around such vents.

Nothing in nature remains the same forever. Cores of ice taken from the Arctic reveal that the area once had a temperate climate. The fossils found in the ice cores suggest that all sorts of vegetation grew where now a wholly different kind of vegetation exists in the harsh climate of the Arctic Circle. Evidence also substantiates the theory that the polar climate is still changing, once again becoming warmer. This warming has implications for the entire world: If the polar ice caps begin to melt at a rapid rate, low-lying areas of the world may become flooded and uninhabitable. Studying the Arctic Ocean and the circumpolar regions will raise awareness that everything in nature is connected and that no alteration in nature, whether natural or human-made, is without consequences.

—*R. Baird Shuman*

Further Reading

Alley, Richard B. *The Two-Mile Time Machine. Ice Cores, Abrupt Climate Change and Our Future* Princeton University Press, 2014. Describes the analytical methods and interpretations of physical clues of past climates trapped in ice that is thousands of years old.

American Museum of Natural History. *Ocean*. Dorling Kindersley Limited, 2006. Discusses the geology, circulation, climate, and physical characteristics of the ocean. Covers marine biology and ocean chemistry. Includes a discussion of icebergs and polar ocean circulation. An excellent starting point for anyone learning about oceans and marine ecology. Includes images on each page, an extensive index, a glossary, and references.

"Arctic Marine Conservation is Not Prepared for the Coming Melt" *ICES Journal of Marine Science* August 2017. https://www.researchgate.net/publication/306300142_How_predictable_is_the_timing_of_a_summer_ice-free_Arctic_PREDICTING_A_SUMMER_ICE-FREE_ARCTIC. Accessed 2 May 2018.

"Arctic Sea Ice Annual Minimum Ties Second Lowest on Record." *NASA*, 15 Sept. 2016, www.nasa.gov/feature/goddard/2016/arctic-sea-ice-annual-minimum-ties-second-lowest-on-record. Accessed 18 Oct. 2016.

Austin, B. *Arctic Basin: Results from the Russian Drifting Stations.* Springer, 2010. Discusses the results obtained by manned research stations on the drift ice of the high Arctic. Describes the meteorological, oceanographic, and geophysical observations.

Ballard, Robert D., and Malcolm McConnell. *Explorations: My Quest for Adventure and Discovery Under the Sea.* New York: Hyperion, 1995. Presents fascinating information about the very deep sea, correcting many previously held notions about it.

Bischof, Jens. *Ice Drift, Ocean Circulation and Climate Change.* Springer, 2001. Presents and discusses the concept of ice rafting, in which large pieces of Arctic ice floes break away and drift like rafts on the ocean currents, their movement adding to our knowledge of past and present conditions of oceanic circulation.

Broad, William J., and Dimitri Schidlovsky. *The Universe Below Discovering the Secrets of the Deep Sea.* Paw Prints, 2008. Provides useful information about the very deep sea. Explores the notion that there is little life in the very deep sea. Demonstrates that all sorts of life that have hitherto been undetected inhabit the abyssal depths, sometimes enduring high temperatures from hot flows that would kill most known forms of life.

Byers, Michael. *Who Owns the Arctic? Understanding Sovereignty Disputes in the North.* Vancouver: Douglas & McIntyre Publishers, 2009. Covers many issues of Arctic sovereignty. An especially topical subject since global warming is rapidly freeing up access to Arctic Ocean resources.

Lemke, Peter, and Hans-Werner Jacobi. *Arctic Climate Change: The Acsys Decade and Beyond.* New York: Springer Science+Business Media, 2012. Addresses a number of major topics related to climate change in the Arctic brought on by increasing global temperatures with respect to the role of the Arctic in the global climate system.

Serreze, Mark C., and Roger G, Barry. *The Arctic Climate System.* Cambridge University Press, 2014. Provides a comprehensive, up-to-date assessment of the Arctic climate system for researchers and advanced students.

Stein, Rüdiger, and Robie W Macdonald. *The Organic Carbon Cycle in the Arctic Ocean.* Springer Berlin, 2013. Various topics are discussed in relation to the Arctic Ocean carbon dynamics. Covers dissolved organic matter, particulate organic carbon, productivity and growth rates, benthic carbon cycling, and organic carbon burial rates. Summarizes the Arctic carbon cycle and compares it to global cycling.

U.S. Department of the Interior, Minerals Management Service. *Programmatic Environmental Assessment: Arctic Ocean Outer Continental Shelf Seismic Surveys.* U.S. Department of the Interior Minerals Management Service, Alaska OCS Region, 2006. Provides an overview of seismic surveys and the exploration of the Alaskan continental shelf. Includes alternative scenarios for surveys and their evaluation. Addresses the environmental impact of such surveys.

Wadhams, Peter *A Farewell to Ice. A Report from the Arctic.* Oxford University Press, 2017. Discusses the role of Arctic and Antarctic ice shields, and how feedback systems function. Special attention is given to the potential for the release of quantities of methane from polar sea floors and Arctic permafrost as Earth's average temperature increases.

ATLANTIC HEAT CONVEYOR

FIELDS OF STUDY

Oceanography; Hydroclimatology; Hydrometeorology; Climate Modeling; Fluid Dynamics; Heat Transfer; Bioclimatology

SUMMARY

The Atlantic heat conveyor transfers heat from the equator northward to the polar region of the Atlantic Ocean, through the Atlantic Meridional Overturning Circulation (MOC), commonly called "the Gulf Stream." Because Earth is basically spherical in shape, incoming solar radiation heats the surface of the planet unevenly, with the equator receiving a greater amount of heat per surface area than the polar regions. This uneven heating creates temperature gradients that drive atmospheric currents (wind) and surface currents in the ocean.

Thermohaline Circulation

This map shows the pattern of thermohaline circulation also known as "meridional overturning circulation." This collection of currents is responsible for the large-scale exchange of water masses in the ocean, including providing oxygen to the deep ocean.

PRINCIPAL TERMS

- **overturning circulation:** essentially the equivalent of an oceanic convection current, in which warmer, less dense water moves at the ocean's surface in one direction, sinks as it becomes colder, and moves back through the ocean at lower depths.
- **Atlantic Meridional Overturning Circulation (AMOC):** the official designation of what is commonly called the Gulf Stream.

OCEANIC CURRENTS

In the Atlantic Ocean, the Gulf Stream transports heat from the equator to the polar region as a surface current. The Gulf Stream is a western boundary current that originates in the Gulf of Mexico and travels north along the east coast of the United States and Newfoundland. Near 50° north, the Gulf Stream splits into two branches: the North Atlantic drift (or North Atlantic current) and the Azores current. The North Atlantic drift (NAD) flows northeastward toward northern Europe, while the Azores current flows east toward the Azores and then south as the Canary current. The Gulf Stream is a large, fast-moving current, transporting between 30 million and 150 million cubic meters of water per second. This flow of water transports approximately 1.4 petawatts (1.4 10^{15} watts) of heat per year. As the Gulf Stream moves north and mixes with cooler surface water from the poles, it releases this heat.

This "Atlantic heat conveyor" may be responsible for maintaining a milder climate in northern Europe, as compared to Newfoundland, which is located at the same latitude. However, this idea has been challenged by some scientists, who hypothesize that atmospheric heat transport is more important in maintaining a mild climate in northern Europe than is oceanic heat transfer.

SIGNIFICANCE FOR CLIMATE CHANGE

The Atlantic heat conveyor delivers a large amount of heat to the Arctic region from the equator, making it an important component of the global climate system and of any changes in that system. One consequence of global warming is the melting of glaciers in the Arctic, which would lead to a large influx of fresh water into the surface waters of the Greenland Sea. This cap of fresh water has the potential to serve as a barrier to the northward flow of the Gulf Stream, thereby blocking the Atlantic MOC and disrupting the heat conveyor. Thus, according to this conceptual model, projected global warming would actually lead to a localized cooling of the Northern Hemisphere, particularly Europe.

In 2005, scientists from the National Oceanography Center presented data to suggest that the Atlantic MOC (AMOC) had slowed during the late twentieth century. These data were subsequently challenged by other scientists, who presented different data sets that showed no MOC slowdown. Despite the controversy regarding whether such slowdown has occurred, computer models consistently indicate that a shutdown of the Atlantic heat conveyor could lead to a lesser warming or even cooling in the Northern Hemisphere as a result of global warming. Geologic evidence suggests that the Younger Dryas, a time of global cooling that lasted from 12,800 to 11,500 years before present, may have been caused by MOC collapse in the Atlantic because of a large influx of fresh water to the North Atlantic from the emptying of glacial Lake Agassiz.

The hypothesis of the Atlantic heat conveyor has been criticized by scientists from the Lamont-Doherty Earth Observatory, who believe that the atmospheric transport of heat and the slow response time of the ocean are the main reasons for Europe's mild climate. Additionally, they point to long waves in the atmosphere, created as air masses flow around the Rocky Mountains, as a third potential cause of a mild European climate. According to this hypothesis, global warming will not stop the Atlantic heat conveyor, nor will it lead to a cooling of Europe. This view, however, is not widely held among climate scientists or policy makers. In its Fourth Assessment Report, released in 2007, the Intergovernmental Panel on Climate Change (IPCC) found that there was insufficient evidence to determine if there were trends—either weakening or increasing—in the magnitude of the Atlantic heat conveyor. However, in the 2009–2010 season, it was observed that the AMOC. There is a basic difficulty in describing and predicting climate effects due to the AMOC because the actual state and stability of that "heat conveyor" are not well known or understood.

—Anna M. Cruse

FURTHER READING

Alley, Richard B. *The Two-Mile Time Machine. Ice Cores, Abrupt Climate Change and Our Future.* Princeton University Press, 2014. Describes the analytical methods and interpretations of physical clues of past climates trapped in ice that is thousands of years old.

Broecker, Wally *The Great Ocean Conveyor. Discovering the Trigger for Abrupt Climate Change.* Princeton University Press, 2010. Describes the global oceanic overturning circulation system and how its disruption by climate change may affect the climate of Earth.

Gornitz, Vivien *Rising Seas. Pest, Present, Future.* Presents an overview of sea level rise due to climate change and describes how rising sea levels may affect the stability of the ocean heat conveyor system.

Jones, E. Peter and Anderson, Leif G. "Is the Global Conveyor Belt Threatened by Arctic Ocean Fresh Water Outflow?" Chapter in Dickson, Robert R., Meincke, Jens and Rhines, Peter, eds. *Arctic-Subarctic Ocean Fluxes. Defining the Role of the Northern Seas in Climate.* Springer Science + Business Media B.V., 2008. Provides a detailed analysis of the interaction of Arctic waters with southerly ocean waters and the potential effects of those interactions on the global conveyor system.

Schmittner, Andreas, Chiang, John C.H. And Hemming, Sidney R., eds. *Ocean Circulation. Mechanisms and Impacts.* John Wiley & Sons, 2013. Provides a thorough description of the effects on Earth's climate due to large "turnover" circulation currents, particularly the Atlantic conveyor system.

ATLANTIC MULTIDECADAL OSCILLATION (AMO)

FIELDS OF STUDY

Oceanography; Thermodynamics; Climate Modeling; Statistics; Fluid Dynamics; Hydroclimatology; Heat Transfer

SUMMARY

The Atlantic Multidecadal Oscillation (AMO) is an example of a cyclical or semicyclical pattern of climate change that repeats on a timescale on the order

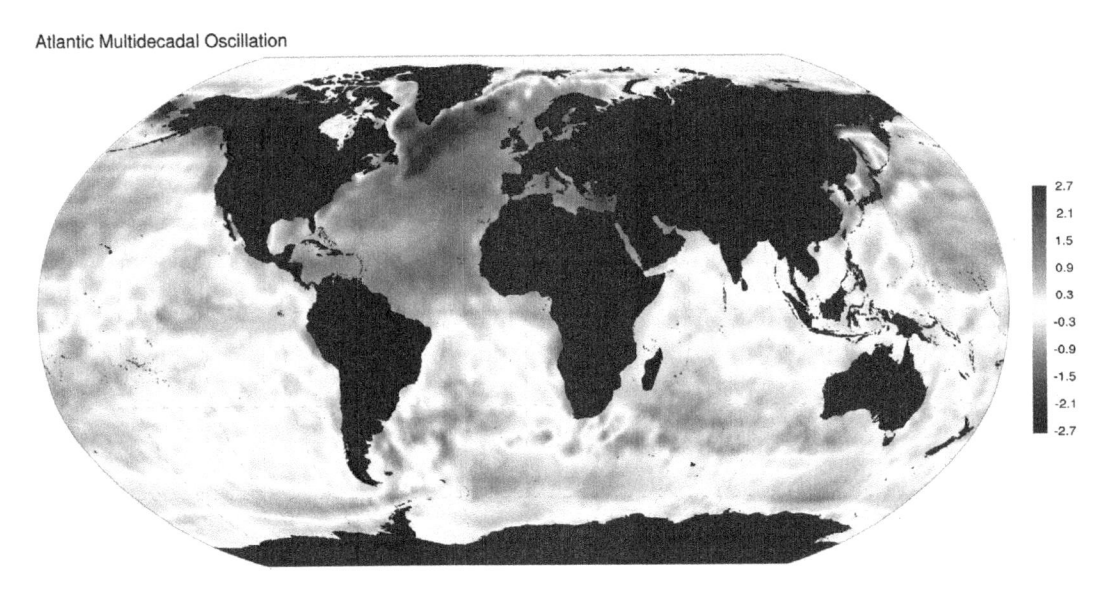

Atlantic multidecadal oscillation spatial pattern obtained as the regression of monthly HadISST sea surface temperature anomalies (1870-2013). By Giorgiogp2 - Own work, CC BY-SA 3.0.

of several decades and affects Earth's climate on a global scale. The AMO is based on sea surface temperatures in the North Atlantic Ocean between the equator and 70° north latitude. Generally, the AMO is computed as a detrended ten-year running mean of these sea surface temperatures and represents variability across the entire North Atlantic basin.

PRINCIPAL TERMS

- **multidecadal oscillation:** an alternation spanning several decades between warm and cool periods in oceanic water flows.
- **thermohaline circulation:** the movement of distinct currents within ocean waters, determined by temperature and salinity.

CHARACTERISTICS OF THE AMO

The AMO exhibits a long-term, quasi-cyclic variation at timescales of fifty to seventy years. Modeling studies reveal that multidecadal variability in the North Atlantic Ocean is dominated by this single mode of sea surface temperature variability. The range in AMO values between warm and cold extremes is only about 0.6°C; however, because the North Atlantic Ocean is so large, even small differences in sea surface temperatures represent extremely large exchanges in energy between the ocean and atmosphere.

The predominant hypothesis is that the AMO is primarily driven by the thermohaline circulation. This hypothesis is supported by both instrumental and climate-model studies. The thermohaline circulation is the large-scale ocean circulation that moves water among all of the world's oceans and is driven by density differences in ocean water caused by heat and freshwater fluxes. When the thermohaline circulation is fast, warm water is moved from tropical areas into the North Atlantic Ocean and the AMO enters a warm phase. When the thermohaline circulation is slow, warm water is not readily moved into the North Atlantic Ocean and the AMO enters a cool phase. During the twentieth century, the AMO was in a warm phase from 1926 through 1963, and it was in cool phases from 1905 through 1925 and 1964 through 1994. In 1995, it entered another warm phase.

Research has suggested that the North Atlantic Ocean may provide information that explains significant amounts of multidecadal climate variability. For example, when the AMO is in a warm phase, the likelihood of drought in North America increases, and when the AMO is in a cool phase, the likelihood of drought in North America decreases. Analysis of major U.S. droughts during the last century indicates that North Atlantic Ocean surface temperatures were warm during the 1930s and 1950s droughts, as well as during the dry period that began in the late 1990s. In contrast, the rainy conditions in the western United States in both the early (1905–1920) and late (1965–1995) twentieth century were associated with cool North Atlantic Ocean surface temperatures. The AMO also has been linked to the occurrence of hurricanes in the North Atlantic Ocean. During warm phases of the AMO, hurricanes are more frequent, whereas when the AMO is in a cool phase hurricane frequency decreases. As might be expected from this cyclic pattern of variability of the AMO, western North America in the twenty-plus year period following 1995 has experienced prolonged dry weather, while hurricane frequency in the Atlantic has increased, with at least two such storms reaching Britain and beyond in late 2017.

Significance for Climate Change

Because the AMO is an important mode of multidecadal climate variability, there are a number of important implications of the AMO for the study of climate variability and change. Understanding what portion of climate variability is due to multidecadal variability, such as that driven by the AMO, allows the discrimination of anthropogenic changes in climate from natural variability. Knowledge of multidecadal climate variability, such as that indicated by the AMO, also has implications for defining and potentially estimating risks in agriculture, water resources, public health, and nature. Variability of the AMO also has an effect on global temperature, and the beginning of a new warm phase of the AMO in 1995 may have contributed to the strong warming of global temperatures in the years immediately following.

Some scientists suggest that if the AMO shifts into a cool phase, the cooling of North Atlantic Ocean surface temperatures may reduce the amount of global warming and may result in a leveling of global temperatures for about a decade. In addition, research has indicated that North Atlantic Ocean surface temperatures may have predictability on the order of a decade or longer, which has important implications for climate forecasting.

The actual physical mechanisms that explain the associations between the North Atlantic Ocean and global climate are not well understood, but several possible mechanisms have been recognized. North Atlantic Ocean surface temperatures may affect Northern Hemisphere atmospheric circulation, such that the frequency of zonal versus meridional atmospheric flow is modulated. Decadal-to-multidecadal variability of North Atlantic Ocean surface temperatures may be aliasing for low-frequency or lagged variations of the tropical oceans. North Atlantic Ocean surface temperatures may be influencing the location and strength of subtropical high pressures. Finally, the North Atlantic Ocean may also be modulating the strength and variability of tropical Pacific Ocean surface temperatures.

—*Gregory J. McCabe*

Further Reading

Anderson, David E., Goudie, Andrew S. and Parker, Adrian G. *Global Environments Through the Quaternary. Exploring Environmental Change.* 2nd ed., Oxford University Press, 2013. Provides concise descriptions of several ocean oscillations in relation to their cyclic effects on climatic conditions.

Barry, Roger G and Hall-McKim, Eileen A. *Essentials of the Earth's Climate System.* Cambridge University Press, 2014. Describes many aspects of the influence of ocean currents and oscillations as drivers of the state of Earth's climate.

Knight, Jeff R., Folland, Chris K. and Scaife, Adam A. "Climate Impacts of the Atlantic Multidecadal Oscillation." *Geophysical Research Letters* 33, no. 17 (September, 2006). Describes some of the climate effects of variations in the AMO.

Maslin, Mark. *Global Warming: A Very Short Introduction.* Oxford University Press, 2009. Provides a concise general introduction to the major factors that define the climate of Earth.

Schmittner, Andreas, Chiang, John C.H. And Hemming, Sidney R., eds. *Ocean Circulation. Mechanisms and Impacts.* John Wiley & Sons, 2013. Provides a thorough description of the effects on Earth's climate due to large "turnover" circulation currents, particularly the Atlantic conveyor system.

Sutton, R. T., and D. L. R. Hodson. "Climate Response to Basin-Scale Warming and Cooling of the North Atlantic Ocean." *Journal of Climate* 20 (2007): 891–907. Discusses the use of a climate model to assess the effects of basin-wide warming and cooling of the North Atlantic Ocean.

Zhang, Rong, Delworth, Thomas L. and Held, Isaac M. "Can the Atlantic Ocean Drive the Observed Multidecadal Variability in Northern Hemisphere Mean Temperature?" *Geophysical Research Letters* 34, no. 2 (January, 2007). Discusses a link between the AMO and Northern Hemisphere temperature variability.

ATLANTIC OCEAN

FIELDS OF STUDY

Oceanography; Physical Geography; Ecosystem Management; Earth System Modeling; Ecology; Heat Transfer; Thermodynamics; Engineering; Environmental Sciences; Environmental Studies; Waste Management; Meteorology; Climate Modeling; Fluid Dynamics; Hydroclimatology; Hydrometeorology; Atmospheric Science; Bioclimatology; Climate Classification; Climate Zones; Hydrology; Petroleum Refining

SUMMARY

The Atlantic Ocean separates the North and South American continents from Europe and Africa. With an area, excluding its dependent seas, of about 83 million square kilometers, it is second in size only to the Pacific Ocean. The dependent seas add another 23 million square kilometers to the Atlantic's total size.

PRINCIPAL TERMS

- **continental drift:** the gradual movement of continental landmasses across the Earth's surface driven by convection processes in the mantle
- **continental shelf:** the part of the sea floor that is generally gently sloping and extends beneath the ocean from adjacent continents
- **continental slope:** the part of the continental shelf that drops off sharply toward the ocean's floor
- **equator:** an imaginary line, equidistant from the North and South Poles, around the middle of the planet
- **estuary:** an area where the mouth of a river broadens as it approaches the sea, characterized by the mixing of freshwater and saltwater
- **lagoon:** a long, narrow body of saltwater that is separated from the ocean by a bank of sand
- **tectonic plates:** large segments of the Earth's crust, affected by the movement of magma in the underlying mantle layer
- **tidal range:** the difference in water depth between high and low tides

LOCATION AND ORIGIN

The Atlantic Ocean, with an average depth of 3,300 meters and a maximum depth of 8,380 meters, touches the eastern coastlines of North and South America and the western coastlines of Europe and Africa. The Atlantic Ocean is often divided into the North Atlantic and the South Atlantic, although the entire oceanic mass may properly be considered a single ocean. The North Atlantic extends from the equator north to the Arctic Ocean. The South Atlantic, that part of the ocean south of the equator that extends as far as Antarctica, meets the Pacific Ocean at Cape Horn, the southernmost tip of South America, and the Indian Ocean at the Cape of Good Hope, the southernmost tip of Africa.

More than 200 million years ago, Earth's land surface consisted of a single supercontinent, referred to as Pangaea, surrounded by an enormous sea, called Panthalassa. A large, shallow bay known as the Tethys Sea also protruded into the supercontinent from Panthalassa. The landmass, like contemporary landmasses, was a large, solid rock plate that floated on the mantle layer of molten rock, surrounded by a relatively thin crust of solidified magma. Over the intervening millions of years, convection currents in the mantle drove Pangaea to break apart into several smaller segments and become the present-day continents. The magmatic currents have driven, and are still driving, the continental masses and tectonic plates across the face of the planet. Looking at a map of the modern world and regarding it as a huge jigsaw puzzle, it is easy to visualize how the western coast of

The Atlantic Ocean off Cape Cod, Massachusetts.

Africa fits into the eastern coast of the Americas and how the coastlines of other geographical areas fit neatly into those facing them.

As Pangaea split apart, the Tethys Sea essentially became enclosed by land as continental drift continued over millions of years, forming part of the present-day Arctic Ocean and Mediterranean Sea. The Pacific Ocean is the remnant of Panthalassa that exists today. The main rift in Pangaea separated large tectonic plates that have since formed North and South America, Asia, Africa, Europe, India, and Australia. The ever-widening space between the separating continents filled with water to form what is now the Atlantic Ocean. Current data regarding the movement of the continents and tectonic plates reveal that the North and South American continental structure and the Europe-Africa-Asia-Australia chain continue to move away from each other as though pivoting about a point in the North Atlantic Ocean approximately at the location of Iceland, such that the greater the distance from that location, the faster the tectonic plate is moving. Thus Europe and the easternmost coast of Canada are separating at the slowest rate, while Australia continues to move in a northeasterly direction at a relatively fast rate.

Two major tectonic plates, each a segment of the Atlantic Ocean floor, are moving apart annually at the average rate of 1.3 centimeters from a volcanically active line that runs approximately down the middle of the Atlantic Ocean, called the Mid-Atlantic Ridge. Where volcanic activity has been intense, molten rock has cooled into volcanic cones that sometimes protrude above the water's surface to create islands. However, in the Atlantic, unlike in the Pacific (where islands are scattered throughout the ocean), more islands are found close to shore than toward the middle of the ocean. The rate of movement of the Atlantic Ocean sea floor away from the Mid-Atlantic Ridge is such that the entire sea floor is replaced about every 100,000 years, precluding the formation of mid-Atlantic islands. The few exceptions are actually the exposed peaks of underwater volcanic mountains, some of which, such as Mount Tiede in the Canary Islands, extend more than 3,600 meters above sea level.

Ancient people named the Atlantic Ocean after the mythological giant Atlas, who is said to have carried the world on his shoulders. Many ancients, believing that earth was flat, feared that anyone sailing far enough into the ocean would eventually fall over the edge of the Earth into an abyss. This belief inhibited early explorers, although even in ancient times some skeptical adventurers sailed far into the ocean and returned safely to shore. Incomprehensibly, a significant and slowly increasing number of people in the present day hold belief in a flat Earth, despite all evidence to the contrary.

Continental Shelf and Coastline

Close to shore, the Atlantic Ocean is generally shallow, seldom exceeding a depth of 545 meters. Waters cover a shelf of rock that, in most of the North Atlantic, extends between 65 and 80 kilometers from the coastline into the ocean. At the end of this shelf, the ocean drops precipitously to a depth of about 3,940 meters, gradually leveling off into what is designated the abyssal plain. The continental shelf off the coast of Africa is much narrower than the continental shelf off the coast of North America.

The coastlines that border the Atlantic Ocean have been forming for millions of years, shaped largely by the motion and activity of the water that laps the shore. The rockiest areas are found in the northern parts of the North Atlantic and in the South Atlantic. The level of the ocean has risen considerably over the past 10,000 years as glaciers have melted. With the end of the Ice Age, huge quantities of water poured down rivers into the oceans. The point at which these rivers meet the ocean is called an estuary. Among the deepwater estuaries are the Chesapeake Bay in the United States and the Falmouth Estuary in Great Britain, both of which are sufficiently deep to accommodate large ships.

Less rocky portions of coastline are often characterized by long stretches of sandy beaches that have been formed over the years as the moving waters of the ocean have pulverized rocks and shells, transforming them into sand. Sometimes sandbars form near the coast. These may be long expanses of sand, often permanently covered by water, while others are below the water's surface only at high tide. Some sandbars that are permanently above water may be broad enough to be inhabited, as are the Outer Banks of North Carolina, although at times devastating floods occur in communities built on sandbars. Sandbars endanger ships that venture too close to them, sometimes causing them to founder. Such ships may float off a sandbar as the water rises at high tide, but in many cases this does not happen, and the ship, trapped on the sandbar, must be hauled off. Most seaports employ pilots whose duty it is to guide incoming ships through deep channels away from dangerous sandbars or rock formations beneath the surface of the water. Lagoons sometimes form between the coast and sandbars that are not permanently covered with water. These are salty bodies of water in which marsh grasses and other vegetation grow. Many of them are rich in the microorganisms that fish need for their survival. Extensive fish and bird populations cluster around lagoons.

No ocean is fed by more rivers than the Atlantic. These rivers have, over hundreds of thousands of years, carried silt and sand toward the ocean they feed. These deposits have, in many places, built up to form deltas, roughly triangular areas whose broad base fronts the ocean. Some deltas, particularly those of the Amazon and Niger Rivers, are hundreds of miles wide with channels running through them. Often in tropical areas, mangrove swamps flourish in the deltas, the roots of the mangrove trees being covered with saltwater at high tide to provide them with the nutrients they require for their growth. Fish and birds flourish in mangrove swamps.

Tides and Trade Winds

Ocean tides are generally predictable. High tide and low tide each occur twice in any twenty-four-hour period. Tidal ranges, the depth variations between high tide and low tide, are most pronounced on continental shelves and in deepwater bays. In some parts of the world, the tidal range is dramatic. In the Canadian province of Nova Scotia, the Bay of Fundy has a tidal range of more than 12 meters, one of the largest in the world.

Gentler tides are observed as one approaches the equatorial regions, although there can be dramatic tidal activity in such areas during hurricanes and tropical storms, which have traditionally occurred in August through October in the Northern Hemisphere and February through April in the Southern Hemisphere. There is considerable concern at present that global climate change is causing more powerful hurricanes and extending the hurricane season. Hurricanes usually form in tropical areas but can move rapidly into more temperate zones where, if they strike land, they can result in injury, death, and substantial destruction. In some cases, such storms erode entire beaches and completely inundate waterside property. The Atlantic coastline in many places is shrinking rapidly as the result of water erosion from such storms.

Predictable wind currents, much like jet streams in the upper atmosphere, blow across the Atlantic Ocean. The trade winds near the equatorial areas blow in a westward direction. Such winds made it possible for early explorers to sail from European ports to the Americas. For their return trips, explorers depended upon winds to the north of the areas where the trade winds blow called the "westerlies," which blow from west to east. Along the equator is a narrow area where waters are calm and where there is virtually no wind. In this area, referred to as the doldrums, sailors can drift languidly for long periods of time with no wind to propel them.

Atlantic Currents

Ocean water is never still. Much ocean water moves in one direction in predictable flows, in essence forming rivers through the ocean. Such rivers are called ocean currents. The warm tropical currents near the equator push the waters of the Atlantic Ocean westward toward North and South America, moving clockwise in the Northern Hemisphere and counterclockwise in the Southern Hemisphere.

The northern equatorial current runs from just north of the equator along the northeastern coast of South America and on toward the southern coast of North America. There, in the ocean east of the United States, the Gulf Stream runs through the Atlantic, flowing in a northerly direction. The Labrador Current that flows from the Arctic Ocean meets the Gulf Stream off Newfoundland, causing temperature variations that result in high humidity and dense fogs. North of Labrador, the Gulf Stream divides into the North Atlantic Drift, which flows north toward Greenland, and the Canaries Current, which flows southeast through the Atlantic off the west coast of Africa. These clockwise currents flow in what is termed the North Atlantic Gyre.

South of the equator, the southern equatorial current moves counterclockwise off South America's east coast, proceeds south to the Brazilian Current, which then veers southeast to meet the Benguela Current that flows from just north of Antarctica to the Tropic of Capricorn. These counterclockwise currents are referred to as the South Atlantic Gyre.

The oceanic rivers that form the currents of the North and South American Gyres have a profound effect upon climate. The Gulf Stream, which brings warm, tropical waters north at the rate of some 130 kilometers per day, makes far northern areas such as Greenland and Iceland warm enough for human habitation. Its southeasterly branch gives Great Britain a more temperate climate than one would expect at such latitudes. The northern regions of the United States are considerably warmer than they would be if the Gulf Stream did not flow along their coasts. Even areas within the Arctic Circle feel the effects of the Gulf Stream, which keeps Russia's Arctic port of Murmansk free of ice throughout the year.

The South Atlantic Ocean's Benguela Current propels cold waters from the Antarctic Ocean north along Africa's southwestern coast, keeping its temperatures much more temperate than would be expected in such latitudes. Cold air, however, does not hold moisture well, so when warm, humid winds from Africa's southwestern coast strike the cold air over the Benguela Current, they cause rain to fall over the ocean. The result is that the whole southwestern coast of Africa is arid. The South Atlantic Ocean there is bordered by desert.

Food Chain

The world's oceans are teeming with life. In a somewhat hierarchical arrangement called the food chain, the sea's larger creatures feed on the smaller ones. At the base of the food chain are microscopic organisms called plankton, which form the basis for life in oceans. Plankton, which exist in both animal and vegetable form, are the basic diet of many aquatic creatures and hence of all ocean life. Plankton cannot live without sunlight, so they are found close to the water's surface. Plankton depend upon nutrients borne in the water for sustenance, and such nutrients are usually found in coastal areas that are not very deep, such as mangrove swamps, where plankton often flourish. Some of the minerals the plankton require are washed into the ocean through estuaries that carry silt from inland areas.

Other minerals come from deep in the sea, carried to the surface by ocean currents. In the Atlantic, the most abundant plankton are found in the extreme northern and southern reaches of the ocean, as well as off the west coast of Africa. These areas have a high concentration of the nutrients that are abundant in deep, cool waters and on which plankton feed. The presence of large amounts of plankton in these areas results in a wealth of other aquatic creatures, most of which live near the surface of the ocean where plankton are found, although exploration of the very deep oceans has revealed a remarkable amount of sea life in parts of the ocean so deep that it was previously thought that nothing could survive there.

The Atlantic has an intriguing population of creatures that range from among the smallest on Earth, such as plankton, to the largest, such as whales. Most of the fish caught by commercial fisherman for human consumption—notably sardines, smelt, cod, halibut, mackerel, hake, sole, and anchovies—swim in large schools near the ocean's surface to harvest the plankton and smaller fish that constitute their diets. Larger fish, such as sharks, whales, swordfish, and sailfish, are fast-moving and live on a diet of smaller fish. They are often found in the cooler waters of the Atlantic, where plankton are abundant enough to nourish the small fish on which larger fish feed.

Some of the sea's creatures are extremely mobile. Whales travel from the polar regions to areas thousands of kilometers away for spawning. Dolphins swim fast enough to cover 245 kilometers in a single day. Conversely, some shellfish, such as mussels, attach themselves to formations in the sea and are immobile, protected from predators by their thick shells. Some forms of sea life are protected by camouflage that allows them to blend in with the surrounding area. Others are able to emit noxious jets of fluid into the waters around them when they are threatened.

Considerable bird life is associated with the Atlantic Ocean. Most of the birds that depend upon the Atlantic for their food supply can fly, but some of them, such as the cormorant and the puffin, can swim as well. They dive into the water to catch their prey and can swim to catch it. Penguins, found in the Antarctic, cannot fly and can barely walk, but they swim as well as many fish. Cranes and other long-legged birds flourish in shallow waters, where they eat tiny sea creatures that they strain from the water and mud through their beaks.

In the middle of the North Atlantic is a large, relatively calm area of water known as the Sargasso Sea, whose surface is covered with huge fields of nutritious seaweed. The Sargasso Sea has become a spawning ground for eels that migrate to it during mating season from Europe and North America.

SEASIDE SETTLEMENTS

Throughout history, humans have tended to settle beside the sea or along rivers. The transportation opportunities provided by such locations still make them desirable places for settlement. Such areas usually offer a temperate climate as well. The ancient cultures of Rome and Greece grew up on the shores of the Mediterranean and the Aegean, two of the dependent seas of the Atlantic Ocean. The Iberian Peninsula borders on the Atlantic to the west, as do France and Great Britain. Advanced civilizations have flourished on the shores of the Atlantic, particularly in Europe and Africa, since long before recorded history.

The islands of the Atlantic, as products of violent volcanic activity that created large, hilly outcroppings, were slow to develop. The exceptions are the Canary Islands and Madeira, both of which became necessary stopover points for explorers sailing from Europe to the Americas. In recent times, considerable development of the Atlantic islands has taken place in such areas as the Caribbean. The Falkland Islands off the southeastern coast of South America have a stable, permanent population, as do such island enclaves as Prince Edward Island and Newfoundland in the north.

Shipping and commerce have been the backbone of the economy in most of the areas that have developed along the Atlantic coastline. Raw materials are brought into Atlantic port cities such as New York, Philadelphia, Rio de Janeiro, and Lisbon to feed the manufacturing industries of those countries. The fishing industry has flourished for many centuries along the Atlantic coastline. Tourism has also become a major economic factor in the more temperate and scenic regions on the Atlantic coast.

Trade patterns were substantially altered with the opening in 1896 of the Suez Canal, which connects the Mediterranean Sea with the Indian Ocean, and with the opening in 1914 of the Panama Canal, which links the Atlantic and Pacific Oceans. These strategic canals opened up a great deal of trade worldwide and overcame the necessity of plying the turbulent and dangerous waters around Cape Horn and the Cape of Good Hope in order to deliver goods to distant markets.

—*R. Baird Shuman*

FURTHER READING

American Museum of Natural History. *Ocean*. Dorling Kindersley Limited, 2006.

Ballard, Robert D., and Malcolm McConnell. *Explorations: My Quest for Adventure and Discovery Under the Sea*. Hyperion, 1995.

Bertness, Mark D. *Atlantic Shorelines: Natural History and Ecology*. Princeton University Press, 2007.

Broad, William J. *The Universe Below: Discovering the Secrets of the Deep Sea*. Simon and Schuster, 1997.

Goni, G. J., and Paola Malanotte-Rizzoli. *Interhemispheric Water Exchange in the Atlantic Ocean*. Elsevier B.V., 2003.

Morozov, Eugene G., Alexander N. Demidov, Roman Y. Tarakanov, and Walter Zenk. *Abyssal Channels in the Atlantic Ocean: Water Structure and Flows*. Springer, 2010.

Stevenson, R. E., and F. H. Talbot, eds. *Oceans*. Time-Life, 1993.

Trujillo, Alan P., and Harold V. Thurman. *Essentials of Oceanography*. Prentice-Hall, 2010.

Ulanski, Stan L. *The Gulf Stream: Tiny Plankton, Giant Bluefin, and the Amazing Story of the Powerful River in the Atlantic*. University of North Carolina Press, 2008.

Voo, Rob van der. *Paleomagnetism of the Atlantic, Tethys and Iapetus Oceans*. Cambridge University Press, 2005.

Waterlow, Julia. *The Atlantic Ocean*. Raintree Steck-Vaughn, 1997.

ATMOSPHERE

FIELDS OF STUDY

Atmospheric Science; Bioclimatology; Thermodynamics; Environmental Sciences; Environmental Studies; Meteorology; Climate Modeling; Fluid Dynamics; Oceanography; Physical Geography; Earth System Modeling; Heat Transfer

SUMMARY

The composition of the atmosphere (excluding water vapor) below an altitude of 80 kilometers is about 78 percent nitrogen and 21 percent oxygen by volume (76 percent nitrogen, 23 percent oxygen by mass). The remaining 1 percent includes all other dry gases, chiefly argon, carbon dioxide, neon, helium, krypton, hydrogen, and ozone. Water vapor, the most variable constituent of the atmosphere, typically occupies between 0 percent and 4 percent of the atmospheric volume. This mixture of gases is commonly referred to as "air."

PRINCIPAL TERMS

- **mean free path:** the average distance traveled by a gas molecule or other free-moving particle between collisions with other molecules or particles.
- **diffusion:** the natural movement of particles through a medium, controlled only by natural vibrations and collisions between particles.
- **evapotranspiration:** the process by which water is transferred from land to air by evaporation from ground-based surfaces and by transpiration from plants and animals.

CONSTITUENTS OF AIR

The two principal constituents of air are greatly dissimilar in their chemical properties. While oxygen is an extremely active chemical, reacting with many substances, nitrogen reacts only under limited conditions. The inert nature of nitrogen is due to the very high strength of the bond between the two nitrogen atoms in the nitrogen molecule, and this is believed to be the reason it came to be the atmosphere's most abundant constituent. Volcanic outgassing in Earth's early history is the likely source of its present atmosphere. Though nitrogen is a minor component of volcanic emissions, the lack of chemical reactions able to remove it from the atmosphere allowed its concentration to grow dramatically over time. Photosynthesis and, to a lesser degree, photodissociation of water by ultraviolet light from the sun are believed to account for atmospheric oxygen.

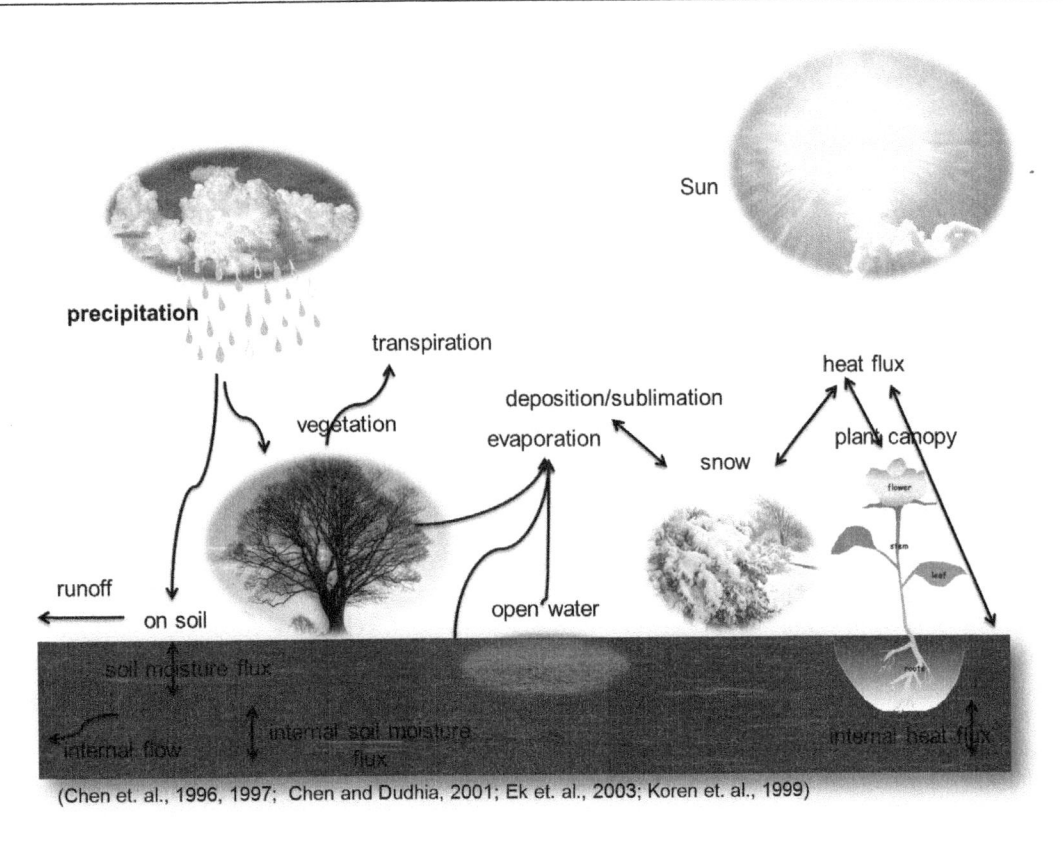

A graphical representation of the global hydrological exchanges.

Carbon dioxide, a principal constituent of volcanic emissions, is also released into the atmosphere by the acid decomposition of carbonate rock, from oceanic sources, as a product of respiration, and from combustion of fossil fuels. Argon, far more abundant in Earth's atmosphere than any of the other noble (inert) gases, is a by-product of the radioactive decay of an isotope of potassium. Helium is also mainly a by-product of radioactive decay.

VERTICAL STRUCTURE

The atmosphere has a well-defined lower boundary at the surface of the planet, but extends indefinitely away from the Earth. At 30,000 kilometers molecules are no longer effectively held in orbit by gravity. Within that distance the atmosphere can be thought of as a series of layers. However, the layering is far subtler than what may be found in, for example, a geologic formation. The most common method of demarcating layers is to examine the average change of temperature as a function of elevation. Earth's surface, warmed by the absorption of solar radiation, conducts heat into the lowest portion of the atmosphere. This lowest layer, known as the troposphere, extends to about 10 kilometers above the surface and is characterized by temperatures that decrease with height. Virtually all the phenomena that are commonly referred to as "weather" occur in the troposphere. The average density of air at sea level is about 1.225 kilograms per cubic meter. Because air is a compressible fluid, air density decreases logarithmically with height. Half the mass of the atmosphere lies below about 5.5 kilometers. Approximately 80 percent of the atmosphere's mass is found in the troposphere.

Between 10 and 50 kilometers, temperatures increase with increasing altitude in the layer known as the stratosphere. The warming of air in this layer is accounted for by the heat released as ozone molecules absorb ultraviolet wavelengths of solar radiation. Ozone concentration is at a maximum in this layer. Historically, it was thought that there was little exchange of air between the troposphere and stratosphere, except during volcanic and atomic explosions, because temperature profiles such as that found in the stratosphere typically suppress mixing. However, the occurrence of human-made chlorofluorocarbons (CFCs) in the stratosphere demonstrated that exchange does take place. The presence of CFCs in the stratosphere is detrimental to ozone and serves as an ozone sink that has no compensating source.

Temperatures once again decrease with increasing height between 50 and 80 kilometers in the mesosphere. The troposphere and stratosphere together account for about 99.9 percent of the atmosphere's mass. The mesosphere contains about 99 percent of the remaining mass.

The thermosphere is situated above the mesosphere and extends indefinitely away from the Earth. Temperatures once again increase with height in this layer and can reach 500 to 2,000 Kelvin depending upon the amount of solar activity. However, temperatures begin to take on a different meaning at these altitudes owing to the relatively small number of molecules and the large mean free path between collisions.

The tops of these four layers are known as the tropopause, stratopause, mesopause, and thermopause, respectively. Temperatures typically remain constant for a few kilometers at the interface of the layers. A feature of note at the tropopause is the jet stream, an especially swift current of air.

The atmosphere can also be partitioned vertically based on how uniformly mixed its constituents are. Turbulent processes in the atmosphere below about 80 kilometers keep the constituents in the lower atmosphere well mixed. This region is known as the homosphere. Air sampled near both the top and bottom of the homosphere will contain nearly equal percentages of each constituent gas, although the densities of the samples will be markedly different. Above 80 kilometers, the vertical mixing of constituents is controlled by molecular diffusion, allowing them to separate by mass, with the lightest gases (hydrogen and helium) present at the highest levels. This region is known as the heterosphere. Sunlight in the heterosphere is more intense than sunlight that penetrates to the homosphere because little filtering has taken place. As a result, ionization occurs in the heterosphere, and this ionization affects the transmittable range of commercially broadcast radio signals that are redirected by the ionized molecules.

BALANCE OF ENERGY

The Sun is the source of nearly all of the energy the atmosphere receives. Minor amounts of energy are contributed by lightning and Earth's internal heat sources. There is a global balance between the solar radiation that heats the atmosphere and the terrestrial radiation emitted to space. However, the balance does not hold for individual latitudes. The complex geometry of a spheroid planet precessing about its rotational axis which is also tilted with respect to the plane of its elliptical orbit about the Sun results in an imbalance between absorbed and emitted radiation. Over the course of a typical year, the tropical region of the Earth between about 37° north and 37° south latitude receives more energy from the Sun than what is regionally emitted back to space. Progressing towards the north and south poles from this region, Earth radiates to space more energy than it receives from the Sun.

As a result of the regional imbalance of energy, there is a continuous transport of energy in the atmosphere and the oceans from the tropical latitudes, where there is a surplus of energy, to the polar latitudes, where there is a deficit. If this transport did not occur, the tropics would continually warm while the polar latitudes would grow colder year after year. The transport of energy by winds and weather systems is most apparent in the middle latitudes of the planet across the interface between the regions of surplus and deficit. In the lower atmosphere the principal forms of the energy are internal energy (associated with the temperature of the air) and latent energy (associated with the phase of water). In the case of the latter, the evaporation of ocean water in the tropics transforms internal energy into latent energy. Water vapor, being a gas and thus highly mobile, is transported away from the tropics and may subsequently condense to form clouds or dew. Condensation releases an amount of energy equal to that used in evaporation. Evaporation and condensation are first-order processes in Earth's heat budget. In addition, they play key roles in Earth's hydrologic cycle. This cycle purifies and redistributes the planet's single most important compound and the resource without which life would not exist.

The Hydrologic Cycle

Though there are approximately 1.3 billion cubic kilometers of water on Earth, about 97 percent of this is salt-bearing ocean water rather than fresh water. Evaporation of ocean water into the atmosphere, its transport by weather systems, and the subsequent condensation in clouds provide life's most precious resource, fresh water, to the continents. The evaporation of water from the oceans and evapotranspiration over land, the transport of water in the atmosphere, and its eventual return to the oceans are collectively known as the hydrologic cycle.

Over the continents, precipitation exceeds evaporation, while the reverse is true over the oceans. Some of the water vapor added to the atmosphere by evaporation from the oceans is transported to the continents, where it combines with water vapor from evapotranspiration, condenses, and falls as precipitation. Some of this precipitation percolates into and becomes part of underground aquifers, or groundwater. Some of the precipitation is returned to the ocean by runoff in rivers. Water vapor is also transported from over the continents to over the oceans in the atmosphere. Generally, water evaporated in one location is not the same water that precipitates on that location. Water vapor is usually transported hundreds or even thousands of kilometers from its source. For example, the majority of water that falls as precipitation on the portion of the United States east of the Rocky Mountains is evaporated off the Gulf of Mexico. Evaporation off the Indian Ocean is the source of the precipitation for the wet Indian monsoon. The hydrologic cycle is rarely completed on a local scale.

Observations indicate that rain and snowfall on the continents is well in excess of the runoff from these same areas. Only about 20 percent of the precipitation that falls on land is returned to the ocean by runoff. While some of the remaining precipitation is stored underground in permeable rock and other types of aquifers, the majority of the excess is transported back to the oceans by air masses. Cold, dry air masses moving towards the equator over land areas are warmed and moistened by evapotranspiration from the surfaces over which they pass. Studies of the change in moisture content of continental polar air moving towards the equator over the Mississippi River drainage basin in the United States indicate that these air masses can remove, by evapotranspiration, a quantity of water equal to nine times the average discharge of the Mississippi River. The hydrologic cycle is subject to great disruptions under conditions of short-term or long-term climate change. Examples of such disruptions include floods and droughts.

Resources from the Atmosphere

The atmosphere is a ready source of several gases used in industry and other applications. The industrial use of gases obtained from the atmosphere began in the early years of the twentieth century. The separation of the constituents of air is basically a three-step process. First, impurities are removed. Second, the purified air is liquefied by compression and refrigeration. Third, the individual components are separated by distillation, making use of the fact that each liquefied component boils at a different temperature.

Air separation plants produce oxygen, nitrogen, and argon for delivery in both the gaseous and liquid phases. The total mass of the atmosphere is about 5.27×10^{18} kilograms. Given the percentage, by mass, of nitrogen (76 percent) and oxygen (23 percent) in the atmosphere, there are in theory about 1.2×10^{18} kilograms of oxygen and 4.0×10^{18} kilograms of nitrogen available for separation and use.

Gases from the atmosphere are used by the steel industry in the cutting and welding of metals. Other user communities include the aerospace industry, chemical companies, and the medical industry. Liquid nitrogen is used in applications requiring extreme cold. The inert nature of gaseous nitrogen and argon makes them useful for flushing air out of systems when one also needs to prevent chemical reactions from occurring. The atmosphere also provides a source of argon, neon, krypton, and xenon and is the only known source of several of the rare gases.

The Atmosphere and Human Health

In addition to being a resource itself, the atmosphere has direct and indirect effects on many other resources and on human health. Examples of aspects dependent on atmospheric conditions include the resistance of crops to disease and insects; the health and productivity of forests; milk, wool, and egg production; and meat quality. Biometeorology, also known as bioclimatology, is the branch of atmospheric science concerned with the effects of weather and climate on the health and activity of human beings.

Deaths from heart attacks and heart disease increase when the human body experiences great thermal stress, as in extreme heat or cold or when temperature changes abruptly. Deaths tend to peak in winter in colder climates and in summer in warmer climates.

An example of the devastating effect high temperature can have on human health is the European heat wave of 2003. Temperatures varied from country to country, but France reported seven days that exceeded 40°C. More than 50,000 people died throughout Europe as a result of the aberrant climate. In Switzerland, where temperatures reached 41°C, flash floods occurred because of melting glaciers. The European agricultural industry suffered extensive losses because of this heat wave: In the wake of severe climate, wheat production fell by 20 percent in France and grapes ripened prematurely.

—*Alan C. Czarnetzki*

Further Reading

Fabian, Peter and Dameris, Martin *Ozone in the Atmosphere. Basic Principles, Natural and Human Impacts.* Springer, 2014. This book provides an advanced discussion of the role of ozone in Earth's environment, its interaction with the many other gaseous components if Earth's atmosphere, and how it is affected by many natural and unnatural factors.

Frederick, John E. *Principles of Atmospheric Science.* Jones and Bartlett 2008. Provides a concise introduction to atmospheric science beginning from the basic constituents and structure of the atmosphere.

Lutgens, Frederick K., Edward J. Tarbuck, Tasa, Dennis G. and Herman, Redina. *The Atmosphere: An Introduction to Meteorology.* 14th ed. Pearson Education, 2018. An updated version of this textbook, in loose-leaf format, provides students with a general overview of the structure and properties of Earth's atmosphere.

Mbane Biouele, Cesar *Earth's Atmosphere Dynamic Balance Meteorology.* Scientific Research Publishing, 2014. Without becoming overly technical, this book discusses the structure of the atmosphere and many of the physical phenomena that occur there.

Novák, Viliam *Evapotranspiration in the Soil-Plant-Atmosphere System.* Springer, 2012. Provides detailed information about the movement of water in the evapotranspiration segment of the hydrologic cycle.

Spellman, Frank R. *The Science of Air. Concepts and Applications.* 2nd ed. CRC Press, 2009. This book is a comprehensive resource covering the components, dynamics, interactions and uses of air as a resource.

Wells, Neil C. *The Atmosphere and Ocean. A Physical Introduction.* 3rd ed. John Wiley & Sons, 2012. This book provides a complete overview of the major principles mentioned in this article.

ATMOSPHERE AND CLIMATE CHANGE

Fields of Study

Atmospheric Science; Atmospheric Chemistry; Bioclimatology; Oceanography; Thermodynamics; Environmental Sciences; Environmental Studies; Meteorology; Climate Modeling; Fluid Dynamics; Hydroclimatology; Hydrometeorology; Earth System Modeling; Heat Transfer

Summary

The atmosphere is a gaseous layer surrounding Earth. It is a mixture of several components. The total weight of Earth's atmosphere is about 5.08 quadrillion tonnes. Its existence makes life on Earth possible, and minor changes in its composition have the potential to produce major changes in Earth's climate, and may even destroy life on Earth. The atmosphere's various physical properties affect many aspects of human existence. Its optical properties make the sky blue and create rainbows and auroras. It carries sound waves, making aural communication possible. It also propagates heat, and seasonal changes in atmospheric thermal properties result in cooler and warmer temperatures at

ATMOSPHERIC STRUCTURE

Because of Earth's gravity, most atmospheric molecules are distributed very close to the Earth's surface. More than 90 percent of Earth's atmosphere lies within 32 kilometers of the planet's surface, and the lower atmosphere begins to shade into the upper atmosphere and outer space at an altitude of about 80 kilometers. This distance is only a fraction of the Earth's radius, which is about 6,400 kilometers.

The atmosphere's density and pressure decrease steadily with altitude. Altitudinal variations in temperature, on the other hand, are more complex and define a layered structure within the atmospheric mass. The temperature decreases with altitude to about 10–16 kilometers from the surface. From that height to about 20 kilometers, the temperature remains relatively constant. From 20 to 45 kilometers, the temperature increases with height, and at altitudes of 45–50 kilometers it again remains constant. Above 50 kilometers, the temperature again begins to decrease with altitude, continuing until reaching a height of about 80 kilometers. From that height to about 90 kilometers, temperature is constant. From 90 kilometers to the end of Earth's atmosphere, temperature increases with altitude, though "temperature" in that extremely rarefied medium has a different meaning than it does nearer to Earth's surface. Based on these altitudinal temperature variations, scientists divide the atmosphere into sections: the troposphere, tropopause, stratosphere, stratopause, mesosphere, mesopause, and thermosphere.

The atmosphere can also be classified according to its composition. In the lower atmosphere (below 80 kilometers), weather convection and turbulent mixing keep the composition of air fairly uniform. This area is therefore known as the homosphere. Above the homosphere, collisions among atoms and molecules are infrequent, and the air is not homogeneously mixed. This layer is therefore called the heterosphere. Furthermore, above the stratosphere, large concentrations of ions and free electrons exist, in a layer known as the ionosphere.

THE ENERGY BUDGET AND THE GREENHOUSE EFFECT

The Sun is Earth's ultimate energy source. Solar energy reaches Earth as solar radiation, or sunlight. During daytime, the surface of Earth is warmed by the Sun, while during nighttime it cools by radiating

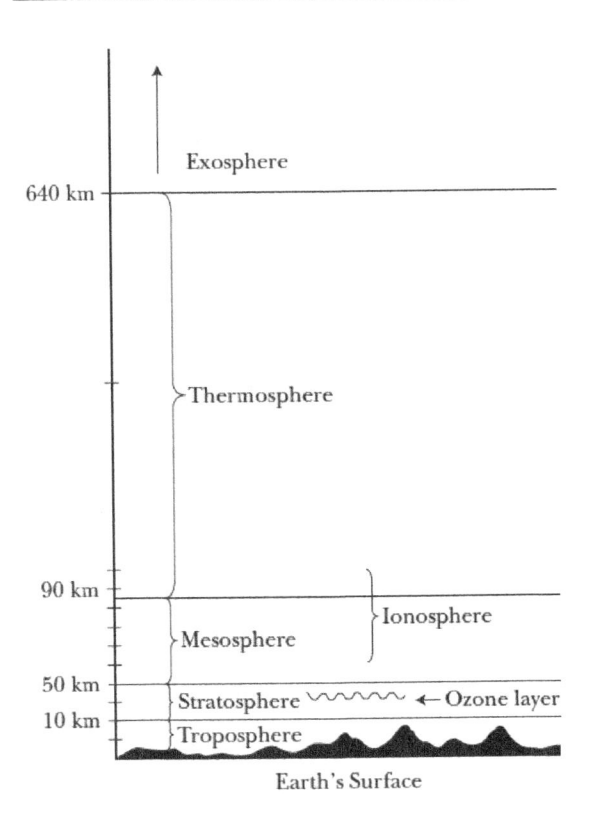

different times of year. More generally, the weather is a function of the atmosphere. Wind, precipitation, humidity, and storm systems are all atmospheric phenomena.

PRINCIPAL TERMS

- **condensation:** the transformation of a substance from its gaseous to its liquid state, accompanied by a release of heat
- **convection:** motion in a fluid that results in the transport and mixing of the fluid's physical properties, such as heat
- **coupled atmosphere-ocean models:** computer simulations of alterations in and interactions between Earth's atmosphere and oceans
- **El Niño and La Niña:** periodic warming and cooling of the eastern tropical Pacific Ocean that affects global weather patterns
- **greenhouse gases (GHGs):** atmospheric trace gases that allow sunlight to reach Earth's surface but prevent heat from escaping into space
- **latent heat:** the heat released or absorbed by a change of state, such as condensation

away as heat energy that it has acquired from sunlight during the day. The sunlight that Earth receives is a form of electromagnetic radiation with a relatively short wavelength, called "shortwave radiation." When Earth cools at night, the heat it releases into space is another form of electromagnetic radiation with a relatively long wavelength. This "longwave radiation" is also known as infrared radiation (IR).

If Earth did not have an atmosphere, it would simply receive shortwave solar radiation during the day and give out longwave radiation at night. At equilibrium, meaning that the amount of energy received is equal to the amount of energy given out, Earth would be balanced at an equilibrium temperature that is much lower than is the actual temperature on Earth with the atmosphere. Some scientists have estimated it at about -18°C. Earth's surface temperature is thus much warmer than it would be without an atmosphere.

This difference in temperature is a result of what has been termed the greenhouse effect. Shortwave solar radiation can mostly penetrate Earth's atmosphere and reach the surface. Infrared radiation, by contrast, is intercepted by the atmosphere, and some of this longwave radiation is reflected back to Earth's surface, increasing the equilibrium temperature of the planet. A new equilibrium temperature is reached that depends on the extent of this energy feedback system, which in turn depends on the amounts and kinds of greenhouse gases in the atmosphere. The present strength of the greenhouse effect and the warmth of Earth's equilibrium temperature are thus dependent on the specific composition and properties of the atmosphere. Changes in atmospheric chemistry and composition can significantly alter Earth's energy balance and equilibrium temperature.

ATMOSPHERIC COMPOSITION, CHEMISTRY, AND GHGS

The atmosphere is composed of nitrogen (about 78 percent by volume), oxygen (21 percent), and various trace gases (totaling about 1 percent). If all trace gases were removed, these percentages for nitrogen and oxygen would remain fairly constant up to an altitude of about 80 kilometers. At the planet's surface, there is an approximate balance between the output and input of these minor gas components.

The two most plentiful components of the atmosphere, nitrogen and oxygen, are of significance to life on Earth. Humans and animals cannot live without oxygen. By contrast, the trace noble gases, such as argon, neon, and helium, are essentially entirely inert chemically. Water vapor is distributed inconsistently in the lower atmosphere; its concentration varies greatly from place to place and from time to time, and it can constitute from 0 to 4 percent of local air. This variable concentration is one reason that water vapor is so important in influencing Earth's weather and climate.

Water vapor provides the main physical substance of precipitation. Its condensation into the liquid state releases the large amount of energy (latent heat) necessary to initiate powerful and violent storms. Water vapor is also a greenhouse gas (GHG). It strongly absorbs longwave radiation and reemits this radiation back toward Earth's surface, acting to increase the temperature of the atmosphere. Clouds, which are generated from water vapor, also play an extremely important role in climate and climate change.

Another very important GHG is carbon dioxide (CO_2). Observations indicate that the concentration of CO_2 in the atmosphere has been rising steadily for more than a century. The increase of CO_2 concentration indicates that CO_2 is entering the atmosphere at a greater rate than its rate of removal. This rise is largely attributable to the burning of fossil fuels, such as coal and oil. Deforestation also contributes to the increase in atmospheric CO_2 concentration by removing a significant fraction of the mechanism by which CO_2 is removed from the atmosphere through photosynthesis. Estimates project that by sometime in the second half of the twenty-first century, CO_2 levels will be twice as high as they were early in the twentieth century. Other GHGs include methane, nitrous oxide, and chlorofluorocarbons (CFCs).

Ozone (O_3) is another important gas for Earth's weather and climate. At Earth's surface, O_3 is a major air pollutant, and it is closely monitored for its effects on air quality. However, at upper levels (about 25 kilometers high), O_3 shields Earth from harmful ultraviolet solar radiation. For this reason, the loss of O_3 high in the atmosphere as a consequence of human activity has become a serious global-scale issue. Ozone depletion by anthropogenic activity, the injection of large quantities of CFCs into the atmosphere over many years, resulted in the formation of a hole in the ozone layer over Antarctica. Finally, aerosols, including particulate matter, are also important constituents of the atmosphere, affecting weather formation, air quality, and climate change.

Weather and Climate

Atmospheric conditions can generally be classified as either weather or climate. Weather is a particular atmospheric state at a given time and place. Climate is an average of weather conditions at a given area over a period of time. Weather includes many atmospheric phenomena of different scales, including middle latitude cyclones (extratropical cyclones), hurricanes (tropical cyclones), heavy rains and floods, mesoscale convective systems, thunderstorms, rain, snow and tornadoes. Climate includes atmospheric conditions that are millennial, centennial, decadal, interannual, or seasonal. For example, global warming can occur on a centennial or longer timescale, an El Niño or La Niña episode will generally occur on an interannual timescale, and seasonal changes occur on relatively short timescales of months, weeks, or days.

Global Warming and Climate Change

Earth's climate has changed constantly over its history. The planet has experienced many cold periods, as well as several warm periods. For example, about ten thousand years ago, the Earth cooled during a period known as the Younger Dryas, when the average global temperature was about 3°C colder than it is today. However, about six thousand years ago, the Earth reached the middle of an interglacial period, known as the Mid-Holocene Maximum. The temperature then was 1°C higher than today's norm. Some of the natural mechanisms contributing to this climate variability include plate tectonics, volcanic activities, ocean circulations, variations in Earth's orbit, and solar variability.

The rapid warming that has occurred in the past hundred years seems to coincide with the socioeconomic development and industrialization of humankind, particularly since the beginning of the period known as the Industrial Revolution. During the past two centuries, human societies have depended heavily on burning fossil fuels to produce usable energy. As a result, increasing amounts of CO_2 have been added to the atmosphere. Global temperatures and CO_2 levels evince a consistent upward trend during the same period. Therefore, many scientists believe that human activity has contributed to climate change as global warming.

Various coupled atmosphere-ocean models have projected that Earth will experience an average temperature increase over the course of the twenty-first century of between 1.4° and 5.8°C. Some possible consequences of this global warming include higher extreme maximum and lower extreme minimum temperatures as weather patterns become more widely variable, more hot days and heat waves, fewer cold days and frost days, more intense precipitation events, more summer drying and drought, increased tropical cyclone intensity, increased Asian summer monsoon precipitation variability, intensified droughts and floods associated with El Niño events, and increased intensity of midlatitude storms. Global warming will also exert some profound effects on many other social and environmental issues. For example, the distribution of water resources and farming may be changed by future warmer climates. Sea-level rise due to the increased warming will have significant effects on the coastal areas of many countries. Loss of Arctic sea-ice will also have geopolitical and economic consequences.

Context

The atmosphere is central to almost all aspects of human existence, constituting not only the source of vital oxygen but also the medium of movement, sound, and weather. It is also the most variable component of Earth's climate system. Because that system is so complex and interconnected, changes in the atmosphere will inevitably result in changes to the rest of the system, some of which are extremely difficult to predict. At the same time, feedback from other components of Earth's environment may exert a significant influence on the atmosphere, including producing positive and negative feedback loops that help alter or maintain Earth's climate.

—*Chungu Lu*

Further Reading

Ahrens, C. Donald. *Essentials of Meteorology: An Invitation to the Atmosphere.* 5th ed. Belmont, Calif.: Thomson Brooks/Cole, 2008. Widely used introductory textbook on atmospheric science; covers a wide range of topics on weather and climate.

Barker, John Roger, et al. *Advances in Atmospheric Chemistry.* World Scientific, 2017.

Dessler, Andrew *Introduction to Modern Climate Change.* 2nd ed. Cambridge University Press, 2016. Provides a good overview of the science and sociology of climate change.

Frederick, John E. *Principles of Atmospheric Science.* Jones and Bartlett 2008. Provides a concise introduction to atmospheric science beginning from the basic constituents and structure of the atmosphere.

Gornitz, Vivien *Rising Seas. Past, Present, Future.* Presents an overview of sea level rise due to climate change and describes how rising sea levels may affect the stability of the ocean heat conveyor system.

Lutgens, Frederick K., Edward J. Tarbuck, Tasa, Dennis G. and Herman, Redina. *The Atmosphere: An Introduction to Meteorology.* 14th ed. Pearson Education, 2018. An updated version of this textbook, in loose-leaf format, provides students with a general overview of the structure and properties of Earth's atmosphere.

Neelin, J. David *Climate Change and Climate Modeling.* Cambridge University Press, 2011. Discusses how climate data is used to identify climate change and to produce predictive models of future climate states.

Wadhams, Peter *A Farewell to Ice. A Report from the Arctic.* Oxford University Press, 2017. Discusses the role of Arctic and Antarctic ice shields, and how feedback systems function. Special attention is given to the potential for the release of quantities of methane from polar sea floors and Arctic permafrost as Earth's average temperature increases.

ATMOSPHERE: EVOLUTION

FIELDS OF STUDY

Geochemistry; Earth System Modeling; Analytical Chemistry; Inorganic Chemistry; Atmospheric Chemistry; Photochemistry; Thermodynamics; Climate Modeling; Process Modeling; Fluid Dynamics; Atmospheric Science; Physical Geography

SUMMARY

The chemical composition of the atmosphere has changed significantly over the 4.6-billion-year history of Earth. The composition of the atmosphere has been influenced by a number of processes, including the "outgassing" of volatile materials originally trapped in Earth's interior during its formation; the geochemical cycling of carbon, nitrogen, hydrogen, and oxygen compounds between the surface, the ocean, and the atmosphere; and the evolution of life.

PRINCIPAL TERMS

- **chemical evolution:** the synthesis of amino acids and other complex organic molecules, as the precursors of living systems, by the action of atmospheric lightning and solar ultraviolet radiation on atmospheric gases
- **photosynthesis:** the biochemical synthesis of glucose and molecular oxygen from carbon dioxide and water by chlorophyll-containing organisms in the presence of sunlight
- **prebiotic:** relating to the period of time before the appearance of life on Earth
- **primordial solar nebula:** an interstellar cloud of gases and dust that condensed by the action of gravitational forces to form the bodies of the solar system about 5 billion years ago
- **solar ultraviolet radiation:** biologically lethal solar radiation in the spectral interval between approximately 0.1 and 0.3 micron (1 micron = 0.0001 centimeter)
- **T Tauri stars:** a class of stars that exhibits rapid and erratic changes in brightness
- **volatile outgassing:** the release of the gases and liquids, such as argon, water vapor, carbon dioxide, and nitrogen sulfur, trapped within Earth's interior during its formation

VOLATILE OUTGASSING

About 5 billion years ago, a cloud of interstellar gas and dust, called the primordial solar nebula, began to condense under the influence of gravity. This condensation led to the formation of the sun, moon, Earth, the other planets and their satellites, asteroids, meteors, and comets. The primordial solar nebula was composed almost entirely of hydrogen gas, with a smaller amount of helium, still smaller amounts of carbon, nitrogen, and oxygen, and still smaller amounts of the rest of the elements of the periodic table. About the time that the newly formed Earth attained its approximate present mass, gases that were released from the planet's interior could be retained by Earth's gravity instead of escaping into space, thus forming a gravitationally bound atmosphere.

It is believed that the atmospheres of the other terrestrial planets, Mars and Venus, also formed in this manner. The release of gases and other volatiles in this manner is called volatile outgassing. The period of extensive volatile outgassing may have lasted for many tens of millions of years. The outgassed volatiles or gases had roughly the same chemical composition as present-day volcanic emissions: 80 percent water vapor by volume, 10 percent carbon dioxide by volume, 5 percent sulfur dioxide by volume, 1 percent nitrogen by volume, and smaller amounts of hydrogen, carbon monoxide, sulfur, chlorine, and argon.

The water vapor that outgassed from the interior eventually reached the saturation point, which is controlled by the atmospheric temperature and pressure. Once the saturation point was reached, the atmosphere could not hold any additional gaseous water vapor. Any new outgassed water vapor that entered the atmosphere would have precipitated out of the atmosphere in the form of liquid water. The equivalent of several cubic kilometers of liquid water released from Earth's interior in gaseous form precipitated out of the atmosphere and formed the oceans. Only small amounts of water vapor remained in the atmosphere, ranging from a fraction of a percent to several percent by volume, depending on atmospheric temperature, season, and latitude.

The outgassed atmospheric carbon dioxide, being somewhat water soluble, dissolved in the newly formed oceans and subsequently formed carbonic acid through its reaction with water. Once formed, carbonic acid can dissociate into ions of hydrogen, bicarbonate, and carbonate. The carbonate ions reacted with ions of calcium and magnesium in the ocean water, forming first insoluble carbonate salts, which precipitated out of the ocean and accumulated as seafloor carbonate sediments, eventually accumulating in sufficient quantities to form beds of carbonate rock. Most of the outgassed atmospheric carbon dioxide formed carbonates, leaving only trace amounts of gaseous carbon dioxide in the atmosphere (about 0.035 percent by volume). Sulfur dioxide, the third most abundant component of volatile outgassing, was chemically transformed into other sulfur compounds, including sulfuric acid and other sulfates in the atmosphere. Eventually, the sulfates formed atmospheric aerosols and gravitated out of the atmosphere to settle on the surface.

The fourth most abundant outgassed compound, nitrogen, is almost completely chemically inert in the atmosphere and thus was not readily transformed in large quantities, as was sulfur dioxide. Only minor amounts of nitrogen would be converted to various oxides by the action of lightning, to be trapped as nitrogen oxide salts in minerals. Unlike carbon dioxide, nitrogen is relatively insoluble in water and, unlike water vapor, does not condense out of the atmosphere. For these reasons, nitrogen built up in the atmosphere to become its major constituent (78.08 percent by volume). Accordingly, outgassed volatiles led to the formation of Earth's atmosphere, oceans, and the earliest, prebiotic carbonate rocks.

CHEMICAL EVOLUTION

It has been demonstrated in laboratory experiments that molecular nitrogen, carbon dioxide, and water vapor in the early atmosphere would have been acted upon by solar ultraviolet radiation and atmospheric lightning. In the process, molecules of formaldehyde and hydrogen cyanide were chemically synthesized in the early atmosphere. These molecules were precipitated and diffused out of the atmosphere into the oceans. In the water, formaldehyde and hydrogen cyanide entered into chemical reactions that eventually led to the chemical synthesis of amino acids—the building blocks of proteins in living systems. The synthesis of amino acids and other compounds from nitrogen, carbon dioxide, and water vapor in the atmosphere is called chemical evolution. Chemical evolution preceded and provided the material for biological evolution.

For many years, it was thought that the early atmosphere was composed of ammonia, methane, and hydrogen rather than of carbon dioxide, nitrogen, and water vapor. Experiments show, however, that ammonia and methane are chemically unstable and are readily destroyed by both solar ultraviolet radiation and chemical reaction with the hydroxyl radical, which is formed from water vapor.

In addition, ammonia is very water soluble and is readily removed from the atmosphere by precipitation. Hydrogen, the lightest element, is readily lost from a planet by gravitational escape. Thus, an early atmosphere composed of methane, ammonia, and hydrogen would be very short lived, unless these gases were produced at a rate equal to their destruction or loss rates (an equilibrium state). These gases are also known to be extremely efficient "greenhouse gases," even more effective than carbon dioxide; their presence in the primordial atmosphere in any substantial amount would have maintained an extraordinarily high atmospheric temperature, by which many of the materials that were formed would thermally decompose. Today, methane and ammonia are very minor components of the atmosphere, at concentrations of 1.7 parts per million by volume and 1 part per billion by volume, respectively. Both gases are produced by microbial activity at the ground surface, and methane is released during coal mining and oil production, and from seafloor accumulations of methane hydrate. Clearly, microbial activity and microbes were nonexistent during the prebiotic phase of the planet, in the time before life existed on Earth.

The atmospheres of the outer gas giant planets Jupiter, Saturn, Uranus, and Neptune all contain quantities of hydrogen, methane, and ammonia. It is believed that the atmospheres of these planets, unlike the atmospheres of the terrestrial planets Earth, Venus, and Mars are captured remnants of the primordial solar nebula resulting from the greater ability of the gravitational fields of those large planets to capture such light materials, preventing them from being drawn toward the sun. Because of the outer planets' great distance from the sun and their very low temperatures, hydrogen, methane, and ammonia are stable and long-lived constituents of their atmospheres. This is not true of hydrogen, methane, and ammonia in Earth's atmosphere.

Some have suggested that at the time of its formation, Earth may have also captured a remnant of the primordial solar nebula as its very first atmosphere. Such a captured primordial solar nebula atmosphere would have been composed of mostly hydrogen (about 90 percent) and helium (about 10 percent), the two major elements of the nebula. Even if such an atmosphere had surrounded the very young Earth, it would have been very short lived. As the young sun went through the T Tauri phase of its evolution, very strong solar winds (the supersonic flow of protons and electrons from the sun) associated with that phase would have quickly dissipated this remnant atmosphere. In addition, there is no geochemical evidence to suggest that early Earth ever possessed a primordial solar nebula remnant atmosphere.

Evolution of Atmospheric Oxygen

There is microfossil evidence for the existence of fairly advanced anaerobic microbial life on Earth by about 3.8 billion years ago. The ability to carry out photosynthesis evolved in one or more of these early microbial species. Through photosynthesis, the organism utilizes water vapor and carbon dioxide in the presence of sunlight and chlorophyll to form glucose and molecular oxygen. The glucose molecules are subsequently used by the organism for food and for biopolymerization into starches and celluloses. The production of oxygen by photosynthesis was a major event on Earth and eventually transformed the composition and chemistry of the early atmosphere as oxygen built up to become the second most abundant constituent of the atmosphere (20.90 percent by volume). It has been estimated that atmospheric oxygen reached only 1 percent of its present atmospheric level 2 billion years ago, 10 percent of its present atmospheric level about 550 million years ago (at the beginning of the Paleozoic), and its present atmospheric level as early as 400 million years ago.

The evolution of atmospheric oxygen had important implications for the evolution of life. Because molecular oxygen is a very effective oxidizing agent and would have been harmful to existing anaerobic life-forms, the presence and buildup of oxygen required the evolution of respiration and aerobic organisms. Accompanying and directly controlled by the buildup of atmospheric oxygen were the origin and evolution of atmospheric ozone, which is chemically formed from oxygen. The evolution of atmospheric ozone resulted in the shielding of Earth's surface from biologically lethal solar ultraviolet rays.

The development of the atmospheric ozone layer and its accompanying shielding of Earth's surface permitted early life to evolve such that it could leave the safety of the oceans and go ashore for the first time in the history of the planet. Prior to the evolution of the atmospheric ozone layer, early life was restricted to a depth of several meters below the ocean surface. At this depth, the ocean water offered shielding from solar ultraviolet radiation. Theoretical computer calculations indicate that atmospheric ozone provided sufficient shielding from biologically lethal ultraviolet radiation for the evolution of non-marine organisms once oxygen reached about one-tenth of its present atmospheric level.

VENUS AND MARS

Calculations indicate that the atmospheres of Venus and Mars also formed as a consequence of volatile outgassing of the same gases that led to the formation of Earth's atmosphere; water vapor, carbon dioxide, and nitrogen. In the case of Venus and Mars, however, the outgassed water vapor may never have existed in the form of liquid water in quantities comparable to those on Earth. Because of Venus's closer distance to the sun (108 million kilometers versus 150 million kilometers for Earth), its lower atmosphere was too hot to permit the outgassed water vapor to condense out of the atmosphere. Thus, the outgassed water remained in gaseous form in the atmosphere and, over geological time, was decomposed by solar ultraviolet radiation into molecular hydrogen and oxygen. The very light hydrogen gas quickly escaped from the atmosphere of Venus, and the heavier oxygen combined with surface minerals to form a highly oxidized surface. In the absence of liquid water on the surface of Venus, the outgassed carbon dioxide remained in the atmosphere and built up to become the overwhelming constituent of Venus's atmosphere, about 96 percent by volume. The outgassed nitrogen accumulated to make up about 4 percent by volume of the atmosphere of Venus.

The present-day carbon dioxide and nitrogen atmosphere of Venus is massive. Its atmospheric surface pressure is about 90 times greater than the surface pressure of Earth's atmosphere. If the outgassed carbon dioxide in Earth's atmosphere had not been dissipated via dissolution in the oceans and carbonate formation, the planet's surface atmospheric pressure would presumably be about 70 atmospheres, with carbon dioxide accounting for about 98 to 99 percent of the atmosphere and nitrogen about 1 to 2 percent. Thus, the atmosphere of Earth would closely resemble that of Venus. The carbon dioxide-rich atmosphere of Venus causes a very significant greenhouse temperature enhancement, giving the surface of Venus a temperature of about 750 kelvins, which is hot enough to melt lead. The surface temperature of Earth is only about 288 kelvins, a range at which water can exist in equilibrium between all three phases, as solid, liquid, and gas.

Like Venus, Mars has an atmosphere, though extremely thin, composed primarily of carbon dioxide (about 95 percent by volume) and nitrogen (about 3 percent by volume). The total atmospheric surface pressure of Mars is only about 7 millibars (1 atmosphere is equivalent to 1,013 millibars). There appears to be large quantities of outgassed water in the form of ice or frost below the surface of Mars, but in the absence of liquid water, the outgassed carbon dioxide has remained in the atmosphere. The smaller mass of the atmosphere of Mars compared to the atmospheres of Venus and Earth may be attributable to the smaller mass of the planet and, accordingly, the smaller mass of gases that could have been trapped in the interior of Mars during its formation. In addition, the amounts of gases trapped in the interiors of Venus, Earth, and Mars during their formation apparently decreased with increasing distance from the sun. Venus appears to have trapped the greatest amounts of gases and was the most volatile-rich planet. Earth trapped the next greatest amounts, and Mars trapped the smallest amounts.

STUDY OF EARTH'S ATMOSPHERE

Information about the origin, early history, and evolution of Earth's atmosphere comes from a variety of sources. Information on the origin of Earth and other planets is based on theoretical computer simulations, with ever-increasing empirical data input from celestial observation. These computer models simulate the collapse of the primordial solar nebula and the formation of the planets. Astronomical observations of what appears to be the collapse of interstellar gas clouds and the possible formation of planetary systems have provided new insights into the computer modeling of this phenomenon. Information about the origin, early history, and evolution of the atmosphere is based on theoretical computer models of volatile outgassing, the geochemical cycling of the outgassed volatiles, and the photochemistry of the outgassed volatiles.

The process of chemical evolution, which could lead to the synthesis of organic molecules of increasing complexity as the precursors of the first living systems on the early Earth, is studied in laboratory experiments in which mixtures of gases simulating Earth's hypothetical early atmosphere are energized by solar ultraviolet radiation and atmospheric lightning. The resulting products are analyzed by chemical techniques. The results of such experiments suggest that a key parameter affecting atmospheric photochemical reactions, chemical evolution, and the theoretical origin of life was the flux of solar ultraviolet radiation on the early Earth. Astronomical measurements of the ultraviolet emissions from young sun-like stars have provided important information about the probable ultraviolet emissions from the sun during the early history of the atmosphere.

Geological and paleontological studies of the oldest rocks and the earliest fossil records have provided important information on the evolution of the atmosphere and the transition from an oxygen-deficient to an oxygen-rich atmosphere. Studies of the biogeochemical cycling of the elements have provided important insights into the later evolution of the atmosphere. Thus, studies of the origin and evolution of the atmosphere are based on a broad cross-section of science, involving astronomy, geology, geochemistry, geophysics, and biology as well as atmospheric chemistry.

SIGNIFICANCE

Studies of the origin and evolution of Earth's atmosphere have provided new insights into the processes and parameters responsible for global change. Understanding the history of the atmosphere provides a sound basis for better understanding its future. Today, several global environmental changes are of national and international concern, including the depletion of ozone in the stratosphere and increasing global temperatures caused by the buildup of greenhouse gases in the atmosphere. The study of the evolution of the atmosphere has provided new insights into the biogeochemical cycling of elements between the atmosphere, biosphere, land, and ocean. Understanding this cycling is a key to understanding environmental problems. Studies of the origin and evolution of the atmosphere have also provided new insights into the origin of life and the possibility of life outside Earth.

—Joel S. Levine

FURTHER READING

Ackerman, Steven A., and John A. Knox. *Meteorology: Understanding the Atmosphere.* 3rd ed. Jones and Bartlett Learning, 2012. Provides an overview of the atmosphere and atmospheric phenomena, beginning from the evolution of the early terrestrial atmosphere. Suitable for university-level readers.

Ahrens, C. Donald. *Essentials of Meteorology: An Invitation to the Atmosphere.* Brooks/Cole Cengage Learning, 2012. Covers various topics in weather and the atmosphere. Discusses topics such as tornadoes and thunderstorms, acid deposition and other air pollution topics, humidity and cloud formation, and temperature.

Archer, David. *The Global Carbon Cycle.* Princeton University Press, 2010.

Berner, Robert A. *The Phanerozoic Carbon Cycle: CO_2 and O_2.* Oxford University Press, 2004. Discusses climate and atmosphere of the Paleozoic, Mesozoic, and Cenozoic eras. Also covers aspects of weathering and erosion on the carbon cycle. Suited to undergraduates. Contains references for each chapter and indexing.

Canfield, Donald Eugene *Oxygen. A Four Billion Year History.* Princeton University Press, 2014. This highly readable book presents a comprehensive overview of atmospheric oxygen and discusses many questions about the origin and significance of atmospheric oxygen.

Denny, Mark. *Making the Most of the Anthropocene—Facing the Future.* Johns Hopkins University Press, 2017.

Glikson, Andrew Y. *Evolution of the Atmosphere, Fire and the Anthropocene Climate Event Horizon.* Springer, 2014. In this book can be found a detailed discussion of the relationship of oxygen and other atmospheric gases with regard to their natural cycles and the effect of human activities.

Harmon, Russel S. and Parker, Andrew *Frontiers in Geochemistry.* John Wiley & Sons, 2011. Presents the existence and history of atmospheric oxygen and its formation in a geochemical context, drawing from several academic sources.

Marshal, John, and R. Alan Plumb. *Atmosphere, Ocean and Climate Dynamics: An Introductory Text.* Elsevier Academic Press, 2008. An excellent introduction to atmospheres and oceans. Discusses the greenhouse effect, atmospheric structure, oceanic and atmospheric circulation, and climate change. Suitable for advanced undergraduates and graduate students with some background in advanced mathematics.

Rhodes, Frank H.T. *Earth. A Tenant's Manual.* Cornell University Press, 2012. In this book the author presents a scientific view of the past, present and future of Earth's atmosphere and its composition.

Voronin, P., and C. Black. "Earth's Atmosphere as a Result of Coevolution of Geo- and Biospheres." *Russian Journal of Plant Physiology* 54 (2007): 132–136. Covers the evolution of the atmosphere's composition and factors altering the gas composition. Provides background content on photosynthesis and chemolithotrophy. Highly technical. Appropriate for readers with a strong chemistry or geology background.

ATMOSPHERE: GLOBAL CIRCULATION

FIELDS OF STUDY

Atmospheric Science; Fluid Dynamics; Heat Transfer; Environmental Sciences; Environmental Studies; Physical Geography; Climate Modeling; Meteorology; Oceanography; Hydroclimatology; Hydrometeorology; Earth System Modeling

SUMMARY

The general circulation of Earth's atmosphere involves the large-scale movements of significant portions of air in the atmosphere. Variations in surface temperatures produce pressure gradients that combine with the Coriolis force to circulate most of the air in the atmosphere. This involves the Hadley circulation, which moves air in the Northern and Southern Hemispheres in three huge convection cells each, and the Walker circulation along the equatorial belt, which produces the El Niño phenomenon when it oscillates.

PRINCIPAL TERMS

- **adiabatic:** the effect of changing the temperature of a gas or other fluid solely by changing the pressure exerted on it, without the input or removal of heat energy
- **anticyclone:** a general term for a high-pressure weather system that rotates clockwise in the Northern Hemisphere and counterclockwise in the Southern Hemisphere
- **Coriolis force:** a non-Newtonian force acting on a rotating coordinate system; on Earth, this causes objects moving in the Northern Hemisphere to be deflected toward the right and objects moving in the Southern Hemisphere to be deflected toward the left due to Earth's rotation
- **cyclone:** a general term for a low-pressure weather system that rotates counterclockwise in the Northern Hemisphere and clockwise in the Southern Hemisphere
- **geostrophic wind:** a wind resulting from the balance between a pressure gradient force and Coriolis force; the flow produces jet streams and is perpendicular to the pressure gradient force and the Coriolis force
- **pressure gradient force:** a wind-producing force caused by a difference in pressure between two different locations

DRIVING FORCES

The general circulation of the atmosphere operates as a heat engine driven by the uneven distribution of solar energy over the surface of the planet. Atmospheric circulation transfers some of this energy from regions where it is abundant to regions where it is scarce. This energy transfer reduces the difference in surface temperature between the equatorial regions and the polar regions, between the oceans and the continents, and between continental interiors and coastal regions. The existence of an atmosphere, with water vapor, carbon dioxide, and other greenhouse gases, keeps the average temperature at the surface considerably warmer than it would otherwise be. The motions of the atmosphere cool the warmer regions and warm the cooler regions, smoothing out the extremes of temperatures.

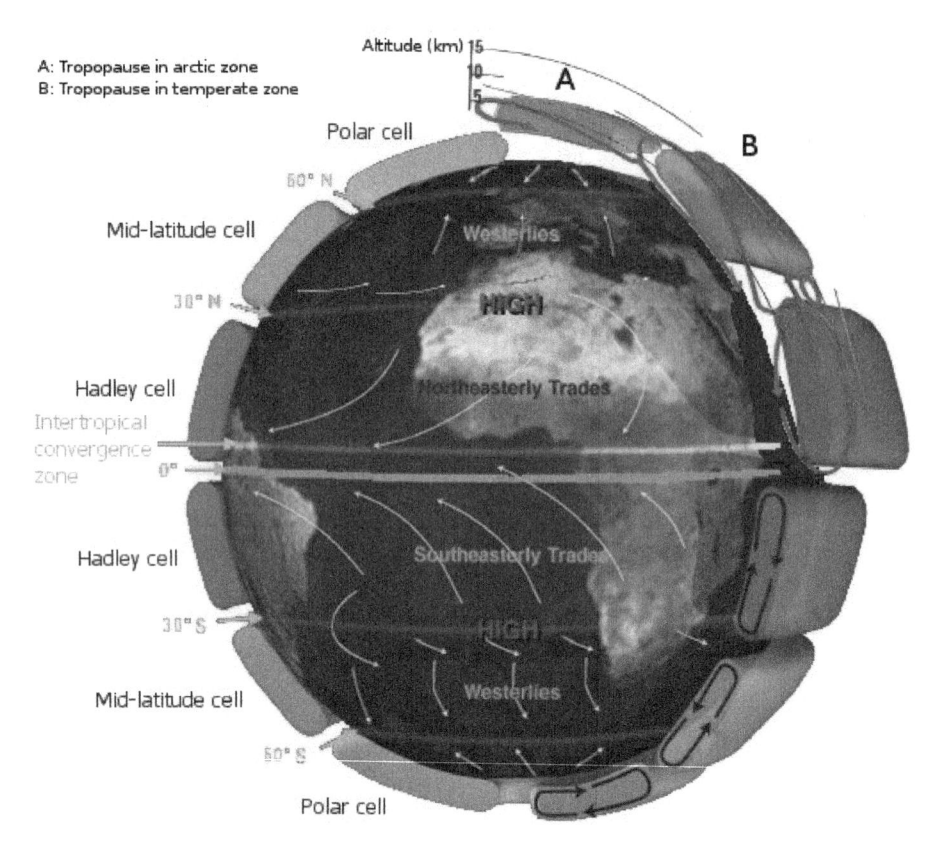

Global circulation of Earth's atmosphere displaying Hadley cell, Ferrell cell and polar cell.

The sun is so far from Earth that its rays of light can be thought of as being parallel when they strike the planet. Because Earth is a sphere, the areas warmed by this light vary with latitude. During the equinoxes, the sun is directly over the equator. A beam of sunlight with a square cross-section of 1 meter on each side is perpendicular to Earth's surface and will illuminate an area of 1 square meter at the equator. However, as the distance from the equator increases, the angle of incidence of that beam of sunlight increases in accord with the circumference of the planetary surface. At a latitude of 45 degrees north (near Ottawa, Ontario, for instance), that same 1 square meter beam of sunlight will be spread out over 1.4 square meters of the surface. At a latitude of 60 degrees north (near Anchorage, Alaska), it will be spread out over 1.7 square meters.

The 23.5-degree tilt of Earth's axis alters this in a cyclic manner over the course of a year. At the Northern Hemisphere summer solstice, 45 degrees north has a 1 square meter of sunlight spread out over 1.07 square meters, and 60 degrees north has it spread out over 1.24 square meters. At the Northern Hemisphere winter solstice, 45 degrees north has its 1 square meter of sunlight spread out over 2.73 square meters, while 60 degrees north has its spread out over 8.83 square meters. The intensity of solar radiation at these two latitudes differs by a factor of 1.16 in the summer and 1.22 in spring and fall, but 3.23 in the winter. Other choices for latitudes would yield similar results. This helps to explain why the temperature contrasts between northern and southern states in the United States is so much greater during the winter months than during the summer ones. It also explains some of the seasonal differences in general atmospheric circulation.

Major Convection Cells

Air above a warm surface absorbs heat and expands, in accordance with the gas law equation of physical chemistry,

$$PV = nRT.$$

This equation specifies the direct relationship between the temperature, pressure, and volume of a gas, so that as temperature increases, the pressure the gas exerts against its surroundings and the volume that it occupies also increase. Because the mass of the heated air has not changed, the expanded air is less dense than the air surrounding it, and it will rise, for the same reason that a hot air balloon rises. The rising air displaces air above it, pushing that air away with a pressure gradient force.

To understand how temperature gradients at the surface produce pressure gradient forces aloft, consider two adjacent columns of air that initially contain the same amount of air, one of which is warmer than the other. The warmer one expands and reaches higher above the surface. Because the mass of air in both columns is initially the same, the air pressure at the surface is also the same beneath both columns. Next, consider the air pressure halfway up the cooler column. This level is beneath one-half of the air in the cool column, but it is beneath more than one-half of the air in the warm column. The air pressure at this elevation in the warm column is therefore greater than the air pressure at this elevation in the cool column. This difference in pressure is a pressure gradient and will cause air to flow aloft from the warm column toward the cool column. As it does so, the total mass in the warm column will decrease, and the total mass in the cool column will increase; therefore, the air pressures at the surface will no longer remain the same. Beneath the cool column, the air pressure at the surface will be greater than beneath the warm column, and this pressure gradient will cause air to flow along the surface from the cool column toward the warm column. Eventually, a steady-state flow will result with an elevation that separates the two directions of air flow. Not surprisingly, above this elevation the total mass in each column will be the same.

If Earth were not rotating, displacements of the air initially above the equator would be toward the poles aloft and toward the equator at the surface; because it is rotating, however, an additional effect called the Coriolis force must be considered. A point on the equator is a distance of one Earth radius away from Earth's rotation axis, whereas a point at one of the poles is directly on the rotation axis. As a parcel of air moves to the north from the equator, it is moving closer to the rotation axis. It inherited a certain angular momentum from when it was on the equator, and conservation of this angular momentum requires it to move somewhat to the east. In contrast, a parcel moving to the south in the Northern Hemisphere is moving farther from the rotation axis, and conservation of its angular momentum requires it to move a bit to the west. In either case, in the Northern Hemisphere, the Coriolis force causes objects to move to the right of the direction that they would normally be moving; in the Southern Hemisphere, it causes things to move to the left. This effect is only significant when there is very little frictional interaction between the moving object and Earth, so it is extremely important in the circulation of the oceans and atmosphere.

Air being displaced from its location over the equator will initially go due north or due south. The Coriolis force deflects it more and more until, in the vicinity of 30 degrees north or south latitude, it will be moving due east. At high altitudes, this is a geostrophic wind, a wind from the west (called a westerly) resulting from a poleward-directed pressure gradient force being deflected 90 degrees by the Coriolis force. The fastest elements of this flow form the subtropical jet stream. No longer moving toward the poles, this high-altitude air is now considerably cooler and denser than it was when first heated at the surface of Earth. As a consequence, it descends. As it gets lower, the additional air pressure it experiences causes it to warm up. Bicycle pumps illustrate this adiabatic effect, as they get hot when the air within them is pressurized without the input of heat energy.

The amount of water vapor that can be contained in air varies with temperature. This is commonly observed as water condenses out of cooler air at night to produce dew or frost. The air rising over the equator is initially warm and heavily laden with water. As it cools, this water vapor condenses, eventually producing the intense rainfall essential for equatorial rainforests. Losing its entrained water vapor warms the air, enhancing its ascent. Later, after moving to the northeast or southeast, this air descends, warming up so that it once again absorbs water vapor. The regions of Earth's surface near 30-degree latitudes are characterized by intense evaporation that produces deserts such as the Sahara and the Kalahari in Africa.

Because the descending air at 30-degree latitudes is denser than average, the air pressure at the surface beneath it will be higher than average. Similarly, the air pressure beneath the rising air column at the equator is lower than average. This difference in pressure causes winds to blow across the surface, from 30 degrees north or south latitude toward the equator. The Coriolis force also affects this flow so that these winds are deflected; by the time they reach the equator, they are coming out of the east. These easterly winds are called the trade winds, a name given because they were favorable for sailing ships bound to the East for trade.

HADLEY CELLS

The movements of air already described connect to form two circulating cells, one in the Northern Hemisphere and one in the Southern Hemisphere. These are called Hadley cells, named after George Hadley (1685-1768), who theorized them in 1735.

The convection responsible for the Hadley cells is not seen as clearly elsewhere on Earth. However, four additional cells, two in each hemisphere, have long been a part of the theoretical development of meteorology. One, called the Ferrel cell, after William Ferrel (1817-1891), who proposed it in 1856, also descends at about 30 degrees latitude and presumably ascends at latitudes of about 60 degrees. The other, also proposed by Ferrel and called the Polar cell, ascends at latitudes of about 60 degrees and descends at the poles. The surface winds produced by these cells return air to the region around 60 degrees. This returning air is deflected by the Coriolis force, producing westerlies between 30 degrees and 60 degrees, and Polar easterlies at higher latitudes.

Much of the general circulation of Earth's atmosphere can be explained by this six-cell model, and it continues to be used in many elementary meteorology and earth-science courses and textbooks. As a theoretical tool, it is useful and easily grasped, and it yields insights about global systems that are generally accurate. Air does descend at the poles and at 30 degrees latitude, and returns to the vicinity of 60 degrees latitude, but the situation there is not as simple as the rising heated air at the equator. The boundary between the Polar cell and the Ferrel cell, called the Polar front, has opposing surface winds, not converging ones; high-altitude winds over this front do not simply diverge as they do above the equator.

FRONTS AND CYCLONES

During World War I, dirigibles were used to drop bombs on locations in England. To avoid assisting this tactic, the global system of weather data gathering was suspended. Subsequently, Norway's fishing fleet was put at risk from the weather, and scientists there developed theoretical models to make up for the lack of distant weather data. Led by Vilhelm Bjerknes (1862-1951) and Halvor Solberg (1895-1974), this group of meteorologists was called the Bergen school. Relying on more closely spaced but effectively synchronous data, they developed the concept of fronts and proposed the cyclone model for storm genesis and evolution.

The cyclone model grew out of observations of storm systems in the middle latitudes. Such a system has a low-pressure region near its center and a pattern of winds moving in a concentric fashion around this low. This is the result of surface winds trying to move into the low-pressure region but being deflected by the Coriolis force. In the Northern Hemisphere, these cyclonic winds move around the low in a counterclockwise direction, whereas in the Southern Hemisphere, they move clockwise. Often, observations revealed a consistent pattern of temperature gradients, precipitation bands, and surface wind configurations. The entire storm system usually moved from west to east and evolved in similar ways from its initial genesis, to being fully developed with maximum winds, to fading out and disappearing. As this evolution occurred, the interactions between air masses of different temperatures and with different moisture content followed reasonably consistent patterns.

The meteorologists of the Bergen school saw that the shear zone between the westerlies of the warm, moist air to the south and the easterlies of the cold, dry air to the north was an important element in generating these storms. A line connecting the various cyclonic disturbances in the middle latitudes defined the Polar front—not as a simple, smooth surface, but one with major excursions to the north and south, much like a meandering river.

Air above the poles is cooler, denser, and therefore more compressed than the warmer air in the middle latitudes. This produces a pressure gradient aloft that is directed toward the pole. This would cause poleward movement, except that the Coriolis force deflects such movement, again producing a westerly geostrophic wind, the fastest part of which is called the Polar jet stream. This is the jet stream referred to on weather maps and in forecasts in the United States and Canada. As already described, the temperature gradients at the surface are greater at higher latitudes, and hence the pressure gradients and velocities of the Polar jet stream are greater than those for the subtropical jet stream. In addition, because the surface temperature gradients are greater in winter than in summer, the velocity and significance of the Polar jet stream are also greater in the winter.

As the Polar jet stream races around the globe at velocities of about 125 kilometers per hour in the winter and 60 kilometers per hour in the summer, instabilities develop that deflect it into a meandering path. As a meander develops, the range of temperatures over which it moves increases, causing higher winds and even greater meandering. Eventually, portions of its path may be nearly north-south, bringing warmer air to higher latitudes and cooler air to lower ones. This diminishes the temperature gradients, causing the meanders to shrink until nearly east-west flow is reestablished. Called Rossby waves, after Carl-Gustav Rossby (1898-1957), who described them in 1939, these meanders form a path that resembles a very blunt, rounded star with three to six points centered on the pole. Moving along such a path, the jet stream speeds up at some places and slows down at others. Speeding up decreases air pressure aloft, while slowing down increases it. Cyclonic disturbances tend to form beneath places where the air pressure aloft is reduced and to move along tracks that lie beneath the jet stream. As the disturbances evolve, fronts develop, precipitation occurs, and warm air is transported to higher altitudes and then to higher latitudes.

El Niño/Southern Oscillation

The surface components of the Hadley cells move toward the equator. The Coriolis force deflects the flow, so that by the time these winds reach the equator, they are coming out of the east. Ocean currents, driven by their own geostrophic flows, are quite similar, with major east-to-west flows near the equator. The water moved by these ocean currents is warm surface water, and its transport to the west results in a buildup of such waters in the western part of the Pacific and, to some extent, the Atlantic. With warmer sea-surface temperatures to the west and upwelling of cooler water on the eastern side of the Pacific, yet another temperature gradient exists of sufficient scale to influence general circulation patterns.

The convection cell in this case is called a Walker cell, named after Sir Gilbert Walker (1868-1958), who identified it in 1924. The conditions needed for its development are neither constant nor periodic. Every three to five years, because of factors not yet well understood, this circulation breaks down. Generally coupled with a decrease in the strength of the trade winds, the Walker cell reverses its orientation: Instead of ascending air in the west, with sufficient rainfall to support equatorial rainforests in Indonesia, the air ascends over the eastern regions of the Pacific, bringing sometimes intense rainfall to regions that are otherwise deserts in Peru. Called El Niño by oceanographers and the Southern Oscillation by atmospheric scientists (often abbreviated as ENSO), this reversal has dramatic effects on weather patterns.

Usually lasting between twelve and eighteen months, the ENSO has been identified in historic and geologic data sets. Droughts in Africa, floods in the American West, and other phenomena appear to be related to the sea-surface temperature in the equatorial Pacific. Certainly one of the more interesting aspects of the ENSO is its effect on the other aspects of general atmospheric circulation.

Significance

Understanding general atmospheric circulation is essential for accurate, useful weather predictions. Knowledge of this circulation permits meteorologists to estimate how various parcels of air will move, how pressures will change, and how and where precipitation will occur. These estimates, in turn, permit them to project further into the future and improve the accuracy of their predictions.

By recognizing which variables are most likely to alter the general atmospheric circulation, and being able to guess how these variables might have been different in the past, geologists and climatologists can make more informed models about ancient weather patterns and climate evolution. The Himalayan Mountains and Tibetan Plateau are comparatively recent features on Earth, for example, and their presence has dramatically altered circulation patterns. Better understanding of past climates will help assess the influence of anthropogenic inputs, such as carbon dioxide and chlorofluorocarbons (CFCs), and should serve to guide public policy.

—*Otto H. Muller*

FURTHER READING

Ahrens, C. Donald. *Essentials of Meteorology: An Invitation to the Atmosphere.* 6th ed. Cengage Learning, 2012. Written in a style that promotes visualization of the concepts being discussed. Treatment of general atmospheric circulation is straightforward and easily grasped. Presents few equations, but includes lists of terms and questions for thought.

Barry, Roger G., and Richard J. Chorley. *Atmosphere, Weather, and Climate.* 9th ed. Routledge, 2010. Covers the subject of general atmospheric circulation in a thorough but not technically challenging style. Presents a vast number of figures and black-and-white line drawings to illustrate its points. Suitable for advanced high school students.

Chang, Julius *General Circulation Models of the Atmosphere.* Elsevier, 2012. The book covers the fundamentals of general circulation models and their application in atmospheric studies.

Donner, Leo, Schubert, Wayne and Somerville, Richard, eds. *The Development of Atmospheric General Circulation Models. Complexity, Synthesis and Computation.* Cambridge University Press, 2011. A curated collection of articles describing the historical development of general circulation models and their development since then.

Lutgens, Frederick K., Edward J. Tarbuck, Tasa, Dennis G. and Herman, Redina. *The Atmosphere: An Introduction to Meteorology.* 14th ed. Pearson Education, 2018. An updated version of this textbook, in loose-leaf format, provides students with a general overview of the structure and properties of Earth's atmosphere.

Marshal, John, and R. Alan Plumb. *Atmosphere, Ocean and Climate Dynamics: An Introductory Text.* Elsevier Academic Press, 2008. An excellent introduction to atmospheres and oceans. Covers the greenhouse effect, convection and atmospheric structure, oceanic and atmospheric circulation, and climate change. Suitable for advanced undergraduates and graduate students with some background in advanced math.

Schneider, Tapio, and Adam H. Sobel, eds. *The Global Circulation of the Atmosphere.* Princeton University Press, 2007. Collects papers from the 2004 conference at California Institute of Technology. Covers large-scale atmospheric dynamics, storm-tracking dynamics, and tropical convection zones. A strong understanding of advanced mathematics is required. Best suited for graduate students and professionals.

Vallis, Geoffrey K. *Atmospheric and Oceanic Fluid Dynamics: Fundamentals and Large-scale Circulation.* Cambridge University Press, 2006. Begins with an overview of the physics of fluid dynamics to provide foundational material on stratification, vorticity, and oceanic and atmospheric models. Discusses topics such as turbulence, baroclinic instabilities, wave-mean flow interactions, and large-scale atmospheric and oceanic circulation. Best suited for graduate students studying meteorology or oceanography.

Wells, Neil. *The Atmosphere and Ocean: A Physical Introduction.* 3rd ed. John Wiley & Sons, 2012. The atmosphere and oceans are both fluids circulating on a rotating planet, intimately and profoundly influencing each other. Treats atmospheric and oceanic interactions in a readable, yet thorough, manner. Presents numerous quantitative concepts and equations with graphs or figures that make them understandable to readers with little technical background.

ATMOSPHERE: STRUCTURE AND THERMODYNAMICS

FIELDS OF STUDY

Atmospheric Science; Atmospheric Chemistry; Thermodynamics; Fluid Dynamics; Climate Modeling; Hydroclimatology; Hydrometeorology; Meteorology; Bioclimatology; Oceanography; Physical Geography; Earth System Modeling

SUMMARY

An atmosphere is a layer of gases that surrounds a planetary surface. Earth's current atmosphere is a complex, dynamic system that interacts closely with and controls the surface environment. Atmospheric thermodynamics involves the process by which energy from the sun is absorbed and deposited on Earth, including its oceans and atmosphere. Early life-forms substantially altered Earth's atmosphere, and humans continue to do so.

PRINCIPAL TERMS

- **adiabatic:** characterizing a process in which no heat is exchanged between a system and its surroundings
- **air drainage:** the flow of cold, dense air downslope in response to gravity
- **chlorofluorocarbons (CFCs):** chemicals in which chlorine and fluorine replace one or more of the hydrogen atoms in the molecular structure of the corresponding hydrocarbon
- **cosmic rays:** high-energy atomic nuclei and subatomic particles, as distinct from electromagnetic radiation
- **environmental lapse rate:** the general temperature decrease within the troposphere; the rate is variable but averages approximately 6.5°C per kilometer
- **exosphere:** the outermost layer of Earth's atmosphere
- **greenhouse effect:** a planetary phenomenon in which the atmosphere absorbs and retains more heat radiation than it passes back out into space
- **growing degree-day index:** a measurement system that uses thermal principles to estimate the approximate date when crops will be ready for harvest
- **heterosphere:** a zone of the atmosphere at an altitude of 80 kilometers, including the ionosphere, made up of rarefied layers of oxygen atoms and nitrogen molecules
- **homosphere:** a major zone of the atmosphere below the heterosphere whose chemical makeup is consistent with the proportions of nitrogen, oxygen, argon, carbon dioxide, and trace gases at sea level; includes the troposphere, stratosphere, and mesosphere
- **infrared radiation:** electromagnetic radiation with frequency in the range of 10^{13} to 10^{14} Hertz (Hz)
- **ionosphere:** the layer of ionized gases in Earth's atmosphere, starting about 50 to 100 kilometers above the surface of the planet (between the thermosphere and the exosphere)
- **latent heat:** the energy absorbed or released during a change of physical state
- **mesosphere:** the extremely rarefied atmospheric layer at altitudes from 50 to 80 kilometers above the surface, characterized by rapid decreases in temperature
- **net radiative heating:** the driving force for atmospheric thermodynamics, essentially the difference between heat entering the atmosphere due to solar heating and heat leaving the atmosphere as infrared radiation
- **photodissociation:** the condition in which light energy absorbed by a molecule is sufficient to dissociate the bonds between atoms in the molecule, typically caused by light in the ultraviolet range
- **radiational cooling:** the cooling of Earth's surface and the layer of air immediately above it by a process of radiation and conduction
- **stratosphere:** the atmospheric zone 20 to 50 kilometers above the surface that contains the functional ozone layer
- **temperature inversion:** a condition in which a region of warmer occupies a position above its normal location, causing air temperature to increase with increasing elevation from Earth's surface
- **thermosphere:** the atmospheric zone extending from 80 to 480 kilometers in altitude, and containing the ionosphere
- **troposphere:** the lowest atmospheric layer, extending from Earth's surface to an altitude of about 18 kilometers, containing 90 percent of the total mass of the atmosphere, marked by considerable turbulence and a decrease in temperature with increasing altitude
- **ultraviolet light:** electromagnetic radiation having a frequency in the range of 10^{15} to 10^{17} Hz

ATMOSPHERIC CONTENT

"Atmosphere" usually refers to the layer of gases that covers Earth. Although most planets have atmospheres of some sort, Earth's atmosphere is unique among those known in this solar system in that it contains a substantial amount of oxygen and supports the equilibrium existence of water in solid, liquid, and vapor phases.

The atmosphere of Earth contains 78 percent nitrogen and 21 percent oxygen; the remainder consists of 0.9 percent argon, 0.03 percent carbon dioxide, and traces of hydrogen, methane, nitrous oxide, and inert gases. In addition, the atmosphere carries varying amounts of water vapor and aerosol particles such as dust and volcanic ash, depending on local and global events. As early as 3.5 billion years ago, primitive algae-like life-forms emerged and fed on the carbon dioxide by using photosynthesis to metabolize carbon dioxide and water molecules to form simple sugars and more complex polysaccharides. Photosynthesis over time gradually brought about a drastic change in the ratios of gases in the atmosphere, eventually producing the current carbon dioxide concentration of only 0.03 percent and free molecular oxygen concentration of 20 percent.

A Dynamic System

Gases are compressible, and they absorb or transmit varying amounts of electromagnetic radiation. Because of these characteristics, the atmosphere is not static but is instead a highly dynamic system. Phenomena such as weather and climate are short- and long-term events involving the exchange of energy and transport of mass within the atmosphere and also between the solid Earth, liquid oceans, and space.

Links between solar activities (especially sunspot cycles), weather, and climate have been sought for decades, but associations between them remain inconclusive. Interactions between the oceans and the air, however, are more substantial, and therefore more readily discernible. Winds drive waves and affect ocean currents; in return, the oceans act as a heat source or sink for the atmosphere. The most famous interaction is the El Niño/Southern Oscillation event. El Niño ("the child"), which usually happens around Christmas, is an upwelling of cold water off the Pacific coast of South America that occurs every two to seven years. Besides having a disastrous effect on the fishing industry, it is associated with changes in circulation and precipitation patterns in the atmosphere over the Pacific basin.

Temperature

The atmosphere has no clear upper boundary, but is generally considered to extend to an altitude of about 300 kilometers, where it responds more to electromagnetic effects and acts less like a fluid body. It can be described in three major characterizations: temperature, chemistry, and electrical activity.

Temperature changes with altitude and, as there is less overlying gas with increasing altitude, pressure and exposure to radiation also affect temperature. The lowest region of the atmosphere, enclosing virtually all life and weather on Earth, is the troposphere, extending to an altitude of 8 to 18 kilometers. The name is taken from the Greek word *tropo*, meaning "turn," and refers to the fact that this region turns with the solid earth. The bottom of the troposphere is the boundary layer, where the atmosphere interacts directly with the surface of the planet. The boundary layer is often turbulent, as moving air masses (winds) encounter and flow around or over obstructions and exchange heat with the ground or water. Temperature and pressure in the troposphere decrease at about 2°C per kilometer until the top of the troposphere, the tropopause, is reached. Life becomes increasingly difficult to maintain with altitude (humans may require additional oxygen above 4 kilometers and must wear pressure suits above 10.6 kilometers). Some 90 percent of the mass of the atmosphere is contained below the tropopause.

At the tropopause, temperature reaches a minimum of about -50°C, then rises again as one enters the stratosphere, to peak at about 15°C at the stratopause at an altitude of about 50 kilometers. Above this level, in the mesosphere, temperature declines again to a low of -60°C at the mesopause, at an altitude of 85 kilometers. Only 1 percent of the atmosphere is in the mesosphere and above; 99 percent lies below. Particles from atomic nuclei to meteors generally are destroyed in the mesosphere. Nuclei, or cosmic rays, encountering gas molecules in the mesophase will be shattered into secondary and tertiary particles. Most meteors are heated by friction and vaporize when they encounter the mesosphere. Above the mesosphere, the thermosphere (the hot atmosphere) extends to approximately 300 kilometers in altitude, and temperatures soar to between 500 and 2,000°C, depending on solar activity. Because the atmosphere is so thin, however, the total heat present is minuscule. Finally, beyond the thermosphere is the exosphere (outer layer), which extends from 300 kilometers to the solar wind (also called the interplanetary medium).

Solar and Infrared Radiation

Atmospheric thermodynamics can be described as the process by which energy from sunlight is absorbed by matter on Earth's surface, in the oceans, and in the atmosphere. This energy must be returned to space in the form of infrared radiation in order for the planet to maintain its stable cyclic range of ambient temperatures. Even though solar radiation and infrared radiation must be approximately balanced on a global level, they are frequently out of balance on a local level, accounting for the ambient weather changes seen on a day-by-day basis.

Absorption of solar energy is most concentrated on Earth's surface, particularly in tropical regions, whereas most of the infrared radiation going out to space originates in the middle troposphere and is more evenly distributed between equatorial and polar regions. Time-averaged temperature distribution is maintained by the system by transporting heat from regions where solar heating dominates over infrared cooling to regions where radiative cooling is able to dominate. In this way, the atmosphere transports heat from the ground to the upper troposphere, and the atmosphere and the oceans work together to transport heat toward the Arctic and Antarctic poles from the equatorial belt.

Net Radiative Heating

Net radiative heating, the difference between heat gained by absorption of solar energy and heat lost due to infrared radiation, is the driving force for atmospheric thermodynamics. Solar radiation reaches the top of the atmosphere, where about 30 percent of it is reflected back into space. The remaining radiation is absorbed. About 69 percent of solar absorption happens at the surface, and about 40 percent of that leaves the surface as net infrared radiation. This serves to leave a net surface radiative heating of about 150 watts per square meter. On a global and annual average, the net radiative heating of Earth's surface is equivalent to the net radiative cooling of the atmosphere; normal thermodynamics within the atmosphere thus maintains an appropriate balance between the two. A prolonged imbalance between these two systems would result in global warming, which could melt the polar ice caps and raise the levels of the oceans.

Over the oceans, most of the energy from net radiative heating of the surface is used in the process of water evaporation. Energy is removed from the surface as latent heat; when water vapor condenses, it deposits almost all of its latent heat into the air rather than into the condensed water, which returns to the surface as precipitation. The result is that heat is transported from the surface to the air, where the potential for condensation takes place. On a global average, about 70 percent of the net radiative heating of the surface is removed by latent heat flux, and the remaining 30 percent leaves the surface by conduction of the sensible heat to the overlying air. The atmosphere thus experiences diabatic heating or cooling from four ongoing processes: latent heating, sensible heating at the surface, solar absorption, and infrared heating or cooling.

Temperature Inversion

The general temperature decrease within the troposphere is called the "environmental lapse rate." The occurrence of shallow layers where temperatures actually reverse this normal pattern and get warmer with increasing height is known as a "temperature inversion." Temperature inversions are often seen in cities located at high altitudes with a low partial pressure of oxygen, and in basins surrounded by mountains that serve to block winds and trap industrial pollutants. A "greenhouse effect" inversion causes a reversal of an ecosystem's normal atmospheric temperature gradient, thermally altering harmful air-borne chemical compounds and enhancing their negative effects on organisms living below. A notable example is the strong eastern wind that blows toward Denver, Colorado, and traps a brown cloud of pollutants against the Rocky Mountains. This requires the daily broadcast of air-quality reports on radio and television and leads to the frequent calls for weak and elderly residents to stay indoors during hotter parts of the day. Trapped carbon monoxide, coming mainly from automobile tailpipes via the incomplete combustion of gasoline, combines quickly with hemoglobin, the oxygen-carrying compound in the blood in humans and animals. By taking up binding sites on the hemoglobin molecule, carbon monoxide impairs oxygen delivery to the tissues. Elderly persons with heart disease are at special risk through any restrictions in oxygen delivery.

The thermosphere, which does not have well-defined limits, exhibits another increase in temperature as a result of the absorption of very short-wavelength solar energy by atoms of oxygen and nitrogen. Although temperatures rise to values in excess of 1,000°C in this outermost layer, it is difficult to compare the temperature in this layer with that seen on Earth's surface. Temperature is directly proportional to the average speed at which molecules are moving, and because gases within the thermosphere are moving at very high speeds, ambient temperature remains very high. However, the gases in this region are so sparse that only a very small number of these fast-moving air molecules collide with foreign bodies, thus causing only a very small amount of energy to be transferred. For this reason, the temperature of a satellite orbiting Earth in the thermosphere is determined by the amount of solar radiation it absorbs and not directly by the temperature of the surrounding environment. Thus, if an astronaut in the space shuttle were to expose his or her hand to ambient space, it would not feel hot.

Ground inversions occur frequently and commonly extend upward for 100 meters or more. They can develop from several different causes. Their primary cause is radiative cooling, which occurs under clear skies at night. During the day, the ground stores thermal energy from solar radiation. After sunset, the ground surface radiates the stored heat energy, thereby cooling the ground surface. Energy from the relatively warm air is physically conducted into the radiationally cooled surface of the ground as the two are in contact, thereby cooling the air immediately above it. Only about the first 100 meters experiences a temperature decrease as a result of radiative cooling. Such induced temperature inversions are more likely to occur on nights with clear skies than on nights with cloudy skies. Calm wind conditions or light breezes also are more conducive to developing a ground inversion. Ground inversions are more likely to develop and to last longer in cold climates because snow and ice reflect sunlight, and the small amount of heat absorbed is utilized in the melting process, thus cooling the surface rapidly and producing an inversion.

A second mechanism for the formation of an inversion is the result of a phenomenon called air drainage. On cold nights over rolling topography, denser cold air responds to gravity and moves downslope to collect in local depressions. Continued cooling can cause inversions to extend over larger areas both vertically and horizontally if the vertical cooling extends above the summits of the rolling terrain. Evidence of the initial development of a ground inversion often is heavy dew or frost. Frequently, ground fogs occur in association with inversions because of cooling of the air. This is particularly true in an air drainage situation where fogs first appear in depressions at the surface.

A third way in which ground inversions are formed is through the movement of a warm air mass into a region. A warm current of air may move over a cool ground surface or a cooler layer of surface air. The lower portions of the air mass are cooled, and stable or nonturbulent conditions result, producing an inversion.

The frequency of ground inversions varies across the United States (frequency being expressed as a percentage of the total time a region has inversions). Most ground inversions occur at night in winter. Summer inversions are less frequent than winter inversions, but they do occur.

CHEMISTRY

From a chemical standpoint, the atmosphere is divided into two major realms: the homosphere and the heterosphere. The homosphere—which overlaps the troposphere, stratosphere, and mesosphere—has a chemical makeup essentially identical to sea-level proportions of nitrogen, oxygen, and trace gases, even though the absolute numbers of atoms and molecules drop sharply. With increasing altitude, some important differences start to appear. Ozone becomes an important constituent of the atmosphere in this realm. Ozone is formed in the stratosphere by short-wavelength ultraviolet sunlight splitting apart oxygen molecules by photodissociation. These free oxygen atoms then form ozone with oxygen molecules. Molecular nitrogen also is dissociated.

Although the stratosphere and mesosphere (sometimes treated together as the middle atmosphere) are quite tenuous compared to the troposphere, gases in them form an optically dense layer that absorbs or reflects short-wavelength ultraviolet and X-ray radiation that would be damaging to life on the surface. Ozone is especially important with regard to the absorption of ultraviolet radiation. Nevertheless, ozone is quite fragile and can be destroyed by chlorofluorocarbons (Freons and related compounds) used for several years as spray can propellants and refrigeration system coolants. Studies indicate that these gases migrate upward in the atmosphere and chemically remove thousands of times their own mass in ozone molecules before they are broken down after several decades or even centuries. This is believed to have led to the formation of an ozone hole over the South Polar region and the appearance of a similar, but less pronounced, effect over the Arctic. Atomic oxygen becomes more common in the mesosphere. In the heterosphere, the gas mixture changes drastically, and hydrogen and helium become dominant.

Electrical Activity

In terms of electrical activity, there are the regions designated as the neutral atmosphere and the ionosphere. The neutral atmosphere, below 50 kilometers in altitude, is largely devoid of electrical activity other than lightning, which might be regarded as localized "noise." Above 50 kilometers, atoms and molecules are ionized largely by sunlight (ultraviolet radiation in particular) and, to a lesser extent, by celestial X-ray sources, collisions with other atoms, and geomagnetic fields and currents. Although the ionosphere as a whole is electrically neutral, it comprises positive (ion) and negative (electron) elements that conduct currents and respond to magnetic disturbances. The ionosphere starts at about 50 to 100 kilometers in altitude and extends outward to more than 900 kilometers as it gradually merges with the magnetosphere and its components. It is sometimes called the Heaviside layer after Oliver Heaviside (1850-1925), who predicted a layer of radio-reflecting ionized gases. The ionosphere is divided into C, D, E, and F layers, which in turn are subdivided (F1, for example).

The ionosphere is one of the most active regions of the atmosphere and one of the most responsive to changes in solar activity. Ions and electrons in the ionosphere form a mirror-like layer that reflects radio waves. Radio waves are absorbed by the lower (D) region of the ionosphere (which also reaches down into the mesosphere). The D-layer dissipates at night in the absence of solar radiation, allowing radio waves to be reflected by the F-layer at higher altitudes, thus causing radio "skip." These effects vary at different wavelengths. Intense solar activity can alter the characteristics of the ionosphere and make it unreliable as a radio reflector, either through the input of high-energy radiation or by the injection of particles carried by the solar wind.

Atmospheric Light Displays

Such particle injections would go unnoticed but appear as the aurora borealis and aurora australis (the Northern and Southern Lights, respectively). Earth's magnetic field shields the planet from most charged radiation particles. At the polar regions, however, where the magnetic field lines are vertical (rising from the surface), the environment is magnetically open to space. Many charged particles from space or from the solar wind are "funneled" into the polar regions along the lines of force of the planetary magnetic field. When the particles strike the atmosphere, they surrender their energy as light with spectral lines unique to the electrochemical interactions taking place. These auroral displays generally take place at 120 to 300 kilometers in altitude, with some occurring as high as 700 kilometers.

The aurora is the best-known atmospheric light display. Other "dayglow" and "nightglow" categories are caused by lithium, sodium, potassium, magnesium, and calcium at altitudes from about 60 to 200 kilometers. These metals may be introduced by meteors as they are vaporized upon entering the atmosphere. A layer of hydroxyl radicals causes an infrared glow at about 100 kilometers, and dull airglows are caused by poorly understood effects at 100 to 300 kilometers in altitude.

Jet Streams and Waves

Although the principal division of the atmosphere is vertical, there are horizontal differences related to latitude and to weather. Two major phenomena that affect the atmosphere are the jet streams and waves. The jet stream is a high-speed river of air moving at about 10 to 20 kilometers in altitude and at 100 to 650 kilometers per hour. Its location plays a major role in the movements of larger air masses that make up weather fronts in the troposphere.

More than twenty wave phenomena take place in the atmosphere in response to different events. The three principal categories are gravity, Rossby, and acoustic. Gravity waves are not associated with relativity but with vertical oscillations of large air masses causing ripples, like a bottle bobbing in a pond. Rossby (or planetary) waves are associated with the wavelike distribution of weather systems. Acoustic waves are related to sound.

Atmospheric Study Tools

The earliest types of instruments used for atmospheric study remain among the most important. Barometers, thermometers, anemometers, and hygrometers provide the most immediate records of atmospheric change and warnings of impending

73

events. Vertical profiles of atmospheric conditions are obtained by transporting such instruments up into the air using balloons and suborbital rockets. The term "sounding rocket" comes from the earliest days of atmospheric study, when scientists were "sounding" the ocean of air just as they would the ocean of water: Small charges were attached to balloons, and the time sound took to reach the ground was a crude measure of atmospheric density. Balloon-borne instrument packages continued to be called radiosondes ("radio sounders").

Instrumentation carried aboard spacecraft is of a different nature. Many of the most revealing devices have been spectrometers of various types that analyze light reflected, emitted, or adsorbed by the atmosphere. Absorption of ultraviolet light by the atmosphere led to the discovery of the hole in the Antarctic ozone cover; drops in absorption meant that ultraviolet light was passing through, rather than being returned to space (typically, such measurements also require observation of the solar ultraviolet output). Optical instruments usually are most effective when they view the atmosphere "edge-on" so as to increase the brightness of the signal (somewhat like viewing a soap bubble at the edges). Atmospheric studies can be difficult when viewing straight down, because the weak signals from airglow and other effects are washed out by the brighter glow of Earth or the stellar background. Special techniques can be employed. The U.S. space shuttle has twice carried sensors designed to monitor carbon monoxide pollution in the atmosphere. Gas cells containing carbon monoxide at different pressures acted as filters that blocked all signals but the wavelengths corresponding to carbon monoxide at the same pressure (that is, altitude) as that in the cell.

The most powerful tools used in studying the atmosphere have been the weather satellites deployed to observe the atmosphere from geostationary orbit (affording continuous views of half a hemisphere) and from lower polar orbits. Images from these satellites reveal the circulation of the atmosphere by the motion of clouds. Other sensors (called sounders) provide temperature profiles of the atmosphere at various altitudes.

The most extensive analyses of the atmosphere have been carried out by the Atmosphere Explorers, the Orbiting Geophysical Observatories, the Atmosphere Density Explorers, and the Dynamics Explorers. Operating in the upper reaches of the atmosphere, these spacecraft have enabled determination of the structure and composition of the atmosphere and the changes it experiences with seasons and solar activity. The more sensitive chemical assays, however, have been conducted by instruments carried aboard the manned Spacelab 1 and 3 missions of the NASA space shuttle program. An Imaging Spectrometric Observatory carried on Spacelab 1 produced highly detailed emission spectra of the atmosphere between 80 and 100 kilometers in altitude. Atmospheric Trace Molecules Observed by Spectroscopy (ATMOS), on Spacelab 3, measured the altitude ranges of some thirty chemicals and identified five, such as methyl chloride and nitric acid, in the stratosphere, where previously they were only suspected.

COMPARATIVE PLANETOLOGY

Comparative planetology analyzes the differences and similarities between and among the planets. Earth, Venus, and Mars are used most often in comparative atmospheric studies. These three "terrestrial" planets are similar in size and in general chemistry but totally different in environment, largely because of their different atmospheres. Venus has a dense atmosphere composed largely of carbon dioxide and topped by clouds of sulfuric acid, which has led to surface temperatures of 900°C and to normal atmospheric pressures ninety times greater than those of Earth. The circulation pattern, though, is unaltered by precipitation and oceans and thus can be used as a model in studying Earth. Efforts to understand how Venus became a "runaway greenhouse" have suggested a similar scenario for Earth. Mars, in contrast, has a tenuous atmosphere composed of carbon dioxide and traces of water vapor and oxygen. Studies of Mars focus on how its climate and atmosphere evolved and whether it was once Earthlike.

ATMOSPHERIC ALTERATION BY LIFE-FORMS

The atmosphere as it currently exists is a relatively recent phenomenon brought about by the gradual alteration of the environment by life-forms. Awareness of this global alteration is helping humans understand the effects they are having on the environment over a relatively short time span—essentially since the onset of the Industrial Revolution. The widespread use of fossil fuels and the burning of forests to clear land for agriculture have converted the carbon that plants spent billions of years converting into solid carbon compounds back into gaseous carbon dioxide. Furthermore, the plants that were "sinks," or absorbers, of carbon dioxide are available in lesser quantities to liberate oxygen. Sulfur compounds are naturally introduced by volcanoes, biological decay, and oceanic processes, but large quantities have been added by industrial processes, including coal burning. One product, sulfur dioxide, combines with water vapor at low altitudes to form sulfuric acid. At high altitudes, it can also alter the ozone layer and the terrestrial radiation balance. In addition, the ratios of nitrogen compounds are altered by combustion and by widespread use of nitrogen-based fertilizers; these products, too, have an adverse effect on ozone.

The immediate concern is not that the oxygen supply will be depleted, although that is a credible, long-term possibility, but that the increased amounts of carbon dioxide in the atmosphere will cause a greenhouse effect. In the greenhouse effect, long-wavelength (infrared) radiation emanating from the soil or ground is absorbed by the atmosphere and retained. Glass serves this purpose for a greenhouse by retaining the radiation-warmed air within the interior of the structure and so increasing the interior temperature. Carbon dioxide has the same effect in Earth's atmosphere, but it functions by absorbing the infrared energy and emitting a significant portion of it back into the atmosphere and toward the ground rather than out into space. Other human-made gases that enhance the greenhouse effect are nitrous oxide, methane (which is also produced naturally), and chlorofluorocarbons (which also deplete ozone, thus allowing more radiation to enter). Because so little is known about causes and effects in this field, there are uncertainties in predicting what will happen. It is expected, however, that increases in the carbon dioxide content of the atmosphere will raise global temperatures and that such a rise in temperature will shift weather patterns and cause large portions of the polar ice caps to melt, thus flooding coastal regions.

Agricultural Applications

There are many practical applications of the knowledge of thermodynamic data, one of which is regularly utilized in agriculture to estimate the approximate date when crops will be ready for harvest.

The growing degree-day index estimates the number of growing degree-days for a particular crop on any given day as the difference between the daily mean temperature and the minimum temperature required for growth of a particular crop. For example, the minimum growing temperature for corn is 10°C (50 degrees Fahrenheit), which means that on a day when the mean temperature is 24°C (75 degrees Fahrenheit), the number of growing degree-days for sweet corn is estimated at fourteen. Starting with the onset of the growth season, the daily growing degree-day values are added. If 1,111 growing degree-days, sometimes called "heat units," are needed for corn to mature in a particular region, the corn should be ready to harvest when the total number of growing degree-days reaches that figure.

—Daniel G. Graetzer
—Dave Dooling

Further Reading

Ahrens, C. Donald. *Essentials of Meteorology: An Invitation to the Atmosphere.* Brooks/Cole Cengage Learning, 2012. Discusses various topics in weather and the atmosphere, including tornadoes and thunderstorms, acid deposition and other air pollution topics, humidity and cloud formation, and temperature.

Ahrens, C Donald, and Robert Henson. *Meteorology Today: An Introduction to Weather, Climate, and the Environment.* Cengage, 2019. Discusses global climates, climate change, and classification. A thorough and useful text designed for college-level students taking an introductory course on atmospheric science, yet readable for anyone with an interest in meteorology.

Lutgens, Frederick K., Edward J. Tarbuck, Tasa, Dennis G. and Herman, Redina. *The Atmosphere: An Introduction to Meteorology.* 14th ed. Pearson Education, 2018. An updated version of this textbook, in loose-leaf format, provides students with a general overview of the structure and properties of Earth's atmosphere.

Marshal, John, and R. Alan Plumb. *Atmosphere, Ocean and Climate Dynamics: An Introductory Text.* Elsevier Academic Press, 2008. An excellent introduction to atmospheres and oceans. Discusses the greenhouse effect, convection and atmospheric structure, oceanic and atmospheric circulation, and climate change. Suited for advanced undergraduates and graduate students with some background in advanced mathematics.

Stacey, Frank D., and Paul M. Davis. *Physics of the Earth*. 4th ed. Cambridge University Press, 2008. Discusses Earth's atmosphere in Chapter 2. An appendix provides information on thermodynamics. Well organized, with additional mathematics and physics concepts geared to graduate students. Includes many appendices and student exercises.

Vallis, Geoffrey K. *Atmospheric and Oceanic Fluid Dynamics: Fundamentals and Large-scale Circulation*. Cambridge University Press, 2006. Begins with an overview of the physics of fluid dynamics to provide foundational material on stratification, vorticity, and oceanic and atmospheric models. Discusses topics such as turbulence, baroclinic instabilities, wave-mean flow interactions, and large-scale atmospheric and oceanic circulation. Best suited for graduate students studying meteorology or oceanography.

Zdunkowski, Wilfred, and Andreas Bott. *Thermodynamics of the Atmosphere: A Course in Theoretical Meteorology*. Cambridge University Press, 2004. Written for graduate students and researchers in meteorology and related sciences. Assumes a significant background in mathematics in the discussion of thermodynamic principles as they relate to atmospheric phenomena.

ATMOSPHERIC AND OCEANIC OSCILLATIONS

FIELDS OF STUDY

Oceanography; Atmospheric Science; Thermodynamics; Heat Flow; Fluid Dynamics; Environmental Sciences; Environmental Studies; Meteorology; Climate Modeling; Process Modeling; Mathematics; Hydroclimatology; Hydrometeorology; Bioclimatology; Climate Classification; Climate Zones; Physical Geography; Earth System Modeling

SUMMARY

The oceans' effect on developing weather patterns has long been known and taken into consideration in weather prediction. Atmospheric and oceanic patterns fluctuate over a one- to twenty-year course. These fluctuations, or oscillations, create major climate change, such as the well-known El Niño Southern Oscillation (ENSO) in the tropical Pacific, which affects weather across the globe.

PRINCIPAL TERMS

- **conveyor belt current:** a large cycle of water movement that carries warm waters from the North Pacific westward across the Indian Ocean, around southern Africa, and into the Atlantic, where it warms the atmosphere, then returns to a deeper ocean level to rise and begin the process again
- **Ekman spiral:** water movement in lower depths that occurs at a slower rate and in a different direction from surface water movement
- **solar radiation:** transfer of energy from the sun to Earth's surface, where it is absorbed and stored
- **trade winds:** winds that blow steadily toward the equator; north of the equator, trade winds blow from the northeast, whereas south of the equator they blow from the southeast
- **upwelling:** the process by which colder, deeper ocean water rises to the surface and displaces surface water

AIR AND OCEAN INTERACTION

Currents in the water and the air carry heated water toward cooler climates, where the heat is released, helping to regulate temperature on northern continents. As heat is released, currents in the ocean and atmosphere return the cooled water to warmer climates, and the process continues in an endless cycle. These exchanges are an important way of recycling energy through the ocean-atmosphere system, which tends to balance land temperatures and climate response. Ocean currents, which keep water in constant motion, are affected by Earth's rotation, the sun's energy, wind, and the salinity (salt content) and temperature of the ocean. Just as there are a series of air streams, pressures, and currents in the various levels of the atmosphere, the ocean also contains a similar network of circulation patterns and pressure zones.

Some currents that flow on and in the deep sea, such as the Gulf Stream, push through the waters much as rivers flow through land. By coupling the sciences of meteorology and oceanography, several major ocean currents have been detected using peak evaporation levels caused by warm, dry air in the subtropics. Strong evaporation throughout the year off the coast of South Africa characterizes the presence of the Agulhas Current, strong evaporation off the eastern United States in January characterizes the Gulf Stream, and strong evaporation throughout the year in the northeast Atlantic characterizes the North Atlantic Drift. These currents all have generally higher surface temperatures in relation to overlying air, especially in winter, and are frequently accompanied by strong winds. Approximately 40 percent of the total heat transported from the Southern to the Northern Hemisphere is through the action of surface ocean currents.

Another type of ocean current, called an upwelling, brings deeper, colder waters to the surface. This colder water replaces warmer surface waters that are pushed away by the strong trade winds. Upwellings carry with them large amounts of nutrients that nourish the plankton, which make up the base of the food chain. Such currents are an excellent example of how atmospheric currents and ocean currents interact: Without the trade winds to push surface waters away, colder water could not surface.

Ocean currents called Ekman spirals also demonstrate how the atmosphere affects ocean waters. Winds drive surface water in the same direction that the wind blows. Water just below the surface moves more slowly due to friction and the moving water is slightly deflected to the right by the Coriolis Effect. Therefore, water at lower depths moves even more slowly, and in the opposite direction of the flow of surface water. This causes the water to flow in a downward motion called the Ekman spiral. Because of the Ekman spiral, surface water actually flows at a 90-degree angle to the wind flowing out to sea along coastlines, thus creating the opportunity for upwellings.

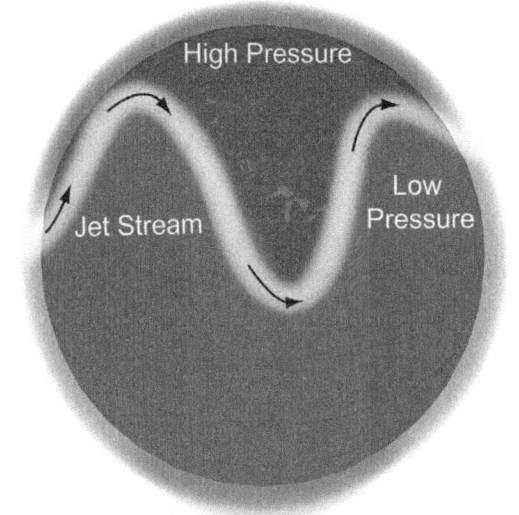

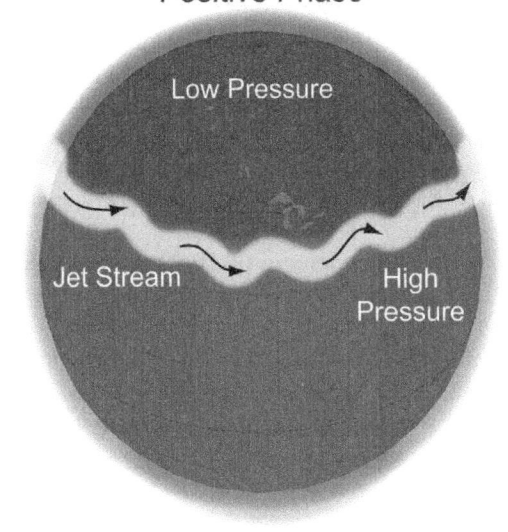

CLIMATE SHIFTS AND GLOBAL IMPACT

Changes in ocean temperature and air currents influence weather, and weather changes over time

Arctic Oscillation

determine climate. As warm-water masses move toward the coasts, they bring with them atmospheric moisture that causes rainfall. When air currents shift, as when strong trade winds die down or reverse course, areas that normally get little rain may be flooded, and areas that normally get a lot of rain may have droughts.

The best-known climate shift has occurred for centuries. It was named El Niño (Spanish for "Christ child" or "the little boy") by Peruvian fisherman because it occurs during the Christmas season. Peru is known for its anchovy fisheries, and El Niño hampers this important harvest. Normally, strong trade winds push warmer surface waters away from Peru's coast so upwelling can occur. This upwelling brings rich nutrients from the lower ocean waters, and Peruvian fishermen reap strong harvests. During El Niño years, which cycle approximately every seven years and may last up to two years, trade winds weaken, and warmer surface waters remain along Peru's coast. Upwellings do not take place, surface waters heat up, and the anchovy harvest suffers.

The effects of El Niño do not end with the Peruvian anchovy fisheries, however. The phenomenon triggers a climatic ripple effect that disrupts weather patterns around the globe. Unusual numbers of storms may rage across North America. Drought conditions have been experienced in northeastern Brazil, southeastern Africa, and western Pacific islands. El Niño effects have also caused unusually wet springs in the eastern United States and elsewhere.

The 1982–1983 El Niño is estimated to have caused more than $8 billion in damage. The severe 1997–1998 El Niño, in turn, is thought to have caused more than $15 billion in damage. It was not until the 1982 El Niño, however, that meteorologists and oceanographers began to seriously study the phenomenon to learn ways to predict both its approach and the severe weather it often causes. Researchers learned that a fluctuating wind pattern known as the Southern Oscillation, which is the same fluctuation that causes the trade winds to weaken, triggers El Niño. Researchers also learned that the ocean-current pattern is dependent upon the wind pattern. Together, this climatic event is known as the El Niño Southern Oscillation (ENSO).

ENSO

Meteorologists and oceanographers continue to learn all they can about ENSO and other atmosphere-ocean oscillations around the globe. The ENSO pattern was the first to be studied, and many weather forecasting models have been created to aid the study of atmosphere-ocean oscillations. These computer models have also helped researchers study ENSO historical trends for the past five hundred years. Many have come to believe that ENSO events may have contributed to plagues and other similar disasters throughout history.

ENSO patterns fluctuate between the extremes of heated tropical waters associated with El Niño and a colder weather-front pattern known as La Niña ("the little girl"). This oscillation pattern takes about seven years to cycle, and each extreme (hot or cold) in the pattern may last up to two years. ENSO researchers have also linked the onset of El Niño with other climate events in surrounding ocean waters. Just before El Niño begins warming the Pacific Ocean, the tropical Indian Ocean warms. Warming then appears in the tropical Pacific, triggering ENSO, which causes warm winds to blow over South America. About nine months after this occurs, the circulation of the tropical Atlantic changes and the waters there begin warming.

The study of the ENSO pattern has led researchers to discover a major ocean circulation pattern called the conveyor belt. This conveyor belt of water is thought to have existed for the several million years since the continents have occupied their current positions in the oceans. It serves to connect the major bodies of water making up the global ocean. Its significance was overlooked until oscillation patterns were studied. The conveyor belt current circulates heated ocean water from the tropical Pacific Ocean to the North Pacific Ocean, turning clockwise around the Pacific, to pass westward between Australia and Malaysia across the Indian Ocean, around the southern tip of Africa, and up into the North Atlantic Ocean. As the surface waters cool in the North Atlantic Ocean, they sink to a lower ocean level and return eastward across the Atlantic Ocean, around southern Africa, across the Indian Ocean, passing south of Australia, and back to the tropical waters of the Pacific Ocean.

Some researchers are concerned that if this conveyor belt current were to stop or slow down, heat would build up in the Southern Hemisphere, while the Northern Hemisphere would experience a severe drop in temperature. Such an occurrence may have caused the last ice age, and research findings continue to stress the importance of this conveyor belt current to maintaining global climate. Temperatures of surface water change in direct relation to changes in this conveyor belt current, the mass flow of which is estimated to equal that of the Amazon River a hundred times over, delivering a heat load to the upper North Atlantic equivalent to one-fourth of the solar energy that reaches the surface in that region.

NAO AND PADO

The study of the ENSO pattern and the use of computer modeling to forecast weather revealed other atmospheric-oceanic oscillations, such as the North Atlantic Oscillation (NAO), which fluctuates on a twenty-year time scale. The atmospheric behavior of NAO has long been known to meteorologists. It is typically a low-pressure, counterclockwise wind circulation that centers over Iceland. This weather pattern contrasts with a high-pressure, clockwise circulation near the Azores Islands off the coast of Portugal. Strong winds blow west to east between these two weather centers. NAO seesaws between these two centers, and strong winds drive heat from the Gulf Stream across Eurasia during high-index years, producing unusually mild winters there. At the other extreme, air pressure builds up over Iceland, weakening the warming winds and delivering bitter winters to Europe and Greenland.

Oceanographers have concluded the ocean must be the cause of the unique atmospheric patterns created by NAO. Because the ocean has a huge capacity for storing heat and reacts at a slower rate than the atmosphere, it must provide the input for atmospheric patterns operating in the same mode year after year. The source of the oceanic oscillator is a pipeline of warm water fed by the Gulf Stream. It takes twenty years to complete one cycle, thus setting the timing for the long-term swings of NAO. Researchers are unsure, however, how the temperature of the waters in the pipeline actually triggers NAO.

The discovery of the Pan-Atlantic Decadal Oscillation (PADO) resulted from a theory that tropical oscillations in the Atlantic actually extended beyond the tropical region. PADO covers an area of more than 11,000 kilometers extending from the southern Atlantic to Iceland. Meteorologists proposed the existence of PADO, claiming it fluctuates on a ten- to fifteen-year time scale. PADO consists of east-to-west bands of water spanning the Atlantic Ocean. The bands alternate between warmer and cooler water and are accompanied by changes in atmospheric circulation. An oscillation to the other extreme reverses the temperature variance. It is also believed that PADO is triggered by NAO.

The study of ENSO, NAO, and PADO has led many researchers to theorize that such oscillations are linked in a chain that circles the global ocean. The Arctic Oscillation, which reverberates from the far northern Atlantic Ocean to the far northern Pacific Ocean, is associated with the NAO, especially in winter. The two phases that characterize the Arctic oscillation correlate precisely to those of the NAO. Varying from day to day and decade to decade, the oscillation is a natural response of the atmosphere to the complex interactions of the ocean-atmosphere system.

The Pacific Decadal Oscillation (PDO) is believed to be another link in the chain of atmosphere-ocean oscillations. This high-latitude oscillation was first noted in 1977 when the northern Pacific Ocean cooled dramatically. An atmospheric low-pressure center off the Aleutian Islands intensified and shifted eastward. This brought more frequent storms to the West Coast of North America and warmed Alaska, while Florida experienced periodic winter freezes. A multitude of other environmental changes also resulted. Researchers have been able to pinpoint other shifts occurring in 1947 and 1925 and have also identified close ties between PDO and ENSO events.

SIGNIFICANCE

The greatest impact of the study of atmosphere-ocean oscillations has been the joining of two separate fields of science. As meteorologists and oceanographers research the manner in which atmospheric and oceanic patterns rely on each other and ultimately influence climate, they have become more dependent on coupled research.

Computer simulations, sometimes referred to as "oceans in a box," have become more sophisticated as researchers from both fields collaborate and combine data. Climate scientists began using computer simulations in the mid-1980s to study winds in the atmosphere to determine how they stir ocean currents and alter pressure patterns in the atmosphere that feed back on the ocean again. Joint research between oceanographers and climatologists extends knowledge about the complex interactions of atmosphere-ocean oscillations and their influence upon each other and the climate.

—Lisa A. Wroble

FURTHER READING

Bigg, Grant R. *The Oceans and Climate.* Cambridge Univ. Press, 2006. Written by a noted professor of environmental sciences. Details atmospheric and oceanic circulation patterns, and discusses their influence on meteorological developments. Describes the influence of the atmosphere and the ocean on each other and demonstrates how this interaction influences major ocean-atmosphere oscillations.

Christopherson, Robert W., and Ginger H Birkeland. *Geosystems: An Introduction to Physical Geography.* Pearson, 2018. Presents a discussion of atmospheric and oceanic oscillations as fundamental systems of the global environment. Includes numerous references and links to online resources. Accessible to general readers.

Easterbrook, Don. *Evidence Based Climate Science: Data Opposing CO_2 Emissions as the Primary Source of Global Warming.* Elsevier, 2011. Presents an evidence-based analysis of the scientific data concerning climate change and global warming. Authored by eight of the world's leading climate scientists who refute the claims embraced by proponents of CO_2 emissions as the cause of global warming. Includes comprehensive citations and references, as well as an extensive bibliography.

Garrison, Tom S. *Oceanography: An Invitation to Marine Science.* Brooks/Cole, Cengage Learning, 2010. Discusses oceanic currents, circulation, and oscillations. Builds a story from the formation of oceans, through waves and tides, to economics and conservation of the ocean, its inhabitants, and its resources. Provides abundant diagrams that aid readers from the layperson to advanced undergraduates.

Houghton, John. *Global Warming: The Complete Briefing.* 4th ed. Cambridge University Press, 2009. Provides an overview of the evidence for global warming. Several chapters focus on the climate system and seasonal forecasting. Offers accessible coverage of the ENSO system. Includes discussions of how ENSO influences weather and the potential for improved weather modeling based on research of ENSO.

Marshal, John, and R. Alan Plumb. *Atmosphere, Ocean and Climate Dynamics: An Introductory Text.* Elsevier Academic Press, 2008. Offers an excellent introduction to atmospheres and oceans. Discusses the greenhouse effect, convection and atmospheric structure, oceanic and atmospheric circulation, and climate change. Best suited for advanced undergraduates and graduate students with some background in advanced mathematics.

McKinney, Frank. *The Northern Adriatic Ecosystem: Deep Time in a Shallow Sea.* Columbia University Press, 2007. Covers the paleogeography of the Adriatic Sea. Discusses the succession of the ecosystem as the sea's geography changed. Discusses oceanography topics such as circulation and sedimentation. Topics are thorough and logically ordered, making this book accessible to undergraduates.

Michaels, Patrick J., and Robert C. Balling. *Climate of Extremes: Global Warming Science They Don't Want You to Know.* Cato Institute, 2009. Presents a view of climate change as the natural and inevitable consequence of oceanic and atmospheric oscillations, instead of the apocalyptic result of anthropogenic effects.

Sarachik, Edward S., and Mark A Cane. *The El Niño-Southern Oscillation Phenomenon.* Cambridge University Press, 2010. Offers a comprehensive discussion of ENSO and other oceanic-atmospheric processes. Covers research measurements, models, and predictions of future occurrences. Provides many diagrams, appendices, a reference list, and index.

TAO Project Office of NOAA/Pacific Marine Environmental Laboratory. *Upper Ocean Heat Content and ENSO.* National Oceanic and Atmospheric Administration (NOAA) Web site (http://www.pmel.noaa.gov/tao). Researchers at NOAA and other research institutions work together to track and model current oscillations, including ENSO and NAO. More information about their findings and other ocean-atmosphere interactions are available through links from their Web site.

Vallis, Geoffrey K. *Atmospheric and Oceanic Fluid Dynamics: Fundamentals and Large Scale Circulation.* Cambridge University Press, 2006. Begins with an overview of the physics of fluid dynamics to provide foundational material on stratification, vorticity, and oceanic and atmospheric models. Discusses topics such as turbulence, baroclinic instabilities, wave-mean flow interactions, and large-scale atmospheric and oceanic circulation. Best suited for graduate students studying meteorology or oceanography.

Woods Hole Oceanographic Institution (WHOI) Web site (http://www.whoi.edu/index.html). Offers general information about the ocean, research, education, and resources, including video animations. WHOI researchers work to understand the complexities of the ocean; atmosphere-ocean interaction is just one component of their research.

ATMOSPHERIC BOUNDARY LAYER

FIELDS OF STUDY

Fluid Dynamics; Physical Geography; Earth System Modeling; Meteorology; Climatology; Earth System Modeling; Atmospheric Science; Thermodynamics; Environmental Sciences; Environmental Studies; Oceanography; Heat Transfer

SUMMARY

The atmospheric boundary layer (ABL) is the lowest 10–20 percent of the troposphere, the lowest layer of the atmosphere. Its contact with a planetary surface directly influences its behavior. A large percentage of the mass and energy of the atmosphere is localized within the ABL.

PRINCIPAL TERMS

- **fluid:** the state of matter characterized by the ability to flow and conform to the shape of its container
- **friction:** resistance to relative motion caused by contact at a shared surface and influenced by surface irregularities
- **urban heat island:** a spot on Earth's surface that is significantly warmer than the surrounding area as a result of human alterations to the landscape, primarily cities.

CHARACTERISTICS OF THE ABL

Also known as the planetary boundary layer (PBL), the atmospheric boundary layer (ABL) is the lowest 10–20 percent of the troposphere, the lowest layer of the atmosphere. Its contact with a planetary surface directly influences its behavior. The ABL contains a disproportionately large amount of the mass and the kinetic energy of the atmosphere. It is the most dynamically active of the layers of Earth. The phrase "boundary layer" originates in the study of boundary layers in fluid flows: The boundary layer is the layer of fluid that is most influenced by friction with the Earth's surface. A defining characteristic of the ABL is turbulence caused by thermal convection, due to thermal buoyancy, and wind shear, due to frictional forces. The atmospheric boundary layer has three layers: the surface layer, the core, and the entrainment layer, also called the capping inversion layer.

SIGNIFICANCE FOR CLIMATE CHANGE

The atmospheric boundary layer is important for ensuring that Earth's atmospheric composition remains relatively homogeneous throughout, despite external heat and energy inputs. Consequently, it is important for ensuring that life can be sustained on Earth. ABL considerations are particularly important in the area of the urban environment. The development of large cities has changed the ABL in those areas, resulting in surface heating, urban heat islands, and artificial boundary layers have developed that trap pollutants. Calculations of the boundary layer can help architects develop urban environments in such a way as to minimize impact on the boundary layer. ABL research can help improve weather forecasts, especially long-term forecasts and forecasts of longtime climate models. Without a properly constituted boundary layer, Earth would lose the unique conditions that make it hospitable to human existence.

The ABL is important meteorologically in the area of assessing convective instability. The entrainment zone (at the top of the ABL) acts as a lid on rising air parcels attributable to temperature inversion. If that entrainment layer is broken, capped air parcels can rise freely, resulting in vigorous convection that produces severe thunderstorms. When sunlight enters the atmosphere, a part of it is immediately reflected

This grey glowing cloud over the city of Berlin is due to light scattered off of suspended aerosols below the planetary boundary layer (PBL). The sky above the PBL appears dark because the air is cleaner.

back to space. The remainder penetrates the atmosphere, and the Earth's surface absorbs it. This energy is then reemitted by the Earth back into the atmosphere in the form of longwave, or infrared, radiation. Carbon dioxide and water molecules absorb this energy, and then emit much of it back toward Earth and into the atmosphere again. This delicate exchange of energy between Earth's surface and the atmosphere is what keeps the average global temperature from changing drastically from year to year. When this exchange is disrupted, climate problems result.

—*Victoria Price*

Further Reading

Çengel, Yunus A., and John M. Cimbala. *Fluid Mechanics: Fundamentals and Applications.* McGraw-Hill, 2010.

Frederick, John E. *Principles of Atmospheric Science.* Jones and Bartlett 2008. Provides a concise introduction to atmospheric science beginning from the basic constituents and structure of the atmosphere.

Lutgens, Frederick K., Edward J. Tarbuck, Tasa, Dennis G. and Herman, Redina. *The Atmosphere: An Introduction to Meteorology.* 14th ed. Pearson Education, 2018. An updated version of this textbook, in loose-leaf format, provides students with a general overview of the structure and properties of Earth's atmosphere.

Metcalfe, Sarah and Derwent, Dick *Atmospheric Pollution and Environmental Change.* Routledge, 2014.

Schimel, David *Climate and Ecosystems.* Princeton University Press, 2013. One of the "Princeton Primers on Climate" series, this book describes the interaction mechanisms of various ecosystems and climate.

ATMOSPHERIC CHEMISTRY AND CLIMATE CHANGE

FIELDS OF STUDY

Atmospheric Chemistry; Atmospheric Science; Photochemistry; Environmental Chemistry; Geochemistry; Oceanography; Organic Chemistry; Inorganic Chemistry; Physical Chemistry; Thermodynamics; Environmental Sciences; Environmental Studies; Waste Management; Meteorology; Climate Modeling; Process Modeling; Mathematics; Fluid Dynamics; Chemical Kinetics; Hydroclimatology; Hydrometeorology; Physical Geography; Spectroscopy; Particle Physics

SUMMARY

The significance of the Earth's atmosphere is vastly disproportionate to its size. Although its thickness relative to Earth's sphere is comparable to an apple's skin, it is essential for life. It was not until the eighteenth century that scientists began to understand the role of atmospheric gases such as oxygen and carbon dioxide (CO_2) in plant and animal life, and it was not until the end of the nineteenth century that scientists grasped the details of how soil microorganisms utilized atmospheric nitrogen to create compounds necessary for the health of plants and animals. Throughout the twentieth century, climatologists, atmospheric chemists, and others gathered information about how such anthropogenic gases as CO_2, methane, and nitrous oxide were increasing Earth's greenhouse effect and elevating the planet's average global temperature. This enhanced greenhouse effect fosters climate changes that are potentially so devastating that some scholars have called climate change the most important issue of the twenty-first century.

PRINCIPAL TERMS

- **anthropogenic:** resulting from human activities
- **chlorofluorocarbons (CFCs):** compounds of chlorine, fluorine, and carbon, popularly known by the trade name Freon
- **greenhouse effect:** result of atmospheric trace gases that allow high-energy sunlight to reach the terrestrial surface but absorb low-energy heat that is radiated back
- **greenhouse gases (GHGs):** tropospheric gases such as carbon dioxide, methane, and water vapor that cause the greenhouse effect
- **ozone layer:** a stratospheric region containing relatively high concentrations of triatomic oxygen (ozone) that prevents much ultraviolet solar radiation from reaching Earth's surface
- **primary air pollutants:** harmful substances that are emitted directly into the atmosphere
- **secondary air pollutants:** harmful substances that result from the reaction of primary air pollutants with principal atmospheric components
- **stratosphere:** an atmospheric region extending from about 17 to 48 kilometers above the Earth's surface
- **troposphere:** an atmospheric region extending from the Earth's surface to about 17 kilometers high over equatorial regions and to about 8 kilometers high over polar regions

Chemical Composition of the Earth's Atmosphere

Approximately three-quarters of Earth's air mass is located in the troposphere, and dry air in this region is approximately 78 percent nitrogen, 21 percent oxygen, and 1 percent argon and other gases, by volume. The troposphere also contains trace amounts of many other gases, such as methane, various nitrogen oxides, ammonia, sulfur dioxide, and ozone, arising from both natural and anthropogenic sources. Human activities have not changed the concentrations of the two major gases in the atmosphere, but scientific evidence accumulated over the past century indicates that human beings, particularly in advanced industrialized societies, are effectively changing the concentrations of certain trace gases. Examples of these include CO_2, methane, nitrous oxide, carbon monoxide, chlorofluorocarbons (CFCs), and sulfur dioxide. Some of these atmospheric trace gases, particularly CFCs, are synthetic or artificial products that have been used as refrigerants and aerosols. Others, such as CO_2 and sulfur dioxide, are produced by the burning of fossil fuels. Agricultural practices are also significant sources of gases such as methane and nitrous oxide.

Although Earth's stratosphere contains much less gaseous matter than the troposphere, the ratio of nitrogen to oxygen is very similar to that of the troposphere. It differs markedly from the troposphere, however, in its concentrations of water vapor and ozone. Stratospheric water-vapor concentrations are only about one-thousandth of tropospheric concentrations, but ozone concentrations are much higher in the stratosphere. Ozone is localized in a layer ranging from about 15 to 35 kilometers above Earth's surface. This ozone layer, whose molecules are created when oxygen interact with high-energy solar radiation, prevents about 95 percent of the Sun's ultraviolet radiation from reaching Earth's surface, where it would damage and ultimately destroy living organisms. Accordingly, the ozone layer also prevents the conversion of tropospheric oxygen to ozone, which, in the lower atmosphere, is a dangerous air pollutant.

Chemical Reactions in the Troposphere

Besides being home to such major gases as nitrogen and oxygen, the troposphere contains hundreds of other distinct molecular species, formed by and undergoing myriad chemical reactions driven by sunlight, some of which have an influence on climate change. Because oxygen is such a reactive species, many of these reactions are oxidations, and some scientists see these reactions as constituting a low-temperature combustion system. Fueling this combustion are chemicals released from both natural and artificial sources. For example, methane enters the troposphere in large amounts from swamp and bog emissions, termites, and ruminant animals. Human activities contribute a large number of organic compounds, and CO_2 and water are the end results of their oxidation. Both of these are potent greenhouse gases (GHGs).

Global Sulfur Emissions by Source and Latitude

Atmospheric chemical inputs can vary greatly by region, as illustrated by the table below listing the vastly different sources of atmospheric sulfur in different parts of the globe.

Latitude	Anthropogenic %	Marine %	Terrestrial %	Volcanic %	Biomass Burning %
90° south	0	0	0	0	0
75° south	0	80	0	19	1
58° south	2	97	0	0	1
45° south	22	72	0	9	1
28° south	67	28	0	1	4
15° south	21	47	1	22	10
0°	21	39	1	33	7
15° north	40	30	1	19	1
28° north	85	6	0	8	1
45° north	88	4	0	7	1
58° north	86	3	0	10	1
75° north	30	40	0	23	7
90° north	0	0	0	0	0

Source: Pacific Marine Environmental Laboratory, National Oceanic and Atmospheric Administration.

Atmospheric chemists have also been attempting to work out in detail the influence of chemical radicals on tropospheric gases. Radicals can be charged or neutral atoms or parts of molecules that have single electrons available for reaction with other chemical species. Such charged groups of atoms as the hydroxyl radical (composed of one hydrogen and one oxygen atom) play an important role in the daytime chemistry of the troposphere, and the nitrate radical (composed of one nitrogen atom and three oxygen atoms) is the dominant nighttime oxidant of atmospheric organic chemical species. Fossil-fuel combustion is a significant contributor to tropospheric pollution. Particulates such as soot were a factor in some "killer smogs," and scientists have recently discovered that particulates contribute to global dimming, a lessening of sunlight's ability to penetrate the atmosphere and reach Earth's surface. Sulfur dioxide, which is produced by the combustion of coal and oil, can be a primary air pollutant, since it is toxic to living organisms as well as damaging to buildings. It can also be a secondary air pollutant, because it reacts with water vapor to create sulfuric acid, which is an acid-rain component, causing harm to various lifeforms, including trees and fish.

CHEMICAL REACTIONS IN THE STRATOSPHERE

Just as in the lower atmosphere, chemical reactions in the upper atmosphere exhibit great variety, and some of these reactions have an important influence on climate change. Over the past decades, the chemical species that has received the most attention has been ozone. Scientists paid heightened attention to the chemical reactions in the ozone layer when, in the late 1980s, a "hole" was discovered in this layer above the Antarctic. During the 1970s scientists had found a threat to the ozone layer when they worked out the reactions between chlorine-containing radicals and ozone. These reactions changed ozone molecules into diatomic oxygen molecules, thus weakening the ability of the ozone layer to protect Earth's surface from high-energy solar radiation.

A primary source of these catalytic, chlorine-containing species turned out to be CFCs. General Motors had developed and introduced various CFCs in 1930, and they proved to be successful as refrigeration and air-conditioning coolants, as well as being effective as aerosol propellants. Because of the widespread and accelerating use of CFCs, the tropospheric concentrations of these chemicals increased from the 1930s to the 1970s, when Mexican chemist Mario Molina and American chemist F. Sherwood Rowland showed that CFCs, although seemingly inert in the troposphere, became very reactive in the stratosphere. There, ultraviolet radiation split the CFCs into highly reactive radicals that, in a series of reactions, promoted the debilitation of the protective ozone shield.

The exhaust from aircraft and spacecraft also helped deplete stratospheric ozone. Despite attempts, such as the Montreal Protocol (1987), to reduce concentrations of CFCs and other ozone-depleting chemicals in the atmosphere, the Antarctic ozone hole continued to grow in the 1990s and early twenty-first century. This meant that countries near Antarctica began experiencing higher levels of ultraviolet solar radiation. Current data as of 2017, however, indicate that the ozone layer has recovered somewhat as a result of the elimination of CFCs and similar products such as brominated solvents from general use following the institution of the Montreal Protocol and subsequent efforts.

Atmospheric Chemistry and Global Climate Change

Humans tend to be most aware of weather, a local area's short-term temperature and precipitation variations. Scientists such as atmospheric chemists tend to concentrate on climate, or a large region's long-term variations in temperature, precipitation, and cloud cover. Because of discoveries revealing the extreme complexity of chemical reactions in the atmosphere, atmospheric chemistry has become a profoundly interdisciplinary field, depending on new facts and ideas found by physicists, meteorologists, climatologists, oceanographers, geologists, ecologists, and other scientists.

Paleoclimatologists have studied changes in Earth's atmosphere over hundreds of millions of years, while other environmental and atmospheric chemists have focused on such pivotal modern problems as global warming. These studies have led to research aimed at understanding the causes of global warming and the development of theories to explain existing data. Particularly useful has been computerized modeling of Earth's atmosphere, through which experiments can be performed to help scientists understand likely future effects of climate change. These theoretical predictions have placed pressure on various governments to make important changes in policy, such as taxing fossil-fuel use to motivate reductions in GHG emissions.

Atmospheric chemists have come to realize that the goal of their research on global climate change is to understand the relevant chemical species in the atmosphere, their reactions, and the role of anthropogenic chemicals, especially GHGs, in relation to climate change and global warming. Many atmospheric chemists believe that a greenhouse effect beyond what would normally exist is a certainty, and they are also highly confident that human activities generating GHGs are a significant element in the recent rise in average global temperatures. Less certain are predictions about the future.

Computer models developed to synthesize and test theories about the complex chemical interactions in the troposphere and stratosphere necessarily involve assumptions and simplifications. For example, the numbers of chemical compounds and their reactions have to be reduced to formulate even a crude working model of Earth's atmosphere. Despite these problems, many environmental chemists, building on what they are most sure of, have played an important part in several countries in determining governmental policies as they relate to global climate change.

CONTEXT

Atmospheric chemists' discoveries have had a major influence on how environmentalists and other scientists understand the seriousness, interrelated nature, and complexity of atmospheric problems. Many atmospheric chemists educate their students and the public about issues relating to global climate change, while others have been carefully monitoring the changes in the Earth's atmosphere. They have also participated in international discussions and agreements about controlling GHG emissions, developing substitutes for CFCs, and passing local and international laws that would lessen the likelihood of some catastrophic scenarios predicted by various computer models. Just as the many components and reactions in the atmosphere make a full understanding of these complexities very difficult, so, too, environmental chemists find themselves in an even more complex milieu in which they have to integrate their understanding with those of other scientists, industrialists, and government officials in both developed and developing countries. Therefore, though global climate change is, at root, a physical and chemical issue, to solve the problem of global climate change will require an integrated, multidisciplinary, and international approach that, though daunting, appears to be increasingly necessary.

—*Robert J. Paradowski*

FURTHER READING

Websites:

National Aeronautics and Space Administration (NASA) "NASA Ozone Watch" <https://ozonewatch.gsfc.nasa.gov/>

Books:

Jacob, Daniel J. *Introduction to Atmospheric Chemistry.* Princeton University Press, 2007. Undergraduate textbook written by a Harvard professor; provides an overview of the new and rapidly growing field of atmospheric chemistry. Illustrations and index.

Möller, Detlev. *Chemistry of the Climate System.* 2nd ed. De Gruyter, 2014. A more advanced book providing detailed information about the chemistry of atmospheric gases and air-borne constituents, using a problem-based approach.

National Academy of Sciences. *The Future of Atmospheric Chemistry Research. Remembering Yesterday, Understanding Today, Anticipating Tomorrow.* National Academies Press, 2016. This report from the National Academy of Sciences offers a significant account of the state of atmospheric chemistry research.

Seinfeld, John H. and Pandis, Spyros N. *Atmospheric Chemistry and Physics. From Air Pollution to Climate Change.* 3rd ed. John Wiley & Sons, 2016. A complete introduction to the structure and chemistry of the atmosphere.

ATMOSPHERIC CIRCULATION

FIELDS OF STUDY

Fluid Dynamics; Mathematics; Atmospheric Science; Heat Transfer; Thermodynamics; Environmental Studies; Environmental Sciences; Earth System Modeling; Climate Modeling; Meteorology; Process Modeling; Hydroclimatology; Hydrometeorology; Climate Zones; Climate Classification; Oceanography; Computer Science

SUMMARY

Atmospheric circulation is the large-scale movement of air that distributes heat from tropical to polar latitudes across the surface of Earth. The global wind patterns are guided by three distinct convection cells known as the Hadley cell, Ferrel cell, and Polar cell. These cell currents transport heat by circulating air at various latitudes and that extend from Earth's surface to the upper boundary of the troposphere. The troposphere is the lowest layer of the atmosphere and extends from Earth's surface upward to approximately 15 kilometers (9.3 miles) above the surface, where it is separated from the stratosphere by an area of temperature inversion known as the tropopause. Because the troposphere is where nearly all of the world's weather conditions originate, it is important for scientists to understand atmospheric circulation patterns in order to more accurately predict climate

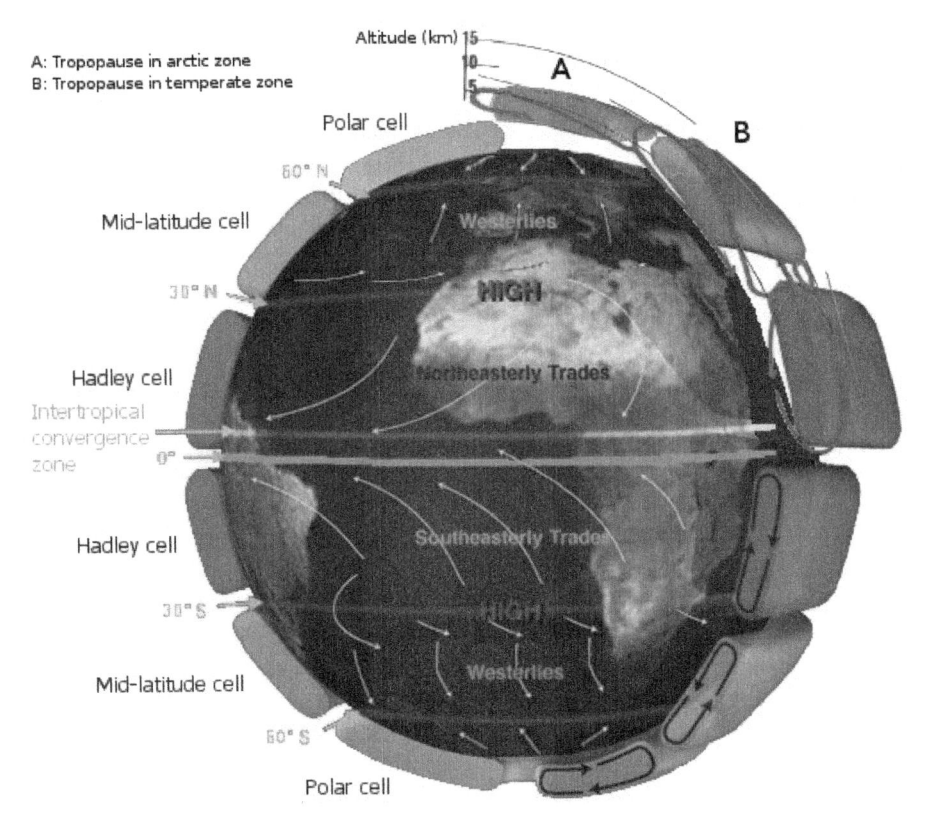

Atmospheric circulation diagram, showing the Hadley cell, the Ferrel cell, the Polar cell, and the various upwelling and subsidence zones between them.

conditions that can affect everything from crop production to transportation safety.

PRINCIPAL TERMS

- **Coriolis Effect:** an apparent force acting on a rotating coordinate system; on Earth this causes things moving in the Northern Hemisphere to be deflected toward the right and things moving in the Southern Hemisphere to be deflected toward the left
- **eastern trade winds:** winds blowing in a northeasterly or southeasterly direction near the equator as a result of the Coriolis Effect and the Hadley circulation, so named because they favored sailing ships traveling eastward on trade missions
- **geostrophic current:** a current resulting from the balance between a pressure gradient force and the Coriolis Effect; the current moves horizontally and is perpendicular in direction to both the pressure gradient force and the Coriolis Effect
- **pressure gradient:** a difference in pressure that causes fluids (both liquids and gases) to move from regions of high pressure to regions of low pressure
- **wind-driven circulation:** the surface currents on the ocean that result from winds and geostrophic currents

Background

English meteorologist George Hadley (1685–1768) was interested in finding out why sailors encountered westerly winds at the midlatitudes and easterly winds, known as the trade winds, closer to the equator. In 1735, Hadley described atmospheric circulation as a massive version of a huge sea breeze in which warm air rises over the equator and sinks over the poles and is moved directionally along latitudinal lines as a result of the rotation of the Earth. His is regarded as the first attempt to describe how weather patterns combine and interact to produce a general circulation of the atmosphere. In recognition, the largest of the three convention cells was named after Hadley.

The Hadley cell lies nearest to the equator, stretching north and south from the equatorial line to approximately 30 degrees latitude. Within the Hadley cell, warm air rises from along the equator and flows toward the poles within the troposphere before cooling and descending in the subtropics. Near the surface, trade winds blow toward the equator in a westward direction and often develop into thunderstorms as they rise near the equator, in what is called the Inter-Tropical Convergence Zone. The rising warm air from the equator circulates toward higher latitudes and then sinks at approximately 30 degrees latitude, creating high-pressure regions over the world's subtropical oceans and deserts.

The Ferrel cell, named in honor of nineteenth-century American meteorologist William Ferrel (1817–1891), represents the midlatitude segment of the Earth's atmospheric circulation, ranging between 30 and 60 degrees north and south latitude. Air circulation in the Ferrell cell is opposite the flow in the Hadley cell. In the Ferrel cell, air near the surface flows toward the poles in an eastward direction, and air at higher altitudes flows toward the equator in a westward direction. The prevailing winds in this cell, known as the westerlies because they originate in the west and flow eastward, are more susceptible to passing weather systems—particularly subtropical highs—than the prevailing winds in the Hadley and Polar cells and can change direction abruptly.

The Polar cell lies at the farthest distance from the Earth's equator, extending from 60 degrees latitude to the North and South Poles. These are the smallest and weakest of the atmospheric circulation cells. Air in the Polar cell rises at lower latitudes and moves toward the poles through the troposphere. When this circulating air reaches the pole, it has cooled significantly and descends, traveling along the surface back toward the equator in a westward direction. The prevailing winds in this cell are known as the polar easterlies.

OVERVIEW

Atmospheric circulation occurs when pressurized air moves around the globe in convection cells, with warmer, denser air rising from the surface and cooler, less dense air descending from the troposphere. Air circulation is also driven by movement from dense, high-pressure areas to low-pressure areas.

The vertical and horizontal air movements come together to influence climate and weather conditions in the various parts of the world. For example, land is heated more quickly than water during the daytime hours due to the differences in the specific heat capacity of land and water. Therefore, the air above the land becomes warmer and rises (vertical movement), adding to the atmospheric pressure. Horizontal air flow then moves the pressurized air into lower-pressure areas over the sea, creating less air mass over the land. The cycle perpetuates when the pressurized air over the sea makes its way to the lower-pressure atmosphere near the land, where it heats up again and continues the rotation. The flow is reversed in the evening—land loses heat more quickly than water—creating an opposite current of circulating air.

Air moves through the atmosphere under the influence of pressure gradients that propel it from high-pressure areas to low-pressure areas. Horizontal winds that travel long distances appear to follow curved trajectories because of the eastward rotation of the Earth. The specific arc is a result of air's speed of movement and its latitude. For example, a mass of air that is flowing from the equator toward the pole appears to be deflected because the air is moving faster to the east at the equator than its destination at the pole. This is because a stationary object at the equator completes a path of approximately 40,000 kilometers in one day because of the Earth's rotation, while an object located at 60 degrees latitude travels only half that distance in the same time. This force is known as the Coriolis Effect in honor of the French physicist Gustave-Gaspard Coriolis (1792–1843), who described the phenomenon in 1835.

Knowing how air pressure and Earth's natural forces relate to air movement is essential for making predictions and preparations with respect to regional climates, ocean currents, storm systems, and wind behavior that can impact the safety, well-being, and livelihoods of people all over the world.

—*Shari Parsons Miller, MA*

FURTHER READING

Barry, Roger G., and Richard J. Chorley. *Atmosphere, Weather, and Climate.* Routledge, 2009.

Henderson-Sellers, Ann, and Kendal McGuffie. "Atmospheric Composition Change: Climate-Chemistry Interactions" *The Future of the World's Climate: A Modeling Perspective.* Elsevier, 2012. 309–66.

Leroux, Marcel. *Dynamic Analysis of Weather and Climate: Atmospheric Circulation, Perturbations, Climatic Evolution.* Springer, 2014.

Philander, S. George. "Investigating Atmospheric Circulation." *Our Affair with El Niño: How We Transformed an Enchanting Peruvian Current into a Global Climate Hazard.* Princeton University Press, 2005, 177–88.

Satoh, Masaki. *Atmospheric Circulation Dynamics and General Circulation Models.* Springer, 2014.

Seager, Sara. "Atmospheric Circulation." *Exoplanet Atmospheres: Physical Processes.* Princeton University Press, 2010. 211–28.

Vallis, Geoffrey K. *Atmospheric and Oceanic Fluid Dynamics: Fundamentals and Large-Scale Circulation.* Cambridge University Press, 2006.

ATMOSPHERIC DYNAMICS

FIELDS OF STUDY

Atmospheric Science; Atmospheric Chemistry; Fluid Dynamics; Thermodynamics; Heat Transfer; Oceanography; Hydrology; Physical Geography; Photochemistry; Meteorology; Climate Modeling; Hydroclimatology; Hydrometeorology; Bioclimatology; Climate Zones; Climate Classification; Earth System Modeling

SUMMARY

In the eighteenth century, Edmond Halley (1656–1742), for whom Halley's comet is named, charted the monsoons and the trade winds, making the first known meteorological map. In an effort to understand the trade winds, Halley correctly surmised that the Sun-warmed air over the equator would rise high into the atmosphere and then flow toward the poles. He further supposed that the air would cool off and sink at the poles, and then return to the equator as a surface wind, but he could not explain why the trade winds came from the northeast, or even the east, instead of from the north.

PRINCIPAL TERMS

- **air parcel:** a theoretical house-sized volume of air that remains intact as it moves from place to place
- **Coriolis Effect:** in the Northern Hemisphere, the westward deflection of southward-moving air and the eastward deflection of northward-moving air—caused by Earth's rotation
- **stratosphere:** the atmospheric region just above the tropopause and extending up about 50 kilometers
- **tropopause:** the transition region between the troposphere and the stratosphere
- **troposphere:** the lowest layer of the atmosphere—in which storms and almost all clouds occur, extending from the surface to an altitude of 8 to 15 kilometers

THE ROLE OF THE CORIOLIS EFFECT

Earth is about 40,000 kilometers in circumference at the equator, and it rotates once in twenty-four hours. Thus, at the equator the land, sea, and air are rushing eastward at nearly 1,700 kilometers per hour. At 45° north latitude, by contrast, Earth is only about 28,000 kilometers in circumference, so a point located at that latitude travels eastward at only about 1,200 kilometers per hour.

Consider a parcel of air at rest with respect to the land at the equator. Suppose that it is filled with red smoke so that its location is easily seen. Now let the parcel move northward; because of its eastward momentum, it will also be moving eastward with respect to the land north of the equator. This eastward deflection of northward-moving air parcels is called the Coriolis Effect and is named for Gaspard-Gustave de Coriolis (1792–1843), who studied it in 1835. Also as a result of the Coriolis Effect, if an air parcel at rest with respect to the land at some point north of the equator begins moving southward toward the equator, it will be deflected westward, because its

Hurricane Michelle, in November, 2001.

eastward momentum will be less than that of the ground below it.

In 1735, some fifty years after Halley's surmises, George Hadley (1685-1768) proposed the first workable explanation for the direction of the trade winds by referring to the Coriolis Effect. Hadley believed that air parcels heated at the equator rose high and then migrated north to the pole. Cooled during the journey, the parcel would sink to the surface and head back south. Because of the Coriolis Effect, the southward moving air would be deflected westward. This would explain the trade winds north of the equator, and this proposed air circulation route was called a Hadley cell.

The American meteorologist William Ferrel (1817-1891) pointed out that atmospheric dynamics could not actually be that simple, since the prevailing winds at midlatitudes are westerlies, not easterlies such as the trade winds. Ferrel suggested that the Hadley cell extended only to about 30° latitude north of the equator, where the cooled air sank and returned to the equator. The Ferrel cell lies between about 30° north latitude and 60° north latitude. Air rises at 60°, flows southward, cools and descends at 30°, and flows northward and from the west near the ground—hence the westerlies. The Polar cell extends from 60° to the pole, with air rising at 60° and sinking at the pole. The cells of the Southern Hemisphere mirror those of the Northern Hemisphere.

JET STREAMS

Where the Ferrel cell meets the Polar cell, the temperatures and pressures of the air masses are generally different. These differences give rise to winds blowing north from the Ferrel cell toward the Polar cell, but this wind is soon deflected eastward by the Coriolis Effect. Hemmed in between the Polar and Ferrel cells, the wind becomes the polar jet stream—a river of air 160 to 500 kilometers wide, 1 kilometer deep, and generally 1,500 to 5,000 kilometers long. Several discontinuous segments of the jet stream together might come close to circumnavigating the Earth. These segments wax and wane over time and sometimes disappear completely.

The polar jet stream forms at the tropopause, 7 to 12 kilometers above sea level. (The tropopause is the transition region between the troposphere below and the stratosphere above.) The speed of the jet stream averages 80 kilometers per hour in the summer and 160 kilometers per hour in the winter, but it can reach speeds of up to 500 kilometers per hour. While it generally flows eastward, sometimes it also meanders hundreds of kilometers south and then back north.

A second jet stream, the subtropical jet stream, occurs between the Hadley and the Ferrel cells, but since the tropopause is higher there, this jet stream is between 10 and 16 kilometers above ground level. This jet stream tends to form during the winter, when temperature contrasts between air masses are the greatest. Other low-level jet stream segments may form near the equator. The jet streams of the Southern Hemisphere mirror those of the Northern Hemisphere. Studies show that jet streams help carry carbon dioxide (CO_2) from where it is produced to other parts of the world.

There are practical reasons for studying jet streams. Jet streams influence the paths of storms lower in the atmosphere, so meteorologists must take them into account in their forecasts. Pilots flying from Tokyo to

Los Angeles can cut their flight times by one-third if they can use the jet stream for a tailwind. About just one percent of the energy of the world's jet streams could satisfy all of humanity's current energy needs, if it were possible to tether wind-powered generators within them.

Oscillations

In 1899 India was stricken by a severe famine and drought because the monsoons failed. In response, Sir Gilbert Walker (1868–1958) headed a team at the Indian Meteorological Department using statistical analysis on weather data from the land and sea looking for a link to the monsoons. They eventually found a link between the timing and severity of the monsoons and the air pressure over the Indian Ocean and over the southern Pacific Ocean. The team found that high pressure over the Pacific meant low pressure over the Indian Ocean, and vice versa. Walker named this alternating of pressures the Southern Oscillation and linked it with the monsoon. It has since been linked to other weather phenomena.

Normally, there is a large region of high pressure in the Pacific just off the coast of South America. The trade winds near the surface blow westward from this high-pressure region to a low-pressure region over Indonesia. The winds pick up moisture as they cross the Pacific and deliver it in the monsoons over Indonesia, India, and so forth. Energy from the condensation of moisture heats the air and causes it to rise higher; then, the air flows back eastward to the South American coast, cools, and sinks to complete the Walker cycle.

Just after Christmas, a warm current flows south by the coasts of Ecuador and Peru. In some years, that current is stronger and warmer, and then it brings beneficial rains to the South American coast, a kind of "Christmas gift." The event is called El Niño (little boy) with reference to the Christ Child. The event is also called the El Niño-Southern Oscillation, or ENSO, and its affects weather on a global scale. The trade winds weaken, and warm water surges eastward to the South American coast. Air rises as it is heated by the warm coastal waters, and air comes from the west to replace the rising air. As a result, the trade winds are reversed and blow eastward. The winds that carried moisture from the Indian Ocean toward the equator are weakened, so the monsoons are weakened. Low pressure develops over the Indian Ocean and pulls the subtropical jet stream south. The displaced jet stream brings more rain to East Africa and drought to Brazil. Central Asia, the northwestern United States, and Canada experience heat waves, while Central Europe experiences flooding.

After a few years, the El Niño event weakens and things return to normal—except that two-thirds of the time nature overshoots "normal," and La Niña (little girl) appears. The west-blowing trade winds return but are much stronger than normal. Cool water rises from the deep and forms a cool region off the west coast of South America. Colder-than-normal air blows over the Pacific Northwest and over the northern Great Plains, but the rest of the United States enjoys a milder winter. The Indian monsoons strengthen, and the subtropical jet stream returns to its normal position. Eventually, things quiet down and return to normal.

Since El Niño begins when warm water collects off the western coast of South America, global warming will probably increase the frequency and intensity of El Niño. The accompanying heat waves, flooding rains, and droughts will likely be more severe.

Fronts and High-Pressure Air Masses

A weather front is the boundary between two air masses of different densities. They normally differ in temperature and humidity. Cold air is denser than warm air, so the air of an advancing cold front wedges beneath and lifts warm air. This upward motion produces low pressure along the front, and as the air lifts and cools, moisture condenses and forms a line of clouds or showers along the front. Light rain may begin 100 kilometers from the front, with heavy rain beginning 50 kilometers from the front.

Cold fronts generally come from the north and head south, while warm fronts generally come from the south and head north. The leading edge of an advancing warm front takes an inverted wedge shape, such that the high cirrus clouds that mark the approach of the front may be hundreds of kilometers ahead of the front's ground-level location. The cloud base continually lowers as the front approaches, and with enough moisture present rain may extend 300 kilometers in front of the ground-level front. Fronts are the principal cause of nonrotating storms.

Sinking air forms a high-pressure area at the surface. This air is usually dry, having delivered its moisture elsewhere, rendering the sky clear. If there were any warm, moist air, it would be prevented from rising by the descending air of the high-pressure area, so it could not form clouds and rain. Many of Earth's deserts form where the circulating air of the Hadley cell descends. Any high-pressure area that remains in place for a long time can cause drought. Winds blowing outward from the high-pressure area will be turned by the Coriolis Effect in a direction established by a simple rule: In the Northern Hemisphere, if one stands so that the low-pressure area is on one's left, the wind will be at one's back. In the Southern Hemisphere, the wind would be at one's front.

HURRICANES AND LOW-PRESSURE AIR MASSES

Rising air forms a low-pressure area near Earth's surface. Higher-pressure air from outside the area will flow toward the low-pressure area (like water running downhill), but it will be turned by the Coriolis Effect and slowly spiral inward. This air will eventually be caught up in the rising air currents, will cool off as it rises, will be heated as its moisture condenses, and will then rise higher and contribute to the updraft. At this point, the low-pressure area has become a storm. If its rotation speed is over 120 kilometers per hour, the storm has become a hurricane. Global warming models predict that as CO_2 levels increase, the number of hurricanes occurring annually may not increase, but the average intensity of such storms will increase.

CONTEXT

More than a century ago, Sir Gilbert Walker began the process of describing the world's weather in one comprehensive model. The weather in each part of the world is tied in some fashion to the weather elsewhere, through jet streams, circulation cells, or movement of air masses, in accord with the nature of fluids. As Earth's global climate undergoes changes, the interrelationship of the planet's weather systems becomes more important than ever, since the climate change will affect not only individual weather patterns and events but also the way in which those individual events affect one another and Earth's weather generally.

—*Charles W. Rogers*

FURTHER READING

Ahrens, C. Donald. *Essentials of Meteorology: An Invitation to the Atmosphere.* 5th ed. Belmont, Calif.: Thomson Brooks/Cole, 2008. Widely used introductory textbook on atmospheric science; covers a wide range of topics on weather and climate.

Frederick, John E. *Principles of Atmospheric Science.* Jones and Bartlett 2008. Provides a concise introduction to atmospheric science beginning from the basic constituents and structure of the atmosphere.

Gill, Adrian E. and Donn, William L. *International Geophysics Series Vol. 30. Atmosphere-Ocean Dynamics.* Elsevier, 2016. This book provides a comprehensive overview of the physical and mathematical principles describing the interaction of the atmosphere and oceans of Earth.

Iribarne, Julio V. *Atmospheric Physics.* Springer, 2013. As a short introduction to the field of atmospheric dynamics, this book attempts to provide a comprehensive yet elementary survey of the terrestrial atmosphere and how it functions.

Lutgens, Frederick K., Edward J. Tarbuck, Tasa, Dennis G. and Herman, Redina. *The Atmosphere: An Introduction to Meteorology.* 14th ed. Pearson Education, 2018. An updated version of this textbook, in loose-leaf format, provides students with a general overview of the structure and properties of Earth's atmosphere.

ATMOSPHERIC INVERSIONS

FIELDS OF STUDY

Atmospheric Science; Atmospheric Chemistry; Photochemistry; Spectroscopy; Thermodynamics; Fluid Dynamics; Heat Transfer; Environmental Sciences; Environmental Studies; Meteorology; Climate Modeling; Process Modeling; Hydroclimatology; Hydrometeorology; Physical Geography; Ecosystem Management; Earth System Modeling; Ecology; Toxicology; Pathology

SUMMARY

Air temperature normally decreases with height above Earth's surface. An atmospheric inversion is defined in meteorology as a scenario in which air temperature increases with height. To understand its importance it is necessary to consider the concept in light of the broader structure of the atmosphere.

PRINCIPAL TERMS

- **adiabatic:** referring to thermodynamic processes that occur without a net change in heat content
- **anthropogenic:** deriving from human sources or activities
- **inversion:** reversal of the normal order
- **stratosphere:** an atmospheric region extending from about 17 to 48 kilometers above Earth's surface
- **troposphere:** the lowest layer of Earth's atmosphere extending from the surface to the tropopause, the transition layer to the mesosphere, characterized by decreasing temperature with increasing altitude

STRUCTURE OF THE ATMOSPHERE

The vertical structure of the atmosphere is characterized by four broad regions; in order of increasing altitude, these are the troposphere, stratosphere, mesosphere, and thermosphere. On average, temperatures decrease with height in the troposphere and mesosphere, and increase with height in the stratosphere and thermosphere. The atmosphere's vertical temperature profile is important because it affects the ability of air to move vertically (that is, to rise and fall) in the atmosphere. Vertical motions are generally permitted in regions where air temperatures decrease with height, and they are suppressed in regions where air temperatures increase with height. The latter condition, known as a temperature inversion, creates a layer in the atmosphere that has the ability to limit, or cap, vertical mixing. The stratospheric inversion is a prime example; it caps vertical motions associated with storms in the troposphere.

Temperature inversions play an important role in trapping anthropogenic (human-caused) pollutants near the Earth's surface, leading to the formation of smog and reduced air quality in many metropolitan areas. Temperature inversions also play an important role in the formation of severe thunderstorms and mixed precipitation (such as freezing rain and sleet).

Smaller-scale temperature inversions that occur within the troposphere may have significant impacts on air quality at urban and regional levels because they trap polluted air masses close to the ground.

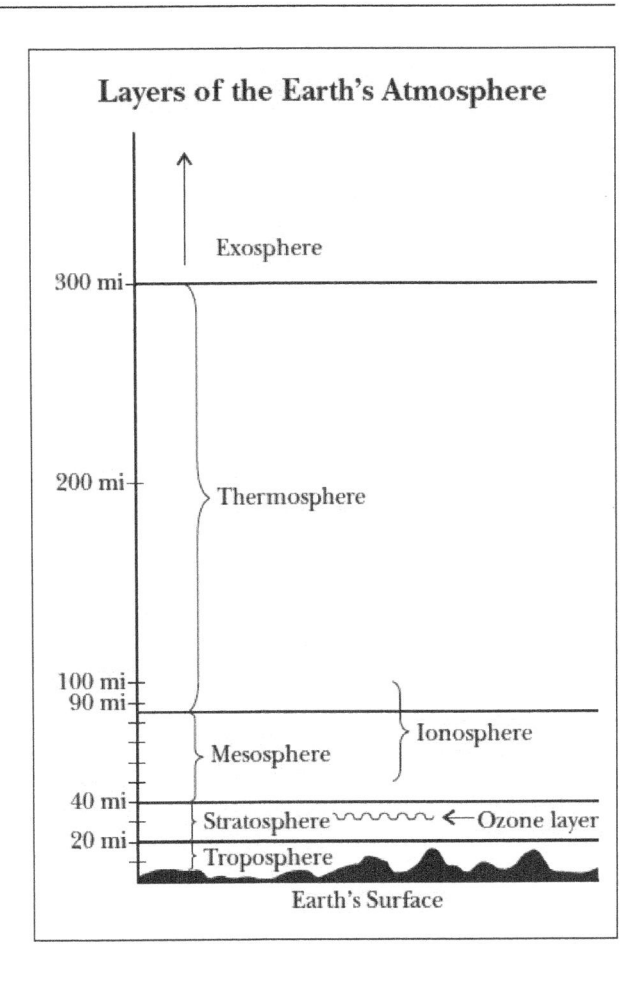

A common example of this is a radiation inversion, which often develops during winter months on days with limited wind and low humidity. These conditions allow the ground to cool quickly, thereby causing the air immediately above the ground to cool more than the air aloft. The warmer air aloft forms a stable layer, or cap, over the atmosphere below it. Radiation inversions are an air-quality concern because they often set up during the evening traffic rush hour and trap associated pollutants close to the surface. In general, radiation inversions affect small geographic regions and dissipate during the early morning, when solar radiation raises temperatures near the surface.

A larger-scale subsidence inversion forms when a high-pressure system causes air to subside and warm adiabatically (without gaining or losing heat) over a broad region. Different rates of subsidence between air near the surface and air aloft allow the air aloft to warm to a greater degree, setting up a large-scale temperature inversion. In contrast to radiation inversions, subsidence inversions have the potential to persist for long periods of time. This scenario is common in many metropolitan areas that are known for poor air quality, particularly Los Angeles, California. In Los Angeles, the impact of temperature inversions is magnified by the fact that the city is surrounded on three sides by mountains, which limit horizontal mixing and promote higher concentrations of air pollutants in the region.

The pollutants that build up in urban areas as the result of temperature inversions have important long-term health implications and, at times, can prove to be serious short-term threats. For example, in 1948 toxic air conditions that resulted from subsidence inversion in the western Pennsylvania town of Donora caused the deaths of twenty residents. A similar incident known as the London smog disaster killed nearly four thousand residents of London, England, in 1952.

—*Jeffrey C. Brunskill*

FURTHER READING

Aguado, Edward, and James E. Burt. *Understanding Weather and Climate.* 5th ed. Upper Saddle River, N.J.: Pearson Education, 2010.

Ahrens, C Donald, and Robert Henson. *Meteorology Today: An Introduction to Weather, Climate, and the Environment.* Cengage, 2019.

ATMOSPHERIC OPACITY

FIELDS OF STUDY

Observational Astronomy; Theoretical Astrophysics

SUMMARY

Atmospheric opacity is the degree to which electromagnetic radiation penetrates the layers of atmosphere surrounding a celestial body. Electromagnetic radiation in the form of light or radio waves reaches the surface in differing amounts, depending on the type of wave, its frequency and energy, and the level of opacity of the surrounding atmosphere. Atmospheric opacity affects surface conditions on Earth and provides information about the potential for life on other celestial bodies.

PRINCIPAL TERMS

- **electromagnetic waves:** the classical form of electromagnetic radiation, produced when electric and magnetic fields come together and interact; can be in the form of radio waves, microwaves, infrared, optical, ultraviolet, x-rays, or gamma rays, depending on their frequency, energy, and wavelength
- **greenhouse effect:** an atmospheric phenomenon in which certain molecular components of the atmosphere capture infrared radiation from a planet's surface and prevent it from radiating back out into space
- **greenhouse gases (GHGs):** atmospheric gases whose molecules absorb infrared radiation from a planet's surface and then emit a significant portion of it back into the atmosphere, most commonly carbon dioxide, methane, water and ozone.
- **opacity:** the degree to which a substance or object lets various forms of electromagnetic radiation pass through it.

PROPERTIES OF THE ATMOSPHERE

Earth's atmosphere is a complex, layered field of gases, water vapor, and dust surrounding the planet. It provides and protects the conditions necessary for life. The atmosphere, contained by Earth's gravity, holds in the oxygen and other gases that make up breathable air and allows the light and heat of the sun to reach Earth's surface. It also deflects or absorbs most harmful forms of electromagnetic waves, such as ultraviolet, gamma rays, and x-rays.

Earth is not the only celestial object with an atmosphere. While all atmospheres serve as filters for light, heat, and radiation, each differs in the amount of waves it allows to reach the surface. The degree to which an atmosphere does this is called atmospheric opacity. Opacity is affected by both the conditions of the atmosphere and the properties of the

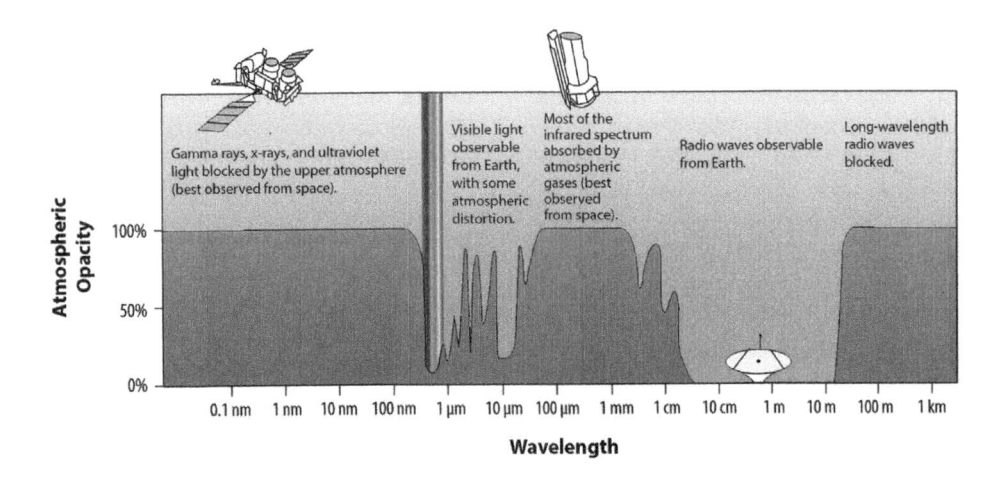

Percentage of atmospheric opacity for various electromagnetic wavelengths. Particular wavelengths cannot pass through the atmosphere while others can penetrate the atmosphere to varying degrees. Gamma rays, x-rays, and ultraviolet light are blocked by the upper atmosphere. Visible light is observable from Earth with some atmospheric distortion. Most of the infrared spectrum is absorbed by atmospheric gases. Radio waves are fully observable from Earth. Long-wavelength radio waves are blocked by the atmosphere.

electromagnetic waves. Relevant atmospheric conditions include temperature, cloud cover, and amount of water vapor.

Early Atmospheric Study

Scientists have been learning about atmospheric opacity for as long as they have been studying Earth's atmosphere. The earliest known experiments in the seventeenth and eighteenth centuries were limited to studying temperature and air pressure on Earth's highest mountains. The invention of hot air balloons in the late eighteenth century allowed scientists to study the higher levels of the atmosphere, but the effects of extreme cold and air pressure on humans hampered these experiments. At very high altitudes, the scientists got frostbite and lost consciousness. French meteorologist Léon Teisserenc de Bort (1855–1913) became one of the first to launch an unmanned weather balloon. Instruments attached to the balloon allowed him to record information at altitudes higher than could be safely reached by humans.

In 1901, Italian physicist and inventor Guglielmo Marconi (1874–1937) proved that one layer of the atmosphere, known as the ionosphere, reflects radio waves. He used this property to bounce the first wireless radio signal across the Atlantic Ocean, from England to Canada. The next year, de Bort presented his research on the upper atmosphere, identifying the troposphere and stratosphere. In 1932, American physicist Karl Guthe Jansky (1905–1950) found that the static heard in radio signals is caused by radio waves from deep in the Milky Way galaxy. This was the first proof that radio waves could also pass through the atmosphere.

Improved technology in the twentieth century enabled closer study of atmospheric layers and their properties. The development of spaceflight in the late 1950s and 1960s represented a huge leap forward. Scientists began using satellites and special telescopes to study Earth's atmosphere, both from space and from within its layers.

Several key research instruments, including the International Space Station (ISS), orbit in the part of Earth's atmosphere known as the thermosphere. The outer boundary of this layer is 690 kilometers (429 miles) above Earth's surface. Other satellites and telescopes have allowed scientists to study the atmospheres of other planets.

Properties of Electromagnetic Radiation

The nature of electromagnetic waves is determined by their frequency and wavelength. These properties are related through the following wave equation:

$$c = \lambda f$$

In this equation, c represents the speed of light, λ is wavelength, and f is frequency. Wavelength is the distance between a point in one wave and the same point in the following wave, normally measured in meters. Frequency is the number of wave cycles per unit time. The International System of Units (SI) unit of frequency is the hertz (Hz), equal to one cycle per second.

The different types of electromagnetic waves, in order of decreasing wavelength and increasing frequency, are radio waves, microwaves, infrared, optical or visible light, ultraviolet, x-rays, and gamma rays. All electromagnetic waves travel at the speed of light in a vacuum. Radio and optical waves are the easiest to study because they can pass readily through the atmosphere. Studying other forms of electromagnetism requires specially calibrated filters, cameras, and telescopes.

EFFECTS OF ATMOSPHERIC OPACITY
The opacity of an atmosphere affects the conditions on the surface below. Earth's atmosphere filters the amount of ultraviolet radiation that reaches the planet and helps keep Earth at temperatures suitable to sustain life. First it allows the warming rays of the sun to reach the surface and then it absorbs the heat energy that the surface emits as infrared radiation, or heat. Certain molecular components of the atmosphere called 'greenhouse gases, or GHGs, absorb some of that heat and emit it back into the atmosphere instead of back to space. This serves to maintain the average temperature of Earth's atmosphere significantly higher than it would be otherwise. The effect of retaining some of the heat in this manner is what is known as the greenhouse effect.

Opacity also impacts radio-wave transmission. One of the highest levels of Earth's atmosphere, the ionosphere, is so named because it is made up of ions and free electrons. When atoms or molecules collide with ultraviolet and x-rays, they lose electrons, becoming positively charged ions. Together, the free electrons and positive ions form a plasma. A plasma is a state of matter that consists of unbound positively and negatively charged particles but is electrically neutral as a whole, because the positive and negative charges balance out. It was the ionosphere that bounced Marconi's radio waves across the ocean.

The ionosphere also has sublayers that change seasonally and even daily. When there is no sunlight, one of the layers disappears and another two merge. This allows radio waves to travel farther at night than in the daylight.

IMPORTANCE OF ATMOSPHERIC OPACITY
The opacity of Earth's atmosphere is important to sustain human life, and humans may be just as important to atmospheric opacity. Some researchers think that the gases produced by certain human activities, such as farming, burning fossil fuels, and maintaining landfills, cause changes in the atmosphere's opacity that enhance the greenhouse effect. This could increase the average surface temperature on all or part of Earth's surface. Scientists differ on how drastic these changes could be, what their overall effects may be, and how fast those effects might occur.

Scientists also study the atmospheric opacity of other planets. NASA's Viking program, which sent two orbital probes to Mars in the 1970s, included studies of the depth and opacity of Mars's atmosphere. Further studies have been done on other planets in an effort to determine if their atmospheric opacity allows for the conditions necessary to sustain life.

BIBLIOGRAPHY
"A Blanket around the Earth." *National Aeronautics and Space Administration.* California Inst. of Technology, n.d. Web. 27 Apr. 2015.
"Atmosphere." *National Geographic.* National Geographic Soc., n.d. Web. 27 Apr. 2015.
"Electromagnetic Radiation." *Cosmos: The SAO Encyclopedia of Astronomy.* Swinburne U of Technology, n.d. Web. 27 Apr. 2015.
"The Electromagnetic Spectrum. *National Aeronautics and Space Administration.* NASA, Mar. 2013. Web. 27 Apr. 2015.
"History of Discovery of the Atmosphere." *UCAR Center for Science Education.* U Corporation for Atmospheric Research, n.d. Web. 27 Apr. 2015.
"Karl Jansky and the Discovery of Cosmic Radio Waves." *National Radio Astronomy Observatory.* Assoc. U, 16 May 2008. Web. 27 Apr. 2015.
Randall, David A. *Atmosphere, Clouds, and Climate.* Princeton: Princeton University Press, 2012.

ATMOSPHERIC OXYGEN

FIELDS OF STUDY

Atmospheric Chemistry; Atmospheric Science; Analytical Chemistry; Thermodynamics; Environmental Chemistry; Biochemistry; Physical Chemistry; Photochemistry; Organic Chemistry; Environmental Studies; Oceanography; Waste Management; Chemical Kinetics

SUMMARY

Oxygen exists in Earth's atmosphere almost entirely as the reactive diatomic molecular species O_2. It is a by-product of photosynthesis and is believed to have accumulated in the primordial atmosphere as a result of photosynthesis occurring in cyanobacteria in the oceans. Algae and green plants are also important contributors of atmospheric oxygen. In photosynthesis, organisms take up carbon dioxide (CO_2) from the air and water and minerals from the soil, and, using the energy of sunlight, they convert these substances into chemical compounds from which they build their cellular constituents. The oxygen released in the process is used in respiration by animals, including humans, as well as by plants and many other organisms, to retrieve the energy stored in chemical bonds. Carbon dioxide is given off as the by-product of respiration.

PRINCIPAL TERMS

- **aerobic:** taking place in or requiring the presence of oxygen
- **anaerobic:** taking place in or requiring the absence of oxygen
- **endosymbiont:** an organism living inside a cell or body of another organism
- **eukaryote:** an advanced cell, containing a nucleus and other membrane-bound organelles
- **fermentation:** the biochemical generation of energy from sugars and organic compounds derived from sugars in the absence of oxygen
- **organic matter:** carbon-containing compounds produced by life processes
- **photosynthesis:** the process occurring in green plants for the production of glucose from carbon dioxide and water in the presence of chlorophyll and sunlight

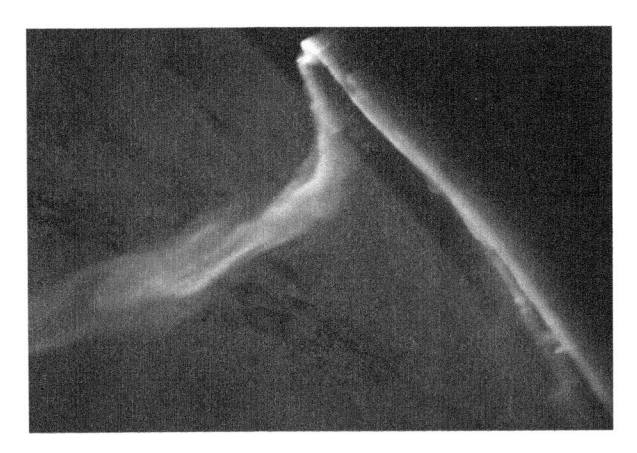

Aurora during a geomagnetic storm that was most likely caused by a coronal mass ejection from the Sun on 24 May 2010, taken from the ISS. (NASA)

- **prokaryote:** a primitive cell (bacterium), lacking a nucleus and other membrane-bound organelles
- **respiration:** the biological generation of energy through biochemical reactions involving molecular oxygen (O_2)

The Early Atmosphere and Photosynthesis

When Earth was formed, some 4.5 billion years ago, the atmosphere contained almost no free oxygen. This early atmosphere probably consisted of CO_2, water vapor, nitrogen, hydrogen, and trace gases. Earth's first living things, bacteria, appeared around 3.8 billion years ago, in the oceans. These early prokaryotes were unable to photosynthesize. They probably derived their nutrients from chemical compounds in seawater, and their energy through fermentation. Some prokaryotes developed the ability to photosynthesize, and, about 2.8 billion years ago, one group of these, the cyanobacteria, began to produce oxygen as a by-product. These microbes still exist today.

Effects of Atmospheric Oxygen on Life

Oxygen reached appreciable levels in the atmosphere by about 2.3 billion years ago. It changed the course of evolution. Because it is a highly reactive gas, oxygen was a deadly poison to many of the earliest bacteria, but some bacteria harnessed it to break the chemical bonds in their food, yielding energy. Thus, they developed respiration, an aerobic pathway for energy production. Respiration is more efficient than fermentation. It became life's predominant energy-producing pathway.

The build-up of oxygen in the atmosphere aided life in another way through the formation of the planet's ozone layer. Ozone (O_3) is produced by the interaction of the Sun's ultraviolet radiation with O_2 gas. The ozone layer shields terrestrial life from about 95 percent of the ultraviolet radiation that comes to Earth from the Sun. Without this shield, the total ultraviolet radiation would effectively sterilize Earth's surface and render the land surface uninhabitable for both plants and animals.

About 1.5 billion years ago, the accumulation of abundant free oxygen was accompanied by the evolution of a new, more complex cell type, the eukaryote. Those eukaryotes that acquired photosynthetic endosymbionts became algae. By 600 million years ago, multicellular eukaryotic forms had arisen, and the groups of organisms that dominate the globe today, the animals, plants and fungi, became established.

Atmospheric Oxygen and the Geological Carbon Cycle

The countervailing processes of photosynthesis and respiration help stabilize the atmospheric concentrations of oxygen and CO_2 on timescales of single years to tens of thousands of years. Over longer timescales spanning millions of years, recycling of Earth's crust regulates carbon exchange between rocks, oceans, and the atmosphere and affects the balance of atmospheric gases. As plant matter and sulfur in rocks and sediments are alternately buried and oxidized, oxygen is added to the atmosphere or removed from it. The abundance of CO_2 is generally inversely related to the abundance of oxygen in these processes.

The concentration of oxygen in Earth's atmosphere is about 21 percent. The remainder of the atmosphere consists mainly of the relatively unreactive gas nitrogen, at about 78 percent. The other atmospheric components are trace gases, including CO_2, at about 0.035 percent, and water vapor, which fluctuates in concentration. Although a slight decrease in atmospheric oxygen has been recorded during recent decades, attributable to the burning of fossil fuels, the reservoir of atmospheric oxygen is so large that significant change cannot readily occur. Conversely, atmospheric CO_2 is present in such small concentrations that minor changes make a large proportional difference.

There is evidence that, in the distant past, the atmospheric oxygen concentration was not completely stable. About 540 million years ago, oxygen composed about 15 percent of the atmosphere. Then, around 300 million years ago, it reached about 35 percent, as a result of the conquest of the land by plants and the attendant increase in global photosynthesis. Oxygen concentration also increased by the burial of forests in coal swamps, which sequestered vast amounts of organic carbon from the atmosphere. Under normal circumstances, the trees would decay, and their carbon would be oxidized to CO_2 and water, but the anaerobic conditions in the swamps prevented this from happening. By around 200 million years ago, the atmospheric oxygen concentration had plummeted back to about 15 percent, accompanied by a major extinction event at the Permian-Triassic boundary.

Context

Oxygen's existence has been known only since 1774, when the English scientist Joseph Priestley (1733–1804) isolated the gas and noted its special properties, including its ability to make a flame burn especially brightly. In Priestley's day, the atmosphere was just beginning to be understood as a collection of individual gases, rather than the single, uniform "air" it had been supposed to be. Priestley was also the first to note that plants and animals exist in a reciprocal relationship, mediated by the gases of photosynthesis and respiration. It took nearly two centuries following Priestley's discovery of oxygen for the origin of this gas to be understood. Only in the 1960s did the idea that photosynthesis was responsible for the build-up of atmospheric oxygen over the eons become widely accepted by scientists.

—*Jane F. Hill*

Further Reading

Archer, David. *The Global Carbon Cycle*. Princeton University Press, 2010.

Broda, E *The Evolution of the Bioenergetic Processes*. Elsevier, 2014. The author gives a thorough description of the respiration and fermentation processes, and their respective roles in regard to prokaryotes and eukaryotes.

Canfield, Donald Eugene *Oxygen. A Four Billion Year History*. Princeton University Press, 2014. This highly readable book presents a comprehensive overview of atmospheric oxygen and discusses many questions about the origin and significance of atmospheric oxygen.

Glikson, Andrew Y. *Evolution of the Atmosphere, Fire and the Anthropocene Climate Event Horizon*. Springer, 2014. In this book can be found a detailed discussion of the relationship of oxygen and other atmospheric gases with regard to their natural cycles and the effect of human activities.

Harmon, Russel S. and Parker, Andrew *Frontiers in Geochemistry*. John Wiley & Sons, 2011. Presents the existence and history of atmospheric oxygen and its formation in a geochemical context, drawing from several academic sources.

Rhodes, Frank H.T. *Earth. A Tenant's Manual*. Cornell University Press, 2012. In this book the author presents a scientific view of the past, present and future of Earth's atmosphere and its composition.

Spellman, Frank R. *The Science of Air. Concepts and Applications*. 2nd ed. CRC Press, 2009. This book is a comprehensive resource covering the components, dynamics, interactions and uses of air as a resource.

ATMOSPHERIC PHYSICS

FIELDS OF STUDY

Aeronomy; Acoustics; Applied Physics; Atmospheric Dynamics; Mathematics; Atmospheric Chemistry; Climatology; Cloud physics; Computer Modeling; Engineering; Environmental Physics; Fluid Dynamics; Fluid Mechanics; Meteorology; Physics; Radiation; Remote Sensing; Statistics; Statistical Mechanics; Spatial Statistics; Thermodynamics

SUMMARY

Atmospheric physics is a subfield of atmospheric science that studies the physics principles underlying atmospheric phenomena. Atmospheric physicists study the flow of energy as cold and warm air masses collide and as fast and slow rivers of air interact. Mathematics is an integral part of studying these processes, as physics itself depends on mathematics. In atmospheric physics, the goal is to mathematically model, and therefore predict, what occurs in the atmospheric layers that surround Earth and other planets. Statistics and computers play an important role in studying the atmosphere, as does chemistry; the chemical makeup of the atmosphere affects its activity. To gather data, the field depends on the advanced design and manufacture of sensing devices.

PRINCIPAL TERMS

- **albedo:** a value between zero and one representing the fraction of incident light reflected from a body such as Earth
- **anticyclonic:** the clockwise rotation of an air mass in the Northern hemisphere, counterclockwise in the Southern hemisphere
- **cyclonic:** the counterclockwise rotation of an air mass in the Northern hemisphere, clockwise in the Southern hemisphere
- **Coriolis Effect:** the effect by which an object moving in a straight line across a rotating surface is seen to follow a curved path
- **longwave radiation:** electromagnetic radiation in the infrared range, having wavelengths longer than those of the visible spectrum, associated with heat and heating
- **shortwave radiation:** electromagnetic radiation in the ultraviolet range, having wavelengths shorter than those of the visible spectrum, associated with the transport of energy capable of promoting chemical reactions

BASIC PRINCIPLES

Historically, atmospheric research has been two-pronged. Physical meteorology studies what is seen and heard in the atmosphere, such as cloud formation, rainfall, lightning, tornadoes, and other tangible phenomena. The dynamics side of atmospheric research studies large-scale atmospheric motions, such as those that are hundreds of kilometers or many days long. This includes frontal systems, tropical storms, jet streams, and related effects. Also central to atmospheric physics are air pressure, density, and the capture of water, especially from oceans.

Atmospheric physics

The Visible Infrared Imager Radiometer Suite (VIIRS) on the United States' newest Earth-observing satellite. VIIRS collects radiometric imagery in visible and infrared wavelengths of the Earth's land, atmosphere, and oceans.

These two historical approaches are best merged in an interdisciplinary approach. The study of climate change, for instance, involves advanced dynamics, chemistry, and radiation research. The mathematics of physics, especially fluid flow equations, remains essential to this study. For example, scattering theory uses mathematics to understand the behavior of atmospheric matter scattering in particles and/or waves. The study of atmospheric physics also uses mathematics to model wave propagation, or the ways that waves travel.

The application of physics to atmospheric studies has grown in recent decades. The late twentieth and early twenty-first centuries brought advances in atmospheric physics thanks to satellites and computers. The gaseous envelope of atmosphere that makes life on earth possible is never fully at rest, so much of scientists' knowledge of it still relies on Newton's laws of motion, formulated in the seventeenth century. The constant motion of air makes it difficult to capture facts without the application of continuum mechanics, which studies the behavior of solids and fluids as entire masses rather than separate particles. Joseph-Louis Lagrange (1736–1813) was the next scientist after Isaac Newton (1643–1727) to make an in-depth study of mechanics. Lagrange's work enabled the development of mathematical physics, and his approach has become widespread ever since. In the Lagrangian description of fluid motion, a physicist examines a parcel or several parcels of air to learn how properties transform and interact as a single system within those bounds and with the environment beyond the parcels.

CORE CONCEPTS
The atmosphere is a complicated mixture of gaseous materials, water, and solid matter. A relatively thin layer compared to the mass of the planet, it has far more horizontal than vertical movement. The atmosphere, like the ocean, has tides. For the most part, they are created through daily heating by the sun's radiation, solar gravitation, and molecular resonance, in which a molecule vibrates among several alternate structures. Though solar radiation has an effect on atmospheric physics, gravity, compressing most of the air to within 16 kilometers of the surface, has more effect. The pressure exerted by gravity causes air to become denser at sea level than in the mountains or on airplanes. Density decreases exponentially with altitude, and pressure changes in the troposphere create flows of air mass. These air masses are also affected by the rotation of the planet. Planetary rotation impacts the upper atmosphere too, by creating waves that affect the movement of heat, chemicals, and aerosols.

ATMOSPHERIC COMPOSITION
Earth's atmosphere contains mostly nitrogen and oxygen. The remaining gases, including carbon dioxide (CO_2) and ozone (O_3), make up only 1 percent of the air. One of the most important components of the atmosphere is water, occurring both in molecular form and in fine aerosols. Aerosols are tiny solid or liquid particles suspended in gas. The air carries natural aerosols, coming from volcanic ash, sea spray, pollen, and other sources; the air also carries human-produced aerosols.

Ozone is made in the stratosphere when solar rays collide with O_2 molecules, imparting the energy needed to separate the O_2 molecule into two separate O atoms. These can then combine with other O_2 molecules, resulting in the formation of O_3 (ozone). Stratospheric ozone is a good thing, as it protects

Schematic for Global Atmospheric Model

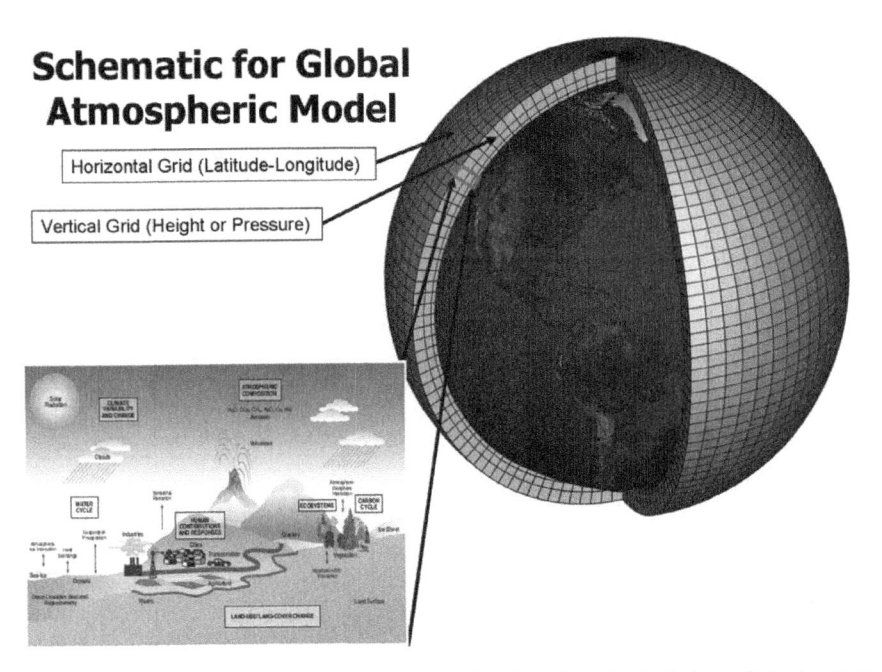

Climate models are systems of differential equations based on the basic laws of physics, fluid motion, and chemistry. Atmospheric models calculate winds, heat transfer, radiation, relative humidity, and surface hydrology.

Earth from harmful ultraviolet rays coming from the sun. Harmful ozone is a type of surface pollution, which happens in the troposphere when sunlight and heat trigger chemical reactions involving nitrogen oxides typically created by human activity, and volatile organic chemicals. Harmful ozone can make breathing difficult for people with respiratory ailments.

WATER MOVEMENT

Water movement in the air is a central concern for meteorologists. Water vapor, considered only a trace of what makes up the atmosphere, is found almost exclusively in the troposphere. It is produced at the surface and in the tropics, drying out increasingly toward the upper troposphere and the poles. Convection, or the movement of molecules within fluids, builds cells (clouds) vertically near the tropics, but most cloud motion is horizontal. As trained physicists, weather forecasters need to understand the basic elements of motion. For example, speed differs from velocity. Speed describes how fast in time an object moves (e.g., kilometers per hour), whereas velocity describes both the speed and particular direction of an object's movement. Acceleration, in units per second, and force, including the force of gravity, are part of understanding cloud physics and other parts of the atmosphere. Earth's rotation adds centripetal (inward-pulling) and centrifugal (outward-pushing) forces. In addition, the Coriolis Effect causes atmospheric masses to sway with relation to the ground, as viewed from above. Northern Hemisphere air masses swerve right, and Southern appear to go left; thus, cyclones spin clockwise north of the equator, and counterclockwise south of it.

THERMODYNAMICS

The study of thermodynamics focuses on the relationship between energy and work and the transfer of energy between systems. One aspect of thermodynamics in atmospheric physics is thermal equilibrium, in which a transfer of heat has occurred between two systems to the point at which there is no longer an exchange of energy; the systems' temperatures are now the same. Another thermal effect of the atmosphere is that it makes water change states. Water can be liquid, gas, solid, or plasma. The discipline of thermodynamics supplies the necessary formulae and conceptual mathematics to study the phase changes of water, such as how vapor transforms into water or ice. An understanding of thermodynamics also helps physicists to calculate the condensation of vapor, which must occur faster than evaporation in order to form liquid water. These are just two basic examples of the applications of thermodynamics to atmospheric physics.

Atmospheric Zones

Earth's atmosphere is considered to be about 500 hundred kilometers thick, though gravity concentrates most of it to within 16 kilometers of the surface of Earth. The atmosphere also does not end at a particular height. Rather, it continuously thins until it has merged with space. It is helpful to understand Earth's atmosphere by dividing it into different zones.

The thermosphere is the highest layer of Earth's atmosphere. Temperature within it increases with altitude, because a smaller amount of gas is absorbing a large amount of solar radiation. The top portion of the thermosphere, right before space begins, is called the exosphere. The exosphere extends from about six hundred kilometers high up to 1,600 kilometers or more, depending on solar activity. The second part of the thermosphere is the ionosphere, which stretches up from approximately 80 kilometers high from Earth's surface. The ionosphere contains a large amount of ions which, when impacted by solar wind, cause aurora light displays. Solar radiation is so powerful here that it breaks electrons free, producing ions out of molecules. Radio waves are reflected here, making radio communication possible.

Below the thermosphere is the mesosphere, the coldest layer of earth's atmosphere, stretching from about 50 kilometers to 80 kilometers high. The air here, while thin, is thick enough that it can burn up meteoroids and cause meteor showers to be visible. The mesosphere is too high for airplanes and weather balloons, but too low for orbital satellites.

The next layer, the stratosphere, stretches between approximately 16 and 50 kilometers high. The stratosphere's temperature increases the higher it goes, due to heating ozone. This is where "good" ozone acts as a protective shield, intercepting ultraviolet rays from the sun and preventing their energy from penetrating the lower atmospheric layers.

The zone closest to Earth's surface and where most of Earth's atmosphere is concentrated is called the troposphere. This is where weather happens. The sun warms the surface, and then the surface warms air masses, which rise as cool air masses fall. The troposphere cools with gains in altitude; its vertical instability is one major cause of weather. Between the troposphere and the stratosphere is the tropopause, which acts as a soft boundary. High-altitude jet streams race along this level. The height of the tropopause is affected by vertical instability from convection, meaning that it lies closer to the cold of the poles than the warmth of the equator. The tropopause is very dry and caps the weather zone; just above it is the best place for commercial airplanes to fly.

Applications Past and Present
Remote Sensing

On October 4, 1957, the Soviet Union launched *Sputnik I*, the first earth-orbiting, human-built satellite; this started the space race of the 1950s and 1960s. Since the launch of *Sputnik I*, satellites have come a long way. Remote sensor satellites, basically platforms carrying sensors that orbit Earth and send back data, have become so common that debris associated with those no longer functioning—space junk—is a navigational hazard. Even so, remote sensing has revolutionized communications, intelligence gathering, weather forecasting, and atmospheric research. NOAA, the National Oceanic and Aeronautic Administration, operates satellites that track storms, gather ocean temperatures, take innumerable data readings, and make these readings available around the clock to scientists in the field. Every satellite reports to an earth base, where scientists receive the data, process it, and make it available to others. The U.S. Office of Satellite Products and Operations, OSPO, can provide data on such topics as atmospheric profiles, rain, clouds, wind, ozone, aerosols, and the radiation budget.

The satellites that remote sensing technology rides on are platforms that exist in the harsh conditions of space. Their data gathering cameras or other sensors are dependent on power systems and antennae, and none of it can be maintained once the unit is in orbit. Even when a satellite is performing well, it can be thrown off course by gravitational forces and will lose altitude over time. Correcting for position is often done with gas canisters; the usable life of the satellite may depend less on its advanced technology than on how many gas canisters can be packed onto the platform. Despite this difficulty, satellites provide information that cannot be obtained in other ways. Scientists continually work toward the improvement of satellite technology so that more accurate and detailed data, about the weather and other atmospheric concerns, can be provided.

Meteorology
Meteorology, the study of the phenomena in the atmosphere collectively known as weather has been part of atmospheric physics since long before the satellite age. It includes such phenomena as changes in air pressure, moisture, temperature, and wind direction. Since antiquity, for example, people with rheumatism have been said to feel pressure changes in their joints. "Weather" in the atmospheric physics community is a fluid term, but an important focus. Weather can be defined as air flow patterns kilometers long, occurring in the troposphere over the course of days rather than weeks. Meteorologists can track weather patterns, given that they show some regularity over months and in annual cycles. However, weather can still be random on any given day and generally remains unpredictable. Beyond meteorology, other fields of study and their corresponding data sets, such as ocean temperature change, impinge on atmospheric research. Atmospheric physics is part of the overall study of climate change. The field therefore must keep advancing so that civilization-impacting effects such as droughts, hurricanes, and the effects of climate change can be predicted and appropriate planning can take place.

Flight
As society demands increasingly fuel-efficient airplanes and effective military aircraft, aerodynamic design depends on increasing knowledge of atmospheric physics. Wings of aircraft "fly" because the speeding up of airflow over the top surface of a certain shape creates a difference in air pressure between the top and bottom surfaces. Orville and Wilbur Wright were among the first test this concept when their *Flyer* rose into the air above Kitty Hawk, North Carolina, in 1903. Pilots know that air pressure differences and motions cause winds and turbulence, requiring skill and experience to negotiate. Aircraft must withstand greater atmospheric temperatures the faster they fly. The space shuttle *Columbia*, for example, was destroyed in 2003 by hot gases that penetrated its exterior in one spot that was under intense pressure during reentry through the atmosphere. Such challenges concerning altitude and speed are a significant part of the importance of atmospheric physics to aircraft design.

Acoustics
Sound waves are one type of atmospheric wave that physicists study. Acoustics, the effect of sound in a certain space, affects modern living in a number of ways. One is noise control. In an increasingly noisy world, the design and manufacture of materials that can, for example, separate highway noise from homes or appliance noise from one apartment to the next, are important. Acoustics affects architecture in transmitting sound as well. Public arenas, such as concert and lecture halls, represent Lagrangian parcels, each with their own needs. Acoustics is also important in industrial and military diagnostics, as high powered engines can be tested for imperfections by listening for a rattling sound. Finally, the study of acoustics is needed for the ever-better transmission of communications over wireless areas, as these areas steadily grow.

Climate Change
Weather occurs over large spaces and in short amounts of time. Climate, however, occurs across the planet over decades and centuries. The study of climate change is a relatively new field, in the sense that its effects are more pressing than they have been in past centuries. The United States Environmental Protection Agency (EPA) reported that the first decade of the twenty-first century had a planet-wide average temperature rise of 0.6°C, or 1.4°F. The summer of 2012 saw widespread drought in the United States' grain-growing regions, resulting in record-high prices for corn and soybeans. These staple crops are used for more than just food; ethanol, certain plastics, and other products are made with corn and soy. Climate change has also caused storms to rage worldwide and with increased ferocity. Atmospheric carbon dioxide and other greenhouse gases are widely held to be causes of global warming. The study of atmospheric physics, therefore, is used in most efforts made by industry or government to address climate change and to secure the well-being of future generations.

Energy Balance and Thermal Equilibrium
The first law of thermodynamics states that in a closed system, such as the atmosphere, energy can never be created nor destroyed, but it does transition from one form to another. Additionally, the amount of work done by a system is equal to the amount of energy available. According to this law, the atmosphere will heat or cool internal to itself to the degree equal to heat supplied by the sun, minus work done by the atmosphere. That work is the radiation of excess heat back out into space. If Earth simply absorbed heat from the sun, it would be like Venus, a boiling ball. Instead, Earth's atmosphere accepts the sun's shortwave radiation and emits longwave, or infrared radiation (IR).

A portion of shortwave radiation, known as the albedo, is reflected from ground features such as ice and snow and from clouds. Clouds play an important role in Earth's thermal equilibrium because they are highly reflective. Heat bouncing off the ground can sometimes hit the bottoms of clouds. When this happens, the clouds will still reflect albedo, but this time back to Earth's surface. In addition to this complexity, aerosol particles help form clouds by providing places for water vapor to coagulate. They also absorb IR from anywhere and scatter shortwave solar rays. Aerosols' effect on thermal equilibrium has been documented throughout history in association with volcanic ash emitted from massive eruptions. However, the full effects of aerosols are believed to be more complex than scientists yet know, since the amounts and kinds of aerosols in a given atmospheric parcel are often impacted by human activity.

To achieve equilibrium, incoming solar energy is distributed throughout the atmosphere as Earth turns. Outgoing longwave radiation (OLR) is complicated by the greenhouse effect, in which the troposphere turns half the OLR back to Earth's surface to bounce again. The troposphere is a heat sink—meaning that it absorbs heat—that both drives atmospheric circulation and acts as a blanket to trap heat and make life on Earth possible. As atmospheric chemical and water vapor levels change, atmospheric physicists are needed to determine the effects this will have.

SOCIAL CONTEXT AND FUTURE PROSPECTS
Climate Change
The issue of climate change is often approached from two areas: forcings and feedbacks. Forcings are what impact or initiate change, whether anthropogenic (caused by human activity) or natural. Aerosols are considered forcing agents. Some types of aerosols initiate atmospheric cooling, while others initiate warming. Other forcings include anthropogenic changes in the atmosphere's composition, changes in land use, volcanic eruptions, alterations in solar output, and long-term changes of Earth's orbital parameters. Though research into these effects has been conducted since the beginning of the twenty-first century, much more needs to be done.

The second area that scientists watch closely is called feedback. Feedbacks are the results of forcings. They are either positive, by increasing warming, or negative, by decreasing warming. Feedbacks, which have been studied more than forcings, come from water vapor, albedo, atmospheric lapse-rate (the rate at which temperature decreases upward), and clouds. For example, an increase in greenhouse gases may increase the amount of bright, low-level clouds. This, in turn, leads to less absorption of the sun's rays and thus an increase solar radiation hitting Earth. The Intergovernmental Panel on Climate Change (IPCC) has called for research into clouds as a top priority. Studies of precipitation feedback have been in the earliest stages of mathematical modeling; models vary so widely that more atmospheric researchers who focus in cloud physics and precipitation are needed.

Communications
Storms in the ionosphere disrupt radio, electronic navigation, and GPS (Global Positioning Systems). Modern living has become increasingly dependent on electronic navigation and GPS. Satellites that serve cell phone communications depend on electronic navigation to stay balanced, stay in orbit, and keep away from space junk. Large-scale agriculture has become more dependent on tractor-mounted GPS to plant and maintain crops. Better knowledge of Earth's ionosphere will matter more and more as human reliance on technology, whether large or small, continues to increase into the future.

Air Quality
More people live in densely-populated areas than ever before. As the world's population grows, concerns about urban air quality also grow. Ground-level ozone is one pollutant associated with industries

and automobiles. Another pollutant is the particular matter of urban atmospheric aerosols, such as dust and ash. Particulate matter is one of the most harmful pollutants for people. While scientists have been working to lower ozone-creating emissions, controlling dust is also important. Atmospheric physicists who understand the motion of particles at the very lowest levels of the troposphere are needed for this purpose.

—*Michael P. Auerbach, MA*

FURTHER READING

Dessler, Andrew E., and Edward A. Parson. *The Science and Politics of Global Climate Change: A Guide to the Debate.* Cambridge University Press, 2010. Introduction to the issue of climate change, including atmospheric chemistry and other atmospheric properties research. Also discusses the potential future of research in the area of climate change.

Frederick, John E. *Principles of Atmospheric Science.* Jones & Bartlett, 2008. Introductory text describing the various fields of atmospheric sciences, including atmospheric chemistry, atmospheric physics, and climatology. Describes techniques and research methods utilized in modern climate and atmospheric research.

Houghton, John T. *The Physics of Atmospheres.* 3rd ed. Cambridge University Press, 2002. Revised textbook with chapters on topics such as remote sensing, numerical modeling, climate change, chaos, and predictability.

NASA Goddard Institute for Space Studies. National Aeronautics and Space Administration, 2012. Web. 21 Aug. 2012. Describes a variety of current research programs in the environmental sciences, physics, and atmospheric chemistry. Also contains descriptions of using atmospheric physics in the study of climate change and global warming.

Spencer, Roy. *Global Warming.* Roy Spencer, 2012. Web. 23 Aug. 2012. Spencer, a climatologist, used to work for NASA. His website offers alternative ideas to the common view that global warming is caused by human activity.

ATMOSPHERIC SCIENCES

FIELDS OF STUDY

Atmospheric Chemistry; Environmental Chemistry; Geochemistry; Physical Chemistry; Photochemistry; Thermodynamics; Environmental Sciences; Environmental Studies; Waste Management; Physical Sciences; Meteorology; Climate Modeling; Statistics; Process Modeling; Mathematics; Fluid Dynamics; Hydroclimatology; Hydrometeorology; Bioclimatology; Climate Classification; Climate Zones; Oceanography; Physical Geography; Ecosystem Management; Ecology; Heat Transfer; Computer Science; Software Engineering; Particle Physics; Mass Spectrometry; Spectroscopy

SUMMARY

Atmospheric sciences include the fields of physics and chemistry and the study of the composition and dynamics of the layers of air that constitute the atmosphere. Related topics include climatic processes, circulation patterns, chemical and particulate deposition, greenhouse gases, oceanic temperatures, interaction between the atmosphere and the ocean, the ozone layer, precipitation patterns and amounts, climate change, air pollution, aerosol composition, atmospheric chemistry, modeling of pollutants both indoors and outdoors, and anthropogenic alteration of land surfaces that in turn affect conditions within the ever-changing atmosphere.

PRINCIPAL TERMS

- **anthropogenic:** deriving from human sources or activities
- **chlorofluorocarbons (CFCs):** chemical compounds with a carbon backbone and one or more chlorine and fluorine atoms
- **greenhouse effect:** an atmospheric phenomenon in which certain molecular components of the atmosphere capture infrared radiation from a

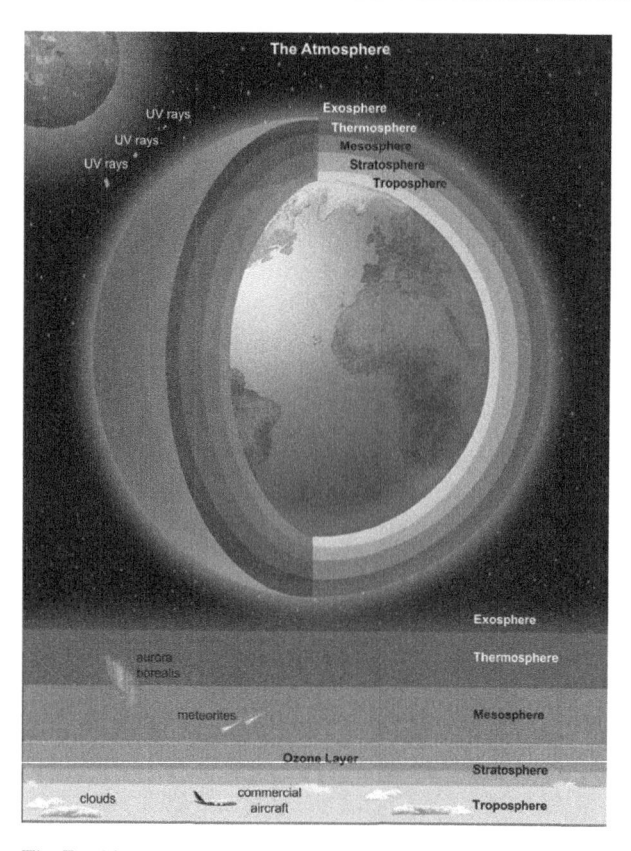

The Earth's atmosphere is divided into five layers.

- planet's surface and prevent it from radiating back out into space
- **greenhouse gases (GHGs):** atmospheric gases whose molecules absorb infrared radiation from a planet's surface and then emit a significant portion of it back into the atmosphere, most commonly carbon dioxide, methane, water and ozone.
- **longwave radiation:** electromagnetic radiation in the infrared range, having wavelengths longer than those of the visible spectrum, associated with heat and heating
- **shortwave radiation:** electromagnetic radiation in the ultraviolet range, having wavelengths shorter than those of the visible spectrum, associated with the transport of energy capable of promoting chemical reactions

DEFINITION AND BASIC PRINCIPLES

Atmospheric sciences is the study of various aspects of the nature of the atmosphere, including its origin, layered structure, density, and temperature variation with height; natural variations and alterations associated with anthropogenic impacts; and how it is similar to or different from other atmospheres within the solar system. The present-day atmosphere is in all likelihood quite dissimilar from the original atmosphere. The form and composition of the present-day atmosphere is believed to have developed about 400 million years ago in the late Devonian period of the Paleozoic era, when plant life developed on land. This vegetative cover allowed plants to take in carbon dioxide and release oxygen as part of the photosynthesis process.

The atmosphere consists of a mixture of gases that remain in place because of the gravitational attraction of Earth. Although the atmosphere extends about 100,000 kilometers above Earth's surface, the vast proportion of its gases (97 percent) is located in the lower 30 kilometers. The bulk of the atmosphere consists of nitrogen (78 percent) and oxygen (21 percent). The last 1 percent of the atmosphere contains all the remaining gases, including an inert gas (argon), which accounts for 0.93 percent of the 1 percent, and carbon dioxide (CO_2), which makes up a little less than 0.04 percent. Carbon dioxide has the ability to absorb longwave radiation leaving Earth and shortwave radiation from the Sun; therefore, any increase in carbon dioxide in the atmosphere has profound implications for global warming.

BACKGROUND AND HISTORY

Evangelista Torricelli (1608-1647), an Italian physicist, mathematician, and secretary to Galileo (1564-1642), invented the barometer, which measures barometric pressure, in 1643. The first attempt to explain the circulation of the global atmosphere was made in 1686 by Edmond Halley (1656-1742), an English astronomer and mathematician. In 1735, George Hadley (1685-1768), an English optician, described a pattern of air circulation that became known as a Hadley cell. In 1835, Gustave-Gaspard Coriolis (1792-1843), a French engineer and mathematician, analyzed the movement of air on a rotating Earth, a pattern that became known as the Coriolis Effect. In 1856, William Ferrel (1817-1891), an American meteorologist, developed a model of hemispheric circulation of the atmosphere that became known as a Ferrel cell. Christophorus Buys Ballot (1817-1890), a Dutch meteorologist, explained the relationship between the distribution of pressure, wind speed, and direction in 1860.

Manned hot-air balloon flights beginning in the mid-nineteenth century facilitated high-level observations of the atmosphere. For example, in 1862, English meteorologist James Glaisher (1809–1903) and English pilot Henry Coxwell (1819–1900) reached 29,000 feet, at which point Glaisher became unconscious and Coxwell was partially paralyzed so that he had to move the control valve with his teeth. In 1902, Léon Teisserenc de Bort (1855–1913) of France was able to determine that air temperatures begin to level out at 39,000 feet and actually increase at higher elevations. In the twentieth century, additional information about the upper atmosphere became available through radio waves, rocket flights, and satellites.

How It Works

A knowledge of the basic structure and dynamics of the atmosphere are a necessary foundation for understanding applications and practical uses based on atmospheric science.

Layers of the Atmosphere

The heterosphere and the homosphere form the two major subdivisions of Earth's atmosphere. The uppermost subdivision (heterosphere) extends from about 80 kilometers above Earth's surface to the outer limits of the atmosphere at about 100,000 kilometers. Nitrogen and oxygen, the heavier elements, are found in the lower layers of the heterosphere, and lighter elements such as hydrogen and helium are found at the uppermost layers of the atmosphere. The homosphere (or lowest layer) contains gases that are more uniformly mixed, although their density decreases with height. Some exceptions to this statement occur with the existence of an ozone layer at an altitude of 20 to 50 kilometers and with variations in concentrations of carbon dioxide, water vapor, and air pollutants closer to Earth's surface.

The atmosphere can be divided into several zones based on decreasing or increasing temperatures as elevation increases. The lowest zone is the troposphere, where temperatures decrease from sea level up to an altitude of 16 kilometers in equatorial and tropical regions and up to an altitude of 6 kilometers at the poles. This lowermost zone holds substantial amounts of water vapor; aerosols that are very small; and light particles that originate from volcanic eruptions, desert surfaces, soot from forest and brush fires, and industrial emissions. Clouds, storms, and weather systems occur in the troposphere.

The tropopause marks the boundary between the troposphere and the next higher layer, the stratosphere, which reaches an altitude of 50 kilometers above Earth's surface. Circulation in this layer occurs with strong winds that move from west to east. There is limited circulation between the troposphere and the stratosphere. However, manned balloons, certain types of aircraft (Concorde and the U-2), volcanic eruptions, and nuclear bomb tests are able to break through the tropopause and enter the stratosphere.

The gases in the stratosphere are generally uniformly mixed, with the major exception of the ozone layer, which is found at an altitude range of 20 to 50 kilometers above Earth. This layer is extremely important because it shields life on Earth from the intense and harmful ultraviolet radiation from the Sun. The ozone layer has been diminishing because of the release of chlorofluorocarbons (CFCs), organic compounds containing chlorine, fluorine, and carbon, used as propellants in aerosol sprays and in synthetic chemical compounds used for refrigeration purposes. In 1978, the use of CFCs in aerosol sprays was banned in the United States, but they are still used in some refrigeration systems. Other countries continue to use CFCs, which eventually get into the ozone layer and result in ozone holes of considerable size. In 1987, members of the international community took steps to reduce CFC production through the Montreal protocol, and by 2003, the rate of ozone depletion had began to slow down. However, in the period from 2008 to 2018, monitoring stations have detected a steady increase in the atmospheric content of some banned CFCs, indicating that someone somewhere is producing the banned substance in bulk quantities once more, despite the global ban. Although the manufacture and use of CFCs can be controlled, at least in principle, natural events that are detrimental to the ozone layer cannot be prevented. For example, the 1991 eruption of Mount Pinatubo in the Philippines reduced the ozone layer in the midlatitudes nearly 9 percent.

Temperatures decrease with elevation at the stratopause, where the mesosphere layer begins at about 50 kilometers and continues to an altitude of about 80 kilometers. The mesopause at about 80 kilometers marks the beginning of the thermosphere, where the density of the air is very low and holds minimal

107

amounts of heat. However, even though the atmospheric density is minimal at altitudes above 250 kilometers, there is enough atmosphere to have a drag effect on spaceships.

Atmospheric Pressure
The gas molecules in the atmosphere exert a pressure due to gravity that amounts to about 15 pounds per square inch on all surfaces at sea level. As the distance from Earth gets larger, in contrast to the various increases and decreases in atmospheric temperature, atmospheric pressure decreases at an exponential rate. For example, air pressure at sea level varies from about 28.35 to 31.01 inches of mercury, averaging 29.92 inches. The pressure at the top of Mount Everest at 20,029 feet can get as low as 8.86 inches. This means that each inhalation of air at this altitude is about one-third of the pressure at sea level, producing severe shortness of breath.

Earth's Global Energy Balance
Earth's elliptical orbit about the Sun ranges from 146.4 million kilometers at perihelion (closest point to the Sun) on January 3 to 151.2 million kilometers at aphelion (furthest from the Sun) on July 4, averaging 148.8 million kilometers. Earth intercepts only a tiny fraction of the total energy output of the Sun. Upon reaching Earth, part of the incoming radiation is reflected back into space, and part is absorbed by the atmosphere, land, or oceans. Over time, the incoming shortwave solar radiation is balanced by a return to outer space of longwave radiation.

Earth-Moon Differences
Scientists believe that the moon's surface has a large number of craters formed by the impact of meteorites. In contrast, there are relatively few meteorite craters on Earth, even though, based simply on its size, Earth is likely to have been hit by as many or even more meteorites than the Moon. This notable difference is attributed to Earth's atmosphere, which burns up incoming meteorites, particularly small ones (the Moon does not have an atmosphere). Larger meteorites can pass through Earth's atmosphere, but their impact craters may have been filled in or washed away over millions of years. Only the more recent ones, such as Meteor Crater in northern Arizona, with a diameter of about 1,450 meters (about 1.45 kilometers) and a depth of about 220 meters, remain easily recognizable.

Air Masses
Different types of air masses within the troposphere, the lowest layer of the atmosphere, can be delineated on the basis of their similarity in temperature, moisture, and to a certain extent, air pressure. These air masses develop over continental and maritime locations that strongly determine their physical characteristics. For example, an air mass starting in the cold, dry interior portion of a continent develops thermal, moisture, and pressure differences that can be substantially different from an air mass that develops over water. Atmospheric dynamics also allow air masses to modify their characteristics as they move from land to water and vice versa.

Air mass and weather front terminology were developed in Norway during World War I. Norwegian meteorologists were unable to get weather reports from the Atlantic theater of operations; consequently, they developed a dense network of weather stations that led to impressive advances in atmospheric modeling that are still being used.

The Radiation Budget
The incoming solar energy that reaches Earth is primarily in the shortwave (or visible) portion of the electromagnetic spectrum. Earth's energy balance is attained by about one-third of this incoming energy being reflected back to space and the other two-thirds leaving Earth as outgoing longwave radiation. This balance between incoming and outgoing energy is known as Earth's radiation budget. The decades-long, ongoing National Aeronautics and Space Administration (NASA) program known as Clouds and Earth's Radiant Energy System (CERES) is designed to measure how much shortwave and longwave radiation leaves Earth from the top of the atmosphere.

Clouds play a very important role in the global radiation balance. For one thing, they constantly change over time and in type. Some clouds, such as high cirrus clouds found near the top of the troposphere at 40,000 feet, can have a substantial impact on atmospheric warming. Accordingly, the value of CERES is based on its ability to observe if human or natural changes in the atmosphere can be measured even if they are smaller than large-scale energy variations.

Greenhouse Effect

Selected gases in the lower parts of the atmosphere trap heat and then radiate some of that heat back to Earth. If there was no natural greenhouse effect, Earth's overall average temperature would be close to -18°C, rather than the existing +15°C.

The burning of coal, oil, and gas makes carbon dioxide (CO_2) the major greenhouse gas, accounting for nearly half of the total amount of heat-producing gases in the atmosphere. Before the Industrial Revolution in Great Britain in the mid-eighteenth century, the estimated level of carbon dioxide in the atmosphere was about 280 parts per million by volume (ppmv). Estimates for the natural range of carbon dioxide for the past 650,000 years range from 180 to 300 ppmv. All these values are less than the 391 ppmv recorded in January 2011. In 2013, the Mauna Loa Observatory reported that the global concentration of carbon dioxide in the atmosphere had reached 400 ppmv for the first time in history, and that number was subsequently surpassed by 2016. Carbon dioxide levels have been increasing since 2000 at a rate of 1.9 ppmv each year. The radiative effect of carbon dioxide accounts for about one-half of all the factors that affect global warming. Estimates of carbon dioxide levels at the end of the twenty-first century range from 490 to 1,260 ppmv.

The second most important greenhouse gas is methane (CH_4), which accounts for about 14 percent of all global warming factors. The origin of this gas is attributed to the natural decay of organic matter in wetlands, but anthropogenic activity—rice paddies, manure from farm animals, the decay of bacteria in sewage and landfills, and biomass burning (both natural and human induced)—results in a doubling of the amount of this gas over what would be produced solely by wetland decay.

Chlorofluorocarbons (CFCs) absorb longwave energy (warming effect), but they also have the ability to destroy stratospheric ozone (cooling effect). The warming radiative effect is three times greater than the cooling effect. CFCs account for about 10 percent of all global warming factors. Tropospheric ozone from air pollution and nitrous oxide (N_2O) from motor vehicle exhaust and bacterial emissions from nitrogen fertilizers account for about 10 percent and 5 percent, respectively, of all global warming factors.

Several kinds of human actions lead to a cooling of Earth's climate. For example, the burning of fossil fuels results in the release of tropospheric aerosols, which acts to scatter incoming solar radiation back into space, thereby lowering the amount of solar energy that can reach Earth's surface. These aerosols also lead to the development of low and bright clouds that are quite effective in reflecting solar radiation back into space.

APPLICATIONS AND PRODUCTS

Atmospheric science is applied in many ways. It is used to help people better understand their global and interplanetary environment and to make it possible for them to live safely and comfortably within that environment. By using the principles of this field, researchers, engineers, and space scientists have developed a vast number of applications. Among the most important are those used to track and predict weather cycles and climate.

Remote Sensing Techniques

Oceans cover about 71 percent of Earth's surface, which means that large portions of the world do not have weather stations or places where precipitation can be measured with standard rain gauges. To provide more information about precipitation in the equatorial and tropical parts of the world, NASA and the Japan Aerospace Exploration Agency initiated the Tropical Rainfall Monitoring Mission (TRMM) in 1997. The orbit of the TRMM satellite monitors Earth between 35 degrees north and 35 degrees south latitude. The goal of the study is to obtain information about the extent of precipitation, along with its intensity and length of occurrence. The major instruments on the satellite are radar to detect rainfall, a passive microwave imager that can acquire data about precipitation intensity and the extent of water vapor, and a scanner that can examine objects in the visible and infrared portions of the electromagnetic spectrum. The goal of data collection is to obtain the necessary climatological information about atmospheric circulation in this portion of Earth to develop better mathematical models for determining large-scale energy movement and precipitation.

Geostationary Satellites

Geostationary operational environmental satellites (GOES) enable researchers to view images of the planet from what appears to be a fixed position above Earth. The satellites are actually circling the globe at a speed that is in step with Earth's rotation. This means that a satellite at an altitude of 35,520 kilometers will make one complete revolution in the same twenty-four hours and direction that Earth is turning above the equator. At this height, the satellite is in a position to view nearly one-half of the planet at any time. On-board instruments can be activated to look for special weather conditions such as hurricanes, flash floods, and tornadoes. On-board instruments are also used to make precipitation estimates during storm events.

Doppler Radar
Doppler radar was first used in England in 1953 to pick up the movement of small storms. The basic principle guiding this type of radar is that back-scattered radiation frequency detected at a certain location changes over time as the target, such as a storm, moves. A transmitter is used to send short but powerful microwave pulses. When a foreign object (or target) is intercepted, some of the outgoing energy is returned to the transmitter, where a receiver can pick up the signal. An image (or echo) from the target can then be enlarged and shown on a screen. The target's distance is revealed by the time that elapses between transmission and return. The radar screen can not only indicate where the precipitation is taking place but also reveal the intensity of the rain by the amount of the echo's brightness. In short, Doppler radar has become a very useful device for determining the location of a storm and the intensity of its precipitation and for obtaining good estimates of the total amount of precipitation.

Responses to Climate Change
Since the 1970s, many scientists have pointed out the possibility that human activity is having more than a short-term impact on the atmosphere and therefore on weather and climate. Although much debate continues on the full impact of human activities and greenhouse gas emissions, the atmospheric sciences have led to conferences, United Nations conventions, and agreements among nations on ways that human beings can alter their behavior to halt or at least mitigate the possibility of global climate change. The impact of these agreements, still in their infancy, remains unknown, as does the overall effect of human activity on weather and climate (the models for which are highly complex). However, the insights contributed by the atmospheric sciences to the overall debate on whether climate change is primarily anthropogenic (human caused), and whether global warming is actually taking place, have caused many nations and individuals to modify their attitudes toward human relationships with the global environment, resulting in national and intergovernmental changes in policies concerning carbon emissions, as well as personal decisions ranging from the consumption of "green" building materials to the purchase of vehicles fueled by noncarbon sources of energy.

SOCIAL CONTEXT AND FUTURE PROSPECTS
Climate change may be caused by both natural internal/external processes in Earth-Sun system or by human-induced changes in land use and the atmosphere. Article 1 of the United Nations Framework Convention on Climate Change (entered into force March, 1994) states that the term "climate change" should refer to anthropogenic changes that affect the composition of the atmosphere rather than natural causes, which should be referred to "climate variability." An example of natural climate variability is the global cooling of about 0.3°C in 1992-1993 that was caused by the 1991 eruption of Mount Pinatubo in the Philippines. The 15 million to 20 million tonnes of sulfuric acid aerosols ejected into the stratosphere reflected incoming radiation from the sun, thereby creating a cooling effect. Many suggest that the above-normal temperatures experienced in the first decade of the twenty-first century provide evidence of climate change caused by human activity. Based on a variety of techniques that allow scientists to estimate the temperature in previous centuries, the years 2005, 2010, and 2014 were the warmest in the last thousand years. A 2009 article published by the American Geophysical Union suggests that human intervention in Earth systems has reached a point where the Holocene epoch of the past 12,000 years is becoming a new Anthropocene epoch in which human systems have become primary Earth systems rather than simply influencing natural systems.

Numerous observations strongly suggest a continuing warming trend. Snow and ice have retreated from areas such as Mount Kilimanjaro in Tanzania, which at 5,860 meters is the highest mountain in Africa. Glaciated areas in Switzerland also provide evidence of this warming trend. The Special Report on Emission Scenarios issued in 2001 by the Intergovernmental Panel on Climate Change (IPCC) examined the broad spectrum of possible concentrations of greenhouse gases by considering the growth of population and industry along with the efficiency of energy use. The IPCC computer climate models were used to estimate future trends. For example, a global temperature increase of 19.55°C to 21.78°C by the year 2100 is an IPCC standard estimate.

—*Robert M. Hordon, PhD*

FURTHER READING

Christopherson, Robert W., and Ginger H Birkeland. *Geosystems: An Introduction to Physical Geography.* Pearson, 2018.

Coley, David A. *Energy and Climate Change: Creating a Sustainable Future.* John Wiley & Sons, 2008.

Ellis, Erle C., and Peter K. Haff. "Earth Science in the Anthropocene: New Epoch, New Paradigm, New Responsibilities." *EOS, Transactions, American Geophysical Union,* vol. 90, no. 49, 2009, p. 473.

Gautier, Catherine. *Oil, Water, and Climate: An Introduction.* Cambridge UP, 2008.

Lutgens, Frederick K., Edward J. Tarbuck, Tasa, Dennis G. and Herman, Redina. *The Atmosphere: An Introduction to Meteorology.* 14th ed. Pearson Education, 2018. An updated version of this textbook, in loose-leaf format, provides students with a general overview of the structure and properties of Earth's atmosphere.

Strahler, Alan. *Introducing Physical Geography.* 5th ed., John Wiley & Sons, 2011.

Wolfson, Richard. *Energy, Environment, and Climate.* W. W. Norton, 2008.

ATMOSPHERIC STRUCTURE AND EVOLUTION

FIELDS OF STUDY

Atmospheric Science; Atmospheric Chemistry; Geochemistry; Biochemistry; Bioclimatology; Oceanography; Inorganic Chemistry; Photochemistry; Climate Modeling; Hydroclimatology; Physical Geography; Earth System Modeling

SUMMARY

The atmosphere is an ocean of gases held to the Earth and compressed by gravity. If greenhouse gases (GHGs) were newly introduced into the atmosphere, less energy would leave the Earth than strike it, so the planet would warm until those two rates balanced. (Infrared radiation from a warmer Earth has a shorter wavelength and is therefore more likely to escape into space.) GHGs in the atmosphere keep Venus 500°C, Earth 35°C, and Mars 7°C warmer than each planet would be without its atmosphere. Without the greenhouse effect, much of Earth would be permanently covered with snow and ice.

PRINCIPAL TERMS

- **albedo:** percentage of incident light Earth reflects back into space

- **greenhouse gases (GHGs):** gases that allow sunlight to pass through to the ground but trap, at least partially, the infrared radiation that would otherwise escape into space
- **ice age:** a period during which the average global temperature is reduced, allowing sea ice and glaciers to cover a significant fraction of Earth's surface
- **late heavy bombardment:** a period about 3.9 to 4.0 billion years ago, when Earth was pummeled by debris from space at 1000 times the normal rate, heating the atmosphere and melting the crust
- **Milankovitch cycles:** recurring time periods during which the shape of Earth's orbit, the tilt of its axis, and the occurrence of its farthest distance from the Sun all change

EARTH'S FIRST ATMOSPHERE

If Earth's atmosphere had formed along with Earth itself, its composition would be expected to reflect the relative abundance of solar elements extant at the time and present in gases heavy enough to be retained by Earth's gravity. Hydrogen and helium are the two most abundant gases, but Earth's gravity is not strong enough to hold them, so they

Solar Abundances of the Elements

Element	Solar Nebula Abundance per Hydrogen Atom
Hydrogen	1
Helium	0.16
Oxygen	0.00089
Neon	0.0005
Carbon	0.0004
Nitrogen	0.00011
Silicon	0.000032
Magnesium	0.000025
Sulfur	0.000022
Argon	0.0000076

gradually escape into space. Oxygen is the next most abundant gas, but it is so chemically active that, without plant life to replenish it, it would soon disappear from Earth's atmosphere. Neon is next in solar abundance, is chemically inert, and is has sufficient mass to be retained by Earth's gravity. The fact that it is not the most abundant gas in Earth's present atmosphere provides evidence that Earth's primordial atmosphere escaped into space, probably during a flare-up of the young Sun or during the late, heavy bombardment following the planet's formation.

Scientists believe that the current atmosphere consists of gases released by volcanic activity as rocks and minerals within Earth were heated by the energy from radioactive decay. Water vapor is the most abundant volcanic gas and presumably condensed to form the oceans. Water is also a significant component of comets and meteors, and additional water would have been acquired when such extraterrestrial objects impacted the early planet. The next most abundant volcanic gases are carbon dioxide (CO_2), nitrogen (N_2), and argon (Ar). Large quantities of CO_2 are removed from the atmosphere as it dissolves in the oceans, where most of it eventually combines with calcium and oxide ions, and eventually forms carbonate rocks (limestone). Sulfur dioxide and sulfur trioxide are chemically active and do not stay in the atmosphere very long due to reaction with atmospheric water to form sulfuric acid and precipitating as acid rain. Thus, if the Earth's atmosphere came from volcanoes, it should be dominated by nitrogen, followed by CO_2 and then argon. The atmospheric CO_2 would soon be depleted as it dissolved in the oceans. The atmospheres of Venus and Mars are richer in CO_2 than is Earth's atmosphere, because those planets have no oceans to remove their CO_2.

In the absence of oxygen, iron dissolves in water, but when oxygen is dissolved in the same water, iron oxide precipitates out and sinks to the bottom. Ore deposits of this iron compound first appeared about 2.6 billion years ago and peaked about 1.8 billion years ago. This probably reflects the increasing abundance of plants that released oxygen into the atmosphere and the increasing atmospheric concentration of oxygen that resulted. Earth's atmospheric oxygen concentration reached a maximum of 30 percent 300 million years ago and a minimum of about 12 percent 200 million years ago. It now essentially stable at about 21 percent.

Composition of Air and of Volcanic Gases

Gas	Air (Percent by Volume)	Volcanic* Gas (Percent by Volume)
N_2 (nitrogen)	77	5.45
O_2 (oxygen)	21	
H_2O (water vapor)	0.1 to 2.8	70.8
Ar (argon)	0.93	0.18
CO_2 (carbon dioxide)	0.033	14.07
Ne (neon)	0.0018	
CH_4 (methane)	0.00015	
NH_3 (ammonia)	0.000001	
SO_2 (sulfur dioxide)		6.4
SO_3 (sulfur trioxide)		1.92
CO (carbon monoxide)		0.4
H (hydrogen)		0.33

*Kilanea volcano, Hawaii.

THE FAINT EARLY SUN PARADOX

Stars such as the Sun become approximately double in brightness over their lifetimes as normal stars. The Sun is thus already about 30 percent brighter than it was when it first became a normal star, 4.56 billion years ago. Based on contemporary values of Earth's albedo and its atmosphere, Earth should not have had liquid water before about 2 billion years ago. There is, however, abundant geological evidence that liquid water has existed on Earth for at least 3.8 billion years. Mars exhibits a similar paradox, since it seems to have had abundant surface water 3.8 billion years ago.

If the Sun had been born 7 percent more massive, the Earth would have been warm enough for liquid water. Because at normal rates the Sun can have lost only 0.05 percent of its mass since it was born, however, this is probably not the case. A sufficiently large amount of CO_2 in the atmosphere could have made the early Earth warm enough for water, but geological evidence for that much atmospheric CO_2 is lacking. However, Philip von Paris and his colleagues at the Aerospace Centre in Berlin have developed a plausible computer model that allows the early Earth to have developed liquid water with only 10 percent of the CO_2 previously thought necessary, an amount not ruled out by geological evidence.

STRUCTURE OF THE ATMOSPHERE

The atmosphere has settled into a series of layers, one on top of the other, like the layers of an onion. The lowest layer of the atmosphere is called the troposphere. It extends from the ground up to about 15 kilometers over the equator and slants down to just 8 kilometers in height over the poles. The word "troposphere" is based on the Greek word *tropos*, which means "turning" or "mixing." Storms are a result of the mixing that takes place in this layer, so storms and almost all clouds are confined to the troposphere. In the troposphere, temperature decreases by about 8°C with each 1 kilometer of altitude. Tourists are often surprised to find that the South Rim of the Grand Canyon, in Arizona, is 12°C cooler than the inner gorge, 1.5 kilometers below.

The atmospheric layer above the troposphere is the stratosphere (from the Latin *stratum*, meaning "horizontal layer"). It extends from the troposphere up to an altitude of about 50 kilometers. Commercial airliners usually cruise in the lower stratosphere, where the reduced air density produces less drag and where they are above clouds and storms. The ozone layer, which protects Earth life from most of the Sun's ultraviolet radiation, is located in the stratosphere, mostly between 20 and 40 kilometers high.

Above the stratosphere lies the mesosphere (from the Greek *mesos,* meaning "middle"). The mesosphere extends from about 50 kilometers up to 80 or 90 kilometers in altitude. Incandescent trails may be observed in the mesosphere. These trails result when meteoroids the size of sand grains or pebbles strike Earth from space: They are heated by the friction of their swift passage through the mesosphere until they vaporize and disintegrate.

The thermosphere (from the Greek *thermos,* meaning "heat") begins about 90 kilometers above the ground and extends upward about 500 kilometers higher. Gas molecules of the thermosphere absorb solar energy and convert it into kinetic energy (energy of motion). The speed of a molecule is a measure of its temperature, and speeds corresponding to temperatures of up to 15,000°C are expected in thermosphere gas molecules. While that may seem hot, an unprotected person in the thermosphere would soon freeze, because such gas molecules are few and far between, making the average temperature of this layer extremely cold. Auroras form in the lower thermosphere when high-energy particles from the Sun and other stars impact with air atoms.

The International Space Station (ISS) orbits in the thermosphere, about 340 kilometers above Earth's surface. The atmosphere there is thin enough that drag on the ISS is small, but not zero. The advantage to this location is that, although it must periodically be reboosted by a supply ship while in use, when the ISS is finally abandoned it will slowly and naturally deorbit itself.

Beginning in the lower thermosphere, air atoms are far enough apart that when ultraviolet light from the Sun drives electrons away from their parent atoms, they do not recombine for some time. The clouds of free electrons (or ions) that thus form can reflect radio waves. This region is the ionosphere.

Glacial and Interglacial Periods

Years Before Present (thousands)	Temperature Deviation*	Description
0.2 to 0.4	-1	Little Ice Age (Northern Europe)
0.8 to 1.2	+1	climatic optimum
20 to 500	+2 and -6	4 cold periods and 4 warm periods
Years Before Present (millions)		
3 to 10	+1 to +2	warm period
10 to 25	+3	Antarctica thawing
25 to 35	+1	Antarctica glaciation
35 to 80	up to +6	Eocene optimum
260 to 360	-2 to -5	Karoo glaciation
420 to 450	+2 to -5	Andean-Saharan glaciation
635 to 800	-5	Cryogenian glaciation
2,400 to 2,100	-5	Huronian glaciation

*Rough global averages; deviation is from 17° Celsius, expressed in ° Celsius.

Depending upon the intensity of incident sunlight, the ionosphere can extend throughout the thermosphere and up into the exosphere. The exosphere begins 500 to 600 kilometers above Earth's surface and extends out to about 10,000 kilometers, where it shades into interplanetary space.

Earth's Greenhouse Gases

Clouds and water vapor together are responsible for roughly 80 percent of Earth's greenhouse warming. People can affect the amount of water vapor in the atmosphere locally by activities such as deforestation and agricultural irrigation, but the global amount is determined by Earth's vast oceans. The global amount of atmospheric water vapor remains essentially constant over time, so most atmospheric scientists conclude that it cannot be responsible for recent warming. Water vapor also plays a major role in feedback loops. For example, if Earth warms, more water will evaporate from the oceans, becoming vapor. More water vapor will produce more clouds, and more clouds blanketing Earth will keep it warmer, but more clouds will also reflect more sunlight back into space, keeping Earth cooler. Experts theorize that the warming effect would be greater than the cooling effect, but it is nevertheless apparent that global warming is quite a complex phenomenon in which water vapor is just one of several factors.

Most of the rest of Earth's greenhouse effect is accounted for by CO_2. The amount of atmospheric CO_2 is increasing by about 0.4 percent per year, with perhaps the majority of that increase coming from transportation, power generation, and other human activities. It also comes from volcanoes and burning vegetation, and is removed from the atmosphere by plants and by being dissolved in the oceans. In the oceans, it may combine with calcium to make limestone. It is also removed from the ocean by sea creatures that use calcium carbonate to make their shells.

Methane is present in the atmosphere in only trace amounts, but, molecule for molecule, it is many times more effective than CO_2 as a GHG. Atmospheric methane has increased by 11 percent since 1974. Significant amounts of it are released by coal and oil production, cattle and sheep, swamps and rice paddies, jungle termites and the decomposition of permafrost terrains and of methane hydrates on the sea floor.

Ice Ages

Life on Earth has survived many cooling and warming cycles in the past, but scientists do not fully understand the causes of climate change. Different factors dominate changes at different times. For example, increased atmospheric CO_2 from volcanoes may warm Earth enough to end an ice age, but in other cases atmospheric CO_2 might not increase until centuries after the climate warms. A few of the other factors to consider are Milankovitch cycles, solar intensity, albedo, and the ability of the oceans to absorb CO_2 and convert it to limestone.

Context

At one other time, humanity stood on the verge of having the power to change the climate. When world stockpiles of nuclear weapons were at their highest levels, some people wondered if an all-out nuclear war would result in nuclear winter—months or years of freezing temperatures and little sunlight—followed by ten thousand years of nuclear spring—temperatures several degrees above normal and dangerous levels of ultraviolet light reaching the ground. Humans are again on the verge of being able to change the climates of Earth, and not enough is known to predict all of the results of changing the climate or of taking actions designed to avoid changing it.

—*Charles W. Rogers*

Further Reading

Canfield, Donald Eugene *Oxygen. A Four Billion Year History.* Princeton University Press, 2014. This highly readable book presents a comprehensive overview of atmospheric oxygen and discusses many questions about the origin and significance of atmospheric oxygen.

Catling, David C. and Kasting, James F. *Atmospheric Evolution on Inhabited and Lifeless Worlds.* Cambridge University Press, 2017. A complete and comprehensive treatment of the evolution of Earth's atmosphere, with application to the study of other planets.

Condie, Kent C. *Earth as an Evolving Planetary System.* 3rd ed. Elsevier, 2016. Provides a complete description of Earth's internal and external structures with regard to the evolution of the planet and its atmosphere.

Hewitt, C.N. And Jackson, A.V., eds. *Atmospheric Science for Environmental Scientists.* John Wiley & Sons, 2009. An introductory textbook of atmospheric science, discusses many aspects of the structure and evolution of Earth's atmosphere.

Sánchez-Lavega, Agustín *An Introduction to Planetary Atmospheres.* CRC Press, 2011. A comprehensive resource discussing many aspects of atmospheric science and evolution.

AVERAGE WEATHER IN CLIMATE STUDIES

SUMMARY

The term "average," from a statistical point of view, denotes the arithmetic mean of a set of numbers taken from a sample or a representative population. That is, the average rainfall for the month of January for a specific place is the average of the actual January rainfalls for a period of time. In most cases, this use of "average" in weather and climate studies and applications makes sense, but in some cases it can give an erroneous impression of the average weather.

For example, to determine the average January rainfall for the city of Tauranga, New Zealand, one must first decide which January rainfalls to average. In Tauranga, rainfalls have been measured from several sites for the periods 1898–1903, 1905–1907, and 1910–2009. For a correct analysis, data for all the observation sites must first be carefully adjusted so that they apply to the current recording site. When this is done, the average for all of these January months is 88 millimeters. This average includes a rainfall of 268 millimeters in January, 1989 (the wettest January recorded), and a rainfall of only 1 millimeter in January, 1928 (the driest January recorded). The average in this case is therefore of some use and reveals something about the January rainfalls over a period of one hundred years in Tauranga, but it is not an exhaustive description, especially if one seeks to determine whether Tauranga is getting wetter or drier.

For other weather elements, such as temperature, sunshine, cloudiness, and wind, the situation is similar: The average for a particular period, such as a month, season, year, decade, or set of years, is simply the average of the values found in a series of observations.

SIGNIFICANCE FOR CLIMATE CHANGE

As noted, in most cases, this use of the average in weather and climate studies and applications makes sense, but in some cases it can give an erroneous and sometimes distorted view of the average weather. For example, in the case of temperatures, a problem has arisen in climate change discussions, because the average temperature is traditionally determined by climatologists by taking the average of the highest (maximum) and the lowest (minimum) temperatures for a particular day. While this practice produces useful information, especially when comparing the average temperatures of, say, Chicago with those of Bangkok, a difficulty arises when the daytime and nighttime temperatures are important. For example, in a continental climate, such as that of Moscow, the difference between the daytime and nighttime temperature is significant, whereas in a tropical climate, such as that of Singapore, the day-to-night temperature difference is relatively small.

Average weather is generally reflected by measurements of average rainfall, average day- and nighttime temperatures, average sunshine, and so on. However, when one considers changes in

Severe thunderstorms and tornadoes swept across the Midwest and Appalachians on March 2, 2012. Meteorologist Jeff Masters described the outbreak as a result of warm, wet air from the Gulf of Mexico mixing with cold, dry air aloft.

climate—small changes in the average weather over time—it becomes necessary to take particular note of the period involved and the specific weather element being measured. For example, climate change may cause a particular place to be wetter in the winter and drier in the summer. To measure such a change, one would need to assess the changes in the summer rainfall and the winter rainfall over a period of time of at least thirty years and ideally one hundred years. Such an analysis might show that during the first fifty of the last one hundred years, winter rainfalls were lower than those of the following fifty years, whereas summer rainfalls were higher. One must therefore treat all climate data with a degree of caution and be careful always to compare apples with apples.

For example, if one considers again the city of Tauranga, New Zealand, as a typical example, temperature observations have been taken there since 1913. Various observation sites have been used, and data from all sites have been adjusted to reflect what the temperature would have been if all observations had been taken from the same site. With these adjustments, the highest average monthly temperatures for each month in Tauranga have occurred in January, 1935; February, 1928; March, 1916; April, 1938; May, 1916; June, 2002; July, 1916; August, 1915; September, 1915; October, 1915; November, 1954; and December, 1940. The highest annual average daily maximum temperature was 20.4° Celsius, recorded in 1916, followed by 20.2° Celsius in 1928, 20.1° Celsius in 1914, and 20.0° Celsius in 1998. These observations show that, despite the indications of global warming in some parts of the world, not all areas are the same, and memories can be deceiving.

—*W. J. Maunder*

B

BAROMETRIC PRESSURE AND BAROMETRY

FIELDS OF STUDY
Fluid Dynamics; Atmospheric Science; Physical Sciences; Meteorology; Physical Geography

SUMMARY
Barometric pressure is a measure of atmospheric pressure as recorded by a scientific instrument known as a barometer. Barometric pressure readings have been made since the eighteenth century and remain useful today in forecasting weather. These readings allow meteorologists to identify the areas of high and low pressure that are integral parts of weather events.

PRINCIPAL TERMS

- **altimeter:** scientific instrument that measures the altitude of an object above a fixed level
- **aneroid barometer:** device that uses an aneroid capsule composed of an alloy of beryllium and copper to measure changes in external air pressure
- **atmospheric pressure:** force exerted on a surface by the weight of air above that surface; measured in force per unit area
- **barograph:** a graph that records atmospheric pressure in time
- **barometer:** device for measuring atmospheric pressure; some are water-based, some use mercury or an aneroid cell, and some create a line graph of atmospheric pressure
- **high-pressure area:** region in which the atmospheric pressure is greater than that in the areas around it; represented by H on weather maps
- **low-pressure area:** region where the atmospheric pressure is lower than that in surrounding areas; represented by L on weather maps
- **mercury barometer:** glass tube of a minimum of 84 centimeters (33 inches), closed at one end, with a mercury-filled pool at the base; the weight of the mercury creates a vacuum at the top of the tube; mercury adjusts its level to the weight of the mercury in the higher column
- **meteorology:** the study of changes in temperature, air pressure, moisture, and wind direction in the troposphere; the interdisciplinary scientific study of the atmosphere
- **water-based barometer:** also known as a storm glass or Goethe barometer, a device with a glass container and a sealed body half full of water; also has a spout that fills with more or less water depending upon atmospheric conditions and their forces

Digital barometric pressure sensor BMP085, Bosch Sensortec, used for altimeters, for example.

BAROMETRIC PRESSURE DEFINED
Barometric pressure, also known as atmospheric pressure, is the measure of the amount of pressure the atmosphere exerts on the surface of the earth at a given point in time. Barometric pressure is the accumulated weight of the air, influenced by the force of gravity, above the point being measured.

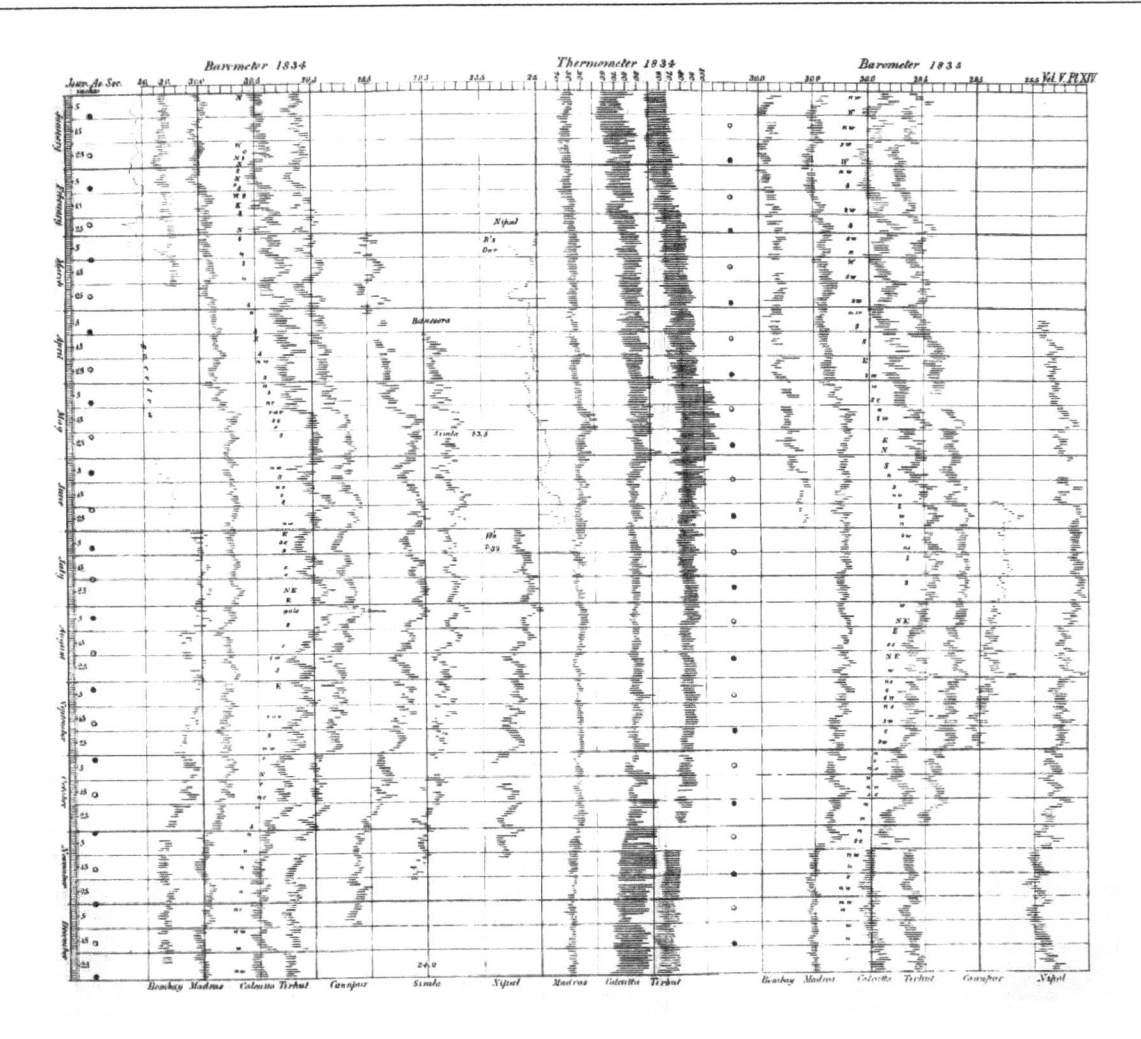

Chart of barometer and thermometer measurements.

When the temperature is cold and the air is dense, the air pressure will be higher than it is when the temperature is warm and the air is less dense. This occurs because the denser, heavier air exerts more downward pressure than warmer, lighter air.

An active interest in atmospheric pressure can be traced to Galileo Galilei and Evangelista Torricelli in the seventeenth century. Galileo wondered how long a straw could be and still remain useful when moving fluid up that straw. The question was based on actual observations of pumps used to bring water from beneath the ground. The pumps were effective up to about 10 meters (33 feet). Beyond that height, the pumps failed. Torricelli began exploring an understanding of this phenomenon.

Having observed that water could not reach a height of more than 10 m, Torricelli calculated that if atmospheric pressure were the cause, mercury, with a greater density, would not be able to attain a height of more than 74 centimeters (29 inches). Through experimentation, he had removed the air from the top of a glass straw by pumping it out; he correctly concluded that the force of the atmosphere had caused the water to rise to a height of 10 m and the mercury to rise to a height of 74 cm.

Torricelli noticed that the height of the column of mercury changed from day to day. He also noticed that if he traveled to areas that ranged in altitude, the column of mercury changed in height depending upon the altitude. He theorized that the weight of the atmosphere was changing but was unable to identify how or why.

Rene Descartes was next to investigate atmospheric pressure. In 1647, he added numbers to a Torricelli tube so that the readings from day to day could be accurately recorded. Blaise Pascal posited that air became thinner as one goes higher in the atmosphere and proved this in 1648. In 1666, Robert Boyle was the first to describe the Torricelli tube as a barometer.

By the eighteenth century, mercury barometers were an important part of the equipment of ocean-going vessels, in large part because of the work of Edmund Halley and John Locke in proving a correlation between barometer readings and weather conditions. To this day, barometric pressure readings play a vital role in predicting the weather. They help track the flow of air from high-pressure areas to low-pressure areas.

BAROMETRY DEFINITION AND BASIC PRINCIPLES

Barometry is the science of measuring the pressure of the atmosphere. Derived from the Greek words for "heavy" or "weight" (*baros*) and "measure" (*metron*), it refers generally to the measurement of gas pressure. In gases, pressure is fundamentally ascribed to the momentum flowing across a given surface per unit time, per unit area of the surface. Pressure is expressed in units of force per unit area. Although pressure is a scalar quantity, the direction of the force due to pressure exerted on a surface is taken to be perpendicular and directed onto the surface. Therefore, methods to measure pressure often measure the force acting per unit area of a sensor or the effects of that force. Pressure is expressed in newtons per square meter (pascals), in pounds per square foot (psf), or in pounds per square inch (psi). The pressure of the atmosphere at standard sea level at a temperature of 288.15 kelvin (K) is 101,325 pascals, or 14.7 psi. This is called 1 atmosphere. Mercury and water barometers have become such familiar devices that pressure is also expressed in inches of water, inches of mercury, or in torrs (1 torr equals about 133.3 pascals).

BAROMETER TYPES AND MEASURING BAROMETRIC PRESSURE

The initial weather-forecasting barometer, the 'Torricellian tube' was invented by Evangelista Torricelli (1608 – 1647) and measured the height of a water column that the pressure of air would support, with a vacuum at the closed top end of a vertical tube. This barometer is an absolute pressure instrument. Atmospheric pressure is obtained as the product of the height, the density of the barometric liquid, and the acceleration because of gravity at the Earth's surface. The aneroid barometer uses a partially evacuated box the spring-loaded sides of which expand or contract depending on the atmospheric pressure, driving a clocklike mechanism to show the pressure on a circular dial. A barograph is an aneroid barometer mechanism adapted to plot a graph of the variation of pressure with time, using a stylus moving on a continuous roll of paper. The rate of change of pressure helps weather forecasters to predict the strength of approaching storms. All barometers determine the weight of the atmosphere, according to the force of gravity, pressing down upon them.

Widely used in the eighteenth century, mercury barometers are seldom used outside a lab today. Mercury barometers rely upon columns of mercury in glass tubes marked with a scale, and the height of the mercury column is noted. As the pressure of the atmosphere becomes greater, the column of mercury rises and the high pressure is noted. As the pressure of the atmosphere becomes less, the column of mercury falls and low pressure is noted. By recording readings at the same time and in the same place each day, the readings can be compared to note the trends in pressure. These trends can be used to forecast the weather.

Aneroid barometers came into wide use in the mid-nineteenth century but do not include mercury. Instead they include a bellows that contracts and expands with the pressure of the atmosphere. When the pressure is great, the pin measuring the barometric pressure moves to a higher position and indicates high pressure. When the pressure is less, the pin measuring the barometric pressure moves to a lower position and indicates low pressure. As is the case with the mercury barometer, the trend in readings can be used to forecast the weather. A very simple aneroid barometer can be easily constructed by confining an inflated balloon within a container that is open at one end and attaching the stem of the balloon to an indicating device.

In the 1970's, devices based on the aneroid barometer principle were developed, in which the deflection of a diaphragm caused changes in electrical capacitance that then were indicated as voltage changes in a circuit. In the 1980's, piezoelectric materials were developed, enabling electrical voltages to be created from changes in pressure. Micro devices based on these largely replaced the more expensive but accurate diaphragm-based electromechanical sensors. Digital signal processing enabled engineers using the new small, inexpensive devices to recover most of the accuracy possessed by the more expensive devices. Electronic barometers, in use today, rely upon internal sensors and electronic circuits to measure and display atmospheric pressure. These barometers are very accurate and often can display graphs of recent readings. Their displays are digital and generally include other readings, such as temperature, which are of use.

Influence of Gravity on Barometric Pressure

Gravity plays a part in barometric pressure readings because gravity determines the weight of the atmosphere and therefore the pressure it exerts at any particular point within the atmosphere. If a person envisions himself or herself in what Torricelli referred to as an ocean of air, that person is at the bottom of that "ocean." The pressure is not felt because it comes from all sides and effectively cancels itself out. However, the force of gravity plays a part in the force of the atmosphere.

If a person goes to a location 1,524 m (5,000 ft) above sea level, that individual will find that the atmosphere is always lighter than it is at a location at or below sea level. Because of this, adjustments are made to bring all readings to sea level. In this way, forecasters can note meaningful differences between locations that vary in altitude. It also is useful to track changes in barometric pressure at one location because this allows the forecaster to note the trends for that specific location.

Barometric Pressure and the Weather

Barometric pressure readings are integral to weather forecasting, as air flows from high-pressure areas into low-pressure areas because of the difference in atmospheric pressure. This flow of air is what brings a weather event.

The air flowing into low-pressure areas is doing so because of the force of gravity as it pulls upon the denser air. The air flowing from the high-pressure area does so in the form of winds. These winds are subject to the rotation of the earth and to the Coriolis force. The Coriolis force, or effect, describes the effect of an inertial force and demonstrates that an object appears to deviate from its path when viewed in a coordinate system. In truth, the object is not deflected; however, the wind is observed to move in a counterclockwise direction around a low-pressure area and clockwise around a high-pressure area in the Northern Hemisphere. (The reverse is true in the Southern Hemisphere.) The weather conditions experienced will be the result of the winds, temperatures, humidity, and other factors in play. By noting the trend in the barometric pressure, forecasters have an idea of what type of weather is developing and approaching and can tailor their observations to those expectations when creating their forecasts.

—*Gina Hagler*
—*Narayanan M. Komerath, PhD*

Bibliography

Burch, David. *The Barometer Handbook: A Modern Look at Barometers and Applications of Barometric Pressure.* Starpath, 2009. A thorough discussion of the uses and importance of the barometer. Includes worldwide average monthly pressures and their standard deviations. Provides a firm understanding of the importance of this simple instrument in weather prediction.

Frederick, John E. *Principles of Atmospheric Science.* Jones and Bartlett, 2008. A complete introduction to atmospheric science. Presents the fundamental scientific principles and concepts related to the earth's climate system.

Gillum, Donald R. *Industrial Pressure, Level, and Density Measurement.* 2d ed. Research Triangle Park, N.C.: International Society for Automation, 2009. Teaching and learning resource on the issues and methods of pressure measurement, especially related to industrial control systems. Contains assessment questions at the end of each section.

Randall, David A. *Atmosphere, Clouds, and Climate.* Princeton University Press, 2012. An overview of the major atmospheric processes. Covers the function of the atmosphere in the regulation of energy flows and the transport of energy through weather systems, including thunderstorm and monsoons. Examines obstacles in predicting the weather and climate change.

Salby, Murray L. *Physics of the Atmosphere and Climate.* Cambridge University Press, 2012. Provides an integrated treatment of the earth-atmosphere system and its processes. Begins with first principles and continues with a balance of theory and applications as it covers climate, controlling influences, theory, and major applications.

Teague, Kevin Anthony and Gallicchio, Nicole *The Evolution of Meteorology. A Look Into the Past, Present and Future of Weather Forecasting.* Wiley-Blackwell, 2017. A comprehensive presentation of weather forecasting that includes descriptions of barometers and other devices in context.

C

CARBON AND CARBON DIOXIDE EQUIVALENT

FIELDS OF STUDY

Atmospheric Chemistry; Chemical Kinetics; Analytical Chemistry; Thermodynamics; Atmospheric Science; Environmental Sciences; Environmental Studies; Waste Management; Meteorology; Climate Modeling; Statistics; Mathematics; Heat Transfer

SUMMARY

In contrast to the CE, the CO_2e uses the functionally equivalent CO_2 concentration as the reference (the amount of CO_2 that would have the same GWP). CE is usually given for a specified timescale, effectively expressing the time-integrated radiative forcing. This also differentiates CE from CO_2e, which describes the instantaneous radiative forcing.

PRINCIPAL TERMS

- **greenhouse effect:** a natural process by which water vapor, carbon dioxide, and other gases in the atmosphere absorb heat and reradiate it back to Earth
- **greenhouse gases (GHGs):** gases (or vapors) that trap heat in the atmosphere by preventing or delaying the passage of long-wavelength infrared radiation from Earth's surface out to space
- **infrared radiation:** radiation with wavelengths longer than those of visible light, felt by humans and animals as heat

CARBON VS CARBON DIOXIDE

Various greenhouse gases (GHGs) such as methane, nitrous oxide, chlorofluorocarbons, and carbon dioxide (CO_2) have different potential effects on global warming. Of these, CO_2 is the most common. Greenhouse gases function by absorbing infrared radiation emitted from the surface of Earth, preventing it from continuing on out into space, and subsequently radiating at least half of that heat energy back into the atmosphere and toward Earth's surface. In order to compare GHGs based on their global warming potential (GWP), carbon dioxide (CO_2) is chosen as the reference gas. The carbon dioxide equivalent (CO_2e) for a given amount of GHG is defined as the amount of CO_2 that would have the same GWP measured over a specified time period, usually one hundred years. The carbon equivalent (CE), similar to carbon dioxide equivalent (CO_2e), is a metric used to quantify how much global warming a given quantity of gas can cause, using the functionally equivalent concentration of carbon, converted to the amount of carbon dioxide (CO_2) that would have the same GWP as the reference.

SIGNIFICANCE FOR CLIMATE CHANGE

The CO_2e metric provides a universal standard measure by which to evaluate and compare the global warming effects of emissions of various GHGs, as well as to calculate the total effects of the GHGs present in Earth's atmosphere. Those uses of CO_2e have important implications for decisions about climate-change mitigation. On one hand, CO_2e measurements of GHG emissions make it easy to compare the various impacts of different plans to prevent such emissions. The fact that the CO_2e of methane is 21 times that of CO_2 helps determine whether to focus on methane or CO_2 reduction in a given context.

On the other hand, the calculation of total effect of GHGs expressed in CO_2e concentration can provide a way to compare existing concentrations of GHGs in the atmosphere with theoretical critical threshold levels. Recent studies have confirmed that one such threshold is represented by a global temperature rise of 2°C during the twenty-first century.

123

The Orbiting Carbon Observatory, which was lost during launch on February 24, 2009, was designed to be the first spacecraft to study Earth's atmospheric carbon dioxide, a main cause of global warming.

In order to prevent surpassing this threshold, the total concentration of all GHGs must be less than 450 parts per million in CO_2e. However, GHG concentration is already beyond that level.

In addition to facilitating global climate study, the CO_2e metric as a reference for evaluating GHGs also plays an important role in emissions trading. Various national and international regulatory structures allow polluters that reduce their emissions below a certain level to trade emissions credits (the right to pollute) to other entities whose emissions are still above the maximum permitted level. The existence of a universal standard of emission measurement, CO_2e, allows these credit systems to function much more effectively and efficiently.

The CE of a gas or a mixture of gases is a metric measure that can be used to compare the GWP of greenhouse gases (GHGs) for the purposes of analysis, computer modeling, and reporting, as was done, for example, in such international treaties as the Kyoto Protocol and the Montreal Protocol. The CE of a gas is calculated by multiplying the mass (in metric tons) by the GWP of the gas and is expressed for a specified timescale. CE values can be converted to CO_2e units simply by multiplying the CO_2 by 3/11 (the ratio of the molecular weight of carbon to that of CO_2). Time is an important factor when comparing CE or GWP values, as they are functions of the time periods over which these values are calculated.

The United Nations climate change panel, known as the Intergovernmental Panel on Climate Change (IPCC), an intergovernmental organization charged with evaluating the risk of climate change caused by human activity, expresses CE in billions of tonnes of CO_2e. In industry, CE units are often expressed in millions of metric tons of CO_2e; for vehicles, units are in grams of CO_2e per kilometer. For example, the GWP for methane gas over one hundred years is 25; the equivalent value for nitrous oxide is 298. This means that the emission of 1 million metric tons of methane or nitrous oxide has the equivalent ability to contribute to global warming as 25 million or 298 million metric tons of CO_2, respectively. Over the same time span, the hydrofluorocarbons fluoroform (CHF_3, HFC-23) and 1,1,1,2-Tetrafluoroethane ($C_2H_2F_4$, HFC-134a), shown to have large negative effects on the environment, have GWPs of 14,800 and 1,430, respectively.

—*Rena Christina Tabata*
—*To N. Nguyen*

FURTHER READING

Bradley, Raymond S. *Paleoclimatology: Reconstructing Climates of the Quaternary.* Academic Press, 2015. A definitive introduction to the science of paleoclimatology, and the use of climate proxies.

Hyman, Andrew. *Principles of Paleoclimatology.* Callisto Reference, 2017.

Mann, Michael E. *The Hockey Stick and the Climate Wars. Dispatches from the Front Lines.* Columbia University Press, 2014. This very readable book lays out the science behind global climate change and identifies the conflict between climate change proponents and deniers.

Reay, David *Greenhouse Gas Sinks.* CAB International, 2007. Covers the importance of carbon dioxide, methane and nitrous oxide as greenhouse gases, and describes the function of sinks for each gas.

Trabalka, J.R. And Reichle, D.E., eds. *The Changing Carbon Cycle. A Global Analysis.* Springer Science + Business Media, 2016. The various chapters in this book present scientific discussions of as many aspects of the study of atmospheric carbon dioxide.

CARBON DIOXIDE IN EARTH'S ATMOSPHERE

FIELDS OF STUDY

Atmospheric Chemistry, Atmospheric Science; Heat Transfer; Environmental Chemistry; Geochemistry; Oceanography; Photochemistry; Environmental Sciences; Environmental Studies; Bioclimatology; Waste Management; Meteorology; Climatology; Climate Modeling; Chemical Kinetics

SUMMARY

Carbon is the principal building block of life on Earth; all plants and animals are composed largely of carbon. Carbon dioxide (CO_2) emission is a normal part of the respiration cycle. It is also emitted from the burning of plant or animal material. The burning of fossil fuels for energy causes CO_2 to be emitted into the atmosphere, and forest clearing also results in a net increase in atmospheric CO_2. Once in the atmosphere, CO_2 acts as a greenhouse gas (GHG), retaining heat in the atmosphere and contributing to global warming. Presently, CO_2 is the most common GHG in the atmosphere, and its atmospheric concentration is increasing. The proper regulation of human activities that exacerbate this situation is therefore the focus of a good deal of attention.

PRINCIPAL TERMS

- **anthropogenic:** produced as a result of human activities
- **fossil fuel:** combustible materials formed by pressure on plant and animal material over time that are now used as fuels, including coal, oil and natural gas
- **greenhouse gases (GHGs):** gases that act to retain heat within Earth's atmosphere, tending to increase average global temperatures

FOSSIL FUELS

Fossil fuels are the remnant of plant and animal matter that has collected over many millions of years, slowly being converted to coal, crude petroleum and natural gas. All fossil fuels are composed primarily of carbon, with small amounts of other elements reflecting their presence in the proteins and other tissues of their once-living sources. The burning of fossil fuels constitutes the major anthropogenic contribution of CO_2 in the atmosphere. Approximately 57 percent of GHGs consist of CO_2 generated by burning fossil fuels. Another 17 percent is due to the decay of biomass and deforestation. The Intergovernmental Panel on Climate Change (IPCC) has stated with a high degree of confidence that human activities since 1750 have contributed to the addition of these and other GHGs to the atmosphere. When Charles Keeling began measuring the accumulation of CO_2 in the atmosphere from an observatory atop Mauna Loa, Hawaii, in 1958, his measurements indicated that the concentration of

U.S. Anthropogenic GHG Emissions, 2006

Greenhouse Gas	Amount (kilograms of CO_2 equivalent)	Percent of Total GWP
Energy-related carbon dioxide	5,825.5	82.3
Other carbon dioxide	108.8	1.5
Methane	605.1	8.6
Nitrous oxide	378.6	5.4
All other Kyoto gases*	157.6	2.2

Data from U.S. Energy Information Administration.
*Hydrofluorocarbons, perfluorocarbons, and sulfur hexafluoride.

CO_2 in the atmosphere was 315 parts per million. By 2005, the measured concentration of CO_2 was 379 parts per million, which exceeded the natural range of atmospheric CO_2 levels over the last 650,000 years. The CO_2 content of the atmosphere, derived from ice-core data, has varied over time by 10 parts per million around a mean value of 280 parts per million.

Industrialization was initially fueled by coal in the 18th and 19th centuries. Coal continued to be a major source of energy for much of the twentieth century, but oil and natural gas also were used in increasing amounts as energy sources. Gasoline, derived from oil, fueled the growth of automobile culture in many parts of the world in the twentieth century. One thing that all of these energy sources have in common is that they are fossil fuels. Combustion of fossil fuels produces CO_2, along with smaller quantities of other GHGs. Much of industrial civilization in the early twenty-first century continues to be powered by fossil fuels.

Energy sources that do not generate CO_2 are becoming available, and more efficient energy sources and automobiles are being built that generate less CO_2. However, it is nonetheless true that each atom of carbon that is used as fuel produces a corresponding molecule of CO_2, and CO_2 will continue to be a by-product of industrial society for years to come. Controlling the impact of CO_2 on the global climate will require different approaches to energy generation and consumption, as well as new technologies such as carbon sequestration that remove carbon from the atmosphere.

Land Use

Human use of the land has led to the accumulation of CO_2 (and methane, another GHG) in the atmosphere for the past several thousand years. The growth of population in the last two hundred years has magnified this impact. Clearing forested land for agriculture has generally led to the burning of much of the cleared vegetation. As noted above, combustion of carbon-based entities produces CO_2 as a by-product. The decay of biomass, often generated by agriculture, also produces CO_2, although the biomass produced annually was also formed annually by removing the same amount of CO_2 from the atmosphere. As societies have increased their use of metals over time, metal smelting, first using wood for its energy source, then coal, has also contributed CO_2 to the atmosphere.

Some of the anthropogenic CO_2 has been fixed in the oceans and wetlands as peat, rather than in the atmosphere. As wetlands are drained for other uses, however, this carbon sink is diminished, so that more carbon enters the atmosphere. Population pressure and the demands for agricultural products drive land clearing and wetland degradation in many parts of the world. Land clearing continues to increase dramatically in some parts of the world, such as the Amazon basin in Brazil.

Who Produces CO_2?

The industrial nations of the world have been the major producers of CO_2 over time. As each nation has industrialized, it has begun burning fossil fuels extensively, as well as clearing land for agriculture. Industrialized countries continue to generate most of the world's CO_2, either directly through energy generation and automobile use or indirectly by their demand for agricultural products from other areas, which leads to further land clearing and tillage in those areas and by the additional transportation required. Nations that are rapidly industrializing, such as China and India, are doing so largely through the use of fossil fuels. Although the United States is currently the largest consumer of fossil fuels, China is poised to move into second place and will probably surpass the United States by 2030 as a producer of GHGs. Even less-industrialized nations make extensive use of fossil fuels.

Dealing with CO_2 emissions in the future will require concerted efforts by industrialized countries. Less industrialized nations cannot be expected to forgo the economic progress generated by industrialization, but they too will have to manage their carbon footprints over time.

Context

The major driver of what is called the greenhouse effect is CO_2. Some estimates of the amount of CO_2 that will be in the atmosphere by 2100 are as high as 1,000 parts per million if emissions grow unchecked. Such a concentration could produce a global temperature increase of 5°C or more, a change not seen for several million years. Even if the growth of CO_2 in the atmosphere is brought somewhat under control, its concentration could still reach 440 parts per million, which could produce a temperature increase of as much as 3°C. Controlling the rate of carbon emissions into the atmosphere without harming economic well-being will be a challenge. Failure to control the growth of carbon emissions will lead to what several authorities consider to be a much different and undesirable life for much of the planet. The production of CO_2, which has been the hallmark of economic progress, may lead to economic decay if it is not checked.

—*John M. Theilmann*

BIBLIOGRAPHY

Alley, Richard B. *The Two-Mile Time Machine. Ice Cores, Abrupt Climate Change and Our Future* Princeton University Press, 2014. Describes the analytical methods and interpretations of physical clues of past climates trapped in ice that is thousands of years old.

Archer, David and Pierrehumbert, David, eds. *The Warming Papers. The Scientific Foundation for the Climate Change Forecast.* John Wiley & Sons, 2011. The role of atmospheric carbon dioxide and its role in climate regulation are discussed in a number of contexts in this book.

Conkling, Philip W, et al. *The Fate of Greenland: Lessons from Abrupt Climate Change*. MIT Press, 2013.

Houghton, John. *Global Warming: The Complete Briefing*. 4th ed. New York: Cambridge University Press, 2009. Excellent, comprehensive discussion of global warming and the role of CO_2 that includes extensive references. Houghton is the former chairman of the Scientific Assessment Working Group of the IPCC.

MacKenzie, Fred T. and Lerman, Abraham *Carbon in the Geobiosphere. Earth's Outer Shell*. Springer, 2006. This book presents descriptions and analyses of the role of carbon dioxide in Earth's history.

Trabalka, J.R. And Reichle, D.E., eds. *The Changing Carbon Cycle. A Global Analysis*. Springer Science + Business Media, 2016. The various chapters in this book present scientific discussions of as many aspects of the study of atmospheric carbon dioxide.

CARBON EQUIVALENT

SUMMARY

The carbon equivalent (CE), similar to carbon dioxide equivalent (CO_2e), is a metric measure used to quantify how much global warming a given quantity of gas can cause (global warming potential, or GWP), using the functionally equivalent concentration of carbon, converted to the amount of carbon dioxide (CO_2) that would have the same GWP as the reference. In contrast to the CE, the CO_2e uses the functionally equivalent CO_2 concentration as the reference (the amount of CO_2 that would have the same GWP). CE is usually given for a specified timescale, effectively expressing the time-integrated radiative forcing. This also differentiates CE from CO_2e, which describes the instantaneous radiative forcing.

SIGNIFICANCE FOR CLIMATE CHANGE

The CE of a gas or a mixture of gases is a metric measure that can be used to compare the GWP of greenhouse gases (GHGs) for the purposes of analysis, computer modeling, and reporting—as was done, for example, in such international treaties as the Kyoto Protocol and the Montreal Protocol. The CE of a gas is calculated by multiplying the mass (in metric tons) by the GWP of the gas and is expressed for a specified timescale. CE values can be converted to CO_2e units simply by multiplying the CO_2 by 3/11 (the ratio of the molecular weight of carbon to that of CO_2). Time is an important factor when comparing CE or GWP values, as they are functions of the time periods over which these values are calculated.

The United Nations climate change panel, known as the Intergovernmental Panel on Climate Change (IPCC), an intergovernmental organization charged with evaluating the risk of climate change caused by human activity, expresses CE in billions of metric tons of CO_2e. In industry, CE units are often expressed in millions of metric tons of CO_2e; for vehicles, units are in grams of CO_2e per kilometer. For example, the GWP for methane gas over one hundred years is 25; the equivalent value for nitrous oxide is 298. This means that the emission of 1 million metric

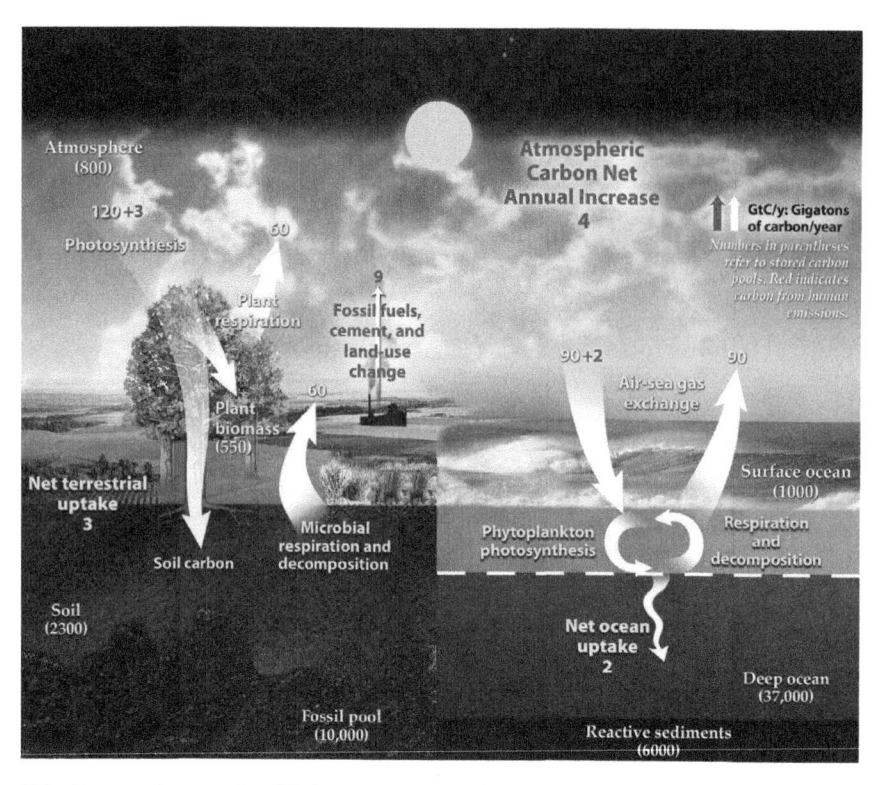

This diagram shows a simplified representation of the contemporary global carbon cycle. Changes are measured in gigatons or carbon per year (GtC/y). Numbers in parentheses refer to stored carbon pools.

tons of methane or nitrous oxide has the equivalent ability to contribute to global warming as 25 million or 298 million metric tons of CO_2, respectively. Over the same timespan, the hydrofluorocarbons fluoroform (HFC-23) and 1,1,1,2-Tetrafluoroethane (HFC-134a)—shown to have large negative effects on the environment—have GWPs of 14,800 and 1,430, respectively.

—*Rena Christina Tabata*

CARBON-OXYGEN CYCLE

FIELDS OF STUDY

Atmospheric Chemistry; Biochemistry; Environmental Chemistry; Geochemistry; Photochemistry; Environmental Sciences; Bioclimatology; Oceanography; Physical Geography; Ecology

SUMMARY

The carbon-oxygen cycle is the process by which oxygen and carbon are cycled through Earth's environment. The cycle includes phenomena such as photosynthesis, which produces oxygen by producing the sugar glucose from carbon dioxide and water, and respiration, which uses oxygen to break glucose down into carbon dioxide and water. The cycle is critical for the homeostasis of the environment.

PRINCIPAL TERMS

- **biosphere:** life on Earth, and the area it inhabits
- **carbon sequestration:** the process of storing carbon in a stable state to negate carbon's effects on climate
- **combustion:** reactions by which oxygen and organic materials become carbon dioxide, water and other oxides

- **decomposition:** process by which organic matter is broken down into its most basic components by microorganisms
- **greenhouse effect:** an atmospheric phenomenon in which certain molecular components of the atmosphere capture infrared radiation from a planet's surface and prevent it from radiating back out into space
- **homeostasis:** the condition in which all components of a multicomponent system are in dynamic balance with each other so that the overall state of the system remains constant
- **hydrosphere:** water on Earth, and its area
- **pedosphere:** the soil
- **photosynthesis:** the process occurring in green plants for the production of glucose from carbon dioxide and water in the presence of chlorophyll and sunlight
- **respiration:** process by which organisms break down glucose and other sugars for energy

THE CARBON-OXYGEN CYCLE

The carbon-oxygen cycle involves the flow of carbon and oxygen through the environment. Carbon is recycled between carbon dioxide and carbon compounds in life-forms. The cycle includes the transfer mechanisms of the carbon compounds and oxygen in the biosphere and in the atmosphere, hydrosphere, and pedosphere. Climatological research is focusing on the rates of processing and the homeostasis of the system to determine carbon dioxide release and capture.

The carbon-oxygen cycle is the result of the storing of carbon from atmospheric and dissolved carbon dioxide as organic material through photosynthesis. The carbon is subsequently released through respiration, combustion, or decomposition, all of which combine the carbon in the organic material with atmospheric or dissolved oxygen as carbon dioxide (CO_2). The CO_2 so released into the atmosphere is reabsorbed by photosynthetic organisms such as cyanobacteria, some protists, and green plants, thus completing the cycle and starting the process anew.

The cycle demonstrates how different parts of Earth's environment and climate are interrelated. For example, analysis of the flow of material from rivers into seas allows scientists to consider the impact of biologic materials. Similarly, the release of

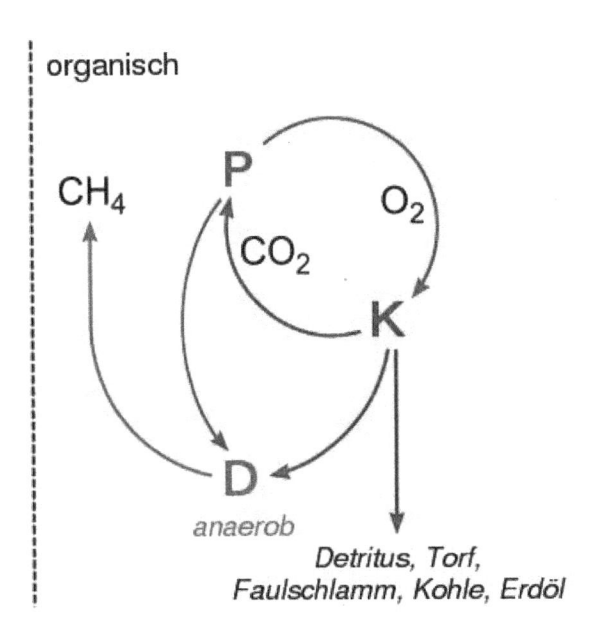

Anaerobic carbon oxygen cycle (P = producer, K = consumer, D = decomposer).

the gases into the atmosphere and the capturing of carbon by plants allow scientists to see the effect of land use on climate.

By examining environmental factors such as land use or propensity for fire in a given area, one can get an idea of how a given region will impact other areas and can estimate the total amount of gases it might contribute to the atmosphere. Additionally, environmental factors provide hints as to effective methods of managing or sequestering carbon. Forests and trees are typical examples of carbon sequestration.

Based on the carbon-oxygen cycle, it is known that trees take in a certain amount of carbon over time, that the trees will produce living space for some number of animals, and that those animals will produce a certain amount of carbon dioxide throughout their lives and when they die and decompose. Also factored is whether a forest is prone to fires. By looking at such factors, one can get an idea of how much carbon dioxide a forest can remove from the atmosphere and for how long, and if there is anything humans can do to make that forest more efficient.

Most of the oxygen released into Earth's atmosphere is produced by oceanic plankton, whereas forests of all kinds absorb huge amounts of carbon as sites of oxygen turnover.

PROCESSES

Photosynthesis is a set of chemical reactions that occur in green plants and in some bacteria and some protists. Photosynthesis uses energy from the sun to convert carbon dioxide and water into the sugar glucose and oxygen in the presence of chlorophyll.

The process itself is the interaction of light-dependent reactions and the Calvin or dark cycle. The light-dependent reactions separate the hydrogen from the oxygen in water and are thus the oxygen-producing portion of the cycle and prepare reactants needed for the other cycles. The Calvin or dark cycle turns the carbon dioxide and light cycle products into sugar. The light cycle runs directly on captured photons, whereas the Calvin cycle does not depend directly upon sunlight.

Along with rates of release from combustion, which is the conversion of organic material to carbon dioxide and water vapor, it is also necessary to factor in rates of respiration and decomposition. Respiration is how organisms get energy from organic compounds, such as sugars. Plants and animals use aerobic respiration, which evolved after oxygen became prevalent in the atmosphere. Aerobic respiration occurs in special organelles of eukaryotic cells called ribosomes. Here sugars are broken down into carbon dioxide and water in a process known as the Krebs or citric acid cycle.

Decomposition releases various greenhouse gases, as decomposing organisms are often excellent habitats for anaerobic life-forms. Anaerobic respiration releases such gases as carbon dioxide, hydrogen sulfide, and methane. Thus the total impact of the disruption of an environment, such as the burning of a forest, is not only in the combustion of the trees but also in the drop in photosynthesis and decomposition of any detritus.

HISTORY OF THE CYCLE

The understanding of the carbon-oxygen cycle has existed since the eighteenth century, after the discoveries of Antoine Lavoisier (1743 – 1794), who found that respiration was a process similar to burning, and of Joseph Priestley (1733 – 1804), who discovered oxygen and who also conceptualized the relationship between blood and air. The full mechanics of the carbon-oxygen cycle and its implications, however, were discovered by scientists in subsequent centuries.

The workings of the chemical reactions of photosynthesis were discovered in the 1940's, and research continues into the role of quantum physics in the light-harvesting stage of photosynthesis. With the growing awareness of climate change, more work is being done to examine the cycle.

Current research often takes samples from the area in question, be they soil samples to estimate bacterial carbon production or air samples to measure carbon levels at a given site. These samples are then analyzed together to get a picture of how each part is affecting the overall carbon-oxygen balance for that area. The carbon data are often mixed with data on other climatological factors, such as albedo, humidity, and temperature, to build a picture of the effects on climate.

ATMOSPHERIC COMPOSITION AND THE GREENHOUSE EFFECT

The carbon-oxygen cycle affects climate because it affects Earth's atmosphere. Ignoring water vapor, which is variable, it is known that the standard dry atmosphere is 78.09 percent nitrogen, 20.95 percent oxygen, 0.93 percent argon, and 0.039 percent carbon dioxide; the rest of the atmosphere comprises various trace gases. Water vapor composes, on average, about 1 percent of the atmosphere.

The carbon-oxygen cycle plays an important role in maintaining homeostasis in the atmosphere. The cycle "cycles" air and carbon to maintain equilibrium. The gases in the atmosphere and their relative amounts have an important impact upon climate. Some gases absorb infrared rays reflected or emitted from the earth's surface and send them back to Earth and into the lower atmosphere, thus trapping the energy and raising Earth's overall average temperature. This process is known as the greenhouse effect.

Gases that are best at absorbing and emitting in the infrared range are known as greenhouse gases and include water vapor, carbon dioxide, methane, nitrous oxide, and ozone. Water vapor and carbon dioxide make up the largest contribution to the effect, with methane and ozone playing a smaller role. Many of these gases are produced in industrial reactions such as combustion. Combustion from cars and factories and from other human activities since the start of the Industrial Revolution in the 18th century have been major factors in anthropogenic, or human-caused, climate change. It is here that the carbon-oxygen cycle comes into play. There is a maximum rate at which a particular environment can cycle the carbon dioxide into oxygen, meaning that excess carbon dioxide builds up in the atmosphere, enhancing the greenhouse effect.

By studying the plants and animals that are part of the process, scientists can find the effects of increased temperature on those components. The environmental impact, or increased environmental stress, leads to less capacity to handle the increased carbon dioxide. This creates a positive feedback loop that further increases the temperature. Increased global temperature could cause increased cloud cover, which would increase Earth's albedo. Increased temperature also could spawn algal blooms in the oceans, which would absorb carbon dioxide and thus reach a new equilibrium. Such events are indicative of a large amount of environmental damage.

The greenhouse effect is not wholly negative, however. It is actually required for life to exist on this planet. In the absence of greenhouse gases, Earth would have a significantly colder climate than it does now. The average temperature of the planet in the absence of the naturally-occurring greenhouse effect has been estimated to be a balmy -18°C. Additionally, Earth's atmospheric composition has varied throughout its history. Earth's early atmosphere was very different from that of today. It consisted mostly of water vapor, carbon dioxide, hydrogen sulfide, and ammonia. All of the greenhouse gases of the time kept the average Earth temperature fairly high.

Once photosynthetic life evolved about 3.5 billion years ago, the oxygen it produced began to break down the methane and carbon dioxide in the atmosphere, and the level of oxygen began to increase. This resulted in what is called the oxygen catastrophe, in which many of the non-aerobic organisms at the time became extinct, leaving those that adapted to the new situation by developing aerobic respiration. This development had a great impact upon later life, as aerobic respiration became far more efficient.

The oxygen catastrophe also may have caused the Huronian glaciation event, which occurred most likely as a result of greenhouse gas depletion by photosynthetic life-forms. The event was one of the longest glaciation periods in Earth's history, lasting some 300 million years. It and the later Snowball Earth periods that likely caused the evolution of multicellular life and the Cambrian explosion were ended only by the buildup of greenhouse gases from volcanic eruptions and by the weathering of rocks.

Since this time, greenhouse gas levels have oscillated by period; for example, during the Devonian period, carbon dioxide was higher and oxygen was lower than present levels. In the Cretaceous period, both carbon dioxide and oxygen levels were higher. During both periods, the average temperature was higher than it is today.

The Permian-Triassic mass extinction event of 250 million years ago, in which a shutdown of the carbon dioxide cycle may have been responsible for the greatest mass extinction in Earth's history, is a dramatic example of how the carbon-oxygen cycle figures into climate. While the exact causes remain unknown, the mass extinction is thought to have begun with the eruption of the Siberian Traps, a massive set of volcanoes that released massive quantities of carbon dioxide and debris into the atmosphere. This largest eruptions in known history are believed to have caused a small period of global cooling, which stressed many environments. As a result, many trees died before they could process the carbon dioxide that was added to the atmosphere. Evidence seems to show that global warming set in around the peak of the extinction, and it is thought that the oceans became saturated with carbon dioxide, to the point in which there was a degassing event. The seas released their carbon in huge clouds that swept over the land, suffocating all nearby creatures. Massive erosion suggests that the climate became arid and the interior of continents were more or less lifeless. While another event on this scale is unlikely, it provides a strong example of the kind of phenomena that can arise from the disruption of the carbon-oxygen cycle.

GREENHOUSE GAS EMISSIONS

More than 3 million tonnes of carbon dioxide are released into the atmosphere each year. Compounding that are ongoing releases of other gases such as methane, which has twenty-five times the effect of carbon dioxide as a greenhouse gas in a one-hundred-year period.

The largest sources of anthropogenic emissions are deforestation and industrial technology, including factories, cars, and cement making (which releases carbon dioxide). However, agriculture also releases greenhouse gases. Cattle, for example, are estimated to release 16 percent of anthropogenic methane, although the effect of millions of cattle is unlikely to be very different from that of the millions of bison that previously roamed the prairies of North America naturally.

Humans have always released carbon dioxide into the environment. Various indigenous North American cultures burned forests annually to encourage habitats for large game animals. Also, much of Europe has been deforested for agricultural space and shipbuilding. However, the emissions from industrial and postindustrial technology are on a wholly greater scale. Equally important is that the carbon being released now has not been part of the carbon-oxygen cycle since it was stored in the form of coal and petroleum than 250 million years ago. Thus, with the combination of environmental disruption and greenhouse gas emission, recent activities are more akin to a sustained volcanic eruption than to any other event on Earth.

As the system moves further from homeostasis, the compensatory mechanisms, such as increased plant growth, are overloaded or are not given the chance to function, as is the case with heavy deforestation. This forms positive feedback, pushing the environment further from equilibrium. Something as simple as planting additional trees can act to forestall the harmful trend, however. An important point is that the carbon in a living tree has not been expelled; it remains stored as long as the tree lives. Once the tree dies the carbon dioxide is released, either slowly through natural decomposition of the cellulose that makes up the wood, or rapidly through combustion of the wood.

Carbon Sequestration

Carbon sequestration involves storing carbon in a stable state to negate carbon's harmful effects on the environment. As mentioned, trees serve as reservoirs for carbon. Many efforts focus on manipulating the carbon-oxygen cycle to increase the amount of carbon stored. Means of sequestering carbon range from genetically modifying plants so that they carry out a more efficient form of photosynthesis, to timing forest harvesting and replanting to ensure peak carbon storage, as in the Canadian lumber industry. Other forms of carbon sequestration under consideration include injecting carbon into worked out coal seams; sinking carbon to the bottom of the ocean through algae growth, where it would presumably eventually become petroleum or coal once more in time; producing certain carbon-absorbing compounds; and developing processes that utilize carbon dioxide as a raw material. With proper manipulation of the carbon-oxygen cycle through methods such as carbon sequestration, many of the worst effects of climate change can be mitigated.

—*Gina Hagler*

Further Reading

Bashkin, V. N., and Robert W. Howarth. *Modern Biogeochemistry.* Kluwer Academic, 2002. An excellent detailed description of the carbon-oxygen cycle, its precise mechanics, means for quantitative analysis, the development of the system, and its relationship to other cycles in the environment. Most suitable for college students and other readers with some background in chemistry and biology.

Kirby, Richard R. *Ocean Drifters: A Secret World Beneath the Waves.* Studio Cactus, 2010. Focuses on plankton, the life-forms responsible for most of the oxygen release on Earth. Includes enlarged photographs. Recommended for general readers.

Kondratyev, K. Y., V. F. Krapivin, and Costas A. Varotsos. *Global Carbon Cycle and Climate Change.* Springer, 2003. Detailed analysis of the relationship of the carbon-oxygen cycle to climate change. Includes the effects of other greenhouse gases and how it relates with Earth's environment. Recommended for those with some background in the subject.

Le´vêque, C. *Ecology from Ecosystem to Biosphere.* Science, 2003. Contextualizes the carbon-oxygen cycle in its ecological role, which allows readers to see how the cycle fits with the larger functioning of the ecosystem and biosphere. Suitable for general readers.

Vallero, Daniel A. *Environmental Biotechnology: A Biosystems Approach.* Academic, 2010. Examines the carbon-oxygen cycle in the context of environmental biotechnology. Systems based, it deals with the larger scale and interactions of the cycle.

Walker, Sharon, and David McMahon. *Biochemistry Demystified.* McGraw-Hill, 2008. An accessible introduction to the processes behind respiration and photosynthesis that underlie the carbon-oxygen cycle. Written for beginners. Also covers biochemistry.

CARBON SINK

FIELDS OF STUDY

Atmospheric Science; Atmospheric Chemistry; Geochemistry; Oceanography; Bioclimatology; Environmental Sciences; Environmental Studies; Ecology; Tree Ring Analysis; Process Modeling; Physical Geography; Ecosystem Management; Earth System Modeling

SUMMARY

A sink is a process or mechanism that removes or absorbs carbon dioxide (CO_2) and other greenhouse gases (GHGs) from the atmosphere. The 1992 U.N. Framework Convention on Climate Change (UNFCCC) defines a sink as any process, activity, or mechanism that removes or absorbs a GHG, aerosol, or precursor of a GHG from the atmosphere.

PRINCIPAL TERMS

- **anthropogenic:** produced as a result of human activities
- **fossil fuel:** combustible materials formed by pressure on plant and animal material over time that are now used as fuels, including coal, oil and natural gas
- **greenhouse gases (GHGs):** gases that act to retain heat within Earth's atmosphere, tending to increase average global temperatures
- **photosynthesis:** the biochemical process occurring in green plants by which atmospheric carbon dioxide and water are converted into glucose in the presence of chlorophyll and sunlight

How Sinks Function

For a process to be a sink, in accord with the UNFCCC definition, it must be a net sink. That is, if it also releases GHGs into the atmosphere, it must remove more GHGs than it emits. Otherwise, it would count as a source of GHG emissions. The most common sinks are carbon sinks, which include oceans and terrestrial ecosystems such as forests, grasslands and soils. Carbon sinks remove carbon from the atmosphere, primarily as carbon dioxide, and store it. In the latter respect, they are also referred to as reservoirs, since the carbon stored by them can also be released from them at a later time.

In mid-February, the crew of the R/V Roger Revelle came nose-to-nose with an Antarctic iceberg. NOAA researchers are participating in a research effort cosponsored by NOAA and the National Science Foundation.

Forests, grasslands and oceanic algae act as carbon sinks through the process of photosynthesis. Ocean waters act as a sink when atmospheric CO_2 dissolves in ocean surface waters and is stored there. This continues until the surface waters are saturated, at which point the rate of CO_2 uptake declines. The CO_2 remains near the surface until the oceans turn over, which happens in cycles of about one thousand years. During this continuous overturning process, the surface waters move downward, carrying with them the dissolved carbon. This enables the oceans to continue to absorb carbon from the atmosphere.

Significance for Climate Change

The most important anthropogenic GHG is CO_2, and the amount of CO_2 in the atmosphere has increased by about 35 percent in the industrial era, mainly by human activities. However, the increase in CO_2 in the atmosphere is less than the increase in CO_2 emissions. This is because, of the approximately 545 billion tonnes of carbon released into the atmosphere by human activity from 1870 to 2014, only about 40 percent has remained in the atmosphere. The rest has been absorbed by land and ocean carbon sinks. Without these carbon sinks, the amount of CO_2 in the atmosphere, the principal cause of anthropogenic climate change, would be considerably higher than it is currently.

Terrestrial ecosystems of soil and vegetation currently act as a net global sink for carbon. They are, however, also potentially major sources of GHG emissions. Over the years, deforestation has contributed an estimated 30 percent to GHG concentration. When trees are cut down, the atmosphere is affected in two ways: The trees, which also act as reservoirs of CO_2, are enabled to release the stored carbon into the atmosphere, and the CO_2 that would otherwise have been removed from the atmosphere by these trees remains in the atmosphere. These two effects of deforestation on the whole tend to increase the warming of the climate.

It is thought that modifying carbon sinks to enhance their carbon uptake would have the effect of slowing the rate of climate change. By reducing the rate of deforestation or by planting more trees (afforestation or reforestation), Earth's carbon sinks can be expanded. Under the Kyoto Protocol's clean development mechanism (CDM), afforestation and reforestation projects are eligible to earn "carbon credits" that are used to offset a portion of the industrial restrictions on CO_2 emissions. This provides an incentive to increase the planting of trees and hence to enhance Earth's reserve of carbon sinks.

—*Tomi Akanle*

Further Reading

Ashton, Mark S., Tyrrell, Mary S., Spalding, Deborah and Gentry, Bradford, eds. *Managing Forest Carbon in a Changing Climate.* Springer Science + Business Media, 2012. Provides a close examination of the role of forests and forest management in the regulation of atmospheric carbon dioxide.

Cohen, Stewart J. and Waddell, Melissa W. *Climate Change in the 21st Century.* McGill-Queens University Press, 2009. Discusses many aspects of climate change, with emphasis on the role and function of natural and artificial greenhouse gas sinks.

Nelleman, Christian, Corcoran, Emily, Duarte, Carlos M., Valdes, Luis, DeYoung, Cassandra, Fonseca, Luciano and Grimsditch, Gabriel, eds. *Blue Carbon. The Role of Healthy Oceans in Binding Carbon.* UNEP/Earthprint, 2009. This "rapid response report" examines the role of effective management of ocean ecosystems as carbon sinks for mitigating climate change.

Rackley, Stephen A. *Carbon Capture and Storage.* 2nd ed., Elsevier/Butterworth-Heinemann, 2017. An in-depth presentation of the methodologies of artificial carbon sink technologies.

Reay, David *Greenhouse Gas Sinks.* CAB International, 2007. Covers the importance of carbon dioxide, methane and nitrous oxide as greenhouse gases, and describes the function of sinks for each gas.

Reay, David *Nitrogen and Climate Change: An Explosive Story.* Springer, 2015. Discusses the role of nitrogen in climate change and of the oceans as a sink for nitrogen oxides and other greenhouse gases.

Sarmiento, Jorge L. and Gruber, Nicolas *Ocean Biogeochemical Dynamics.* Princeton University Press, 2013. Describes the principal mechanisms by which the oceans act as a sink for carbon and other materials.

CATASTROPHIST-CORNUCOPIAN DEBATE

FIELDS OF STUDY

Climatology; Ecology; Environmental Sciences; Environmental Studies; Physical Sciences; Atmospheric Science; Earth System Modeling

SUMMARY

Two schools of thought dominate the worldviews of the climate discussion, and of society in general. One sees a future in which the inevitable result of increasing population and the associated demands on the environment is the catastrophic failure of human society. The other sees a future in which technology provides the means to solve all problems facing this planet.

PRINCIPAL TERMS

- **catastrophist:** someone who holds that human society leads to or advances by sudden drastic events rather than by a smooth evolutionary process
- **Malthusian:** related to the inevitability of the negative future of technological societies
- **Promethean:** related to the inevitability of a positive future for technological societies

Optimism and Pessimism

Many people are pessimistic about Earth's future when they consider the myriad environmental problems facing today's world. In the early nineteenth century, Thomas Robert Malthus (1766–1834) predicted a dismal future of overpopulation and mass starvation. Neo-Malthusian environmentalists foresee a catastrophic future in which too many humans battle for ever-dwindling resources, leading to vice, misery, and the collapse of civilization.

On the other hand, optimists believe that technology is a cornucopia that, like the mythical horn of plenty, will provide an abundance of ingenious new cures for the world's environmental problems. This Promethean environmentalism, named for the titan who gave humans the gift of fire in ancient Greek mythology, argues that past innovative technologies have repeatedly averted predicted disasters with new inventions. It asserts that this historic pattern of progress and abundance will continue indefinitely into the future.

Thomas Malthus.

Significance for Climate Change

According to catastrophists, the disasters accompanying global warming are inevitable unless drastic changes in human society and behavior are implemented immediately. Even then, they hold that it may already be too late to avoid an impending doom. Although excessive dwelling on future disasters can become a self-fulfilling prophecy, if nothing is done the feared consequences are more likely to occur. It is not unreasonable for environmentally concerned people to feel moral indignation over abundant excesses, abuses, and needless waste, but attempting to shock or shame people into altering such behaviors is often futile. Progressive, positive action is seldom motivated by fear alone. If the predicted disasters fail to materialize, the public may assume a false sense of complacency.

On the other hand, prometheans tend to emphasize historical precedents for the solution of environmental problems. Although this approach is comfortable, it can lead the public into a false sense of security by causing people to expect technology to fix all problems without any changes to the human behaviors that have caused the problems. Blind faith in technology then becomes an excuse for continuing behavior that exacerbates existing problems.

A balanced viewpoint exists between these two extremes. This worldview recognizes that there are serious environmental problems facing the world, but asserts that obstacles can be conquered when faced openly and creatively. Such a viewpoint embraces the cornucopian belief that solutions to all problems are possible, but it also embraces the catastrophist belief that the cooperation of nearly every human society and individual will be necessary to achieve those solutions. It rejects the inevitability of environmental doom, but it also rejects the belief that anonymous scientists or inventors will solve problems on humanity's behalf, a belief that excuses individuals from working to solve those problems themselves.

—*George R. Plitnik*

FURTHER READING

Epstein, Alex *The Moral Case for Fossil Fuels.* Portfolio/Penguin, 2014. An unabashed and unapologetic discussion of the inevitable victory of technological advances in controlling climate change.

Jouzel, Jean, Lorius, Claude and Raynard, Dominique *The White Planet. The Evolution and Future of Our Frozen World.* Princeton University Press, 2013. Examines the past, present and future roles of ice on Earth, concluding that climate change and global warming will ultimately have disastrous consequences for human societies on Earth regardless of technological advances.

Marris, Emma *Rambunctious Garden: Saving Nature in a Post-Wild World.* Bloomsbury Publishers, 2011. Describes as an "optimistic book," the author promotes an approach of living with climate change and allowing wild growth to expand into all available niches, rather than attempting to maintain natural wilderness areas in their pristine state by technological intervention to control climate change.

Yuen, Eddie "The Politics of Failure Have Failed: The Environmental Movement and Catastrophism." Chapter in Lilley, Sasha, McNally, David, Yuen, Eddie and Davis, James *Catastrophism: The Apocalyptic Politics of Collapse and Rebirth*, PM Press, 2012. A philosophical analysis of the use of catastrophic "scare tactics" in an effort to effect political change in regard to climate change.

CHLOROFLUOROCARBONS

FIELDS OF STUDY

Atmospheric Chemistry; Organic Chemistry; Photochemistry; Physical Chemistry; Environmental Chemistry

SUMMARY

Developed in the 1930s as refrigerants, chlorofluorocarbons (CFCs) gained rapid acceptance in the 1940s. New uses were found for them as aerosol propellants, blowing agents, solvents, fire suppressants, and inhalation anesthetics. Production climbed, reaching as high as 566,591 tonnes in the United States by 1988. In 1971, it was shown that CFCs had accumulated in the atmosphere, and by 1974 a relationship was demonstrated between atmospheric CFCs and depletion of the ozone layer. Over the next twenty years, manufacture and use of CFCs were drastically reduced to protect the ozone layer from further harm.

PRINCIPAL TERMS

- **Freon:** trade name for CFCs made by DuPont chemical company
- **greenhouse effect:** a phenomenon in which gas molecules within Earth's atmosphere absorb infrared heat energy from the surface and retain it in the atmosphere, preventing it from radiating away into space
- **halocarbons:** the general family of compounds that includes CFCs, HCFCs, HFCs, and other molecules in which carbon atoms are bonded to halogen atoms

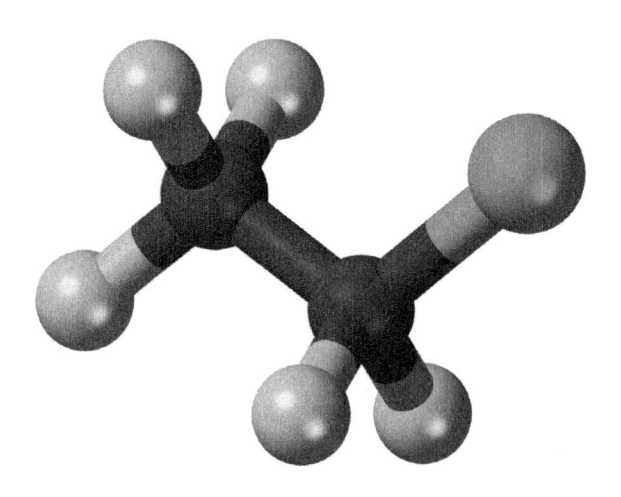

Ball-and-stick model of the chloropentafluoroethane molecule, a CFC refrigerant, now phased out due to ozone layer damage. Color code for image: black is carbon (C); yellow-green is fluorine (F); green is chlorine (Cl).

- **halon:** a compound containing bromine, carbon, chlorine, and fluorine
- **Kyoto Protocol:** a 1997 international agreement to limit greenhouse gas emissions
- **Montreal Protocol:** a 1987 international agreement to phase out the manufacture and use of ozone-depleting chemicals, especially CFCs
- **ozone layer:** the portion of the Earth's stratosphere (from 10–50 kilometers altitude) where ozone has formed and absorbs dangerous ultraviolet radiation from the Sun

Fate of Halocarbons in the Atmosphere

Chlorofluorocarbons (CFCs), although denser than air, readily mix throughout the atmosphere and eventually reach the stratosphere (10–50 kilometers in altitude). Although CFCs have low chemical reactivity (and hence long lifetimes) in the lower atmosphere, in the stratosphere they encounter and absorb energetic ultraviolet radiation, resulting in the liberation of free chlorine atoms. These chlorine atoms can act as catalysts for the destruction of stratospheric ozone molecules. Because this process is catalytic, each individual chlorine atom can lead to the destruction of thousands of ozone molecules. Other volatile chlorine compounds such as chloromethane, dichloromethane, chloroform and carbon tetrachloride can also liberate destructive chlorine atoms. Bromine atoms are also destructive of ozone.

Bromine-containing compounds include the halons, used as fire suppressants and inhalation anesthetics, and methyl bromide, a soil fumigant and natural product of sea organisms.

In 1971, James E. Lovelock, using a sensitive detector, found traces of CFCs in air samples from different parts of the world. F. Sherwood Rowland and Mario Molina realized the destructive potential of CFCs with regard to ozone and raised concern among industrialists and politicians that led to the signing of the Montreal Protocol in 1987. Because of the long lifetimes of CFCs, however, even in the absence of further production many years will be required for the existing pollutants to dissipate and allow the ozone layer to recover. This is time would be greatly extended in the event that the restrictions agreed to in the Montreal Protocol were to be ignored, and unfortunately this is not only a possibility, but has become a reality. Environmental scientists reported in May, 2018, that a 25 percent increase in emissions of the compound CFC-11 has been detected since 2012, indicating that the compound is being manufactured somewhere in quantities well beyond those reported to the United Nations under the provisions of the Montreal Protocol. Known by its chemical name of trichlorofluoromethane (CCl_3F), CFC-11 was once widely used as a component of insulating foams before its use and mass production were banned.

Substitutes for CFCs

Phasing out CFCs meant that substitutes were needed for applications in refrigeration, air conditioning, and aerosols. The ideal substitute would be nontoxic, nonflammable, noncorrosive (like a CFC), and of suitable physical properties (such as boiling point and heat of vaporization), but without the ability to destroy ozone like the CFCs. Attention naturally focused on the related compounds hydrochlorofluorocarbons (HCFCs) and hydrofluorocarbons (HFCs). The HCFCs, although they contain chlorine, tend to be destroyed by chemical reactions in the lower atmosphere before they reach the ozone layer. HFCs, which contain no chlorine, have negligible ozone destructiveness even should they reach the stratosphere. In addition to adopting these new compounds, the existing stocks of CFCs in abandoned equipment had to be trapped and either recycled or disposed of in an environmentally acceptable manner, rather than simply being vented into the atmosphere as they always had been.

CONTEXT

The impact of CFCs and their related compounds is mainly on the ozone layer, because of their catalytic effect. CFCs are also potent greenhouse gases but are present in the atmosphere at such low levels (0.1–0.5 parts per billion) that their contribution to the total greenhouse effect is very small. Loss of ozone in the stratosphere also affects temperature in complex ways, warming some parts of the atmosphere and cooling others. Nevertheless, the sheer numbers of individual halocarbons, even at low levels individually, add up to a warming potential that is worth controlling. Atmospheric levels of substances controlled by the Montreal Protocol have declined a great deal. Not surprisingly, the levels of their substitutes have risen. This trade-off is good for the ozone layer, but it leaves much to be accomplished in regard to global warming.

—*John R. Phillips*

FURTHER READING

Andersen, Stephen O. and Sarma, K. Madhavi *Protecting the Ozone Layer. The United Nations History.* UNEP/Earthscan, 2012. In this book, the authors provide a detailed history and discussion of the effects of chlorofluorocarbons and related compounds on the stratospheric ozone layer, and of the efforts made to curtail the damage.

Benedick, Richard Elliot *Ozone Diplomacy. New Directions in Safeguarding the Planet.* Harvard University Press, 2009. This enlarged edition provides a deep historical account and analysis of the meshing of science and diplomacy in addressing the problem of depletion of the ozone layer by chlorofluorocarbons.

Cumberland, John H., Hibbs, James R. and Hoch, Irving, eds. *The Economics of Managing Chlorofluorocarbons. Stratospheric Ozone and Climate Issues.* Routledge, 2016. An extensive examination of the history, uses and environmental effects of chlorofluorocarbons.

Lambright, W. Henry *NASA and the Environment. The Case of Ozone Depletion.* NASA SP-2005-4538, 2005. An interesting case study of the problem of ozone depletion, subdivided into eight recognizable stages.

VanLoon, Gary W. and Duffy, Stephen J. *Environmental Chemistry. A Global Perspective.* 4th ed., Oxford University Press, 2017. A well-composed text book that presents a good discussion of chlorofluorocarbons and related compounds.

CHLOROFLUOROCARBONS AND RELATED COMPOUNDS

FIELDS OF STUDY

Atmospheric Chemistry; Organic Chemistry; Photochemistry; Physical Chemistry; Thermodynamics; Heat Transfer; Chemical Engineering; Environmental Sciences; Environmental Studies; Waste Management; Fluid Dynamics; Chemical Kinetics; Ecosystem Management; Earth System Modeling

SUMMARY

Developed in the 1930s as refrigerants, chlorofluorocarbons (CFCs) gained rapid acceptance in the 1940s. New uses were found for them as aerosol propellants, blowing agents, solvents, fire suppressants, and inhalation anesthetics. Production climbed, reaching as high as 566,591 tonnes in the United States by 1988. In 1971, it was shown that CFCs had accumulated in the atmosphere, and by 1974 a relationship was demonstrated between atmospheric CFCs and depletion of the ozone layer. Over the next twenty years, manufacture and use of CFCs were drastically reduced to protect the ozone layer from further harm.

Hydrofluorocarbons (HFCs) are a family of organic chemical compounds composed entirely of hydrogen, fluorine, and carbon. They are generally colorless, odorless, and chemically unreactive gases at room temperature. HFCs fall under the broader classification of haloalkanes. While HFCs do not harm the ozone layer, they contribute to global warming as greenhouse gases (GHGs) and are considered one of the major groups of high global warming potential (HGWP) gases.

PRINCIPAL TERMS

- **alkanes:** the class of organic compounds that consist solely of carbon and hydrogen atoms in fixed proportions
- **Freon:** trade name for CFCs made by DuPont chemical company
- **global warming potential (GWP):** a measure of the effectiveness of a particular gas as a greenhouse gas relative to carbon dioxide; a GWP of 140 indicates the gas retains 140 times the amount retained by the same mass of CO_2
- **greenhouse effect:** a phenomenon in which gas molecules within Earth's atmosphere absorb infrared heat energy from the surface and retain it in the atmosphere, preventing it from radiating away into space
- **haloalkanes:** alkane compounds in which one or more hydrogen atoms have been replaced in the corresponding molecular structure by a halogen atom (F, Cl, Br or I)
- **halocarbons:** the general family of compounds that includes CFCs, HCFCs, HFCs, and other molecules in which carbon atoms are bonded to halogen atoms
- **halon:** a compound containing bromine, carbon, chlorine, and fluorine
- **Kyoto Protocol:** a 1997 international agreement to limit greenhouse gas emissions
- **Montreal Protocol:** a 1987 international agreement to phase out the manufacture and use of ozone-depleting chemicals, especially CFCs
- **ozone layer:** the portion of the Earth's stratosphere (from 10–50 kilometers altitude) where ozone has formed and absorbs dangerous ultraviolet radiation from the Sun

FATE OF HALOCARBONS IN THE ATMOSPHERE

Chlorofluorocarbons (CFCs), although denser than air, readily mix throughout the atmosphere and eventually reach the stratosphere (10–50 kilometers in altitude). Although CFCs have low chemical reactivity (and hence long lifetimes) in the lower atmosphere, in the stratosphere they encounter and absorb energetic ultraviolet radiation, resulting in the liberation of free chlorine atoms. These chlorine atoms can act as catalysts for the destruction of stratospheric ozone molecules. Because this process is catalytic, each individual chlorine atom can lead to the destruction of thousands of ozone molecules. Other volatile chlorine compounds such as chloromethane, dichloromethane, chloroform and carbon tetrachloride can also liberate destructive chlorine atoms. Bromine atoms are also destructive of ozone. Bromine-containing compounds include the Halons, used as fire suppressants and inhalation anesthetics, and methyl bromide, a soil fumigant and natural product of sea organisms.

In 1971, James E. Lovelock, using a sensitive detector, found traces of CFCs in air samples from different parts of the world. F. Sherwood Rowland and Mario Molina realized the destructive potential of CFCs with regard to ozone and raised concern among industrialists and politicians that led to the signing of the Montreal Protocol in 1987. Because of the long lifetimes of CFCs, however, even in the absence of further production many years will be required for the existing pollutants to dissipate and allow the ozone layer to recover. This is time would be greatly extended in the event that the restrictions agreed to in the Montreal Protocol were to be ignored, and unfortunately this is not only a possibility, but has become a reality. Environmental scientists reported in May, 2018, that a 25 percent increase in emissions of the compound CFC-11 has been detected since 2012, indicating that the compound is being manufactured somewhere in quantities well beyond those reported to the United Nations under the provisions of the Montreal Protocol. Known by its chemical name of trichlorofluoromethane (CCl_3F), CFC-11 was once widely used as a component of insulating foams before its use and mass production were banned.

SUBSTITUTES FOR CFCs

Phasing out CFCs meant that substitutes were needed for applications in refrigeration, air conditioning, and aerosols. The ideal substitute would be nontoxic, nonflammable, noncorrosive (like a CFC), and of suitable physical properties (such as boiling point and heat of vaporization), but without the ability to destroy ozone like the CFCs. Attention naturally focused on the related compounds hydrochlorofluorocarbons (HCFCs) and hydrofluorocarbons (HFCs). The HCFCs, although they contain chlorine, tend to be destroyed by chemical reactions in the lower atmosphere before they reach the ozone layer. HFCs, which contain no chlorine, have negligible ozone destructiveness even should they reach the stratosphere. In addition to adopting these new compounds, the existing stocks of CFCs in abandoned equipment had to be trapped and either recycled or disposed of in an environmentally acceptable manner, rather than simply being vented into the atmosphere as they always had been.

THE HYDROFLUOROCARBONS

Many of the HFCs were developed for use in industrial, commercial, and consumer products as alternatives to ozone-depleting substances such as chlorofluorocarbons (CFCs) and hydrochlorofluorocarbons (HCFCs). Common HFCs, in order of atmospheric abundance, include HFC-23 (fluoroform, CHF_3), HFC 134a (C2H2F4, as 1,1,1,2-Tetrafluoroethane, tetrafluoroethane, R-134a, Genetron 134a, or Suva 134a), and HFC-152a ($C_2H_4F_2$, as 1,1-Difluoroethane, difluoroethane, or R-152a). HFC-23 is used in a wide range of industrial processes and is a by-product of Teflon(TM) production. HFC-134a is primarily used as a refrigerant for domestic refrigeration and automobile air conditioners. HFC-152a is commonly used in refrigeration, electronic cleaning products, and automobile applications as an alternative to HFC-134a.

The global warming potentials (GWPs) of HFCs range from 140 (for HFC-152a) to 11,700 (for HFC-23). These GWPs are significantly lower than those of the gases the HFCs are designed to replace. The atmospheric lifetimes of the same two HFCs are just over 1 year and 260 years, respectively. HFCs are one of two groups of haloalkanes targeted in the Kyoto Protocol. HFC emissions are projected to increase in the coming years, as industry continues to strive to decrease CFC and HCFC production.

HFCs are preferred over CFCs and HCFCs, because HFCs lack chlorine. When CFCs are emitted into the atmosphere, chlorine (Cl) atoms contained in those CFCs become disassociated through interaction with ultraviolet light. The resulting free Cl atoms decompose ozone into oxygen, and regenerated Cl atoms go on to degrade more ozone molecules. This reaction continues for the atmospheric lifetime of the Cl atom, which ranges from one to two years. On average, a single Cl atom destroys 100,000 ozone molecules. Thus, the lack of Cl in HFCs makes them a desirable alternative.

Some studies indicate that excessive exposure to HFCs may affect the brain and heart. However, this has only been established for concentrations higher than those found in the atmosphere.

CONTEXT

The impact of CFCs and their related compounds is mainly on the ozone layer, because of their catalytic effect. CFCs are also potent greenhouse gases but are present in the atmosphere at such low levels (0.1–0.5 parts per billion) that their contribution to the total greenhouse effect is very small. Loss of ozone in the stratosphere also affects temperature in complex ways, warming some parts of the atmosphere and cooling others. Nevertheless, the sheer numbers of individual halocarbons, even at low levels individually, add up to a warming potential that is worth controlling. Atmospheric levels of substances controlled by the Montreal Protocol have declined a great deal. Not surprisingly, the levels of their substitutes have risen. This trade-off is good for the ozone layer, but it leaves much to be accomplished in regard to global warming.

—*John R. Phillips*
—*Rena Christina Tabata*

FURTHER READING

Andersen, Stephen O. and Sarma, K. Madhavi *Protecting the Ozone Layer. The United Nations History.* UNEP/Earthscan, 2012. In this book, the authors provide a detailed history and discussion of the effects of chlorofluorocarbons and related compounds on the stratospheric ozone layer, and of the efforts made to curtail the damage.

Benedick, Richard Elliot *Ozone Diplomacy. New Directions in Safeguarding the Planet.* Harvard University Press, 2009. This enlarged edition provides a deep historical account and analysis of the meshing of science and diplomacy in addressing the problem of depletion of the ozone layer by chlorofluorocarbons.

Cumberland, John H., Hibbs, James R. and Hoch, Irving, eds. *The Economics of Managing Chlorofluorocarbons. Stratospheric Ozone and Climate Issues.* Routledge, 2016. An extensive examination of the history, uses and environmental effects of chlorofluorocarbons.

Lambright, W. Henry *NASA and the Environment. The Case of Ozone Depletion.* NASA SP-2005–4538, 2005. An interesting case study of the problem of ozone depletion, subdivided into eight recognizable stages.

VanLoon, Gary W. and Duffy, Stephen J. *Environmental Chemistry. A Global Perspective.* 4th ed., Oxford University Press, 2017. A well-composed text book that presents a good discussion of chlorofluorocarbons and related compounds.

CLIMATE AND THE CLIMATE SYSTEM

FIELDS OF STUDY

Climate Classification; Climate Zones; Bioclimatology; Hydroclimatology; Meteorology; Hydrometeorology; Climate Modeling; Atmospheric Science; Environmental Science; Physical Geography; Oceanography; Tree Ring Analysis; Analytical Chemistry

SUMMARY

Climate is an aggregation of near-surface atmospheric conditions and weather phenomena over an extended period globally or in a given area. It is characterized by statistical means and such variables as air temperature, precipitation, winds, humidity, and frequency of weather extremes. The time period over which a climate is described is typically thirty years, as specified by the World Meteorological Organization (WMO).

PRINCIPAL TERMS

- **biosphere:** the portion of Earth that contains living organisms and ecosystems
- **cryosphere:** the portion of Earth's surface that exists below the freezing point of water, including polar ice caps, sea ice, glaciers, snow caps, and permafrost
- **lithosphere:** the portion of Earth consisting of solid rock

CLASSIFYING CLIMATES

World climate is classified by either the empirical method, focusing on climatic effects, or the genetic method, emphasizing the causes of climatic effects. The empirical Köppen system, based on annual mean temperature and precipitation combined with vegetation distribution, divides world climate into five groups: tropical, dry, temperate, continental, and polar. Each group contains subgroups, depending on moisture and geographical location. The genetic Bergeron, or air-mass, classification system is more widely accepted among atmospheric scientists, as it directly relates to climate formation and origin. Air-mass classification uses the two fundamental attributes of moisture and thermal properties of air masses. Air masses are classified into dry continental (C) or moist maritime (M) categories. A second letter is assigned to each mass to describe the thermal characteristic of its source region: P for polar, T for tropical, and (less widely used) A for Arctic or Antarctic. For example, the dry cold CP air mass originates from a continental polar region. Sometimes, a third letter is used to indicate the air mass being cold (K) or warm (W) relative to the underlying surface, implying its vertical stability.

CLIMATE SYSTEM

In a broad sense, climate often refers to an intricate system consisting of five major components: the atmosphere, hydrosphere, cryosphere, land surface (a portion of the lithosphere), and biosphere, all of which are influenced by various external forces such as Earth-Sun orbit variations and human activities. The atmosphere, where weather events occur and most climate variables are measured, is the most unstable and rapidly changing part of the system. Earth's atmosphere is composed of about 99 percent permanent gases (nitrogen and oxygen) and 1 percent trace gases, such as carbon dioxide (CO_2) and water vapor. All weather and climate phenomena are associated with the trace gases called greenhouse gases (GHG), whose presence in the atmosphere is responsible for maintaining the average global

A composite satellite image depicting Earth's interrelated climate systems.

temperature approximately 18°C higher than it would be in their absence. This enables the dynamic nature of the atmosphere. Long-term increases in GHG concentration tend to warm the climate overall, while day-to-day variations in atmospheric thermal and dynamic structures are responsible for daily weather events.

The hydrosphere comprises all fresh and saline waters. Freshwater runoff from land returning to the ocean influences the ocean's composition and circulations, while transporting a large amount of minerals, chemicals and energy. Because of their great thermal inertia and huge moisture source, oceans regulate the Earth's climate. The cryosphere consists of those parts of the Earth's surface covered by permanent ice in polar regions, alpine snow, sea ice, and permafrost. It has a high reflectivity (albedo), reflecting solar radiation back into space, and is critical in driving deep-ocean circulations.

Land surfaces and the terrestrial biosphere control how energy received at the surface from the Sun is returned to the atmosphere, in terms of heat and moisture. The partitioning between heating and moistening the atmosphere has profound implications for the initiation and maintenance of convection and thus for precipitation and temperature. Marine and terrestrial biospheres have major impacts on the atmosphere's composition through the uptake and release of GHG during photosynthesis and organic material decomposition.

INTERACTIONS AMONG CLIMATE SYSTEM COMPONENTS

The individual components of the climate system are linked by physical, chemical, and biological interactions over a wide range of space and time scales. The atmosphere and oceans are strongly coupled by moisture and heat exchange. This coupling is responsible for the El Niño-Southern Oscillation (ENSO), the North Atlantic Oscillation, and the Pacific Decadal Oscillation, resulting in climate swings on interannual to interdecadal scales. The terrestrial biosphere and atmosphere exchange gases and energy through transpiration, photosynthesis, and radiation reflection, absorption, and emission.

These interactions form the global water, energy, and carbon cycles. The hydrologic cycle leads to clouds, precipitation, and runoff, redistributing water among climate components. Oceans and land surfaces absorb solar radiation and release it into the atmosphere by diffusion and convection. Global carbon and other gas cycles are completed by photosynthesis fixing CO_2 from the atmosphere and depositing it into the biosphere, soil, and oceans as organic materials, which are then decomposed by microorganisms and releasing the CO_2 back into the atmosphere.

Any change or disturbance to the climate system can lead to chain reactions that may reinforce or suppress the initial perturbation through interactive feedback loops. If the climate warms, melting of glaciers and sea ice will accelerate, and the surface will absorb more solar radiation, further enhancing warming. On the other hand, warmer air temperatures result in more moisture in the atmosphere, increasing cloud cover, which increases albedo and reduces the absorbed solar radiation. This leads to cooling, compensating for the initial warming. There exist many such positive and negative feedback mechanisms, which makes the causality of climate change complex.

—*Zaitao Pan*

FURTHER READING

Archer, David. *The Global Carbon Cycle*. Princeton University Press, 2010.

Barry, Roger G. and Hall-McKim, Eileen A. *Essentials of the Earth's Climate System.* Cambridge University Press, 2014. This book describes the many individual components of Earth's climate in order to assemble an all-encompassing understand of the global climate.

Lutgens, Frederick K., Edward J. Tarbuck, Tasa, Dennis G. and Herman, Redina. *The Atmosphere: An Introduction to Meteorology.* 14th ed. Pearson Education, 2018. An updated version of this textbook, in loose-leaf format, provides students with a general overview of the structure and properties of Earth's atmosphere.

Maslin, Mark. *Global Warming: A Very Short Introduction.* Oxford University Press, 2009. Provides a concise general introduction to the major factors that define the climate of Earth.

Saha, Pijushkanti *Modern Climatology.* Allied Publishers Pvt Ltd, 2012. An entry-level textbook that provides an concise, yet thorough, overview of the science of climatology and climate relationships.

Schimel, David *Climate and Ecosystems.* Princeton University Press, 2013. One of the "Princeton Primers on Climate" series, this book describes the interaction mechanisms of various ecosystems and climate.

Schmittner, Andreas, Chiang, John C.H. And Hemming, Sidney R., eds. *Ocean Circulation. Mechanisms and Impacts.* John Wiley & Sons, 2013. Provides a thorough description of the effects on Earth's climate due to large "turnover" circulation currents, particularly the Atlantic conveyor system.

CLIMATE CHANGE AND GLOBAL WARMING

FIELDS OF STUDY

Climatology; Climate Science; Environmental Sciences; Environmental Studies; Oceanography; Meteorology; Hydroclimatology; Hydrometeorology; Thermodynamics; Heat Transfer; Physical Geography

SUMMARY

Climate is characterized by mean air temperature, humidity, winds, precipitation, and frequency of extreme weather events over a period of at least thirty years. Global warming is an example of climate change, and so are increases in the magnitude or frequency of storms, floods and droughts experienced in many parts of the world during the past several decades. Climate change includes both natural variability and anthropogenic changes.

PRINCIPAL TERMS

- **anthropogenic:** caused by human activity
- **climate:** long-term, average, regional or global weather patterns
- **emission scenario:** a set of posited conditions and events, involving climatic conditions and pollutant emissions, used to project future climate change
- **greenhouse gases (GHGs):** trace atmospheric gases that trap heat in Earth's atmosphere, preventing it from escaping into space
- **proxy:** an indirect indicator of past climate conditions
- **weather:** the set of atmospheric conditions obtaining at a given time and place

DEFINING CLIMATE CHANGE

Although climate changes on longer than millennial timescales are natural, it is extremely likely that the global warming since 1950 is anthropogenic, according to the fifth Assessment Report of the Intergovernmental Panel on Climate Change (IPCC) of the United Nations. The United Nations is concerned primarily with anthropogenic climate change, both because it poses a threat to global security and because it can be altered by altering human and governmental behavior. For this reason, the United Nations Framework Convention on Climate Change (UNFCCC) defines climate change as being a change of the climate state that is attributable either directly or indirectly to human activity that changes the composition of the atmosphere on a global scale, and that can be differentiated as occurring in addition to

Ten Indicators of a Warming World

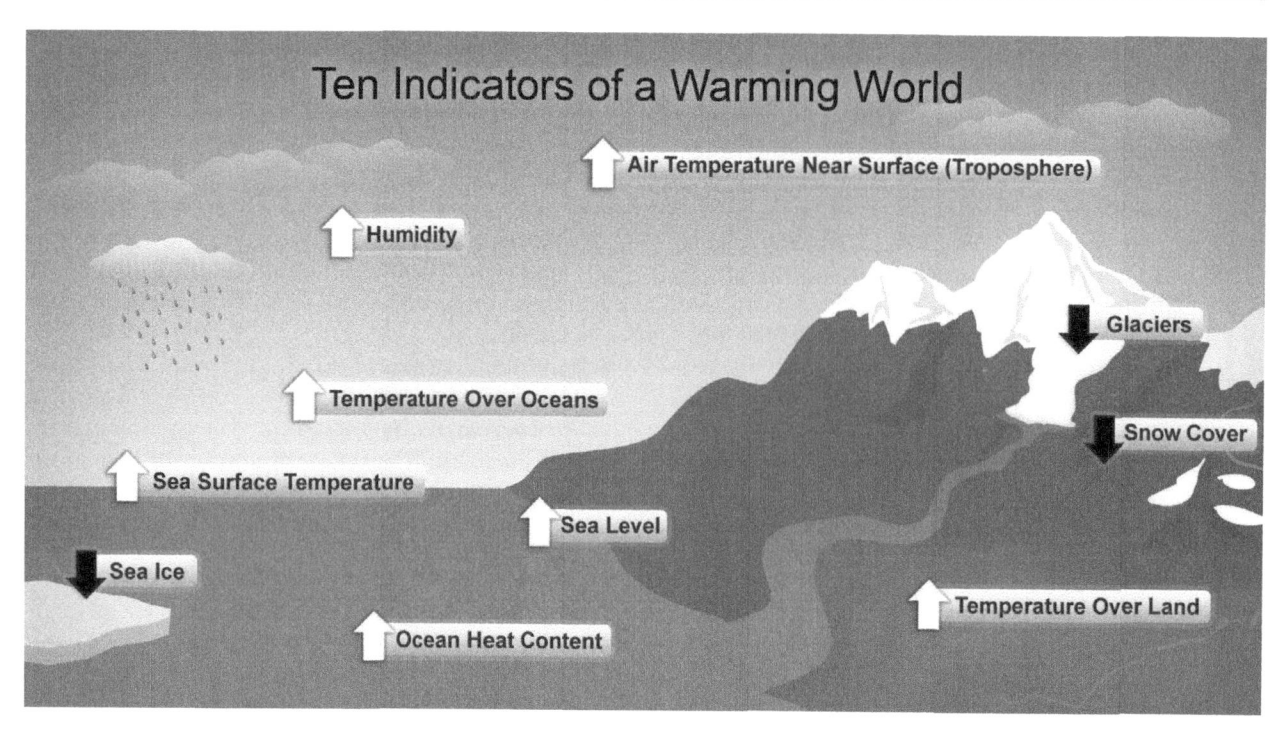

This diagram shows ten indicators of global warming.

naturally-occurring climate variability that has been observed over similar periods of time.

CLIMATE CHANGE DETECTION

Earth's atmosphere is chaotic, and weather can change dramatically in a matter of days or even hours. The temperature in some places may rise or fall by 20°C or more in one day. On the other hand, climate, as the average of years of weather conditions, changes on a much smaller scale. For example, the global mean surface air temperature increased by only 0.6°C during the twentieth century. By the same token, such a seemingly small increase can have extremely significant effects.

The climatic increase in mean surface air temperature is computed from tens of thousands of weather station records spanning decades. The difficulty of ensuring data continuity in time, uniformity in space, and constancy in observational methods poses serious challenges to climatologists. To discern slight trends amid diverging data, scientists use advanced mathematical tools to synchronize all observations, adjust discontinuities, and filter out local influences such as heat island effects.

Modern climate change has generally been observed with in situ thermometers and, later, with remote sensing devices. Paleoclimate change (change before about 1850) is inferred from proxy climate data. Tree rings can provide evidence of temperature and precipitation history for two to three thousand years, while tiny air bubbles trapped in Antarctic ice provide data extending hundreds of thousands of years into the past. Pollen and zooplankton cells in river and sea sediments also contain useful proxy climate data.

Detecting climate change depends on individual variables. Temperature change is the most reliable such variable, because its internal variability is small and it is more widely observed than other variables. Long-term precipitation changes are more difficult to discern, because rain- and snowfall vary so greatly from one year to the next. The intensity and frequency of extreme weather events such as hundred-year floods are even more difficult to detect, because these events are rare, so a significant data set must cover many years.

CLIMATE CHANGE THEORIES

FIELDS OF STUDY

Climatology; Atmospheric Science; Oceanography; Thermodynamics; Heat Transfer; Fluid Dynamics; Climate Modeling; Statistics; Mathematics; Earth System Modeling; Hydroclimatology; Hydrometeorology; Bioclimatology; Climate Classification; Climate Zones; Physical Geography; Observational Astronomy

SUMMARY

Earth's climate is a complex system in constant flux. An understanding of how those changes occur has emerged in recent decades. This insight has allowed for a better understanding of Earth's natural history and has helped to mitigate the dangers of modern climate change.

PRINCIPAL TERMS

- **albedo:** a value between zero and one reflecting the fraction of radiation reflected by a surface
- **carbon-oxygen cycle:** the process by which oxygen and carbon are cycled through Earth's environment
- **cosmic ray:** high-energy subatomic particles that are produced by phenomena in space, such as supernovae
- **eccentricity:** the departure of an ellipse from circularity; less circularity means greater eccentricity
- **El Niño and La Niña:** periodic warming and cooling of the eastern tropical Pacific Ocean that affects global weather patterns
- **greenhouse effect:** a phenomenon in which gas molecules within Earth's atmosphere absorb infrared heat energy from the surface and retain it in the atmosphere, preventing it from radiating away into space
- **obliquity:** the angle of tilt between the earth's rotational axis and an axis perpendicular to the plane of its orbit
- **precession:** a cyclical change in the orientation of the axis of rotation in a rotating body or system
- **proxies:** traces of ancient environments that reveal details, such as climatic data, about those environments
- **sedimentary rock:** rock formed by the repeated deposition of sediment in a body of water or by the layering of material on land

CLIMATE CHANGE THEORIES

An understanding of Earth's climate as a dynamic system originated in the eighteenth century, arising from discoveries in geology and paleontology. Continued research in the twentieth century provided an increasingly complex image of how Earth systems are interrelated. An understanding of future anthropogenic (human-caused) climactic shifts is thus based on an understanding of Earth's climate mechanisms, which are derived from matching historical data with the geologic record.

METHODOLOGY

Climatology employs many methods, including tree ring analysis, ice core sampling, sediment core sampling, and satellite observation. Tree ring analysis, or dendroclimatology, works by examining the rings of trees, which reflect the growing conditions and growth of a tree during its growing season. Rings are wider under favorable growing conditions and narrower under less favorable growing conditions. Dendroclimatology works only to the point at which preserved trees can be found and such trees have provided high-resolution data as far back as 11,000 years ago.

Another common method in climatology is to examine ice cores from glaciers. Ice cores are shafts of ice pulled from an ice cap by drilling into that cap with a core drill. Cores are commonly taken from Greenland and Antarctica, but cores can also be taken from mountaintop glaciers. Additionally, ice cores trap air; however, that air is not always the same age as the ice because rates of compression of snow to ice can be low. Ice core samples have been obtained from as much as 1.5 million years ago. The oldest ice core records have been obtained in Antarctica. Ice cores also can catch useful environmental and climatological proxies, such as iridium, pollen, and volcanic ash.

Another common method for climate data collection is sediment drilling from lakes and sea floors. Like their icy counterparts, these sediment cores retain similar data, but they also can record fossil data. Seafloor cores can theoretically go back 200 million years.

Terrestrial rocks go back to about 3.5 billion years, with some isolated instances being 1 billion years older. Analysis of their constituent materials

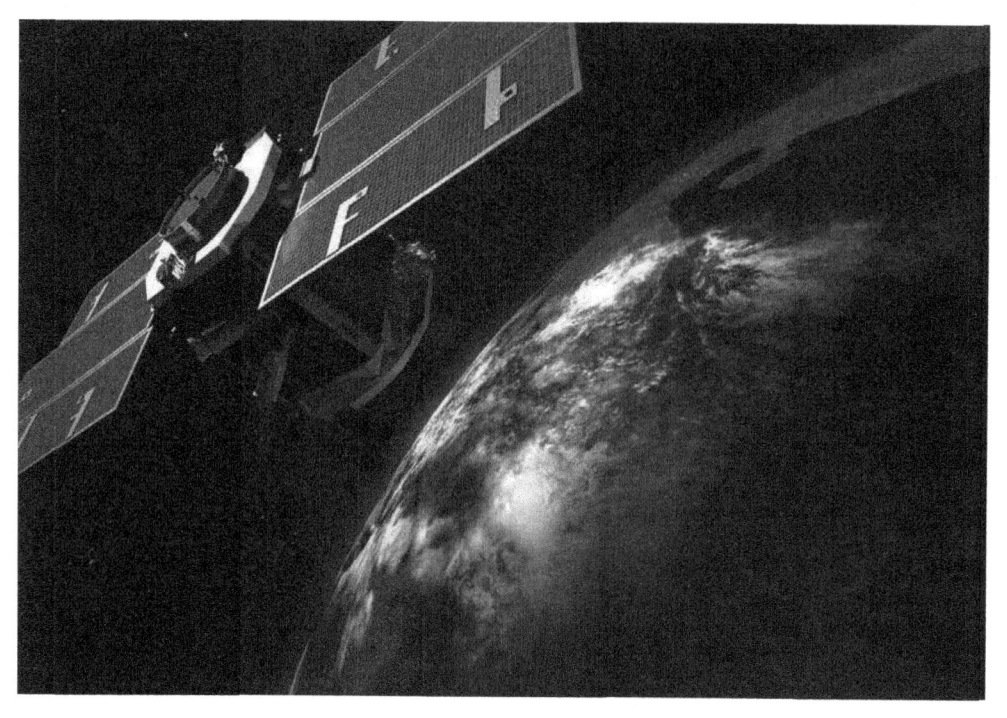

Artitst's conception of NASA's CloudSat satellite, which measures cloud properties to provide information on how they affect weather and climate.

and minerals can give an idea of the environment in which they were formed. Ancient sedimentary rocks can give ideas of water levels and temperatures, while other stones record the shores and flows of lakes and rivers. Along with these markers, the fossil record also often gives an idea of ancient Earth conditions.

To study the current climate, networks such as the Earth Observation System (EOS) of the National Aeronautic and Space Administration gather various sorts of data. The EOS includes more than fifteen satellites, all designed to monitor different aspects of Earth's environment. For example, the satellite *Aqua* observes the water cycle and *CloudSat* measures cloud altitude and properties.

MILANKOVITCH THEORY

The Milankovitch theory is one of the most comprehensive long-term climate change theories available. The theory considers how Earth's movement affects climate. The theory, discovered by a Serbian engineer during World War I, considers the processes of orbital eccentricity, obliquity, and precession, allowing for scientists to explore the impact of these processes' individually and in combination.

Earth's orbit is not a perfect circle; it is an ellipse that varies in eccentricity in time. Changes in the eccentricity of Earth's orbit from gravitational interactions with other planets mean that, on occasion, Earth's orbit is narrower, which results in greater difference in seasons, and at other times, Earth's orbit is more circular, which leads to a lesser difference in seasons. These changes occur because of Earth's proximity to the sun.

The effect of the variance of eccentricity of Earth's orbit can be great. At its most elliptical, Earth receives about 23% more radiation at perihelion (the point closest to the sun) than it does at aphelion (the point farthest from the sun). The eccentricity ranges between about 0.0034 and 0.058. This process takes about 413,000 years. Through interactions with other cycles the process combines to create a cycle of about 100,000 years.

The obliquity of the ecliptic (the tilt of Earth's axis) varies between about 22.1 to 24.5 degrees in about 41,000 years. When the obliquity is greater, the difference in solar radiation received is also greater, meaning a greater difference in the seasons. When the obliquity is lower, less difference occurs between

The Greenhouse Effect

Some solar radiation is reflected by the Earth and the atmosphere.

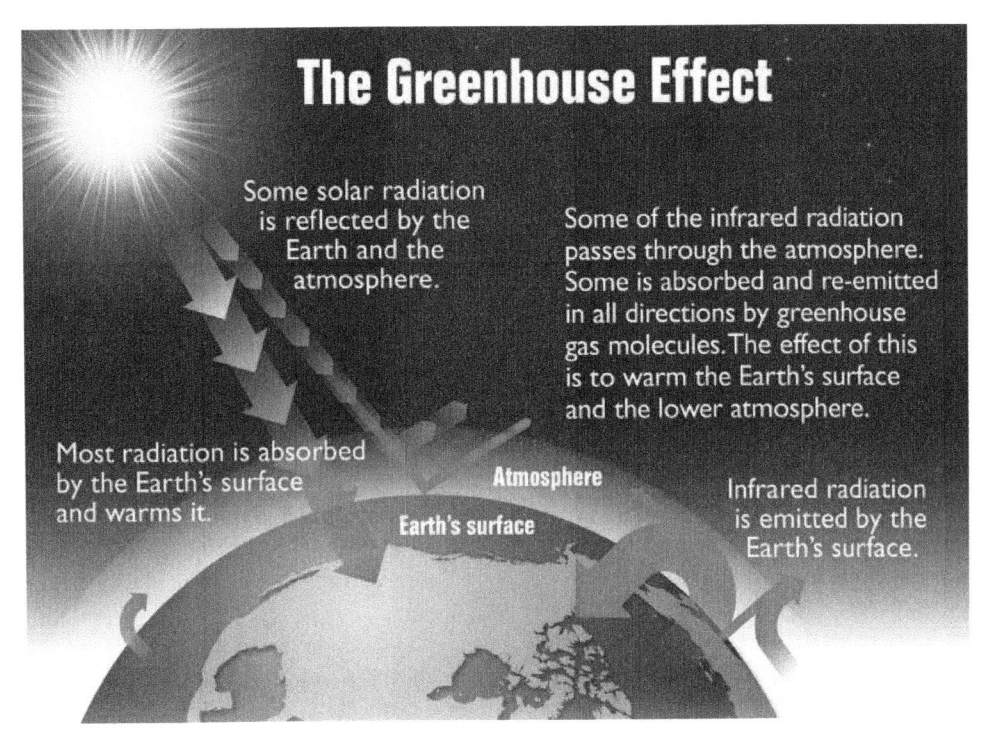

Some of the infrared radiation passes through the atmosphere. Some is absorbed and re-emitted in all directions by greenhouse gas molecules. The effect of this is to warm the Earth's surface and the lower atmosphere.

Most radiation is absorbed by the Earth's surface and warms it.

Infrared radiation is emitted by the Earth's surface.

Illustration of the earth's greenhouse effect.

the seasons. It is thought that low obliquity favors ice ages. The cycle now is heading toward its lowest obliquity and, therefore, toward an ice age. It is expected to reach lowest obliquity in about 8,000 years; however, anthropogenic climate change has mitigated this trend.

The axis of Earth's rotation also precesses in a cycle of about 26,000 years. When Earth's axis tilts toward the sun at perihelion, the hemisphere facing the sun has a greater difference in seasons that year. When the axis is not pointing toward the sun at a solstice, but instead does so at an equinox, less difference in the seasons occurs between the hemispheres.

The orbital ellipse itself also precesses and operates with changes in eccentricity to alter the length of the seasons. Orbital inclination was not an original part of Milankovitch's work but has since been added. Earth's orbit is pulled up and down relative to the plane of the solar system by the gravitational effects of Jupiter. Orbital inclination works on a time scale of about 100,000 years and is thought to have an effect because of the dust clouds and other debris altering the amount of light reaching the earth.

The Milankovitch theory does have its weaknesses: Some trends have been observed but not predicted, and some trends have been predicted but not observed. On the whole, however, the theory fits the observed periodicities of climate.

Cosmic Rays

Another space-based phenomenon that could affect terrestrial climate is cosmic rays. Though disputed, there seems to exist evidence for such a proposition. Cosmic rays ionize the atmosphere and cause increased cloud cover, resulting in a lower temperature. This would be most noticeable on a long-time scale as the earth moves through the spiral arms of the galaxy and their supernovae density. Earth moves through a spiral arm about once every 135 ± 25 million years, causing flux in cosmic rays with a period of 143 ± 10 million years, matching the cycle of 145 ± 8 million years. An absence in ice ages on earth between 2 billion and 1 billion years ago coincides with a drop in star formation, giving the idea further credence.

Greenhouse Gases and the Carbon-Oxygen Cycle

Another theoretical framework for climate shifts comes from examination of the carbon-oxygen cycle, which works in tandem with the Milankovitch cycle and is used to explain some points in which the climatic reaction far exceeds the change in the cycle. Because carbon dioxide a greenhouse gas, it has great influence on the climate. Disruption of this mechanism is one of the main causes of anthropogenic climate change.

A good example of how the carbon-oxygen cycle figures into climate is the *Azolla* event of 49 million years ago. At that time, the Arctic Ocean was warm and closed off, such that it started to collect a layer of freshwater. It is thought that the fresh-water plant *Azolla* grew and covered the Arctic Ocean. After the *Azolla* started to die, it sank to the ocean bottom; the freshwater and seawaters did not mix, which sequestered the carbon. It is estimated that this single phenomenon could have caused an estimated 80% drop in atmospheric carbon dioxide, dramatically reducing the greenhouse effect and turning the climate from a much warmer one to a far cooler one.

GREENHOUSE AND ICEHOUSE EARTH

"Greenhouse Earth" is characterized by high temperatures, which can sometimes reach the polar regions. Eighty percent of Earth's history has featured this climate. The rest of the climate has been "Icehouse Earth." More specifically, Earth is presently in an interglacial era. Interglacials make up 20 percent of the time in the average glacial period.

A third state, called "Snowball Earth," occurred when the entire planet was believed to have frozen over, though there is some debate about whether or not there was an open or seasonally open ocean near the equator. The state was most likely ended by the buildup of greenhouse gases from volcanic activity. It is thought that the transition of "icehouse" to "greenhouse" was caused by plate tectonics.

CONTINENTAL DRIFT

An important factor in determining climate is continental placement because continents shift the workings of various cycles. It is thought that the closing of the Isthmus of Panama, the opening of the Drake Passage, and the opening of the Tasmanian Gateway helped Earth's climate transition from "greenhouse" to "icehouse" by changing the flow of ocean currents. The surrounding seas were open to circulation around Antarctica and less able to circulate at the equator. In the south, this led to the creation of the Antarctic Circumpolar Current, which kept relatively warm water from reaching Antarctica, allowing it to freeze. This led to increased albedo, which led to more freezing. A positive feedback loop like this is also thought to have led to "Snowball Earth."

IMPACT

Increased greenhouse gases has caused a drop in the amount of ozone, a gas that shields the planet from ultraviolet radiation. When the temperature drops, as in an ice age, ozone is depleted. When ozone is depleted, the stratosphere becomes still colder, further depleting the ozone. Ozone also cools the troposphere. Additionally, low-altitude ozone is a greenhouse gas.

An understanding of climate change not only illuminates historical processes, but also gives an idea of what occurs with anthropogenic climate change. Understanding Earth's climate provides warning of climate change and also aids in finding solutions to related climate concerns.

—*Gina Hagler*

FURTHER READING

AghaKouchak, Amir, et al. *Extremes in a Changing Climate: Detection, Analysis, and Uncertainty.* Springer, 2013.

Bartlein, Patrick J., and John A. Matthews. *The Sage Handbook of Environmental Change.* Sage, 2013.

Battarbee, R. W., and H. A. Binney. *Natural Climate Variability and Global Warming: A Holocene Perspective.* Blackwell, 2008.

Bradley, Raymond S. *Paleoclimatology: Reconstructing Climates of the Quaternary.* Academic, 1999.

Cowie, Jonathan. *Climate Change: Biological and Human Aspects.* Cambridge University Press, 2007.

Cronin, Thomas M. *Principles of Paleoclimatology.* Columbia University Press, 1999

Dove, Michael. *The Anthropology of Climate Change: A Historical Reader.* Wiley, 2014

Gornitz, Vivien. *Encyclopedia of Paleoclimatology and Ancient Environments.* Springer, 2009.

Martin, Ronald. *Earth's Evolving Systems: The History of Planet Earth.* Jones, 2013.

Schmidt-Thome, Philipp, and Johannes Klein. *Climate Change Adaptation in Practice: From Strategy Development to Implementation.* Wiley, 2013.

Wilson, Elizabeth J., and David Gerard. *Carbon Capture and Sequestration: Integrating Technology, Monitoring, and Regulation.* Ames: Blackwell, 2007.

CLIMATE FEEDBACK

FIELDS OF STUDY

Climate Modeling; Climate Science; Climate Classification; Climate Zones; Oceanography; Heat Transfer; Physical Geography; Earth System Modeling; Climatology; Atmospheric Science; Physical Chemistry; Atmospheric Chemistry; Photochemistry; Environmental Sciences; Environmental Studies; Observational Astronomy

SUMMARY

Factors that can alter Earth's radiative balance are called "climate forcings" and "climate feedbacks." Understanding climate forcings and feedbacks is at the heart of understanding how greenhouse gases emitted into the atmosphere might affect future temperature trends. Climate feedbacks are particularly important to understand because computer models projecting future warming incorporate assumptions about such feedbacks (especially water vapor) that significantly elevate predicted temperature increases due to greenhouse gas emissions.

PRINCIPAL TERMS

- **aerosols:** tiny particles or liquid droplets suspended in the atmosphere; some, such as sea salt, are natural, while others, such as soot from power plants, are of human origin
- **albedo:** the extent to which an object reflects radiation; the reflectivity of objects with regard to incoming solar radiation
- **climate forcing:** factors that alter the radiative balance of the atmosphere (the ratio of incoming to outgoing radiation)
- **greenhouse gases (GHGs):** gases (or vapors) that trap heat in the atmosphere by preventing or delaying the passage of long-wavelength infrared radiation from Earth's surface out to space
- **infrared radiation:** radiation with wavelengths longer than those of visible light, felt by humans and animals as heat
- **lapse rate:** the change in a variable with height, often used to discuss changes in temperature with altitude in a context of climate change
- **radiative balance:** the balance between incoming and outgoing radiation of a body in space, such as Earth

Balancing the System

Like all planets, Earth is bombarded by radiation from the Sun, stars, and other space-based sources of energy. Eventually, that energy is returned to space. Incoming and outgoing radiation must ultimately balance out, in accord with the first law of thermodynamics, the law of conservation of energy. In the case of Earth, which has a significant atmosphere and varied surface geography, the pathways by which radiation reaches the surface and is radiated back out to space are somewhat convoluted. Some incoming radiation is reflected back to space by clouds and particles in the atmosphere before it reaches the ground, some is reflected back toward space from various surfaces, and some is absorbed and then reradiated away in the form of long-wave, or infrared, radiation. A significant proportion of this reradiated infrared radiation, or heat energy, is retained within the atmosphere by certain atmospheric gases or water vapor before it eventually makes its way back into space. Still, over time, the total Earth-atmosphere system is said to be in radiative balance, an equilibrium system in which there is no net gain or loss of energy with the conditions that exist.

Climate Forcing

One cannot understand climate feedbacks without understanding climate forcings. A climate forcing (technically a "radiative forcing") is something that exerts a direct effect on the radiative balance of the

Glacial iceberg in Argentino lake shows a section which was previously under water, now exposed due to weight shifting as the iceberg melts.

Earth's atmosphere, something that changes the balance of incoming versus outgoing radiation either permanently or transiently. Forcings include incoming solar radiation, the heat-retaining ability of greenhouse gases present naturally in the atmosphere, anthropogenic greenhouse gas emissions and conventional air pollutants, changes in land use that might alter the reflectivity of the Earth's surface, and more. The Intergovernmental Panel on Climate Change (IPCC) identifies nine major radiative forcing components, some of which are considered well understood, and others that are not so well understood. Forcings identified by the IPCC include the greenhouse gases, ozone, stratospheric water vapor, surface albedo, aerosols, contrails, and solar irradiance.

Climate Feedback

Climate feedbacks are secondary changes to the radiative balance of the climate stemming from the influence of one or another climate forcing. Such climate feedbacks may be either positive or negative. A positive feedback would amplify the effect of a change in a given climate forcing, while a negative feedback would damp down or lessen the effect of a change in a given climate forcing. For example, a change in the atmosphere's water vapor content can result in greater cloudiness, which could constitute a positive feedback by tending to retain more heat energy and thus increasing the atmospheric water content. It can also act as a negative feedback by reflecting more of the incoming energy back into space, thus tending to cool the atmosphere and reduce atmospheric water content.

According to the National Research Council, part of the U.S. National Academies of Science, climate feedbacks that primarily affect the magnitude of climate change include clouds, atmospheric water vapor, the lapse rate of the atmosphere (defined as the change in temperature with altitude), the reflectivity (albedo) of ice masses, biological, geological and chemical cycles, and the carbon cycle. Feedbacks that primarily affect temporary responses of the climate include ocean heat uptake and circulation feedbacks. Finally, feedbacks that mostly influence the spatial distribution of climate change include land hydrology and vegetation feedbacks, as well as natural climate system variability.

When used in projecting future temperatures stemming from greenhouse gas emissions, estimates of some climate feedbacks are incorporated into computerized models of the climate system. These climate feedbacks—clouds, water vapor, surface albedo, and the lapse rate—are expected to contribute as much (or more) warming to the atmosphere as changes in the greenhouse gases do by themselves.

Context

The extent to which climate feedbacks might increase or decrease the heat-trapping effects of humanity's greenhouse gas emissions is an important factor in public policy development. If computer models understate the extent of positive feedbacks, future warming could be worse than projected, and actions undertaken to combat climate change might be insufficient to the challenge. On the other hand, if computer models overstate positive feedbacks, or underestimate negative feedbacks, projected future warming scenarios could be too high. In this case, massive resources spent on controlling greenhouse gas emissions could be wasted, leaving society less able to deal with other challenges, environmental or otherwise.

—*Kenneth P. Green*

Further Reading

Goosse, Hugues *Climate System Dynamics and Modeling*. Cambridge University Press, 2015. The individual components and observable mass and energy cycles of Earth are discussed initially, and subsequent chapters analyze and discuss individual feedbacks.

Houghton, John. *Global Warming: The Complete Briefing*. Cambridge University Press, 2009. A leading textbook on climate change, updated to reflect findings of the IPCC's most recent assessment reports.

Neelin, J. David *Climate Change and Climate Modeling*. Cambridge University Press, 2011. A more advanced textbook that describes many of the concepts of atmospheric science and presents a detailed discussion of climate feedbacks in regard to climate change.

North, Gerald, and Tatiana Erukhimova. *Atmospheric Thermodynamics: Elementary Physics and Chemistry*. Cambridge University Press, 2009. A textbook co-authored with a climate modeler deeply steeped in the study of climate forcings and feedbacks. Suitable for an undergraduate audience.

North, Gerald R. and Kim, Kwang-Yul *Energy Balance Climate Models*. Wiley-VCH, 2017. This book provides a mathematics-based presentation of climate phenomena and feedbacks.

Rosenzweig, Cynthia, Rind, David, Lacis, Andrew and Manley, Danielle, eds. *Lectures in Climate Change, Vol. 1. Our Warming Planet. Topics in Climate Dynamics*. World Scientific, 2018. The chapters in this book are transcripts of individual lectures in climate science, with emphasis on the role of climate feedbacks.

CLIMATE MODELS AND MODELING

FIELDS OF STUDY

Climate Modeling; Computer Science; Software Engineering; Climate Studies; Climate Sciences; Climate Classification; Climate Zones; Atmospheric Science; Atmospheric Chemistry; Physical Geography; Thermodynamics; Heat Transfer; Statistics; Process Modeling; Mathematics; Fluid Dynamics; Bioclimatology; Hydroclimatology; Hydrometeorology; Oceanography; Earth System Modeling; Observational Astronomy; Electromagnetism; Particle Physics

SUMMARY

Climate models are the most important tools for quantitatively estimating how natural and anthropogenic forces affect different aspects of the climate system. A climate model is a numerical representation of the climate system, including the physical, chemical, and biological properties of its components and the interactions between those components. It consists of a large number of mathematical equations that quantitatively describe the processes occurring within the climate system. Climate models span the entire range from very simple models that can be executed with a pencil and paper to very complex models that require large supercomputers.

PRINCIPAL TERMS

- **albedo:** the fraction of incident light reflected from a body such as Earth
- **convection:** the cyclic movement of matter in fluids such as air or water, by which warmer, less dense matter rises through cooler matter and then descends elsewhere as its heat energy decreases
- **emissivity:** a measure of the ability to radiate absorbed energy
- **greenhouse gases (GHGs):** atmospheric trace gases that absorb heat and retain it in the atmosphere, preventing it from escaping into space
- **Navier-Stokes equation:** an equation describing the flow of air and other fluids
- **parameterization:** the assignment of measurable factors within a system

HIERARCHY OF CLIMATE MODELS

Climate model equations describe the conservation and transport of atmospheric heat, moisture, and momentum along three directions: latitude, longitude, and vertical altitude. For simplicity, climate models are often averaged along one or more of these directions. Depending on the nature of averaging, the following hierarchy of climate models develops.

ZERO-DIMENSIONAL ENERGY-BALANCE MODELS: The simplest form of climate model is averaged along all three directions. It consists of a single equation describing the balance of incoming and outgoing energy at the top of the atmosphere. These models calculate the temperature of the Earth as a function of incoming solar radiation, Earth's albedo, and emissivity.

The advantage of zero-dimensional models is their simplicity. These are the only type of climate models that can be solved by hand using a simple calculator, or even just pencil and paper. They have been widely used to study how global temperature may change in response to changes in solar radiation during the eleven-year and longer solar cycles, in Earth's albedo resulting from changes in cloud cover or sea-ice extent, and in emissivity.

ONE-DIMENSIONAL MODELS: Radiative-convective (RC) models are averaged along latitude and longitude, but not along the vertical direction. They include the two most important processes of vertical energy transport in the atmosphere: upward and

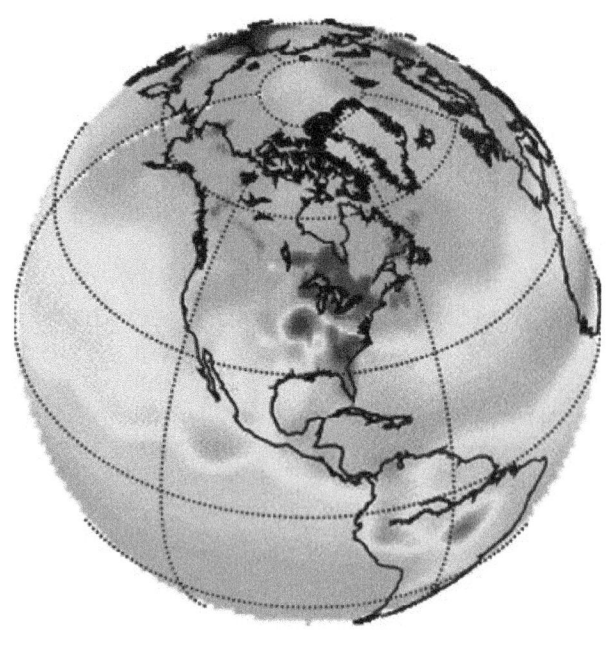

A computer model of global carbon dioxide emissions.

downward transport of radiation and upward transport of heat from Earth's surface by convection. The inclusion of vertical transport is important for climate modeling. Because gases in the atmosphere are unevenly distributed vertically, the impact of climate change varies with altitude. RC models can accurately simulate the vertical profile of temperature and temperature change in the atmosphere.

Energy balance (EB) models, in contrast to RC models, are averaged along the longitudinal and vertical directions. Thus, they may be used to account for equator-pole heat transport that arises because the equator receives more solar energy than do the poles. Unlike RC models, which use very sophisticated and realistic heat transport equations, the transport in EB models is described through relatively simple parameterizations that have limited applicability. EB models have been used to study the mechanisms of poleward heat transport, ice age climates, and ocean-atmosphere interactions.

TWO-DIMENSIONAL MODELS: Two-dimensional RC models are averaged along the longitudinal direction. Thus they combine one-dimensional RC and EB models. They account for vertical and poleward heat transport, the climate system's two most important heat transport processes. These models have been widely used to study general atmospheric circulation patterns such as Hadley cells. Because of their relative simplicity, these models have also been coupled with chemistry and radiation models to study stratospheric chemical-radiative-dynamical interactions.

Two-dimensional EB models average atmospheric properties and processes over the height of the atmosphere to describe the energy balance over Earth's entire surface. These models have limited applicability and have been used to study ice-age climates.

THREE-DIMENSIONAL GENERAL CIRCULATION MODELS

General circulation models (GCMs) are the most complex climate models. They depend on solutions of the full set of Navier-Stokes equations for atmospheric flow and thermodynamic and of microphysical equations for conservation of energy and moisture along all three cardinal directions. The three-dimensional nature of the models allows them realistically to handle the transport of energy, moisture, and momentum in the horizontal as well as vertical dimensions. Hence, GCMs are very useful for investigating regional aspects of climate and climate change.

Sometimes these models are coupled with dynamic ocean models, atmospheric chemistry models, or ecosystem dynamics models. Such multi-model coupling allows scientists to study the interactions between all three components of the climate system (atmosphere, biosphere, and hydrosphere).

The major disadvantage of GCMs, especially coupled atmosphere-ocean GCMs (AOGCMs), is that they are computationally very expensive, and require costly access to the supercomputers needed to run simulations with GCMs. Because of their size and complexity, it takes a long time, often weeks, to complete each simulation. Another drawback of these models is their coarse resolution, although as computer technology and software applications become more advanced the resolution of such models also improves. Each grid cell in a GCM currently spans about 1° of latitude by 1° of longitude. GCMs therefore cannot explicitly resolve fine-scale processes such as boundary-layer turbulence or cumulus clouds. The models do not completely ignore such fine-scale processes, however. Rather, these processes are approximated by subgrid parameterizations.

Earth systems models of intermediate complexity (EMICs) are another very powerful tool in climate studies. These models bridge the gap between complex GCMs and simpler models. EMICs simulate all the physical and dynamical processes contained in GCMs, but they use simpler parameterizations and coarser resolution, making them much faster to run than GCMs. Because of their computational efficiency, EMICs are becoming very popular in the climate policy field. They are widely used to study the feasibility of different adaptive and mitigation policies on climate change.

Context

Climate models in general form the basis for nearly all rational argumentation as to the causes and severity of climate changes of the past and present, as well as the effects of human actions on climate in the future. They are thus crucial to both climate science and climate policy. However, different climate modeling systems, and different deployments of the same systems, have yielded different results. This ambiguity has rendered it difficult thus far to come to definitive conclusions as to the optimal course of action in either the near term or the long term. Applying a hierarchy of climate models can help reduce some of the uncertainty by deploying the most appropriate model for a given purpose, but precise, definitive predictions of climate change remain elusive.

—*Somnath Baidya Roy*

Further Reading

Gettelman, Andrew and Rood, Richard B. *Demystifying Climate Models. A Users Guide to Earth System Models.* Springer Open, 2016. An excellent introduction to climate modeling, this book first gives a basic introduction to Earth's climate, and then individually describes the many parameters that are used to construct climate models.

Goosse, Hugues *Climate System Dynamics and Modeling.* Cambridge University Press, 2015. The individual components and observable mass and energy cycles of Earth are discussed initially, and subsequent chapters analyze and discuss individual feedbacks.

Kaper, Hans and Engler, Hans *Mathematics and Climate.* SIAM, 2013. This book guides the reader through the mathematical basis of climate models, beginning with understanding the use of statistics and continuing through discussions of individual factors affecting the global climate system.

Lloyd, Elisabeth A. and Winsberg, Eric, eds. *Climate Modelling. Philosophical and Conceptual Issues.* Springer, 2018. In this book, individual chapters present discussions geared to the interpretation and understanding of climate data from climate models rather than the models themselves.

Stocker, Thomas *Introduction to Climate Modelling.* Springer, 2011. This book presents a discussion of the basic structures and applications of different types of climate models.

CLIMATE PREDICTION AND PROJECTION

Fields of Study

Climate Science; Climate Modeling; Statistics; Atmospheric Science; Earth System Modeling; Process Modeling; Meteorology; Thermodynamics; Fluid Dynamics; Bioclimatology; Hydroclimatology; Hydrometeorology; Oceanography; Physical Geography; Heat Transfer

Summary

Weather is the atmospheric conditions that are experienced on a daily basis; climate is what produces those conditions in a given geographical area. Climatology has been described as "geographical meteorology." The principal purpose of studying a region's climate is to discover measurable factors that accurately predict, months or years in advance, what the general weather regime will be like in a given area. In making predictions, climatologists first look at established historical patterns, both recorded from first-hand observation and documented through proxies. Increasingly they also investigate whether there are systematic perturbations in established historical patterns, and they also incorporate the effects of events whose probability is low or unknown.

Principal Terms

- **anthropogenic climate change:** changes in overall long-term weather patterns due to human activity

Earth's Energy Budget

Earth's Energy Budget: The earth's energy budget is a model for evaluating the energy gains and losses to the atmosphere.

- **extreme weather events:** natural disasters caused by weather, including floods, tornadoes, hurricanes, drought, and prolonged severe hot and cold spells
- **proxies:** measurable parameters, correlated with climate, that are preserved in the geologic record (for example, oxygen isotope ratios and fossil pollen)

HISTORY

Attempts by humans to forecast climate date back several millennia, at least to the days of ancient Babylon, when astrologers used the motions of the Sun, Moon, and planets to predict whether the coming season would be favorable for agriculture. The practice may be even older. Anthropologist Johannes Wilbert recorded a system of climate prediction among the Warao Indians of Venezuela, a group of primitive Stone Age agriculturalists, based in their religious beliefs. In the Bible, the story of Joseph (c. 1800 BCE) relates how Joseph interpreted Pharaoh's dream of seven lean cattle devouring seven fat cattle as a prediction of impending drought and famine, enabling the Egyptians to prepare in advance.

Climate prediction for agricultural purposes remained a major function of astrology, and later astronomy, from Babylon to the pioneer European astronomer Tycho Brahe (1546–1601) in the sixteenth century. Even today, many successful American farmers follow the *Farmers' Almanac*, with its astrologically based recommendations for planting crops. The practice persists in part because it has some basis in fact: The phases of the Moon, and to a lesser extent the orbits of Jupiter and Saturn, do affect climate.

Interest in a more scientific and systematic approach to climate prediction gained impetus in the late nineteenth century in response to expansion of Europeans into regions that experience more extreme and variable climatic conditions than Europe. The eleventh edition of the *Encyclopedia Britannica* (1911) divides the globe into climatic zones, describes the variability of each, and discusses evidence for a general warming trend, which the author of the encyclopedia article was inclined to dismiss as unproven. A decade later, Sir Gilbert Walker (1868-1958) began publishing his pioneering work on fluctuations in the Indian monsoons, their relationship to periodic famines, and the correlation between them and oscillation of high- and low-pressure areas between the Indian Ocean and the tropical Pacific. This Southern Oscillation, later shown to be linked to the El Niño phenomenon in the eastern Pacific, is the most important determinant of cyclical global weather patterns, and is thought to have been the factor that caused the drought in the story of Joseph, as well as affecting other civilizations throughout the world.

BUILDING A GLOBAL CLIMATE PREDICTION NETWORK

Climate prediction depends upon having large numbers of accurate measurements of many different variables, which can then be correlated mathematically. Correlation is an uncertain process at best, rarely yielding unequivocal results. For example, all of the complex hurricane-predicting machinery of the United States' National Oceanic and Atmospheric Administration (NOAA) produced, as of the beginning of June, 2009, the prediction that the upcoming August-October Atlantic hurricane season would have a 50 percent chance of being average and a 25 percent chance of being above or below average. Vague predictions such as this are one reason that skeptics such as Marcel Leroux, of the University of Adelaide, could plausibly question that scientists have demonstrated any general global warming effect. More recent and improved observations, however, leave little room for doubt that climate change and global warming are real.

Global meteorological monitoring, coordinated through national weather services and the United Nations, involves a network of satellites capable of measuring physical parameters including surface temperatures, wind speeds, cloud cover, and barometric pressure, at points 50 kilometers apart, at hourly intervals. National weather centers are well apprised, for example, of the exact status of El Niño on any given day and how it has been developing, but unless it has recently exhibited extraordinary features, these data give only a general picture of what the climate will do in affected regions.

The main thrust of global climate prediction was, and to a large extent still is, extreme weather events, including occasions of extreme heat or cold, droughts, floods, and cyclonic or anticyclonic storms. Predictions impact disaster preparedness and help nations minimize mortality. Knowing in advance that El Niño is likely to produce drought in Australia and South Africa in a given year helps those countries stockpile grain, devote more acreage to drought-resistant crops, and prepare for wildfires. In the United States and elsewhere, projections for hurricane and tornado activity affect insurance policies and land-use decisions. For extreme weather projections, a decadal time frame is sufficient. Beyond that, only a few cyclical phenomena can be projected with yearly accuracy, and the number of unknown variables becomes too large to allow useful prediction.

CLIMATE PROJECTION

All of the models used in predicting decadal climate variability were developed using historical data and assume that assigned parameters are constant, oscillate in a regular manner around a mean, or are increasing or decreasing at a constant rate. With respect to the carbon dioxide content of the atmosphere, none of these conditions is currently met. However, once a model is developed, climatologists can use a computer simulation to project, for example, how atmospheric carbon dioxide and global temperatures would respond if there were a large increase or decrease in emissions. Because of complexities, uncertainties, and unknown variables, such projections often fall short of reliability.

CONTEXT

With respect to the controversy over global warming, whether it is occurring and to what it is attributable, input from climatologists associated with NOAA and other agencies suffers from distortion between laboratory and the media. When scientists correctly project from their models that a massive eruption of the Yellowstone supervolcano such as occurred 200,000 years ago would produce abrupt catastrophic cooling, completely dwarfing any anthropogenic warming, it implies neither that such an eruption is expected in the near future nor that efforts to curtail emissions and environmental degradation are futile in the face of overwhelming nature, yet that is the lesson many people would derive from such statements. The high resolution, global coverage, and international cooperation among climatology centers ensure that no event or trend of significance escapes attention. Thus, should media attention not translate into regulatory action, it is not the fault of the climatologists.

—*Martha A. Sherwood*

FURTHER READING

Fagan, Brian *Floods, Famines and Emperors. El Niño and the Fate of Civilizations.* Basic Books, 1999. A readable historical examination of how the El Niño—Southern Oscillation (ENSO) phenomenon have affected civilizations on a global scale throughout the past.

Gramelsberger, Gabriele and Feichter, Johann, eds. *Climate Change and Policy. The Calculability of Climate Change and the Challenge of Uncertainty.* A curated collection of articles addressing problems of uncertainty about climate modeling and climate change on an individual basis.

Heymann, Matthias, Gramelsberger and Mahoney, Martin, eds. *Cultures of Prediction in Atmospheric and Climate Science. Epistemic and cultural shifts in computer-based modelling and prediction.* Routledge, 2017. Individual chapters in this book present discussions of climate model design and the use of predictions and projections within political and regulatory discussions.

Visconti, Guido *Problems, Philosophy and Politics of Climate Science.* Springer, 2018. Presents an introduction to some factors affecting global climate, then addresses the practical and philosophical issues arising in climate observation and modeling.

Warner, Thomas Tomkins *Numerical Weather and Climate Prediction.* Cambridge University Press, 2011. A book describing the mathematical basis of climate prediction and projection.

CLIMATE RECONSTRUCTION

FIELDS OF STUDY

Climate Classification; Climate Zones; Climate Modeling; Paleoclimatology; Paleontology; Environmental Sciences; Tree Ring Analysis; Spectroscopy; Optics

SUMMARY

Scientists use a combination of techniques to reconstruct and describe aspects of past climates, such as temperature, precipitation, and atmospheric carbon dioxide (CO_2) concentration, using historical accounts and proxies. These methods include analysis of preserved pollen, tree rings, and ice-core oxygen isotope ratios and CO_2 concentrations, as well as paleobotanical methods.

PRINCIPAL TERMS

- **isotopes:** atoms of the same element that differ by the number of neutrons in their nuclei
- **paleobotany:** the study and identification of plants preserved in the fossil record
- **paleoclimate:** the climate that existed at a particular period in Earth's distant past
- **paleontology:** the study of the flora and fauna that existed in Earth's distant past
- **proxy:** observable factors that retain the effects of past climates

SIGNIFICANCE FOR CLIMATE CHANGE

To develop accurate climate-prediction models, scientists require data that predate modern science, forcing them to rely on historical records and a variety of climate proxies to reconstruct the climate of hundreds, thousands, and even millions of years ago. Climate factors that can be reconstructed include local and global temperatures, precipitation, sea levels and salinity, atmospheric pressure, atmospheric CO_2 concentration, ice volume, and ocean circulation.

Many studies use a combination of proxies in an attempt to minimize error. Data from these proxies

Sources of Climate Reconstruction Data

Years Ago	Available Data Sources
1,000	Written records, tree rings
10,000	Varved and lake sediments
50,000	Lake levels, mountain glaciers
250,000	Polar ice cores
1,000,000+	Ocean sediments, cave deposits

Source: National Ice Core Laboratory, U.S. Geological Survey.

can be combined with modern climate data to create models that infer past climate as well as predict future climate. paleoclimatology and climate reconstruction are important for explaining current ecosystems and understanding factors that affect climate change. Climate reconstruction data also are important for improving models that help predict the effects of possible climate change scenarios, such as the potential effect of increased atmospheric CO_2.

PROXIES IN CLIMATE RECONSTRUCTION

Two of the most crucial sources of paleoclimate data are ice and sediment cores. Deep ice cores preserve atmospheric gases, water, and pollen. Scientists analyze isotope ratios, CO_2 concentrations, and pollen assemblages to infer information about past climates. Perhaps the most famous ice core was taken from Lake Vostok in Antarctica. Data from this core have shown that East Antarctica was colder and drier, and that atmospheric circulation was more vigorous during glacial periods than they are now. Scientists can also correlate atmospheric CO_2 and methane with temperature.

Deep-sea sediment cores provide similar information about past climate through marine microorganisms such as diatoms and foraminifera. The shells and exoskeletons of these tiny creatures incorporate the naturally-occurring carbon and oxygen isotopes as calcium carbonate when they are formed. The isotope ratios are thus preserved in their shells over the passage of time. The proportions of these isotopes allow scientists to infer such factors as past water temperatures, while the community makeup of the microorganisms can be used to make other inferences about their environments. Scientists also take sediment cores from lakes. Charcoal layers in sediment strata can indicate fires, and pollen can also provide climate information.

Pollen is extremely tough and holds up well for millions of years in the fossil record. Pollen assemblages from cores can be used to infer climate by comparison with modern plants and their climate tolerances, although scientists must be careful not to assume that plants today live exactly as their ancestors did. Pollen records are often correlated with records from other sources, such as marine plankton and ice cores, to minimize error. For example, researchers at the Faculte de St. Jerome in France were able to estimate climatic range and variability in the Eemian interglacial period, approximately 130,000 to 120,000 years ago, using pollen analysis in conjunction with other methods. They found that the warmest winter temperatures occurred in the first three millennia of the period, followed by a rapid shift to cooler winter temperatures between 4,000 and 5,000 years after the beginning of the Eemian. After that, annual variations of temperature and precipitation were slight, only 2–4°C and 200–400 millimeters per year. Pollen records can have very high resolution, on the order of a single year when taken from annually deposited lake sediments.

157

Tree-ring analysis is usually employed within the time span of the historical record, although it can also be employed with fossilized trees. The thickness of tree rings is affected by temperature, precipitation, daily sunlight and other environmental factors. Trees grow thicker rings in years with optimal conditions, and thinner rings when conditions are less conducive to the growth of the trees. Scars and burn marks can also be used to identify fires and other events. These events can often be correlated with historical records to establish precise dates. Slices of different trees can also be correlated with one another to construct records stretching back hundreds and even thousands of years. In areas with good tree records, such as the dry American Southwest, tree-ring analysis has extremely fine resolution. In the White Mountains, the bristlecone pine tree chronology goes back nearly ten thousand years, to 7,000 BCE, almost to the end of the last ice age. Bristlecone pine chronologies have been used to recalibrate the carbon 14 dating process. Tree-ring analysis can also provide information about the effects of pollution. Using similar methods with coral, scientists have reconstructed sea surface temperatures and salinity levels for the last few centuries.

Several other methods of climate reconstruction are used to infer temperature and precipitation from millions of years ago. These methods often rely on the fossil record, particularly that of plants. Leaf physiognomy methods rely on physical characteristics of leaves thought to be independent of species in order to estimate precipitation and temperature. For example, leaf-margin analysis compares the ratio of leaves with smooth margins to leaves with toothed margins. Tropical environments have a higher percentage of smooth-margined leaves than do temperate environments. The stomatal index—the ratio of the tiny holes in a given area of a leaf to the overall number—can provide information about atmospheric CO_2. Scientists have used many other proxies to reconstruct aspects of past climates, and new methods are developed every year, further refining understanding of past climates and improving models of future climates.

—*Melissa A. Barton*

FURTHER READING

Alley, Richard B. *The Two-Mile Time Machine. Ice Cores, Abrupt Climate Change and Our Future* Princeton University Press, 2014. Describes the analytical methods and interpretations of physical clues of past climates trapped in ice that is thousands of years old.

Bradley, Raymond S. *Paleoclimatology: Reconstructing Climates of the Quaternary*. Academic Press, 2015. A definitive introduction to the science of paleoclimatology, and the use of climate proxies.

Mackay, Anson, Battarbee, Rick, Birks, John and Oldfield, Frank, eds. *Global Change in the Holocene*. Routledge, 2014. Chapters by several contributors describe Holocene climate reconstruction from the resources of many different climate proxies.

Mann, Michael E. *The Hockey Stick and the Climate Wars. Dispatches from the Front Lines.* Columbia University Press, 2012. In presenting the historical contentions of the exposition of anthropogenic climate change, the role of the application of climate reconstruction methods is discussed at some length.

Speer, James H. *Fundamentals of Tree-Ring Research*. University of Arizona Press, 2010. A thorough presentation of the methodology of tree-rings and their use as environmental and climate proxies.

CLIMATE REFUGEES

FIELDS OF STUDY

Environmental Studies; Environmental Science; Physical Geography; Ecology; Sociology; Anthropology

SUMMARY

Humans, as well as animal and plant species, have always migrated in response to climatic changes, and a rapid global climate change in the twenty-first century would be no exception. In some parts of the world, climate change may make the environment inhospitable or unsuitable for human habitation, leading to exodus of the affected populations. People who migrate for climatic reasons are sometimes referred to as "climate refugees." However, refugees, as defined by the 1951 U.N. Refugee Convention, are those who flee their home country under justified fear of persecution due to their religion, ethnicity, nationality, or social or political affiliation. Migrants fleeing the effects of climate change do not usually fit this definition. Rather, they fall into the broader category of displaced persons, those who migrate internally or internationally to escape intolerable conditions such as civil strife, economic collapse, or land degradation. Such movement is termed "forced migration."

PRINCIPAL TERMS

- **NGO:** acronym for Non-Governmental Organization, generally not-for-profit, charitable and humanitarian in purpose
- **sea-level rise:** generally indicates the gradual increase over time of the average sea level due to the influx of meltwater from glaciers and polar icecaps, but also applies locally in events such as storm surge

DRIVERS OF FORCED MIGRATION

Climate change can cause forced migration in several different ways that render a region unsuitable for the species that exist in that region. "Slow disasters," such as drought, desertification, and glacier loss, may gradually render an area uninhabitable. Alternatively, climate change can also alter regions that have been unsuitable for occupation so that it becomes possible for those species to exist there. This is more applicable to plants and animals, however, and less so for humans. Climate-related natural disasters such as hurricanes and floods can also necessitate permanent migration. Perhaps the most obvious driver of forced migration for humans is sea-level rise, which can obliterate homes, communities, and entire countries. Finally, violent conflict may arise as a secondary effect from climate-induced resource scarcity.

SIGNIFICANCE FOR CLIMATE CHANGE

International law confers refugee status only upon those who cross international borders to flee violent persecution. Climate "refugees" who are internally displaced, and those who are escaping nonviolent conditions, currently have no formal recourse to international aid. However, many countries and organizations have begun to acknowledge the need for a concerted response to climate-induced migration.

Kibumba is a displaced persons camp about forty minutes north of Goma, close to the border with Rwanda. It is built upon the ruins of the camps for Rwandan refugees.

An impending problem with no legal precedent is the disappearance of an entire country due to sea-level rise, leaving the citizens of that country stateless. With projected rates of sea-level rise, several island states will lose their entire territory within the twenty-first century. One such country is Tuvalu, from which citizens have already begun evacuating. Although other countries have no legal obligation to accept Tuvaluan migrants, New Zealand has formally invited them since 2001 (at the rate of 75 per year). Australia has declined to enter into a similar agreement. Tuvalu Overview, a Japanese NGO, has sought to call attention to the plight of Tuvalu and other small island states by documenting the life of each of Tuvalu's ten thousand citizens in photos and stories.

Other climate-related causes of human migration, such as floods, droughts, storms, and land degradation, are even more ambiguous in their legal implications, since these events cannot be definitively attributed to climate change. The Office of the United Nations High Commissioner for Refugees (UNHCR) has expressed concern about the impending problem of populations displaced by climate change, but its mandate does not extend to most of those populations. New legal and humanitarian arrangements may prove necessary. Since attribution of blame will often be difficult or impossible, the best solution may be voluntary aid from nations that have the resources to assist.

The number of people likely to migrate because of climate change is highly uncertain. Predictions of such migration over the next century vary from a few tens of millions of people (mainly due to the direct effect of sea-level rise) to over a billion people (due to drought, crop failure, storms, conflict, and other indirect effects). More accurate estimates are needed to help institutions prepare for the likely consequences.

Despite the great uncertainties surrounding climate-related forced migration, there is a general consensus that developing countries will be the most negatively affected. This is due to their greater climatic vulnerability, their lower adaptive capacity, and the fact that they already host most of the world's displaced persons and refugees. However, vigorous adaptation efforts may be able to prevent the most serious consequences of climate change and reduce the need for migration. There will still be instances of unavoidable migration, such as disappearance of a country below sea level, in which case strengthened institutional and legal frameworks will be needed to help resettle the affected population as promptly and equitably as possible.

—*Amber C. Kerr*

FURTHER READING

Avidan, Kent and Behrman, Simon *Facilitating the Resettlement and Rights of Climate Refugees. An Argument for Developing Existing Principles and Practices.* Routledge, 2108. This book attempts to construct a contextual framework for the issues of climate refugees as a base for the development of principles and policies for nations that will become climate refugee hosts.

Behrman, Simon and Kent, Avidan, eds. *Climate Refugees. Beyond the Legal Impasse?* Routledge, 2018. An in-depth presentation of the social, economic and legal implications of climate change in regard to climate refugees.

McLeman, Robert A. *Climate and Human Migration. Past Experiences, Future Challenges.* Cambridge University Press, 2014. An advanced text that examines why people migrate, then considers each of the large-scale drivers of forced migration.

Pilkey, Orrin, Pilkey-Jarvis, Linda and Pilkey, Keith C. *Retreat from a Rising Sea. Hard Choices in an Age of Climate Change.* Columbia University Press, 2016. This book presents a close look at the impending doom of low-lying coastal cities and human societies as sea levels rise.

Wennersten, John R. and Robbins, Denise *Rising Tides: Climate Refugees in the Twenty-First Century.* Indiana University Press, 2017. Examines the potential for economic and social disaster arising from large numbers of climate refugees.

CLIMATE SENSITIVITY

FIELDS OF STUDY

Thermodynamics; Atmospheric Science; Climate Science; Climate Modeling; Mathematics; Fluid Dynamics; Earth System Modeling; Heat Transfer

SUMMARY

The global climate is a complex system that reacts to changes in its components, such as atmospheric carbon dioxide (CO_2) concentration. A change in the atmospheric concentration of CO_2 may cause a change in the radiation balance of the Earth; such a change is called "radiative forcing." Many changes can cause radiative forcing, including changes in greenhouse gas (GHG) concentration, the output of the Sun, ice cover, and aerosol concentration. In response to a change in the Earth's radiation balance, the planet's temperature will change until global energy balance is restored. How much the global temperature changes depends on internal feedbacks in the Earth's climate system that cause net amplification (positive feedback) or reduction (negative feedback) of the initial radiative forcing. Internal climate feedbacks include changes in water vapor, lapse rate, albedo, and clouds.

PRINCIPAL TERMS

- **aerosol:** fine liquid and solid particles suspended in the atmosphere, typically measuring between 0.01 and 10 microns and either natural or anthropogenic in origin
- **albedo:** a value between zero and one representing the fraction of light reflected from a body such as a planet.

- **greenhouse gases (GHGs):** gases that act to retain heat within Earth's atmosphere, tending to increase average global temperatures
- **negative feedback:** a cause-and-effect mechanism in which the result tends to decrease the effectiveness of the cause
- **positive feedback:** a cause-and-effect mechanism in which the result tends to augment the effectiveness of the cause

Equilibrium Climate Sensitivity

Equilibrium climate sensitivity (ECS) is a useful summary statistic of the behavior of the Earth's climate system. ECS is defined as the change in equilibrium of global average surface temperature in response to a doubling of the atmospheric concentration of CO_2 from preindustrial levels (from 280 parts per million to 560 parts per million). A doubling of atmospheric CO_2 causes radiative forcing of about 3.7 watts per square meter by increasing long-wave radiative absorption by CO_2. ECS is not a simple measure of the amount of thermal energy added to Earth's climate system by the CO_2 alone, because climate changes cause feedback loops in complex combinations. For example, melting ice decreases Earth's albedo, causing the planet's surface to absorb more heat and to further increase thermal energy, a positive feedback loop. At the same time, the elevated temperature tends to increase atmospheric water content and cloud cover, which acts to increase Earth's albedo and lessen the amount of energy that is absorbed, a negative feedback loop. If there were no internal climate feedbacks, then ECS would be about 1.2°C. However, because of internal climate feedbacks, ECS is likely between 2°C and 4.5°C.

ECS allows scientists to assess the change in average global temperature after Earth has reached equilibrium over several thousand years. However, it is computationally intensive to run complex global climate models to equilibrium. Thus, a modified concept termed "effective climate sensitivity" has been developed as an approximation of ECS. Effective climate sensitivity is calculated by estimating the climate feedback parameter at a specific point in time during transient climate conditions (not at equilibrium) using estimates of ocean heat storage, radiative forcing, and surface temperature change. Some studies find that effective climate sensitivity calculations underestimate the true ECS of a given model.

Significance for Climate Change

ECS is used to summarize and compare different climate models, as well as to combine information from models, historical records, and paleoclimate reconstructions. It is immensely important for making, understanding, and reacting to projections of future climate change. Every single degree of temperature difference in ECS can imply vastly different impacts on the planet over the long term.

A large number of studies have estimated ECS using a variety of methods, models, and data. There are four main categories of strategies used to estimate ECS. One method is to estimate ECS directly from observations of past climate changes. Another is to compile expert opinions. A third creates multiple versions of a single climate model by varying its parameters, and then compares the climate simulated by each version with climate observations to determine which is most likely. A final strategy is to combine the results of multiple methods into a single probability distribution. There are two main sources of climate observations used in this type of research: modern instrumental observations (after 1850), and paleoclimate reconstructions over the past thousands or millions of years.

After considering all the available research, the Fourth Assessment Report of the Intergovernmental Panel on Climate Change (IPCC) concluded that ECS is likely (with a probability of greater than 66 percent) to be between 2°C and 4.5°C, very likely (greater than 90 percent probability) to be larger than 1.5°C, and most likely to have a value of about 3°C. The estimated range of ECS has been relatively stable over a thirty-year period: A range of 1.5°—4.5°C was proposed in 1979 in a report by the National Academy of Sciences. Scientists have since improved the certainty of their estimates of the lower limit of climate sensitivity and of the transient climate response. The upper limit of ECS, however, remains difficult to quantify as a result of nonlinearities that cause a skewed probability distribution. The persistence of such large uncertainty in the value of ECS is a significant barrier to narrowing the range of projections of future climate change.

Context

There are several important limitations to the concept of ECS. It is potentially dependent upon the state of the climate system and upon the rate and magnitude of radiative forcing: A doubling of CO_2 versus a halving of CO_2 may not cause the same magnitude of temperature change. Additionally, different forcing mechanisms can have different sensitivities to radiative forcing, so ECS values may be specific to changes in CO_2. Lastly, ECS quantifies equilibrium temperature change over thousands of years, so it does not give direct projections for future climate changes over periods of hundreds of years. A separate summary statistic, transient climate response (TCR), was developed to compare the transient responses of climate models and provide shorter-term projections.

—*Carolyn P. Snyder*

FURTHER READING

Chin, Mian *Atmospheric Aerosol Properties and Climate Impacts.* DIANE Publishing, 2009. This report to the U.S. Congress presents a detailed analysis of the impacts and effects of atmospheric aerosols with respect to controlling climate change.

National Research Council *Climate Stabilization Targets. Emissions, Concentrations, and Impacts Over Decades to Millenia.* National Academies Press, 2011. Offers an extensive analysis of climate forcing factors in the context of climate sensitivity.

Rapp, Donald *Assessing Climate Change. Temperatures, Solar Radiation, and Heat Balance.* 3rd ed., Springer, 2014. This book discusses historical and present factors affecting global climate, with regard to their role in climate sensitivity assessments and global climate models.

Singh, S.N., ed. *Trace Gas Emissions and Plants.* Springer, 2013. This book provides a basic introduction to the mathematical principles of climate sensitivity, assuming only a basic understanding of differential calculus.

CLIMATE VARIABILITY

FIELDS OF STUDY

Climatology; Climate Classification; Climate Zones; Atmospheric Science; Thermodynamics; Environmental Sciences; Environmental Studies; Physical Chemistry; Meteorology; Climate Modeling; Process Modeling; Fluid Dynamics; Hydroclimatology; Hydrometeorology; Bioclimatology; Oceanography; Hydrology; Physical Geography; Earth System Modeling

SUMMARY

The term "climate variability" denotes the inherent tendency of the climate of a specific area to change over time. Climate lag denotes a delay in the climate change prompted by a particular factor. The time period considered would normally be at least fifty years, but a period of at least one hundred years is usually more appropriate. Instrumental climatic observations have been taken in most areas of the world for at least one hundred years, and in some areas for more than two hundred years, and any analysis of climate variability over time should utilize the full record of those observations. Longer time periods, from one thousand years to a geological period, may also be studied using proxies.

PRINCIPAL TERMS

- **climate change:** alterations in long-term meteorological averages in a given region or globally
- **climate fluctuations:** changes in the statistical distributions used to describe climate states
- **climate normals:** averages of a climatic variable for a uniform period of thirty years
- **climatic oscillation:** a fluctuation of a climatic variable in which the variable tends to move gradually and smoothly between successive maxima and minima
- **climatic trend:** a climatic change characterized by a smooth monotonic increase or decrease of the average value in the period of record

CLIMATE

Climate is an abstraction, a synthesis of the day-to-day weather conditions in a given area over a long period of time. The main climate elements are precipitation, temperature, humidity, sunshine, radiation, wind speed, wind direction, and phenomena such

Two views of Briksdalsbreen (The Briksdal glacier) photographed from nearly the same place. The picture on the left has been taken in the end of July in year 2003 and the another picture on the right has been taken on 4th August 2008.

as fog, frost, thunder, gales, cloudiness, evaporation, and grass and soil temperatures. In addition, meteorological elements observed in the upper air may be included where appropriate. The climate of any area may also be described as a statistical analysis of weather and atmospheric patterns, such as the frequency or infrequency of specific events.

In the most general sense, the term "climate variability" denotes the inherent tendency of the climate of a specific area to change over time. The time period considered would normally be at least fifty years, but a period of at least one hundred years is usually more appropriate. Instrumental climatic observations have been taken in most areas of the world for at least one hundred years, and in some areas for more than two hundred years, and any analysis of climate variability over time should utilize the full record of those observations. Longer time periods, from one thousand years to a geological period, may also be studied using proxies.

Magnitude of Climate Variability

The degree or magnitude of climate variability can best be described through the statistical differences between long-term measurements of meteorological elements calculated for different periods. In this sense, the measure of climate variability is essentially the same as the measure of climate change. The term climate variability is also used to describe deviations of the climate statistics over a period of time (such as a month, season, or year) from the long-term statistics relating to the same calendar period. In this sense, the measure of climate variability is generally termed a climate anomaly.

Climate Properties

Three basic properties characterize the climate of an area. Thermal properties include surface air temperatures above water, land, and ice. Kinetic properties include wind and ocean currents, which are affected by vertical motions and the motions of air masses, aqueous humidity, cloudiness, cloud water content, groundwater, lake lands, and the water content of snow on land and sea ice. Finally, static properties include pressure and density of the atmosphere and oceans, composition of the dry air, salinity of the oceans, and the geometric boundaries and physical constants of the system. These three types of properties are interconnected by various physical processes, such as precipitation, evaporation, infrared radiation, convection, advection, and turbulence. The climate is a complex system, and any consideration of climate variations, especially in terms of global warming, must be carefully evaluated.

Climate Variability Over the Last Thousand Years

One thousand years ago, parts of the Earth were warm and dry. The Atlantic Ocean and the North Sea were almost free of storms. This was the time of the great Viking voyages. Vineyards flourished in England.

In contrast, frosts occurred in the Mediterranean area, and rivers such as the Tiber in Rome and the Nile in Cairo occasionally froze. This suggests that a shift occurred in the pattern of large-scale European weather systems. However, by about 1200, the benign climate in Western Europe began to deteriorate, and climate extremes characterized the next two centuries. From about 1400 to 1550, the climate grew colder again, and about 1550 a three-hundred-year cold spell known as the Little Ice Age began. (The term "Little Ice Age" is used differently by different writers. Many use it to refer to the climate cooling from about 1300 to 1850, while others use it for the latter half of that interval, when cooling was greatest, beginning around 1550 or 1600.) Around 1850, the cold temperatures began to moderate, and from about 1900 a relatively steady warming trend, with a few intervening cold periods, occurred in many areas of the world. After the peak warmth of the year 1998 for the world as a whole, global temperatures remained relatively stable, with some cooling trends in a few areas. In the twenty-first century, however, a consistent, but slow, global warming trend has been observed that is most noticeable in the accelerated loss of ice coverage in the Arctic and Antarctic.

Climate Lag

A delay in the climate change can be prompted by a particular factor and can occur when the system is influenced by another, slower-acting, factor. For example, when carbon dioxide (CO_2) is released into the atmosphere, its full effect may not be recognized immediately, because some of the CO_2 may be partially absorbed by ocean waters and subsequently be released much later. The considerable amount of lag in geophysical systems can be seen in the delay between actions that increase or decrease climate forcings (changes that affect the energy balance of Earth) and their consequent impacts on the climate. Lag can be accounted for in several ways: Some occurs because of the length of time it takes for certain chemicals to cycle out of the atmosphere, some results from the effects of warming upon natural cycles, and some comes from the slow pace of oceanic temperature change. Many climate scientists estimate climate lag to be between twenty and thirty years. Thus, in this example, even if all additional carbon emissions were to immediately cease, the climate would continue on as though nothing had happened for two to three decades before the effect of the cessation in CO_2 emissions would be seen.

Context

The climate has always varied and will continue to vary, but it is important to differentiate internal variations, or changes that do not imply instability, with external variations, which are attributable to forcing. Changes in the intensity of seasons or of rainfall may presage global warming, or they may simply be temporary oscillations within an existing system. The longer the time period under discussion, moreover, the more difficult it is to tell where unidirectional climate change begins or ends. The longer some kinds of climate disruption are delayed, and the more climate commitment is built up, the more likely it is that feedback effects will be seen. "Climate commitment" refers to the fact that climate reacts with a delay to influencing factors.

Climate variability, when seen as an inherent characteristic of Earth's atmosphere, can be treated as a reason to accept global changes as natural and beyond human control. When seen as a result of human activity, the same variability can be treated as a call to action to reverse global warming. There are at least three responses to this situation: prevention, mitigation, and remediation. The potential for dangerous feedback effects drives a prevention response, that is, action that reduces the global warming risk. A mitigation response takes a practical approach: Climate disaster is already imminent, and the best prevention efforts may be too little, too late, but there is still the need to reduce the worst of the threats. A remediation approach would not look at ways to change anthropogenic causes and their consequences. For example, efforts might be made to use geoengineering, that is, to alter the core geophysical processes that relate to anthropogenic climate factors.

Evaluations of the limits of natural variability are therefore crucial precursors of evaluations of climate change itself. What will prove to be the correct argument remains for the future, but irrespective of the truth of the two arguments, it is evident that people and societies must adapt to climate changes, and those communities that adapt to the varying climate will be in a better position to withstand the climatic variability of the future.

—*W. J. Maunder*
—*Victoria Price*

FURTHER READING

Broecker, W. S. "Does the Trigger for Abrupt Climate Change Reside in the Ocean or the Atmosphere?" *Science* 300 (2003): 1519–1522. Good example of a discussion on some aspects of the natural causes of climate change.

Broecker, Wally *The Great Ocean Conveyor. Discovering the Trigger for Abrupt Climate Change.* Princeton University Press, 2010. Describes the global oceanic overturning circulation system and how its disruption by climate change may affect the climate of Earth.

Conkling, Philip, Alley, Richard, Broecker, Wallace and Danton, George *The Fate of Greenland: Lessons from Abrupt Climate Change.* MIT Press, 2013.

Mann, Michael E. *The Hockey Stick and the Climate Wars. Dispatches from the Front Lines.* Columbia University Press, 2014. This very readable book lays out the science behind global climate change and identifies the conflict between climate change proponents and deniers.

National Research Council, Division of Earth and Life Sciences *Abrupt Impacts of Climate Change: Anticipating Surprises.* National Academics Press, 2013.

Rashid, Haruna, Polyak, Leonid and Mosely-Thompson, Ellen *Abrupt Climate Change Mechanisms, Patterns and Impacts.* John Wiley & Sons, 2013.

CLIMATE ZONES

FIELDS OF STUDY

Climate Classification; Climate Zones; Climatology; Bioclimatology; Ecology; Physical Geography; Environmental Sciences; Environmental Studies

SUMMARY

Climate, the average weather conditions in a specific region over a period of at least thirty years, is determined by various factors, the most important of which are the amount of precipitation and the temperature of the air. Climate controls the major ecological community types, or biomes; that is, the climate in a given region determines the flora and fauna that will thrive in that region.

PRINCIPAL TERMS

- **biome:** a particular climatic region and the animals and plants that normally inhabit that region
- **climate:** long-term, average, regional or global weather patterns
- **deforestation:** the loss of forest cover in an area by natural or artificial means
- **tundra:** the treeless region between the northern tree line, characterized by low-growing vegetation and permafrost subsoil

THE KÖPPEN CLASSIFICATION SYSTEM

In 1900, Wladimir Köppen (1846–1940), a German climatologist, developed what has become the most widely used system for classifying world climates. The Köppen system identifies five major climate zones: tropical moist climates (A zone), dry climates (B zone), humid middle latitude climates (C zone), continental climates (D zone), and cold climates (E zone). Köppen also used two subgroups to more specifically describe the zones.

Tropical moist climates are characterized by year-round high temperatures and large amounts of rain. Rainfall is adequate all year round, and there is no dry season. This zone is typical of northern parts of South America, central Africa, Malaysia, Indonesia, and Papua New Guinea. The dry climate zone has little rain and a wide range of daily temperatures. There is a dry season in the summer and winter, with a mean annual temperature around 18°C, as in the western United States, northern and extreme southern Africa, parts of central Asia, and most of Australia. The humid middle latitude climate, or temperate zone, has hot-to-warm, dry summers and cool, wet winters, but no dry season as such. Southeastern sections of the United States and South America, westernmost Europe, and the southeast corner of China fit this category. The continental climate zone, in interior regions of large land masses such as Canada and northern Europe and Asia, experience varied seasonal temperatures and moderate rainfall. The cold climate zone, characterized by permanent ice and ever-present tundra, occupies Greenland and the most northerly parts of Asia.

Africa, Present Day

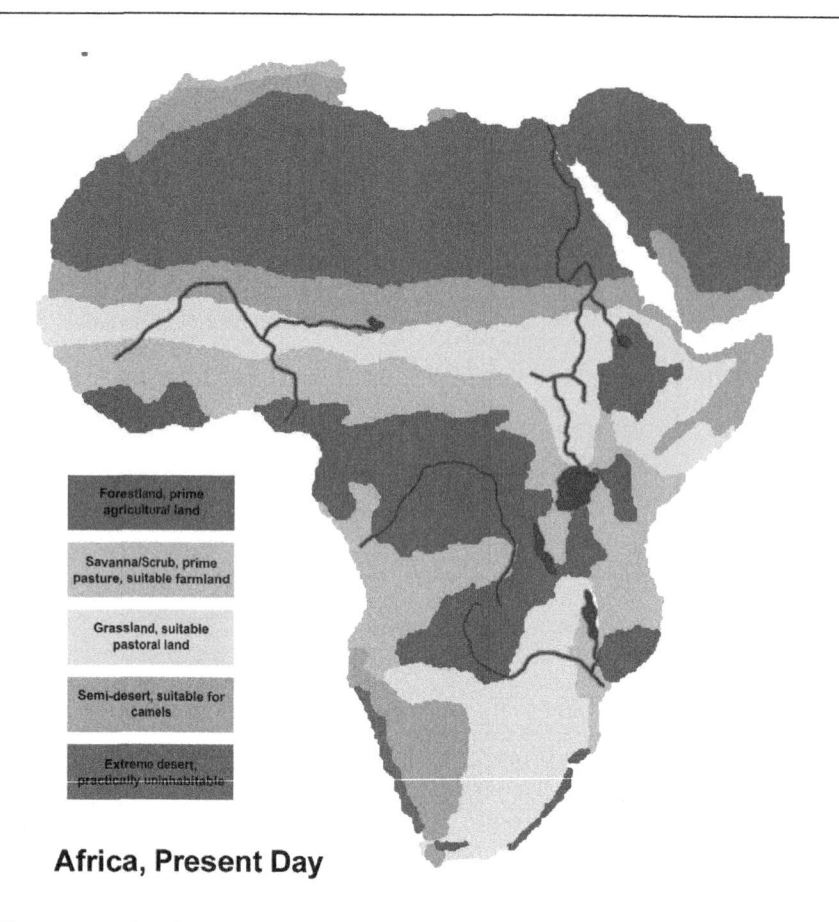

Climate areas based on data and maps from the Oak Ridge National Laboratory Paleovegitation project.

SIGNIFICANCE FOR CLIMATE CHANGE

While acknowledging some unknowns and uncertainty, researchers predict that, if global warming caused by carbon dioxide (CO_2) and other greenhouse gas (GHG) emissions continue at the current rate, recognized climate zones will shift in extent and some of the climate zones recognized in the early twenty-first century could disappear entirely by the end of the century, giving way to new climate zones on up to 39 percent of the world's land surface. Major areas that could be affected are tropical highlands and polar regions. Broad strips of areas labeled tropics and subtropics at the beginning of the twenty-first century could develop new climates that do not resemble any of the zones in categories assigned in the Köppen Climate Classification System. Researchers predict that heavily populated areas such as the southeastern United States, southeastern Asia, parts of Africa such as its mountain ranges, the Amazonian rain forest, and South American mountain ranges are likely to be the most severely affected.

Climate change patterns could affect ecosystems on a global scale. For example, major changes in the forests of North America could result. The ranges of the yellow birch, sugar maple, hemlock, and beech trees are expected to shift northward up to 1,000 kilometers while abandoning entirely their present-day locations. Animals will also be affected, as those from colder climes are forced to move into more northerly regions and those from warmer climes range farther north and south. Temperature and rainfall patterns would change breeding and migration patterns.

For humans, a grave concern is global food production. One climate model predicts that the corn belt in midwestern North America will move northward, even farther into central Canada, and winter wheat may replace corn in parts of the present corn belt. Within several decades, the Swiss Alps could become home to a Mediterranean climate, with wet winters and long, dry, warm summers. Within one hundred years, the climate zone in southern Switzerland may move northward by as much as 500 kilometers. In the western Alps, the climate may come to resemble that found in southern France in the early twenty-first century.

Biodiversity in South Africa is expected to be affected substantially by shifting climate zones: Species will experience extinction on a wide scale; up to half the country will see a climate not known before; succulent karoo, a globally important arid-climate hotspot, and biomes in the *fynbos* (a Mediterranean-climate thicket) will suffer. While the degree of this change remains speculative, availability of food could affect sub-Saharan West Africa, as vegetation zones move southward.

Deforestation will continue in the Amazonian forest areas of South America, and new climates are expected to be created near the equator. Some researchers predict that mountainous areas such as those found in Peru and the Colombian Andes, as well as regions in Siberia and southern Australia, could experience the disappearance of climates completely. Devastation of critical ecosystems and changes in agricultural patterns could severely affect Australia, New Zealand, and the developing island nations of the Pacific, while rising sea levels will eliminate some low-lying island nations from existence. With so many factors still undetermined or speculative, however, it remains to be seen how climate zone changes will play out in the future.

—*Victoria Price*

FURTHER READING

Bonan, Gordon *Ecological Climatology: Concepts and Applications.* 3rd ed., Cambridge University Press, 2015. Provides a concise definitive description of Earth's recognized climate zones within the broader context of climatology.

Maslin, Mark. *Global Warming: A Very Short Introduction.* Oxford University Press, 2009. Provides a concise general introduction to the major factors that define the climate of Earth.

Roggema, Rob *Adaptation to Climate Change: A Spatial Challenge.* Springer, 2009. The chapters of this book present the planned approaches of several European nations for adapting to the challenges that shifting climate zones will present.

Schimel, David *Climate and Ecosystems.* Princeton University Press, 2013. Describes the mutual interaction of climate and ecosystem within biomes.

CLIMATOLOGY

FIELDS OF STUDY

Atmospheric Chemistry; Biochemistry; Environmental Chemistry; Geochemistry; Inorganic Chemistry; Physical Chemistry; Phorochemistry; Thermodynamics; Chemical Engineering; Environmental Sciences; Environmental Studies; Physical Sciences; Meteorology; Process Modeling; Fluid Dynamics; Chemical Kinetics; Hydroclimatology; Atmospheric Science; Oceanography; Hydrology; Physical Geography; Ecosystem Management; Ecology; Spectroscopy

SUMMARY

Climatology deals with the science of climate, which includes the huge variety of weather events. These events change at periods of time that range from months to millennia. Climate has such a profound influence on all forms of life, including human life, that people have made numerous attempts to predict future climatic conditions. These attempts resulted in research efforts to try to understand future changes in the climate as a consequence of anthropogenic and naturally caused activity.

DEFINITION AND BASIC PRINCIPLES

"Weather" pertains to atmospheric conditions that constantly change, hourly and daily. In contrast, "climate" refers to the long-term composite of weather conditions at a particular location, such as a city or a state. Climate at a location is based on daily mean conditions that have been aggregated over periods of time that range from months and years to decades and centuries. Both weather and climate involve

The Greenhouse Effect

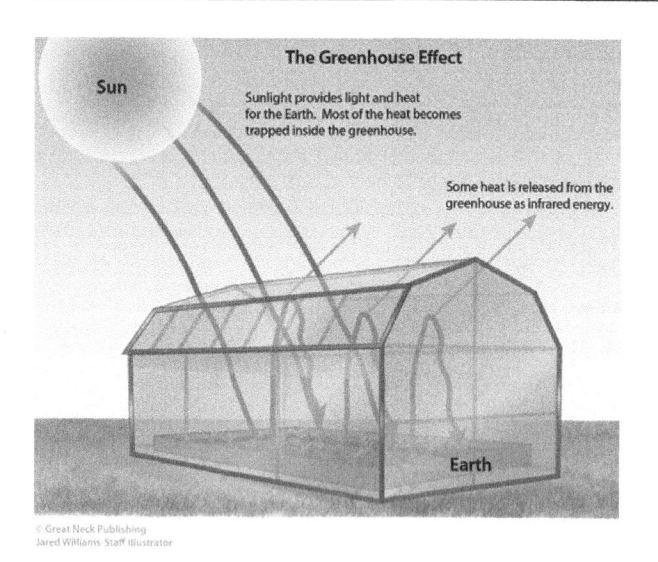

The greenhouse effect refers to the process in which longwave radiation is trapped in the atmosphere and then radiated back to the Earth's surface.

measurements of the same conditions: air temperature, water vapor in the air (humidity), atmospheric pressure, wind direction and speed, cloud types and extent, and the amount and kind of precipitation.

Estimates of ancient climates, going back several thousand years or more, are produced in various ways. For example, the vast amount of groundwater discovered in southern Libya indicates that during some period in the past, that part of the Sahara Desert was much wetter. In ancient Egypt, Nilometers, stone markers built along the banks of the Nile, were used to gauge the height of the river from year to year. They are similar to the staff gauges that are used by the US Geological Survey to indicate stream or canal elevation. The height of the Nilometer reflects the extent of precipitation and associated runoff in the headwaters of the Nile in east central Africa.

Background and History

Measurements of precipitation were being made and recorded in India during the fourth century BCE. Precipitation records were kept in Palestine about 100 BCE, Korea in the 1440s, and in England during the late seventeenth century. Galileo invented the thermometer in the early 1600s. Physicist Daniel Fahrenheit created a measuring scale for a liquid-in-glass thermometer in 1714, and Swedish astronomer Anders Celsius developed the centigrade scale in 1742. Italian physicist Evangelista Torricelli, who worked with Galileo, invented the barometer in 1643.

The first attempt to explain the circulation of the atmosphere around the Earth was made by English astronomer Edmond Halley, who published a paper charting the trade winds in 1686. In 1735, English meteorologist George Hadley further explained the movement of the trade winds, describing what became known as a Hadley cell, and in 1831, Gustave-Gaspard Coriolis developed equations to describe the movement of air on a rotating Earth. In 1856, American meteorologist William Ferrel developed a theory that described the mid-latitude atmospheric circulation cell (Ferrel cell). In 1860, Dutch meteorologist Christophorus Buys Ballot demonstrated the relationship between pressure, wind speed, and direction (which became known as Buys Ballot's law).

The first map of average annual isotherms (lines connecting points having the same temperature) for the northern hemisphere was created by German naturalist Alexander von Humboldt in 1817. In 1848, German meteorologist Heinrich Wilhelm Dove created a world map of monthly mean temperatures. In 1845, German geographer Heinrich Berghaus prepared a global map of precipitation. In 1882, the first world map of precipitation using mean annual isohyets (lines connecting points having the same precipitation) appeared.

Earth's Global Energy Balance.

The Earth's elliptical orbit about the Sun ranges from 91.5 million miles at perihelion (closest to the Sun) in January, to 94.5 million miles at aphelion (furthest from the Sun) in July, averaging 93 million miles. The Earth intercepts about two-billionth of the total energy output of the Sun. Upon reaching the Earth, a portion of the incoming radiation is reflected back into space, while another portion is absorbed by the atmosphere, land, or oceans. Over time, the incoming shortwave solar radiation is balanced by a return to outer space of longwave radiation.

The Earth's atmosphere extends to an estimated height of about six thousand miles. Most of it is made up of nitrogen (78 percent by volume) and oxygen (about 21 percent). Of the remaining 1 percent, carbon dioxide (CO_2) accounts for about 0.0385 percent of the atmosphere. This is a minute amount, but carbon dioxide can absorb both incoming shortwave radiation from the sun and outgoing longwave radiation from the Earth. The measured increase in carbon dioxide since the early 1900s is a major cause for concern as it is a very good absorber of heat radiation, which adds to the greenhouse effect.

Air Temperature

Air temperature is a fundamental constituent of climatic variation on the Earth. The amount of solar energy that the Earth receives is governed by the latitude (from the equator to the poles) and the season. The amount of solar energy reaching low-latitude locations is greater than that reaching higher-latitude sites closer to the poles. Another factor pertaining to air temperature is the fivefold difference between the specific heat of water (1.0) and dry land (0.2). Accordingly, areas near the water have more moderate temperatures on an annual basis than inland continental locations, which have much greater seasonal differences.

Anthropogenic (human-induced) changes in land cover in addition to aerosols and cloud changes can result in some degree of global cooling, but this is much less than the combined effect of greenhouse gases in global warming. The gases include carbon dioxide from the burning of fossil fuels (coal, oil, and natural gas), which has been increasing since the second half of the twentieth century. Other gases such as methane (CH_4), chlorofluorocarbons (CFCs), ozone (O_3), and nitrous oxide (NO_3) also create additional warming effects.

Air temperature is measured at five feet above the ground surface and generally includes the maximum and minimum observation for a twenty-four-hour period. The average of the maximum and minimum temperature is the mean daily temperature for that particular location.

Earth's Available Water

Water is a tasteless, transparent, and odorless compound that is essential to all biological, chemical, and physical processes. Almost all the water on the Earth is in the oceans, seas, and bays (96.5 percent) and another 1.74 percent is frozen in ice caps and glaciers. Accordingly, 98.24 percent of the total amount of water on this planet is either frozen or too salty and must be thawed or desalinated. About 0.76 percent of the world's water is fresh (not saline) groundwater, but a large portion of this is found at depths too great to be reached by drilling. Freshwater lakes make up 0.007 percent, and atmospheric water is about 0.001 percent of the total. The combined average flows of all the streams on Earth—from tiny brooks to the mighty Amazon River—account for 0.0002 percent of the total.

Air Masses

The lowest layer of the atmosphere is the troposphere, which varies in height from ten miles at the equator and lower latitudes to four miles at the poles. Different types of air masses within the troposphere can be delineated on the basis of their similarity in temperature, moisture, and to a certain extent, air pressure. Air masses develop over continental and maritime locations that strongly determine their physical characteristics. For example, an air mass starting in a cold, dry interior portion of a continent would develop thermal, moisture, and pressure characteristics that would be substantially different from those of an air mass that developed over water. Atmospheric dynamics also allow air masses to modify their characteristics as they move from land to water and vice versa.

Air mass and weather front terminology were developed in Norway during World War I. The Norwegian meteorologists were unable to get weather reports from the Atlantic theater of operations; consequently, they developed a dense network of weather stations that led to impressive advances in atmospheric modeling.

Greenhouse Effect

Selected gases in the lower parts of the atmosphere trap heat and radiate some of that heat back to Earth. If there was no natural greenhouse effect, the Earth's overall average temperature would be close to 0 degrees Fahrenheit rather than 57 degrees Fahrenheit.

The burning of coal, oil, and gas makes carbon dioxide the major greenhouse gas. Carbon dioxide accounts for nearly half of the total amount of heat-producing gases in the atmosphere. In mid-eighteenth century Great Britain, before the Industrial Revolution, the estimated level of carbon dioxide was about 280 parts per million (ppm). Estimates for the natural range of carbon dioxide for the past 650,000 years are 180–300 ppm. All of these values are less than the 2015 estimate of 399 ppm. Since 2000, atmospheric carbon dioxide has been increasing at a rate of 1.9 ppm per year. Carbon dioxide emissions accounted for about three-quarters of all the gases that affect global warming as of 2014. Estimates of carbon dioxide values at the end of the twenty-first century range from 490 to 1,260 ppm.

The second most important greenhouse gas is methane (CH_4), which accounts for about 16 percent of all of the global warming factors as of 2014. This gas originates from the natural decay of organic matter in wetlands, but anthropogenic activity in the form of rice paddies, manure from farm animals, the decay of bacteria in sewage and landfills, leaks from natural gas production and distribution, and biomass burning (both natural and human-induced) doubles the amount produced.

Chlorofluorocarbons (CFCs) absorb longwave energy (warming effect) but also have the ability to destroy stratospheric ozone (cooling effect). The warming radiative effect is three times greater than the cooling effect. CFCs accounted for about 2 percent of all of the global warming factors in 2014. Nitrous oxide (N_2O) from motor vehicle exhaust and bacterial emissions from nitrogen fertilizers accounted for about 6 percent of all the global warming factors.

Several human actions lead to a cooling of the Earth's climate. For example, the burning of fossil fuels results in the release of tropospheric aerosols, which acts to scatter incoming solar radiation back into space, thereby lowering the amount of solar energy that can reach the Earth's surface. These aerosols also lead to the development of low and bright clouds that are quite effective in reflecting solar radiation back into space.

Climatology involves the measurement and recording of many physical characteristics of the Earth. Therefore, numerous instruments and methods have been devised to perform these tasks and obtain accurate measurements.

MEASURING TEMPERATURE

At first glance, it would appear that obtaining air temperatures would be relatively simple. After all, thermometers have been around since 1714 (Fahrenheit scale) and 1742 (Celsius scale). However, accurate temperature measurements require a white (high-reflectivity) instrument shelter with louvered sides for ventilation, placed where it will not receive direct sunlight. The standard height for the thermometer is five feet above the ground.

REMOTE-SENSING TECHNIQUES

Oceans cover about 71 percent of the Earth's surface, which means that large portions of the world do not have weather stations and places where precipitation can be measured with standard rain gauges. To provide more information about precipitation in the equatorial and tropical parts of the world, the National Aeronautics and Space Administration (NASA) and the Japanese Aerospace Exploration Agency began a program called the Tropical Rainfall Monitoring Mission (TRMM) in 1997. The TRMM satellite monitors the area of the world between 35 degrees north and 35 degrees south latitude. The goal of the study is to obtain information about the extent of precipitation, its intensity, and length of occurrence. The major instruments on the satellite include radar to detect rainfall, a passive microwave imager that can acquire data about precipitation intensity and the extent of water vapor, and a scanner that can examine objects in the visible and infrared portions of the electromagnetic spectrum. The goal of collecting this data is to obtain the necessary climatological information about atmospheric circulation in this portion of the Earth so as to develop better mathematical models for determining large-scale energy movement and precipitation.

GEOSTATIONARY SATELLITES

Geostationary orbiting earth satellites (GOES) enable researchers to view images of the planet from what appears to be a fixed position. To achieve this, these satellites circle the globe at a speed that is in step with the Earth's rotation. This means that the satellite, at an altitude of 22,200 miles, will make one complete revolution in the same twenty-four hours and direction that the Earth is turning above the Equator. At this height, the satellite is in a position to view nearly half the planet at any time. On-board instruments can be activated to look for special weather conditions such as hurricanes, flash floods, and tornadoes. The instruments can also be used to make estimates of precipitation during storm events.

RAIN GAUGES

The accurate measurement of precipitation is not as simple as it may seem. Collecting rainfall and measuring it is complicated by the possibility of debris, dead insects, leaves, and animal intrusions occurring. Standards were established, although the various national climatological offices use more than fifty types of rain gauges. The location of the gauge, its height above the ground, the possibility for splash and evaporation, its distance from trees, and turbulence all affect the results. Accordingly, all gauge records are really estimates. Precipitation estimates are also affected by the number of gauges per unit area. The number of gauges in a sample area of 3,860 square miles for Britain, the United States, and Canada is 245, 10, and 3, respectively. Although the records are reported to the nearest 0.01 inch, discrepancies occur in the official records. It is important to have a sufficiently dense network of rain gauges in urban areas. Some experts think that five to ten gauges per 100 square miles is necessary to obtain an accurate measure of rainfall.

DOPPLER RADAR
Doppler radar was first used in England in 1953 to pick up the movement of small storms. The basic principle behind Doppler radar is that the backscattered radiation frequency detected at a certain location changes over time as the target, such as a storm, moves. The mode of operation requires a transmitter that is used to send short but powerful microwave pulses. When a foreign object (or target) is intercepted, some of the outgoing energy is returned to the transmitter, where a receiver can pick up the signal. An image (or echo) from the target can then be enlarged and shown on a screen. The target's distance is revealed by the time between transmission and return. The radar screen can indicate not only where the precipitation is taking place but also its intensity by the amount of the echo's brightness. Doppler radar has developed into a very useful device for determining the location of storms and the intensity of the precipitation and for obtaining good estimates of the total amount of precipitation.

CONCLUSION
Climate change may be caused by both natural internal and external processes in the Earth-Sun system and human-induced changes in land use and the atmosphere. The United Nations Framework Convention on Climate Change states that the term "climate change" should refer to anthropogenic changes that affect the composition of the atmosphere as distinguished from natural causes, which should be referred to "climate variability." An example of natural climate variability is the global cooling of about 0.5 degrees Fahrenheit in 1992–1993 that was related to the 1991 Mount Pinatubo volcanic eruption in the Philippines. The 15 million to 20 million tons of sulfuric acid aerosols that were released into the stratosphere reflected incoming radiation from the sun and created a cooling effect. Many experts suggest that climate change is caused by human activity, as evidenced by the above-normal temperatures in the 2000s. Based on a variety of techniques that estimate temperatures in previous centuries, the year 2005 was found to be the hottest in the preceding thousand years.

Numerous observations strongly suggest a continuing warming trend. Snow and ice have retreated from areas such as Mount Kilimanjaro, which at 19,340 feet is the highest mountain in Africa, and glaciated areas in Switzerland. In the Special Report on Emission Scenarios (2000), the Intergovernmental Panel on Climate Change examined the broad spectrum of possible concentrations of greenhouse gases by examining the growth of population and industry along with the efficiency of energy use. The panel estimated future trends using computer climate models. For example, it estimated that the global temperature would increase 35.2–39.2 degrees Fahrenheit by the year 2100.

Given the effect that climate change will have on humanity, many agencies and organizations will be doing research in the area, and climatologists are likely to be needed by a variety of governmental and private entities.

—*Robert M. Hordon, PhD*

BIBLIOGRAPHY
Coley, David A. *Energy and Climate Change: Creating a Sustainable Future.* Hoboken: Wiley, 2008. Print.
Gautier, Catherine. *Oil, Water, and Climate: An Introduction.* New York: Cambridge UP, 2008. Print.
"Global Greenhouse Gas Emissions Data." *EPA,* US Environmental Protection Agency, 9 Aug. 2016, www.epa.gov/ghgemissions/global-greenhouse-gas-emissions-data. Accessed 28 Oct. 2016.
Lutgens, Frederick K., and Edward J. Tarbuck. *The Atmosphere: An Introduction to Meteorology.* 12th ed. Upper Saddle River: Prentice Hall, 2013. Print.
Strahler, Alan. *Introducing Physical Geography.* 6th ed. Hoboken: Wiley, 2013. Print.
Wolfson, Richard. *Energy, Environment, and Climate.* New York: Norton, 2008. Print.

CLOUDS AND GLOBAL CLIMATE

FIELDS OF STUDY

Atmospheric Science; Atmospheric Chemistry; Hydroclimatology; Hydrometeorology; Hydrology; Physical Chemistry; Thermodynamics; Heat Transfer; Photochemistry; Environmental Sciences; Environmental Studies; Meteorology; Climate Modeling; Mathematics; Fluid Dynamics; Chemical Kinetics; Oceanography; Climate Classification; Climate Zones

SUMMARY

Clouds have a profound influence on local and global climate. Scientists have known for many years that clouds are a major component of the greenhouse effect, which makes life possible on Earth. What is not well understood is what effect clouds may have on global warming and cooling trends. Of the many variables that factor into global warming, clouds present the greatest uncertainty in predicting climate change.

PRINCIPAL TERMS

- **aerosols:** fine liquid and solid particles suspended in the atmosphere, typically measuring between 0.01 and 10 microns and either natural or anthropogenic in origin
- **enhanced greenhouse effect:** increased retention of heat in the atmosphere resulting from anthropogenic atmospheric gases
- **global radiative equilibrium:** the maintenance of Earth's average temperature through a balance between energy transmitted by the Sun and energy returned to space
- **greenhouse gases (GHGs):** gases that act to retain heat within Earth's atmosphere, tending to increase average global temperatures
- **negative feedback:** a cause-and-effect mechanism in which the result tends to decrease the effectiveness of the cause
- **positive feedback:** a cause-and-effect mechanism in which the result tends to augment the effectiveness of the cause

CLOUDS

Clouds are composed of microdroplets or ice crystals that form around aerosols in the atmosphere. These microdroplets accumulate until they become large enough to be visible. The many shapes and sizes of clouds are divided into three basic types: cirriform (wispy and transparent), stratiform (layered), and cummuliform (mounded and fluffy). Three altitude divisions are also used for classification: high (more than 5,000 meters above sea level), middle (between 2,000 and 5,000 meters above sea level), and low (less than 2,000 meters above sea level). The terms are combined, with the upper and middle altitudes given the prefixes cirro- and alto- to further differentiate the various cloud types. For example, a midlevel stratus cloud is termed altostratus, and an upper level cirrus cloud is termed altocirrus.

Clouds reflect solar radiation back toward space, much as a mirror reflects light. They also reflect a significant portion of Earth's thermal radiation back toward the surface. Thus, all other factors being equal, nights are warmer and days are cooler when clouds are present. The amount of reflection back to space depends on cloud thickness, with a low of approximately 20 percent for cirrus clouds and a high of 90 percent for cumulonimbus clouds. Clouds contribute significantly to the global radiative balance, or radiative equilibrium, an overall balance between solar radiation received by the Earth and heat reflected back to space. If the Earth absorbed all the radiation it receives, it would be much too hot to sustain life.

GREENHOUSE EFFECT AND FEEDBACK

However, Earth does not absorb all the solar energy it receives, but instead reflects visible radiation and emits infrared radiation. Without an atmosphere, the total return of energy to space would make the Earth much colder than it is. The average surface temperature if there were no atmosphere to provide the thermal buffer that it does, would be approximately –18°C, and the vast majority of Earth would be permanently covered with ice. Clouds and other greenhouse gases (GHGs) absorb some of the energy, trapping it and sending some of it back toward Earth's surface and into the atmosphere where it is absorbed and re-emitted by other GHG molecules. The overall effect is to retain the thermal energy within the atmosphere. This is the greenhouse effect.

"Feedback" is a term that applies to any multipart system in which a change in one part produces

Types of Clouds

Name	Altitude (km)
Altocumulus	2-7
Altostratus	2-7
Cirrocumulus	5-13.75
Cirrostratus	5-13.75
Cirrus	5-13.75
Cumulonimbus	to 2
Cumulus	to 2
Nimbostratus	2-7
Stratocumulus	to 2
Stratus	to 2

Source: National Oceanic and Atmospheric Administration.

a change in the other part, which then affects the original part, and so on. A simple example is a thermostat, which responds to a drop in temperature by turning on a furnace. The furnace raises the air temperature, which then causes the thermostat to turn off the furnace. Feedback systems can affect changes in cloud thickness and prevalence as a response to climate changes, particularly temperature. The greenhouse effect creates a state of equilibrium, and cloud feedback disturbs that equilibrium.

Cloud feedback can be either positive or negative. Positive feedback creates the enhanced greenhouse effect by augmenting the natural equilibrium atmospheric heat retention and acting to raise global temperatures, while negative feedback acts to lower global temperatures. Most clouds provide both positive feedback (transmitting energy down toward the Earth) and negative feedback (transmitting solar radiation back to space). Determining whether the net feedback is positive or negative is a complicated process.

Clouds and Climate Change

Climate change can increase both positive and negative feedback. Studies have shown that cold water produced by the melting polar ice cap causes phytoplankton to release chemicals that produce more and brighter clouds, thereby increasing negative feedback. Aerosols may produce either positive or negative feedback, depending on the source of the aerosols. Volcanic eruptions and pollution from technologically advanced countries, consisting of sulfates and nitrates, generate clouds that produce negative feedback. However, the developing world produces pollution that contains these substances as well as large amounts of black carbon, the by-product of incomplete combustion of carbon-based fuels. Black carbon aerosols generate positive feedback. The net feedback of pollution is very difficult to determine.

A long-held belief is that if the Earth's climate warms, water vapor amounts in the atmosphere will increase, creating more low-level thick clouds that will generate negative feedback. That belief is being called into question by recent studies that have shown that turbulence created by rising warm air currents will actually lead to fewer clouds being formed overall.

A Source of Uncertainty

The impact of clouds on climate change is extremely difficult to model. Most climate modeling systems represent clouds with a small number of variables, masking the subtleties of cloud dynamics. This is due to the extremely complicated mathematics required to model clouds realistically. While analyzing the results of some large-scale climate studies, scientists have concluded that neither the magnitude nor the sign of cloud feedback can be relied upon.

When weather data are entered into climate models, the resulting pictures of cloud cover and thickness often do not match actual conditions. Further complicating matters, studies done with live data collection also show contradictory results. For example, some studies indicate that decreasing cloud cover over China (resulting from large amounts of pollution) may be responsible for increasing temperatures there. However, other studies performed in other parts of the world indicate that temperatures have increased as cloud cover has increased.

CONTEXT

The impact of clouds on global climate cannot be overstated. Clouds are extremely sensitive to fluctuations in solar radiation, Earth's surface temperature, and many other environmental factors, including pollution levels. It is this sensitivity, combined with nearly infinite variations in cloud size, make-up, and altitude, that makes predicting cloud feedback so difficult. Since clouds may either mitigate or increase global warming, it is imperative that scientists intensify their efforts to produce better predictive climate modeling systems, as well as promote studies that analyze data collected in the field. Throughout the world, decisions are being made using climate predictions that may be flawed as a result of the uncertainty presented by cloud behavior. As these decisions will have a considerable impact both economically and sociologically, minimizing the uncertainty presented by clouds in a warming climate may become a priority in environmental agendas.

—*Kathryn Rowberg*
—*Gail Rampke*

FURTHER READING

Andronache, Constantin, ed. *Mixed-Phase Clouds. Observations and Modeling.* Elsevier, 2018. An advanced text in which the basic principles of cloud formation are discussed and analyzed.

Boucher, Olivier *Atmospheric Aerosols: Properties and Climate Impacts.* Elsevier, 2015. For advanced readers, this book discusses the role of atmospheric aerosols in the context of cloud physics.

Heintzenberg, Jost and Charlson, Robert J. *Clouds in the Perturbed Climate System: Their Relationship to Energy Balance, Atmospheric Dynamics and Precipitation.* MIT Press, 2013. Presents and discusses the complex relationship of clouds in the control of Earth's climate.

Lohman, Ulrike, Lüönd, Felix and Mahrt, Fabian *An Introduction to Clouds From the Microscale to Climate.* Cambridge University Press, 2016. A thorough introduction to the formation and function of clouds and their role in determining the climate of Earth.

CONTINENTAL CLIMATE

FIELDS OF STUDY

Physical Geography; Climate Classification; Climate Zones; Climate Modeling; Meteorology; Bioclimatology; Hydrology; Ecosystem Management; Ecology; Atmospheric Science

SUMMARY

Defining a climate is difficult. The accuracy of weather observations and the length of time over which these data have been recorded can seriously affect the statistical understanding of a climate. Additionally, having data to produce statistical averages is no assurance that a climate's classification can accurately describe its conditions.

PRINCIPAL TERMS

- **loess soils:** fine-grained calcareous silt, usually light brown or grey in color, that is easily eroded by wind and water
- **paleoclimate:** the climate that existed during a particular period of time in the geological past
- **Pleistocene:** the geologic period from about 2.5 million to 11,700 years in the past

CONSIDER CHICAGO...

Consider Chicago's humid, continental climate: The average temperature of the city is near 10°C, but the nature of the climate in the region lends itself to extremes. Chicago's daily temperatures are usually

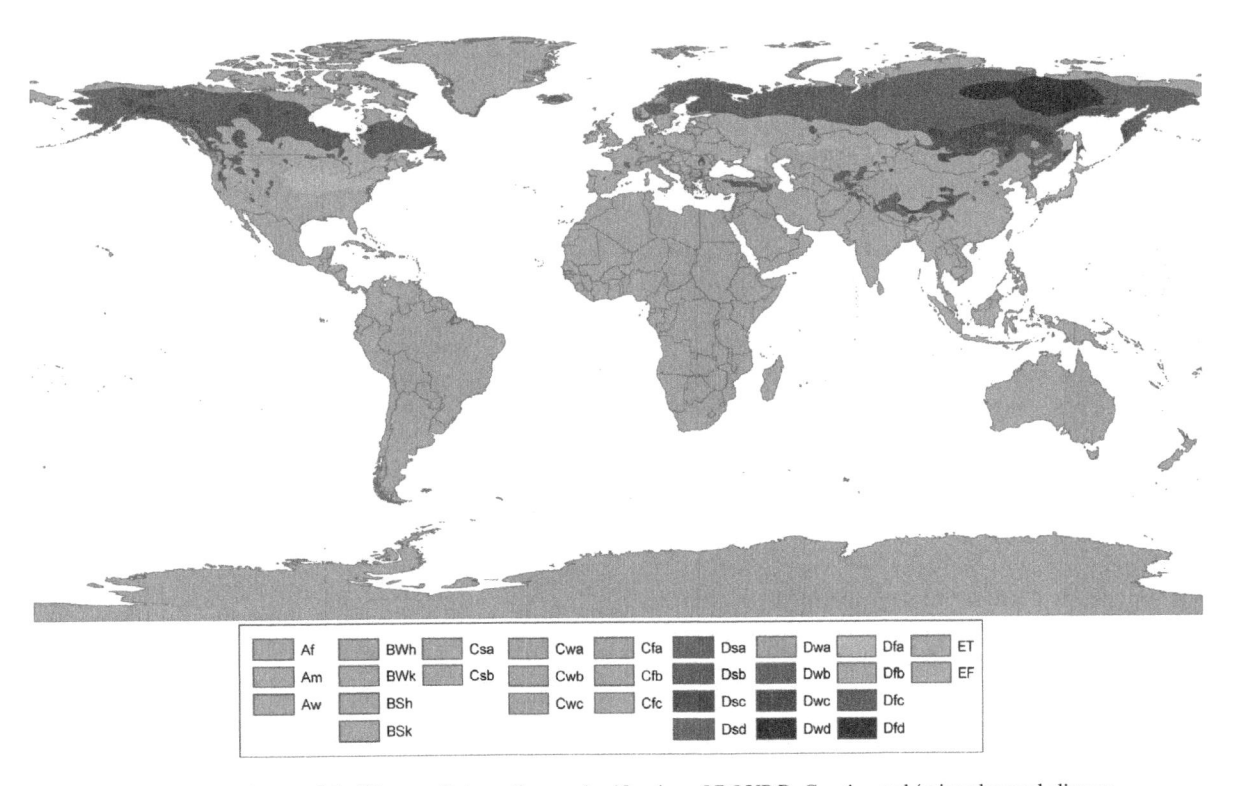

Updated world map of the Köppen-Geiger climate classification. GROUP D: Continental/microthermal climate.

either well above or well below the average. The average precipitation in Chicago is about 84 centimeters per year. This is equal to the precipitation level in Seattle, which has a west coast marine climate. Precipitation in Seattle is mostly rain, however, while Chicago experiences a mix of rain and snow. Rain in Chicago falls mostly in the summer, during violent convective storms. Rain in Seattle continues through the year, with a slight increase during the winter. Seattle's maritime influence moderates the tendency toward convective storms.

Continental climates are characterized by extreme temperatures, and they are unique in that they have a global Northern Hemispheric distribution. In fact, to understand the potential impact that climate change might have on a continental climate, one must first consider its unique geographic character. Humid continental climates are found in North America, Europe, and Asia. In North America, they encompass all of the northeastern quarter of the country, and they are bounded by the humid, subtropical climates of the southern and southeastern United States.

The North American continental climate extends northward into Canada at a line of latitude that marks the extent of viable agricultural production. It extends westward into southern portions of the Prairie Provinces of Canada and north of the middle-latitude, dry climates that prevail just east of the Rocky Mountains. In Europe, the humid continental climate begins directly east of the marine-climate boundaries, along the west coast of the continent. It stretches into Russia, Germany, southern Sweden, and Finland, and is also found in Romania and Poland. Parts of Asia are also included in this climate. Almost all of northern China, the northern part of Japan, and North Korea fall within the humid continental climate zone.

The Köppen classification system places these climates in a group of midlatitude climates. Such climates include the west coast marine climate, the middle latitude dry climate, the humid continental-warm (long) summer climate, and the humid continental-cool (short) summer climate. The term "continental" has significance from the standpoint of the climate's paleoclimatic character. Continental climates exist presently in only the Northern Hemisphere as a result of the movement of tectonic plates, which have concentrated Earth's landmass in the northern half of the globe. The Southern Hemisphere lacks a sufficient continental landmass to form a continental climate. The Southern Hemisphere is more ocean (71 percent) than land. Most important, this region is agriculturally productive. Many soils in the region result from glaciation in the recent Pleistocene, 2 million years ago.

Significance for Climate Change

The continental climate sustains a global "breadbasket." The climate can be divided further by the growing regions of corn (maize) or wheat. An imaginary line running east through the southern half of South Dakota, and as far east as New England, defines the margin zones of the humid continental long (warm) summer climate (to the south of this line) and the humid continental short (cool) summer climate (to the north of it). A similar line divides the climate in Europe and Asia. As the global average temperature rises, the boundary line is expected to shift northward. Thus, global climatic change within this zone has the potential to affect world food supplies.

In a warming scenario affecting the continental climate zone, ancillary problems for farming may emerge. Some models suggest that warming will increase annual precipitation. The humid continental climate averages 76 centimeters of precipitation per year, with snow being the predominant form of precipitation in the winter. Precipitation in this climate zone can vary from about 51 centimeters near drier areas to 127 centimeters near the oceans. An increase in precipitation in the continental climate zone could lead to increased flooding, noxious weeds, and plant diseases. In contrast, some suggest that warming would lead to a longer growing period and a shift of the temperature toward the north.

Conversely, in a global cooling scenario, the continental climate zone might experience shorter summers, and the temperature line between warm and cool summers could move south. Drier conditions and drought would prevail. The region might expect to see more rainfall relative to snowfall, as rain might continue further into the winter months. The continental short (cool) summer climate zone is characterized by cyclonic storms in winter that can bring huge snowfalls.

During the summer, convectional storms, many of them severe with lightning, are normal in continental short (cool) summer climate zones. The temperature average during summer is 24°C. During the winter, average temperatures fall to 11–12°C below freezing. With the influence of cold northern air, it is not uncommon to experience temperature extremes well below –18°C. With increasing temperatures, the possibility for a prolonged period of convective storms might exist. Dry-land crops, such as wheat, would need to be modified to accommodate warmer and moister conditions. More rainfall might increase erosion in already tenuous soils, especially in highly productive löess soils. New farming methods to accommodate these changes would have to be implemented.

—*M. Marian Mustoe*

Further Reading

Aguado, Edward, and James E. Burt. *Understanding Weather and Climate.* 7th ed. Pearson Prentice Hall, 2015. The formation of midlatitude storms is surveyed. Continental climates and the structure of climate are discussed.

Ahrens, C Donald, and Robert Henson. *Meteorology Today: An Introduction to Weather, Climate, and the Environment.* Cengage, 2019. Discusses global climates, climate change, and classification.

Bailey, Robert G. *Ecosystem Geography. From Ecoregions to Sites.* 2nd ed., Springer, 2009. Presents a concise description of the various climate regions of Earth, followed by a more granular analysis of the smaller regions comprising the climate zones, and discusses how climate zones and regions could change as the over all climate changes.

Gruza, George Vadimovich, ed. *Environmental Structure and Function: Climate System, Volumes 1 & 2.* EOLSS Publishers, 2009. A detailed description of atmospheric characteristics and of the various climate zones of Earth.

Maslin, Mark. *Global Warming: A Very Short Introduction.* Oxford University Press, 2009. Provides a concise general introduction to the major factors that define the climate of Earth.

CORIOLIS EFFECT

FIELDS OF STUDY

Physical Sciences; Physical Geography; Meteorology; Climate Modeling; Fluid Dynamics; Atmospheric Science; Oceanography

SUMMARY

The Coriolis Effect is an apparent angular acceleration of a moving object as seen in a rotating system. The acceleration is not a true change in velocity, but an illusion caused by the rotation of the system beneath the moving object.

PRINCIPAL TERMS

- **anticyclonic:** the clockwise rotation of an air mass in the Northern hemisphere, counterclockwise in the Southern hemisphere
- **cyclonic:** the counterclockwise rotation of an air mass in the Northern hemisphere, clockwise in the Southern hemisphere
- **inertia:** the tendency of physical objects to remain at rest or with unchanging speed and direction unless acted upon by an external force
- **inertial frame:** the frame of reference by which speed and distance measured
- **relative motion:** proportional speed and distance of two or more objects moving within the sane inertial frame of reference

RELATIVE MOTION

An unconscious tendency to regard Earth as a fixed frame of reference generates an unrecognized expectation on the part of many people that objects free of forces will move with unchanging direction and speed. That is, they will be unaccelerated. This expectation is consistent with the First Law of Motion, proposed by Sir Isaac Newton (1642–1727) which states that a body in motion will remain in motion, in a constant direction, with constant speed, unless acted upon by an external force.

Newton's First Law of Motion, however, applies only in frames of reference that are themselves not accelerating. These so-called inertial frames must be moving in a constant direction with constant speed. This condition is not met by rotating frames, such as Earth's surface, where every point in the frame (though traveling at a constant speed) is constantly changing direction as it completes a circular path around the axis of rotation.

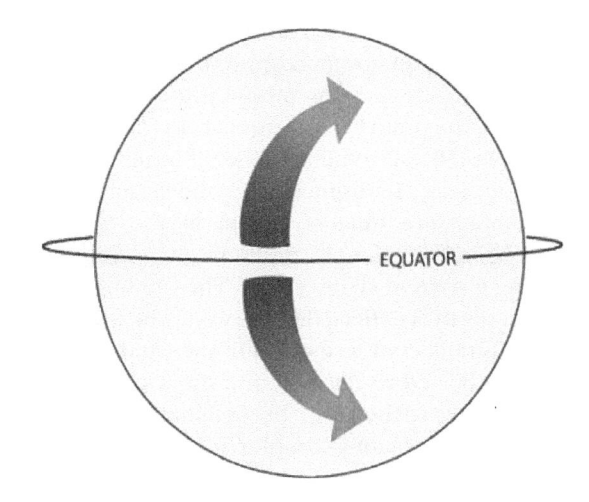

The Coriolis Effect causes moving fluids to be deflected to their right in the Northern Hemisphere and to their left in the Southern Hemisphere.

The farther an object is from the axis of rotation, the faster it will travel on its circular path. An object on the equator, for example, completes a path of 40,000 kilometers in one day, while an object at 60° north latitude travels only half that distance in the same time. Thus, the object at 60° north latitude travels at half the speed of the object on the equator.

Wind and water that leave the equator headed due north carry their equatorial speed with them. As they move north, they pass over territory that is traveling eastward more slowly than they are. As a result, the wind and water move eastward relative to the ground or the seafloor. Conversely, wind and water starting at northern latitudes and moving due south will cross ground that is moving eastward faster than they are; they will move westward relative to the ground or seafloor. This deflection from the original direction of motion is the Coriolis Effect, named for Gustave-Gaspard Coriolis (1792–1843).

SIGNIFICANCE FOR CLIMATE CHANGE

Convection-driven currents carry warm water and air towards the poles from the equator and carry cooler air and water from the polar regions toward the tropics. Both the tropical and the polar currents are deflected to the right, relative to their direction of motion, in the Northern Hemisphere and to the left, relative to their direction of motion, in the Southern Hemisphere. In regions where the currents converge, the deflections merge into circular rotations about the point of convergence. In the Northern Hemisphere, these rotations move counterclockwise; in the Southern Hemisphere, they move clockwise.

A low-pressure weather system in the Northern Hemisphere, for example, draws in air from the surrounding terrain in all directions. The wind flowing in from the north is deflected to the west. The wind from the west is deflected to the south, the wind from the south is deflected to the east, and the wind from the east is deflected to the north. In combination, the winds form a vortex rotating counterclockwise about the center of the low-pressure area. The winds flowing from a high-pressure system, by contrast, create a clockwise vortex. In the Southern Hemisphere, these directions are reversed. Similar effects occur in ocean currents.

The magnitude of the deflection caused by the Coriolis Effect is proportional to the distance from the point of deflection to the rotation axis. For that reason, the Coriolis Effect is most prominent at the equator. It is also proportional to the speed of the currents involved. High winds associated with hurricanes readily display the effect, generating the characteristic circular wind pattern with a calm eye at the center.

The Coriolis Effect establishes the circulation pattern of major storms, trade winds, jet streams, and large-scale ocean currents. All of these convection currents transport thermal energy from the warm tropics to the temperate and polar regions, moderating the global difference in temperatures. The heat carried by the warm waters of the Gulf Stream, for example, keeps Great Britain, Ireland, and the North Atlantic coast of Europe substantially warmer than other regions of the Northern Hemisphere that are located at the same latitude. Air currents also transport large amounts of water evaporated from tropical oceans to temperate and polar regions, where the water precipitates as rain and snow.

The rate at which convection currents transport mass and heat poleward from the tropics is a function of the temperature difference between the two regions. If climate change raises average temperatures in the tropics more than it raises them at the poles, it will create more energetic and powerful currents. If climate change raises polar temperatures more than it raises equatorial temperatures, it will dampen these currents. The resulting effects on the number, type, and destructive power of storms in either case would be complex and difficult to model.

—*Billy R. Smith, Jr.*

FURTHER READING

Cossu, Remo, and Matthew G. Wells. "The Evolution of Submarine Channels under the Influence of Coriolis Forces: Experimental Observations of Flow Structures." *Terra Nova* 25.1 (2013): 65–71. *Academic Search Complete.* Web. 19 Mar. 2015.

Denny, Mark *How the Ocean Works. An Introduction to Oceanography.* Princeton University Press, 2012. This book provides an in-depth examination of the effect of the rotational motion of Earth and the Coriolis Effect.

Dunkel, Ged *Understanding the Jet Stream. Clash of the Titans.* Authorhouse, 2010. A book that describes the role of the Coriolis Effect in the functioning of the jet stream.

Goh, Gahyun, and Y. Noh. "Influence of Coriolis Force on the Formation of a Seasonal Thermocline." *Ocean Dynamics* 63.9/10 (2013): 1083–1092. *Energy & Power Source.* Web. 19 Mar. 2015.

Pidwirny, Michael *Understanding Physical Geography.* Our Planet Earth Publishing, 2017. An entry-level textbook of physical geography that presents a complete description of the Coriolis Effect and how it is driven by Earth's rotation.

CRYOSPHERE

FIELDS OF STUDY

Physical Geography; Environmental Sciences; Oceanography; Earth System Modeling; Environmental Studies; Climate Classification; Climate Zones; Fluid Dynamics; Hydroclimatology; Hydrometeorology

SUMMARY

The cryosphere (from the Greek *kruos*, meaning "extremely cold") refers to those parts of the Earth's surface where temperatures are sufficiently low that water is frozen solid, in the form of either snow or ice. The conditions that freeze the available water within a particular area can be seasonal or can last for years or centuries. The cryosphere includes land covered with snow in the winter; freshwater lakes and river systems that freeze over seasonally; glaciers that freely move about larger water systems and are thus prone to melting and reshaping; and permafrost, or frozen soil and rock that remains frozen year round. The places most associated with the cryosphere are the North and South Poles, but frozen surfaces are found in many high-elevation regions of both the Northern and Southern Hemispheres.

PRINCIPAL TERMS

- **albedo:** the extent to which an object reflects radiation; the reflectivity of objects with regard to incoming solar radiation
- **glaciation:** the environmental process by which a terrain becomes covered by a permanent accumulation of ice
- **glacier:** a permanent covering of ice formed in a cold climate by the slow accumulation of snow that becomes compressed over time to form ice
- **permafrost:** a subsoil layer in which the entrained water remains in a frozen state regardless of the surface temperatures

CRYOSPHERE FORMATIONS

Scientists distinguish two types of formations that make up the cryosphere: land ice and sea ice. Land ice is formed slowly by compressed snow that becomes layers of ice. Land ice is thus freshwater. Perhaps the most familiar examples of land ice are glaciers, great slow-moving ice masses that store at any one time as much as 70 percent of the world's available fresh water. Other examples of land ice are ice shelves where glaciers have extended over open ocean water. Portions of ice shelves break off and head into the open oceans as icebergs. Ice shelves are found in coastal areas of Greenland, northern Canada, northern China, lower South America, southern Australia, and, of course, both poles.

Conversely, the polar oceans, both north and south, are covered with sea ice, or frozen seawater. Sea ice floats on the surface of the water and has an average thickness of 1 meter in the Antarctic and nearly 3 meters in the Arctic. Because this ice exists within a dynamic environment subject to

A Tour of the Antarctic Cryosphere.

temperature changes, wind, and ocean currents, sea ice can be measured by its duration (generally one year or multiyear). Because navigation depends on charting these fluid conditions, climatologists measure the sea ice as it cracks and even splits into huge moving parts, particularly as it inevitably diminishes during the abbreviated summer seasons at the poles. In the most extreme reaches of both poles, sea ice survives summer melting and becomes much thicker, measuring up to 381 centimeters.

SIGNIFICANCE FOR CLIMATE CHANGE

Although seasonal fluctuations in mean temperatures in polar regions are a normal phenomenon and do not affect the general dynamic of the cryosphere, long-term climate shifts resulting from decades of burning fossil fuels have contributed to Earth's rising average air temperature. This global warming, in turn, affects the thousands of square kilometers that make up the cryosphere. As that fragile environment undergoes radical changes over a relatively brief period of time, those changes affect a variety of climate and meteorological conditions around the globe. Interest in the cryosphere has greatly increased over the last generation, as climatologists see this frozen environment as the earliest indicator of rising global temperatures. The National Snow and Ice Data Center at the University of Colorado monitors the cryosphere.

Most dramatically, the diminishing of the snow and ice cover and the shortening of the winter season at the poles means that the planet's natural insulation from the direct bombardment of solar energy is diminishing. The bright surface of the snow and ice of the cryosphere contributes to Earth's albedo and is responsible for reflecting back into space some 70 percent of the energy received from the Sun. As that protection recedes, Earth absorbs more solar energy, adding to an increase in mean air temperature.

The global warming trend causes inland waterways to thaw earlier than they otherwise would, disrupting navigation lines and storm patterns and affecting the ecosystems of indigenous wildlife and plants. Groundwater levels in turn decline, glaciers melt, and scientists must confront the possibility of significant impacts on Earth's water system and the need for global water management. There is cause for concern. Scientists estimate that global sea levels have risen over the last two decades by 7.5 to 10 centimeters, but the loss of significant ice in the endangered Antarctic could raise ocean levels a catastrophic 9 meters in the next century, making the nearly 15 percent of the world's population who live along shorelines climate refugees.

The sea-ice shelves, which protect coastlines in both poles, Alaska, Canada, and Russia from wave erosion, are disappearing, upending peoples who have lived and worked in that difficult environment for centuries. In turn, under the impact of rising air temperatures, the permafrost loses its integrity, a process further complicated by drilling into the rich deposits of fossil fuels that may be found there. But loss of the permafrost has a greater significance. Trapped within its thousands of frozen square kilometers are centuries of decayed plant and animal detritus, and large quantities of toxic materials such as mercury. As the permafrost thaws, the two potent greenhouse gases carbon dioxide and methane are released in great volume to further impact the global air temperature.

—*Joseph Dewey*

FURTHER READING

Archer, David. *The Long Thaw: How Humans Are Changing the Next 100,000 Years of Earth's Climate.* Princeton University Press, 2016. Accessible description of the consequences of cryosphere loss. Pitched to a nonscientific audience, resisting alarmist argument, and generally objective.

Barry, Roger and Gan, Thian Yew *The Global Cryosphere. Past, Present and Future.* Cambridge University Press, 2011. Individual chapters describe each segment of the terrestrial and marine cryospheres, followed by a discussion of the past conditions and future prospects of that global climatic region.

Huggel, Christian, Carey, Mark, Clague, John J. and Kääb, Andreas, eds. *The High-Mountain Cryosphere. Environmental Changes and Human Risks.* Cambridge University Press, 2015. A extensive examination of the global environmental factors that dynamically affect the formation of mountain ice and snow caps and glaciers.

Marshall, Shawn J. *The Cryosphere.* Princeton University Press, 2012. A complete introduction to the physical properties and thermodynamics of ice and snow as the basic matter of the cryosphere and its dynamics.

Michaels, Patrick J. and Balling, Robert C. *Climate of Extremes: Global Warming Science They Don't Want You to Know.* Cato Institute, 2010. Important conservative corrective to growing alarmist projections about crysophere damage. Moderates the predictions and indicates progress in monitoring the cryosphere.

Tedesco, M., ed. *Remote Sensing of the Cryosphere.* Wiley-Blackwell, 2015. Geared to a technical audience and mathematically knowledgeable readers, this book details the basic principles and methodology of remote sensing applied to the cryosphere.

CYCLONES AND ANTICYCLONES

FIELDS OF STUDY

Physical Geography; Fluid Dynamics; Heat Transfer; Thermodynamics; Environmental Sciences; Environmental Studies; Meteorology; Climate Modeling; Process Modeling; Mathematics; Hydroclimatology; Hydrometeorology; Atmospheric Science; Oceanography; Earth System Modeling; Computer Science

SUMMARY

Cyclones and anticyclones are large-scale weather systems with opposite properties. A cyclone is

Tornadoes & Hurricanes

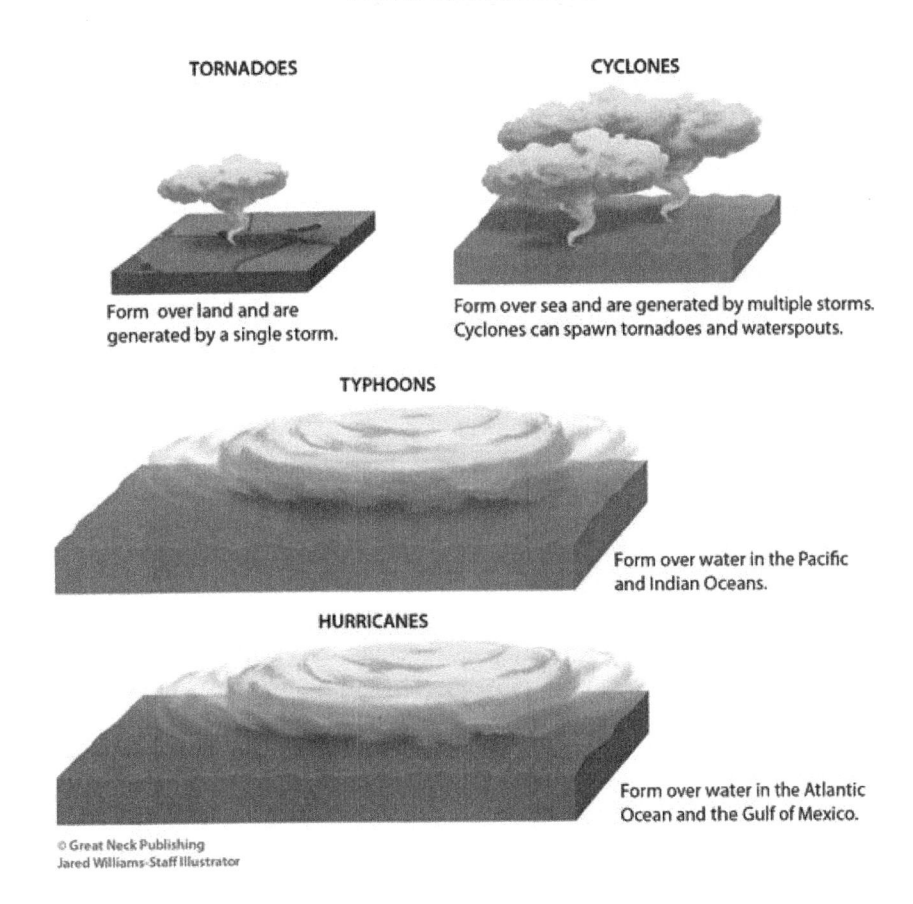

TORNADOES — Form over land and are generated by a single storm.

CYCLONES — Form over sea and are generated by multiple storms. Cyclones can spawn tornadoes and waterspouts.

TYPHOONS — Form over water in the Pacific and Indian Oceans.

HURRICANES — Form over water in the Atlantic Ocean and the Gulf of Mexico.

© Great Neck Publishing
Jared Williams-Staff Illustrator

Tornadoes, Cyclones, and Hurricanes: Tornadoes form over land, cyclones form over seas, typhoons form over the Pacific and Indian Oceans, and hurricanes form over the Atlantic Ocean and Gulf of Mexico.

characterized by a central region of low atmospheric pressure and an anticyclone is characterized by a central region of high atmospheric pressure. Because cyclones are a major cause of stormy weather and anticyclones typically bring good weather, accurate meteorological predictions are greatly informed by an understanding of how these weather systems originate and develop.

PRINCIPAL TERMS

- **convergence:** a tendency of air masses to accumulate in a region where more air is flowing in than is flowing out
- **Coriolis Effect:** the illusion of deflection observed when a body moves through the atmosphere with regard to an individual situated on the moving surface of the Earth
- **cyclogenesis:** the series of atmospheric events that occur during the formation of a cyclone weather system
- **divergence:** a tendency of air masses to spread in a region where more air is flowing out than is flowing in
- **front:** the boundary between two masses of air with different densities and temperatures; usually named for the mass that is advancing (for example, in a cold front, the mass that is colder is moving toward a warmer mass)
- **hurricane:** a cyclone that is found in the tropics (between 23.5 degrees north and south of the

equator) and that has winds that are equal to or exceed 64 knots, or 74 miles per hour
- **isobar:** on a map, a line connecting two or more points that share the same atmospheric pressure, either at a particular time or, on average, in a particular period
- **mid-latitude cyclone:** a synoptic-scale cyclone found in the mid-latitudes (between 30 and 60 degrees north and south of the equator)
- **synoptic scale:** a scale used to describe high- and low-pressure atmospheric systems that have a horizontal span of 1,000 kilometers (621 miles) or more

ATMOSPHERIC PRESSURE AND AIR CIRCULATION

The term *atmospheric pressure* describes the physical force exerted by the weight of the air above a given area on the Earth's surface. Meteorologists map pressure distributions with isobars, which appear as a series of curved lines connecting points that share the same atmospheric pressure. This pressure is measured in units called millibars.

Under average conditions, the atmospheric pressure at sea level is approximately 1,013.2 millibars. Values that are greater than the average sea-level pressure are considered high and those that are lower than average sea-level pressure are considered low. Pressure gradients are horizontal or vertical differences in atmospheric pressure. Pressure gradients create winds because air is constantly moving from areas of high pressure to areas of low pressure, seeking to create equilibrium. In other words, pressure gradients cause air to move perpendicular to isobars.

Pressure gradients, however, are not the only force at work. A phenomenon known as the Coriolis Effect also affects the way air circulates in the atmosphere. The Coriolis Effect is an apparent force that acts on moving objects, such as masses of air, in a rotating system, such as the rotating Earth. The result is that the moving object shifts perpendicular to the axis of this rotation.

To understand the Coriolis Effect, one can imagine trying to throw a ball in a straight line from the North Pole to the equator. Because the Earth is wider at the equator than it is at the poles, points at the equator must travel a greater distance than points at the poles in the same period of time. A ball thrown from the North Pole to the equator would thus appear to bend to the right. Similarly, when a mass of air is moving in the Northern Hemisphere, the Coriolis force appears to deflect that mass toward the right. When a mass of air is moving in the Southern Hemisphere, the Coriolis force appears to deflect that mass toward the left.

The force arising from pressure gradients and the force associated with the Coriolis Effect are roughly equal in magnitude, and at the upper levels of the atmosphere they balance each other to create winds that travel more or less parallel to isobars. Friction, or air resistance, reduces the effects of the Coriolis force at the Earth's surface.

CYCLONES AND THEIR FORMATION

Cyclones and anticyclones are both large-scale weather systems that are shaped by atmospheric pressure gradients, the Coriolis Effect, and surface friction. A cyclone has a central region of low atmospheric pressure with winds circulating around that center. On a weather chart, a cyclone appears as a series of roughly circular or oval isobars; the area inside the innermost isobar is the region of lowest pressure. Isobars that take this particular configuration are known as troughs.

The direction in which a cyclone's winds circulate depends on the hemisphere in which the weather system forms. In the Northern Hemisphere, a cyclone has winds that move in a counterclockwise direction. The reverse is true in the Southern Hemisphere. Because cyclones are a major cause of severe stormy weather, including blizzards and floods, meteorologists creating detailed and accurate weather forecasts pay close attention to how these atmospheric systems originate and develop.

The atmospheric events that take place as a cyclone forms are known collectively as cyclogenesis. Mid-latitude cyclones are cyclones that occur between 30 and 60 degrees north and south of the equator and are about 1,000 kilometers (621 miles) or more in diameter. These typically form at fronts, or the boundaries or transition zones between two masses of air that have different temperatures and densities. At this first stage of cyclogenesis, the heavy cold air and lighter warm air are simply pushing against each other. Because the air masses are not moving, the place where they meet is known as a stationary front.

At the next stage, the cold air—because it is denser and heavier—begins to sink below the light warm air. In turn, the light warm air is forced upward and then over the cold and heavy air mass. Instead of a single stationary front, two fronts are formed: one consisting of the advancing edge of the cold air and one consisting of the advancing edge of the warm air. Because of the Coriolis force, these masses of cold and warm air do not simply exchange places vertically but begin to revolve around each other, turning inward toward the area of low pressure in the center of the rotation. This pattern of winds causes warm air to "pile up" in the center of the cyclone, near the surface. This phenomenon is known as convergence.

When air converges low to the ground, it has nowhere to move but up and out. As the warm-air front rises and expands (diverges), it carries water vapor that cools and condenses into clouds and rain. Different characteristic weather patterns are seen along each front. Brief, intense thunderstorms tend to form along the cold front, and slow, steady rains tend to fall along the warm front. When the cyclone nears its end, the cold front pushes on the warm front so much so that the mass of warm air is entirely separated from the low pressure center. This is known as an occluded front, and it is usually associated with more rainy weather.

ANTICYCLONES

As its name suggests, an anticyclone has properties opposite of those of a cyclone. Whereas a cyclone consists of winds circulating around a center of low atmospheric pressure, an anticyclone has a center of high atmospheric pressure with winds circulating around that center. In the Northern Hemisphere, an anticyclone has winds that move in a clockwise direction; the reverse is true in the Southern Hemisphere.

Like a cyclone, an anticyclone appears on a weather chart as a series of roughly circular or oval isobars; however, in an anticyclone, the area bounded by the innermost isobar is the region of highest pressure. When isobars take this configuration, they are known as ridges.

An anticyclone forms when dense cold-air masses in the upper atmosphere converge, or pile up. When this convergence reaches a high enough level, the air begins to sink to the Earth's surface. As it descends, the air is compressed by increasing pressure and becomes warmer and drier. Anticyclones are generally associated with clear weather.

TROPICAL CYCLONES (HURRICANES)

Tropical cyclones, which are known as typhoons in the western Pacific Ocean and as hurricanes in the Atlantic and eastern Pacific Oceans, are cyclones that form in the tropics (between 23.5 degrees north and south of the equator). They are typically smaller than mid-latitude cyclones but are characterized by extremely high winds, usually exceeding speeds of 119 km (74 mi) per hour.

Because of these winds and the intense thunderstorms, occasional flash floods, and storm surges with which they are associated, tropical cyclones can cause great damage to life and property in coastal areas. Storm surges refer to seawater pushed inland by strong winds.

Tropical cyclones differ from mid-latitude cyclones in not being associated with a front. Instead, this type of cyclone arises when each of a set of specific environmental conditions is present. The ocean waters above which a potential cyclone would form must be at least 27°C (80°F) up to a depth of 46 meters (about 150 feet) or more. This condition makes it natural that tropical cyclones tend to originate relatively near the equator.

The air near the middle of the troposphere, the lowest region of the Earth's atmosphere, must be moist. There must be relatively low vertical wind-shear between the ocean's surface and the upper levels of the troposphere, meaning that wind speeds must not be changing radically as they ascend. Air must be able to cool relatively quickly as it ascends. Finally, there must exist some kind of atmospheric disturbance, or "seedling," such as a trough or an elongated area of low pressure.

If all these conditions are present, a tropical cyclone may begin to form, first with the establishment of a pattern of convection over the ocean. Warm and moist air rises into the atmosphere, cooling and condensing. As the air releases heat, it becomes even lighter, thus powering its own ascent. The water vapor that is released forms the clouds and thunderstorms commonly associated with tropical cyclones. As air moves up and away from the surface, more air rushes in to take its place, creating high winds at the surface. This process creates a self-sustaining cycle that can cause the cyclone to grow and intensify as long as it remains over the water. Usually, however, tropical cyclones begin to dissipate as soon as they move inland because the cycle is broken when the storm system no longer has access to the warm, moist air that moves over the ocean.

—*M. Lee*

FURTHER READING

Ackerman, Steven, and John Knox. "Extratropical Cyclones and Anticyclones." In *Meteorology: Understanding the Atmosphere*. Jones and Bartlett, 2012. This chapter, richly illustrated with color photographs, diagrams, and charts, explains each stage of both weather systems' formations. Includes an outline, summary, key terms, and review questions.

Ahren, C. Donald. "Air Masses, Fronts, and Middle-Latitude Cyclones." In *Essentials of Meteorology: An Invitation to the Atmosphere*. Brooks/Cole, 2011. Notable for being especially clear and well organized, this chapter walks readers through each fundamental concept in a logical order, from air masses and fronts to convergence, divergence, and storm formation. Suitable for beginning students.

Chan, Johnny C. L., and Jeffrey D. Kepert, eds. *Global Perspectives on Tropical Cyclones: From Science to Mitigation*. World Scientific, 2010. Includes chapters on forecasting and modeling tropical cyclones, the effects of climate change on cyclone activity, and approaches to disaster response. Highly technical; best suited to college students and above with some background in meteorology.

De Villiers, Marc. *Windswept: The Story of Wind and Weather*. New York: Walker, 2006. An accessible, scientifically accurate book of popular science that explores the history of human attempts to understand the weather. Contains black-and-white figures and several useful appendices, including two covering tropical cyclone statistics.

Fahy, Frank. *Air: The Excellent Canopy*. Chichester, UK: Horwood, 2009. A well-written introduction to the properties of air and the forces that govern atmospheric circulation, designed for nonspecialists. Explains physics concepts using detailed descriptions and analogies rather than equations.

Longshore, David. *Encyclopedia of Hurricanes, Typhoons, and Cyclones*. New York: Facts on File, 2008. A comprehensive reference book containing about four hundred cross-referenced entries covering the science, history, and cultural significance of severe weather phenomena. Contains black-and-white photographs and other images. Suitable for high school readers and older.

D

DATING METHODS FOR CLIMATE CHANGE

FIELDS OF STUDY

Analytical Chemistry; Atmospheric Chemistry; Biochemistry; Environmental Chemistry; Geochemistry; Inorganic Chemistry; Physical Chemistry; Thermodynamics; Mass Spectrometry; Optics; Spectroscopy; Crystallography; Tree Ring Analysis

SUMMARY

Because the geological and climatological history of Earth began long before recorded history, scientific dating methods are necessary to determine when many climatic events occurred. For example, such methods could be used to determine when glacial deposits formed or when a boulder was dropped on top of those deposits by a melting glacier.

PRINCIPAL TERMS

- **cosmogenic isotope:** an isotope, possibly radioactive, produced when a cosmic ray strikes the nucleus of an atom
- **decay constant:** a measure of the radioactivity of an isotope, determined with a Geiger counter
- **half-life:** the time needed for half of a quantity of a radioactive isotope to decay; it is calculated from the decay constant of the specific isotope, not measured directly
- **isotopes:** atoms of an element that have the same number of protons in their nuclei, but different numbers of neutrons; many elemental isotopes are radioactive
- **primordial isotope:** an isotope that has been present on Earth since the planet formed 4.5 billion years ago
- **varve:** an annual layer in a sediment, usually the result of seasonal variation in inputs

PRIMORDIAL ISOTOPES

When Earth formed, it inherited an inventory of radioactive elements that have been decaying ever since. The decay constant for a particular isotope can be determined by measuring the rate at which disintegrations occur in a sample of known mass. Half-lives are calculated from decay constants. Known half-lives of radioactive isotopes enable scientists to determine the age of objects that contain those isotopes. For example, water moving through the ground will often dissolve small amounts of uranium. A stalagmite may form from this water in a cave as the water evaporates, incorporating any uranium-238 (^{238}U) present. The uranium will decay to produce thorium-234 (^{234}Th). Thorium is insoluble in water, so it can be assumed that the stalagmite initially contained no thorium. Thorium, too, is radioactive, and may decay into ^{234}U, which decays to ^{230}Th, which is also radioactive. Using the decay constants and the amounts of ^{238}U, ^{234}U, and ^{230}Th present in a specimen, the amount of time that has passed since the uranium came out of solution can be calculated. This technique is limited to ages less than 500,000 years.

COSMOGENIC ISOTOPES

Cosmic rays are subatomic particles traveling at very high velocities. When they strike the nucleus of an atom, they can eject nucleons, altering the identity of the atom. An atom of nitrogen-14 (^{14}N), for instance, might become carbon-14 (^{14}C), or atoms of silicon or oxygen might become beryllium-10 (^{10}Be) or aluminum-26 (^{26}Al). All of ^{14}C, ^{10}Be, and ^{26}Al are radioactive, and their decay constants are known, so they provide a means of dating organic material and the surfaces of boulders.

On Earth, ^{14}C is generally created only as a result of cosmic ray bombardment in the atmosphere, so only atmospheric carbon replenishes its ^{14}C level.

185

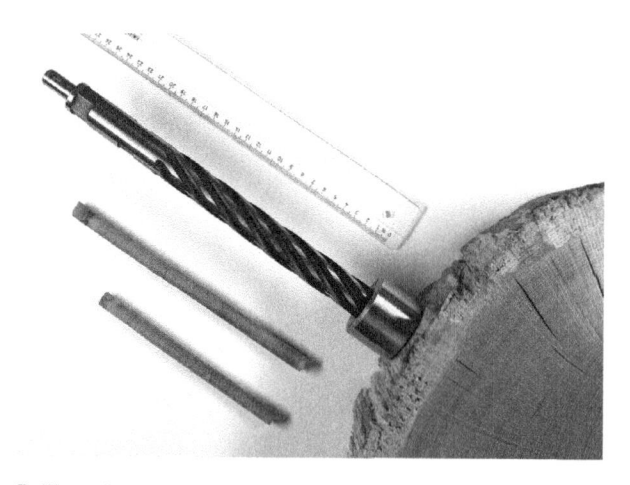

Drill to take samples for dendrochronology from trees.

Non-atmospheric ^{14}C decays over time without replenishment. An organism will interchange carbon with the atmosphere while it is alive, maintaining a relatively constant ratio of ^{14}C to carbon-12 (^{12}C), but once it dies that interchange will cease and the ratio will decrease. By assuming a historically constant ratio of ^{14}C to ^{12}C in the atmosphere (and thus in living organisms) and by comparing that ratio to the ratio in a sample of tissue from a deceased organism, it is possible to determine how many half-lives of ^{14}C have passed since the organism died. The assumption of a constant ratio is known to be invalid, but it will produce the same errors in all samples, giving the same results for samples of the same age.

If the goal of analysis is to compare different samples with one another and there is little need for actual calendar years, the ages of samples are often reported in ^{14}C years. To convert results accurately to calendar years, corrections are made using calibration curves derived from other dating techniques that may produce different calibrated ages from the same ^{14}C age. The effective limit of this technique is about forty-five thousand years.

Cosmic rays also cause reactions in the outer layers of quartz-rich rocks. The isotopes ^{10}Be and ^{26}Al accumulate in these layers at small but relatively constant rates. These isotopes are produced slowly, at a rate of about 100 atoms per gram of rock per year, requiring accelerator mass spectrometry (AMS) techniques to detect them. Cosmic rays do not penetrate solids by more than a few meters, so the exposure age of a surface can reveal when glacial ice melted away above that surface.

NONRADIOMETRIC METHODS

DENDROCHRONOLOGY. Dendrochronology is a method for determining the age of wood by counting and examining annual tree rings. The thickness of a given ring in a tree is determined by environmental factors obtaining during the year in which the ring was formed. Such factors as sunlight, temperature and rainfall affect the rate of growth and overall health of trees. As a result, patterns of ring thickness in trees that were alive at the same time in the same area tend to resemble one another. Matching patterns of ring thickness between trees of known and unknown age can thus provide evidence that the trees were alive at the same time. The reliability of this method has been extended back to about ten thousand years.

VARVES. Just as trees have annual growth cycles, so do sediments deposited in lakes in regions near glaciers. In the summer, rains bring coarse sediments into the lake. In the winter, fine clays have time to settle out. The banded sediments that result from this seasonal alternation are called varves. Just as with tree rings, patterns of thick and thin layers can be correlated in different varved sequences. Some sequences cover more than thirteen thousand years.

LICHENOMETRY. Lichens grow at fairly constant rates in a given area. In any particular area, rocks covered by larger lichens have surfaces that have been exposed longer than have the surfaces of rocks covered by smaller lichens of the same type. By calibrating measurements using tombstones and other objects of known age, the absolute exposure age of lichen-covered surfaces can be estimated.

CONTEXT

Understanding climate change requires knowledge of Earth's climatological history, which in turn requires methods capable of dating events of climatic significance over the last few million years. As technology has improved, the precision and accuracy of these methods has increased dramatically, and the size of the samples required for accurate dating has decreased by orders of magnitude.

Scientists looking at isotope-ratio records in marine sediment cores have sometimes found that different radiometric techniques indicate different dates for the same climatic excursion. As the climatic excursions were found to be global and strongly correlated with known astronomical cycles, it became possible to determine their age with greater accuracy, validating some results over others. This correlation with astronomical cycles could also be used to calibrate radiometric dating methods, just as counting tree rings was used to calibrate ^{14}C dating methods. As methods continue to be developed, it becomes possible to date specific geologic and climatic events, such as the encroachment or retreat of ice from a given area, the speed of uplift of a surface, or the rate of development of a valley, with ever greater accuracy.

—Otto H. Muller

Further Reading

Alley, Richard B. *The Two-Mile Time Machine. Ice Cores, Abrupt Climate Change and Our Future* Princeton University Press, 2014. Describes the analytical methods and interpretations of physical clues of past climates trapped in ice that is thousands of years old.

Bradley, Raymond S. *Paleoclimatology: Reconstructing Climates of the Quaternary*. 3rd ed., Academic Press, 2015. Provides a solid introduction to the theoretical basis of multiple dating methods and their application.

Ruddiman, William F. *Earth's Climate Past and Future*. 3rd ed., W. H. Freeman, 2014. This elementary college textbook has several sections concerning dating methods, their limitations, errors, and resolution. Illustrations, figures, tables, maps, bibliography, index.

Wagner, Günther A. *Age Determination of Young Rocks and Artifacts. Physical and Chemical Clocks in Quaternary Geology and Archaeology*. Springer, 2011. Provides very readable descriptions of the more robust dating methods used by climate scientists and archaeologists, and the manner in which those methods are interpreted.

Walker, Mike *Quaternary Dating Methods*. John Wiley & Sons, 2013. Based on atomic structure, the chapters discuss the methodology of radiometric dating and the interpretation of annually-banded strata to determine dating chronologies.

DEEP OCEAN CURRENTS

Fields of Study

Oceanography; Physical Geography; Fluid Dynamics; Earth System Modeling

Summary

Deep ocean currents, the dynamics of which are not yet well understood, involve significant vertical and horizontal movements of seawater. They distribute oxygen- and nutrient-rich waters throughout the world's oceans, thereby enhancing biological productivity, and play an important role in global heat transport that determines weather and climate.

Principal Terms

- **benthic:** relating to organisms that inhabit the floors of lakes and seas
- **benthonic:** synonymous with benthic, relating to organisms that inhabit the benthos, or ocean floor region
- **Coriolis effect:** the effect by which an object moving in a straight line across a rotating surface is seen to follow a curved path
- **saltation:** the process by which grains of sand and other particulate matter are made to move in a leap-frog manner within a moving fluid such as water or air, resulting in the formation of characteristic structures ranging in size from small ripples to giant sand dunes
- **thermohaline circulation:** saline ocean water circulation currents that are driven by differences of both of heat and relative density

Evidence for Existence

Deep-sea or deep-ocean currents involve vertical as well as horizontal movements of seawater and are generated by density differences in water masses that result in the sinking of colder, denser water to the bottom of the ocean. For many years, however, most oceanographers refused to accept the

World Ocean Currents

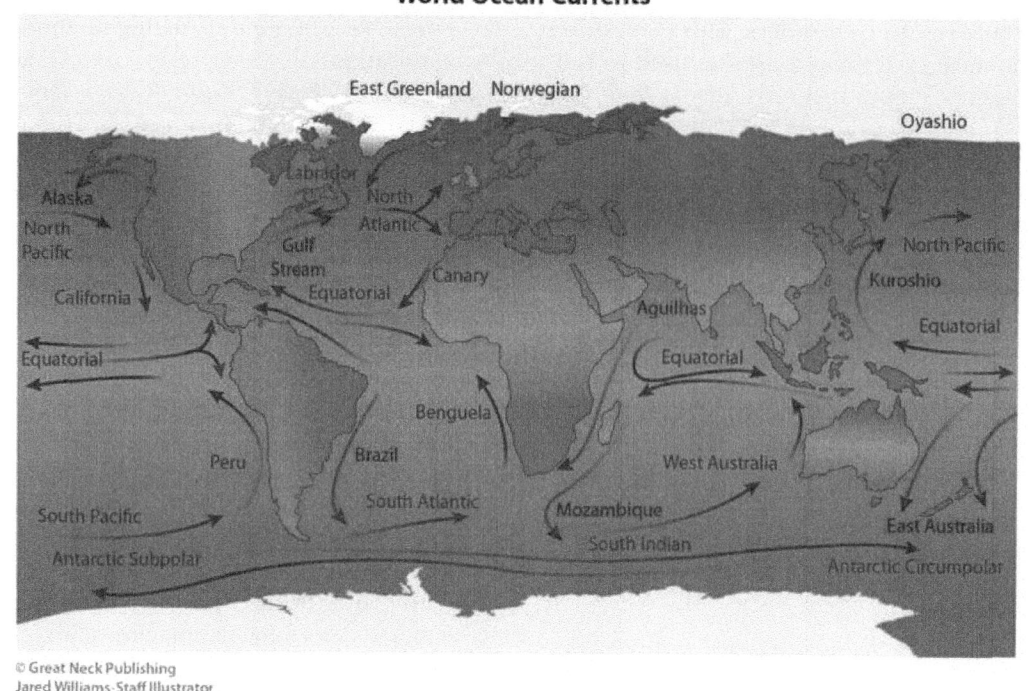

Ocean Systems: World ocean currents are warm and cold. They move through every ocean in the world.

presence of these currents. Even when the Deep Sea Drilling Project, an international effort to drill numerous holes into the ocean floor, was initiated, most researchers envisioned the deep sea as a tranquil environment characterized by sluggish or even stationary water. More recently, however, oceanographers and marine geologists have accumulated abundant evidence to suggest the opposite: that the deep sea can be a very active area in which currents sweep parts of the ocean floor to the extent that they affect the indigenous marine life and even physically modify the sea floor.

In the 1930s, Georg Wust (1890–1977) argued for the likelihood that the ocean floor is swept by currents. Furthermore, he suggested that these currents play an important role in the transport of deep-sea sediment. Wust's ideas were not widely accepted at first, but in the 1960s, strong evidence for the existence of deep-sea currents began to accumulate. In 1961, for example, oceanographers detected deep-sea currents moving from five to ten centimeters per second in the western North Atlantic Ocean. These researchers also determined that the currents changed direction over a period of one month.

In 1962, Charles D. Hollister (1936–1999), while examining cores of deep-sea sediment drilled from the continental margin off Greenland and Labrador, noted numerous sand beds that showed evidence of transport by currents. The nature of these deposits suggested to Hollister that they did not accumulate from turbidity currents, dense sediment-water clouds that periodically flow downslope from nearshore areas. Moreover, it appeared to Hollister that the sand was transported parallel to the continental margin rather than perpendicular to it, as might be expected of sediment transported by a turbidity current. He argued that the sand beds in the cores were transported by, and deposited from, deep-sea currents moving along the bottom of the ocean parallel to the continental margin. Since then, extensive photography of the ocean floor has provided direct evidence for the existence of deep-sea currents. Such evidence includes smoothing of the sea floor; gentle deflection, or bending, of marine organisms attached to the sea floor, as though they were standing in the wind; sediment piled into small ripples by saltation; and local scouring of the sea floor.

THERMOHALINE CIRCULATION

Essentially all earth scientists now agree that the deep-sea floor is swept by rather slow-moving (less than two centimeters per second) currents. The driving force behind these currents, and all oceanic currents for that matter, is energy derived from the sun. Differential heating of the air drives global wind circulation, which ultimately induces surface ocean currents. The vertical circulation of seawater, and thus the generation of deep-sea currents, is controlled by the amount of solar radiation received at a point on Earth's surface. This value is greatest in equatorial regions; there, the solar radiation heats the surface water that lies within the upper three hundred to one thousand meters of the ocean. As this water is heated, it begins to move toward the poles along paths of wind-generated surface circulation, such as the Gulf Stream current of the northwestern Atlantic Ocean.

The cold waters that compose the deep-sea currents originate in polar regions. There, minimal solar-radiation levels produce cold, dense surface waters. The density of this water may also be increased by the seasonal formation of sea ice, ice formed by the freezing of surface water in polar regions. When sea ice forms, only about 30 percent of the salt in the freezing water becomes incorporated into the ice. The salinity and density of the nearly freezing water beneath the ice are therefore elevated. This cold, and therefore denser, saline seawater slowly sinks under the influence of gravity to the bottom of the ocean, where it moves slowly toward the equator where it eventually warms and rises to replace the surface water that has moved polewards. Deep-sea circulation driven by temperature and salinity variations in seawater is termed "thermohaline circulation" and is much slower than surface circulation; the cold, dense water generated at the poles moves only a few kilometers per year. After moving along the bottom of the ocean for anywhere from 750 to 1,500 years, the cold seawater rises to the surface in low-latitude regions to replace the warm surface water, which, as noted above, moves as part of the global surface circulation system back to the polar regions.

Thermohaline circulation and related deep-sea currents are commonly affected by the shape of the ocean floor. Although sinking cold seawater seeks the deepest route along the sea floor, deep-sea currents may be blocked by barriers. The Mid-Atlantic Ridge, the large volcanic ridge effectively bisecting the Atlantic Ocean basin, may interfere with or even prevent the movement of water from the bottom of the western Atlantic to the eastern Atlantic. Conversely, the funneling of deep-sea currents through narrow passages or gaps in seafloor barriers will lead to an increase in the velocity of the current. Once beyond the passage, however, the current spreads and velocity is reduced. Because both air and water are fluids, these effects and behaviors are entirely analogous to those of winds produced in the atmosphere by air-density differences resulting from the uneven distribution of solar energy over Earth's surface.

CORIOLIS EFFECT

The circulation pattern of deep-sea currents is controlled to a large extent by Earth's rotation. The Coriolis effect, the apparent force achieved by Earth's rotation that causes particles in motion to be deflected to the right in the Northern Hemisphere and to the left in the Southern Hemisphere, induces deep-sea currents to trend along the western margins of the major oceans. Thus, water sinking from sources in the North Atlantic Ocean and moving south toward the equator will be deflected to the right, causing it to run along the western side of the North Atlantic. Similarly, north-directed deep-sea currents generated by the sinking of cold water from the Antarctic region will also be deflected to the western margin of the Atlantic.

The Coriolis effect guides deep-sea currents along bathymetric contour lines, lines on a map of the ocean floor that connect points of equal depth. Deep-sea currents that have a tendency to move parallel to the bathymetric contours are known as bottom currents. Barriers to flow may locally deflect deep-sea currents from the bathymetric contours; nevertheless, bottom currents are most conspicuous along the western margins of the major oceans.

SHORT- AND LONG-TERM CONTROLS

The formation of the cold seawater required to set deep-sea currents in motion can itself be considered in terms of short- and long-term controls. Seasonal sea-ice formation is probably the most important process in the production of the north-flowing water generated at the south polar region, or the Antarctic bottom water (AABW). Velocities of the AABW are highest in March and April, that period of the year when sea-ice production in the ocean surrounding

189

Antarctica is greatest. During Southern Hemisphere summers, however, the sea ice melts, and there is an increase in the freshwater flux to the ocean from the continent. Both of these factors reduce the salinity and therefore the density of the seawater, thereby decreasing AABW production.

Many oceanographers and marine geologists have argued that long-term variations in the production of the cold, dense bottom water required to generate deep-sea currents may be related to global climatic changes. More specifically, deep-sea currents appear to be most vigorous during glacial periods, when sea-ice production is enhanced and the sea ice remains on the ocean surface for a greater proportion of the year. Nevertheless, there is also evidence to suggest that the velocities of deep-sea currents in the North Atlantic Ocean were much lower during the most recent glacial periods than they were during the times between glacial phases. Much more work is required to gain a more complete understanding of long-term controls on deep-sea currents.

Measurement Tools and Techniques

The most common methods for the study of deep-sea currents include direct measurement of current velocities, bottom photography, echo sounding, and the sampling of ocean-floor sediment. The speed and direction of deep-sea currents have been determined by the use of free-fall instruments, such as the free-instrument Savonics rotor current meter. This device is dropped unattached into the ocean, where it records current velocities and directions over a period of several days. It then returns automatically to the surface of the ocean, at which time a radio transmitter directs a ship to its position. Other current-measuring devices can be suspended at various depths in the ocean from fixed objects, such as buoys or light ships, to monitor currents for long periods. One such anchored meter measures the flow of water past a fixed point. Flowing water causes impeller blades, similar to the blades of a fan, to rotate at a rate proportional to the current's speed. In addition, the blades cause the meter to align with the current's direction. Electrical signals indicating the direction and speed of the current are transmitted by radio or cable to a recording vessel. Current velocities of less than one centimeter per second can be detected by this meter.

To get the most complete picture of the variability of the ocean, a combination of various measurement techniques with remote sensing may be employed. Such a multidimensional approach may involve the measurement of current velocity, pressure (a measure of depth), water temperature, and water conductivity (a measure of salinity). These data can be transmitted via satellite to a land station or even directly to a computer.

Additional Study Methods

Perhaps the most persuasive evidence for the existence of deep-sea currents and their influence on the ocean bottom has been gained through bottom photography. Sediment waves, or ripples, apparently formed by the saltation of sediment carried by deep-sea currents, along with evidence of current-induced scour of the ocean floor, were first photographed in the Atlantic Ocean in the late 1940s. Since then, the technology of bottom photography has advanced greatly. Bottom photography permits detailed study of some of the smaller features on the ocean floor apparently formed by deep-sea currents. The bending of benthonic, or benthic, organisms (marine organisms that live attached to the ocean floor) in the flow of the current is a particularly intriguing example of the phenomena recorded by this technique.

Echo-sounding studies of the sea floor have yielded abundant information on ocean-floor features that are either formed or modified by deep-sea currents. Notable among these are very long ridges in the North Atlantic evidently constructed from sediment carried by deep-sea currents. In echo sounding, a narrow sound beam is directed from a ship vertically to the sea bottom, where it is reflected back to a recorder on the ship. The depth to the sea floor is determined by multiplying the velocity of the sound pulse by one-half the amount of time it takes for the sound to return to the ship. The depths to the ocean floor are recorded on a chart by a precision depth recorder, which produces a continuous profile of the shape of the sea floor as the ship moves across the ocean.

Sediment transported and deposited by deep-sea currents can be studied directly by actually sampling the ocean floor. Sampling of these deposits is best accomplished by the use of various coring devices capable of recovering long vertical sections, or cores, of seafloor sediment. Sediment recovery is achieved by forcing the corer, a long pipe usually with an inner plastic liner, vertically into the sediment. The simplest coring device, the gravity corer, consists of a pipe with a heavy weight at one end. This type of corer will penetrate only two to three meters into the sea floor. The piston corer, used to obtain longer cores, is fitted with a piston inside the core tube that reduces friction during coring, thereby permitting the recovery of eighteen-meter or longer cores. Analysis of the sediment recovered from the ocean floor by these and other coring devices reveals much information about small-scale features formed by deep-sea currents.

Importance to Life on Earth

Because cold bottom-water masses often are nutrient-rich and contain elevated abundances of dissolved oxygen, deep-sea currents are extremely important to biological productivity. There are areas of Earth's surface where nutrient-rich cold bottom waters rise to the ocean surface. These locations, known as upwelling ecosystems or areas of upwelling, are generally biologically productive and therefore are important food sources. Especially pronounced upwelling occurs around Antarctica. Bottom waters from the North Atlantic upwell near Antarctica and replace the cold, dense, sinking waters of the Antarctic.

The great amount of time required for seawater to circulate from the surface of the ocean to the bottom and back again to the surface has become an important practical matter. If pollutants are introduced into high-latitude surface waters, they will not resurface in the low latitudes for hundreds of years. This delay is particularly important if the material is rapidly decaying radioactive waste that may lose much of its dangerous radiation by the time it resurfaces with the current. The introduction of toxic pollutants into a system as sluggish as the deep-sea circulation system, however, means that they will remain in that system for prolonged periods. Nations must, therefore, be concerned with the rate at which material is added to this system relative to the rate at which it might be redistributed at the surface of the ocean by wind-induced surface circulation. The multinational Geochemical Ocean Sections (GEOSECS) program, introduced as part of the International Decade of Ocean Exploration and carried out from 1970 to 1980, attempted to better assess the problem of how natural and synthetic chemical substances are distributed throughout the world's oceans. The program yielded abundant information regarding the movement of various water masses and, among other things, the distribution of radioactive material in the oceans. For example, GEOSECS demonstrated that tritium produced in the late 1950s and early 1960s by atmospheric testing of nuclear weapons had been carried to depths approaching five kilometers in the North Atlantic Ocean by 1973.

—Gary G. Lash

Further Reading

Broecker, Wally. *The Great Ocean Conveyor: Discovering the Trigger for Abrupt Climate Change.* Princeton University Press, 2010.

Garrison, Tom S. *Oceanography: An Invitation to Marine Science.* 8th ed. Brooks, 2013.

Trujillo, Alan P., and Harold V. Thurman. *Essentials of Oceanography.* 11th ed. Prentice, 2014.

DEGLACIATION

SUMMARY

Deglaciation is the uncovering or exposure of a land surface that was previously covered by glacial ice. It results from ice melting or subliming (transforming from solid directly into vapor). Deglaciation, therefore, accompanies the end of a glacial stage. As deglaciation occurs, several processes take effect. Among these processes are meltwater stream flow, development of meltwater lakes, addition of water to the world's oceans (raising the global sea level), exposure of the land, and rebound of the land (lifting of the land's elevation as a result of the removal of the weight of the overlying ice). In addition, faunal and floral changes accompany deglaciation in response to changes in landscape and ecology.

Glaciers have a seemingly infinite capacity to entrain and transport sedimentary material, from tiny clay particles to giant rock boulders. Glaciers move

GLACIAL DEPOSITION

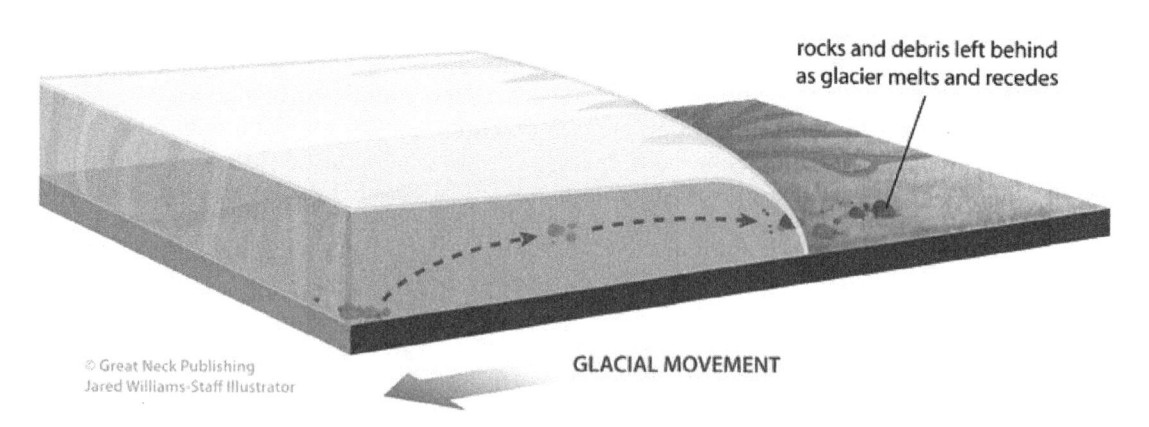

Glacial Deposition: Various substances like rocks, soil, and debris are broken down and redeposited as a nearby glacier moves along.

these materials within and upon the ice, but when they melt all this sediment is deposited in the area where the ice melts. For this reason, areas that have experienced deglaciation are typically covered by glacially transported sediment. Alternatively, glaciers may sweep an area clean of loose material, creating deglaciated areas of bare bedrock. Where deglaciation has formed modern shorelines, those shorelines tend either to be laden with glacial sediment or to present bare bedrock to the waves. In some places, rebound has lifted the land along the modern shore, forming sea cliffs.

Deglaciation accompanies the transition from a glacial stage to an interglacial or warm stage. There have been several such transitions over the past two million years. In addition, deglaciation—to a lesser extent—accompanies minor warming events that occur during glacial stages. Prior to the current epoch of glacial and interglacial stages, the Earth experienced several periods during which glaciers episodically covered large parts of its surface. There have been at least four such glacial periods during the past one billion years.

Significance for Climate Change

Deglaciation accompanies climate change and can be a cause of climate change. The geological record indicates that, when deglaciation commences, there is typically a climatic turn toward global warming. In other words, after climatic warming initiates deglaciation, the deglaciation itself can create a positive feedback loop engendering further warming. Glaciers are highly reflective of sunlight, contributing to Earth's albedo (the percentage of sunlight reflected back into space from the planet's surface). As glaciers melt, white ice and snow are replaced with darker surface elements. Earth's albedo decreases, and more solar radiation is absorbed and retained by the planet. As the planet warms, more glaciers melt, the albedo decreases even further, and the process continues. Melting glaciers also contribute water to lakes and oceans, which help retain atmospheric heat. Rising sea levels due to glacial meltwater contribute to global climate change as well.

Using the modern deglaciation as an example, loss of glacial ice cover on the land has had and continues to have profound consequences for global climate change. For example, release of water locked up in glaciers has affected the amount of water in the oceans as well as on land, in rivers and streams, and in the form of groundwater. This has affected coastal and interior ecosystems, which are dependent upon water for life. Changing patterns in the distribution of water cause both climates and ecosystems to change.

Deglaciation, as compared to glaciation, can be a relatively rapid process, once the melting triggers feedback mechanisms that increase its pace. The rapid nature of this change has led to disequilibrium conditions on land, such as unstable slopes, high gradients in rivers and streams, and unstable lakes that drain catastrophically. In the biotic realm, rapid deglaciation has led to mass death and mass extinction among plant and animal groups, as well as mass migration of animal populations.

Deglaciation leaves behind profound physical effects of ice movement, such as landscapes altered by the erosive forces of massive ice sheets, depositional landforms created by sediment released from melting ice, and lakes created by the meltwater—including waters from marooned blocks of ice that melt long after the main glacial mass is gone. The resulting altered landscapes generally have low levels of vegetation (at least initially), as well as areas of low elevation where water can accumulate. This type of landscape has a higher capacity to retain radiated heat from the Sun and therefore also contributes to atmospheric warming.

—David T. King, Jr.

BIBLIOGRAPHY

Alley, R. B., and P. U. Clark. "Deglaciation of the Northern Hemisphere." *Annual Reviews* 28 (2000): 149–182. One of the best references for a complete understanding of the current deglaciation, its chronology, and its overall effects. Well illustrated.

Ehlers, J., and P. L. Gibbard, eds. *Quaternary Glaciations: Extent and Chronology*. 3 vols. San Diego, Calif.: Elsevier, 2004. Massive, comprehensive compendium of studies of the glaciations and deglaciations of the Quaternary period. Includes five CD-ROMS of digital maps and other supporting material.

Stanley, Steven. *Earth Systems History*. 3d ed. New York: W. H. Freeman, 2009. Lays out the history of Earth, including the factors affecting global climates and climate change, over the vastness of geological time. Selected chapters focus on climate change in terms of glaciation and deglaciation.

DROUGHT

FIELDS OF STUDY

Climatology; Meteorology; Environmental Sciences; Environmental Studies; Climate Modeling; Physical Geography; Hydrometeorology; Hydroclimatology; Atmospheric Science; Climate Zones; Climate Classification; Hydrology; Earth System Modeling

SUMMARY

Drought is an unusually long period of below-normal precipitation. It is a relative rather than an absolute condition, but the end result is a shortage of water for plant growth, affecting the people who live in that region and beyond. Drought is particularly disastrous for farmers and the practice of agriculture.

PRINCIPAL TERMS

- **adiabatic:** a change of temperature within the atmosphere that is caused by compression or expansion without transfer of heat into or out of the system
- **desertification:** the relatively slow, natural conversion of fertile land into arid land or desert
- **evapotranspiration:** the combined water loss to the atmosphere from both evaporation and plant transpiration
- **Palmer Drought Index:** a widely adopted quantitative measure of drought severity that was developed by W. C. Palmer in 1965
- **potential evapotranspiration:** the water needed for growing plants, accounting for water loss by evaporation and transpiration
- **Sahel:** the semiarid southern fringe of the Sahara in West Africa that extends from Mauritania on the Atlantic coast to Chad in the interior
- **soil moisture:** water that is held in the soil and that is therefore available to plant roots
- **subsidence:** in meteorology, the slow descent of air that becomes increasingly dry in the process

IMPACT OF DROUGHT

Droughts have had enormous impacts on human societies since ancient times. The most obvious effect is crop and livestock failure, which have caused famine and death through thousands of years of human history. Drought has resulted in the demise of some ancient civilizations and, in some instances, the

Drought   Principles of Climatology

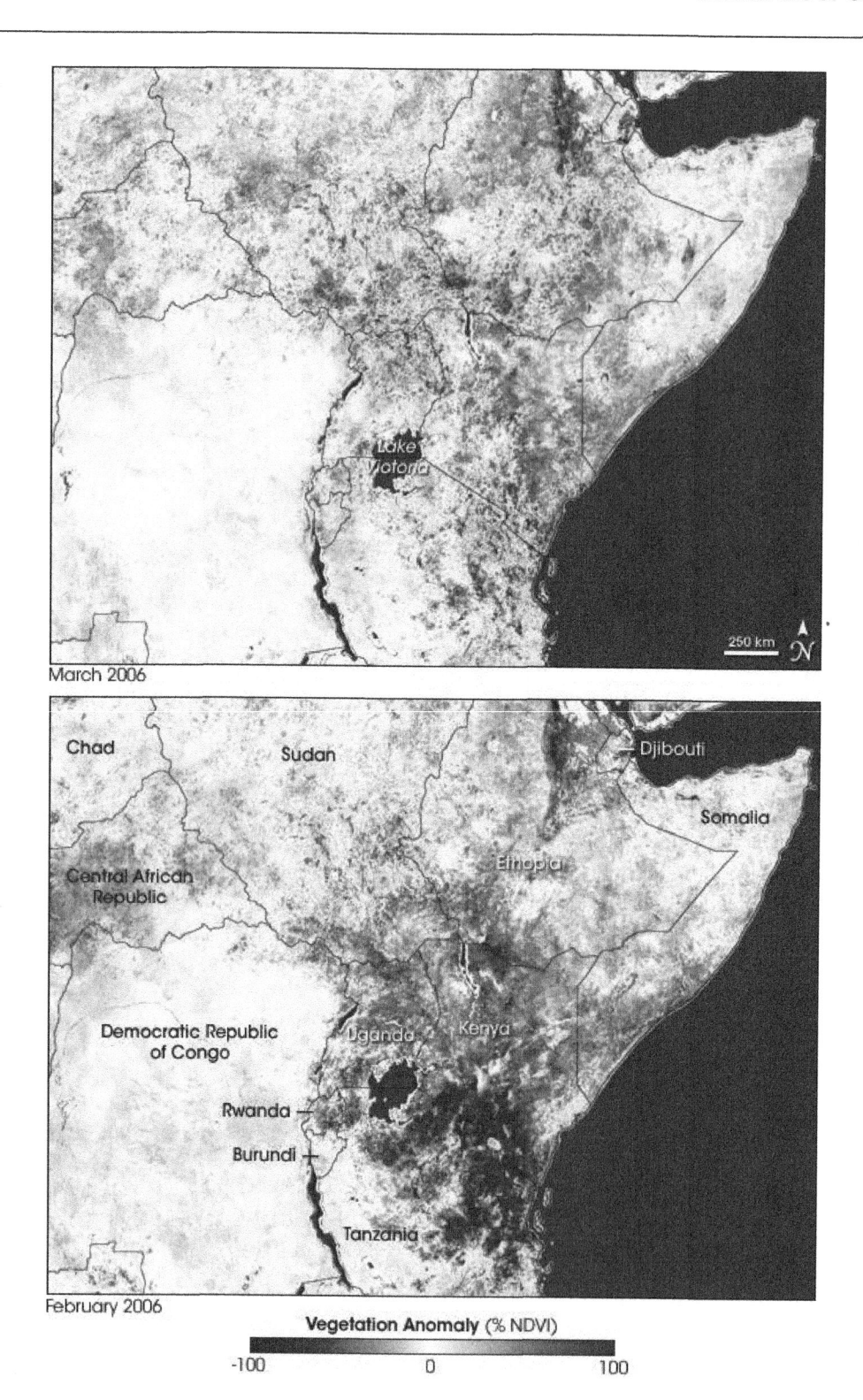

March 2006

February 2006

Vegetation Anomaly (% NDVI)

-100   0   100

vegetation anomalies in East Africa in February and March 2006, visualizing the drought in the region at this time forced mass migration of large numbers of people. Water is so critical to all forms of life that a pronounced shortage could, and has, decimated whole populations.

The effects of drought are profound. The dry conditions in the Great Plains of North America in the early 1930's in conjunction with extensive and improper farming activities resulted in the creation of the Dust Bowl, which at one point covered more than 200,000 square kilometers, or an area about the size of Nebraska. During the early 1960's, a severe drought affected the Mid-Atlantic states. Parts of New Jersey experienced sixty consecutive months of below-normal precipitation, so depleting local water supplies that plans were actively considered to bring rail cars of water into Newark and other cities in the northern part of the state, as the reservoirs that usually supplied the region were practically dry. The Sahel region south of the Sahara in West Africa had a severe drought beginning in the late 1960's and continuing into the early 1970's, creating an enormous negative impact on the local population, livestock, and vegetation. Hundreds of thousands of people starved, thousands of animals died, and many tribes were forced to migrate south to areas of more reliable precipitation. Large areas of eastern Africa in the Sudans and Ethiopia continue to experience an on-going state of drought.

Drought Characteristics

Almost all droughts occur when slow-moving air masses that are characterized by subsiding air movements dominate an area. Often, the air comes from continental interiors where the amount of moisture that is available for evaporation into the atmosphere is very limited. When these conditions occur, the potential for precipitation is low for a number of reasons. First, the humidity in the air is already low, as the continental air mass is distant from maritime (moist) influences. Second, air that subsides undergoes adiabatic heating at the rate of 10°C per 1,000 meters.

The term "adiabatic" refers to a change of temperature within a gas (such as the atmosphere) that occurs as a result of compression (descending air) or expansion (rising air), without any input or extraction of heat from external sources. For example, assume that air at a temperature of 0°C is passing over the Sierra Nevada in eastern California at an elevation of 3,500 meters. As the air descends and reaches Reno, Nevada, at an elevation of 1,200 meters, the higher atmospheric pressure found at lower elevations results in compression and heating at the dry adiabatic rate of 10°C per 1,000 meters, yielding a temperature in the Reno area of 23°C. Thus, adiabatic heating from subsiding air masses results in a decline in relative humidity and an increase in moisture-holding capacity. In addition, the movement of air under these conditions is usually unfavorable for vertical uplift and the beginning of the condensation process. The final factor that reduces precipitation potential is the decrease in cloudiness and corresponding increase in sunshine, which in turn leads to an increase in evapotranspiration demands that favor soil moisture loss.

Another characteristic associated with droughts is that once established, they appear to persist and even expand into nearby regions, resulting in desertification in extreme cases. This tendency is apparently related to positive feedback mechanisms. For example, the drying out of the soil influences air circulation and the amount of moisture that is then available for precipitation farther downwind. At the same time, the atmospheric interactions that lead to unusual wind systems associated with droughts can induce surface-temperature variations that, in turn, lead to further development of the unusual circulation pattern. Thus, the process builds on itself, causing the drought to both last longer and intensify. The situation persists until a major change occurs in the circulation pattern in the atmosphere.

Many climatologists concur with the concept that precipitation is not the only factor associated with drought. Other factors that demand consideration include moisture supply, the amount of water in storage, and the demand generated by evapotranspiration. Although the scientific literature is replete with information about the intensity, length, and environmental impacts of drought events, the role of individual climatological factors that can increase or decrease the severity of a drought is not fully understood.

Drought Identification

Research in drought identification has been changing over the years. Drought was once considered solely in terms of precipitation deficit. Although that lack of precipitation is still a key atmospheric component of drought, sophisticated techniques are now used to assess the deviation from normal levels of the total environmental moisture status. These techniques have enabled investigators to better understand the severity and length of drought events, as well as the extent of the affected area.

Drought has been defined in a number of ways. Some authorities consider it to be merely a period of below-normal precipitation, while others relate it to the likelihood of forest fires. Drought is also said to occur when the yield from a specific agricultural crop or pasture is significantly less than expected. It has also been defined as a period when soil moisture or groundwater decreases to a critical level.

Drought was identified early in the twentieth century by the U.S. Weather Bureau as any period of twenty-one or more days when precipitation was 30 percent or more below normal. Subsequent examination of drought events that were identified by this method revealed that soil moisture reserves were often elevated during these events to the extent that there was sufficient water to support vegetation. It was also determined that the amount of precipitation preceding the drought event was ample or even heavy. Thus, it became apparent that precipitation should not be used as the sole measure to identify drought. Subsequent research has shown that the moisture status of an area is affected by additional factors.

Further developments in drought identification during the middle decades of the twentieth century began to focus on the moisture demands that are associated with evapotranspiration in an area. Evaporation is primarily the process by which liquid water becomes water vapor at the surface and enters the atmosphere. To a lesser extent, this also includes the conversion of "solid water," as ice and snow, either directly by sublimation or through first melting into an intermediate liquid state. Transpiration refers to the loss of moisture by plants to the atmosphere. Although evaporation and transpiration can be studied and measured separately, it is convenient to consider them in applied climatological studies as the single process of evapotranspiration.

There are two ways to define evapotranspiration. The first is actual evapotranspiration, which is the actual or real rate of water-vapor return to the atmosphere from the earth and vegetation; this process could also be called "water use." The second is potential evapotranspiration, which is the theoretical rate of water loss to the atmosphere if one assumes continuous plant cover and an unlimited supply of water. This process could also be called "water need," as it indicates the amount of soil water needed if plant growth is to be maximized. Procedures have been developed that enable one to calculate the potential evapotranspiration for any area from monthly mean temperature and precipitation values.

Some drought-identification studies have focused on agricultural drought, looking at the adequacy of soil moisture in the root zone for plant growth. This procedure involved the evaluation of precipitation, evapotranspiration, available soil moisture, and the water needs of plants. The goal of this research was to determine drought probability based on the number of days when soil moisture storage is reduced to zero.

Evapotranspiration was also used by the Forest Service of the U.S. Department of Agriculture when it developed a drought index to be used by fire-control managers. The purpose of the index was to provide a measure of flammability that could create forest fires. This index has limited applicability to nonforestry users, as it is not effective for showing drought as an indication of total environmental stress.

PALMER DROUGHT INDEX

One of the most widely adopted drought-identification techniques was developed by W. C. Palmer in 1965. The method, which became known as the Palmer Drought Index, defines drought as the period of time, usually measured in months or years, when the actual moisture supply at a given location is consistently less than the climatically anticipated or appropriate supply of moisture. The calculation of this index requires the determination of evapotranspiration, soil moisture loss, soil moisture recharge, surface runoff, and precipitation. The Palmer Drought Index values range from approximately +4.0 for an extremely wet moisture status class to -4.0 for extreme drought. Normal conditions have a value close to 0. Positive values indicate varying stages of abundant moisture, whereas negative values indicate varying stages of drought.

Although the Palmer Drought Index is recognized as an acceptable procedure for incorporating the role of potential evapotranspiration and soil moisture in magnifying or alleviating drought status, there have been some criticisms of its use. For example, the method produces a dimensionless parameter of drought status that cannot be directly compared with other environmental moisture variables, such as precipitation, which are measured in units (centimeters, millimeters) that are immediately recognizable. In addition, the index is not especially sensitive to short drought periods, which can affect agricultural productivity.

In order to address these shortcomings, other researchers use water-budget analysis to identify deviations in environmental moisture status. The procedure is similar to the Palmer method inasmuch as it incorporates the environmental parameters of precipitation, potential evapotranspiration, and soil moisture. However, the moisture status departure values are expressed in the same units as precipitation and are therefore dimensional. Drought classification using this index method ranges from approximately 25 millimeters for an above-normal moisture status class to -100 millimeters for extreme drought. The index would be close to 0 for normal conditions.

SIGNIFICANCE

Drought is invariably associated with some form of water shortage, yet many regions of the world have regularly occurring periods of dryness. Three different forms of dryness have a temporal dimension described as perennial, seasonal, and intermittent. Perennially dry areas include the major deserts of the world, such as the Sahara, Arabian, and Kalahari. Precipitation in these areas is not only very low but also very erratic. Seasonal dryness is associated with regions where the bulk of the annual precipitation comes during a few months of the year, leaving the rest of the year without rain or other precipitation. Intermittent dryness is associated with those instances where the overall precipitation is reduced in humid regions or where the rainy season in seasonally dry areas does not occur or is shortened.

The absence of precipitation when it is normally expected creates variable problems. For example, the absence of precipitation for one week in an area where daily precipitation is the norm would be considered a drought. In contrast, it would take two or more years without any rain in parts of Libya in North Africa for a drought to occur. In those areas that have one rainy season, a 50 percent reduction in precipitation would be considered a drought. In regions that have two rainy seasons, the failure of one could lead to drought conditions. Thus, the word "drought" is a relative term, as it has different meanings in different climatic regions.

User demands also influence drought definition. Distinctions are often made among climatological, agricultural, hydrologic, and socioeconomic drought. Climatological, or meteorological, drought occurs at irregular periods of time, usually lasting months or years, when the water supply in a region falls far below the levels that are typical for that particular climatic regime. The degree of dryness and the length of the dry period are used as the definition of drought. For example, drought in the United States has been defined as occurring when there is less than 2.5 millimeters of rain in a forty-eight hour period. In Great Britain, drought has been defined as occurring when there are fifteen consecutive days with less than 0.25 millimeter of rain for each day. In Bali, Indonesia, drought has been considered as occurring if there is no rain for six consecutive days.

Agricultural drought occurs when soil moisture becomes so low that plant growth is affected. Drought must be related to the water needs of the crops or animals in a particular place, since agricultural systems vary substantially. The degree of agricultural drought also depends on whether shallow-rooted or deep-rooted plants are affected. In addition, crops are more susceptible to the effects of drought at different stages of their development. For example, inadequate moisture in the subsoil in an early growth stage of a particular plant will have minimal impact on crop yield as long as there is adequate water available in the topsoil. However, if subsoil moisture deficits continue, then the yield loss could become substantial.

Hydrologic drought definitions are concerned with the effects of dry spells on surface flow and groundwater levels. The climatological factors associated with the drought are of lesser concern. Thus, a hydrological drought for a particular watershed is said to occur when the runoff falls below some arbitrary value. Hydrological droughts are often out of phase with climatological and agricultural droughts and are also basin-specific; that is, they pertain to the particular watersheds that they affect.

Socioeconomic drought includes features of climatological, agricultural, and hydrological drought and is generally associated with the supply and demand of some type of economic good. For example, the interaction between farming (demand) and naturally occurring events (supply) can result in inadequate water for both plant and animal needs. Human activities, such as poor land-use practices, can also create a drought or make an existing drought worse. The Dust Bowl in the Great Plains and the Sahelian drought in West Africa provide ready examples of the symbiotic relationship between drought and human activities.

In a sense, droughts differ from other major geophysical events such as volcanic eruptions, floods, and earthquakes because they are actually nonevents. They result from the absence of events (precipitation) that should normally occur. Droughts also differ from other geophysical events in that they often have no readily recognizable beginning and take some time to develop. In many instances, droughts are only recognized when plants start to wilt, wells and streams run dry, and reservoir shorelines recede.

There is wide variation in the duration and extent of droughts. The length of a drought cannot be predicted, as the irregular patterns of atmospheric circulation are not fully known and remain unpredictable. A drought ends when the area receives sufficient precipitation and water levels rise in the wells and streams. Because the severity and areal extent of a drought cannot be predicted, all that is really known is that they are an integral part of the overall natural system and that they will continue to occur.

—Robert M. Hordon

Further Reading

Ahrens, C. Donald. *Meteorology Today: An Introduction to Weather, Climate and the Environment.* 8th ed. Thomson Brooks-Cole, 2007. One of the most widely used and authoritative introductory textbooks for the study of meteorology and climatology. Explains complex concepts in a clear, precise manner. Provides numerous images and diagrams.

Bryson, Reid A., and Thomas J. Murray. *Climates of Hunger.* University of Wisconsin Press, 1977. An interesting account of the profound effect of climate on human societies going back to ancient times. Discusses major climate changes and droughts for various regions in the world in separate chapters. Treatment is nonmathematical and suitable for senior-level high school students and above.

Climate, Drought, and Desertification. World Meteorological Organization, 1997. Deals with the climatic factors, such as drought, that lead to desertification worldwide. Includes color illustrations.

Dixon, Lloyd S., Nancy Y. Moore, and Ellen M. Pint. *Drought Management Policies and Economic Effects in Urban Areas of California, 1987-92.* RAND, 1996. Examines the impacts of the 1987-1992 drought in California on urban and agricultural users. Assesses the effects of the drought on a broad range of residential, commercial, industrial, and agricultural water users.

Fisher, R. J. *If Rain Doesn't Come: An Anthropological Study of Drought and Human Ecology in Western Rajasthan.* Manohar, 1997. Focuses on the human ecology associated with India and other countries prone to drought. Examines the factors involved with drought and its durations, as well as drought relief tactics and evaluation of such programs. Illustrations, maps, and references.

Frederiksen, Harald D. *Drought Planning and Water Resources Implications in Water Resources Management.* World Bank, 1992. Contains two papers on drought planning and water-use efficiency and effectiveness. Deals with policy and program issues in water-resources management from the perspective of the World Bank.

Hidore, John, John E. Oliver, Mary Snow, and Richard Snow. *Climatology.* 3d ed. Merrill, 1984. Discusses all aspects of climatology. Contains numerous black-and-white illustrations and maps. Suitable for college-level students. Although quantitative measures are included, no particular mathematical background is necessary.

Knight, Gregory, Ivan Raev, and Marieta Staneva, eds. *Drought in Bulgaria.* Ashgate Publishing Limited, 2004. Presents the occurrence of drought in Bulgaria as a case study to provide information on prospective future global conditions. Discusses the geological, economic, and ecological effects of drought in Bulgaria from 1982 to 1994. Concludes with policy and conservation recommendations for the future.

Mainguet, Monique. *Aridity: Drought and Human Development.* Springer-Verlag, 2010. Discusses global and local aridity, drought, and associated changes in vegetation and hydrology. The second half of the text focuses on anthropogenic impact. Includes references and multiple indexes.

Lutgens, Frederick K., Edward J. Tarbuck, and Dennis Tasa. *The Atmosphere: An Introduction to Meteorology.* 11th ed. Prentice Hall, 2009. An excellent introduction and description of the atmosphere, meteorology, and weather patterns. Suitable for readers new to the study of these subjects. Color illustrations and maps.

Wilhite, Donald A., ed. *Drought and Water Crises: Science, Technology, and Management Issues.* CRC Press, 2005. Covers water management practices currently used in areas around the world. Discusses monitoring, drought planning, and water conservation policies. Provides case studies that discuss many issues of and solutions to drought. Concludes with a chapter discussing the future of water conservation.

Wilhite, Donald A., and William E. Easterling, with Deborah A. Wood, eds. *Planning for Drought: Toward a Reduction of Societal Vulnerability.* Westview Press, 1987. An extensive collection of thirty-seven short chapters on drought covering a wide range of topics from the climatological to the institutional. Provides a good background to the many issues pertaining to drought, its social impacts, governmental response, and adaptation and adjustment. Suitable for college-level students.

Workman, James G. *Heart of Dryness: How the Last Bushmen Can Help Us Endure the Coming Age of Permanent Drought.* Walker Publishing Company, 2009. Introduces the reader to the occurrence of drought and its impact on humans, animals, vegetation, and land. Describes in detail the issues faced and created by humans in desiccated regions through the eyes of the people facing it presently.

E

EARTH-SUN RELATIONS

FIELDS OF STUDY
Astronomy; Observational Astronomy; Spectroscopy; Photochemistry; Thermodynamics; Heat Transfer; Environmental Sciences; Physical Sciences; Meteorology; Climatology; Climate Modeling; Climate Classification; Oceanography; Physical Geography; Earth System Modeling; Computer Science; Particle Physics; Electromagnetism

SUMMARY
The sun is the center of the solar system and the supporter of all life on Earth. It affects the weather, climate, and seasons; it drives the food chain through the process of photosynthesis; and it provides heat, light, and energy. Ultimately, the demise of the sun will bring about the end of Earth.

PRINCIPAL TERMS

- **aphelion:** the time at which the distance between Earth and the sun is smallest; generally occurs on one of the first days of January, two weeks after the December solstice
- **aurora:** a glowing light display resulting from charged particles from solar wind being pulled into Earth's atmosphere by Earth's magnetic field; most often visible near Earth's North and South Poles
- **equinox:** a twice-a-year occurrence during which the tilt of Earth's axis is such that Earth is not tilted toward or away from the sun; the center of the sun is directly aligned with Earth's equator
- **geomagnetic storm:** the effect of variations in solar wind's interactions with Earth's atmosphere; can result in communications disruptions and auroral displays in lower than usual latitudes
- **nuclear fusion:** atomic nuclei join together to form a heavier nucleus; in the sun and other main-sequence stars, hydrogen is fused to form helium
- **perihelion:** the time at which the distance between Earth and the sun is largest; generally occurs on one of the first days of July, two weeks after the June solstice
- **solar cycle:** an approximately eleven-year-long cycle of varying solar activity; solar cycles are tracked based on the visibility of sunspots
- **solar wind:** a stream of charged particles that the sun's atmosphere ejects into space, where it can interact with the magnetic fields of planets
- **solstice:** a twice-a-year occurrence during which the sun appears at its highest point in the sky (once a year as seen from the North Pole and once a year as seen from the South Pole)
- **sunspot:** a cooler area on the sun's surface that appears darker than the surrounding area; a zone of decreased temperature resulting from the complex shape of the sun's magnetic field

OVERVIEW
The sun was born about 4.57 billion years ago out of the collapse of a portion of a stellar nursery, a molecular cloud of enormous density and size. The giant, hot sphere of plasma and gas is the center of the solar system and the supporter of all life on Earth.

The sun accounts for approximately 99.86 percent of the mass of the solar system; its total mass is 2×10^{30} kilograms (about 330,000 times the mass of Earth). Its diameter is roughly 1.4 million kilometers (more than one hundred times that of Earth), and its surface temperature is estimated to be 5,778 kelvins (5,505°C or 9,940°F), while its core temperature is estimated at 1.571×10^7 kelvins. Meanwhile, on Earth, the maximum surface temperature is around 331 kelvins. Three-fourths of the sun is made of hydrogen and nearly one-fourth is made of helium. Trace amounts of heavier elements make up the rest.

The sun is the closest star to Earth and thus the brightest to appear in the sky. While the distance between the sun and Earth is constantly changing based on Earth's 365.25-day orbit around the sun and based on Earth's twenty-four-hour rotation around its own axis, the mean distance between the sun and Earth is 149.6 million km (93 million miles), a distance that defines one astronomical unit. It takes light from the sun eight minutes and nineteen seconds to cover this distance and reach Earth. For comparison, consider the 1.3 seconds that it takes light to reach Earth from Earth's moon, or, on the other end of the spectrum, the one hundred thousand years it takes for light to cross the Milky Way galaxy.

The sun affects Earth in a multitude of ways. Its complex magnetic field causes a range of effects known as solar activity, including sunspots and solar flares, and these phenomena in turn cause space weather in Earth's atmosphere, leading to beautiful visual effects (auroras) but troublesome disruptions of communications and electricity. Space weather also influences Earth's climate, weather, and seasons. All of this solar activity varies on a roughly eleven-year-long solar cycle. While these effects are mostly within Earth's upper atmosphere, the sun also directly affects life at the surface.

Sunlight drives photosynthesis in plants, algae, and some bacteria, allowing them to convert sunlight into food; without this process, the entire food chain would collapse. Humans also both enjoy and suffer direct effects from the sun, which can damage human eyes and skin while also providing sanitizing ultraviolet rays and driving vitamin D production. Solar power can influence the future of technology. Earth's ultimate demise will also be brought about by the sun.

Effects of the Sun's Magnetic Field and the Solar Cycle

The sun releases energy in two main forms: electromagnetic radiation and charged particle emission. Through the activities of the sun's magnetic field and the different ways in which these types of energy are released and moved, Earth experiences a variety of effects.

The sun's strong magnetic field has a complex spiral shape. The star's high temperature means that it is composed solely of gases and plasma, a composition that allows it to rotate faster around its equator than at the poles; this differential rotation twists the magnetic field into its distinctive form. Ultimately, it is the twists in the magnetic field that lead to such solar activity as sunspots and solar flares.

When solar activity is at its maximum in the eleven-year-long solar cycle, the magnetic field flips direction. The sun's magnetic field exerts its influence well beyond the sun itself because the solar wind, a plasma stream of charged particles, carries it through the solar system, altering the magnetic fields of the planets.

Varying solar activity, such as sporadic corona eruptions (coronal mass ejections), can alter the activity and intensity of the solar wind, resulting in geomagnetic storms on Earth. These storms cause beautiful glowing lights in Earth's atmosphere called auroras, particularly aurora borealis (the northern lights) and aurora australis (the southern lights). On the negative side, though, the storms also interrupt communications on Earth, disrupting radio signals and electric transmissions and interfering with compass-based navigation, among other effects.

The frequency of geomagnetic storms roughly aligns with the sunspot cycle, which in turn is determined by the solar cycle. Sunspots are visual representations of the sun's magnetic field; when the sun's magnetic field lines emerge from within the sun, they create magnetic loops in the photosphere (the sun's visible surface). The temperature is lower within these loops, appearing as the dark patches called sunspots. Other solar activity includes solar flares and solar plumes.

The solar cycle, which usually takes eleven years, is marked by a solar maximum (at which time solar activity is at its highest level) and, conversely, a solar minimum. During a solar maximum in 1859, for example, a geomagnetic storm was so intense that the northern lights could be seen as far south as Rome. Throughout history, several prolonged solar minimums have been observed, during which Earth temperatures were at an all-time low. One such period, known as the Maunder Minimum, spanned the second half of the seventeenth century and into the beginning of the eighteenth century.

Earth-sun relationship is partially driven by the solar cycle, during which solar activity is always varying. Earth-sun relationship also is strongly influenced by Earth's orbit around the sun and by Earth's rotation around its own axis, both of which directly affect the distance between Earth and the sun. Perihelion (when the distance between Earth and sun is shortest) occurs once a year, in early January, and aphelion (when the distance is longest) occurs once a year, in early July. These generally occur two weeks after the December solstice and June solstice, respectively.

The solstices are the two points in the year when the sun appears highest in the sky, either in the north or in the south, because of the tilt of Earth's axis. The northern (or summer) solstice occurs in June, and it corresponds with summer in the Northern Hemisphere and winter in the Southern Hemisphere. Conversely, the southern (or winter) solstice occurs in December, during winter in the Northern Hemisphere and summer in the Southern Hemisphere. Similarly, two equinoxes occur each year, one in late March and one in late September, when the tilt of Earth's axis is such that the sun is aligned in the same plane as the equator. The equinoxes are hallmarks of spring and autumn.

The big picture is that variations in the sun's magnetic field cause different types of space weather, which affects Earth's upper atmosphere, ultimately influencing Earth's climate, weather, and seasons; the degree to which space weather occurs remains a topic of ongoing research. The whole relationship is also influenced by the ever-changing distance between Earth and the sun.

Photosynthesis and Health Effects of the Sun

The sun's effects on Earth are not limited to climate control; the sun is also entirely responsible for the food chain (through photosynthesis), and it has a number of other health effects, both positive and negative.

Photosynthesis is a process that occurs in plants, algae, and some bacteria. In basic terms, the process involves sunlight reacting with carbon dioxide and water to yield sugar (food) and oxygen. To elaborate somewhat, the first stage of photosynthesis, light-dependency, begins with an organism trapping and storing sunlight. The sunlight's energy is then converted into chemical energy, stored in the molecules ATP and NADPH.

The second stage does not require any more light; the already-created ATP and NADPH drive the reduction of carbon dioxide to glucose (sugar) and other useful organic molecules. Not only does photosynthesis serve as the basis for the food chain, but it also maintains the necessary oxygen levels in the atmosphere because oxygen is a waste product of the process.

The human health implications of sunlight, particularly the ultraviolet (UV) radiation in sunlight, are both positive and negative. On the positive side, human bodies synthesize vitamin D from sunlight; a deficiency in vitamin D can lead to the thinning and softening of bones, and some research has suggested a link between vitamin D deficiency and seasonal affective disorder (SAD), a seasonal depression.

On the negative side, too much UV radiation can lead to skin aging, skin cancer, and eye damage, such as cataracts and age-related macular degeneration. Looking at the sun through binoculars without a proper UV filter can cause instant retinal damage and even permanent blindness. It is particularly dangerous to look at the sun directly or with an optical aid during an eclipse.

Solar Power

Much as light energy can be converted to food by photosynthesis, it also can be converted to electricity. The harnessing of solar power began on a commercial scale in the 1970s and 1980s, and it continues to gain prevalence, showing promise for the future because of its numerous advantages over other power sources. Most importantly, light energy is renewable: Sunlight will not be depleted, at least not while the sun is still in roughly its current form for the next several billion years.

Solar power also is a clean energy; it does not release greenhouse gases or pollute the ocean, as does oil. Additionally, solar cells are long-lasting and require little maintenance. Although renewable, sunlight is not always available. Storage solutions are important because solar energy cannot be collected at night.

There are two major methods of solar power: first, the use of photovoltaics, which involves the use of the photoelectric effect of sunlight to convert light energy directly into an electric current; and second, the use of concentrated solar power, which involves the use of lenses or mirrors to focus sunlight into a small beam. This focused light is then converted to heat, which then powers a steam turbine or other heat engine to ultimately generate electricity. Both technologies are under development, and the hope is that costs will decrease and efficiency will increase in the coming decades.

THE END OF EARTH

Just as the sun has supported more than 4 billion years of life on Earth, it is ultimately the sun that will bring about the end of Earth. The sun is a main-sequence (or dwarf) star, a type of star marked by fusion of hydrogen to helium within its core.

It is estimated that Earth's sun will exist about 10 billion years in its main-sequence form before it has fused all of its hydrogen; it has already covered nearly one-half of that time span. This means that in about 5 billion years it will enter its next phase of life, as a red giant.

The outer layers of the red giant sun will expand as its core contracts and heats up, and this overall decrease in mass will likely loosen its gravitational pull on the solar system, allowing the planets to move outward. The sun's radius will be greater than the current distance between the sun and Earth, and it is unclear whether the looser pull will allow Earth's orbit to move far enough away or if Earth will be swallowed up by the sun. Even if Earth does escape being enveloped by the sun, the heat will still likely be enough to boil off all of Earth's water and destroy much of its atmosphere. Even before any of this occurs, though (perhaps 1 billion years from now), it is quite possible that life on Earth will cease to exist because of rising surface temperatures of the sun (and thus Earth).

—*Rachel Leah Blumenthal*

FURTHER READING

Alexander, David. *The Sun.* Greenwood Press, 2009. Part of the Greenwood Guides to the Universe series, this easy-to-understand text gives readers the necessary background on the sun before delving into a comprehensive look at recent discoveries and future research goals.

Alley, Richard B. *Earth: The Operators' Manual.* Norton, 2011. Through storytelling, this book illuminates the history of humankind's use of energy while also discussing the negative effects this has had on Earth. The book also delves into current and future alternative and renewable energy options, such as solar power.

Crosby, Alfred W. *Children of the Sun: A History of Humanity's Unappeasable Appetite for Energy.* Norton, 2006. More historical than scientific, this book will nonetheless be interesting for science-minded readers who wish to put scientific knowledge of the sun in its broader historical and cultural context, particularly with regard to its energy potential for humanity.

Eddy, John A. *The Sun, Earth, and Near-Earth Space: A Guide to the Sun-Earth System.* National Aeronautics and Space Administration, 2009. Provides a comprehensive but easy-to-understand look at the sun's structure, function, and relationship with Earth.

Lang, Kenneth R. *Sun, Earth, and Sky.* Springer, 2006. This beautiful text uses a multitude of illustrations and images to illuminate the workings of the sun and its relationship with Earth. It is easy to understand and perfectly suited for students new to the subject.

McFadden, Lucy-Ann Adams, Paul Robert Weissman, and T. V. Johnson. *Encyclopedia of the Solar System.* Academic Press, 2007. From the origins of the solar system to modern-day planetary exploration, this comprehensive text covers topics about the sun, solar wind, and the relationship between the sun and Earth.

ECOSYSTEMS' INFLUENCE ON CLIMATE

FIELDS OF STUDY

Heat Transfer; Environmental Sciences; Physical Sciences; Meteorology; Climatology

SUMMARY

An ecosystem is a functional system, encompassing all organisms (plants, animals, and microorganisms) and all elements of the nonliving physical environment that interact together in a given area. Organisms extract chemical elements (including water, carbon dioxide, and nutrients) as substrates from the physical environment, using these substrates for their own survival, growth, and reproduction. Physical processes and chemical reactions in the environment are catalyzed by organisms so as to influence

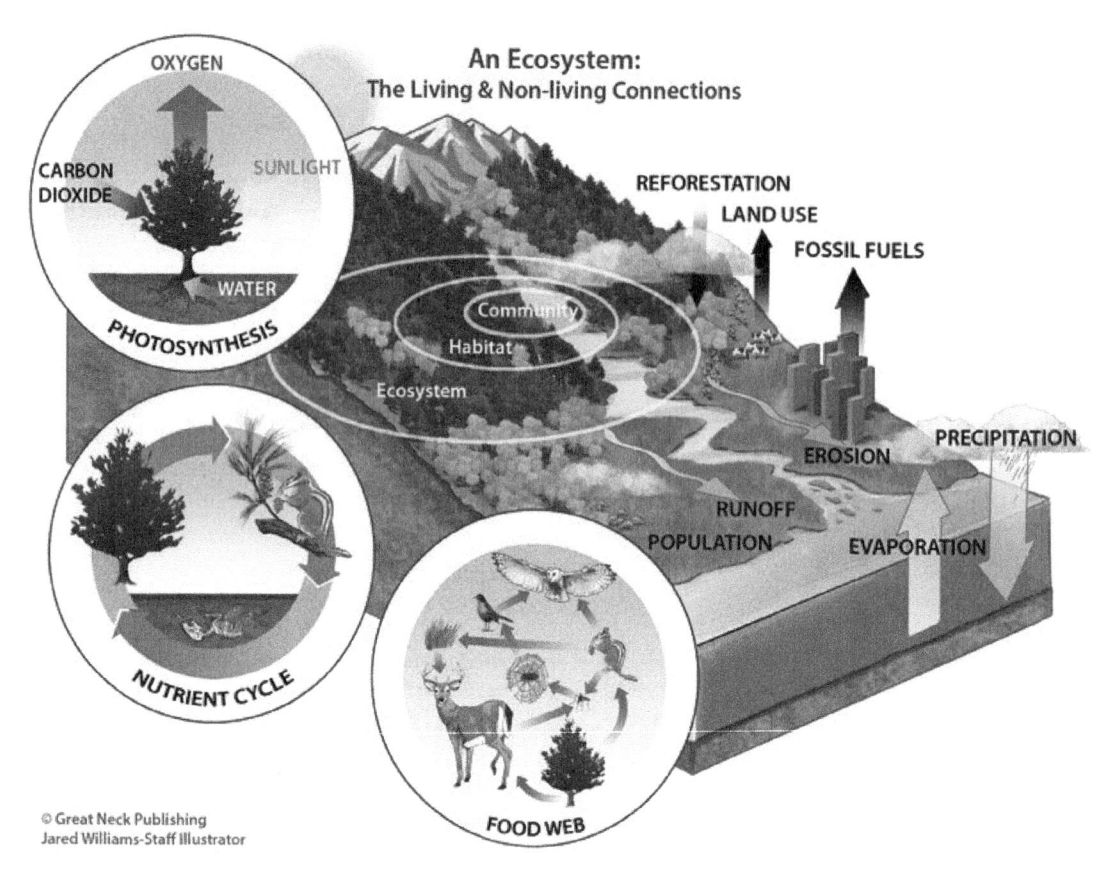

Ecosystems are complex systems involving all of the interactions among living organisms and the environment.

energy balance and to form biogeochemcial cycles of carbon, water, and other elements within the system. Ecosystems can be bounded on various scales, from a microcosm to the entire planet.

PRINCIPAL TERMS

- **albedo:** the percentage of the solar radiation of all wavelengths reflected by a body or surface
- **biome:** a geographically defined area of similar plant community structure shaped by climatic conditions
- **canopy:** the upper layers of vegetation or uppermost levels of a forest, where energy, water, and greenhouse gases are actively exchanged between ecosystems and the atmosphere
- **evapotranspiration:** processes through which water on surfaces or in plants is lost to the atmosphere
- **greenhouse gases (GHGs):** atmospheric gases that trap heat, preventing it from escaping into space
- **latent heat flux:** the flux of thermal energy from land surface to the atmosphere that is associated with evaporation and transpiration of water from ecosystems
- **photosynthesis:** a metabolic pathway that absorbs inorganic carbon dioxide from the atmosphere and converts it to organic carbon compounds using sunlight as an energy source
- **respiration:** metabolic reactions and processes to convert organic compounds to energy that release CO_2 as a by-product
- **sensible heat flux:** the flux of thermal energy that is associated with a rise in temperature
- **stoma:** a pore in the leaf and stem epidermis that is used for gas exchange

CLIMATE AND GEOGRAPHICAL DISTRIBUTIONS OF ECOSYSTEMS

Various types of ecosystems exist on Earth, including ocean ecosystems, land ecosystems, and freshwater ecosystems on a broad scale. Within land ecosystems, vegetation displays different patterns, forming different ecosystems at regional scales, such as forests, deserts, grasslands, and croplands. Except for artificial ecosystems, patterns of natural ecosystems are primarily shaped by climate conditions (such as temperature and precipitation). Along a precipitation gradient from wet to dry regions, ecosystem types change from forests, woodlands, and grasslands to deserts. Along a temperature gradient from the equator to the polar region, ecosystems vary from tropical forests, subtropical forests, temperate deciduous forests, temperature mixed forests, and boreal forests to tundra. In polar climate zones with average temperatures below 10° Celsius in all twelve months of the year, ecosystems include tundra and ice cap in Antarctica and in inner Greenland. Thus, climate and other physical environmental characteristics determine the distribution of ecosystems on the globe.

ECOSYSTEM RESPONSES TO CLIMATE CHANGE

Ecosystems are very sensitive to changes in temperature, atmospheric carbon dioxide (CO_2), and precipitation. Rising atmospheric CO_2 primarily stimulates carbon influx, leading to increases in carbon sequestration and thus potentially mitigating climate change. Rising atmospheric CO_2 concentration has relatively minor impacts on canopy energy balance and water exchange at the surface. Climate warming influences ecosystem feedback related to climate change in several ways, such as exchange of greenhouse gases (GHGs), surface energy balance, and water cycling. It is generally assumed that warming affects carbon release more than carbon uptake, leading to net carbon loss from land ecosystems to the atmosphere. Temperature also affects phenology and length of growing seasons, nutrient availability, and species composition. All these processes influence carbon balances, potentially leading to the net carbon uptake from the atmosphere and negative ecosystem feedback to climate warming.

Increasing temperature also stimulates evapotranspiration, resulting in cooler land surfaces in wet regions and thus negative feedback to climate change. The ecohydrological feedback to climate warming via altering land surface energy balance is weak in dry regions. Altered precipitation regimes (that is, alterations in amount, seasonality, frequency, and intensity) under climate change modify ecosystem carbon cycling, energy balance, and water exchange with the atmosphere. Increased precipitation, for example, usually stimulates plant productivity and ecosystem carbon uptake from the atmosphere. Decreased precipitation generally causes land surfaces to be warmer and generates a higher albedo than does ambient precipitation. Impacts of altered precipitation seasonality, frequency, and intensity are complex and region-specific. In addition, precipitation regimes have long-term impacts on soil development, nutrient availability and vegetation distributions, which can be different from short-term impacts of precipitation on ecosystem processes. Moreover, climate change involves a suite of changes in temperature, precipitation, and GHGs. Those global change factors can interactively influence ecosystem processes and their feedback to climate change.

ECOSYSTEM REGULATION OF CLIMATE CHANGE VIA ENERGY BALANCE

Land surface energy balance influences the climate system by causing fluctuations in temperature, winds, ocean currents, and precipitation. The surface energy balance, in turn, is determined by fractions of absorbed, emitted, and reflected incoming solar radiation. One of the key parameters to determine the energy balance at the land surface is albedo, which regulates differences between the amount of absorbed shortwave radiation (input) and the outgoing longwave radiation (output). Different vegetation covers have different albedo values. When land use and land cover changes occur due to either climate change or anthropogenic activities, land surface energy balance is altered. Overgrazing, for example, may increase albedo. As a consequence, evapotranspiration decreases with associated decline in energy and moisture transfer to the atmosphere. In general, vegetation absorbs more solar energy, transpires more water, drives more air circulation, and results in more local precipitation in a region with low than high albedo. Thus, ecosystems influence energy balance in the atmosphere and feed back to climate change.

ECOHYDROLOGICAL REGULATION OF CLIMATE CHANGE

Water vapor exchange at the land surface significantly affects climate dynamics at local, regional, and global scales. Ecosystems receive water input via precipitation and lose water via evapotranspiration. Plant vegetation is the primary regulator of evapotranspiration. Thus, types of ecosystems significantly affect energy and water transfers from ecosystems to the atmosphere.

Because water transpired through leaves comes from the roots, rooting systems play a critical role in ecohydrological regulation of climate. Woody encroachment to grasslands, for example, can accelerate the ecosystem hydrologic cycle and then influence climate dynamics because trees usually have deep taproots to take up water from deep soil layers. Conifer forests can transpire water from the soil to the atmosphere in early spring and late fall and have longer seasons of transpiration than deciduous forests. Conversion of grasslands to winter wheat croplands accelerates evapotranspiration in winter and early spring when wheat actively grows and grasses are dormant. However, evapotranspiration is lower in fallow fields after wheat harvest than in grasslands in summer and fall. In addition, rooting systems are highly adaptive to climate change. When climate warming increases soil temperature and water stress, plants grow more roots to take up water. The adaptive rooting systems can significantly regulate climate change.

CARBON-CLIMATE FEEDBACK

Ecosystems can regulate climate change via changes in uptake and releases of GHGs. The GHGs involved in ecosystem feedbacks to climate change include CO_2, methane (CH_4), nitrous oxide (N_2O), and ozone (O_3). Their uptakes and releases are modified by changes in temperature, precipitation, atmospheric CO_2 concentration, land use and land cover changes, and nitrogen deposition. For example, ecosystems absorb CO_2 from the atmosphere by photosynthesis and release it back to the atmosphere via respiration. Photosynthetically fixed carbon from the air is converted to organic carbon compounds. Some of the carbon compounds are used to grow plant tissues while others are used for plant respiration. Plant tissues die, adding litter to soil. Litter is partly decomposed by microorganisms to release CO_2 back to atmosphere and partly incorporated to soil organic matter. The latter can store carbon in soil for hundreds and thousands of years.

Many factors and processes can alter the carbon cycles and then influence carbon-climate feedback. For example, deforestation usually results in net release of carbon from ecosystems to the atmosphere, enhancing climate change. Rising atmospheric CO_2 usually stimulates plant growth and ecosystem carbon sequestration, mitigating climate change. Climate warming can stimulate both photosynthesis and respiration. Most models assume that respiration is more sensitive than photosynthesis to climate warming and predict a positive feedback between terrestrial carbon cycles and climate warming. Field experiments, however, suggest much richer mechanisms driving ecosystem responses to climate warming, including extended growing seasons, enhanced nutrient availability, shifted species composition, and altered ecosystem-water dynamics. The diverse mechanisms likely define more possibilities of carbon-climate feedbacks than projected by the current models.

CONTEXT

Ecosystems are basic units of the biosphere. The latter is the global ecological system integrating all living organisms and their interaction with the lithosphere, hydrosphere, and atmosphere. Biosphere-atmosphere interactions occur via exchanges of energy, water, and GHGs in ecosystems. Specifically, ecosystems interact with the atmosphere via emission and absorption of GHGs so as to influence energy balance in the atmosphere; variations in albedo to influence the amount of heat transferred from ecosystems to the atmosphere; and changes in evapotranspiration to cool the land surface, to influence water vapor dynamics, and to drive atmospheric mixing. In addition, ecosystems can influence climate dynamics by changes in production of aerosols and surface roughness and coupling with the atmosphere. Thus, understanding ecosystem processes that regulate energy balances, water cycling, and carbon and nitrogen dynamics is critical to Earth-system science.

—*Yiqi Luo*

BIBLIOGRAPHY

Chapin, F. Stuart, III, et al. "Changing Feedbacks in the Climate-Biosphere System." *Frontiers in Ecology and the Environment* 6 (2008): 313–320. Provides an overview of interrelationships between ecosystems and the climate system in terms of energy balance, water cycling, and greenhouse gas release and uptake in ecosystems.

Chapin, F. Stuart, III, Harold A. Mooney, and Pamela Matson. *Principles of Terrestrial Ecosystem Ecology.* New York: Springer, 2002. Written by three prominent ecologists, this text provides a good introduction for beginners in ecosystem ecology. Comprises four major sections: context, mechanisms, patterns, and integration.

Field, C. B., D. B. Lobell, and H. A. Peters. "Feedbacks of Terrestrial Ecosystems to Climate Change." *Annual Review of Environmental Resources* 32 (2007): 1–29. Reviews major ecosystem processes that potentially result in either negative or positive feedbacks to climate change. Discusses regional differences between those ecosystem-climate feedback processes.

Luo, Y. Q. "Terrestrial Carbon-Cycle Feedback to Climate Warming." *Annual Review of Ecology Evolution and Systematics* 38 (2007): 683–712. Provides an overview of modeling results and experimental evidence regarding carbon cycle-climate change feedback. Summarizes major regulatory mechanisms underlying the carbon-climate feedbacks, including extended growing seasons, enhanced nutrient availability, shifted species composition, and altered ecosystem-water dynamics in response to climate warming.

EL NIÑO-SOUTHERN OSCILLATION AND GLOBAL WARMING

FIELDS OF STUDY

Oceanography; Heat Transfer; Fluid Dynamics; Atmospheric Science; Meteorology; Climate Modeling; Statistics; Earth System Modeling; Hydroclimatology; Hydrometeorology; Physical Geography

SUMMARY

El Niño is a cyclic buildup of warm sea surface water in the central and eastern equatorial Pacific Ocean, adjacent to the Peruvian coast. The cycle's opposite, cooling phase is called La Niña. The Southern Oscillation (SO) is a "seesaw" of air pressure and air circulation between the eastern Pacific and the Indonesian region. The terms El Niño and La Niña are used to denote the extremes of the oscillation. The broader term El Niño-Southern Oscillation (ENSO) describes the range of atmospheric and oceanic processes and their accompanying changes. Although ENSO is based in the tropics, it influences weather throughout the Northern Hemisphere and possibly the globe as a whole. Because of its effects on global temperature averages, ENSO is often discussed in connection to global warming.

PRINCIPAL TERMS

- **Hadley cell:** an atmospheric circulation system of air rising near the equator, flowing poleward, descending in the subtropics, and then flowing back toward the equator
- **La Niña:** the cooling half of the cycle of which El Niño is the warming half
- **ocean-atmosphere coupling:** the interaction between the sea surface and the lower atmosphere that drives many patterns and changes in Earth's weather systems
- **paleoclimates:** climates of the distant past
- **Walker circulation:** an atmospheric circulation pattern in the Pacific and elsewhere in which hot, moist air rises, travels eastward, cools and dries, descends, and returns westward

CHARACTERISTICS OF ENSO

Definitions of ENSO vary, but a common aspect of El Niño is the irregular warming of sea surface water off the coasts of Ecuador, northern Peru, and occasionally Chile. This warming is linked to irregular changes in air pressure at sea level across the Pacific Ocean. During El Niño conditions, the westward-flowing trade winds slacken. During La Niña conditions, by contrast, the westward-flowing trade winds are stronger than normal. A common measure of the SO is the Southern Oscillation Index (SOI), which is usually based on changes in sea-level air pressure at locations on opposite sides of the tropical Pacific. The most common basis for the SOI is the mean sea-level air pressure difference between Tahiti

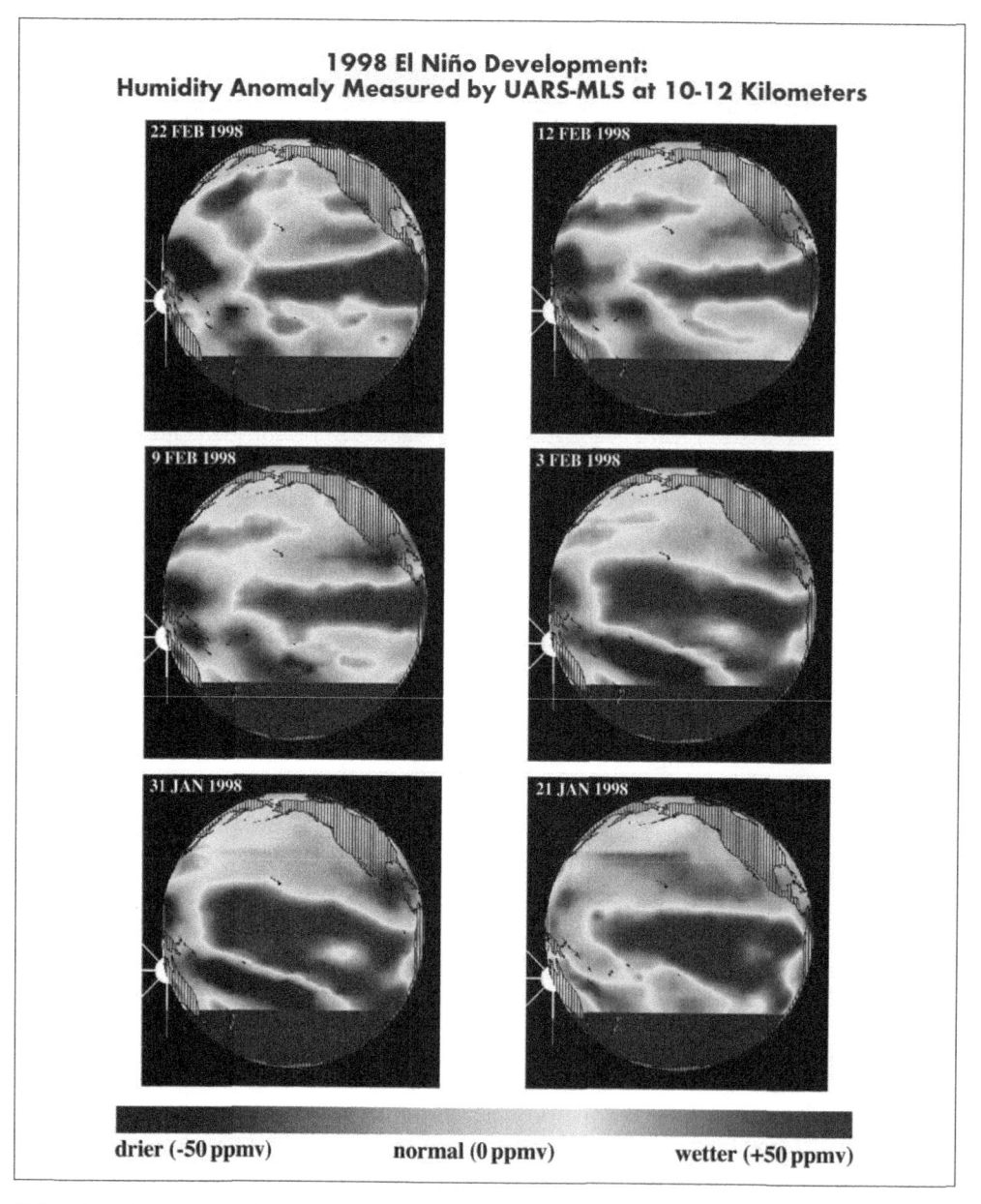

**1998 El Niño Development:
Humidity Anomaly Measured by UARS-MLS at 10-12 Kilometers**

drier (-50 ppmv)   normal (0 ppmv)   wetter (+50 ppmv)

This series of images shows the development of water vapor over the Pacific Ocean during the El Niño event of January and February, 1998.

and Darwin, Australia, expressed as the long-term difference of their monthly pressures.

Normally, there is a low-pressure zone of warm air in the western Pacific and a high-pressure zone of cool air in the eastern Pacific. This pressure differential drives a loop of warm air from over the western Pacific that rises just east of Indonesia, travels eastward, and descends over the eastern Pacific. The loop then flows in a westerly direction at the surface back toward the west. The strength of this circulation, known as the Walker circulation, is heavily influenced by the seesaw-like sea-level pressure differences between the eastern and western Pacific. The pattern is named after Gilbert Walker

(1868–1958), whose work led to the discovery of the Southern Oscillation.

The formation and breakdown of the Walker circulation cell is reflected in the pressure difference across the Pacific. When pressure is low in Tahiti, it is high in Darwin, and vice versa. This periodic yearly-to-decadal seesaw of atmospheric and oceanic circulation is the SO. When the SOI is negative, sea surface temperatures are warmer than usual in the eastern equatorial Pacific, off the coats of Ecuador and northern Peru (and occasionally Chile). A negative SOI is associated with El Niño conditions. When the SOI is positive, sea surface temperatures are cooler than usual in the eastern equatorial Pacific. A positive SOI is associated with La Niña conditions. During La Niña events, the east-west movement of air in the Walker circulation is enhanced with well-defined and vigorous rising and sinking branches.

Links to Global Climate Change

During El Niño events, there is an increase in Hadley cell circulation, the circulation cell of air that rises over the equator and descends in the subtropical latitudes on both sides of the equator. A more vigorous overturning of the Hadley cell circulation leads to an increase in heat transfer from tropical to higher latitudes in both hemispheres. Often, the climatic effects of increased Hadley circulation can be seen globally as above-average temperatures and extreme precipitation. Various other climatic consequences of ENSO events have also been reported. Warmer ocean waters have bleached coral reefs and vigorous atmospheric circulation has driven ocean currents northward, warming the Arctic Ocean and decreasing the amount of sea ice there.

The longer-term relationship between ENSO and global climate change is not fully understood and continues to be studied. While records since the 1980s indicate that El Niños have become more frequent and La Niñas less frequent, there is not enough data to effectively interpret the trend. However, scientists increasingly suspect that global warming causes more extreme and more volatile El Niño events, at least in the short term. For example, the El Niño of 2015 was considered the most powerful since record keeping began. Regardless of the exact interrelation, most scientists are convinced that ENSO and climate change are inextricably linked.

This satellite image depicts the average variation in sea surface temperatures, which alter in El Niño years.

Assumptions about ENSO and Anthropogenic Global Warming

The Intergovernmental Panel on Climate Change (IPCC) has commented on the possibility of connections between ENSO and anthropogenic increases in atmospheric greenhouse gases. Based on the output of complex but unvalidated global climate models, the IPCC indicates that as temperatures increase, the average Pacific climate could more consistently emulate El Niño conditions. However, the IPCC accepts that some climate models point to a more La Niña-like response to global warming, because Hadley cell circulation may decrease with increasing global temperature. Paleoclimatic studies support the view that global warming is likely to foster weaker and less frequent El Niño events. However, the level of scientific uncertainty and the existence of conflicting results render prediction of future climate conditions unreliable.

Some climate change skeptics even claim that the natural temperature variation created by ENSO is responsible for the observed trend of increasing global average temperatures, discounting the influence of anthropogenic climate change. However, virtually all mainstream scientists reject this point of view and accept the influence of humans on global warming. For example, while scientists from NASA and the U.S. Oceanic and Atmospheric Administration (NOAA) acknowledge that El Niño played a role in the record-setting high temperature average in 2015, they suggested that the record would have been reached regardless of the presence of ENSO that year. It is generally accepted that ENSO's influence on temperature averages plays out on a relatively short scale, while global warming, including that caused by human activity, is an independently observable and longer-scale phenomenon. Still, scientists remain interested in the interconnection between ENSO and climate change and continue to investigate the subject.

CONTEXT

Owing to the complexity and uncertainty surrounding global climate and climate change, neither the mean annual values of ENSO nor the interannual variability of ENSO can be reliably simulated in global climate models. Despite this, projections have been made about future trends in precipitation extremes linked to ENSO. The phenomenon has a noticeable influence on mean global temperature, and shifts in temperature are consistent with shifts in the SOI, but the relationship between temperature and ENSO effects has not been consistently strong. On one hand, strong El Niño events create significant spikes in mean global air temperature. On the other hand, there is evidence that long-term warming depresses El Niño activity. For this reason, the mutual effects of climate change and ENSO upon each other remain difficult to predict.

—C. R. de Freitas

FURTHER READING

Websites:

Cropper, Thomas. "Did El Niño Drive the Record Heat of 2015?" *Niskanen Center*. Niskanen Center, 25 Jan. 2016. Web. 1 Feb. 2016.

"NASA, NOAA Analyses Reveal Record-Shattering Global Warm Temperatures in 2015." *Goddard Institute for Space Studies*. NASA, 20 Jan. 2016. Web. Accessed 18 Oct. 2016.

Books:

Fagan, Brian *Floods, Famines and Emperors. El Niño and the Fate of Civilizations*. Basic Books, 1999. A readable historical examination of how the El Niño—Southern Oscillation (ENSO) phenomenon have affected civilizations on a global scale throughout the past.

Grove, Richard and Adamson, George *El Niño in World History*. Palgrave/Macmillan, 2018. Examines historical climate conditions that can be attributed to the El Niño phenomenon, and the effects on contemporary human societies.

Li, Tim and Hsu, Pang-chi *Fundamentals of Tropical Climate Dynamics*. Springer, 2018. Although just one chapter of this book is devoted to the ENSO phenomenon, most of the other chapters include a conversation about the interaction of the ENSO and a number of other phenomena such as the monsoons and other large-scale oscillation phenomena.

Sarachik, Edward S., and Mark A. Cane. *The El Niño-Southern Oscillation Phenomenon*. Cambridge University Press, 2018. The book offers a thorough description of the history and mechanics of the two components of ENSO, before delving into a technical analysis of the phenomenon.

ENHANCED GREENHOUSE EFFECT

SUMMARY

Greenhouse gases (GHGs) in the atmosphere—such as carbon dioxide (CO_2), methane, nitrous oxide, water vapor, and ozone—absorb infrared radiation from the Sun and reradiate some of it at the surface, warming Earth's atmosphere. The average temperature of the atmosphere has been estimated to be more than 30° Celsius warmer than it would be without these gases. The natural greenhouse effect occurs when this process is the result of nonhuman activities; the enhanced greenhouse effect denotes increases in the effect caused by GHGs emitted into the atmosphere by human activities. The enhanced greenhouse effect leads to anthropogenic (human-caused) climate change.

SIGNIFICANCE FOR CLIMATE CHANGE

The concentrations of CO_2 and, to a lesser extent, other GHGs have gradually increased in the

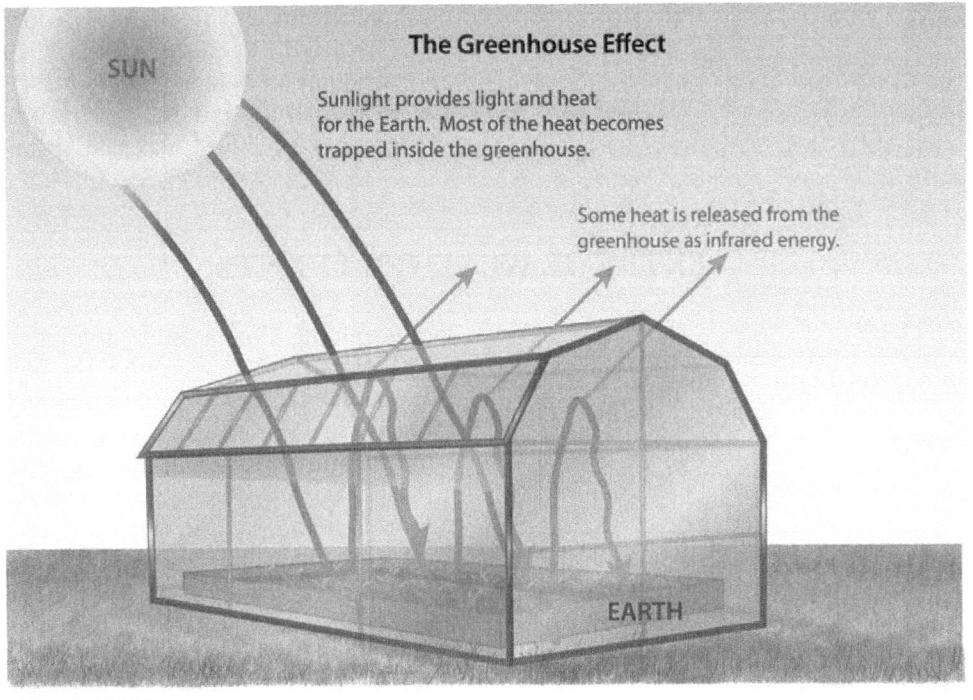

Greenhouse Effect: The glass walls and glass ceiling of a greenhouse trap the sun's heat, making the structure very hot inside.

atmosphere, especially during the twentieth century. For instance, the CO_2 content of the atmosphere in the Hawaiian Islands has increased from 313 parts per million in 1960 to 375 parts per million in 2005. Arctic ice-core samples indicate that the CO_2 content of the atmosphere has also gradually increased over longer timescales. Much of this increase in atmospheric CO_2 concentration appears to be due to human activity, although the importance of human activity relative to natural processes such as in volcanism is not clear. It is known, however, that CO_2 is released to the atmosphere by human activity, such as the burning of fossil fuels (petroleum, natural gas, and coal). Deforestation of tropical and other forests—such as the Amazon rain forest in Brazil—produces a great deal of CO_2 as the plants decay, and it also reduces an extremely important carbon sink, increasing the amount of CO_2 that remains in the atmosphere rather than being converted to biomass and oxygen. CO_2 is also liberated in cement production.

A continued increase in GHGs will likely cause a continued increase in the average temperature of the atmosphere. The greatest increase in temperature will likely be over polar landmasses. For example, a doubling of the amount of CO_2 in the atmosphere has been predicted to cause an average increase of 3° to 4° Celsius at high northern latitudes, resulting in much less snow and ice. Summer conditions might last an extra two months with a correspondingly shorter winter.

Warmer air holds more water vapor than cooler air, so global warming will likely cause the evaporation rate to increase. This increase in evaporation may result in increased drought, desertification, and water shortages in some regions. Although water availability may decrease, greater levels of atmospheric CO_2 will mean more CO_2 is available to drive photosynthesis, so some plants may benefit from this increase. Warmer temperatures at higher latitudes should also allow some crops such as wheat to be grown further to the north than at the present.

—*Robert L. Cullers*

BIBLIOGRAPHY

"The Enhanced Greenhouse Effect." *Joint Nature Conservation Committee*, 4 July 2008, http://jncc.defra.gov.uk/page-4389. Accessed 31 Jan. 2017.

Zillman, John, and Steven Sherwood, reviewers. "The Enhanced Greenhouse Effect." *Nova*, Australian Academy of Science, 28 Nov. 2016, http://www.nova.org.au/earth-environment/enhanced-greenhouse-effect. Accessed 31 Jan. 2017.

EXTREME WEATHER EVENTS

FIELDS OF STUDY

Oceanography; Heat Transfer; Atmospheric Science; Meteorology; Climate Modeling; Statistics; Earth System Modeling

SUMMARY

Extreme weather events are weather systems that become abnormally severe and have high impacts on human life, property, and the environment. In the last hundred years, especially in the latter half of the twentieth century, global surface temperature has experienced a rapid increase. This rapid warming may cause significant changes in Earth's weather patterns, including affecting the frequency and intensity of extreme weather events.

PRINCIPAL TERMS

- **cyclone:** a storm system that rotates about a low pressure area
- **drought:** a long period of no or scarce precipitation
- **extratropical cyclone:** a cyclone originating and subsisting outside the tropics
- **heat wave:** an extended period of abnormally high temperatures
- **hurricane:** a cyclone originating in the tropics
- **severe thunderstorms:** mostly summer convective storms involving microscale rotating winds
- **tornado:** a narrowly focused, funnel-shaped violent windstorm
- **wildfire:** spontaneously ignited, naturally occurring fire

HEAVY PRECIPITATION AND FLOODS

One of the possible consequences of global warming is that it will enhance evaporation and transpiration of water vapor from oceans, rivers, lakes, and vegetated lands. Temperature is the main factor determining the moisture-holding capacity of air. The higher the temperature, the more moisture an air parcel can hold. Although global warming may not occur uniformly across the globe, the averaged temperature increase in the atmosphere should give rise to a corresponding increase of average humidity. More water content in the atmosphere will increase the probability of heavy precipitation, which can lead to floods.

Another global warming–related factor that influences heavy precipitation and flooding is a convection-related increase in the severity of storms. In addition to increased temperatures, global warming may involve a higher extension of Earth's troposphere. Both of these processes can lead to more and stronger convection, which can in turn produce more violent convective storms and heavier precipitation.

DROUGHTS

Since global warming is not uniform, a strong warming can occur in some parts of the world while other parts of the world cool. In response to such non-uniform conditions, different patterns of atmospheric circulation can be realized in different parts of the globe. As a result, in some areas air may be enriched by moisture, and in other areas the air may lose moisture, depending on the general atmospheric circulation patterns and moisture transport in each region. Therefore, while global warming may increase the intensity and frequency of heavy precipitation and floods in some locations, it may also increase the severity and frequency of drought conditions in others.

HEAT WAVES

Global warming increases air temperature in both average and extreme contexts. That is, it results in an increase not only in average temperature but also

Years of Greatest Accumulated Cyclone Energy (ACE)

Year	ACE	Year	ACE	Year	ACE
2005	248	1995	227	1933	213
1950	243	2004	224	1961	205
1893	231	1926	222	1955	199
				1887	182

Note: ACE is equal to the sum of the squares of the maximum sustained wind speed of each tropical storm measured in knots every six hours

in daily, monthly, and yearly maximum temperatures. The increase in temperature extremes suggests an increase in the number of hot days as well. Therefore, global warming will most likely increase both the severity and frequency of heat waves. Such heat waves, along with drought, may increase the occurrence of wildfires.

Although wildfire is not a weather phenomenon, its occurrence is closely related to weather conditions. In particular, warm temperatures and low humidity are the two necessary conditions for wildfires. Because global warming can generate warm surface temperatures and frequent drought conditions, it will increase the likelihood of wildfires.

Tropical Storms

One of the necessary conditions for tropical-storm formation is high sea surface temperature (SST). Tropical storms typically develop over the ocean when SSTs exceed 26 to 27 degrees Celsius. Based on this criterion, recent global warming trends seem to suggest an increase in the number of tropical storms. Furthermore, tropical storms derive energy from latent heat brought by water vapor evaporated from oceans. Higher SSTs promote greater evaporation of water into the atmosphere. This factor suggests that future warm climates may also produce more powerful hurricanes and typhoons. However, the question of whether global warming would cause an increase in storm frequency is unresolved, and different studies have produced conflicting results.

Extratropical Cyclones

Unlike tropical storms, extratropical cyclones derive energy from a non-uniform temperature distribution, or a temperature contrast between locations in the northern and southern latitudes. In meteorology, such a condition is called a "temperature gradient." Strong temperature gradients will generate unstable atmospheric conditions, which will initiate large-scale cyclones. These cyclones typically occur in the cool season, and they are enforced by upper-level jet streams and also characterized by surface fronts.

A general consensus exists that global warming will decrease temperature gradients and also decrease the intensity of jet streams. The combined effect of these changes would tend to decrease the intensity and frequency of extratropical cyclones. However, some scientists argue that, because global warming tends to increase humidity, extratropical cyclones may gain extra energy from latent heat flux due to water vapor condensation.

Severe Thunderstorms and Tornadoes

Severe thunderstorms and tornadoes are convective-scale and microscale weather systems. They are different from extratropical cyclones, which are forced by large-scale temperature gradients. Thunderstorm development strongly depends on convection, which is influenced by surface radiative heating, convergence of surface flows, and topographic forcing. In a future warm climate, increase of global surface temperatures may provide favorable conditions for convection to occur. The number of both annual tornado sightings and annual tornado warning days increased over the second half of the twentieth century, but the number of the most severe tornadoes (F2-F5) exhibited a slight decrease.

CONTEXT

A 2007 report by the Intergovernmental Panel on Climate Change (IPCC) predicted that global warming would likely increase the number and frequency of extreme weather events. Despite the difficulty of attributing a particular weather event to climate change, a growing number of researchers sought to do just that. Between 2010 and 2016, the American Meteorological Society analyzed 131 scientific articles on selected extreme weather events and concluded in 2016 that human-caused climate change effects made an extreme weather event more likely or more severe in about 65 percent of cases. Increased extreme weather events are expected to generate profound impacts on global socioeconomic development. Such impacts include, among many other things, increased risks to human life and health, increased property and infrastructure losses, business income and tax revenue losses, increased cost and pressure on government's disaster relief and mitigation resources, and increased costs of private insurance.

The year 2017 was the costliest year ever recorded in the United States for extreme weather events, totalling an estimated $306 billion in damage, of which three hurricanes accounted for $265 billion, wildfires $18 billion, and tornadoes, hail storms, severe storms, flooding, and a freezing event the remainder. Low-income members of society were disproportionately affected by those disasters.

—Chungu Lu

BIBLIOGRAPHY

Ahrens, C. Donald. *Essentials of Meteorology: An Invitation to the Atmosphere.* 5th ed. Belmont, Calif.: Thomson Brooks/Cole, 2008. Widely used introductory textbook on atmospheric science; covers a wide range of topics on weather and climate.

Berke, Jeremy. "The US Spent More Money on Weather Disasters in 2017 Than Any Year on Record—Here's the Final Total." *Business Insider,* 8 Jan. 2018, www.businessinsider.com/weather-disasters-damage-cost-record-306-billion-in-2017-2018-1. Accessed 22 Feb. 2018.

Diffenbaugh, N. S., R. J. Trapp, and H. Brooks. "Does Global Warming Influence Tornado Activity?" *Eos* 89, no. 53 (2008). Discusses the impact of global warming on tornado activity in the United States.

Herring, Stephanie C., et al., editors. "Explaining Extreme Events of 2016 from a Climate Perspective." *Bulletin of the American Meteorological Society,* vol. 99, no. 1, Jan. 2018, pp. S1–S157, www.ametsoc.net/eee/2016/2016_bams_eee_low_res.pdf. Accessed 22 Feb. 2018.

Intergovernmental Panel on Climate Change. *Climate Change, 2007—Synthesis Report: Contribution of Working Groups I, II, and III to the Fourth Assessment Report of the Intergovernmental Panel on Climate Change.* Edited by the Core Writing Team, Rajendra K. Pachauri, and Andy Reisinger. Geneva, Switzerland: Author, 2008. Comprehensive overview of global climate change published by a network of the world's leading climate change scientists under the auspices of the World Meteorological Organization and the United Nations Environment Programme.

Lowrey, Annie. "The Most Expensive Weather Year Ever." *The Atlantic,* 20 Dec. 2017, www.theatlantic.com/business/archive/2017/12/expensive-weather-storms/548579. Accessed 22 Feb. 2018.

Lutgens, Frederick K., and Edward J. Tarbuck. *The Atmosphere.* 10th ed. Upper Saddle River, N.J.: Pearson Prentice Hall, 2007. Introductory textbook that covers a wide range of atmospheric sciences.

Tebaldi, C., J. M. Arblaster, K. Hayhoe, and G. A. Meehl. "Going to the Extremes: An Intercomparison of Model-Simulated Historical and Future Changes in Extreme Events." *Climate Change* 79: 185–211 (2006). Discusses model-based assessment of global warming's effects on extreme climate conditions.

F

FABRY QUANTIFIES OZONE IN THE UPPER ATMOSPHERE

FIELDS OF STUDY

Spectroscopy; Photochemistry; Analytical Chemistry; Environmental Chemistry; Thermodynamics; Meteorology; Mathematics; Chemical Kinetics; Atmospheric Science; Electromagnetism; Particle Physics

SUMMARY

Experiments performed in the eighteenth century showed that air is a mixture composed of different substances rather than a single element as had long been supposed. Early study of the major reactive gases of the atmosphere, oxygen, nitrogen, and carbon dioxide, was followed by investigations of other components.

Charles Fabry's quantification of the amount of ozone in an atmospheric column led to the discovery of the ozone layer.

Date: January 17, 1913
Locale: Marseilles, France

KEY FIGURES

Charles Fabry: (1867–1945), French physicist
Frederick Alexander Lindemann: (1886–1957), British scientist and head of the Clarendon Laboratory
G. M. B. Dobson: (1889–1976), English upper-atmosphere meteorologist
F. W. P. Götz: (1891–1954), Swiss meteorologist

PRINCIPAL TERMS

- **ozone:** a triatomic molecule consisting of three oxygen atoms, its structure and electron distribution rendering the molecule highly reactive
- **ozone layer:** a stratospheric region containing relatively high concentrations of triatomic oxygen (ozone) that prevents much ultraviolet solar radiation from reaching Earth's surface
- **spectroscopy:** study and use of the interactions of light with atoms and molecules, enabling the analysis of molecular structures and properties
- **ultraviolet:** electromagnetic radiation having wavelengths shorter than those of the visible spectrum, associated with the transport of energy capable of promoting chemical reactions

THE METHOD OF DISCOVERY

Ozone is a notable gas because of its sharp odor. Impure ozone is produced easily by a spark in oxygen and can collect in the atmosphere near large electric motors. Curious about the possible concentration of ozone at the Earth's surface, Walter Noel Hartley (1845–1913) devised chemical procedures for collecting and testing for ozone in the laboratory in 1881. Because the concentration of ozone is extremely low, the volume of air required for such tests is large, so Hartley needed to apply spectroscopy to detect ozone in the higher atmosphere.

The science of spectroscopy was at an early stage of development in the late nineteenth and early twentieth centuries. No good light detectors were available other than photographic film. Although film images were all black-and-white, scientists recognized that the dispersion of light into different colors caused by a prism made the different colors fall onto different places on a film. The relative attenuation of light at each wavelength, corresponding to each position on a film, is characteristic of a particular substance. It can be used to identify the presence and quantity of a substance in a mixture such as air.

By studying films, researchers learned that ozone absorbs both visible and ultraviolet light. Virtually

French physicist Charles Fabry (1867-1945).

complete atmospheric absorption of sunlight at wavelengths below 300 nanometers in the ultraviolet was noticed first by Alfred Cornu (1841–1902). Hartley identified this with ozone, thus absorptions in this range are termed the Hartley bands. Other examples of absorptions include Chappuis bands, in the region of visible light, and Huggins bands, in the ultraviolet above 300 nanometers. The Chappuis bands are responsible for the blue color of ozone.

Because surface measurements at high and low altitudes showed little difference in the wavelength of cutoff of sunlight, Hartley suggested that much ozone existed high in the atmosphere. Charles Fabry determined how much ozone existed at that level. Fabry was a physicist whose life's work involved studies of optics. With Alfred Pérot, he invented an interferometer, which produces a succession of light and dark rings on a screen caused by the interference of different colors of light. The instrument is used to measure short distances with unmatched precision. Fabry developed an interest in astronomy while he was working with his two brothers: Eugène, a mathematician, and Louis, an astronomer. Fabry applied his interferometer to the study of the different wavelengths of light received from the Sun and stars. The use of different wavelengths was the key to quantitative determination of ozone in the atmosphere.

The intensity of the radiation reaching the Earth's surface is a function of the intensity of the sunlight reaching the top of the atmosphere, the path length of the light through the atmosphere and thus the angle from the zenith (vertical), the amount of dust and molecular scattering of the light on its way through the atmosphere, and the absorption of light by gas molecules such as ozone. All of these except dust scattering and path length vary with wavelength.

Working with Henri Buisson (1839–1944), Fabry carefully measured the absorption coefficients of ozone bands at different wavelengths in the laboratory. The researchers used this information to pick two wavelengths, with a known difference in absorption, at which to view sunlight. By measuring at several angles from the zenith, they could make adjustments for the other variations and determine the total amount of ozone in a vertical column of the atmosphere. The unit in which the result has commonly been reported is the thickness of the ozone if it were held in a pure layer at normal, sea-level atmospheric pressure and temperature. A thickness of 0.01 millimeter is known as the Dobson unit. Fabry suggested that about 5 millimeters of ozone exist in an atmospheric column.

In 1920, Fabry and Buisson modified their instrument to reduce the amount of stray sunlight reaching the photographic film. They made repeated measurements of ozone and checked several pairs of absorptions to ensure that it was indeed ozone they were measuring. They found its concentration steady at 3 millimeters. In discussing their results, Fabry and Buisson showed remarkable insight. They noted that ozone probably forms in the upper atmosphere because of absorption of solar radiation and that the maximum concentration of ozone might be at a height of 40 kilometers (about 24.9 miles).

From his own study of meteor tracks in 1921, Frederick Alexander Lindemann (who later became Viscount Cherwell) suggested that the atmosphere above 30 kilometers (about 18.6 miles) is much warmer than had been supposed. He believed that absorption of solar radiation by ozone is responsible for the warmth. As it was already known that the temperature of the stratosphere changes with weather patterns, there was some speculation that changes in ozone concentrations might be responsible for these patterns.

Soon thereafter, G. M. B. Dobson established a program at Oxford University for the monitoring of atmospheric ozone. Using a spectrograph he built based on the design of Fabry and Buisson, Dobson collected a series of measurements from a hill outside Oxford, England, in 1925. From these the annual variation in atmospheric ozone was first seen: a maximum in the spring and a minimum in the fall. In the next year and a half, Dobson distributed a set of six spectrographs throughout Europe to investigate the relation between ozone and weather patterns. His finding of higher concentrations of ozone behind cyclones (large wind systems rotating around regions of low pressure) and ahead of anticyclones (systems around regions of high pressure) further reinforced interest in ozone measurement.

In 1928 and 1929, the same instruments were sent to sites throughout the world to examine gross variations in ozone with latitude and location. Concentrations were found to be higher and far more variable at higher latitudes. Later measurements suggest a maximum at roughly 60 degrees latitude, but with considerable variation.

In Dobson's early work, all exposed photographic plates were returned to Oxford University for development and analysis. This practice removed a potential source of inconsistency. After 1930, the measurement program continued until World War II, with specially designed spectrophotometers using photomultipliers as detectors in place of film. These were particularly advantageous for measurements at low light levels, such as on cloudy days and early or late in a day. These took on great importance after it was found that information about the altitude distribution of the ozone could be gleaned from such measurements.

One of the people who produced exposures for Dobson was F. W. P. Götz, who was working with Dobson to establish the vertical distribution of ozone. Based on variation of their measurements with zenith angle, they tentatively concluded that ozone might be concentrated in a layer near 50 kilometers (about 31.1 miles) above the Earth's surface, in rough agreement with the suggestion of Fabry and Buisson.

Because Götz's measurements were made in summer at the far northern island of Spitsbergen, Norway, the Sun remained close to the horizon for relatively long periods. Light intensity drops steadily at all wavelengths as the angle of the Sun from the zenith increases, because beams must pass through a greater thickness of atmosphere. Götz noted, however, that the ratio of intensity at a short versus a long wavelength went through a minimum near 85 degrees, then, contrary to expectation, increased for observations made overhead while the Sun was approaching the horizon.

Light measured in this way has all been scattered. The higher the altitude of scattering, the smaller the opportunity to be absorbed. The longer-wavelength light is effectively scattered from lower altitudes, so its intensity keeps decreasing rapidly as the Sun sets. Shorter-wavelength light is scattered above the ozone layer, and so its intensity decreases only slowly. One can determine the altitude range of the ozone layer by measuring this *Umkehr*, or reversal effect. A concentration maximum occurs at roughly 25 kilometers (15.5 miles), decreasing rather sharply at higher altitudes and more slowly toward the Earth's surface.

The simplicity of the measurement, and the ability to use the instruments already in service for determining total ozone in a column, made this the most common approach to locating the ozone layer until much later in the twentieth century, when satellite data became widely available. A scattering of balloon and rocket measurements confirmed Götz's conclusion and suggested some variations of ozone concentration that ground-based measurements could not detect.

One instructive improvement on Dobson's work was based on measurement of the changes in absorption of light by ozone with change in temperature. At the low temperatures of the upper atmosphere, background absorption is reduced, thus one can use wavelengths at absorption bands of ozone to determine ozone concentration; measurement of background absorption gives ozone temperatures. Such data confirm a region of high temperatures above 35 kilometers (21.7 miles), which is caused by ozone absorption.

Sydney Chapman (1888–1970) proposed the first theoretical model of upper-atmospheric chemistry after attending an informal conference during which Fabry, Dobson, and others presented results of their studies. Oliver R. Wulf (1897–1987) and Lola S. Deming (?–?) demonstrated in 1936 that ozone is produced through photochemical dissociation of oxygen. They also contended that, although ozone is unstable, once formed in the region near its concentration maximum, it can exist long enough to drift lower in the atmosphere, where its concentration is affected by cyclonic wind patterns.

Research concerning ozone in the late twentieth century focused more on its consumption than on its formation. In 1950, Marcel Nicolet (1912-1996) examined the role of radical hydroxyl in the reaction chemistry of ozone. In 1970, Paul J. Crutzen (1933-) further modified the picture to include the catalytic and stoichiometric reactions with nitrogen oxides. In the late 1970s and 1980s, researchers raised concerns about the catalytic destruction of ozone by chlorine radicals, especially at high latitudes. Such destruction has the potential to increase the intensity of ultraviolet radiation reaching the Earth, changing the temperature patterns of the stratosphere and thus changing weather patterns around the world.

SIGNIFICANCE

Fabry and Buisson's publication of their findings on January 17, 1913, stimulated little other work regarding ozone, and such research was curtailed sharply during World War I. The experiment was clever, but the result was initially merely a curiosity. The first quantitative determination of ozone in the atmosphere was not Fabry's most notable achievement as a physicist.

The paper that Fabry and Buisson published in 1921 had a greater effect. The results were more reliable because of improvements on the instrument and more repetitions of measurements. More important, Lindemann and Dobson could apply the conclusion and discussion immediately to explain their own meteor data. Fabry and Buisson's paper gave Dobson a basis for believing that an extended series of measurements of atmospheric ozone was not only possible but also potentially useful.

Once started on a program of global monitoring for ozone, Dobson found that he could use his early results to convince granting bodies that money invested in such measurements was well spent. The monitoring results that Dobson reported in 1930 served as a framework on which all later measurements were based. Dobson's conclusions proved to be erroneous only in a few points, demonstrating his careful attention to equipment and experimental design.

The studies on ozone helped form or correct meteorologists' views of the atmosphere and thus helped keep the fields of atmospheric physics and meteorology on a sound footing. For example, the study of ozone offered a means to measure gas temperatures above the altitudes reached by most balloons. The most impressive feature of ground-based ozone study is the long-term and global nature of the information. It was not until rocket observations were made in the late 1950s and 1960s that detailed investigation of the stratosphere was possible. Without a comprehensive database already in place, sensible choices and interpretations of experiments would not have been possible.

Because investigation of matter in the upper atmosphere is so difficult, scientists have applied a great deal of effort to the development of mathematical models of atmospheric chemistry and physics. These models consider all the different kinds of molecules and their reactions in attempting to understand what is happening. This work requires actual measurements with which to check the models. Although data gathered by satellites and high-altitude airplane flights have improved researchers' understanding of ozone distribution, the scale provided is not fine enough to allow the examination of ozone movements near the tropopause. Further improvement of atmospheric models requires even more ozone monitoring, so that scientists can understand local variations as well as differences between widely scattered points.

—*James A. Carroll*

FURTHER READING

Christie, Maureen. *The Ozone Layer: A Philosophy of Science Perspective.* Cambridge University Press, 2001. Presents the history of human knowledge about stratospheric ozone in a manner accessible to lay readers. Addresses basic issues of both real-world science and the philosophy of science. Includes figures, references, and index.

Craig, Richard A. *The Edge of Space: Exploring the Upper Atmosphere.* Doubleday, 1968. Discusses the various methods of measuring atmospheric ozone. Written at a time when understanding of ozone chemistry was far less developed than it is in the twenty-first century, but still useful. Based on an important book in aeronomy, *The Upper Atmosphere: Meteorology and Physics*, which Craig published in 1965, this volume is aimed at a general audience.

Dobson, G. M. B. "Forty Years' Research on Atmospheric Ozone at Oxford: A History." *Applied Optics* 7 (March, 1968): 387–405. Dobson's personal recollection of the development of the global network for ozone monitoring presents a fascinating account of the scientific work. Notes the unexpected observation in 1956 of low concentrations of ozone until late in the Antarctic spring—the phenomenon, intensified in subsequent years by the presence of radical chlorine from chlorofluorocarbons, now known as the ozone hole.

Gribbin, John. *The Hole in the Sky: Man's Threat to the Ozone Layer.* Rev ed. Bantam Books, 1988. Details concerns about the destruction of the ozone layer. Probably the most balanced of the books published on this topic in the late 1980s.

_____, ed. *The Breathing Planet.* Basil Blackwell, 1986. Collection of short articles originally published in the English journal *New Scientist* includes a section on ozone. Chapter titled "Monitoring Halocarbons in the Atmosphere" suggests the difficulties and uncertainties of monitoring substances in the atmosphere.

Kerr, J. B., I. A. Asbridge, and W. F J. Evans. "Intercomparison of Total Ozone Measured by the Brewer and Dobson Spectrophotometers at Toronto." *Journal of Geophysical Research* 93 (September, 1988): 11129–11140. Plots of careful ozone measurements and discussion of factors recognized as affecting such measurements over long periods at one site demonstrate, to the seriously suspicious, the variability that makes spotting trends in stratospheric ozone concentration so difficult.

Parson, Edward Anthony. *Protecting the Ozone Layer: Science and Strategy.* Oxford Univ. Press, 2010. Comprehensive technical discussion of efforts to protect the ozone layer undertaken through international cooperation. Chapter 2 is devoted to a review of early stratospheric science. Includes notes, references, and index.

FLOODS

FIELDS OF STUDY

Fluid Dynamics; Physical Geography; Engineering; Mathematics; Environmental Sciences; Ecosystem Management; Earth System Modeling; Meteorology; Climate Modeling; Process Modeling; Hydroclimatology; Atmospheric Science; Hydrology; Ecology; Waste Management

SUMMARY

Floods are extreme conditions of flowing water. They generally occur because of inordinate amounts of rainfall or snowmelt, but floods may also result from other causes, such as dam failures or volcanic eruptions. Floods play a major role in shaping river systems, and their occurrence is critical to the human use of riparian lands.

PRINCIPAL TERMS

- **discharge:** the volume of water moving through a given flow cross-section in a given unit of time
- **flash floods:** rises in water level that occur unusually rapidly, generally because of especially intense rainfall
- **flood:** a rising body of water that overtops its usual confines and inundates land not usually covered by water
- **hydrology:** the branch of science dealing with water and its movement in the environment
- **jökulhlaup:** a flood produced by the release of water sequestered by a glacier, most often due to the failure of some type of glacial dam or to subglacial volcanic activity
- **monsoon:** a seasonal, reversing pattern of wind between warm ocean bodies and landmasses
- **recurrence interval:** the average time interval in years between occurrences of a flood of a given magnitude in a measured series of floods
- **runoff:** that part of precipitation that flows across the land and eventually gathers in surface streams

CAUSES OF FLOODS

Floods involve excessive or extremely large flows of water in rivers and streams. Some technical hydrological definitions of floods refer to stages, or heights, of water above some reference level, such as the banks of a river channel. In practice, however, floods can be thought of as any extreme flow of water that exceeds

The Great Mississippi River Flood of 1927. Flooded power plant at Oswego, Ks., on the Neosho River. April 23, 1927 - river stage 25.4 feet.

the normal ability of the surrounding terrain to contain the flow, often damaging or threatening life and property.

Floods may be caused by a variety of physical factors, including dam failures and the subsidence of land. The most common kind of flood occurs when excessive precipitation and physical factors of the land combine to produce maximum runoff of water. Precipitation can yield runoff directly, or water from melting snow can produce the flow. Very intense flash floods are usually associated with heavy, short-duration rainfall from thunderstorms. Such rainfall easily overwhelms the infiltration capacity of the ground, and water rapidly runs off from the local drainage area or watershed into adjacent stream channels. If enough water concentrates in the channel to exceed the carrying capacity of the channel, it will constitute a flood.

Physical factors on the land surface also determine the rate of concentration of floodwater in stream channels. For example, less permeable soils will allow water to run off faster, as will a lack of vegetation. Artificial enhancement of runoff occurs when slopes are covered by impermeable materials. This situation commonly occurs in construction, when the natural ground surface is replaced by buildings and pavement, resulting in extreme enhancement of runoff. In cities, the yield of floodwater from paved surfaces may be ten times greater than the same storm's yield under natural conditions. In this way, human construction tends to exacerbate flooding problems in urban areas.

SEDIMENT AND BEDROCK

As the flow of water increases in a stream channel, it is usually associated with considerable sediment. If such waterborne sediment also composes extensive deposits adjacent to the stream channel, the river is termed alluvial. An alluvial river commonly has a floodplain of deposited sediment adjacent to the channel. The floodplain is really an intimate part of the river system, as sediment is added to it every time the river rises above its banks. As the water rises, the river's depth rapidly increases, causing an increased ability to transport sediment that is eroded from the bed and banks. If the stream is appropriately loaded with sediment, it will be deposited when the banks are overtopped, and the width of the flow greatly increases. If not enough sediment is supplied by erosion, however, the increasingly energetic flood flows will be erosive, attacking the banks, widening the channel, and restoring the appropriate sediment load to the stream.

When the bed and banks of a river are composed of bedrock, these adjustments of sediment load to flow energy cannot occur. Because the bedrock can be very resistant to erosion, the energy level in the flow can rise spectacularly without being damped by sediment. The excess energy of such sediment-impoverished floods goes into the development of turbulence. Turbulence at high energy levels takes on an organized structure of powerful vortices that produce immense pressure changes. These pressure effects may be sufficient to erode the bedrock boundary by a "plucking" action.

A famous geological controversy once surrounded the problem of bedrock erosion by great floods. In the 1920s, University of Chicago geologist J. Harlen Bretz (1882–1981) proposed that immense tracts of eroded basalt bedrock in Washington State had been created by a catastrophic glacial flood. Bretz subsequently showed that the fascinating landforms of that region, known as the Channeled Scablands, were created when a great lake impounded by glacial ice burst. The lake, glacial Lake Missoula, had been more than six hundred meters deep at its ice dam. It took Bretz nearly fifty years to convince his many critics that catastrophic flooding could explain all the bizarre features of the Channeled Scablands. Geologists now know that the physics of catastrophic flooding is completely consistent with Bretz's observations. Missoula flood flows moved at depths of one hundred to two hundred meters and velocities of twenty to thirty meters per second. The power (rate of energy expenditure) per unit area for such flows is thirty thousand times that for a normal river, such as the Mississippi, in flood. The reason for such immense flow power is that the Missoula flooding occurred on very steep slopes, giving the water great potential energy when the dam burst.

The Missoula floods occurred during the last ice age, more than twelve thousand years ago. There are, however, modern examples of glacial floods called jökulhlaups. In Iceland, jökulhlaups occur where glaciers overlie active volcanoes. Volcanic heat releases water that is stored in subsurface reservoirs by melting the overlying ice. Lakes may also form adjacent to the ice masses. Because ice is less dense than water, such juxtapositions are inherently unstable. When the pressure is high enough, the water may lift the ice dam and burst out from beneath the glacier. The jökulhlaups move house-sized boulders and transport immense quantities of sediment.

DISCHARGE AND RECURRENCE INTERVAL
The volume of water released by a flood per unit of time is termed its discharge. This quantity is the magnitude of the flow that will potentially inundate an area. The chances of experiencing floods of different magnitudes are expressed in terms of frequency. Large, catastrophic floods have a low frequency, or probability of occurrence; smaller floods occur more often. The probability of occurrence for a flood of a given magnitude can be expressed as the odds, or percent chance, of the recurrence of one or more similar or bigger floods in a certain number of years. Analyses of flood magnitude and frequency are achieved by measuring floods and statistically analyzing the data. Results are expressed in terms of the probability of a given discharge being equaled or exceeded in any one year. The reciprocal of this probability is the return period, or recurrence interval, of the flooding, expressed as a number of years.

The concept of a recurrence interval is sometimes confusing; it is simply a statistical measure, based on historical measures of the flood frequency of a river flow, of the likelihood of a flood of a given magnitude occurring over a one-year period. A flood that has a probability of occurrence in one year of 0.1, or 10 percent, is called a ten-year flood, because the one-year probability (0.1) multiplied by 10 is equal to 1. By extension, a larger flood that has a probability of occurrence of 0.01, or 1 percent, in a single year is called a hundred-year flood, because 0.01 multiplied by 100 is equal to 1. Note that these numbers do not preclude several such floods occurring in a given period; for example, two or more floods of the one-hundred-year magnitude could occur in the same year. Such an event, while unlikely, does have a small probability of occurring. In fact, in August 2017, before the "thousand-year flood" caused by Hurricane Harvey in Houston, Texas (0.001 or 0.1 percent probability of occurring in a single year), meteorologist Eric Holthaus (1981–) reported that the Houston area had seen four hundred-year floods since May, 2015.

Flood magnitude-frequency relationships vary immensely with climatic regions. In the humid-temperate regions of the globe, such as the northern and eastern United States, streamflow is relatively continuous. Even rare floods are not appreciably larger than more common floods. A result of this relationship is that stream channels have developed in size to convey the relatively common, moderate-sized floods with maximum efficiency. In contrast, the more arid regions of the southwestern United States have immensely variable flood responses. Stream channels may be dry most of the time, filling with water only after rare thunderstorms. The flash floods that characterize these streams may also be highly charged with sediment; indeed, the sediment may so dominate the flow that the phenomenon is called mudflow or debris flow rather than streamflow. Because extreme events dominate in these environments, the stream channels have developed in size according to these rare, great floods.

MONSOONS AND ANCIENT FLOODS
In tropical areas, some of the greatest known rainfalls are produced by tropical storms and monsoons. Monsoons are seasonal wet-to-dry weather patterns driven by atmospheric pressure changes over the

oceans and continents, which result in alternations between dry periods that inhibit vegetation growth and immensely wet periods that facilitate runoff. Tropical rivers thus have a pronounced seasonal cycle of flooding, and some floods may be immense. These rivers also show channel-size development in accordance with rare, great flows. Some of the most populous places on Earth are situated on seasonal tropical rivers, such as the Ganges and Brahmaputra Rivers in India and Bangladesh. Immense tragedies have occurred when a particularly severe monsoon or tropical storm has produced especially great floods. In 1974, monsoon-related flooding in Bangladesh killed 2,500 people. This pales in comparison to a flood in 1970 that took 500,000 lives and another in 1991 that killed more than 100,000 people. India's National Disaster Management Authority has estimated that an average of 1,600 lives are lost annually due to flooding, with the greatest loss of life—11,316 people—having occurred in 1977.

Studies of ancient floods (paleofloods) through geological reconstruction of past discharges show that monsoons and other flood-generating systems have varied in the past. Between about ten thousand and five thousand years ago, floods were apparently much more intense in many world areas on the boundaries between the tropics and the midlatitude deserts. These intense floods may have been related to long-term glacial-to-interglacial cycles. Because of modern increases in atmospheric carbon dioxide and other greenhouse gases, it appears quite likely that tropical floods may again become more intense, if this process has not already begun. Such a situation could have grave consequences for flood-prone tropical countries.

Study of Floods

Hydrologists study floods by measuring flow in streams. Measurements are taken at stream gauges, locations at which mechanical devices are used to record the water level, or stage, of the river. To transform these stage measurements into discharge values, the hydrologist must perform a rating of the stream gauge. This is accomplished by measuring velocities in the stream channel during various flow events. A velocity meter with a rotor blade calibrated to the flow rate is used for this purpose. When the average measured velocity of the stream channel is multiplied by the cross-sectional area of the channel, the result is the discharge for that flow event. When several flows at different stages are measured, the data are used to generate a rating curve for the gauge. This curve shows discharges corresponding to any stage measured at the gauge.

The discharge values obtained at a stream gauge are collected over many years, constituting a record that can then be used in flood-frequency analysis. Several statistical procedures can be employed to plot the flood experience and to extrapolate to ideal values of the ten-year flood, the one-hundred-year flood, and so on. The discharges associated with these recurrence intervals are then used as design values for hazard assessment, dam construction, and other flood controls. "Flood control" is probably a misnomer, however, because flows larger than the design floods are always possible. Flood control really involves providing various degrees of flood protection.

Another approach to evaluating extreme flood magnitudes involves careful study of the precipitation values that generate the greatest known floods. By transposing the patterns of known extreme storms to other areas, scientists can, in theory, calculate what the runoff would be from hypothetical great storms. Such calculations involve the use of a rainfall-runoff model. These models are prevalent in hydrology because they can be easily programmed. The models give idealized predictions of how water from a given storm would concentrate. The flood discharge modeled from the assumed maximum rainfall is called the probable maximum flood.

Unfortunately, there are problems inherent to both of these traditional hydrological approaches to flood studies. Both make assumptions in calculating potential flood flows. Another procedure is to study the natural records of ancient floods, or paleofloods, that are preserved in geological deposits or in erosional features on the land. It has been found for some sections of bedrock that nonalluvial rivers act as natural recorders of extremely large flood events. These natural flood gauges can be interpreted only by detailed studies that combine geological analysis with hydraulic calculations of the ancient discharges.

Paleoflood hydrology generates real data on the largest floods to occur in various drainages over several millennia. The data include the floods' ages, or the periods in which they occurred, and their discharges. The information can be used directly in a flood-frequency analysis, or it can be used to assess the probable validity of extrapolations from conventional data on smaller floods. Paleoflood data can also be compared with probable maximum flood estimates. In this way, the expense of overdesign and the danger of underdesign can be avoided in flood-related engineering projects.

SIGNIFICANCE

People have lived with floods since the beginning of civilization. The first great civilizations on Earth developed along the fertile but flood-prone valleys of rivers such as the Nile, the Tigris and the Euphrates, the Indus, and the Yangtze and the Huang He. Various means of coping with floods have been documented since the biblical accounts of Noah. Early societies merely avoided zones that tradition told them were hazardous. It has only been in the modern era that large cities have systematically developed on immense tracts of flood-prone lands. Thus the natural process of flooding has become an unnatural hazard to humans.

Floods can have immense consequences for both human life and infrastructure. In the 1930s, a national U.S. program began to respond to such problems by constructing large dams. Despite, or perhaps because of, this expensive effort, flood damage to life and property is much greater today than it was in the 1930s. A single tropical storm system, Hurricane Agnes, killed 128 people overall and generated nearly $3 billion in flood damage to the northeastern United States in 1972; in 2011, the flooding of the Ohio, Mississippi, and Missouri Rivers in April and May, plus the flooding caused by Hurricane Irene and Tropical Storm Lee in August and September, caused a total of 108 deaths and more than $8 billion in damages in the United States; and New Orleans may never fully recover from the flooding caused by Hurricane Katrina in 2005, which resulted in as many as 1,464 deaths and $70 billion in damages in that city alone. The flooding caused by Hurricane Harvey in 2017 was estimated to have caused between $65 billion and $190 billion worth of damage in Texas, the upper range of which would make it the costliest natural disaster in U.S. history, while Hurricane Irma, which followed just days after, inflicted an additional $50 billion to $100 billion in damages in the southeastern United States.

Elsewhere, flooding is a particularly serious issue in South Asia, where monsoon season is an annual occurrence and flood-control measures are largely limited to Dutch-style embankments that do not take into account the greater sediment present in the major rivers. In 2007, unusually heavy rains during monsoon season caused thirty-seven major floods throughout several South Asian countries, including India, Bangladesh, Pakistan, and Nepal, killing more than 4,000 people and displacing or otherwise affecting more than 45 million. In summer 2017, flooding in India, Bangladesh, and Nepal caused more than 1,200 deaths and destroyed close to a million houses, affecting more than 41 million people.

In addition to the dangers posed by flooding itself, the aftermath of a major flood can pose serious health risks to residents. Remaining floodwater may be contaminated with sewage and chemicals, and that contamination can be transferred to flood-damaged homes. Mold, which grows quickly in warm, damp conditions, can cause or trigger respiratory issues. Standing water attracts mosquitoes, increasing the risk of mosquito-borne illnesses. In addition, the stress, anxiety, and depression caused by the event can lead to serious mental health issues for survivors.

One trend in flood management has been to manage flood-hazard zones with multiple approaches that respond to the nature of the flood risk. The river is treated as a whole integrated system, rather than as individual segments, for engineering design. Management alternatives for this system are not limited solely to structural controls, such as dams and levees. Instead, options are considered for land-use adjustment. Flood-prone lands can be used for parks, greenbelts, and bikeways instead of industrial warehouses, stores, and housing. Even when construction must be done on floodplains, it may be possible to make provision for flood risks. Warehouses, for example, can be organized for rapid transfer of materials to second stories or to temporary, safe storage sites. Such adjustments require accurate and timely warning systems that involve measuring rainfall in headwater areas and rapidly predicting the flood consequences to downstream sites at risk.

There is a general need to educate the public that floodplains are a natural part of rivers. Living on a floodplain is really choosing to play a game of "floodplain roulette," in that it is known with certainty that a flood will happen, but it is not known when that flood will happen. For this reason, the most accurate and reliable methods of evaluating flood magnitudes and frequencies are necessary. The choices made for land use on floodplains cannot be based merely on idealized theories of how floods behave.

—*Victor R. Baker*

FURTHER READING

Aldrete, Gregory S. *Floods of the Tiber in Ancient Rome.* Johns Hopkins University Press, 2007.

American Society of Civil Engineers, Hurricane Katrina External Review Panel. *The New Orleans Hurricane Protection System: What Went Wrong and Why.* ASCE, 2007. *ASCE Library,* doi:10.1061/9780784408933. Accessed 6 Sept. 2017.

Baker, Victor R., editor. *Catastrophic Flooding: The Origin of the Channeled Scabland.* Ross, 1981.

Baker, Victor R., et al. *Flood Geomorphology.* Wiley, 2010.

Bedient, Philip B., et al. *Hydrology and Floodplain Analysis.* 2019.

Christensen, Jen. "The Hidden Health Dangers of Flooding." *CNN,* 31 Aug. 2017, www.cnn.com/2017/08/27/health/health-consequences-flood-waters/index.html. Accessed 22 Sept. 2017.

Comerio, Mary C. *Disaster Hits Home: New Policy for Urban Housing Recovery.* University of California Press, 1998.

Dasgupta, KumKum. "Why India Is Failing to Minimise Monsoon Flood Destruction." *Hindustan Times,* 26 July 2017, www.hindustantimes.com/analysis/why-india-is-failing-to-minimise-monsoon-flood-destruction/story-4qC6DuVWacseb-WBwnvLbjI.html. Accessed 22 Sept. 2017.

Freitag, Bob, et al. *Floodplain Management: A New Approach for a New Era.* Island Press, 2009.

Holthaus, Eric. "A Texas-Size Flood Threatens the Gulf Coast, and We're So Not Ready." *Grist,* 23 Aug. 2017, grist.org/article/a-texas-size-flood-threatens-the-gulf-coast-and-were-so-not-ready/. Accessed 22 Sept. 2017.

Irfan, Umair. "The Stunning Price Tags for Hurricanes Harvey and Irma, Explained." *Vox,* 18 Sept. 2017, www.vox.com/explainers/2017/9/18/16314440/disasters-are-getting-more-expensive-harvey-irma-insurance-climate. Accessed 22 Sept. 2017.

Knighton, David. *Fluvial Forms and Processes: A New Perspective.* Rev. and updated ed., Hodder Arnold, 1998.

Lui, Kevin. "Severe Flooding in South Asia Has Caused More Than 1,200 Deaths This Summer." *Time,* 29 Aug. 2017, time.com/4921340/south-asia-floods-india-mumbai-bangladesh-nepal/. Accessed 22 Sept. 2017.

O'Loughlin, Karen Fay, and James F Lander. *Caribbean Tsunamis: A 500-Year History from 1498–1998.* Kluwer Academic Publishers, 2010.

Orsi, Jared. *Hazardous Metropolis: Flooding and Urban Ecology in Los Angeles.* University of California Press, 2004.

Prothero, Donald R. *Catastrophes! Earthquakes, Tsunamis, Tornadoes, and Other Earth-Shattering Disasters.* Johns Hopkins University Press, 2011.

Samenow, Jason. "Harvey Is a 1,000-Year Flood Event Unprecedented in Scale." *The Washington Post,* 31 Aug. 2017, www.washingtonpost.com/news/capital-weather-gang/wp/2017/08/31/harvey-is-a-1000-year-flood-event-unprecedented-in-scale/. Accessed 22 Sept. 2017.

Sampath, Nikita. "Mismatched Flood Control System Compounds Water Woes in Southern Bangladesh." *New Security Beat,* Environmental Change and Security Program, Woodrow Wilson International Center for Scholars, 3 Jan. 2017, www.newsecuritybeat.org/2017/01/mismatched-flood-control-system-compounds-water-woes-southern-bangladesh/. Accessed 22 Sept. 2017.

Schneider, Bonnie. *Extreme Weather: A Guide to Surviving Flash Floods, Tornadoes, Hurricanes, Heat Waves, Snowstorms, Tsunamis and Other Natural Disasters.* Palgrave Macmillan, 2012.

Sene, Kevin. *Flash Floods: Forecasting and Warning.* Springer, 2013.

Sene, Kevin. *Flood Warning, Forecasting and Emergency Response.* Springer, 2008.

Singh, V. P. *Hydrology of Disasters.* Springer, 2011.

Srivastava, Sanjay, et al. *South Asian Disaster Report 2007.* SAARC Disaster Management Centre, 2008. *SAARC Disaster Management Centre,* saarc-sdmc.nic.in/sdr_p.asp. Accessed 6 Sept. 2017.

Warner, Jeroen Frank, et al., editors. *Making Space for the River: Governance Experiences with Multifunctional River Flood Management in the U.S. and Europe.* IWA Publishing, 2013.

FLUID DYNAMICS

FIELDS OF STUDY

Fluid Dynamics; Mathematics; Thermodynamics; Engineering; Physical Sciences; Statistics; Process Modeling; Atmospheric Science; Oceanography; Hydrology; Physical Geography

SUMMARY

Fluid dynamics is an interdisciplinary field concerned with the behavior of gases and liquids in motion. An understanding of fluid dynamic principles is essential to the work done in aerodynamics. It informs the design of air and spacecraft. An understanding of fluid dynamic principles is also essential to the field of hydromechanics and the design of oceangoing vessels. Any system with air, gases, or water in motion incorporates the principles of fluid dynamics.

PRINCIPAL TERMS

- **compressible fluid:** a fluid that can be made to occupy a smaller volume by applying external pressure
- **fluid:** the state of matter characterized by the ability to flow and conform exactly to the shape of its container
- **incompressible fluid:** a fluid whose volume does not change in response to increased pressure
- **viscosity:** a property of fluids determined by intermolecular attractive forces which in turn determines the resistance of the fluid to motion

Definition and Basic Principles

Fluid dynamics is the study of fluids in motion. Air, gases, and water are all considered to be fluids. When the fluid is air, this branch of science is called aerodynamics. When the fluid is water, it is called hydrodynamics.

The basic principles of fluid dynamics state that fluids are a state of matter in which a substance cannot maintain an independent shape. A fluid will take the shape of its container, forming an observable surface at the highest level of the fluid when it does not completely fill the container. Fluids flow in a continuum, with no breaks or gaps in the flow. They are said to flow in a streamline, with a series of particles following one another in an orderly fashion in parallel with other streamlines. Real fluids have some amount of internal friction, known as viscosity. Viscosity is the property that causes some fluids to flow more readily than others. It is the reason that molasses flows more slowly than water at room temperature.

Fluids are said to be compressible or incompressible. Water is an incompressible fluid because its density does not change when pressure is applied. Incompressible fluids are subject to the law of continuity, which states that fluid flows in a pipe are constant. This theory explains why the rate of flow increases when the area of the pipe is reduced and vice versa. The viscosity of a fluid is an important consideration when calculating the total resistance on an object.

The point where the fluid flows at the surface of an object is called the boundary layer. The fluid "sticks" to the object, not moving at all at the point of contact. The streamlines further from the surface are moving, but each is impeded by the streamline between it and the wall until the effect of the streamline closest to the wall is no longer a factor. The boundary layer is not obvious to the casual observer, but it is an important consideration in any calculations of fluid dynamics.

Most fluids are Newtonian fluids. Newtonian fluids have a stress-strain relationship that is linear. This means that a fluid will flow around an object in its path and "come together" on the other side without a delay in time. Non-Newtonian fluids do not have a linear stress-strain relationship. When they encounter shear stress, their recovery varies with the type of non-Newtonian fluid.

A main consideration in fluid dynamics is the amount of resistance encountered by an object moving through a fluid. Resistance, also known as drag, is made up of several components, all of which have one thing in common: they occur at the point where the object meets the fluid. The area can be quite large, as in the wetted surface of a ship, the portion of a ship that is below the waterline. For an airplane, the equivalent is the body of the plane as it moves through the air. The goal for those who work in the field of fluid dynamics is to understand the effects of fluid flows and minimize their effect on the object in question.

BACKGROUND AND HISTORY

Swiss mathematician Daniel Bernoulli (1700-1782) introduced the term "hydrodynamics" with the publication of his book *Hydrodynamica* in 1738. The name referred to water in motion and gave the field of fluid dynamics its first name, but it was not the first time water in action had been noted and studied. Leonardo da Vinci (1452-1519) made observations of how water flows in a river and was the one who realized that water is an incompressible fluid and that for an incompressible fluid, V = constant. This law of continuity states that fluid flow in a pipe is constant. In the late 1600s, French physicist Edme Mariotte (1620-1684) and Dutch mathematician Christiaan Huygens (1629-1695) contributed the velocity-squared law to the science of fluid dynamics. They did not work together, but they both reached the conclusion that resistance is proportional not to velocity itself but to the square of the velocity.

Sir Isaac Newton 1642-1727) put forth his three laws of motion in the 1700s. These laws play a fundamental part in many branches of science, including fluid dynamics. In addition to the term hydrodynamics, Bernoulli's contribution to fluid dynamics was the realization that pressure decreases as velocity increases. This understanding is essential to the understanding of lift. Leonhard Euler (1707-1783), the father of fluid dynamics, is considered by many to be the preeminent mathematician of the eighteenth century. He is the one who derived what is today known as the Bernoulli equation from the work of Daniel Bernoulli. Euler also developed equations for inviscid flows. These equations were based on his own work and are still used for compressible and incompressible fluids.

The Navier-Stokes equations result from the work of French engineer Claude-Louis Navier (1785-1836) and British physicist George Gabriel Stokes (1819-1903) in the mid-nineteenth century. They did not work together, but their equations apply to incompressible flows. The Navier-Stokes equations are still used. At the end of the nineteenth century, Scottish engineer William John Macquorn Rankine (1820-1872) changed the understanding of the way fluids flow with his streamline theory, which states that water flows in a steady current of parallel flows unless disrupted. This theory caused a fundamental shift in the field of ship design because it changed the popular understanding of resistance in oceangoing vessels. Laminar flow is measured today by use of the Reynolds number, developed by British engineer and physicist Osborne Reynolds (1842-1912) in 1883. When the number is low, viscous forces dominate. When the number is high, turbulent flows are dominant.

American naval architect David Watson Taylor (1864-1940) designed and operated the first experimental model basin in the United States at the start of the twentieth century. His seminal work, *The Speed and Power of Ships* (1910), is still read. Taylor played a role in the use of bulbous bows on vessels of the navy. He also championed the use of airplanes that would be launched from naval craft underway in the ocean.

The principles of fluid dynamics took to the air in the eighteenth century with the work done by aviators such as the Montgolfier brothers, Joseph (1740-1810) and Jacques (1745-1799), and their hot-air balloons and the parachute of French physicist Louis-Sébastien Lenormand (1757-1837). It was not until 1799, when English inventor Sir George Cayley (1771-1857) designed the first airplane with an understanding of the roles of lift, drag, and propulsion, that aerodynamics came under scrutiny. Cayley's work was soon followed by the work of American engineer Octave Chanute (1832-1910). In 1875, he designed several biplane gliders, and with the publication of his book *Progress in Flying Machines* (1894), he became internationally recognized as an aeronautics expert.

The Wright brothers, Orville (1871-1948) and Wilbur (1867-1912), are rightfully called the first aeronautical engineers because of the testing they did in their wind tunnel. By using balances to test a variety of different airfoil shapes, they were able to correctly predict the lift and drag of different wing shapes. This work enabled them to fly successfully at Kitty Hawk, North Carolina, on December 17, 1903.

German physicist Ludwig Prandtl (1875-1953) identified the boundary layer in 1904. His work led him to be known as the father of modern aerodynamics. Russian scientist Konstantin Tsiolkovsky (1857-1935) and American physicist Robert Goddard (1882-1945) followed, and Goddard's first successful liquid propellant rocket launch in 1926 earned him the title of the father of modern rocketry.

All of the principles that applied to hydrodynamics as the study of water in motion also applied to aerodynamics, the study of air in motion. Together these principles constitute the field of fluid dynamics.

How It Works
When an object moves through a fluid such as gas or water, it encounters resistance. The amount of resistance depends upon the amount of internal friction in the fluid (the viscosity) as well as the shape of the object. A torpedo, with its streamlined shape, will encounter less resistance than a two-by-four that is neither sanded nor varnished. A ship with a square bow will encounter more resistance than one with a bulbous bow and V shape. All of this is important because with greater resistance comes the need for greater power to cover a given distance. Since power requires a fuel source and a way to carry that fuel, a vessel that can travel with a lighter fuel load will be more efficient. Whether the design under consideration is for a tractor trailer, an automobile, an ocean liner, an airplane, a rocket, or a space shuttle, these basic considerations are of paramount importance in their design.

Applications and Products
Fluid dynamics plays a part in the design of everything from automobiles to the space shuttle. Fluid dynamic principles are also used in medical research by bioengineers who want to know how a pacemaker will perform or what effect an implant or shunt will have on blood flow. Fire flows are also being studied to aid in the science of wildfire management. Until now the models have focused on heat transfer, but new studies are looking at fire systems and their fluid dynamic properties. Sophisticated models are used to predict fluid flows before model testing is done. This lowers the cost of new designs and allows the people involved to gain a thorough understanding of the trade-off between size and power, given a certain design and level of resistance.

Careers and Course Work
Fluid dynamics plays a part in a host of careers. For careers in any area of climate and meteorology, an understanding of fluid dynamics in the behavior of the atmosphere is essential. The interactions and movements of high and low pressure cells within the atmosphere, for example, are determined primarily by fluid dynamic principles, while the importance of fluid dynamics in describing the motion of surface and deep ocean currents should be readily apparent. Naval architects use fluid dynamic principles to design vessels. Aeronautical engineers use the principles to design aircraft. Astronautical engineers use fluid dynamic principles to design spacecraft. Weapons are constructed with an understanding of fluids in motion. Automotive engineers must understand fluid dynamics to design fuel-efficient cars. Architects must take the motion of air into their design of skyscrapers and other large buildings. Bioengineers use fluid dynamic principles to their advantage in the design of components that will interact with blood flow in the human body. Land-management professionals can use their understanding of fluid flows to develop plans for protecting the areas under their care from catastrophic loss due to fires. Civil engineers take the principles of fluid dynamics into consideration when designing bridges. Fluid dynamics also plays a role in sports: from pitchers who want to improve their curveballs to quarterbacks who are determined to increase the accuracy of their passes.

Students should take substantial course work in more than one of the primary fields of study related to fluid dynamics (physics, mathematics, computer science, and engineering), because the fields that depend upon knowledge of fluid dynamic principles draw from multiple disciplines. In addition, anyone desiring to work in fluid dynamics should possess skills that go beyond the academic, including an aptitude for mechanical details and the ability to envision a problem in more than one dimension. A collaborative mind-set is also an asset, as fluid dynamic applications tend to be created by teams.

Social Context and Future Prospects
The science of fluid dynamics touches upon a number of career fields that range from sports to bioengineering. Anything that moves through liquids such as air, water, or gases is subject to the principles of fluid dynamics. The more thorough the understanding, the more efficient vessel and other designs will be. This will result in the use of fewer resources in the form of power for inefficient designs and help create more efficient aircraft and launch vehicles as well as medical breakthroughs.

—*Gina Hagler, MBA*

FURTHER READING

Anderson, John D., Jr. *A History of Aerodynamics and Its Impact on Flying Machines.* Cambridge University Press, 1997.

Carlisle, Rodney P. *Where the Fleet Begins: A History of the David Taylor Research Center.* Naval Historical Center, 1998.

Çengel, Yunus A., and John M. Cimbala. *Fluid Mechanics: Fundamentals and Applications.* McGraw-Hill, 2010.

Darrigol, Olivier. *Worlds of Flow: A History of Hydrodynamics from the Bernoullis to Prandtl.* Oxford University Press, 2005.

Eckert, Michael. *The Dawn of Fluid Dynamics: A Discipline Between Science and Technology.* Wiley-VCH, 2006.

Fernando, H J S. *Handbook of Environmental Fluid Dynamics.* Taylor & Francis, 2013.

Ferreiro, Larrie D. *Ships and Science: The Birth of Naval Architecture in the Scientific Revolution, 1600–1800.* MIT Press, 2007.

Mahan, A. T. *The Influence of Sea Power upon History, 1660–1783.* 1890. Reprint. Noble Books, 2004.

FOG AND REGULATION OF THE GLOBAL CLIMATE

FIELDS OF STUDY

Atmospheric Science; Atmospheric Chemistry; Meteorology; Hydroclimatology; Physical Geography; Oceanography; Heat Transfer; Thermodynamics; Environmental Sciences

SUMMARY

Fog is a cloud that forms near the ground. Fog forms as the temperature of air falls below its dew point and condensation occurs. Fog also forms and dissolves rapidly depending on whether the temperature is higher or lower than the dew point. Fogs are named based on the specific process through which the humidity has reached its saturation point, causing the fog to form.

PRINCIPAL TERMS

- **adiabatic:** referring to thermodynamic processes that occur without a net change in heat content
- **longwave radiation:** electromagnetic radiation in the infrared range, having wavelengths longer than those of the visible spectrum, associated with heat and heating
- **radiative cooling:** the process of emitting energy as longwave radiation to balance the solar energy absorbed as shortwave radiation
- **shortwave radiation:** electromagnetic radiation in the ultraviolet range, having wavelengths shorter than those of the visible spectrum, associated with the transport of energy capable of promoting chemical reactions
- **thermal inertia:** a statement of the rate at which a body of matter approaches thermal equilibrium with its surroundings

TYPES OF FOG

Radiation fog forms as air temperature falls below the dew point by a rapid radiative cooling through a clear, calm night. The cool and heavy air containing this type of fog often settles in low-lying areas, such as mountain valleys and low spots in fields. Most inland fogs are radiation fogs. Advection fog usually forms in coastal regions as warm air blows over a cold surface and loses heat to the underlying surface. Upslope fog forms as air blows over mountain slopes, becomes cool adiabatically, and reaches saturation.

Evaporation fog occurs as a cold, dry air blows over warm, moist surfaces. Thermal and humidity gradients between the two entities facilitate rapid evaporation from the surface. As the warmer, moist air from the ground mixes with cold, dry air above, it reaches saturation and forms fog. Steam-like fog that forms over lakes during late fall and early winter is a typical form of evaporation fog.

Early morning radiation fog on a cool morning in Texas.

SIGNIFICANCE FOR CLIMATE CHANGE

Fog, as a surface-level cloud, has roughly the same climatic effects as low-level clouds. Fog dictates air and surface temperature patterns by moderating the amount of radiation that enters and leaves the surface. High moisture content imparts fog with great thermal inertia. Thus, once a fog has formed, it will prevent air temperature from changing. In the morning, fog's high albedo allows less incoming solar radiation to reach the surface, by reflecting more of it back out into space. Thus, morning fogs keep air temperatures low, allowing only gradual increases until the fog evaporates. As a result, fogs decrease daytime high temperatures.

In the night, fog absorbs outgoing longwave radiation and prevents air temperature from dropping as it would on a clear night. Fog also potentially warms the air near the surface, because condensation is a warming process due to the release of energy as water vapor returns to the liquid state. Overall, however, fog has a greater cooling effect than warming effect on Earth. As global warming continues, greater incidences of fog and lower clouds are expected, because warmer air can contain more water vapor. This increased fog may affect humans by reducing visibility and by trapping air pollutants to form smog. Fog can also produce precipitation, in the form of rain drizzle and light snow.

—*Jongnam Choi*

FURTHER READING

Ahrens, C. Donald. *Essentials of Meteorology: An Invitation to the Atmosphere.* 5th ed. Belmont, Calif.: Thomson Brooks/Cole, 2008. Widely used introductory textbook on atmospheric science; covers a wide range of topics on weather and climate.

Alaby, Michael *Fog, Smog, and Poisoned Rain.* Infobase Publishing, 2014. An entry-level book that describes the characteristics of fog formations and their effects.

Denny, Mark *Making Sense of Weather and Climate. The Science Behind the Forecasts.* Columbia University Press, 2017. An informative yet entertaining book that looks at many readily-visible aspects of climate effects such as fog from a question and answer perspective.

Koračin, Darko and Dorman, Clive E., eds. *Marine Fog: Challenges and Advancements in Observations, Modeling and Forecasting.* Springer, 2017. A curated collection of articles focusing on the phenomenon of marine fog.

FORESTS AND THE MITIGATION OF ANTHROPOGENIC CLIMATE CHANGE

FIELDS OF STUDY

Ecology; Ecosystem Management; Climatology; Bioclimatology; Climate Modeling; Environmental Sciences; Environmental Studies; Hydroclimatology; Physical Geography; Earth System Modeling; Tree Ring Analysis

SUMMARY

Eight thousand years before the present, before the rise of human civilization, one-half of the Earth's surface was covered by forest. In the intervening years, 20 percent of previously forested land has lost its forest cover, resulting in the loss of a major carbon

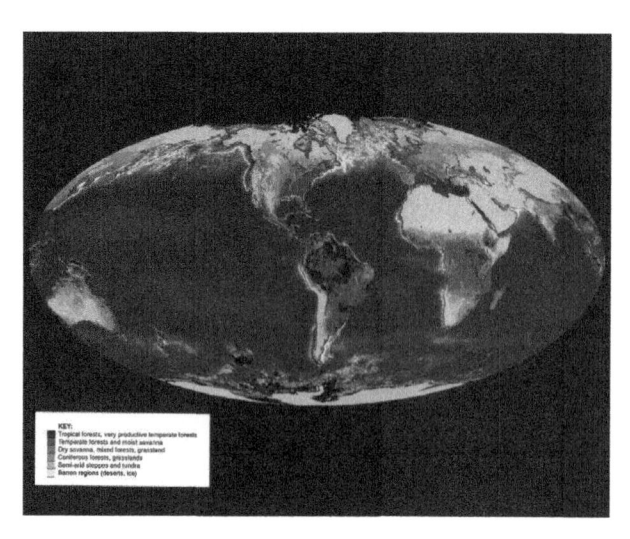

An image produced by NOAA's Polar Orbiting Environmental Satellite, showing the distribution of Earth's forests.

sink. One of the easiest ways for humans to prevent further carbonization of the atmosphere is to prevent further deforestation and to maintain existing forests intact.

PRINCIPAL TERMS

- **afforestation:** creating forests on lands that was not previously forested
- **carbon sink:** plant growth, mineral formation and synthetic activities that act to remove carbon dioxide and other carbon sources from the environment
- **dormancy:** the portion of the year during which no growth occurs
- **growing season:** the portion of the year during which photosynthesis occurs
- **reforestation:** planting trees to replace forests that have been eliminated either naturally or by human action

DISTRIBUTION OF FORESTS

The forests of the world fall into three primary zones: tropical, temperate, and boreal. These zones are defined by their climate and precipitation. Trees require a minimum of 25 centimeters of rainfall per year to grow; land that, by virtue of its location, does not attract the minimum rainfall annually will not support trees. Such land will be either grassland, shrubland, or desert. Trees also require a frost-free part of the year for a growing season. Generally, the longer the growing season, the greater the annual growth.

The tropical rain forests on either side of the equator support tree growth that continues virtually all year long. About half of the world's forests are tropical forests, and they contain the greatest biological diversity of species. Because trees through photosynthesis convert atmospheric carbon to contained carbon, the half of the world's forests in tropical areas are of vital importance.

The temperate forest does not grow all year long. The falling temperatures reach a point at which growth is no longer supported, and the trees go into dormancy for part of the year. The temperate zone has a less varied collection of tree species than the tropics, but it contains a mix of coniferous and broadleaved species, the latter of which generally lose their leaves during the dormant period.

The forest growing farthest to the north is the boreal forest, composed very largely of coniferous species. It grows more slowly than the other forests, because its growing season is the shortest, and it exists in dormancy for a major part of each year. While the true polar regions may have enough precipitation to support trees, their growing season is too short for trees to grow enough to support the needles that provide respiration.

TWENTIETH CENTURY CHANGES

The conversion of formerly forested land to other land uses was greater in the twentieth century than in any previous century. This conversion was driven in large part by the growth in human populations in all parts of the world, but especially during the latter half of the century, primarily in the tropical regions. Between 1950 and 2000, the world's population more than doubled, from 2.5 billion to 6 billion, and the developing world, largely located in the same area where tropical forests are found, contained three-fourths of that larger population. Since forests provide the largest carbon sink, the loss of many forests, especially the tropical forests, is mainly responsible for the rising proportion of carbon in the atmosphere.

The chief driver of deforestation during the twentieth century has been the demand for additional agricultural land to support the growing population. The conversion of forestland to agricultural land adds to atmospheric carbon in several ways. In many cases, the trees that are cut down to release the land for agriculture are burned, and the carbon stored in them is released into the atmosphere. Second, the carbon contained in forest soils over time is also released into the atmosphere when the trees are no longer there to prevent its release. Third, the trees that had been on the land are no longer there to capture future carbon as they grow.

Beside the need for agricultural land to grow crops to feed the growing world population, that population has experienced a change in dietary demand, particularly to include meat products. Some of the land freed up by deforestation is converted not to cropland but to pasture land. Specialized crops such as sugar or soybeans have a market price that poor populations seek to realize by converting forestland to agricultural land. In many areas, the production of specialized crops with major markets in the developed world, such as rubber, has also promoted deforestation.

Locations of Deforestation

The loss of forestland to agriculture has occurred in most parts of the world adapted to tropical forests. The Amazonian forest in Brazil and the forests in central America have been subject to important depletion since the mid-twentieth century. The deforestation has also been substantial in Southeast Asia, notably on many of Indonesia's islands, as well as in Malaysia and Thailand. The forest in Africa has been less affected.

The temperate forest that was heavily deforested in the nineteenth century has started to bounce back as urbanization and the mechanization of agriculture have reduced the demand for agricultural land. Thus where the population is heavily urbanized, former agricultural land is being gradually reforested, as in the United States and Europe.

Much of the world's sawtimber comes from the coniferous trees in the temperate forest and from the coniferous trees that cover the boreal forest. The Russian forest constitutes one-fifth of the total forestland of the world, but in recent years much cutting to supply sawtimber to the developed world has depleted some of that forest. Combined with softwood coming from Canada's boreal forests, these two sources have supplied a major portion of the dimensional lumber used by the developed world for the construction of houses. Lumber production peaked in the United States in 1906, at 46 billion board feet, but since then lumber production supplied by U.S. forests has gone down. However, wood remains the third most productive commodity in world trade, behind petroleum and natural gas. A substantial proportion becomes fuelwood.

Context

At the Rio Conference in 1992, 158 countries agreed to try to prevent further deforestation. It has, however, proved difficult to accomplish partly because market forces are working against it, and partly because definition has proved elusive. The Kyoto Protocol of 1999 identified "reforestation" and "afforestation" as processes that could mitigate deforestation, but compensation to those who carry out such measures has not found widespread acceptance. No way has yet been found to value existing forests such that their preservation could be financially rewarded.

—*Nancy M. Gordon*

Further Reading

Humphreys, David. *Logjam: Deforestation and the Crisis of Global Government*. Earthscan, 2006. Highly critical of the failure of international organizations to take on the task of preventing deforestation.

Malmesheimer, R. W., et al. "Preventing GHG Emissions Through Avoided Land-Use Change." *Journal of Forestry* 106, no. 3 (April/May, 2008). This special issue is devoted entirely to the Society of American Foresters' task force report on forest management solutions for mitigating climate change.

Stern, Nicholas. *The Economics of Climate Change: The Stern Review*. New York: Cambridge University Press, 2007. This famous report tackles the charges of those who claim that the economic costs of attempting to prevent climate change are too great.

Williams, Michael. *Deforesting the Earth: From Prehistory to Global Crisis: An Abridgment*. University of Chicago Press, 2006. This massive account by an author already well known for his careful evaluation of forest history is definitive.

FRONTS AND CLIMATE CHANGE

FIELDS OF STUDY

Climatology; Climate Modeling; Environmental Sciences

SUMMARY

A front is a band of low-pressure systems and marks the transition from one weather regime to another. It is typically formed at the boundary of two distinct air masses. In most cases, fronts are associated with a type of large-scale weather system called a midlatitude cyclone, which has a low-pressure center and causes winds to blow cyclonically (that is, in a counterclockwise direction). Midlatitude cyclones are the largest weather systems on Earth and generate most of the winter storms over the midlatitude continents. A front is a part of the midlatitude cyclone system, which trails a band of low-pressure air extending outward from the low-pressure center of the cyclone. Therefore, various weather systems, such as thunderstorms, heavy precipitation, snowstorms, and tornadoes, are also formed along the frontal band.

PRINCIPAL TERMS

- **anticyclonic:** the clockwise rotation of an air mass in the Northern hemisphere, counterclockwise in the Southern hemisphere
- **cyclonic:** the counterclockwise rotation of an air mass in the Northern hemisphere, clockwise in the Southern hemisphere
- **midlatitudes:** the latitudes north and south of the equator between the Tropics of Cancer and Capricorn

AIR MASSES AND FRONTS

Different air masses are characterized by different physical properties of the atmosphere, such as density, temperature, pressure, winds, and moisture. Because a front is a line-like demarcation that separates two different air masses, the atmosphere exhibits different physical properties on either side of a front. For example, air can change from warm to cold or from cold to warm, winds can blow from northerly to southerly or from westerly to easterly, and air can vary from dry to moist or from moist to dry across the frontal zone. Based on the movement of the frontal band and the temperature and humidity differentials, fronts can be classified as cold fronts, warm fronts, stationary fronts, and occluded fronts.

A cold front is formed when a cold and dry air mass advances and replaces a warm and moist air mass. In this case, the cold and dry air pushes and undercuts the warm and moist air ahead of it. The temperature will generally decrease in an area where a cold front is passing through. Because the cold front will lift the warm and moist air, clouds and precipitation can form at or behind the cold front.

A warm front is formed when a warm and moist air mass advances and replaces a cold and dry air mass. In this case, the warm and moist air pushes and overrides the cold and dry air, and the cold and dry air retreats. The temperature will generally rise in an area where a warm front is passing through. Because of the overriding of warm and moist air over the cold and dry air ahead of it, clouds and precipitation typically form ahead of a warm front.

An occluded front forms when a cold front catches up to and overtakes a warm front. In this case, a warm and moist air sector between the cold and warm fronts disappears, causing a complete convection of warm air in the storm center. This stage marks the full maturity of a midlatitude cyclone. Further dynamic and thermodynamic supports for the storm no longer exist, and the storm will dissipate from this time on.

A stationary front can form when a cold and a warm front move in opposite directions. When they

A squall line marks the leading edge of a cold front.

meet, they can be locked in location. The cold and warm air mix together, so that there is no dominant overtake and apparent movement from either warm or cold air. This kind of situation often arises when fronts interact with the surface topography beneath them.

Significance for Climate Change

Fronts are important weather systems affecting the daily lives of people. They mainly occur in middle latitudes, where large landmasses and dense human populations are located. A midlatitude cyclone, fronts, an upper-level jet stream, and specific storm tracks are all related, one to another, and constitute a complete synoptic weather system (a weather system that can be analyzed on a weather map). Thus, a change in one part of the atmospheric environment will result in a change of the entire weather system. Studies show that global warming tends to widen the tropics and extend the troposphere vertically. There are many consequences of these changes. One of them is a poleward shifting of future jet streams. This shift would cause climatologic locations for midlatitude cyclones, fronts, and storm tracks to change accordingly.

The current global warming trend may also suggest a decrease of surface temperature gradient, since many observations and atmospheric model simulations indicate that greater warming tends to occur in the colder regions. Since the horizontal temperature gradient is the key mechanism for the development of midlatitude cyclones, global warming might decrease the occurrence and intensity of midlatitude cyclones and associated fronts. On the other hand, because global warming tends to increase water content in the atmosphere, midlatitude cyclones may derive more energy from latent heat release and become more violent. There are no definite answers so far for how midlatitude cyclones and fronts are affected by these competing mechanisms.

Finally, frontal dynamics provides an important mechanism to cause convection and to form clouds. Precipitation related to fronts is a major process for removing water from the midlatitude atmosphere. A potential change in frontal climatology in a future warm climate, regardless of whether it is an increase or decrease, will result in redistribution of snow and rain, changing the distribution of Earth's hydrosphere, especially in middle and high latitudes.

—*Chungu Lu*

Further Reading

Ahrens, C. Donald. *Essentials of Meteorology: An Invitation to the Atmosphere.* 5th ed., Thomson Brooks/Cole, 2008. One of the most widely used introductory books on atmospheric science; covers a wide range of topics on weather and climate.

Archer, Cristina L., and Ken Caldeira. "Historical Trends in the Jet Streams." *Geophysical Research Letters* 35, no. 24 (2008). Reports an investigation of the change of location of the jet stream in response to global warming.

Lutgens, Frederick K., Edward J. Tarbuck, Tasa, Dennis G. and Herman, Redina. *The Atmosphere: An Introduction to Meteorology.* 14th ed. Pearson Education, 2018. An updated version of this textbook, in loose-leaf format, provides students with a general overview of the structure and properties of Earth's atmosphere.

O'Hare, Greg, et al. *Weather, Climate and Climate Change Human Perspectives.* Lord, Taylor & Francis, 2014. Discusses the mass and energy effects of air masses and their interaction, before focusing on the roles and description of fronts.

G

GENERAL CIRCULATION MODELS (GCM)

FIELDS OF STUDY

Atmospheric Science; Oceanography; Fluid Dynamics; Heat Transfer; Physical Geography; Statistics; Mathematics; Hydroclimatology; Earth System Modeling; Computer Science; Software Engineering

SUMMARY

Atmosphere and ocean dynamics are the result of the interactions of many different cycles and processes. These are represented in a General Circulation Model (GCM) by complex mathematical equations. Each component is assigned parameters that represent different measurable aspects of that component. Solutions of GCM calculations are used to make predictions of climatic events.

PRINCIPAL TERMS

- **anthropogenic:** deriving from human sources or activities
- **greenhouse gases (GHGs):** atmospheric gases that trap heat within a planetary system rather than allowing it to escape into space; principally carbon dioxide, methane, water and ozone.
- **parameterization:** assigning of measurable factors that simplify the description of complex phenomena that are otherwise difficult to quantify, to enable predictions to be made from the system model

MODELING GCMs

The dynamics and physics that govern the atmosphere and oceans are the result of many cycles and processes interacting over time. A general circulation model (GCM) is a collection of mathematical equations that represent these cycles and processes using computational algorithms that require powerful computational power, typically a supercomputer. In general, the components of the model, such as the atmosphere and ocean, are modeled separately and interact only at their boundaries, such as, for instance, at the interface between the atmosphere and the ocean. GCMs include equations to solve the dynamics (motions) of the atmosphere and ocean, as well as equations and parameterizations to represent atmospheric, oceanic, sea-ice, and terrestrial physics. Processes represented by parameterizations combine equations with proportionality constants, correlations, and lookup tables based on observational and experimental data.

In order to be solved on a computer, the equations and variables incorporated in a GCM must be rendered discrete, which is typically accomplished by dividing up the atmosphere and ocean on a grid and defining the key variables such as temperature and humidity at defined points on this grid. Using variable values and the governing and parameterization equations, one can compute new variable values to a relatively short time in the future. This integration process is repeated to generate longer time sequences. Validation studies that compare the results of the model computation to experimental and observational data are required to acquire confidence in the accuracy of the simulation.

SIGNIFICANCE FOR CLIMATE CHANGE

GCMs are the fundamental tools for predicting the future evolution of Earth's climate and the impact that anthropogenic greenhouse gas (GHG) emissions or other environmental factors will have on temperatures, rainfall, and other climatic conditions in the future. GCMs arose from efforts at numerical weather prediction (NWP). The English mathematician Lewis Fry Richardson (1881–1953) is credited with making the first NWP calculation when he

Joshua Trees in the Inyo Mountains above Eureka Valley in Inyo County, California. Recent climates, and General Circulation Model results of future climates, portray this area as being suitable for the survival and expansion of Joshua trees.

attempted to predict the weather six hours into the future during World War I. Because he lacked computers, the calculations took six months to complete, but the basic approach was the same that would be used four decades later in NWP. The first working GCM is attributed to Norman A. Phillips (1923–), who completed a two-layer hemispheric atmospheric model in 1955. The late 1960s and the 1970s saw the introduction of coupled atmosphere-ocean models and the first use of a GCM to study the effects of carbon dioxide (CO_2) and pollutants in the atmosphere. From that point forward, GCMs were recognized as critical components in the study of climate change.

Four-component models (comprising atmosphere, ocean, sea ice, and land) are often referred to as atmosphere-ocean general circulation models (AOGCMs). The highest-resolution AOGCM grids have horizontal graduations of 1° (roughly 100 kilometers) and up to sixty vertical layers in the atmosphere. A typical use of these models takes an average of current conditions as the initial conditions then integrates the GCM computationally through future decades, while different parameters and boundary conditions are changed to account for different scenarios. Based on the results of these computations, the likely range of future temperatures and other future climatic conditions is determined to the extent possible given the accuracy of the model and the related inputs.

GCM simulations are also used to compute the magnitude of anthropogenic versus natural forcing of the environment in order to quantitatively isolate the human contribution to climate change. They are also used to assess the future impact of plans to reduce global GHG emissions. Key statements about global warming, such as the anticipated rise in temperature and sea level and changes in precipitation and storm patterns due to the release of anthropogenic GHGs over the next century, are the product of general circulation models.

Context

Given their importance in the debate over global warming, GCMs are subject to much scrutiny. These models are not complete, with some physical processes generally not implemented in the models and other representations being the source of argument and uncertainty. Critical points of contention include the parameterization of cloud-radiation feedbacks and cloud microphysics, including precipitation; the impact of changes in incoming solar radiation; and the general variability of results when using different GCMs. A notable missing process in most GCMs is an explicit numerical treatment of the carbon cycle, causing most global warming studies to rely on predetermined changes in atmospheric GHG content. Ongoing work to improve modeling of physical processes, achieve higher-resolution simulations on more powerful computing platforms and increase the use of validation studies will generate the mechanisms by which these concerns will be addressed. Even with their limitations, GCMs remain the primary tool for analyzing the future of Earth's climate.

—*Raymond P. LeBeau, Jr.*

Further Reading

Chang, Julius *General Circulation Models of the Atmosphere*. Elsevier, 2012. The book covers the fundamentals of general circulation models and their application in atmospheric studies.

Donner, Leo, Schubert, Wayne and Somerville, Richard, eds. *The Development of Atmospheric General Circulation Models. Complexity, Synthesis and Computation*. Cambridge University Press, 2011. A curated collection of articles describing the historical development of general circulation models and their development since then.

Glover, David, Jenkins, William J. and Doney, Scott C., *Modeling Methods for Marine Science* Cambridge University Press, 2011. This book provides the basic methodology for the development of general circulation models from first principles, introduced with MATLab.

Hamilton, Kevin and Ohfuchi, Wataru, eds. *High Resolution Numerical Modelling of the Atmosphere and Ocean.* 2nd ed., Springer, 2008. An advanced reference, this book is a curated collection of articles describing the many aspects of different model systems for GCMs, the mathematical basis of those models and the problems associated with those models.

Houghton, John. *Global Warming: The Complete Briefing.* 4th ed. New York: Cambridge University Press, 2009. This overview of global warming by a leading scientist in the field includes an introductory discussion of climate modeling. Figures, references, glossary, index.

GLACIATIONS

FIELDS OF STUDY

Atmospheric Science; Oceanography; Fluid Dynamics; Physical Geography; Computer Science

SUMMARY

Glaciations presented a challenge to climate science. With ice sheets today only on Greenland and Antarctica, huge continental glaciers were difficult even to imagine. As the idea took root, however, evidence for glaciations throughout geologic history became apparent. What controls the timing of glaciations is now largely understood, but how they grow or retreat—and why—continue to be important subjects of research.

PRINCIPAL TERMS

- *glaciation:* the advance of a continental ice sheet
- *ice age:* a period of time during which major continental ice sheets advanced and retreated
- *interglacial:* the warm period between glaciations
- *marine isotope stage:* half of a glacial cycle, as identified in the oxygen isotope data from ocean cores
- *Milanković cycle:* period of variation in Earth's orbital parameters, including axial inclination, climatic precession, and orbital eccentricity

GLOBAL FACTORS

The temperature of the surface of the Earth is a function of how much energy it receives from the Sun and how much of this energy is radiated back into space. The atmosphere plays an important role, as its clouds and aerosols limit how much light energy reaches the surface and its greenhouse gases (GHGs) limit how much infrared energy escapes. Geologic evidence suggests that average global surface temperatures have remained within a limited range over the past two billion years, and GHGs, particularly carbon dioxide (CO_2), may have been responsible for this consistency.

CO_2 is produced by volcanism and removed by weathering. Weathering is temperature dependent, so it represents a negative feedback system: Higher temperature results in more weathering, which removes more CO_2, resulting in lower temperatures, or vice versa. Although over periods measured in hundreds of millions of years this feedback loop seems to have maintained a relatively constant temperature,

The Great Aletsch Glacier (Grosser Aletschgletscher), Valais. Looking toward Gross Wannenhorn.

its effects are not instantaneous, and perturbations have occurred. When these perturbations produce colder conditions, ice ages can result.

Evidence of a number of ice ages has been identified in rocks more than one billion years old. Little is known about these ice ages, as data are sparse and difficult to interpret. Between 750 and 550 million years ago, there were several major ice ages, usually lumped together and called Snowball Earth, as there is evidence that glaciers then existed at sea level near the equator. Although they are of academic interest, at the time these ice ages occurred, there were no land plants, and the atmosphere had far less oxygen, so it is not clear that efforts to understand them will help in the study of contemporary climate change.

Other ice ages, including a short one 440 million years ago and the Permo-Carboniferous ice age between 325 and 240 million years ago, provide insight into what conditions are required for ice ages. These occurred when the land on which the continental glaciers formed was over the South Pole. As continents were also over the South Pole in the time period between these two ice ages and Antarctica sat over the South Pole for 90 million years before its current glaciers formed, this location seems to be a necessary but not a sufficient condition for ice age formation.

PLEISTOCENE GLACIATIONS

About 50 million years ago, the Earth began to cool. Deep-ocean temperatures gradually dropped from 13° Celsius to the present 1° Celsius. Study of ocean sediment cores has identified some minor ice ages around 40 million years ago, another ice age formed the East Antarctica ice sheet 34 million years ago, and a third ice age around 13 million years ago left evidence in Alaska. Orbital factors probably influenced these advances, perhaps by affecting the carbon cycle, but the details remain obscure.

Starting about three million years ago, the climate developed two different states. Since then, it has oscillated between them, causing glaciations—with continental glaciers extending down to the 40° north parallel of latitude—and interglacials such as the current period, some with temperatures even higher than those of the present. Geologists identified glacial deposits, determined that some were on top of others, and gave names to each, but the names were not standardized internationally. The name of the most recent glaciation is Valdaian on the Russian Plains, Devensian in Britain, Weichselian in Scandinavia, Würm in the Alps, and Wisconsinan in North America. The deposits of one glacier are easily removed by subsequent glaciers, so there might be some for which no evidence remains on the continents.

Oxygen isotope ratios of marine sediments, which are not removed by later glaciations, show that there have actually been around fifty glaciations. Because these ratios indicate how much water is tied up in ice, this record is now seen as the best way of delineating when an advance ended and a retreat began. Each marine isotope stage (MIS) is given a number (odd ones for interglacials, even ones for glaciations) starting with the most recent. Interglacial MIS 103 occurred 2,580,000 years ago.

Once all these additional glaciations were known, analysis showed that astronomical cycles controlled their timing, as had been suggested by James Croll in 1864 and Milutin Milanković in 1941. These results are so robust that geologists now use these cycles to calibrate the more recent part of the geologic time scale.

That Milanković cycles control the timing of glaciations is beyond dispute. How they do so, however, is not well understood. Feedback systems in the oceans, the atmosphere, the biosphere, or perhaps elsewhere are needed to amplify the tiny signal produced astronomically. Identifying and understanding these systems will help scientists understand current climate change.

CONTEXT

Skeptics who wonder if the climate is changing should consider glaciations. Doomsayers, who fear the human race is threatened by global warming, should also consider glaciations. Much of geologic history concerns things that happened so long ago that it is easy to dismiss or ignore them, but glaciations are recent history. *Homo erectus* was on the planet and using fire twenty or so glaciations ago. All of human evolution has taken place as glaciers ebbed and flowed. Human migration to North America occurred 14,600 years ago, just after the peak of the last glaciation. Ice sheets then, although they were getting smaller, still covered most of Maine and northern parts of New York, Vermont, and New Hampsire. They are not there now. Climate changes.

—*Otto H. Muller*

BIBLIOGRAPHY

Imbrie, J., and K. P. Imbrie. *Ice Ages: Solving the Mystery.* Short Hills, N.J.: Enslow, 1979. An excellent history of the efforts to understand what caused the ice ages. Full of personal anecdotes, it gives a sense of the excitement and frustrations experienced by cryologists. Many diagrams, including some now classic ones.

Macdougall, J. D. *Frozen Earth: The Once and Future Story of Ice Ages.* Berkeley: University of California Press, 2004. Easily understood by a general reader, this book explains how glaciations were discovered and understood. Chapter 10 focuses on the role glaciations played in evolution in general, and in human evolution in particular. Photos, diagrams, maps, an annotated bibliography of general-interest books.

Ruddiman, William F. *Earth's Climate Past and Future.* 2d ed. New York: W. H. Freeman, 2008. This elementary college textbook has several sections concerning glaciations and their causes, with particular attention paid to how they can elucidate climate change. Illustrations, figures, tables, maps, bibliography, index.

GLACIERS

FIELDS OF STUDY

Atmospheric Science; Atmospheric Chemistry; Analytical Chemistry; Environmental Chemistry; Geochemistry; Physical Chemistry; Photochemistry; Thermodynamics; Fluid Dynamics; Environmental Sciences; Environmental Studies; Waste Management; Meteorology; Process Modeling; Hydroclimatology; Bioclimatology; Climate Zones; Oceanography; Hydrology; Earth System Modeling; Ecology

SUMMARY

Of the freshwater ice of the world, 99 percent is located in Antarctica and in Greenland: 90 percent of the ice is located in Antarctica (an area of 14 million square kilometers containing 27.6 million cubic kilometers of ice); 9 percent is in Greenland (an area of 1.726 million square kilometers containing 2.85 million cubic kilometers of ice), and 1 percent exists in the glaciers and ice caps scattered throughout the world. The two ice sheets of Antarctica and Greenland represent about 10 percent of the Earth's land area and contain more than three-quarters of its freshwater.

PRINCIPAL TERMS

- **alpine glaciers:** large masses of ice found in valleys, on plateaus, and attached to mountains
- **cryosphere:** the portion of the Earth's surface that is composed of frozen water
- **firn:** the intermediary stage between snow and ice
- **glacial ice:** ice created by the compression of snow, sometimes saturated with meltwater that is refrozen
- **ice caps:** masses of ice covering areas smaller than 50,000 square kilometers
- **ice sheets:** masses of ice covering large landmasses

GLACIAL LANDSCAPE

Geologists can provide useful information about the history of glaciation by studying the geographic features left by glaciers that have long since melted away. When alpine glaciers form, they produce an amphitheater-like depression called a "cirque" at their highest point. When the compression of snow produces ice in a depth of about 20 meters, the ice begins to flow downward due to its own weight along the valleys of former streams. The ice deepens and widens the stream valley to produce a glacial trough. Valleys of tributary streams are filled with ice as well, but the shorter length of the tributary glaciers and the lesser discharge produce less deepening of the channel. Typically, these tributary glaciers erode their channels only down to the current ice surface of the main glacier, resulting in distinctive "hanging valleys" that end in a precipice at their juncture with

Glaciation Time Line

Glaciation	Millions of Years Ago	Era
Karoo	360-260	Paleozoic
Andean-Saharan	450-420	Paleozoic
Cryogenian	800-635	Neoproterozoic
Huronian	2,400-2,100	Paleoproterozoic

the valley produced by the main body of the glacier. The height of these hanging valleys indicates the depth of the ice in the main glacier.

Materials the glacier erodes are pushed to its side, creating lateral moraines, or to its lower end, creating a frontal or end moraine, or accumulate under the glacier, creating a ground moraine, or till. These features identify the greatest breadth (lateral moraines) and furthest extent (terminal moraine) of ancient glaciers. The sediments deposited in glacial lakes called varves, can be used to determine the length of the deposition process. Distinctive small hills called drumlins are formed under an advancing glacier and, because of their teardrop shape, indicate the direction of the ice and water flow. Meltwater accumulates in lakes found in the cirques; these lakes are called tarns. Other lakes occupy the depressions in the glacial troughs. When the glacier retreats, it uncovers a U-shaped valley instead of the V-shaped valley it had prior to the glaciation.

Continental ice sheets transform the landscape differently. When they retreat, they leave a multitude of lakes, many of which are round (kettle lakes). Others are elongated in the direction of the ice flow, such as the Finger Lakes in New York. The Great Lakes system in Canada, beginning with Lake Ontario through Lake Superior and beyond towards the northwest through Lake of the Woods, Lakes Winnipeg and Winnipegosis, Great Slave Lake and Great Bear Lake also reflect the linear movement of ancient continental ice sheets. By reconstructing the history of glaciation from these alterations made to the landscape, glaciologists and geologists have been able to contribute greatly to the understanding of the nature of Earth's climatic changes.

Ice Formation and Location

Glaciers are masses of ice that are produced where the summer temperatures fail to melt the snow that fell during the preceding winter. Over time, this snow is compressed by overlying layers of more recent snow, forcing out some of the air that exists around snow crystals. This air escapes toward the surface; the density of this older snow steadily increases, until the snow turns into ice. At that point, the density of the ice, in grams per cubic centimeter, varies from 0.85 to 0.91. Ice therefore floats on liquid water, whose density is 1. The transformation of snow into ice is slower in polar areas, where compression is the major mechanism at work, because air temperatures remain low all year round, producing very little, if any, meltwater to be refrozen. In temperate climates, ice forms more rapidly than in the polar areas because there are periods of melting when the temperature is above the freezing point of water. The meltwater soaks the snow and refreezes during the next colder period. This process is much faster than compression to achieve the density of ice.

Glaciers can exist at any latitude, even along the equator. They are present on every continent except Australia. The necessary condition for glaciers is that the air temperature remains low enough to prevent melting of the last winter snow. Because temperature decreases with increasing elevation, glaciers are found on high mountains or volcanoes. In Africa,

Mounts Kenya, Kilimanjaro, and Ruwenzori have small glaciers (though they are retreating rapidly). In South America, there are many small tropical glaciers located in the Andes as well. In North America, Europe, and Asia, small glaciers dot the summits of high mountain ranges and volcanoes. Glaciers gain mass, called accumulation, by the deposition of snow in the highest elevation. This high-elevation region of accumulation is called the cirque, generally a large amphitheater-like region at the summit of glaciers.

Glaciers lose mass through processes of ablation, such as melting, sublimation, and calving. Melting takes place in a glacier where the temperature is greater than 0°C. This can occur where the air temperature reaches this value or at the underside of the glacier, where friction of the ice on the ground beneath the glacier causes the temperature to increase, producing meltwater. This water lubricates the underside of the glacier, resulting in faster movement.

The second component of ablation is sublimation. sublimation is the transformation of ice directly into water vapor, without an intermediate liquid stage. In Antarctica, sublimation is a major contributor to the ablation of ice because the ice sheet is affected by very strong winds which enhance the process. The third form of ablation is calving. It involves the breaking of the end of the glacier when it reaches an ocean or a lake. calving produces icebergs—masses of ice that float because of the lower density of ice (0.85–0.91) compared to that of water (1). Calving is responsible for about 40 percent of ablation in Greenland and 80 to 90 percent in Antarctica.

The term "mass balance" refers to the difference in mass between accumulation and ablation. A glacier will grow longer if accumulation is greater than ablation over a period of time. This is called a glacial advance. A glacier will get shorter if the amount of ice removed by ablation exceeds the amount of snow that accumulates in the coldest part, the cirque. When glaciers become shorter they are said to retreat. This does not mean that the ice stops moving downward from the cirque to the front, at the lower end of the glacier. Glacial retreat indicates only that the front of the glacier will be found closer to the cirque.

Today, most glaciers in the world are retreating. Although there are a few exceptions where glaciers are advancing, the worldwide trend is a steady retreat in response to the general rising of the Earth's temperatures.

GLACIOLOGY AND PALEOCLIMATES

The science that studies the cryosphere is called glaciology. One of the first important glaciologists was Louis Agassiz (1807–1873), a native of Switzerland, who studied the glaciers of the Alps in the nineteenth century. Building on Agassiz's work, modern glaciologists discovered that glaciers not only are a good source of information about the global impact of recent environmental changes but also provide valuable data about the long history of climate change on this planet. Modern study of ice cores—cylinders of ice retrieved from glaciers—has shown that ice records the temperatures of the Earth atmosphere at the time the snow fell. The paleotemperatures are inferred from the composition of the water that makes up the ice.

In nature, water is made of two atoms of hydrogen and one atom of oxygen; however, oxygen has three isotopes (elements that have the same number of protons but different numbers of neutrons). The lightest and most abundant of the three is oxygen-16, ^{16}O. Oxygen-18, ^{18}O, is the heaviest but exists in much smaller quantities. Higher Earth temperatures make it easier for the heavier ^{18}O molecule to evaporate, resulting in snow, and therefore glacial ice, that has an increased proportion of it. During glaciations, the lower temperatures lead to a depletion of ^{18}O in the ice of glaciers. The same principle is applied to ice when two isotopes of hydrogen are measured. Because the ice sheets of Antarctica and Greenland are very thick, ice cores obtained by drilling into these ice sheets are very long. They therefore provide a very long record of the climates of the past (known as "paleoclimates"). Based on the cores retrieved so far, scientists have been able to identify a sequence of glacial and interglacial cycles that more than 800,000 years into the past.

Earth has had many ice ages in its 4.5-billion-year history. Most recently, beginning about one million years ago, the Great Ice Age occurred during a time period called the Pleistocene. This glacial period formed an ice sheet in North America, Northern Europe, Northern Asia, and Antarctica that expanded until it reached its maximum extent about twenty thousand years ago. The Pleistocene Glaciation was not uniformly cold; short interglacial periods of warming occurred several times. Finally, about ten thousand years ago, the warming trend continued, melting the ice sheets and uncovering the northern continents. The northern part of Canada became free of ice about six thousand years ago. The mountain glaciers attached to the high mountain ranges of the American West are remnants of the Pleistocene period. In more recent centuries, Earth experienced a shorter period of cold temperatures, called the Little Ice Age, beginning about 1650 and ending approximately in 1850, during which Earth cooled by about 1°C. (The term "Little Ice Age" is used differently by different writers. Many use it to refer to the climate cooling from about 1300 to 1850, while others use it for the latter half of that interval, when cooling was greatest, beginning around 1550 or 1600.) The mountain glaciers that advanced during this period are currently retreating in response to today's higher temperatures.

Context

At their present rate of melting and retreating, glaciers are having a major impact on the populations living in their vicinity. The increase in meltwater can increase the production of hydroelectricity for a short while but at the same time may impact the population's well-being in the very near future by decreasing the amount of available water for irrigation or electricity production when the glaciers will have completely disappeared. Monitoring glacier changes and drawing scientific conclusions about their retreat is not something new. As early as 1894, scientists began cataloging glaciers and their changes. These findings were published by the World Glacier Monitoring Service. Maximum extents of glaciers were computed by using the position of their terminal moraines, and their volumes were estimated by measuring the height of their lower end since it corresponds to the height of ice that used to occupy the glacial trough.

In the 1970s, during the International Hydrological Decade declared by United Nations Educational, Scientific, and Cultural Organization, the Temporal Technical Secretariat for the World Glacier Inventory was created and began making a comprehensive inventory of more than 100,000 glaciers worldwide. Since then, with the help of satellite instruments, it has been determined that there are almost 200,000 glaciers, thousands of whose outlines, retreats, and advances are readily mappable. These measurements allow scientists to rapidly assess the impact of the warming of the Earth on the cryosphere, and their study has proven a valuable tool in monitoring their reaction to the warming of the Earth's atmosphere.

The work of the Intergovernmental Panel on Climate Change and the research conducted for the Fourth International Polar Year (2007–2008) organized by the International Council for Science in conjunction with the World Meteorological Organization focused on understanding the extremely complex relationships between glaciers and climates. One of the goals of the United States National Committee for the International Polar Year was the creation of a network of observation platforms to monitor glaciers in order to provide reliable data by which scientists can able to assess the impact of global warming both on the glaciers themselves and upon the global ecosystem of which they are an essential part.

—Denyse Lemaire
—David Kasserman

Further Reading

Botts, Lee, et al. *The Great Lakes: An Environmental Atlas and Resource Book.* Govt. of Canada, 2002.

Davies, Bethan. "Mapping the World's Glaciers." Antarcticglaciers.org, 25 Nov. 2014. Web. 23 Mar. 2014.

Grotzinger, John P., and Thomas H. Jordan. *Understanding Earth.* 7th ed. Freeman, 2014.

Hambrey, Michael, and Jurg Alean. *Glaciers.* Cambridge University Press, 2006.

Krüger, Tobias. *Discovering the Ice Ages: International Reception and Consequences for a Historical Understanding of Climate.* Brill, 2013.

Rapp, Donald. *Ice Ages and Interglacials: Measurements, Interpretations, and Models.* Springer, 2009.

Ruddiman, William F. *Earth's Climate Past and Future.* 2nd ed. W. H. Freeman, 2008. This elementary college textbook has several sections concerning glaciations and their causes, with particular attention paid to how they can elucidate climate change. Illustrations, figures, tables, maps, bibliography, index.

Stanley, Steven M., and John A. Luczaj. *Earth System History.* Freeman, 2015.

GLOBAL CLIMATE

FIELDS OF STUDY

Climate Science; Climate Classification; Climate Zones; Climatology; Bioclimatology; Hydroclimatology; Climate Modeling; Meteorology; Hydrometeorology; Atmospheric Science; Oceanography; Hydrology; Fluid Dynamics; Earth System Modeling; Heat Transfer; Thermodynamics; Physical Geography

SUMMARY

Climate is a general characterization of long-term weather and environment conditions for a specific location. Several major factors influence climate in a given region, including latitudinal position, the distribution of land and water, and elevation. Ocean currents, prevailing winds, and the positions of high- and low-pressure areas also have significant climatic effects.

PRINCIPAL TERMS

- **Chinook/foehn wind:** a warm, dry wind on the eastern side of the Rocky Mountains or the Alps
- **climate controls:** the relatively permanent factors that govern the general nature of the climate of a region
- **evaporation:** the process by which a liquid changes into a gas
- **Hadley circulation:** an atmospheric circulation pattern in which a warm, moist air ascends near the equator, flows poleward, descends as dry air in subtropical regions, and returns toward the equator
- **Inter-Tropical Convergence Zone (ITCZ):** a low-pressure belt, located near the equator, where deep convection and heavy rains occur
- **Köppen climate classification system:** a system for classifying climate based mainly on average temperature and precipitation
- **monsoon:** a seasonal climate system characterized by wind and precipitation patterns
- **precipitation:** liquid or solid water particles that fall from the atmosphere to the ground
- **rain shadow:** the region on the lee side of a mountain where precipitation is noticeably less than on the windward side
- **subtropical high belt:** a high-pressure belt where warm, dry air sinks closer to the surface
- **transpiration:** the process by which water in plants is transferred as water vapor into the atmosphere

CLIMATE FACTORS

Heterogeneous distributions of heat and water result in rich and varied climates. In particular, the tropical regions receive more energy from solar radiation than they emit in the form of infrared heat. The polar regions, by contrast, receive less energy from the Sun than they emit as heat. As a result, the tropics are regions of heat surplus, while polar regions are deficient in heat. Moreover, because tropical regions include large expanses of ocean, there is more water stored in the tropical atmosphere than is stored in the atmosphere at high latitudes.

The imbalance in the heat and water budget in the tropical and polar regions produces circulation that transports heat and water from and to these regions. These transports are carried out by both atmosphere and oceans. The weakening or strengthening of heat and water transports is an important signal for climate change.

Several methods have been developed to classify global climate. The most widely used method is based on the Köppen classification system. Designed by the German climatologist Wladimir Köppen (1846–1940), this method uses the average annual and monthly temperature and precipitation to describe a global climate for various climate zones. In this method, global climate is divided into six major groups, and each group is divided into subgroups.

TROPICAL MOIST CLIMATE (GROUP A)

Tropical moist climate is typical of most of Earth's tropical regions (from the equator to about 20° north and south latitudes). The climate of these regions is characterized by year-round warm temperatures and abundant rainfall. In the tropics, the annual mean temperature is typically above 18°C, and typical annual average rainfall exceeds 150 centimeters. Tropical moist climate is divided based on rainfall characteristics into three subtypes of climate: tropical wet, or tropical rain-forest climate (Af); tropical monsoon climate (Am); and tropical wet-and-dry climate (Aw).

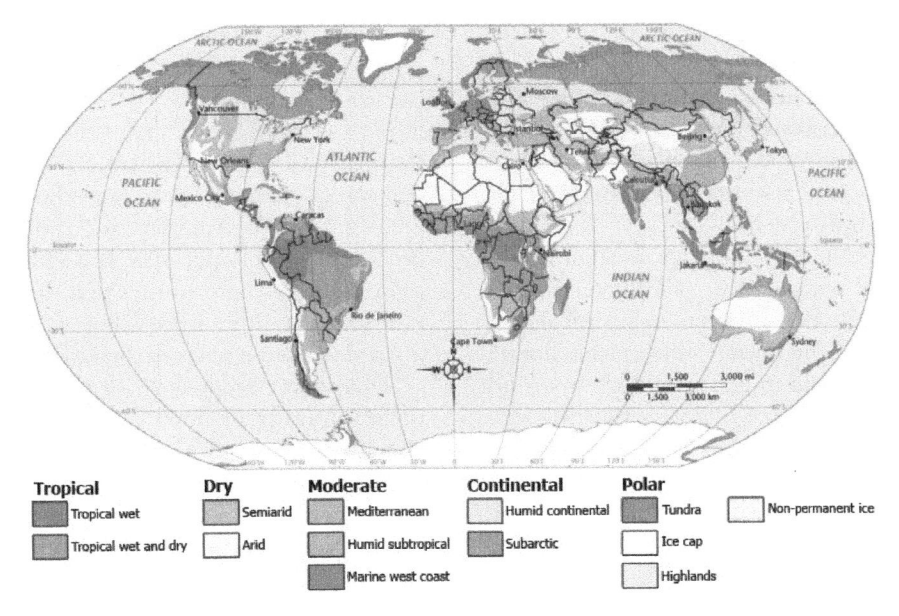

Climate zones of the world.

The tropical rain-forest climate exhibits constant high temperatures and abundant year-round rainfall. As a result, it is marked by dense vegetation, typically composed of broadleaf trees, jungles, and evergreen forests. A large number of diversified plants, insects, birds, and animals inhabit the tropical rain forests. Many lowlands near the equator are in this type of climate, which includes the Amazon River Basin of South America, the Congo River Basin of Africa, and the East Indies, from Sumatra to New Guinea.

Unlike the tropical wet climate, the tropical wet-and-dry climate has distinct wet and dry seasons. Although the annual precipitation usually exceeds 100 centimeters, during the dry season the average monthly rainfall can be less than 6 centimeters. The dry season lasts more than two months. The tropical wet-and-dry climate dominates most of tropical Africa, tropical South America, and South Asia. The variations of dry and wet seasons in these regions are closely associated with the migration of the Inter-Tropical Convergence Zone (ITCZ) in the tropics.

The tropical monsoon climate exists between the tropical rain-forest and wet-dry climates. It has abundant rainfall, in excess of 150 centimeters per year, but the rains do stop briefly, typically for one or two months. Tropical monsoon climate can be seen along the coasts of Southeast Asia and India and in northeastern South America. In contrast to the wet-dry climate, the rain and the pause of rain in these areas are related to monsoonal circulation.

DRY CLIMATE (GROUP B)

Just outside the tropics, most of the continental land located between approximately 20° and 30° latitude in both the Northern and the Southern Hemispheres is in arid or semiarid climates. Precipitation in these areas is scarce most of the year, and evaporation and transpiration exceed precipitation.

The arid climate (BW) is the true desert climate and can be found in the Sahara Desert in Africa, a large portion of the Middle East, much of the interior of Australia, Central Asia, and the west coasts of South America and Africa. These areas are located in the subtropical high belt, which is governed by descending air from the Hadley circulation.

Around the margins of the arid regions, semiarid (BS) areas enjoy a slightly greater rainfall. The light rains of semiarid climes support the growth of short bunch grass, scattered low bushes, trees, and sagebrush. This climate can be found in the western United States, southern Africa, and the Sahel.

THE MOIST SUBTROPICAL MIDLATITUDE CLIMATE (GROUP C)

Most subtropical midlatitude regions are farther poleward from the major dry-climate latitudes.

These areas extend approximately from 25° to 40° latitudes in both the Northern and the Southern Hemispheres. This climate has distinct summer and winter seasons. Winter is mild, with average temperatures for the coldest month of between -3°C and -18°C. The regions in this climate belt are typically humid and have ample precipitation.

There are three major subtypes in the group C climate: humid subtropical (Cfa); west coast marine (Cfb); and dry-summer subtropical, or Mediterranean (Cs). The humid subtropical climate typically presents hot and muggy summers, but mild winters. Summers experience heavy rains, while winters are slightly drier. This climate type can be found principally along the east coasts of continents, such as the southeastern United States, eastern China, southeastern South America, and the southeastern coasts of Africa and Australia.

The west coast marine climate has cool summers and mild winters and produces more precipitation in winter than in summer. The largest area with this climate is Europe. Finally, the dry-summer or Mediterranean climate is distinctively characterized by extreme summer aridity and heavy rains in winter. Countries surrounding the Mediterranean Sea and the U.S. West Coast, including Northern California and Oregon, are in this type of climate.

Moist Continental Climate (Group D)

The moist continental climate is located farther north of the moist subtropical midlatitude climate zone, from 40° to 50° north latitudes. This climate mostly occurs in North America and Eurasia. The general characteristics of the moist continental climate are warm-to-cool summers and cold winters. The average temperature of the warmest month exceeds 10°C, and the coldest month's average temperature generally drops below -3°C. Winters are severe, with snowstorms, blustery winds, and bitter cold. The climate is controlled by a large landmass.

The group D climate is further divided by summer temperature into three major subtypes: humid continental with hot summers (Dfa), humid continental with cool summers (Dfb), and subpolar (Dfc). Both winter and summer temperatures in the Dfa climate are relatively severe. That is, winter is cold and summer is hot. Farther north is the Dfb climate, which experiences long, cool summers and long, cold, windy winters. The subpolar climate presents severely cold winters and short summers. In the subpolar region, moisture supply is limited. Therefore, precipitation is low.

Polar Climate (Group E)

The polar climate exists over the northern coastal areas of North America and Eurasia, Greenland, the Arctic, and Antarctica. It is characterized by low temperatures year-round. Even during the warmest month, the temperature is below 10°C. Precipitation is scarce in these parts of the Earth.

The polar climate is divided into two subtypes: the polar tundra (ET) climate and the polar ice-cap (EF) climate. The tundra climate occupies the coastal fringes of the Arctic Ocean, many Arctic islands, and the ice-free shores of northern Iceland and southern Greenland. In these regions, the ground is permanently frozen to depths of hundreds of meters, a condition known as permafrost. In summer, the temperature can remain above freezing, allowing tundra vegetation to grow in the thin layer of unfrozen soil that develops atop the permafrost layer. The monthly mean temperature under the ice-cap climate is mostly below 0°C. The ice cap occupies the interior ice sheets of Greenland and Antarctica. The growth of plants is prohibited, and the landscape is perpetually covered with snow and ice. Many studies show that these regions are most sensitive to global warming and have experienced rapid snow and ice melting in recent decades.

Highland Climates (Group H)

The distribution of global mountain ranges and plateaus creates another type of climate. Climate in highland regions is unique. Highland climates are characterized by a great diversity of conditions. Because air temperature decreases with altitude, climatic changes corresponding to those from group B to group E will be experienced when ascending mountain slopes. In general, every 300 meters of mountain elevation will correspond to a change of climate type.

In addition to the drop of temperature with increased altitude, orography modifies precipitation and wind patterns in many ways. For example, a mountain's windward slopes typically receive more precipitation than its leeward slopes. Therefore, more dense vegetation grows on the windward slopes of large mountains, such as the western slope of the

Rockies, than on the leeward slopes, such as the eastern slope of the Rockies. Often, the leeward foot of a mountain receives very little precipitation. These areas are often called "rain shadows." Leeward mountain foots are also subject to downslope mountain winds from time to time, especially during winters. These winds are called "Chinook wind" in North America, or "foehn wind" in Europe.

The most prominent highland climate occurs over the Tibetan Plateau, where the average elevation is over 4,000 meters. In North America, highland climates characterize the Rockies, Sierra Nevada, and Cascades ranges. In South America, the Andes range creates a continuous band of highland climate. Many of these mountains and highlands play central roles in monsoonal circulation, an important global climate system in various parts of the world.

Context

Global climate is a complex system that involves interactions among Earth's atmosphere, hydrosphere, cryosphere, lithosphere, and biosphere. In an even larger context, the global climate is just a part of the Sun-Earth system. For a particular place on Earth, the formation of the local climate pattern is dependent upon a set of climate controls. Despite its complexity, the global climate can be classified according to two basic physical parameters: mean temperature and precipitation. Future climate change can be measured and quantified by closely monitoring the change of these parameters in various parts of the world.

—*Chungu Lu*

Further Reading

Ahrens, C. Donald. *Essentials of Meteorology: An Invitation to the Atmosphere.* 5th ed. Thomson Brooks/Cole, 2008. One of the most widely used introductory books on atmospheric science; covers a wide range of topics on weather and climate.

Anderson, Michael *Investigating the Global Climate.* The Rosen Publishing Group, 2011. Geared to younger readers, this book nevertheless provides clear, definitive descriptions of the various segments of the global climate system that are a useful starting point for all levels.

Bridgman, H.A., Oliver, John E., Glantz, Michael, Corveny, Randall S. and Allan, Robert *The Global Climate System: Patterns, Processes and Teleconnections.* Cambridge University Press, 2014. The majority of this book describes the global climate system, while the remainder of the content examines social and economic aspects.

Lutgens, Frederick K., Edward J. Tarbuck, Tasa, Dennis G. and Herman, Redina. *The Atmosphere: An Introduction to Meteorology.* 14th ed. Pearson Education, 2018. An updated version of this textbook, in loose-leaf format, provides students with a general overview of the structure and properties of Earth's atmosphere.

Rodó, Xavier and Comín, Francisco A., eds. *Global Climate. Current Research and Uncertainties in the Climate System.* Springer Science + Business Media, 2013. The chapters in this book represent a curated series of articles that focus on individual segments of the global climate system.

GLOBAL CLIMATE COALITION (GCC)

Fields of Study

Climate Science; Ecosystem Management; Earth System Modeling; Meteorology

Summary

The Global Climate Coalition (GCC) is a now-defunct industry group that worked to set the climate change agenda of the United States. This organization was influential and successful in its endeavors. The GCC dissolved amid increased understanding in the field of climate science and the rise of public opinion in support of public policy to address climate change. The coalition met many of its goals prior to its dissolution.

Principal Terms

- **greenhouse gases (GHGs):** atmospheric gases whose molecules absorb infrared radiation from a planet's surface and then emit a significant portion of it back into the atmosphere, most commonly carbon dioxide, methane, water and ozone.

- **libertarian:** based on the principles of minimal government control and maximal personal and corporate freedom

Mission

The GCC was formed by representatives of multinational corporations reliant on the use of fossil fuels, particularly in the energy and transportation sectors. Prominent members of the GCC included BP, Royal Dutch Shell, DuPont, Ford Motor Company, Daimler Chrysler, Texaco, General Motors, and Exxon (now ExxonMobil). The mission of the GCC was to influence U.S. public policy regarding climate change.

The GCC was organized in reaction to the findings of the Intergovernmental Panel on Climate Change (IPCC). A GCC representative argued at the libertarian think tank the Cato Institute that the IPCC report was flawed and represented the views of pro-climate-change activists. The GCC rejected all government-mandated reductions of greenhouse gas (GHG) emissions in the United States. The GCC was against U.S. participation in the Kyoto Protocol. The coalition argued that climate science and the human impact on climate change were both too uncertain to justify coercive government policy intended to lessen GHG emissions. The GCC conducted lobbying and public relations campaigns to spread the skeptical view of climate change to the public.

Significance for Climate Change

The GCC was an influential factor in setting the U.S. climate change agenda. The GCC pressured the administration under President Bill Clinton to reject command-and-control environmental regulations intended to address climate change. The coalition argued that burdensome regulations designed to address climate change would decrease economic growth and hinder the economic competitiveness of U.S. companies. The GCC was an influential actor in preventing U.S. ratification of the Kyoto Protocol. In 1998, the Senate voted to reject the Kyoto Protocol 95–0.

The demise of the GCC began with the withdrawal of BP, a major multinational oil company. BP broke ranks with the GCC in May, 1997. The GCC lost members as increasing scientific evidence and rising consensus alarmed governments, and people around the world demanded action on climate change. In 2002, the GCC disbanded, stating, "At this point, both Congress and the Administration (Bush II) agree that the United States should not accept mandatory cuts in emissions required by the Protocol." The GCC claimed its mission had been accomplished, so there was no reason to maintain an active presence to shape the climate change agenda.

The skeptical view of climate change found representation in the administration under President George W. Bush. The administration took the view that climate science was uncertain and asked the National Academy of Sciences (NAS) for assistance in determining if IPCC reports had been tampered with by pro-climate-change-policy activists.

Although the GCC has dissolved as an organization, its activities have made a lasting impact in setting climate policy agenda in the United States. Many in the business sector continue in the present day to insist that there is no need to transform the global economy to address climate change. In June, 2008, National Aeronautics and Space Administration climatologist James E. Hansen (1941–) argued that corporate executives who promote contrarian arguments against climate change theory should be held liable for their actions. Hansen contends such activities are "crimes against humanity and nature" and that executives from many fossil-fuel-intensive corporations consciously attempt to confuse the public on the state of understanding and agreement among climate scientists.

—*Justin Ervin*

Further Reading

Pilkington, Ed. "Put Oil Firm Chiefs on Trial, Says Leading Climate Change Scientist." *The Guardian*, Monday, June 23, 2008. Discusses James E. Hansen's return to Congress after he first warned of climate change in 1988. Hansen claims industry groups have been consciously spreading misinformation on climate change that has slowed government reaction to address climate change.

Powell, James Lawrence *The Inquisition of Climate Science*. Columbia University Press, 2012. An exposé of the most prominent deniers of climate change, focusing on their lack of credentials, the extensive funding received from the industrial complex dependent upon carbon-based fuel consumption, and their inability to provide a valid alternate explanation for global warming.

Rahm, Diane *Climate Change Policy in the United States. The Science, the Politics, and the Prospects for Change.* McFarland, 2009. Provides an overview of the issue of global warming and its anthropogenic causes in relation to climate policy in the United States.

Shearer, Christine *Kivalina: A Climate Change Story.* Haymarket Books, 2011. This book presents a view of the "other side" of climate change, the impacts climate-change denial policies have on real people.

Skjaerseth, Jon Birger and Skodvin, Tora *Climate Change and the Oil Industry: Common Problems, Varying Strategies.* Manchester University Press, 2013. While climate change denial still exists, this book describes the internal corporate practices of some major oil companies intended to minimize or eliminate whatever global climate change effects that may result from their activities.

Taylor, Maria *Global Warming and Climate Change. What Australia Knew and Buried...Then Framed a New Reality for the Public.* Australian National University, 2014. This book describes the manner in which political policies and actions in Australia were used to manipulate public knowledge and awareness of the issue of climate change in Australia and around the world.

GLOBAL ECONOMY AND CLIMATE CHANGE

FIELDS OF STUDY

Economics; Environmental Sciences; Statistics; Climate Modeling; Computer Science

SUMMARY

The global economy is the result of a process of increasingly global economic integration that began in the sixteenth century. European powers spread capitalism to colonies that provided cheap labor (including slaves), abundant natural resources, and vital new markets for goods and services. The expansion of European capitalism led to the export of industrialization in the nineteenth and twentieth centuries. The United States worked to reorganize the global economy following the instability of the early twentieth century that culminated with World War II. This reorganization was negotiated at Bretton Woods, New Hampshire, in 1944. The International Monetary Fund (IMF) and the World Bank were created at the Bretton Woods Conference. The General Agreement on Tariffs and Trade (GATT) was signed in 1947. Together, these institutions and agreements were designed to include as much of the world's people and territory as possible under a global capitalist economy. The Soviet Union and China limited this vision at the time. Members of the former Soviet Union and China have since become vital actors in the global economy.

PRINCIPAL TERMS

- **capitalism:** an economic system in which the means of production are privately owned and operated with the goal of increasing wealth
- **commodity:** anything that has commensurable value and can be exchanged
- **economic interdependency:** a state of affairs in which the economic processes of a group of nations are mutually dependent
- **globalization:** the worldwide expansion and consequent transformation of socioeconomic interrelationships
- **industrialization:** the process of transformation from an agrarian society based on animal and human labor to an industrial society based on machines and fossil fuels
- **liberal institutionalism:** a school of thought that focuses on cooperation between countries derived from agreements and organizations
- **neoliberalism:** a school of economic thought that stresses the importance of free markets and minimal government intervention in economic matters

ECONOMIC GLOBALIZATION

The global economy represents a central process of globalization. As such, it increases economic interdependency. Economic globalization is made possible through advances in information and transportation

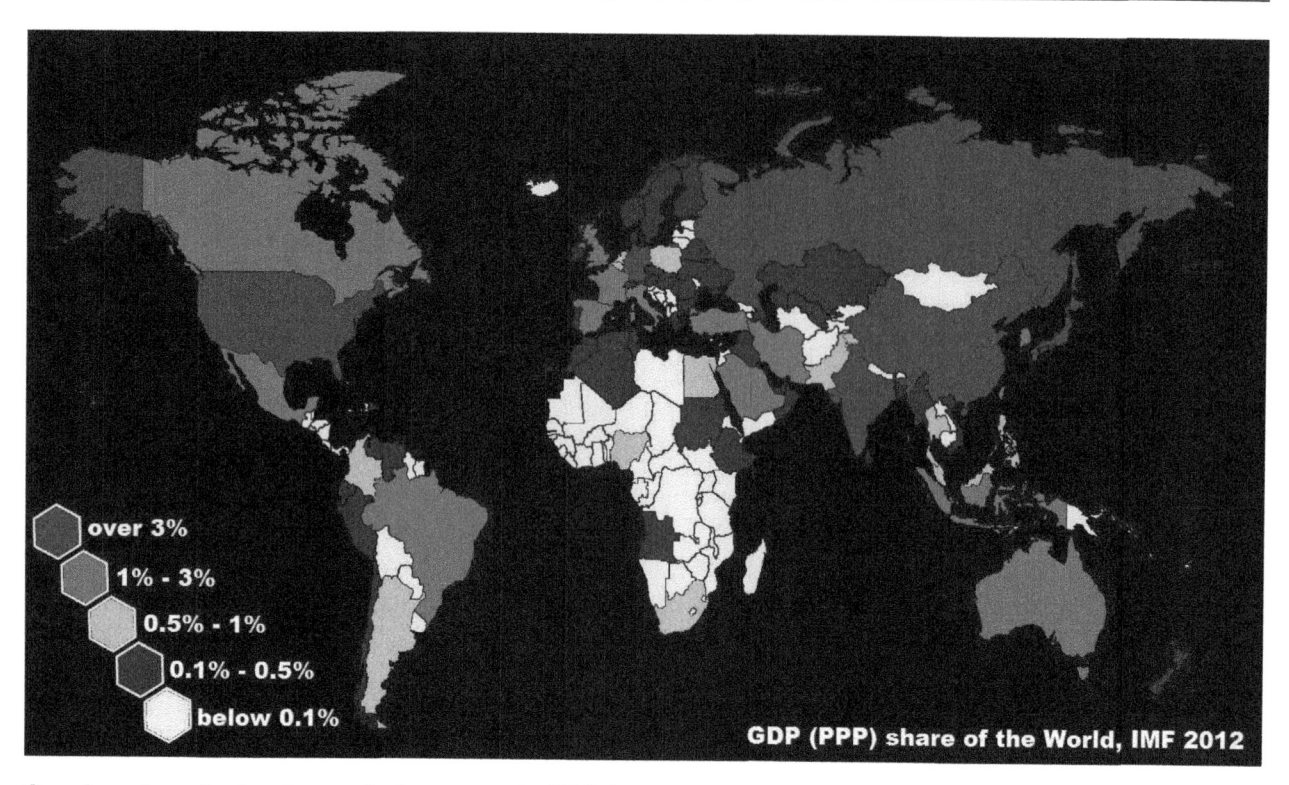

Gross domestic product based on purchasing-power-parity (PPP) share of world total in 2012.

technology. With the rise of technology, many of the world's countries have become interconnected through complex networks of economic production and consumption.

Economic globalization is composed of many actors. Countries legalize rules regarding international economic transactions. Multinational corporations dominate international economic production and the global trade in goods and services. International financial institutions finance the global flow of goods and services and provide money capital for foreign investment.

THE GLOBAL ECONOMY AND CLIMATE CHANGE

The spread and growth of the global economy increases the energy intensity of the world's countries. The increase in the volume of exchange of goods and services, over greater distances, has increased the use of fossil fuels. More people drive more kilometers in automobiles and fly longer distances in airplanes. More people use increasing amounts of electricity, much of which continues to be produced by the combustion of fossil fuels. The on-going expansion of global economic activity thus requires a continuing increase in the consumption of fossil fuel. The consumption of fossil fuel is believed to be a major factor causing climate change.

Economic globalization increases the human transformation of the environment. Deforestation results from the human need for lumber for construction and land for agriculture. Rising living standards lead to increased construction and increasing consumption of meat from cattle raised on land that has been deforested for the purpose of cattle ranching. Deforestation decreases the Earth's ability to remove carbon dioxide from the atmosphere. Increased meat consumption increases the concentration of the greenhouse gas (GHG) methane in the atmosphere.

Problems associated with the global economy and climate change have gained the attention of powerful actors within the global economy. The IMF and the World Economic Forum (WEF) agree that climate change, if left unchecked, is likely to destabilize the global economy. A report, produced by the White House in 2014, announced that climate change would likely cause a 40 percent decrease in global GDP. This could lead to a global economic depression and violent conflict. Climate change could dissolve global economic networks, creating shortages of vital economic inputs, leading to global economic decline.

Political Economic Theory and Climate Change

The impact the global economy has on climate change is addressed by theories of political economy. One's adherence to a particular theory greatly impacts the way the relationship between the global economy and climate change is interpreted, and reflect the catastrophist-cornucopian debate. For example, neoliberal economic theorists argue that global markets will distribute the technologies needed to address climate change, a cornucopian point of view. New technologies, such as wind generators, photovoltaic solar panels, hybrid automobiles, and fuel cells will circulate across the globe under free market capitalism. Liberal institutionalist theorists agree with neoliberals about technological transfer. They argue, however, that the global economy requires active public management to address climate change rather than just reliance on the development of new technologies, the catastrophist point of view. Liberal institutionalists cite the importance of cooperation among countries to address climate change. This cooperation is best realized in the form of international governmental organizations and agreements, such as the Kyoto Protocol.

Some theorists argue the global economy is unsustainable. These theorists propose dramatic transformations for the economy. Ecological economists argue environmental problems such as climate change are symptoms of the Earth no longer being able to assimilate human economic activity. Ecological economists argue that the global economy is unsustainably depleting Earth's natural capital at an ever-increasing rate. This condition cannot last indefinitely, because Earth is a finite system. Ecological economists argue the global economy must attain an optimal scale or face devastating consequences.

Context

The global economy has created unprecedented opportunities and problems for humanity. The global economy has created unprecedented wealth for some, but it has also increased social instability and impoverishment for others while contributing to environmental problems such as climate change. Climate change transcends the ability of individual countries to create solutions. To confront climate change, countries of the world will have to cooperate at unprecedented levels. Rich countries will have to promote policies that help poor countries address climate change. Those living in affluent countries and enjoying energy-intensive lifestyles cannot justify demanding economic sacrifices from those in impoverished circumstances.

Humanity faces a serious economic paradox with climate change. In order to remain stable, the global economy must grow. However, to address climate change, this economic growth must be achieved while diminishing the factors responsible for climate change. Given a global economy that still relies heavily on the combustion of fossil fuels, this will be no easy feat. The global economy and climate change are interconnected but contradictory functions. As the global economy grows, the dangers of climate change increase. As the dangers of climate change increase, global economic growth is threatened. This relationship is critically important and offers no easy solutions.

—*Justin Ervin*

Further Reading

Bahr, Simon *How Does Climate Change Affect Global Economy?* GRIN Publishing, 2017. An essay examining the natural and anthropogenic causes of climate change and the ramifications for the global economy.

Freytag, Andreas, Kirton, John J., Sally, Razeen and Savona, Paolo, eds. *Securing the Global Economy. G8 Global Governance for a Post-Crisis World*. Routledge, 2016. The G* are the eight most affluent nations in the global economy, and this book explores the reasons why the G8 have an ethical responsibility and the ability to address the complex issues of global climate change in order to maintain and develop a stable global economy.

Gilding, Paul *The Great Disruption: How the Climate Crisis Will Transform the Global Economy*. A&C Black, 2012. While not offering more than basic solutions for complex problems related to global climate change, this book nevertheless identifies many of those problems.

Mander, Jerry and Goldsmith, Edward, eds. *The Case Against the Global Economy and For a Turn Towards Localization.* 2nd ed. Routledge, 2014. The argument is made in this book that the globalized economy is fueling anthropogenic climate change with increasingly catastrophic consequences.

Vonnegut, Andrew *Inside the Global Economy; A Practical Guide.* Rowman & Littlefield, 2018. The first eight chapters of this book identify and describe as many major aspects of the global economy. The next five chapters identify, describe and hypothesize how shifts in underlying aspects such as demographics, technology and global climate might affect the global economy.

GLOBAL MONITORING

FIELDS OF STUDY

Atmospheric Science; Atmospheric Chemistry; Spectroscopy; Mass Spectrometry; Optics; Oceanography; Hydrology; Physical Geography; Photochemistry; Environmental Sciences; Environmental Studies; Waste Management; Meteorology; Climatology: Climate Modeling; Statistics; Process Modeling; Hydroclimatology; Hydrometeorology; Bioclimatology; Climate Classification; Climate Zones; Ecosystem Management; Earth System Modeling

SUMMARY

The monitoring of the climate and related variables on global and regional scales is essential if correct analyses of what has happened, and what is happening, to the atmospheric climate are to be made. For example, it is often mentioned, especially by the media, that the climate is getting warmer, or cooler, or drier, or wetter, or more humid, or windier, and so on. What the true situation is, however, cannot be determined without rigorous and ongoing collection and examination of data. For example, are the polar ice areas increasing or deceasing, and are glaciers advancing or retreating? Are these trends different from what happened a few decades ago? The correct monitoring of the global climate and related variables is clearly necessary to advance our knowledge of the true situation.

PRINCIPAL TERMS

- **global:** relating to or encompassing the whole Earth
- **homogeneous climate data:** a sequence of values of a climate variable, such as precipitation, which have been observed under the same or similar conditions and with the same or similar measuring equipment; the combination of climate data from two localities that are near each other is often made when considering climate change
- **monitoring:** the systematic observation of an element such as precipitation, sea surface temperatures, or wind speed; such observations are usually made every six hours and sometimes every three hours, hourly, or (conversely) only once daily
- **observing systems:** systems of collectively gathering and analyzing temperature and other atmospheric observations, particularly (in modern times) through the World Meteorological Organization
- **regional:** relating to areas such as the Pacific, the Atlantic, the tropics, and large land areas, such as North America or Australia

GLOBAL CLIMATE MONITORING SYSTEM

Temperature and other atmospheric observations have been made in many parts of the world for more than a century and in some places for more than two hundred years. Initially these observations were made by single entities, but during the last hundred years, particularly during the last fifty years, there has been an effort to centralize and consolidate these data.

The World Meteorological Organization (WMO) and its forerunner, the International Meteorological Organization (IMO), have, for more than one hundred years, been at the forefront of organizing research on and monitoring the world's climate. In particular, since 1992 the Global Climate Observing System (GCOS) has supported this research

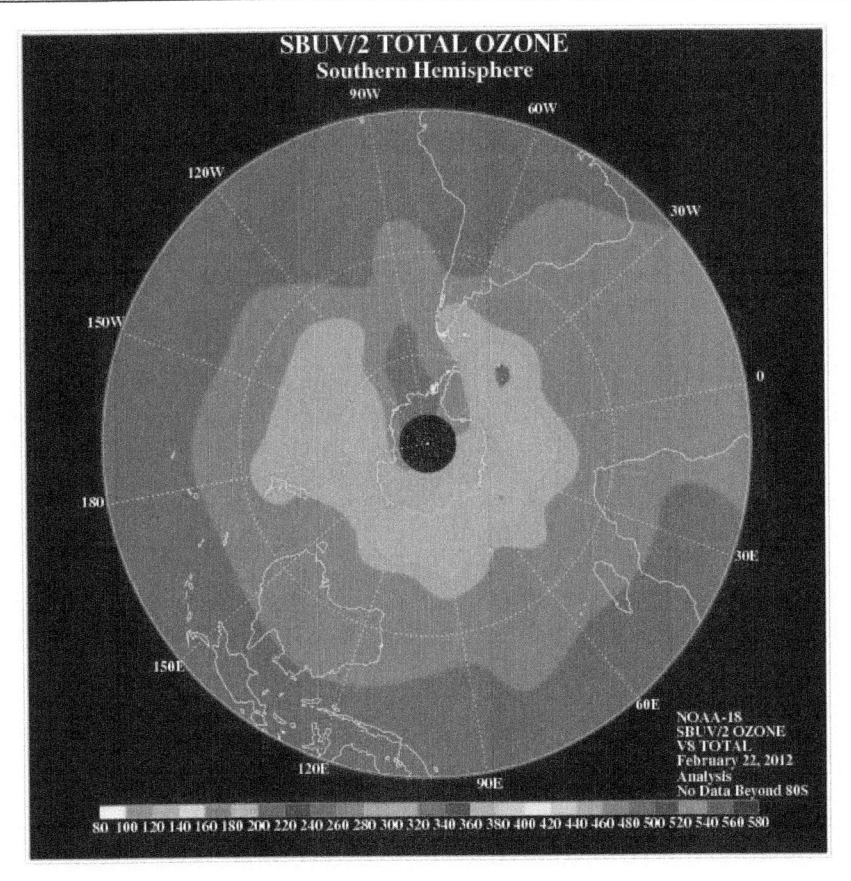

This image shows ozone concentration in the Southern Hemisphere on February 22, 2012.

according to several key principles: operation of historically uninterrupted stations and observing systems should be maintained; high priority for additional observations should be focused on data-poor regions, poorly observed parameters, and regions sensitive to change; and operators of satellite systems for monitoring climate need to sample the Earth system in such a way that climate-relevant (diurnal, seasonal, and long-term interannual) changes can be resolved.

Regional Monitoring

A good example of "regional monitoring" is the South Pacific Sea Level and Climate Monitoring Project (SPSLCMP), which was developed in 1991 as the Australian government's response to concerns raised by member countries of the South Pacific Forum over the potential impacts of human-induced global warming on climate and sea levels in the Pacific region. The first three phases of the project established a network of twelve high-resolution Sea Level Fine Resolution Acoustic Measuring Equipment (SEAFRAME) sea-level and climate-monitoring stations throughout the Pacific.

U.S. National Climate Center

Another example of "regional monitoring" is the Climate Prediction Center (CPC) of the National Weather Service (part of the National Oceanographic and Atmospheric Administration, NOAA). The CPC collects and produces daily and monthly data, time series, and maps for various climate parameters, such as precipitation, temperature, snow cover, and degree days for the United States, the Pacific islands, and other parts of the world. The CPC also compiles data on historical and current atmospheric and oceanic conditions, the El Niño-Southern Oscillation (ENSO) Index, and other climate patterns, such as the North Atlantic and Pacific Decadal Oscillations, as well as stratospheric ozone and temperatures.

U.K. Climate Research Unit

A significant global center for compiling temperature and other climate data sets is the Climate Research Unit (CRU) of the University of East Anglia in the United Kingdom. Some of the data produced are available online, and other sets are available on request. CRU endeavors to update the majority of the data pages at timely intervals. Data sets are available in the following categories: temperature, precipitation, atmospheric pressure and circulation indices, climate indices for the United Kingdom, data for the Mediterranean and alpine areas, and high-resolution grid data sets.

World Data Center System

A good example of a "global center" is the World Data Center (WDC) system, which was created to archive and distribute data collected from the observational programs of the 1957-1958 International Geophysical Year (IGY). Originally established in the United States, Europe, Russia, and Japan, the WDC system has since expanded to fifty-two centers in twelve countries. Its holdings include a wide range of solar, geophysical, environmental, and human dimensions data. The WDC is maintained by Model and Data (M&D), which is hosted at the Max Planck Institute for Meteorology, located in Germany.

Context

In terms of climate change, the monitoring of the climate is extremely important. A number of internationally recognized groups collect, analyze, and publish, mainly through Web sites, climate and other data for various areas of the world, including the globe as a whole, the land as a whole, the ocean as a whole, and regions such as the tropics and polar areas. In addition to the climate variables, a wide variety of other variables are monitored on a global or regional scale, such as the extent of sea ice, sea surface temperatures, carbon dioxide, and methane. All of these values, when correctly analyzed, provide important indicators of what changes have occurred during the past (weeks to many decades) in the broader atmospheric environment. Such information is essential if correct answers to questions relating to the changing climate are to be obtained.

—*W. J. Maunder*

Further Reading

Kininmonth, William. *Climate Change: A Natural Hazard.* Multiscience, 2004. Kininmonth has a career in meteorological and climatological science and policy spanning more than forty-five years. His suspicions that the science and predictions of anthropogenic global warming extend beyond sound theory and evidence were crystallized following the release of the 2001 IPCC assessment report. His book gives information about global and regional monitoring on various scales.

National Aeronautics and Space Administration. "Solar Physics." Available at http://solarscience.msfc.nasa.gov/SunspotCycle.shtml. Presents data on the solar cycle, the Maunder Minimum, and other Earth-Sun interactions.

Singer, S. Fred, and Dennis T. Avery. *Unstoppable Global Warming: Every Fifteen Hundred Years.* Rowman and Littlefield, 2008. This book is dedicated to those thousands of research scientists who have documented evidence of a fifteen-hundred-year climate cycle over the Earth. Refers throughout to various aspects of monitoring of the global climate.

University of Colorado at Boulder. "Sea Level Change." Available at http://sealevel.colorado.edu/results.php. Presents tables, maps, time series, and other data on global sea level.

University of East Anglia. "Climatic Research Unit." Available at http://www.cru.uea.ac.uk/. The CRU presents data, information sheets, and the online journal *Climate Monitor*.

University of Illinois. "The Cryosphere Today." Available at http://arctic.atmos.uiuc.edu/cryosphere/. Offers frequently updated data on the current state of Earth's cryosphere.

U.S. National Space Science and Technology Center. Available at http://www.nsstc.org/. The mission of the NSSTC, an arm of NASA, is "to conduct and communicate research and development critical to NASA's mission in support of the national interest, to educate the next generation of scientists and engineers for space-based research, and to use the platform of space to better understand our Earth and space environment and increase our knowledge of materials and processes."

GLOBAL POSITIONING SYSTEM (GPS)

FIELDS OF STUDY

Engineering; Environmental Studies; Physical Geography; Earth System Modeling

SUMMARY

GPS was originally intended for military applications to accurately determine locations worldwide in all kinds of weather. In the 1980s, the U.S. government made the system available for civilian use. GPS is used as a navigation and positioning tool in transportation, such as fleet cars and commercial trucking, in surveying, and for almost all outdoor recreational activities. In the scientific community, GPS plays an important role in geology, meteorology, wildlife studies, archeology, and many other areas.

PRINCIPAL TERMS

- **geoid:** a dynamic three-dimensional virtual model of Earth in which all points on the "surface" are at sea level
- **ionosphere:** portions of the upper atmosphere consisting of part of the mesosphere, thermosphere, and exosphere; characterized by gas ionization through exposure to solar radiation
- **pseudorandom:** a series of random numbers or code signals generated by a non-random computational method or algorithm
- **trilateration:** identification of a singular point on a curved surface by the intersection of three lines, analogous to triangulation on a plane surface
- **troposphere:** the lowest layer of Earth's atmosphere extending from the surface to the tropopause, the transition layer to the mesosphere, characterized by decreasing temperature with increasing altitude

JUST GPS...PLEASE

The GPS Navigation Satellites Wireless communication Communications global positioning system (GPS) is a satellite-based navigation system comprised of a network of satellites placed into orbit by the U.S. Department of Defense in 1973. GPS was originally intended for military applications to accurately determine locations worldwide in all kinds of weather. In the 1980s, the U.S. government made the system available for civilian use. GPS is used as a navigation and positioning tool in transportation, such as fleet cars and commercial trucking, in surveying, and for almost all outdoor recreational activities. In the scientific community, GPS plays an important role in geology, meteorology, wildlife studies, archaeology, and many other areas. Mathematics was critical in the development of this system and mathematicians work on many ongoing issues, such as precision and error correction.

Global positioning systems have been made available to the private sector but depend on satellites originally placed into orbit for military purposes and require precise calculations.

There are three parts that form the GPS: the space segment (satellites), the
user segment (the receiver), and the control segment (control stations). The control segments are on the geoid (a three-dimensional model of Earth). The first segment of the system consists of a constellation of satellites, orbiting 20,000 kilometers above Earth in 12-hour circular orbits. While the exact number of satellites in operation varies at any given moment, at least six groups of four satellites are necessary to ensure that they can be detected from anywhere on Earth's surface. Each group is assigned a different path, creating six orbital planes that completely surround Earth.

TRILATERATION

The satellites transmit signal information to Earth. GPS receivers take this information and use trilateration to calculate the user's exact location. Each satellite continuously transmits a data stream containing orbit information, equipment status, and the exact time. GPS receivers contain computer chips that then calculate the difference between the time a satellite sends a signal and the time it is received. The unit multiplies this time of signal travel by the speed of travel to get the distance between the GPS receiver and the satellite. Since these are radio waves, the speed used is the speed of light. One satellite gives a sphere on which the receiver sits. Two satellites give two spheres on which the receiver sits. The intersection of two spheres (and they must intersect) is a circle. Adding a third satellite gives the receiver one of two points at which the sphere will intersect

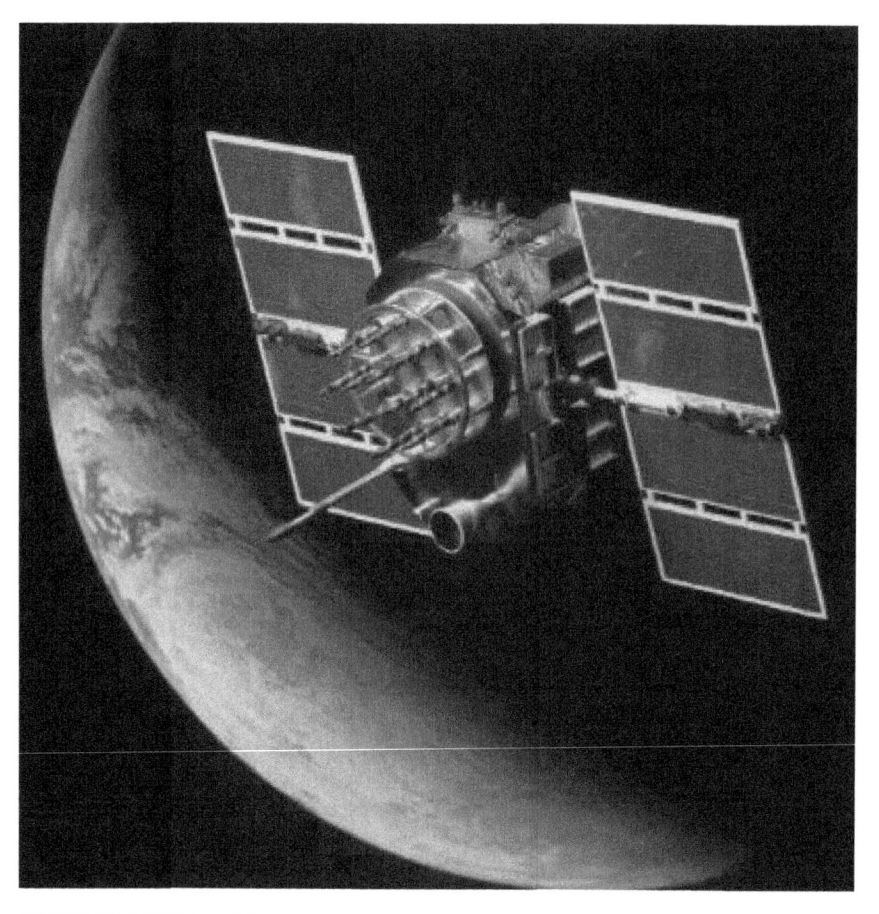

NAVSTAR-2 (GPS-2) satellite.

the circle. Using the geoid as the fourth solid, the receiver fixes the point of location. Despite this, there is still some possibility for error if the clock on the receiver has a slight error. A clock error of only one-thousandth of a second causes a position error of almost 200 miles. The solution is to use geometry. If one more satellite is added, then even if the clock in the receiver is off, it is off for all of the satellites by the same amount. The receiver lies on a line from each of the satellites. If all clocks are exact, then the receiver will sit at the intersection of the lines. However, the error in the receiver clock will cause the lines to intersect in different points, resulting in a polygon surrounding the receiver. The receiver can be calculated to be at the center of this polygon.

GPS Capabilities and Accuracy

A GPS receiver must be locked on to the signal of at least three satellites to calculate the latitude and longitude and to track movement. With four or more satellites, the receiver can determine the user's latitude, longitude, and altitude. Once the user's position has been determined, the GPS unit can calculate other information, such as speed, bearing, track, trip distance, distance to destination, sunrise and sunset times, and more. Most GPS receivers are accurate to within 15 meters on average. Newer GPS receivers often come with wide-area augmentation system (WAAS) capability that can improve accuracy to less than three meters on average. No additional equipment or fees are required to take advantage of WAAS. Users can also get better accuracy with differential GPS (DGPS), which corrects GPS signals to within an average of three to five meters. The U.S. Coast Guard operates the most common DGPS correction service. This system consists of a network of towers that receive GPS signals and transmit a corrected signal by beacon transmitters. In order to get

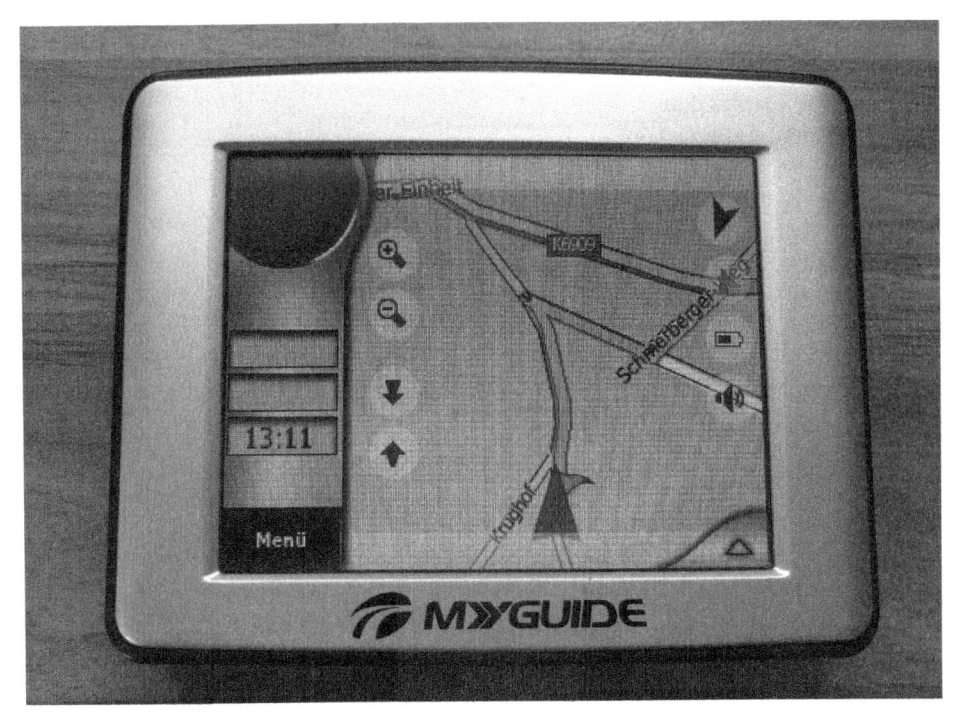

Mobile navigation system.

the corrected signal, users must have a differential beacon receiver and beacon antenna in addition to their GPS.

Possible sources of error include the following:

Ionosphere and Troposphere Delays. Different layers of the atmosphere have different impacts on the speed of the satellite signal through those layers. Mathematicians have been working on creating better models of these atmospheric layers in order to give smaller errors.

Geoid Error. The receiver uses a mathematical model of the surface of Earth, the geoid. Better mathematical models can improve the accuracy as long as they are relatively easy to use in computation.

Signal Multipath. The GPS signal may be reflected off objects, increasing the travel time of the signal, thereby causing errors. Mathematicians are working on developing models to account for multipath based on the relative location of receiver.

Orbital Errors. Inaccuracies in the satellite's reported location are handled by the control segment, which tries to keep each satellite on track.

Number of Satellites Visible. If only three satellites are visible, the receiver gives a position with a warning that it is likely to be very inaccurate.

Satellite Geometry/Shading. Differences in the relative position of the satellites at any given time may cause errors. Ideal satellite geometry exists when the satellites are located at wide angles relative to each other. Poor geometry results when the satellites are located in a line or in a tight grouping.

Intentional Degradation of the Satellite Signal. Selective Availability (SA) is an intentional degradation of the signal previously imposed by the U.S. Department of Defense. SA was intended to prevent military adversaries from using the highly accurate GPS signals. The government turned off SA in May 2000, which significantly improved the accuracy of civilian GPS receivers.

GPS SIGNAL TRANSMISSION

GPS satellites transmit two low-power radio signals, designated "L1" and "L2." Civilian GPS uses the L1 frequency of 1575.42 MHz in the UHF band. A GPS signal contains three different bits of information: a pseudorandom code, ephemeris data, and almanac data. The pseudorandom code is simply an identification code that identifies which satellite is transmitting information. Ephemeris data, which are constantly transmitted by each satellite, contain important information about the status of the satellite (healthy or unhealthy), current date, and time. The almanac data tell the GPS receiver where each GPS satellite should be at any time throughout the day. Each satellite transmits almanac data showing the orbital information for that satellite and for every other satellite in the system.

—David Royster

FURTHER READING

Cooke, D. *Fun with GPS*. ESRI Press, 2005.
Kaplan, Elliot D., and Christopher Hegarty, eds. *Understanding GPS: Principles and Applications*. 2nd ed. Artech House, 2005.
Hinch, Stephen W. *Outdoor Navigation with GPS*. Wilderness Press, 2011.
Levitan, Ben. *GPS Quick Course: Systems, Technology and Operation*. Althos, 2007.

GLOBAL SURFACE TEMPERATURE

FIELDS OF STUDY

Physical Geography; Ecology; Environmental Sciences; Environmental Studies; Earth System Modeling; Climate Modeling; Meteorology; Heat Transfer; Atmospheric Sciences; Ecosystem Management

SUMMARY

Global surface temperature generally indicates the estimated average air temperature at sea level on Earth's surface. Both land-based measurements and sea surface measurements are used, since nearly three-quarters of Earth's surface is water rather than land. Monitoring of the average temperature has indicated a consistent rising trend since the recording of temperatures was begun more than a century ago.

PRINCIPAL TERMS

- **Celsius degree:** the standard unit of temperature measurement on the Celsius scale, for which the freezing and boiling points of pure water are defined as 0° and 100°, respectively
- **global:** relating to or encompassing the whole Earth
- **mean:** the sum of the individual values of a series of measurements, divided by the number of measurements
- **urban heat island:** a spot on Earth's surface that is significantly warmer than the surrounding area as a result of human alterations to the landscape, primarily cities

GLOBAL SURFACE TEMPERATURE IN CONTEXT

Global surface temperature is an estimate of global mean air temperature at Earth's surface, based on thermometer measurements made at land-based weather stations. Because about 70 percent of the Earth's surface is covered by water, land-based data are supplemented by sea-surface-temperature (SST) measurements. Estimates of air temperature are based on the assumption that there is a simple link between the SST and that of the air above. Usually, SST measurements are based on measurements of the temperature of seawater that is taken aboard ships for use as an engine coolant. In the past, SST measurements were made of water taken from buckets tethered to ropes and thrown overboard. Supplementary SST data are gathered from data buoys, small island stations, and shipboard, nighttime measurements of marine air temperatures.

SIGNIFICANCE FOR CLIMATE CHANGE

Three agencies have taken responsibility for the combined global surface temperature record: the Climate Research Unit of the University of East Anglia; the U.S. National Aeronautics and Space Administration's Goddard Institute for Space Studies; and the Global Historical Climate Network of the U.S. National Oceanographic and Atmospheric Administration. To determine global

Deviations in Mean Global Surface Temperature

The following table lists deviations in average global surface temperature from the baseline temperature average set during the period between 1951 and 1980.

Year	Deviation (in 0.01° Celsius)	Year	Deviation (in 0.01° Celsius)
1880	−12	1960	−2
1890	−21	1970	+4
1900	−6	1980	+28
1910	−21	1990	+48
1920	−17	2000	+42
1930	−4	2005	+76
1940	+14	2008	+54
1950	−17		

Data from Goddard Institute for Space Studies, National Aeronautics and Space Administration.

surface temperature changes over time, these agencies locate and analyze anomalous departures from thirty-year temperature averages. These analyses are most commonly based on the area-weighted global average of sea-surface-temperature anomalies and land-surface-air-temperature anomalies. Based on these analyses, the mean global surface temperature of the Earth shows a warming in the range of 0.3°–0.7°C over the past century, or a statistical average of about 0.003°–0.007°C per year. Global data sets from the various agencies show slightly dissimilar trends, as the data are processed in different ways.

Questions arise as to the representative nature of the data on which global surface temperature calculations have been based. These data come from weather stations unevenly distributed over the Earth's surface, mostly on land, close to towns and cities, and predominantly in the Northern Hemisphere. Land use can have significant effects on local climate. The best-documented examples of such effects are urban heat islands, which are significantly warmer than their rural surroundings. Cities replace natural vegetation with surfaces such as concrete and asphalt that can warm them by several degrees Celsius, forming urban "heat islands." Thus, a disproportionate number of weather stations being located in urban environments may skew data regarding global averages. Many weather stations are located at airports, which were originally located in rural areas that have since been developed. Thus, while the data collected at such stations remain reliable measurements of the local, urban environment, they may no longer be equally reliable indicators of global trends.

—*C R de Freitas*

Further Reading

Frederick, John E. *Principles of Atmospheric Science*. Jones and Bartlett 2008. Provides a concise introduction to atmospheric science beginning from the basic constituents and structure of the atmosphere.

Lutgens, Frederick K., Edward J. Tarbuck, Tasa, Dennis G. and Herman, Redina. *The Atmosphere: An Introduction to Meteorology.* 14th ed. Pearson Education, 2018. An updated version of this textbook, in loose-leaf format, provides students with a general overview of the structure and properties of Earth's atmosphere.

Salby, Murray L. *Physics of the Atmosphere and Climate.* Cambridge University Press, 2012. Provides an integrated treatment of the Earth-atmosphere system and its processes. Begins with first principles and continues with a balance of theory and applications as it covers climate, controlling influences, theory, and major applications.

GLOBAL WARMING IN ANTARCTICA

FIELDS OF STUDY

Environmental Sciences; Environmental Studies; Meteorology; Climate Sciences; Climate Modeling; Physical Geography; Ecosystem Management; Ecology; Oceanography; Hydroclimatology; Atmospheric Science; Climate Zones; Climate Classification

SUMMARY

Antarctica, located at and around the South Pole, is the world's fifth largest continent, with a surface area of 12.4 million square kilometers. Approximately 5,500 kilometers wide at its broadest point, it is surrounded by the southern portions of the Atlantic, Pacific, and Indian Oceans. This immense landmass is covered with an ice sheet larger than the continent itself. At its maximum winter extent during the month of July, the ice sheet measures about 14 million square kilometers and contains 30 million cubic kilometers of ice.

PRINCIPAL TERMS

- **Antarctic Peninsula:** a peninsula stretching northward toward South America that contains about 10 percent of the ice of Antarctica
- **East Antarctic ice sheet:** ice sheet located east of the Transantarctic Mountains that stores over 60 percent of the world's total freshwater
- **glacier:** a mass of ice that flows downhill, usually within the confines of a former stream valley
- **ice sheet:** a mass of ice covering a large area of land
- **ice shelf:** a platform of freshwater ice floating over the ocean
- **mass balance:** the difference between the accumulation of snow and the ablation of ice on a given glacial formation
- **sea ice:** frozen ocean water
- **West Antarctic ice sheet:** the smallest, but no less significant, ice sheet in Antarctica, located west of the Transantarctic Mountains

Global Warming Potentials of Major Greenhouse Gases

Greenhouse Gas	Global Warming Potential*
CO_2 (carbon dioxide)	1 (reference)
CH_4 (methane)	25
N_2O (nitrous oxide)	298
HFC-23 (hydrofluorocarbon-23)	14,800
SF_6 (sulfur hexafluoride)	22,800

*One-hundred-year time horizon.

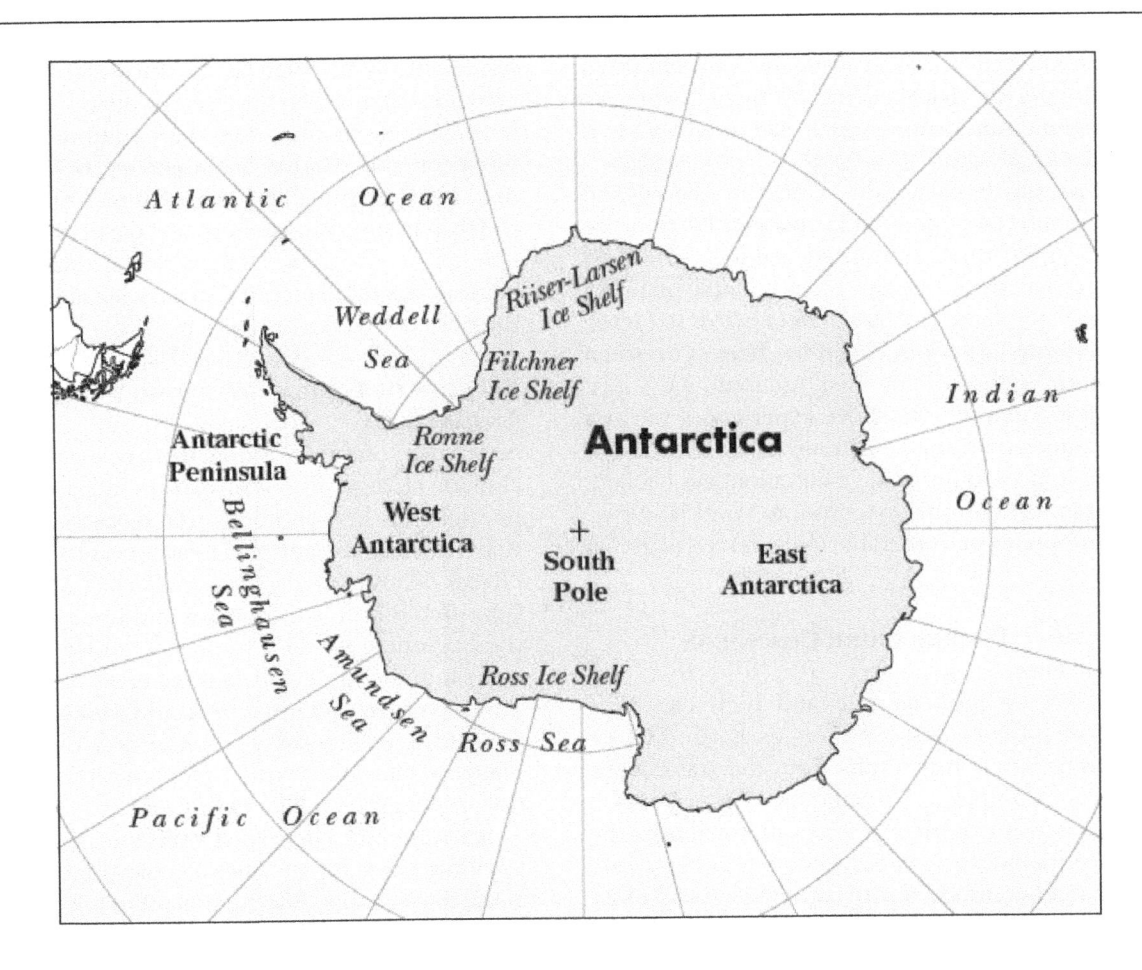

The Antarctic Ice Sheet: Unevenly Divided

More than 98 percent of Antarctica is covered with ice of an average thickness of about 2,100 meters. Most of the 2 percent of the continent not covered by ice is in the Antarctic Peninsula. The Antarctic ice sheet reaches a thickness of almost 5,000 meters at its highest point. From a geologic standpoint, Antarctica is made up of two structural provinces, East Antarctica and West Antarctica. East Antarctica is a stable shield separated from the much younger Mesozoic and Cenozoic belt of West Antarctica. The contact zone between these two provinces is the Transantarctic Mountains and the depression separating the Ross Sea and the Weddell Sea.

The Transantarctic Mountains divide the ice-covered continent into two ice sheets, the largest masses of ice known on Earth. The East Antarctic ice sheet, which is mostly situated in the Eastern Hemisphere, comprises 90 percent of the Antarctic ice. It is surrounded by the southern Atlantic Ocean, the Indian Ocean, and the Ross Sea. The South Pole is located in the East Antarctic ice sheet. The East Antarctic continental landmass on which it rests is close to sea level. Besides its ice sheets, Antarctica has many glaciers, ice streams, and ice shelves. While the East Antarctic ice sheet is dome-shaped, the West Antarctic ice sheet is more elongated along the mountains in the center of the peninsula.

As its name implies, the West Antarctic ice sheet is located in the Western Hemisphere. The northernmost part of the West Antarctic ice sheet protrudes in a peninsula that ends beyond the Antarctic Circle, south of South America; this peninsula is often considered a third, distinct ice sheet, the Antarctic Peninsula ice sheet. The backbone of the peninsula is composed of high mountains, an extension of the Andes mountain range, reaching about 2,800 meters. The northernmost latitude of the peninsula is 63 degrees and 13 minutes south. The largest part of the West Antarctic ice sheet along the Amundsen Sea flows into the Ross Ice Shelf, a platform of floating ice on which the American research station McMurdo is located.

The peninsula glaciers drain into the Weddell and Bellingshausen seas and the Ronne and Filchner ice shelves. Unlike the East Antarctic ice sheet, the West Antarctic ice sheet sits on a continental platform that in some places is 2,500 meters below sea level. It is therefore more influenced by changes in ocean temperatures than is the East Antarctic ice sheet. The West Antarctic ice sheet experiences warmer temperatures than the East Antarctic ice sheet, both because it has a lower average elevation and because the Antarctic Peninsula extends into lower latitudes. It is therefore more vulnerable to the effects of global warming.

MEASURING TEMPERATURE CHANGE IN ANTARCTICA

Because of its high latitude and high elevation, Antarctica is the coldest continent on Earth. The ice that covers Antarctica results from the transformation of snow into ice. The amount of precipitation that Antarctica experiences is not uniform over the entire continent; the coasts, with lower elevation and higher temperatures, record about six times more annual snow accumulation than does the much higher and colder interior, which receives less than three centimeters of water-equivalent precipitation annually. The lowest temperature ever recorded on Earth (-89.2°C) occurred in 1983 at Vostok, a Russian research station in the middle of the East Antarctic ice sheet, where only 166 millimeters of precipitation is received on average per year. (A lower temperature of -93.2°C was recorded in East Antarctica in 2010, but because it was measured by satellite rather than on the ground, the record is not official.)

As Antarctic snow layers are progressively transformed into ice, they preserve evidence of the temperature at the time the snow fell in the form of isotope ratios within the ice. As a result, ice cores may be drilled from the Antarctic ice and examined to obtain a chronology of Antarctic temperatures. Antarctic weather stations, moreover, have been recording temperature and measuring precipitation since the early twentieth century, albeit not in a continuous manner. During and after the International Geophysical Year (1957–58), weather stations were systematically installed at the forty-eight bases created in Antarctica by twelve countries. One of the most challenging tasks that these stations have faced has been the physical maintenance of the devices measuring such a harsh environment. Anemometers, which measure wind speed, are particularly vulnerable to the ferocious katabatic winds that sweep the continent, sublimating the surface ice and damaging the devices that measure them.

EFFECTS OF GLOBAL WARMING IN ANTARCTICA

Antarctica plays an important role in assessing climatic change. Its ice reveals the variation of temperature of the continent over more than 800,000 years. It is also the perfect laboratory for studying the effects of human activities on Earth's atmosphere. Considered hostile to humans and unexplored until the beginning of the twentieth century, Antarctica came into the spotlight when the ozone hole above it was discovered and when the world's longest ice core was retrieved at Vostok. The ozone hole threatens the planet by allowing short-wavelength ultraviolet radiation to penetrate the lower atmosphere.

Scientists are concerned over the potentially calamitous effect of the melting of Antarctic ice. In 2006, for example, Eric Rignot, a French glaciologist working at the University of California, computed that Antarctica had lost 178 billion metric tons of ice, mostly from the Antarctic Peninsula. This would result in a rise in sea level of about 0.5 millimeter. Though the amount is relatively small, the trend it indicated was troubling; the loss had increased from the 102 billion metric tons recorded in 1996. From 1996 to 2006, some glaciers in the west began moving more rapidly toward the sea and thus produced more icebergs, as their terminuses have collapsed into the sea.

In addition, temperatures in Antarctica have notably been increasing in the twenty-first century, with a record high of 17.5 degrees Celsius (63.5 degrees Fahrenheit) being recorded at the tip of the Antarctic Peninsula in March 2015. Research published in the science journal *Nature* in 2016 predicted more dire consequences if global emissions continue to remain at high levels, causing higher temperatures that would intensify the rate of disintegration of the West Antarctic ice sheets. Sea levels would then rise to dangerous levels, with an estimated rise of five to six feet by the year 2100.

In a study published in the journal *Current Biology* in 2017, Matthew Amesbury and colleagues reported that Antarctica is "greening" as a result of increasingly rapid moss growth, caused by rising temperatures and increased meltwater from glaciers. On the northernmost part of Antarctica's mainland, the researchers discovered that in the past fifty years, the two dominant moss species had been growing more than three times faster than before—an average of three or more millimeters per year, compared to previous average annual growth of one millimeter or less. Moss core samples revealed the accelerated growth, as well as the historical sensitivity of moss growth to increases in temperature. As snow and ice cover decreases and vegetation cover increases, more and more heat is absorbed into the ground, which in turn spurs even faster growth. Future warming could lead to a major ecosystem shift that would cause Antarctica to resemble the warmer and greener Arctic.

Measuring Ice Losses

To understand and compute the mass balance of the ice in Antarctica, one must determine when and whether the continent loses or gains mass. Antarctica gains mass when snow falls. The entire continent, which is about 1.5 times the size of the United States, contains only about one hundred weather stations, so it is not easy to estimate with a great degree of certainty how much snow accumulates on it, and the error margin is high. Melting is rare in Antarctica, because the temperature tends to remain below the freezing point year-round. In 2008, Rignot estimated that 99 percent of the ice lost in Antarctica forms icebergs. However, the total picture of ice loss is not completely uniform; while Antarctica is losing ice at a greater rate and the area covered by sea ice in the Arctic has steadily decreased, sea ice in Antarctica has slightly increased, particularly along the East Antarctic ice sheet's coast, a condition to which many climate change skeptics have pointed to in support of their claims. A report published in May 2016 by a team of physicists and NASA scientists indicates, however, that the Antarctic's very cold winds, consistently colder temperatures, thicker snowpack, and the surrounding terrain of the Antarctic Sea and its powerful, circular current all contribute to the growth of Antarctic sea compared to the decrease in Arctic sea ice.

Knowledge of the processes by which the great ice masses of Antarctica grow and shrink is not yet perfect. Improving that knowledge will be of increasing importance, and understanding the climatological mechanisms at work on the continent will be crucial in assessing the impact of human activity on the health and stability of the Earth.

—Denyse Lemaire
—David Kasserman

Further Reading

Alley, Richard B. *The Two-Mile Time Machine. Ice Cores, Abrupt Climate Change and Our Future* Princeton University Press, 2014. Describes the analytical methods and interpretations of physical clues of past climates trapped in ice that is thousands of years old.

"Antarctica Hits Record High Temperature at Balmy 63.5°F." *Reuters*, 1 Mar. 2017, www.reuters.com/article/us-antarctica-temperatures-idUSKBN168417. Accessed 1 June 2017.

Berwyn, Bob. "Why Is Antarctica's Sea Ice Growing While the Arctic Melts? Scientists Have an Answer." *InsideClimate News*, 31 May 2016, insideclimatenews.org/news/31052016/why-antarctica-sea-ice-level-growing-while-arctic-glaciers-melts-climate-change-global-warming. Accessed 1 June 2017.

Donahue, Michelle Z. "Fast-Growing Moss Is Turning Antarctica Green." *National Geographic*, 19 May 2017, news.nationalgeographic.com/2017/05/antarctica-green-climate-moss-environment/. Accessed 1 June 2017.

McGonigal, David, chief consultant. *Antarctica: Secrets of the Southern Continent*. Firefly Books, 2008.

Nghiem, S. V., et al. "Geophysical Constraints on the Antarctic Sea Ice Cover." *Remote Sensing of Environment*, vol. 181, 2016, pp. 281–92, doi:10.1016/j.rse.2016.04.005. Accessed 2 Feb. 2017.

Williams, Rob. "Scientists Record New Coldest Temperature on Earth." *The Independent*, 10 Dec. 2013, www.independent.co.uk/news/science/scientists-record-new-coldest-temperature-on-earth-on-the-east-antarctic-plateau-8995135.html. Accessed 1 June 2017.

GREAT LAKES

FIELDS OF STUDY

Hydrology; Physical Geography; Geology; Ecology; Environmental Sciences; Environmental Studies; Ecosystem Management; Waste Management

SUMMARY

The Great Lakes represent the largest freshwater lake complex on Earth. Created by continental glaciers over the past 18,000 years, these five major lakes (Ontario, Erie, Huron, Michigan, and Superior) and one minor lake (Saint Clair) provide significant resources for Canadians and Americans occupying the surrounding basin.

PRINCIPAL TERMS

- **epilimnion:** a warmer surface layer of water that occurs in a lake during summer stratification; during spring, warmer water rises from great depths, and it heats up through the summer season
- **greenhouse effect:** a natural process by which water vapor, carbon dioxide, and other gases in the atmosphere absorb heat and reradiate it back to Earth
- **isostatic rebound:** a tendency of Earth's continental surfaces to rise after being depressed by continental glaciers, without faulting
- **Pleistocene:** a geologic era spanning about 2 million years that ended about 10,000 years ago, often considered synonymous with the "Ice Age"
- **seiche:** rocking motion of lake level from one end of the lake to the other following high winds and low barometric pressure; frequently, a seiche will follow a storm event
- **storm surge:** a rapid rise in lake level associated with low barometric pressure; the water level is frequently "pushed" above a shoreline on one end of the lake and depressed on the opposite end
- **thermocline:** a well-defined layer of water in a lake separating the warmer and shallower epilimnion from the cooler and deeper hypolimnion
- **wetlands:** areas along a coast where the water table is near or above the ground surface for at least part of the year; wetlands are characterized by wet soils, water-tolerant plants, and high biological production

Geological Development of the Region

The Great Lakes are superlative features on the North American landscape. They make up the largest freshwater lake complex on Earth and represent about 18 percent of the world's freshwater supply. Covering a total area of 245,000 square kilometers, the Great Lakes have a shoreline length of 17,000 kilometers. Lake Superior (82,100 square kilometers), Lake Huron (59,600 square kilometers), and Lake Michigan (57,800 square kilometers) are among the ten largest lakes on Earth.

The rocks forming the foundation of the Great Lakes date back some 600 million years. On the northern and northwestern shore of Lake Superior are remnants of the Canadian Shield, which is composed of igneous rocks of the Precambrian era, more than 1 billion years ago. Following volcanic activity and mountain building during the Precambrian era, the central region of North America was repeatedly covered by shallow tropical seas. At this time, during the Paleozoic era (600 million to 230 million years ago), sediments transported by rivers from adjacent eroding uplands were deposited in a shallow marine environment, and lime, salt, and gypsum precipitated from the seawater. All these soft materials were eventually hardened into sedimentary rock layers such as sandstone, shale, limestone, and halite. A multitude of fauna colonized the submarine environment, including corals, brachiopods, crinoids, and several species of mollusks.

As the layers of sediment accumulated over millions of years, the basin began to subside at its center. The Great Lakes basin structure may be compared to a series of bowls, one stacked on top of another. As viewed from above, only the top bowl is completely visible; however, the rims of the progressively deeper bowls are visible as a number of thin concentric rims along the perimeter of the basin.

The Paleozoic era was followed by the Mesozoic era (230 to 63 million years ago), a time of little deposition. In spite of the great age of the rocks making up the foundation of the Great Lakes, the lakes themselves were created in the relatively recent Pleistocene epoch. Between the 220 million years when the basin's bedrock was deposited and the onset of Pleistocene glaciers, the landscape now occupied by the lakes was occupied by streams. The

Lake Michigan.

streams eroded the softer bedrock to form channels and valleys. The divides between and parallel to the eroded valleys were more resistant to erosion and are represented by higher elevations.

The streams excavated shales, and weaker limestones now occupied the Lake Michigan and Lake Huron basins. An arc, composed of hard dolomitic rock and known as the Niagara Escarpment, extends in a northwesterly direction from Niagara, forming the Bruce Peninsula that separates the east side of Lake Huron from Georgian Bay. The same structure continues across Michigan's upper peninsula, separating Lake Michigan from Green Bay as the Door Peninsula. The ancient stream channels were favored by the glaciers because they were at lower elevations and composed of more erodible bedrock. The linear shape of the lower Great Lakes is clearly related to the initial erosion by streams followed by the continental ice.

Lake Superior is partly located on the Canadian Shield, and its geologic origin is less obvious. East-west faults underlie Lake Superior, and the rocks form a structural sag, or syncline, oriented along the long axis of the lake in an approximately east-west direction.

The glacial origin and development of the Great Lakes is complex for several reasons. Each lake has a unique history and time of formation, making generalizations difficult. For example, the glaciers repeatedly advanced and retreated from many directions, covering and exposing each lake basin. There is abundant evidence regarding the size, elevation, and precise geographical distribution of each ancestral lake; this historical information is documented by coastal landforms such as higher ancient shorelines and relict wave-cut features. Yet the changing of the lakes' outlets to the ocean and reversals of drainage patterns complicate the sequence of events. Furthermore, because of the weight of the ice, the Paleozoic bedrock subsided to a lower level as the continental mass sought to "float" more deeply in the underlying mantle layer. As the glaciers receded, exposing different segments of the basin, the land began to recover its previous level and rise. This process, called isostatic rebound, is active today, causing elevation changes of many fossil shorelines. Although uplift has slowed since the ice exposed the newly created Great Lakes, the process is continuing.

As the ice began to retreat from the region, glacial landforms were deposited. Along many shorelines, moraines composed of fragments of rock, sand, and silt form spectacular bluffs. Along Lake Superior, the ice scraped and removed much of the soil, exposing the bedrock, which now forms high cliffs. Sand eroded from glacial sediment was transported by rivers to the lakes and deposited as beaches. The exposed beach sand was then transported inland by the wind to form coastal dunes.

WEATHER AND CLIMATE

Weather and climate influence several processes occurring in the Great Lakes, including changing lake levels, storm surges and related seiches, and lake stratification and turnover. Through the hydrologic cycle, moisture is evaporated from the lake surfaces and is then returned as precipitation over the water and as runoff from the land. During cooler and wetter years, evaporation is retarded, and more water is contributed to the lakes by excess precipitation, causing lake levels to rise. In warmer and drier years, evaporation increases, and precipitation is retarded, causing lower lake levels. Such changes in water levels are not cyclic and occur over several years. In 1988, Lake Huron and Lake Michigan had record levels of 177.4 meters above sea level. In 1995, the average water level was 176.3 meters above sea level, a difference of 1.1 meters.

Unstable weather conditions generate storms that pass over the region, generally from west to east. When strong winds persist for several hours from a constant direction over a lake, the water level is "pushed" from one side of the lake to the other. This storm surge, accompanied by low atmospheric pressure, may elevate the water level as much as 2 meters along a shoreline in a matter of a few hours.

Gale-force winds on October 30 and 31, 1996, over Lake Erie raised the lake level 1.25 meters at Buffalo, New York. Concurrently, as the water rose at Buffalo, it was lowered in Toledo, Ohio, at the opposite end of the lake, by 2.25 meters. The total difference of water level was 3.50 meters. Following a storm, the level of a lake rocks back and forth as a "seiche" before settling to its normal level.

In turn, the lake waters dramatically affect the local weather. As winter approaches, lake-effect snows commonly occur. The effect is most common in the fall, before the lakes cool and freeze. Cold winds from the north or west pass over the basin, picking up moisture from the relatively warm lakes. The water vapor is then condensed, forming clouds that, in turn, dump heavy snows in coastal zones, especially along eastern Lake Michigan and southern Lake Superior, Lake Huron, Lake Erie, and Lake Ontario. From November 9 to November 12, 1996, 1.2 meters of snow fell along Lake Erie's south shore, paralyzing local communities.

With the exception of Lake Erie, the lake bottoms were scoured by glaciers to depths below sea level. The lakes thus have variable temperatures from the surface to the bottom. As the water temperature changes from season to season, water density is altered. During winter, as ice forms over the lakes, the water beneath the ice remains warmer. As the ice cover breaks up in spring, the deeper warmer water rises, or "turns over," to the surface. It heats up through the summer months, causing stratification of warmer water (called the epilimnion) above colder, denser water. The contrasting water layers are separated by a thermocline, demarcating a rapid temperature transition between the warmer epilimnion and cooler subsurface water.

An issue of concern to scientists regarding the Great Lakes is the impact of the intensification of the greenhouse effect on the lakes' water levels. The increase of greenhouse gases in the atmosphere, especially carbon dioxide, appears to be causing warming at unprecedented rates. Although climatologists differ in their opinions as to the impact of a warmer atmosphere over the Great Lakes, there is general agreement that both evaporation and precipitation will increase and stream runoff will decrease over many years. Based on general circulation models, it appears that lake levels will be from 0.5 meter to 2 meters lower than present levels if the climate continues to warm.

Wetlands

Wetlands along the shorelines of the Great Lakes are significant ecological zones located at the meeting of land and lake. Although many wetlands around the basin have been lost or degraded, the remaining habitat has multidimensional functions as part of both upland and aquatic ecosystems. The wetlands are exposed to both short-term (storm surges and seiches) and long-term changes in water levels that constantly alter the biogeography of these habitats. Because of the state of constant water-level change, or "pulse stability," the distribution and types of wetland plants shift dramatically. Thus, a constant renewal of the flora is occurring. Furthermore, because of the flushing action of the rise and fall of lake levels, peat accumulation in Great Lakes wetlands does not commonly occur, as it does in marine settings.

The coastal wetlands serve significant ecological, economic, and social functions. They provide spawning habitat and nursery and resting areas for many species, including fishes, amphibians, reptiles, ducks, geese and other water birds, and mammals. Largely because of the sport fishing industry, these habitats contribute significant revenue to the surrounding states and to the province of Ontario. Furthermore, pollution control and coastal erosion protection are additional benefits provided by these habitats.

Study of the Great Lakes

The creation and development of the Great Lakes have, in terms of geologic time, occurred relatively recently. Also, modifications such as erosion and deposition of coastal features are continual, active processes. To unravel the events leading to changes in the Great Lakes, scientists use techniques that include varve analysis and radiocarbon dating. A common theme of both techniques is that they express time in numbers of years within a reasonable range of accuracy rather than in a relative or comparative way. Varves consist of alternating light and dark sediment layers deposited in a lake. A light-colored mud is deposited during spring runoff; a dark-colored mud is deposited atop the lighter-colored layer during the following winter, as ice forms and there is less agitation of the lake water. One light and one dark band together represent one year of deposition. Numerous layers can be counted, like tree rings, and the number of years that were required for a sequence to be deposited can be determined. To obtain numerous undisturbed varve layers, researchers use a piston-coring device, which consists of a hollow pipe attached to a cable that is released vertically into the lake. As it falls freely to the bottom, it plunges into the soft sediment. A piston allows the sediment to remain in the pipe as it is raised by an attached cable. The mud can then be extracted from the tube, cut open, and analyzed.

Radiocarbon dating of carbon-rich material such as peat, lime, coral, and even bone material is useful for absolute dating back to about 50,000 years ago. Carbon's abundance in nature, coupled with the youthfulness of the Great Lakes, makes this tool very useful because many glacial, coastal, and sand dune landforms frequently contain some form of carbon suitable for absolute dating.

To map and detect recent changes in the landscape such as coastal erosion or the rate of dune migration, old maps, navigation charts, and aerial photographs are used. Charts and maps of the coastal zone have been available for more than a century, and aerial photographs of the region have been taken since the 1930s. By observing the position of a shoreline on historical sets of detailed aerial photographs over a ten-year period, for example, changes in the shoreline can be detected, and the erosion rate per year can be determined.

Geographic positioning systems can accurately locate the latitude and longitude of a point on a shoreline, store the information, and compare the shoreline position with the position at some future time. Satellite pictures help to detect wetland types and determine acreage; this information can then be compared to a later environmental condition, such as a period of higher lake level, to see if species habitat or acreage have changed.

Because of a geological process known as isostatic rebound, fossil shorelines become uplifted and exposed. Elevations retrieved from older topographic maps reveal how much uplift has occurred. If the age of a relict shoreline can be determined with radiocarbon analysis, the rate of glacial rebound in millimeters per century can then be assessed.

—*C. Nicholas Raphael*

FURTHER READING

Annin, Peter. *Great Lakes Water Wars*. Island Press, 2018.

Bolsenga, S. J., and C. E. Herdendorf. *Lake Erie and Lake St. Clair Handbook*. Wayne State University Press, 1993.

Botts, Lee, et al. *The Great Lakes: An Environmental Atlas and Resource Book*. Govt. of Canada, 2002.

Douglas, R. J. W. *Geology and Economic Minerals of Canada*. Report Number I. Ottawa: Geological Survey of Canada, 1970.

Dowdeswell, Julian A. *Atlas of Submarine Glacial Landforms: Modern, Quaternary and Ancient*. Geological Society, 2016.

Grady, Wayne, et al. *Great Lakes: The Natural History of a Changing Region*. Greystone Books, 2011.

Holman, J. Alan. *In Quest of Great Lakes Ice Age Vertebrates*. Michigan State University Press, 2001.

Le Sueur, Meridel. *North Star Country*. 2nd ed. University of Minnesota Press, 1998.

Spring, Barbara. *The Dynamic Great Lakes*. Independence Books, 2001.

U.S. Environmental Protection Agency. *The Great Lakes: An Environmental Atlas and Resource Book*. Great Lakes Program Office, 1995.

Ver Berkmoes, Ryan, Thomas Huhti, and Mark Lightbody. *Great Lakes*. Lonely Planet Publishing, 2000.

Wooster, Margaret. *Living Waters: Reading the Rivers of the Lower Great Lakes*. Excelsior Editions/State University of New York Press, 2009.

GREENHOUSE GASES AND GLOBAL WARMING

FIELDS OF STUDY

Atmospheric Science; Atmospheric Chemistry; Environmental Chemistry; Photochemistry; Thermodynamics; Heat Transfer; Environmental Sciences; Climate Modeling; Climate Studies; Fluid Dynamics; Chemical Kinetics; Earth System Modeling; Spectroscopy; Mass Spectroscopy

SUMMARY

Greenhouse gases (GHGs) have both natural and anthropogenic sources. They allow sunlight to pass through them and reach Earth's surface, but they trap longwave infrared radiation released by Earth's surface, preventing it from escaping into space. These trace atmospheric gases thus play an important role in the regulation of Earth's energy balance, raising the temperature of the lower atmosphere.

PRINCIPAL TERMS

- **aerosols:** small particles suspended in the atmosphere
- **anthropogenic:** deriving from human sources or activities
- **fossil fuels:** fuels (coal, oil, and natural gas) formed by the chemical alteration of plant and animal matter under geologic pressure over long periods of time
- **global dimming:** reduction of the amount of sunlight reaching Earth's surface
- **global warming:** an overall increase in Earth's average temperature
- **greenhouse effect:** absorption and emission of radiation by atmospheric gases, trapping heat energy within the atmosphere rather than allowing it to escape into space
- **greenhouse gases (GHGs):** atmospheric trace gases that contribute to the greenhouse effect

GHG Sources and Atmospheric Physics

The atmosphere comprises constant components and variable components. It is composed primarily of nitrogen (78 percent) and oxygen (21 percent). Its other constant components include argon, neon, krypton, and helium. Its variable components include carbon dioxide (CO_2), water vapor (H_2O), methane (CH_4), sulfur dioxide (SO_2), ozone (O_3), and nitrous oxide (N_2O). GHG concentrations in the atmosphere have historically varied as a result of natural processes, such as volcanic activity. They have always been a small fraction of the overall atmosphere. However, they exhibit significant effects on the climate despite their low concentrations. Thus, small variations in GHG concentration may have disproportionate effects on Earth's climate. Since the Industrial Revolution, humans have added a significant amount of GHGs to the atmosphere by burning fossil fuels and cutting down trees. Scientists estimate that the Earth's average temperature has already increased by 0.3° to 0.6°C since the beginning of the twentieth century. The variable components affect the weather and climate because they absorb heat emitted by Earth and thereby warm the atmosphere. In addition to the variable natural atmospheric GHGs, anthropogenic halocarbons, other chlorine- and bromine-containing substances, sulfur hexafluoride, hydrofluorocarbons, and perfluorocarbons contribute to the greenhouse effect.

Composed of two oxygen atoms and one carbon atom, CO_2 is a colorless, odorless gas deriving from carbon burning in the presence of sufficient oxygen. It is released to the atmosphere by forest fires, fossil fuel combustion, volcanic eruptions, plant and animal decomposition, oceanic evaporation, and respiration. It is removed from the atmosphere by CO_2 sinks, seawater absorption, and photosynthesis.

Methane is a colorless, odorless, nontoxic gas consisting of four hydrogen atoms and one carbon atom. It is a constituent of natural gas and fossil fuel. It is released into the atmosphere when organic matter decomposes in oxygen-deficient environments. Natural sources include wetlands, swamps, marshes, termites, and oceans. Other sources are the mining and burning of fossil fuels, digestive processes in ruminant animals, and landfills. Methane reacts with hydroxyl radicals in the atmosphere, which break it down in the presence of sunlight, shortening its lifetime.

Nitrous oxide is a colorless, nonflammable gas with a sweetish odor. It is naturally produced by oceans and rain forests. Anthropogenic sources include nylon and nitric acid production, fertilizers, cars with catalytic converters, and the burning of organic matter. Nitrous oxide gas is consumed by microbial respiration in specific anaerobic environments.

Sulfur dioxide is released during volcanic activities, combustion of fossil fuel, transportation, and industrial metal processing. This gas is more reactive than is CO_2, and it rapidly oxidizes to sulfate. It produces acidic gases and acid rain when it reacts with water and oxygen.

Ozone (triatomic oxygen) is a highly reactive, gaseous constituent of the atmosphere. A powerfully oxidizing, poisonous, blue gas with an unpleasant smell, it helps create smog. It is produced in chemical reactions of volatile organic compounds or nitrogen oxide with other atmospheric gases in the presence of sunlight. Oxygen and ozone absorb a critical range of the ultraviolet spectrum, preventing this dangerous radiation from reaching Earth's surface and making possible life on Earth.

Halocarbons, particularly the chlorofluorocarbons, have global warming potentials (GWPs) from

The Greenhouse Effect

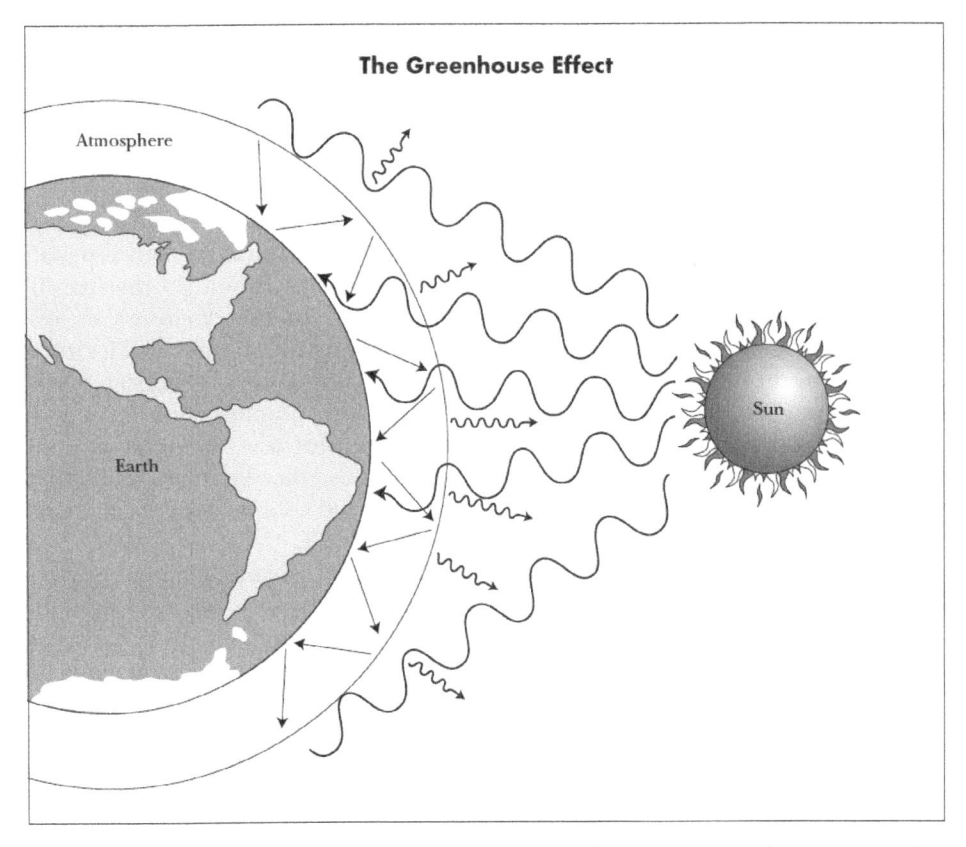

Some heat from the Sun is reflected back into space (squiggled arrows), but some becomes trapped by Earth's atmosphere and reradiates toward Earth (straight arrows), thus heating the planet.

three thousand to thirteen thousand times that of CO_2 and they remain in the atmosphere for hundreds of years. These compounds were commonly used in refrigeration, air conditioning, and electrical systems, but their use has been regulated as a result of their environmental and climatic effects.

Effect on Climate Change

The Working Groups of the Intergovernmental Panel on Climate Change (IPCC) presented a synthesis report in 2007, providing an integrated view of climate change from multiple perspectives. The report observed an increase of global air and ocean temperatures, melting of snows, and rising sea levels. The report estimated the one-hundred-year linear trend of Earth's average temperature between 1906 and 2005 at an increase of 0.74°C, significantly greater than the trend from 1901 to 2000 (0.6°C). The EPA (the Environmental Protection Agency) noted in 2016 that 2000-10 was the warmest decade every recorded. The increase of temperature contributed to changes in wind patterns, affecting extra-tropical storm tracks and temperature patterns.

Global average sea level has risen between 1961 and 2001 at an average rate of 1.8 millimeters per year and between 1993 and 2008 at an average rate of 3.1 millimeters per year. The increase is due largely to melting glaciers and polar ice sheets. Satellite data between 1978 and 2008 show that average annual extent of Arctic sea ice shrank by an average of 2.7 percent per decade. The average summertime extent shrank far more, an average of 7.4 percent per decade.

Increases have been reported in the number and size of glacial lakes and the rate of change in some Arctic and Antarctic ecosystems. Runoff and earlier spring peak discharge in many glacier- and snow-fed rivers have also increased. These increases have in turn had effects on the thermal structure and water quality of the rivers and lakes fed by this runoff. Both marine and freshwater systems have been associated with rising water temperatures and with changes in ice cover, salinity, oxygen levels, and circulation patterns. These ecological changes have affected algal, plankton, and fish abundance.

Precipitation has increased in the eastern parts of North and South America, northern Europe, and northern and central Asia. It has decreased in the Mediterranean and southern Africa. These patterns also have affected algal, plankton, and fish abundance. Globally, since 1970, a greater area of Earth's surface has been affected by drought.

Changes in atmospheric GHG and aerosol concentration, as well as solar radiation levels, affect the energy balance of Earth's climate system. Global GHG emissions increased by 70 percent over preindustrial levels between 1970 and 2004. CO_2 emissions increased by 80 percent, but they began to decline after 2000. The global increase in CO_2 and methane emissions is due to fossil fuel and land use, particularly agriculture.

Coastlines are particularly vulnerable to the consequences of climate change, such as sea-level rise and extreme weather. Around 120 million people on Earth are exposed to tropical cyclone hazards. During the twentieth century, global sea-level rises contributed to increased coastal inundation, erosion, ecosystem losses, loss of sea ice, thawing of permafrost, coastal retreat, and more frequent coral bleaching.

Anticipated future climate-related changes include a rise in sea level of up to 0.6

meter by 2100, a rise in sea surface temperatures by up to 3°C, an intensification of tropical cyclones, larger waves and storms, changes in precipitation and runoff patterns, and ocean acidification. These phenomena will vary on regional and local scales. Increased flooding and the degradation of freshwater, fisheries, and other resources could impact hundreds of millions of people, with significant socioeconomic costs. Degradation of coastal ecosystems, especially wetlands and coral reefs, affects the well-being of societies dependent on coastal ecosystems for goods and services.

Context

In response to global warming, changes are being implemented to reduce GHG emissions. The United Nations Framework Convention on Climate Change prepared the 1997 Kyoto Protocol. Under the protocol, thirty-six states, including highly industrialized countries and countries undergoing transitions to a market economy, entered into legally binding agreements to limit and reduce GHG emissions. Developing countries assumed nonbinding obligations to limit their emissions as well.

In the energy sector, fuel use is slowly transitioning from coal to natural gas and renewable energy (hydropower, solar, wind, geothermal, tidal, wave, and bioenergy). In the transport sector, fuel-efficient, hybrid, and fully electric vehicles are being designed and marketed, and governments are attempting to motivate commuters to use mass-transit systems. More efficient uses of energy, including low-energy light bulbs, day lighting, and efficient electrical, heating, and cooling appliances are being developed and deployed.

Industrial manufacturers have implemented electrical efficiency measures as well, and they have begun recycling, as well as capturing and storing CO_2. Crop and land management techniques have also improved, leading to an increase in soil carbon storage and the restoration of peaty soils and degraded land. Rice cultivation techniques have been improved, and livestock management techniques are being developed to reduce methane and nitrogen emissions. More controversially, dedicated energy crops are being grown to replace fossil fuels.

Afforestation, reforestation, forest management, reduced deforestation, and harvested wood product management are also being geared toward reducing GHG emissions. Forestry products are in use for bioenergy to replace fossil fuels. Improvements are being made in tree species, remote sensing for analyses of vegetation and soil carbon, and mapping of land use. In the waste industry, methane is being recovered from landfills and energy is being recovered from waste incineration. Organic waste is more widely used for composting, wastewater is minimized, and the wastewater produced is treated and recycled. Biocovers and biofilters are being developed to optimize methane oxidation.

—*Ewa M. Burchard*

Further Reading

"Climate Change Indicators: Greenhouse Gases." *EPA*, 6 Oct. 2016, www.epa.gov/climate-indicators/greenhouse-gases. Accessed 2 Feb. 2017.

Dlugokencky, E. J., et al. "Continuing Decline in the Growth Rate of the Atmospheric Methane Burden." *Nature*, 393, 1998, pp. 447–50.

Gore, Al. *An Inconvenient Sequel: Truth to Power.* Rodale, 2017.

Incropera, Frank P. *Climate Change—A Wicked Problem: Complexity and Uncertainty at the Intersection of Science, Economics, Politics, and Human Behavior.* Cambridge UP, 2016.

Lallanila, Marc. "What Is the Greenhouse Effect? *Live Science*, 12 Apr. 2016, www.livescience.com/37743-greenhouse-effect.html. Accessed 2 Feb. 2017. Le Treut, H., et al. "Historical Overview of Climate Change." In *Climate Change, 2007—The Physical Science Basis: Contribution of Working Group I to the Fourth Assessment Report of the Intergovernmental Panel on Climate Change,* edited by Susan Solomon et al. Cambridge UP, 2007.

"A Student's Guide to Global Climate Change." *EPA*, 3 Mar. 2016, www3.epa.gov/climatechange/kids/index.html. Accessed 2 Feb. 2017. Walker, Gabrielle. *An Ocean of Air: Why the Wind Blows and Other Mysteries of the Atmosphere.* Harcourt, 2007.

Ward, Peter Langdon. *What Really Causes Global Warming? Greenhouse Gases or Ozone Depletion?* Morgan James, 2016.

GROUND ICE AND CLIMATE CHANGE

FIELDS OF STUDY

Physical Geography; Environmental Sciences; Environmental Studies; Environmental Chemistry; Waste Management; Ecosystem Management; Ecology; Meteorology; Climate Sciences; Bioclimatology; Climate Zones; Hydrology

SUMMARY

Ground ice is found in cavities, voids, pores, and other openings in frozen or freezing ground. It is a feature of a permafrost region, but the two terms are not interchangeable. A permafrost region is defined as an area where the soil and rocks making up the land remain below the freezing point for two or more consecutive years and can include areas with little or no water content. Permafrost regions account for nearly one-quarter of the landmass of the Northern Hemisphere, and some of these areas have been frozen since the Pleistocene ice age. Ground ice differs from glacier ice as well. Glaciers are created when fallen snow compresses into thick ice masses over long periods of time, and they are capable of river-like movement, albeit typically at a pace so slow as to be essentially imperceptible to the human eye.

PRINCIPAL TERMS

- **permafrost:** soils that remain frozen throughout seasonal changes for a period of two or more years
- **pingo:** a large block of relatively pure water ice that has pushed up through permafrost soil to form a large conical mound
- **reticulate:** resembling a more-or-less regular network of lines and cracks

CHARACTERISTICS OF GROUND ICE

Ground ice is formed once the temperature drops below 0°C, when most of the moisture in the soil freezes. There are two different types of ground ice: structure-forming ice and pure ice. Structure-forming ice, which holds sediment together, comes in multiple varieties, including ice crystals, intrusive ice, reticulate vein ice, segregated ice, and icy coatings on soil particles. Pure, or massive, ice exists primarily toward the surface and is found as ice wedges, massive ice beds, and pingo cores. Massive ice beds are those with a minimum ice-to-soil ratio of 2.5 to 1.

Ancient ground ice, known as fossil ice, is a valuable source of historical information for geologists, paleontologists, and climatologists. Fossil ice preserves organic material and provides a measurable history of sediment and air quality covering hundreds of thousands of years. Cores drilled from fossil ice are studied to learn about the climates of the distant past.

In 2002, the National Aeronautics and Space Administration reported that the Odyssey orbiter had identified ground ice on Mars, a find that was confirmed by the Phoenix lander in June, 2008. The

Strings of ice found in the Adirondack region of upper New York state.

extreme cold and thin atmosphere on Mars would cause surface water to vaporize as dry ice does on Earth, but ground ice buried under the surface remains frozen and should prove invaluable to studies of that planet's history, as well as serving as an analogue for the study of terrestrial permafrost.

Significance for Climate Change

Ground ice and permafrost play crucial roles in the industrial development of a region's energy and mining industries, as well as in the infrastructure needed to support such enterprises. It is possible to build in the presence of ground ice in its structure-forming state, so long as its properties are taken into consideration.

The presence of ground ice is closely linked to a region's climate, especially to the temperature at ground level. Its impact varies based on a variety of factors, including drainage, snow cover, soil composition, and vegetation. Ground ice affects both topography and vegetation, as well as a region's response to changes in climate and population.

In regions where the temperature remains close to freezing, a change of only a few degrees can be enough to cause ground ice to start melting. Changes can occur naturally, with normal temperature fluctuations, or as a result of ground clearing through construction or forest fire. When structure-forming ground ice melts, the surrounding terrain is significantly weakened, creating slope instabilities and thaw settlement.

Scientists predict that as much as 90 percent of the permafrost in the Northern Hemisphere could melt by the end of the twenty-first century, with most of the thaw taking place in the top 3 meters of terrain. Some regions in Alaska and Siberia have already experienced collapsed infrastructure and increased rock fall in the high elevations. The resulting meltwater would eventually reach the oceans, causing sea levels to rise around the world. Since the 1930s, Arctic water runoff has increased by 7 percent, and it is projected to reach a 28 percent increase by 2100.

Permafrost traps and holds approximately 30 percent of Earth's carbon, sequestering it from the atmosphere. Melting ground ice will release the trapped carbon into the atmosphere, contributing to the greenhouse effect, which in turn will cause even more ice to melt.

Coastlines in permafrost regions become more vulnerable to erosion as the ice retreats. The Alaskan island of Shishmaref, home to a native population whose ancestors have lived there for more than four thousand years, lost 7 meters of coastline annually between 2001 and 2006.

—*P. S. Ramsey*

Further Reading

Alley, Richard B. *The Two-Mile Time Machine. Ice Cores, Abrupt Climate Change and Our Future* Princeton University Press, 2014. Describes the analytical methods and interpretations of physical clues of past climates trapped in ice that is thousands of years old.

Jouzel, Jean, Lorius, Claude and Raynard, Dominique *The White Planet. The Evolution and Future of Our Frozen World.* Princeton University Press, 2013. Examines the past, present and future roles of ice on Earth, concluding that climate change and global warming will ultimately have disastrous consequences for human societies on Earth regardless of technological advances.

Margesin, Rosa, ed. *Permafrost Soils*. Springer-Verlag, 2009. A collection of learned articles describing many aspects of Arctic permafrost soils and the effects of global warming on those soils and their content.

Gosnell, Mariana. *Ice: The Nature, the History, and the Uses of an Astonishing Substance*. Chicago: University of Chicago Press, 2007. Examines in detail ice and its impact on Earth and its inhabitants.

Macdougall, J D . *Frozen Earth: The Once and Future Story of Ice Ages*. University of California Press, 2013. A scientific look at the ice ages and their geological impact.

Turney, Chris. *Ice, Mud, and Blood: Lessons from Climates Past*. Macmillan, 2008. Describes the discoveries derived from ice cores and how these discoveries led to a better understanding of paleoclimates.

Wadhams, Peter *A Farewell to Ice. A Report from the Arctic*. Oxford University Press, 2017. Discusses the role of Arctic and Antarctic ice shields, and how feedback systems function. Special attention is given to the potential for the release of quantities of methane from polar sea floors and Arctic permafrost as Earth's average temperature increases.

GULF STREAM

FIELDS OF STUDY

Oceanography; Fluid Dynamics; Physical Geography; Heat Transfer; Thermodynamics; Climate Modeling; Hydroclimatology; Hydrology; Earth System Modeling

SUMMARY

The Gulf Stream is a geostrophic surface current that constitutes the northwestern part of the North Atlantic Gyre. It moves huge quantities of water at remarkably fast velocities across vast distances with many geological, physical, and biological repercussions. By itself, however, it is not responsible for the mild climate of Western Europe.

PRINCIPAL TERMS

- **Coriolis Effect:** an apparent force acting on a rotating coordinate system; on Earth this causes things moving in the Northern Hemisphere to be deflected toward the right and things moving in the Southern Hemisphere to be deflected toward the left
- **geostrophic current:** a current resulting from the balance between a pressure gradient force and the Coriolis Effect; the current moves horizontally and is perpendicular in direction to both the pressure gradient force and the Coriolis Effect
- **gyre:** the major rotating current system at the surface of an ocean, generally produced by a combination of wind-generated currents and geostrophic currents
- **pressure gradient:** a difference in pressure that causes fluids (both liquids and gases) to move from regions of high pressure to regions of low pressure
- **thermohaline circulation:** a mode of oceanic circulation that is driven by the sinking of denser water and its replacement at the surface with less dense water
- **wind-driven circulation:** the surface currents on the ocean that result from winds and geostrophic currents

WATER CIRCULATION

It is convenient to consider any systematic movement of water at sea as being part of either a wind-driven circulation system or a thermohaline circulation system. In the former, the linkage with atmospheric movement is direct, the currents are usually at or near the sea surface, and the velocities of the flows are often in the range of several centimeters per second (or several knots). In thermohaline circulation, the driving force is gravity, which causes denser water to sink and flow to the deepest parts of the sea, and the currents are usually much slower. In the

North Atlantic Ocean, thermohaline circulation occurs on a vast scale as saline surface water gives up its heat, sinks, and eventually flows across the bottom of both the North and South Atlantic basins.

Wind-driven circulation develops a gigantic clockwise circular motion called the North Atlantic Gyre. The most intense flow of this gyre is along its northwestern boundary and is called the Gulf Stream. Because the flow is circular, it does not actually have a beginning or end. As it is usually geographically defined, however, the Gulf Stream begins in the straits of Florida, where water leaving the Gulf of Mexico and the Caribbean joins with water continuing to go around the gyre. There the Gulf Stream moves about 30 million cubic meters per second past any point. The volume of water entrained in this flow continues to increase, and by the time it reaches Cape Hatteras, North Carolina, there are about 85 million cubic meters moving by any point every second. When the Gulf Stream reaches longitude 65 degrees west, off the Grand Banks of Newfoundland, it moves 150 million cubic meters per second.

To put this in perspective, consider that 1 cubic meter of water has a mass of 1,000 kilograms. A 150-pound person has a mass of 68 kilograms. Therefore, 1 cubic meter of water has the mass of about fifteen people. The flow off Cape Hatteras would be equivalent to 1.2 billion people, roughly the population of China, streaming by every second. By the time it gets to the Grand Banks, the flow would be larger by 65 million cubic meters, or 975 million more people. The Gulf Stream truly dwarfs most of nature's other wonders. The total hydrologic cycle, which includes all the rain, snow, sleet, and hail that falls on Earth (oceans plus continents), moves an average of only 10 million cubic meters of water per second.

Coriolis Effect

The secret to maintaining such huge flows of water lies in their circular nature. A virtual "hill" of warm, low-density water sits over the core of the North Atlantic Gyre. Water on or within this hill is driven by gravity or a pressure gradient to move downslope or away from the center of this hill. This moving water is deflected by the Coriolis Effect, which is a consequence of the planet's rotation. In the Northern Hemisphere, the Coriolis Effect causes things moving with a horizontal velocity to move to their right. Therefore, water continuously trying to move out from the center of this hill is deflected to move around the gyre instead. Eventually a balance is achieved between the Coriolis Effect and the pressure gradient forces. As a result, the water does not move away from the center of the hill but instead

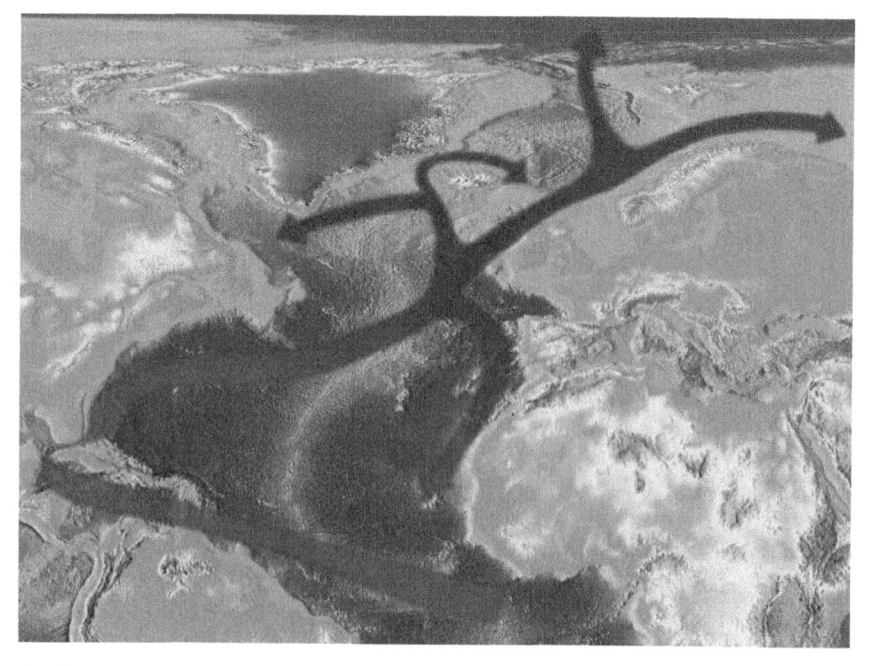

Gulf Stream map.

circles it in a clockwise fashion. This kind of current is called a "geostrophic" current.

Earth's spherical shape means that the Coriolis Effect varies with latitude. One result is that the flow within the North Atlantic Gyre varies with location. The hill is not symmetrical but has its steepest slopes on its northwest edge. Because currents must balance slopes, the gyre is most intense there. The center of the gyre is in the western Atlantic not far from Bermuda. The hill also slopes away to the east, but very gradually, so that the southern flows of the gyre are slow and spread out over a very large region.

Driving the gyre and maintaining the Bermuda High are the winds over the Atlantic Ocean. This subtropical atmospheric feature is a region of descending, dry air near the center of the North Atlantic Ocean. After descending, this air pushes out across the ocean. It, too, is deflected by the Coriolis Effect, developing into somewhat circular, clockwise winds. Near the center of this system, winds are weak, and precipitation is uncommon. Early sailors, faced with long, dry periods of calm, sometimes made their thirsty horses walk the plank. This is the origin of the term "horse latitudes," sometimes used to describe this region.

With little cloud cover and a subtropical latitude, this region receives intense solar radiation that warms the surface water and causes intense evaporation. This causes the water of this area, the Sargasso Sea, to become extraordinarily saline and very warm. Just as oil floats on vinegar in a salad dressing, this warm, and correspondingly less dense, saline water floats on top of the cooler, denser water below.

It is this warm water that forms the hill driving the Gulf Stream. The warm water tries to spread out and flow over cooler surface waters far from the center of the gyre. The Coriolis Effect makes it move around the hill, not down it, and the boundary between this warm, rotating mass of water and the cooler water it is trying to flow over is where the currents are most obvious. This is the Gulf Stream, a distinct boundary between the productive coastal waters, which are green and teeming with life, and the dark blue, nearly lifeless Sargasso Sea waters.

Gulf Stream as a Boundary Current

For decades, schoolchildren have been taught that the mild climate of the British Isles and Western Europe gets its heat from the Gulf Stream. The Gulf Stream is often presented as a river of warm water moving north and then east, eventually to deliver its heat to the European continent. The significance of the Gulf Stream as a source of heat changes when it is recognized for what it is, just a boundary current. It separates the huge hill of warm, saline Sargasso Sea water from the colder surrounding waters.

The Gulf Stream is a very active boundary region with motions driven by the wind and geostrophic currents that surround the huge quantity of surface water, heated by the sun in subtropical high-pressure zones and made especially salty by accompanying evaporation. The Gulf Stream gets all the press, but it is actually this huge quantity of water that conveys heat to Europe.

In the North Atlantic, beyond the northern extent of the circulating gyre, cold winds remove heat from the surface waters. In the process, these winds are warmed and bring pleasant temperatures to Europe. However, by removing heat, these winds cool the saline surface waters until they become dense enough to sink to the bottom of the sea. This is called thermohaline circulation. The sinking waters are replaced by a gradual northward flow of surface water. It is likely that most of this surface water spent time in the North Atlantic Gyre, but its transport northward was independent of that circular motion. If the gyre were to stop tomorrow, the thermohaline circulation would continue, and Europe would stay just as warm as it is today. In fact, some researchers have suggested that the strength of the Gulf Stream actually reduces the warming effects of this thermohaline circulation. If they are correct, were the Gulf Stream to stop, Europe might grow warmer.

Although its role as a heat-delivery system may have been overstated, the Gulf Stream is still an incredibly powerful element of oceanic circulation and, as such, greatly influences the biology and chemistry of the surface ocean. This significance is easily seen where meanders develop on the Gulf Stream, typically beyond Cape Hatteras. Just as meanders can grow and develop on a slow-moving river, they also develop on the Gulf Stream. Whereas a stream meander may form an oxbow lake if the course of the stream closes in upon itself, a meander in the Gulf Stream produces a circulating eddy separated from the rest of the stream when it closes upon itself. These eddies will have a core of warmer, less fertile water and rotate in a clockwise manner if they close off on the southeastern side of a meander. They will have a core of cooler, more fertile water and rotate in a counterclockwise manner if they close off on the northwestern side of a meander. These rings persist for months to a year or more and may establish their own ecosystems during their lifetimes.

The Gulf Stream disperses eggs, seeds, and juvenile and adult organisms. As a chemical agent, it stirs up the surface waters, keeping its warm waters well mixed. As a physical agent, it moves enormous quantities of water. It is clearly a remarkable current and a very important part of the global ecosystem.

STUDY OF THE GULF STREAM
The Gulf Stream is studied by directly measuring the strength of its currents at different depths and locations and by examining the effects of its currents through monitoring the position of floats released within it and designed to stay at particular depths.

Floating current meters can be moored to anchors at the bottom of the sea. Their depth is controlled by the length of the tether keeping them attached to the anchor. They can record data electronically, storing it in computer memory. When a vessel is at the surface, ready to retrieve the meter and its data, it transmits a special coded sound pulse. This instructs the meter to release itself from the tether and rise to the surface, where it transmits a radio signal allowing the vessel to home in on it for recovery. The data are incorporated in complex computer models that tie together the results obtained from hundreds of current meters deployed during overlapping time periods. Snapshots of the current system can then be obtained, and sequences of these snapshots reveal the behavior of the currents over time.

Floating objects can be released at sea and tracked by satellite. To ensure that these objects are being moved by ocean currents rather than surface winds, they usually have a large parachute or sail deployed in the water beneath them. Because the density of seawater increases with depth, floats are often designed to have neutral buoyancy at a particular depth (neither sinking below, nor floating above that depth). A layer of the ocean (the SOFAR channel) acts as a wave guide for sound waves. Floats in this layer can transmit sounds over tremendous distances, permitting them to be tracked efficiently by a small number of surface ships with sonar receivers suspended into this layer.

Because geostrophic currents are driven by the slopes of the ocean's surface, any technique that can measure those slopes can provide valuable insight into the driving forces behind the Gulf Stream. These slopes are very gradual—total dynamic relief over all the world's oceans is about 2 meters—and, consequently, direct measurement is difficult. Satellite techniques, coupled with computer models to filter out waves, tides, and dozens of other confounding effects, are approaching the point where they will be able to measure this topography. Yet this dynamic topography is generally determined indirectly by measuring temperature and salinity as a function of depth and position. These data are used to determine the density of the seawater as a function of depth and position. By assuming that at some depth the horizontal pressure gradients have disappeared, it is possible to reconstruct the differences in the height of the water column needed to accommodate these variations in density. Then the velocities and directions of the resulting currents can be calculated. When the theoretical models are compared with currents measured by moored meters or revealed by the paths of floating objects, the results are in agreement. This gives strong support to the theoretical concepts underlying the study of ocean currents.

Many of the approaches used to study of the Gulf Stream are exercises in applied mathematics. That computed and measured results agree so well is a triumph of geophysical fluid dynamics.

SIGNIFICANCE
The North Atlantic Gyre, which has the Gulf Stream as its northwest boundary current, dominates surface flow in the Atlantic Ocean. Voyages of discovery, exploration, conquest, and exploitation all were affected to some extent by this system and the winds that accompany it.

The Gulf Stream, studied since the eighteenth century, has provided the basis for much of what scientists know of ocean currents. Elaborate mathematical constructs, including the entire concept of geostrophic currents, have been developed to describe, analyze, and comprehend this powerful system.

As scientists have learned more about the Gulf Stream, they have discovered many new areas of study, including branches in the stream, countercurrents at the surface and at depth, and fluctuations in flow and velocity with time and place. Some researchers devote their entire careers to studying just the rings of the Gulf Stream, which contain entire ecosystems. Others investigate the location and strength of the Gulf Stream in the distant past, during and even before the ice ages. There is evidence that at times the current has taken different paths across the continental shelf, perhaps scouring out valleys in the ocean floor in the process.

Scientists' understanding of the dynamics of the planet relies on comprehending the transfer of energy from the equator to the poles. The Gulf Stream is an important component of this transfer. As people have become more aware of the fragility of the environment and become more concerned about issues of global climate change, the role of the Gulf Stream and thermohaline circulation in influencing temperatures in Europe and elsewhere on the planet has taken on a new importance.

—*Otto H. Muller*

FURTHER READING

Allen, P A. *Earth Surface Processes*. Blackwell Science, 2002.

Broecker, Wally. *The Great Ocean Conveyor*. Princeton University Press, 2010.

Colling, Angela. *Ocean Circulation*. 2nd ed. Butterworth-Heinemann, 2001.

MacLeish, William H. *The Gulf Stream*. Houghton Mifflin, 1989.

Segar, Douglas A., and Elaine Stamman, Segar. *Introduction to Ocean Sciences*. Douglas A. Segar, 2012.

Ulanski, Stan L. *The Gulf Stream: Tiny Plankton, Giant Bluefin, and the Amazing Story of the Powerful River in the Atlantic*. University of North Carolina Press, 2008.

Vallis, Geoffrey K. *Atmospheric and Oceanic Fluid Dynamics: Fundamentals and Large-scale Circulation*. Cambridge University Press, 2006.

Voituriez, Bruno. *The Gulf Stream*. UNESCO, 2006.

GYRES AND CLIMATE CHANGE

FIELDS OF STUDY

Oceanography; Physical Geography; Environmental Studies; Environmental Science; Climate Modeling; Process Modeling; Hydrology; Earth System Modeling

SUMMARY

The oceans' gyres are basin-scale circulation patterns in which the net flow of water occurs in a circular pattern around the basin. Each gyre is generally made up of four distinct currents that are driven by wind stresses, and its circulation direction is governed by the Coriolis Effect. The major, subtropical gyres in the Atlantic and Pacific Oceans are located between the equator and approximately 45° north latitude, while the smaller, subpolar gyres lie north of that latitude. The Antarctic Circumpolar Current is a gyre that flows continuously around Antarctica, because there are no landmasses to impede it.

PRINCIPAL TERMS

- **Coriolis Effect:** the effect by which an object moving in a straight line across a rotating surface is seen to follow a curved path
- **decadal:** describing a cyclical event that occur over a period of at least ten years
- **downwelling:** the bulk movement of water from surface levels to lower depths
- **upwelling:** the bulk movement of water from lower depths to surface levels

GYRE CURRENTS

Individual currents that make up a gyre have different characteristics. Because of the Coriolis Effect and resulting differences in sea surface elevation, the western boundary currents (currents that flow on the western side of the ocean basins and flow northward from the equator) are narrow, are relatively deep (up to 1,200 meters), and have velocities of up to 178 kilometers per day. Examples of western boundary currents

Earth's Major Gyres

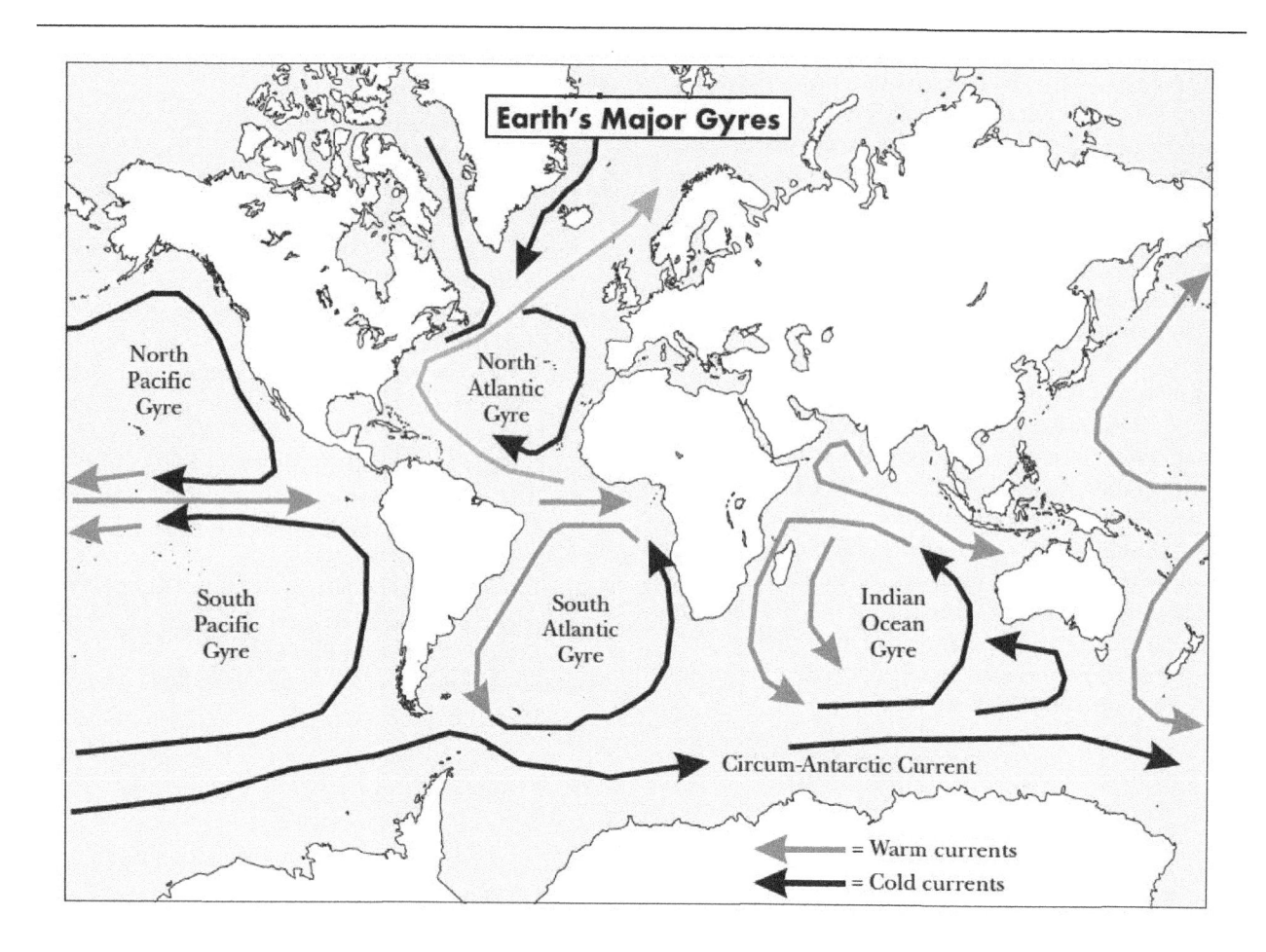

Map showing: North Pacific Gyre, North Atlantic Gyre, South Pacific Gyre, South Atlantic Gyre, Indian Ocean Gyre, and Circum-Antarctic Current. Legend: Warm currents, Cold currents.

include the Gulf Stream in the Atlantic Ocean and the Kuroshio Current in the Pacific Ocean.

Eastern boundary currents have widths of up to almost 1,000 kilometers but are only 500 meters deep. Examples of eastern boundary currents include the California Current in the Pacific Ocean and the Canary Current in the Atlantic Ocean. Offshore Ekman transport associated with the eastern boundary currents leads to upwelling and high levels of productivity in these surface waters. Circulation of the gyres is completed by transverse currents that flow east and west across the ocean basins, connecting the western and eastern boundary currents.

The subtropical gyres are associated with persistent high-pressure regions in the atmosphere, which leads to a net motion of surface water to the center of the gyre—a process known as convergence. These regions of high pressure are associated with low annual rainfall totals, so the salinity of water in the center of the gyres is somewhat elevated. Elevated salinity and convergence lead to downwelling, so the central gyres are regions of low productivity.

SIGNIFICANCE FOR CLIMATE CHANGE

The gyres transfer large amounts of heat from the equator to the poles. For example, the Gulf Stream carries heat north from the Caribbean Ocean, travels along the East Coast of the United States, and then curves eastward toward Europe. The heat released from the Gulf Stream may lead to warmer average temperatures in Europe than those found at similar latitudes in North America. It has been hypothesized that global warming will lead to an increased influx of freshwater to the Arctic Ocean, which could block the northward flow of the warm, salty water of the Gulf Stream. In turn, this could prevent heat transport from the equator and could lead to cooling of the Northern Hemisphere.

The strength of the gyral currents can vary on decadal timescales, such as is observed in the Atlantic

Multidecadal Oscillation (AMO), which manifests as cycles in average sea surface temperatures (SSTs), as lesser or greater amounts of warm water are transported from equatorial regions. The intensity of hurricanes is strongly dependent on SST, with stronger, more frequent hurricanes occurring when SSTs are higher. Thus, if global temperatures increase, the amplitude of the AMO will increase, leading to the circulation of warmer water and increased hurricane intensity.

Decadal variations in gyral flow can also affect the extent of coastal upwelling, salinity, and nutrient concentrations. Scientists from the Georgia Institute of Technology have discovered the existence of the North Pacific Gyre Oscillation. Variations in the mode of the oscillation are thought to cause historical variations in fish populations that are critical to Pacific fisheries. While such oscillations are part of the natural climate cycle, evidence indicates that the amplitude of the oscillations may increase with global warming.

Water in the center of the gyres is isolated from the rest of the world's oceans. An important consequence of this isolation is the low levels of phytoplankton productivity of the water, a situation described as "oligotrophic." The major source of nutrients to support productivity in the central gyres is upwelling of deep waters. However, downwelling, not upwelling, occurs in the central gyres, making these waters the equivalent of large, open-ocean deserts. As global temperatures have increased, the size of these ocean deserts has increased. Scientists from the National Oceanic and Atmospheric Administration and the University of Hawaii have examined a nine-year series of remotely sensed ocean color data from the SeaWiFS satellite and concluded that these open-ocean deserts have expanded by up to 15 percent. The expansion of the oligotrophic regions is consistent with computer models of oceanic vertical stratification in the gyres, but the rate of expansion exceeds all model predictions.

—*Anna M. Cruse*

Further Reading

Broecker, Wally *The Great Ocean Conveyor: Discovering the Trigger for Abrupt Climate Change.* Princeton University Press, 2010. Describes the global oceanic overturning circulation system and how its disruption by climate change may affect the climate of Earth.

Di Lorenzo, Emanuele, et al. "North Pacific Gyre Oscillation Links Ocean Climate and Ecosystem Change." *Geophysical Research Letters* 35, no. 108607 (April, 2008). Overview of the discovery of the North Pacific Gyre Oscillation and its consequences for Pacific ecosystems and fisheries industries.

Open University. *Ocean Circulation.* 2nd ed. Elsevier Butterworth-Heinemann, 2005. This introductory oceanography textbook covers atmospheric circulation, surface currents, and thermohaline circulation. The theory of fluid flow throughout the ocean is developed with a minimum of mathematics (knowledge of algebra and trigonometry is assumed). Illustrations, figures, tables, maps, references, index.

Polovna, Jeffrey J., Evan A. Howell, and Melanie Abecassis. "Ocean's Least Productive Waters Are Expanding." *Geophysical Research Letters* 35, no. 103618 (February, 2008). Presents the SeaWiFS data, showing that the ocean's oligotrophic regions in the subtropical gyres are expanding at rates exceeding those predicted by computer models.

Schmittner, Andreas, Chiang, John C.H. And Hemming, Sidney R., eds. *Ocean Circulation. Mechanisms and Impacts.* John Wiley & Sons, 2013. Provides a thorough description of the effects on Earth's climate due to large "turnover" circulation currents, particularly the Atlantic conveyor system.

Trenberth, Kevin E. "Warmer Oceans, Stronger Hurricanes." *Scientific American* 297, no. 1 (July, 2007): 44–51. Describes how warmer oceans and variations in gyral currents such as the Gulf Stream affect decade-scale circulation patterns and the formation of hurricanes.

H

HADLEY CIRCULATION AND CLIMATE CHANGE

FIELDS OF STUDY

Atmospheric Science; Heat Transfer; Fluid Dynamics; Environmental Sciences; Physical Geography; Earth System Modeling; Meteorology; Climate Modeling; Hydroclimatology; Climate Classification; Climate Zones; Oceanography

SUMMARY

Named for eighteenth century meteorologist George Hadley (1685–1768), who explained how it worked, the Hadley circulation is a loop of air that starts near the equator. The hot air in the equator region rises to the troposphere level and is carried toward the poles. At about 30° latitude, the cooled air drops back to the surface, creating a high-pressure area. The air flows back toward the equator, completing the loop and generating the trade winds. Because Earth is spinning, the air traveling back to the equator moves toward the southwest in the Northern Hemisphere and is called the Northeast Trades. In the Southern Hemisphere, the air movement generates the Southeast Trades. The low-pressure area where the Northeast Trades meet the Southeast Trades and where the hot air rises is called the Inter-Tropical Convergence Zone (ITCZ). The ITCZ does shift with the seasons.

PRINCIPAL TERMS

- **greenhouse effect:** an atmospheric phenomenon in which certain molecular components of the atmosphere capture infrared radiation from a planet's surface and prevent it from radiating back out into space
- **greenhouse gases (GHGs):** atmospheric gases whose molecules absorb infrared radiation from a planet's surface and then emit a significant portion of it back into the atmosphere, most commonly carbon dioxide, methane, water and ozone.
- **solar heating:** heat acquired by absorption of radiant energy from the Sun

THE HADLEY CIRCULATION LOOP

The Hadley circulation is caused by solar heating. Accordingly, the intensity of the Hadley circulation is related to the sea surface temperature. The hot air carries moisture aloft, but as the air rises, it cools and is able to contain less moisture. The moisture condenses, causing the large amount of rainfall in the equatorial region. Nevertheless, the air that continues to travel toward the poles is moist, relatively warm air and drier, cooler air that flows back toward the equator.

The Hadley circulation is one of the major regulators of the Earth's energy budget. It spreads heat collected at the equator to the northern and southern subtropical areas. It also carries heat into the troposphere, where it can radiate into space. Hadley's ascending limb controls the rainfall in the tropical areas, where large amounts of rain occur. The descending limb controls the dryness of the subtropical area. Although the Hadley circulation covers only part of the Earth, it covers the area where a large percentage of the people of the world live.

SIGNIFICANCE FOR CLIMATE CHANGE

A difference in the patterns of the sea surface temperature would force a change in precipitation and cause a change in the Hadley circulation. A change in that system would cause a change in the flow of heat, momentum, and humidity along the meridians. Earth's overall radiative balance, along with the monsoon systems and the ocean circulation, are also affected by a change in Hadley circulation. Different climate models diverge when they project the effect

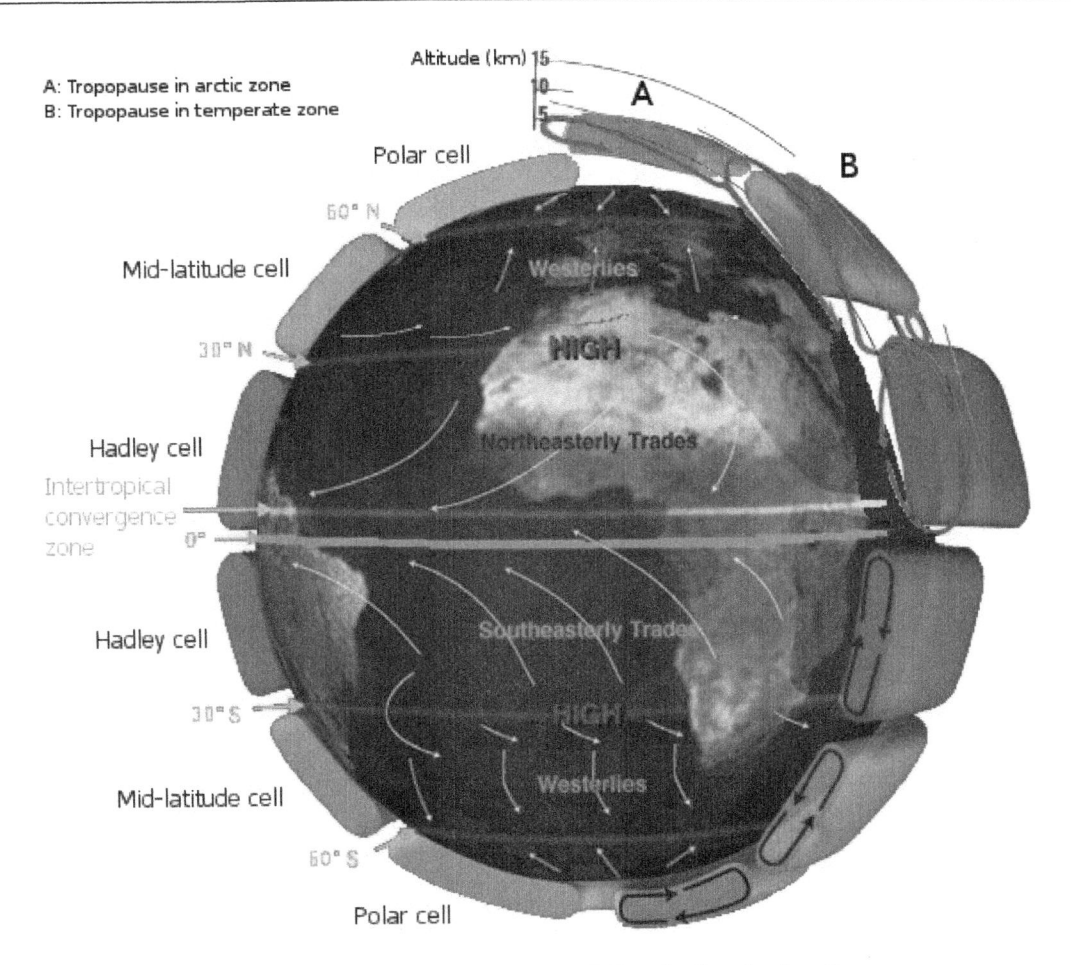

Global circulation of Earth's atmosphere displaying Hadley cell, Ferrell cell, and polar cell.

of an increase in greenhouse gases (GHGs) on the Hadley circulation. Some models indicate an increase of the intensity of the Hadley circulation, causing a more arid subtropical region to develop as the rising GHG concentration causes an increase in sea surface temperature. Other models indicate a weakening of the Hadley circulation.

Since 1950, there has been a more intense Hadley circulation, with a consequential increase in rainfall in the equatorial oceanic region and a drier tropical and subtropical landmass. This increase has accompanied a stronger westerly stratospheric flow and an increase in cyclones in the middle latitudes. The driving force behind these changes has been identified as the warming of the Indo-West Pacific tropical waters. The increased sea surface temperature difference between the winter and summer hemisphere tropics causes a stronger Hadley circulation.

Models indicate that the solar forcing of the increased Hadley circulation is more intense with GHGs than without. It is not clear from the models proposed whether the more intense Hadley circulation is due to a natural fluctuation, is anthropogenic, or is the combined result of natural and human factors. Solar forcing models indicate more evaporation and thus more moisture carried aloft. This causes less cloud cover and more solar heating. Thus, solar forcing seems to represent a positive feedback loop. The more solar heating, the less cloud cover and more solar heating.

Models of past data indicate that the ITCZ may have shifted over time by more than just the annual summer-to-winter position shift. Another shift would cause different areas of Earth to be dry or to be rain-soaked. The extent of the Hadley circulation is influenced by several different factors according to models. One factor that was shown not to be related is the mean global temperature. Just increasing the Earth's temperature will not change the area covered by the Hadley circulation. The Hadley circulation can be likened to El Niño-Southern Oscillation events. Since 1976, the increase in number and strength of El Niño events has caused an increase in the strength of the winter Hadley circulation.

—C. Alton Hassell

FURTHER READING

Diaz, Henry F., and Raymond S. Bradley, eds. *The Hadley Circulation: Present, Past, and Future.* Kluwer Academic, 2004. Essay collection that grew out of a conference examining the Hadley circulation from several different viewpoints. Illustrations, maps, references.

Hadley, G. "Concerning the Cause of the General Trade-Winds." *Philosophical Transactions of the Royal Society of London* 39 (1735): 58–62. George Hadley takes Halley's concept and expands on it.

Halley, E. "An Historical Account of the Trade Winds, and Monsoons, Observable in the Seas Between the Tropicks, with an Attempt to Assign the Physical Cause of the Said Winds." *Philosophical Transactions of the Royal Society of London* 16 (1686): 153–168. The original article explaining the Hadley circulation. Illustrations.

O'Hare, Greg, et al. *Weather, Climate and Climate Change Human Perspectives.* Lord, Taylor & Francis, 2014. Comprehensive review of weather and its causes, including the Hadley cell. Illustrations, references, index.

Watterson, Ian Godfrey, and Edwin K. Schneider. "The Effect of the Hadley Circulation on the Meridional Propagation of Stationary Waves." *Quarterly Journal of the Royal Meteorological Society* 113, no. 477 (July, 1987): 779–813. Based on Watterson's doctoral dissertation of the same name, this article distills one of the only extended studies of Hadley circulation, presenting its most salient insights and conclusions.

HEAT VS. TEMPERATURE

FIELDS OF STUDY

Thermodynamics; Physical Chemistry; Physics; Heat Transfer; Chemical Engineering; Engineering; Environmental Studies; Physical Sciences; Statistics; Process Modeling; Fluid Dynamics; Chemical Kinetics; Atmospheric Science; Oceanography

SUMMARY

The concepts of heat and temperature are intimately related, so these quantities are often confused, and the terms are sometimes improperly used interchangeably by laypersons. Temperature is a measurement of the average kinetic energy of the atoms and molecules of a substance. Heat is the energy that is transferred between two bodies that have different absolute temperatures.

PRINCIPAL TERMS

- **rotation:** refers to the spinning of atoms and molecules about a central axis
- **translation:** refers to the motion of atoms and molecules moving through three-dimensional space
- **vibration:** refers to several different spring-like oscillations that bonds within molecules are capable of undergoing, such as stretching and compression, and asymmetric and symmetric scissoring

HEAT AND TEMPERATURE

The concepts of heat and temperature are closely related, so these quantities are often confused, and the terms are sometimes used interchangeably (but incorrectly) by laypersons. Temperature is a measurement of the average kinetic energy of the atoms and molecules of a substance, due to their physical motions of vibration, rotation and translation. The kinetic energy of an object is calculated according to the following formula:

$$E_K = \tfrac{1}{2} MV^2$$

where E_K is kinetic energy, M is mass, and V is velocity. The kinetic energy of atoms is generally due to

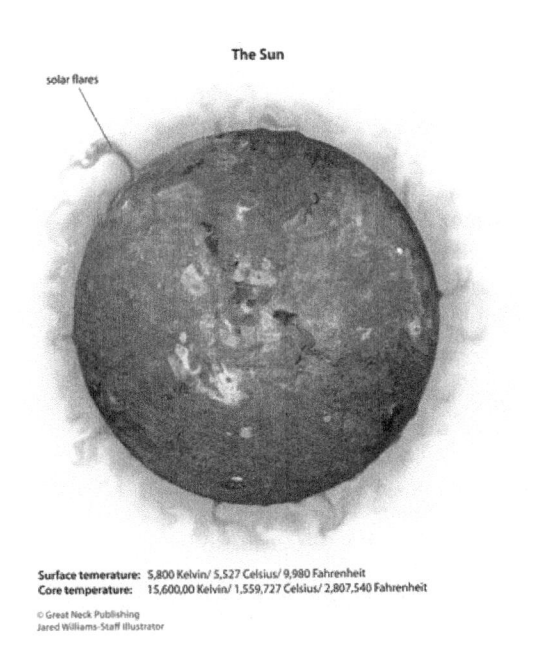

Sun: The sun sends its heat and light to Earth, which is over 150 million kilometers (94 million miles) away.

a combination of their directional velocity and vibrational motion. This energy may be measured relative to a reference standard by a thermometer.

The two most common temperature scales in use are the Fahrenheit and Celsius (formerly centigrade) scales. In the Fahrenheit scale, the freezing point of water is defined as 32° and the boiling point is defined as 212°. In the Celsius scale, the freezing point of water is defined as 0° and the boiling point is defined as 100°. In science and engineering, the Rankine and Kelvin temperature scales are used. These are called absolute temperature scales, because their zero reference level is absolute zero, the lowest temperature that can theoretically exist. At absolute zero, the total kinetic energy of all the atoms in a substance is zero.

Heat is the form of energy transferred across the boundary of an object as a result of a temperature difference across that boundary. Consider what happens across a pane of window glass on a cold winter day. The air inside the heated house is warmer than the air outside. This temperature difference across the window pane causes an energy flow through the glass from the warm air inside to the cold air outside: The outer air next to the glass is warmed, while the air inside the house cools, requiring a heating system to maintain a constant, comfortable temperature.

There are three types of heat transfer: conduction, convection, and radiation. All result from a temperature difference between an object and its surroundings.

Conduction is the primary mechanism of heat transfer in solids. It occurs because of molecular activity in the solid. Convection is the primary mechanism of heat transfer through fluids and results from bulk mixing between fluid layers. Radiation, the third type of heat transfer, is the only mechanism that can transfer heat through a vacuum. When the Sun heats the Earth, there is no solid between them through which heat can be conducted, nor is there is any fluid through which heat can be convected. The temperature difference between the Sun and the Earth still causes heat transfer to the Earth by means of electromagnetic waves. Electromagnetic waves can also transfer energy in the form of light, X rays, or radio waves.

Heat is measured as a function of the temperature change of a substance when heat transfer occurs. The commonly used units for heat are the calorie, the joule, and the British Thermal Unit (BTU). A calorie is the amount of heat required to raise the temperature of one gram of water by one Celsius degree. A BTU is the amount of heat required to raise the temperature of one pound of water by one Fahrenheit degree. One joule is equal to 0.239 calorie.

Significance for Climate Change

Heat transfer to the Earth's oceans and atmosphere is the first step in, and at the very heart of, climate change. Most of the industrial or mechanical processes in which humans engage release heat as a waste by-product. For example, all fossil fuel and nuclear power plants generate waste heat as a by-product of producing power. A typical power plant may have an efficiency of between 35 percent and 50 percent, meaning that this percentage of the stored energy in a fuel is converted to useful energy, while the remaining energy is discharged to the atmosphere or some body of water as waste heat. This heat transfer to the environment causes a temperature increase of the environment. This temperature increase is a contributor to global warming. Even a degree or two increase in the average temperature of the oceans or the atmosphere can cause profound effects upon Earth's biosphere. All combustion engines also dissipate waste heat to the environment, as do heating and cooking appliances. Only power produced by wind, water, or geothermal energy does not directly contribute to global warming, although these types of power production can have other effects.

—*Eugene E. Niemi, Jr.*

FURTHER READING

Çengel Yunus A, et al. *Thermodynamics: An Engineering Approach.* 2019. Sophomore- or junior-level college textbook dealing with heat and energy. Includes a simplified discussion of global warming and the greenhouse effect.

Holman, Jack P. *Heat Transfer.* McGraw-Hill, 2014. College engineering textbook dealing with the various mechanisms of heat transfer.

Thurman, Harold V., and Alan P. Trujillo. *Introductory Oceanography.* Prentice Hall, 2008. Includes a section on air-sea interaction and its relationship to global warming.

HUMIDITY AND GREENHOUSE GASES

FIELDS OF STUDY

Atmospheric Chemistry; Environmental Chemistry; Physical Chemistry; Meteorology; Thermodynamics; Climate Modeling; Hydroclimatology; Atmospheric Science; Heat Transfer

SUMMARY

Humidity is the amount of water present in the atmosphere in the form of vapor. As a gas, water vapor contributes to the local atmospheric pressure in accordance with Dalton's law of partial pressures. If the local partial pressure of water is exactly equal to its vapor pressure, the air is said to be saturated. This state is defined as 100 percent humidity, and the corresponding temperature is water's dew point. Water vapor is the most abundant greenhouse gas (GHG) in Earth's atmosphere, exceeding the amount of CO_2 by a factor of one thousand.

PRINCIPAL TERMS

- **albedo:** a value between zero and one representing the fraction of light reflected from a body such as a planet
- **equilibrium:** the condition in which no net change occurs within a dynamic system, such as when exactly the same amount of water vapor condenses to liquid as liquid water evaporates to water vapor
- **greenhouse effect:** an atmospheric phenomenon in which certain molecular components of the atmosphere capture infrared radiation from a planet's surface and prevent it from radiating back out into space
- **greenhouse gases (GHGs):** gases that act to retain heat within Earth's atmosphere, tending to increase average global temperatures
- **radiation budget:** the balance between solar energy entering Earth's atmosphere from space and the heat energy leaving Earth's atmosphere to pass outward into space
- **saturated:** the condition in which air holds the maximum amount of water vapor for the particular conditions of temperature and pressure
- **supersaturated:** the condition in which the amount of water vapor in air exceeds the maximum amount according to the particular conditions of temperature and pressure

VAPOR PRESSURES AND WATER VAPOR

Humidity is the amount of water present in the atmosphere in the form of vapor. As a gas, water vapor contributes to the local atmospheric pressure in accordance with Dalton's law of partial pressures. In any mixture of gases, the partial pressure of any one component is equal to the total pressure of the mixture multiplied by the fraction of the gas present in the mixture. For example, molecular oxygen constitutes 20 percent of the atmosphere, so the partial atmospheric pressure of oxygen is 20 percent of Earth's total atmospheric pressure. The total pressure is about 1.03 kilograms per square centimeter, so the partial atmospheric pressure of oxygen is about .20 kilograms per square centimeter.

Water normally exists in liquid and solid as well as vaporous form. Its vapor pressure is the pressure at which pure water vapor coexists in equilibrium with either the liquid or the solid state. At equilibrium, the liquid would not evaporate, the solid would not sublimate, and the vapor would not condense. By contrast, if the local partial pressure of water vapor is greater than its vapor pressure, the vapor condenses; if the local partial pressure is less than the vapor pressure, then the liquid evaporates and the solid sublimates. Vapor pressure is not a constant but rather is a function of temperature.

If the local partial pressure of water is exactly equal to its vapor pressure, the air is said to be saturated. This state is defined as 100 percent humidity, and the corresponding temperature is water's dew point. If the vapor pressure of water is equal to the total local atmospheric pressure, the water will evaporate without limit, and the corresponding temperature is water's boiling point. Relative humidity is the ratio, expressed as a percentage, of the local partial pressure of water vapor to the vapor pressure associated with the local temperature.

Humidity can exceed 100 percent, a condition known as supersaturation. In supersaturation, water vapor's partial pressure exceeds the theoretical vapor pressure at that temperature. Condensation cannot take place, however, unless condensation nuclei are present. Water droplets exceeding a certain critical size act as such nuclei, absorbing water vapor and growing; water droplets below the critical size evaporate. If no droplets larger than the critical size exist and no other condensation nuclei are present, then the supersaturated vapor is stable. Fine, dry particles, such as dust or pollutants, also act as condensation nuclei in supersaturated air.

Evaporation is an endothermic process, or one that requires an input of energy in order to occur. The change of phase from liquid to gas takes place at constant temperature. The energy consumed by the process is stored in the water vapor in the form of latent heat of vaporization. When the vapor condenses, all of the latent heat is released, which means that condensation is exothermic. Water has a latent heat of vaporization of 2,256,000 joules per kilogram, an unusually high value for such a simple compound.

Significance for Climate Change

Water vapor is the most abundant greenhouse gas (GHG) in Earth's atmosphere, exceeding the amount of CO_2 by a factor of one thousand. It is transparent to visible radiation but opaque to infrared radiation of 5.25 to 7.5 micrometers in wavelength, which is

Cloud forest. Mount Kinabalu, Borneo.

a lower frequency range than that of visible light. Incoming solar radiation peaks in the visible spectrum. Energy re-emitted by the Earth peaks in the infrared portion of the spectrum. Thus, incoming energy is better able to penetrate water vapor than is outgoing energy.

In order for Earth to maintain thermal equilibrium if there were no humidity in the atmosphere, radiating as much energy back to space as it receives from the Sun, the global average temperature would be much higher than it is in order to accommodate the energy that is currently reflected away as Earth's albedo.. The balance of the amounts of radiation received from the Sun and emitted back into space is called the radiation budget.

No simple statements about the effect of total atmospheric water vapor on climate change are possible, because the atmosphere is a nonequilibrium system. Water vapor resides in the air for fairly short periods before precipitating out as rain, snow, or dew. As a result, the amount of water vapor in the atmosphere itself responds quickly to changes in climate. Water vapor in turn affects Earth's radiation budget and, through it, surface temperatures, closing the loop and generating feedback. Layers of air near the planet's surface are warm enough and close enough to the oceans to stay relatively saturated. Their effect on the radiation budget is small, because they are nearly as warm as the surface itself. Upper layers of the atmosphere are cooler and moistened only by the water vapor that convects or diffuses upward from the layers below. Small quantities of water in these upper layers can have a disproportionate effect on the radiation budget, trapping enough infrared energy to significantly warm the climate.

This feedback is complicated, however, by the presence of water in the air as suspended droplets in the form of clouds and fog. These droplets scatter visible radiation in all directions, preventing an appreciable fraction of the incident solar energy from reaching the ground and contributing to Earth's albedo (the fraction of incident solar energy reflected back into space). Ice and snow have the same effect. Increases in albedo have a cooling effect and act to moderate any global average temperature increases.

—*Billy R. Smith, Jr.*

FURTHER READING

Colman, B. R., and T. D. Potter, eds. *Handbook of Weather, Climate, and Water*. Wiley-Interscience, 2003. Concise and thorough treatment of the hydrologic cycle and its effect on climate. Illustrations, figures, tables, references, index.

Schneider, Tapio, and Adam H. Sobel. *The Global Circulation of the Atmosphere*. Princeton University Press, 2007. Chapter 6, "Relative Humidity of the Atmosphere," may be difficult for the nonexpert to read in depth, but there is value in skimming it to understand the authors' positions on the subject. Figures, tables, and references.

Taylor, F. W. *Elementary Climate Physics*. Oxford University Press, 2014. Readers without significant prior knowledge of climate science should consult this book for help in understanding difficult topics. Illustrations, figures, tables, bibliography, index.

HURRICANES AND TYPHOONS

FIELDS OF STUDY

Atmospheric Science; Heat Transfer; Fluid Dynamics; Meteorology; Climate Modeling; Process Modeling; Hydroclimatology; Hydrometeorology; Earth System Modeling; Spectroscopy; Electromagnetism

SUMMARY

Hurricanes are cyclonic storms that form over tropical oceans. A single storm can cover hundreds of thousands of square kilometers and have interior wind speeds of 65 to 230 knots (74 to 200 miles) per hour near its eye. Destruction is caused by wind damage, as well as by storm surge and subsequent flooding.

PRINCIPAL TERMS

- **condensation:** the process by which water, or any other substance, changes from a vapor state to a liquid state, releasing heat into the surrounding air; this process is the opposite of evaporation, which requires the input of heat
- **Coriolis force:** an apparent force caused by the rotation of the planet, in which objects moving above Earth's surface (such as the wind) deflect to the right in the Northern Hemisphere and to the left in the Southern Hemisphere
- **knot:** a unit of nautical distance equivalent to 1.86 kilometers or 1.15 miles
- **tropical cyclone:** an area of low pressure that forms over tropical oceans, characterized by extreme amounts of rain, a central area of calm air, and winds that attain speeds of up to 300 kilometers per hour rotating counterclockwise in the Northern Hemisphere and clockwise in the Southern Hemisphere
- **tropical depression:** cyclonic thunderstorms with wind speeds from 36 to 64 kilometers per hour
- **tropical storm:** a thunderstorm with cyclonic winds circulating at speeds of 64 to 118 kilometers per hour
- **vortex:** a mass of air, water, or other fluid that spins about a central axis, capable of reaching high velocities

ANATOMY OF A HURRICANE

Hurricanes are huge, swirling storm systems that can cover thousands, and sometimes hundreds of thousands, of square kilometers, averaging about 600 kilometers in diameter. Often called the "greatest

Anatomy of a Hurricane

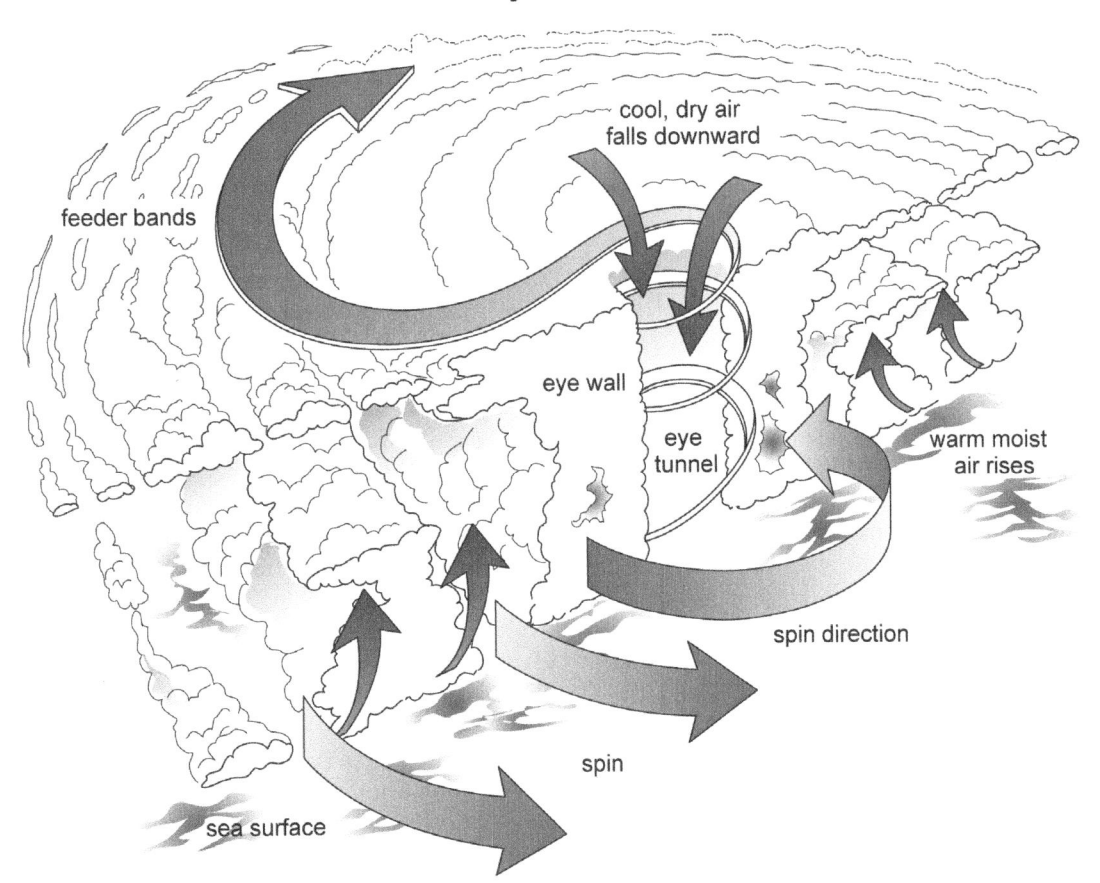

storms on Earth" hurricanes have sustained winds of at least 65 knots (118 kilometers) per hour with maximum wind speeds of 230 knots (320 kilometers) per hour. In the Western Hemisphere, these storms are called hurricanes. They are also referred to as typhoons in the western North Pacific, cyclones in the Indian Ocean, and *baguios* in the Philippines. The swirling motion of these storms is counterclockwise, or cyclonic, in the Northern Hemisphere and clockwise, or anticyclonic, in the Southern Hemisphere.

Hurricanes are as individual and unique as fingerprints. Their behavior is difficult to predict, even when satellites are used to track them. Scientists who have studied decades of hurricane data have found some patterns. Hurricanes evolve primarily in specific areas of the west Atlantic, east Pacific, south Pacific, western north Pacific, and north and south Indian Oceans. They rarely move closer to the equator than 4 to 5 degrees latitude north or south, and no hurricane has ever been known to have crossed the equator. They are more common during the warmer months of the year depending on their ocean of origin. In the Northern Hemisphere, hurricanes are most common from May to September; in the Southern Hemisphere, the hurricane season typically ranges from December to May. More recently it has been found that the hurricane season has been increasing in length, in accord with increased warming of ocean waters.

The reasons for these patterns are that hurricanes need warm surface waters, high humidity, and winds from the same direction at a constant speed in order to form. Cyclonic depressions can only develop in areas where the ocean surface temperatures are more than 24 degrees Celsius; the eye structure, which must be present for the storm to be classified as a hurricane, requires surface temperatures of 26 to 27 degrees Celsius to form. This means that hurricanes will rarely develop above 20-degree latitudes because the ocean temperatures are never warm enough to provide the heat energy needed for their formation. In the Northern Hemisphere, the convergence of air that is ideal for hurricane development occurs above tropical waters when eastward-moving waves develop under the fairly constant force of the trade winds. The region around the equator is traditionally called "the doldrums" because there is no consistent direction to wind flow. Hurricanes, needing wind to form, can be found as close as 4 to 5 degrees away from the equator. At these latitudes, the Coriolis Effect, an apparent deflecting force associated with Earth's rotation, gives the moving air masses the spin necessary to form hurricanes.

In hurricane formation, heat is extracted from the ocean, and warm, moist air begins to rise, forming clouds and causing instability in the upper atmosphere. As the air rises, it spirals inward toward the center of the system. This spiraling movement causes the seas to become turbulent, and large amounts of sea spray are captured and suspended in the rising air. This spray increases the rate of evaporation fueling the storm with water vapor.

As the vortex of wind, water vapor, and clouds spin at a faster and faster rate, the eye of the hurricane forms. The eye, which is the center of the hurricane, is a relatively calm area that has only light winds and fair weather. The most violent activity in the hurricane takes place in the area immediately adjacent to the eye, called the eye wall. It is in the eye wall that the spiraling air rises and cools, while moisture condenses into droplets that form rainbands and clouds. The process of condensation releases latent heat, which causes the air to rise and generate more condensation. The air thus rises rapidly, creating an area of extremely low pressure close to the storm's center. The severity of a hurricane is often indicated by how low the pressure readings are in the central area of the hurricane.

EYE FORMATION

As the air moves higher, up to 15,000 meters, the air is propelled outward in a cyclonic or anticyclonic flow, depending on latitude. However, some of the air is forced inward into the eye. The compression of air in the eye causes the temperature to rise. This warmer air can hold more moisture, and the water droplets in the central clouds will evaporate. As a result, the eye of the hurricane becomes nearly cloud-free.

The temperature is much warmer in the eye, especially in the middle and upper levels, than outside of it. Therefore a large pressure differential develops across the eye wall, which establishes the violence of the storm. Waves of 15 to 20 meters are common in the open ocean because of winds around the eye. Winds in a hurricane are not symmetrical around the eye; when one faces the direction that the hurricane is moving, the strongest winds are usually to the right of the eye. The radius of hurricane-level winds (velocities of 119 kilometers per hour or more) can vary from 15 kilometers away from the eye in small hurricanes to 150 kilometers in large hurricanes. The strength of the wind decreases in relation to its distance from the eye.

Depending on the size of the eye, which can range from 5 to 65 kilometers in diameter, the calm period of blue skies and mild winds can last from a few minutes to hours when the storm hits. The calm is deceiving because it is not the end of the storm but only a temporary lapse until the winds from the opposite direction are encountered.

HURRICANE DAMAGE

Hurricanes, and similar storms that are less intense, are classified by their central pressure and their sustained wind speed. Tropical depressions have wind speeds below 20 to 34 knots per hour, whereas tropical storms have wind speeds of 35 to 64 knots. To be classified as a hurricane, storms must have sustained winds of 65 knots (74 miles) per hour or higher.

The Saffir-Simpson scale further categorizes the intensity of hurricanes into five levels:

Category 1 hurricanes are considered weak, with sustained winds of 63 to 83 knots (118 to 153 kilometers) per hour. They cause minimal damage to buildings but are powerful enough to damage unanchored mobile homes, shrubbery, and trees. Normally they cause coastal road flooding and minor damage to piers.

Category 2 hurricanes have winds of 83 to 95 knots (153 to 178 kilometers) per hour and can easily damage roofs, doors, and windows on buildings. They also cause substantial damage to trees, shrubs, mobile homes, and piers. Flooding of roads and low-lying areas normally occurs with this level of storm.

Category 3 hurricanes are considered to be strong, with winds of 96 to 113 knots (178 to 210 kilometers) per hour. Such storms are sufficient to destroy mobile homes and can cause structural damage to residences and utility buildings. Flooding from this level of hurricane can destroy small structures near the coast, while larger structures normally sustain damage from floating debris. Flooding from this level of hurricane can extend to 15 kilometers or more inland.

- *Category 4* hurricanes are categorized as very strong, boasting winds of 114 to 135 knots (210 to 250 kilometers) per hour. These storms can cause major damage to lower floors of buildings and can cause major beach erosion. Residences often sustain roof structure failure and subsequent rain damage. Land lower than 3 meters above sea level can be flooded by storm surge, which may lead to the mass evacuation of residential areas up to 10 kilometers inland.
- *Category 5* hurricanes are severe, classified by having sustained winds greater than 135 knots (250 kilometers) per hour. These can completely destroy roofs on residential and industrial buildings, and may destroy the buildings themselves. Structures less than 5 meters above sea level can sustain major flood damage to lower floors, and mass evacuations of residential areas 10 to 20 kilometers inland from the shoreline can be required.

Because wind speeds of about 43 knots (80 kilometers) per hour can break tree branches and cause some damage to structures, hurricane winds are considered to be among the most destructive of natural disasters. The wind force applied to an object increases with the square of the wind speed. A building 30 meters long and 3 meters high that has 160-kilometer-per-hour hurricane speed winds blowing against it would have 18,181 kilograms of force exerted against its walls. This is because a 160-kilometer-per-hour wind exerts a force of approximately 195.5 kilograms per square meter. If the wind speed was 138 knots (256 kilometers) per hour, the force against the house would be a 455,000 kilograms. Rain blown by the wind also contributes to an increase of the pressure on buildings.

Hurricane Andrew, which struck the United States in 1992, caused wind gusts of at least 280 kilometers per hour in south Florida and caused an estimated $25 billion in damage, making it the most expensive hurricane to hit the United States to that time. In 2005 Hurricane Katrina caused massive flooding in the city of New Orleans, which is continuing to recover from the effects of that disaster nearly 15 years later. The cost of that recovery, at an estimated $151 billion dwarfs that of Hurricane Andrew.

STORM PATHS

Hurricanes can travel in very sporadic and unpredictable paths. Some will travel in a generally curved path, while others will change course quite rapidly. Their movements can reverse direction, zigzag, veer from the coast back to the ocean, intensify over water, stall, or return to the same area. The paths are affected by pressure systems in the surrounding atmosphere, the prevailing winds, and Earth's rotation. They can also be influenced by the presence of high- and low-pressure systems on the land they encounter. High-pressure areas can act as barriers to hurricanes; if a high is well developed, its outward-spiraling flow will guide the hurricane around its edges. Low-pressure systems tend to attract the hurricane system toward it.

After forming over warm, tropical water, hurricanes generally travel indirectly toward the poles until they lose their energy over cooler waters. An average hurricane can travel from 450 to 650 kilometers per day and more than 1,600 kilometers before it is downgraded to a tropical storm. Rarely, a hurricane can maintain the required wind speed to latitudes as high as 40 degrees.

STORM SURGE

The greatest cause of death and destruction due to a hurricane comes from the rise in sea level as water is pushed ahead of the high winds, a condition known as storm surge. As the hurricane crosses the continental shelf and moves to the coast, the water level may increase as much as 5 to 7 meters. This is caused by the drop in atmospheric pressure at sea level inside the hurricane, which allows the higher air pressure surrounding the hurricane to push up the sea water within the hurricane, in the same way that the air pressure outside of a drinking straw pushes the liquid up within it. At the same time, the winds in front of the hurricane will pile the water up against the coastline. This results in a wall of water that can be up to 7 meters high and 75 to 150 kilometers wide. This wall of water can sweep across the coastline where a hurricane makes landfall. The combination of shallow shore water and a strong hurricane causes the highest surges of water.

If the storm surge arrives at the time of high tide, the water heights of the surge can be increased by an additional 1 meter or more. The height of the storm surge also depends upon the angle at which the storm encounters the mainland. Hurricanes that make landfall at right angles to the coast will cause a higher storm surge than hurricanes that enter the coast at an oblique angle. Often, the slope or shape of the shore and ocean bottom can create a bottleneck effect and cause an even higher storm surge.

Water weighs 1,000 kilograms per cubic meter and thus has considerable destructive power. Storm surge is responsible for 90 percent of deaths in a hurricane. The pounding of the waves caused by the hurricane can easily demolish buildings. Storm surges cause severe erosion of beaches and the destruction of coastal highways. Often, buildings that have survived hurricane winds have had their foundations eroded by the sea surge or have been demolished by the force of the pounding waves. Storm tides and waves in harbors can destroy ships. The saltwater that inundates land can kill existing vegetation, and the residual salt left in the soil makes it difficult to grow new vegetation.

The most destructive hurricane in the twentieth century occurred in November 1970 in what is now known as Bangladesh. Of the 500,000 people who were killed, most of them were victims of storm surge. In 1985, the same area was hit by another hurricane, causing the deaths of 100,000 people.

RAINFALL AND FLOODING
The amount of rainfall received depends on the diameter of the rainband within the hurricane and on the hurricane's speed. One typhoon in the Philippines caused 185 centimeters of rain to fall in a twenty-four-hour period, a world record. In 2016, Tropical Storm Harvey, not classed as a hurricane, stalled over the Gulf of Mexico near Texas and pumped more than 153 centimeters of rain onto the mainland before making landfall as Hurricane Harvey. Heavy rainfall over a small area that has insufficient drainage can cause flash floods or river floods. Flash floods last from thirty minutes to four hours. The excess water overflows streambeds, damaging bridges, underpasses, and low-lying areas. The strong currents associated with flash floods can move cars off the road, wash out bridges, and erode roadbeds.

Floods in an existing river system develop more slowly. It can take two or three days after a hurricane hits before large rivers overflow their banks. River floods cover extensive areas, last one week or more, and destroy both property and crops. The flood waters eventually retreat, leaving buildings and residences full of mud and ruined furnishings. Rain driven by the wind in hurricanes can cause damage to buildings around windows, through cracks, and under shingles.

TORNADOES AND OTHER DANGERS
Hurricanes may also spawn tornadoes. The tornadoes associated with hurricanes are usually about one-half the size and power of tornadoes that form in the Midwest, and are of a shorter duration. The area these tornadoes affect is small, usually only 200 to 300 meters wide and about 2 kilometers long. Despite their smaller size, they can be very destructive. Tornadoes normally occur to the right of the direction of the hurricane's movement. Approximately 94 percent of tornadoes occur within 10 to 120 degrees from the hurricane eye and beyond the area of hurricane-force winds.

Tornadoes associated with hurricanes are most often observed in Florida, Cuba, and the Bahamas, as well as along the coasts of the Gulf of Mexico and the South Atlantic Ocean. In 1961, 26 tornadoes were associated with Hurricane Carla; Hurricane Beulah released 115 tornadoes in 1967. Hurricane Camille, one of the most lethal storms to hit the United States, created more than 100 tornadoes in 1969.

Not all dangers due to hurricanes have their origins in the atmosphere. During hurricanes, snakes are driven from their natural habitats by the high inflow of saltwater. They are strong swimmers and can be found along roads, in buildings, and in high, dry places. Many people who are bitten by snakes during hurricanes have problems receiving medical attention for their bites because of breakdowns in communications and transportation.

—Toby Stewart
—Dion Stewart

FURTHER READING
Ahrens, C Donald, and Robert Henson. *Meteorology Today: An Introduction to Weather, Climate, and the*

Environment. Cengage, 2019. One of the most widely used and authoritative introductory textbooks for the study of meteorology and climatology. Explains complex concepts in a clear, precise manner and supports them with numerous images and diagrams. Discusses hurricanes and the mechanisms that generate them extensively.

Bryant, Edward A. *Natural Hazards*. 2nd ed. Cambridge University Press, 2005. Provides a sound scientific treatment for the educated layperson. Readers should have a basic understanding of mathematical principles. Presents many case studies. Contains photographs, tables, figures, and a glossary of terms.

Emanuel, Kerry. *Divine Winds: The History and Science of Hurricanes*. Oxford University Press, 2005. Covers a wide variety of topics on a number of hurricane events and related research. Presents events in chronological order with related topics scattered between. Written in a popular style, but still contains portions with "hard science." Includes multiple useful appendices, a list of sources, further reading, credits, and indexing.

Fitzpatrick, Patrick J. *Contemporary World Issues: Hurricanes*. 2nd ed. ABC-CLIO, 2006. A reference work that provides background material on the issues, people, organizations, statistics, and publications related to hurricanes.

Mooney, Chris. *Storm World: Hurricanes, Politics and the Battle Over Global Warming*. Harcourt, 2007. Provides a clear discussion of the nature of hurricanes and the perceived changes in their nature that are still hotly debated in regard to their relationship with global warming. Very readable, presenting very technical information in a nontechnical manner.

Prothero, Donald R. *Catastrophes!: Earthquakes, Tsunamis, Tornadoes, and Other Earth-Shattering Disasters*. Johns Hopkins University Press, 2011. Provides a detailed and clear explanation of the many natural and anthropogenic disasters facing our planet. Each chapter is devoted to a different catastrophe, including earthquakes, volcanoes, hurricanes, ice ages, and current climate changes.

Schneider, Bonnie. *Extreme Weather: A Guide to Surviving Flash Floods, Tornadoes, Hurricanes, Heat Waves, Snowstorms, Tsunamis and Other Natural Disasters*. Palgrave Macmillan, 2012. Presents vivid explanations of how, when, and why major natural disasters occur. Discusses floods, hurricanes, thunderstorms, mudslides, wildfires, tsunamis, and earthquakes. Presents a guide of how to prepare for and what to do during an extreme weather event, along with background information on weather patterns on natural disasters.

HYDROGEN POWER

FIELDS OF STUDY

Inorganic Chemistry; Organometallic Chemistry; Physical Chemistry; Thermodynamics; Chemical Engineering; Engineering; Environmental Studies; Process Modeling; Fluid Dynamics; Chemical Kinetics; Atmospheric Science; Heat Transfer; Crystallography; Spectroscopy

SUMMARY

Molecular hydrogen (H_2) is an ideal fuel to be used for transportation, since the energy content of hydrogen is three times greater than that of gasoline and four times greater than that of ethanol. Hydrogen power powered rockets launched by the National Aeronautics and Space Administration (NASA) for many years. Today, a growing number of automobile manufactures around the world are making hydrogen-powered vehicles. Because of depleting supplies and growing demand for oil, H_2 may become an alternative to gasoline.

PRINCIPAL TERMS

- **electrolysis:** an industrial process in which electrical current is used to force the reduction of water molecules (H_2O) to hydrogen (H_2) and oxygen molecules (O_2)
- **greenhouse effect:** an atmospheric phenomenon in which certain molecular components of the atmosphere capture infrared radiation from a

- planet's surface and prevent it from radiating back out into space
- **greenhouse gases (GHGs):** gases that act to retain heat within Earth's atmosphere, tending to increase average global temperatures
- **steam reforming:** an industrial process employing steam and pressure to force conversion of methane to methanol and molecular hydrogen

HYDROGEN AS FUEL

The idea of hydrogen as the fuel of the future was expressed long ago by Jules Verne in his novel *L'Île mystérieuse* (1874-75; *The Mysterious Island*, 1875). However, compared to oil, H_2 is not abundant on Earth. Its atmospheric concentration is only 0.00001 percent, and there is even less of it in the oceans. Though many microorganisms produce H_2 during fermentation, it is such a good source of energy that it is used almost immediately by other microbes. Thus, in order for humans to use hydrogen as fuel, it must be generated using other energy sources.

While molecular hydrogen is rare, the chemical element hydrogen is the most basic and plentiful element in the universe. It also forms two-thirds of the most abundant chemical compound on Earth, water, as H_2O. Therefore, the challenge posed is to find a cost-effective and environmentally friendly way to generate H_2 from water or other chemical compounds. At present, H_2 is obtained mainly from natural gas (methane and propane) via steam reforming. Although this approach is practically attractive, it is not sustainable. Molecular hydrogen can be also produced by electrolysis. In this process, electric energy is employed to split water into H_2 and O_2. The requisite electricity can be obtained using clean, sustainable energy technologies such as wind and solar power. However, the process is not efficient, requiring significant expenditure of energy and purified water.

There are other technological and economic obstacles to hydrogen power. These obstacles include safety issues, as well as the lack of effective solutions for storage and distribution of H_2. Hydrogen has gained an unwarranted reputation as a highly dangerous substance among the general public. Like all fuels, H_2 may produce an explosion, but it has been used for years in industry and earned an excellent safety record when handled properly.

Hydrogen is the lightest chemical, so it has a much lower energy density by volume than do other fuels. As a gas, it requires three thousand times more space for storage than does gasoline. Thus, hydrogen storage, especially in cars, represents a challenge for scientists and engineers. For storage, H_2 is pressurized in cylinders or liquefied in cryotanks at −253°C. Both processes require a significant expenditure of energy and generate large quantities of waste carbon dioxide (CO_2). In most contemporary hydrogen-powered vehicles, H_2 is stored as compressed gas. Because compressed gas cannot be delivered in the same fashion as liquid fuel, gasoline stations and pumps cannot simply be converted into hydrogen stations. Thus, the distribution system for the new fuel would have to be constructed from scratch, requiring considerable monetary investment.

SIGNIFICANCE FOR CLIMATE CHANGE

Fossil fuels generate CO_2, contributing to the greenhouse effect. Switching from fossil fuels to H_2 would eliminate that source of greenhouse gas (GHG) emissions, provided that the new fuel could be produced without carbon-emitting technologies. Burning H_2 for an energy source produces only water as a by-product, and H_2 is also a renewable fuel, since it can be made from water again. Unfortunately, current methods of H_2 production from natural gas also generate CO_2. The ultimate goal is to generate H_2 without emitting GHGs into the atmosphere, perhaps by using wind or solar power.

One promising green method of H_2 production is a biological approach: A great number of

Hydrogen-powered Mazda RX-8.

microorganisms produce H_2 from inorganic materials, such as water, or from organic materials, such as sugar, in reactions catalyzed by the enzymes hydrogenase and nitrogenase. Hydrogen produced by microorganisms is called biohydrogen. The most attractive biohydrogen for industrial applications is that produced by photosynthetic microbes. These microorganisms, such as microscopic algae, cyanobacteria, and photosynthetic bacteria, use sunlight as an energy source and water to generate hydrogen. Hydrogen production by photosynthetic microbes holds the promise of generating a renewable hydrogen fuel using the plentiful resources of solar light and water.

It is possible to use hydrogen to fuel internal combustion engines. Doing so produces at least one GHG, nitrogen oxide, because the burning of hydrogen requires air, which is almost 80 percent nitrogen (N_2). To use hydrogen power in climate-friendly ways, it will be necessary to replace the internal combustion engine with fuel cells, which produces electricity to power vehicles without motors. Fuel cells are like batteries: They generate electricity via chemical reactions between H_2 and O_2. Fuel cells emit water and heat, not CO_2 or other GHGs. In addition, fuel cells are 2.5 to 3 times more efficient in converting H_2 energy than are internal combustion engines.

Hydrogen fuel-cell cars could even provide power for homes and offices if necessary. As of 2015, Toyota was set for a limited commercial release of a hydrogen fuel cell vehicle.

For hydrogen power to become a reality, tremendous research and investment efforts are necessary. The fate of hydrogen power technology will also depend on consumers' willingness to spend money on climate-friendly technologies.

—*Sergei Arlenovich Markov*

FURTHER READING

Ewing, Mark. "Toyota's Hydrogen Fuel Cell Kenworth Can Revolutionize Heavy Transport." *Forbes*, Forbes Magazine, 9 Aug. 2017, www.forbes.com/sites/markewing/2017/08/09/toyotas-hydrogen-fuel-cell-kenworth-can-revolutionize-heavy-transport/#3957fe276e48.

Ogden, Joan. "High Hopes for Hydrogen." *Scientific American* Sept. 2006: 94–99. Print.

Rögner, Matthias. *Biohydrogen*. Walter De Gruyter, 2015.

Service, Robert F. "The Hydrogen Backlash." *Science* 305.5686 (2004): 958–61. Print.

ICE AGES AND GLACIATIONS

FIELDS OF STUDY

Atmospheric Science; Atmospheric Chemistry; Analytical Chemistry; Environmental Chemistry; Geochemistry; Physical Chemistry; Photochemistry; Thermodynamics; Fluid Dynamics; Environmental Sciences; Environmental Studies; Waste Management; Meteorology; Process Modeling; Hydroclimatology; Bioclimatology; Climate Zones; Oceanography; Hydrology; Earth System Modeling; Ecology

SUMMARY

Glaciers are layers of ice that form on Earth's lithosphere where the temperature is sufficiently low to support year-round ice and snow. Extended periods when temperatures drop sufficiently low to support large-scale increases in glaciation are called glacial epochs or ice ages. Earth is now in an ice age that began 2.4 million years ago and has involved twenty or more fluctuations between glacial and interglacial periods. Estimates indicate that Earth is undergoing cycles of glaciation that occur every eleven thousand years.

PRINCIPAL TERMS

- **cryosphere:** the portion of the Earth's surface where the year-round temperature remains constant enough to support permanent ice and snow
- **eccentricity:** variation in the shape of the Earth's orbit around the sun, ranging from circular to elliptical
- **glacial epoch:** an extended period of global temperature reduction and glaciation that generally lasts for millions of years and includes internal glacial and interglacial periods
- **glacial stage:** short-term period of glaciation, generally lasting for less than one million years and alternating with interglacial periods
- **glacier:** buildup of frozen ice on some portion of Earth's lithosphere
- **interglacial:** a period of reduced glacial coverage that alternate with glacials within a global glacial epoch
- **isostacy:** equilibrium between the lithosphere of the Earth and the liquid layer of rock in the inner layers of the strata
- **marine isotope stage:** half of a glacial cycle, as identified in the oxygen isotope data from ocean cores
- **Milankovitch cycle:** period of variation in Earth's orbital parameters, including axial inclination, climatic precession, and orbital eccentricity
- **obliquity:** the tilt of an orbiting body's rotational axis relative to its orbital axis
- **Pleistocene-Quaternary glaciation:** current ice age beginning approximately 2.4 million years ago
- **precession:** the gradual change of a rotating body's axis of rotation

GLACIERS OF MODERN EARTH

The part of Earth in which the temperature is permanently below the freezing point of water (0°C, or 32°F) is known as the cryosphere. Glaciers are large bodies of ice that form within the cryosphere and are considered permanent by the standard of the human life cycle.

Glaciers are formed primarily from layers of snow that have become compacted by the weight of overlying snow and by the pull of gravity to recrystallize into ice. Glaciers today exist primarily in the far Northern and Southern Hemispheres past the snow line, which is the latitudinal mark beyond which the ambient temperature remains below the freezing point of water.

The portion of the Earth covered in permanent ice changes over time according to cycles that affect

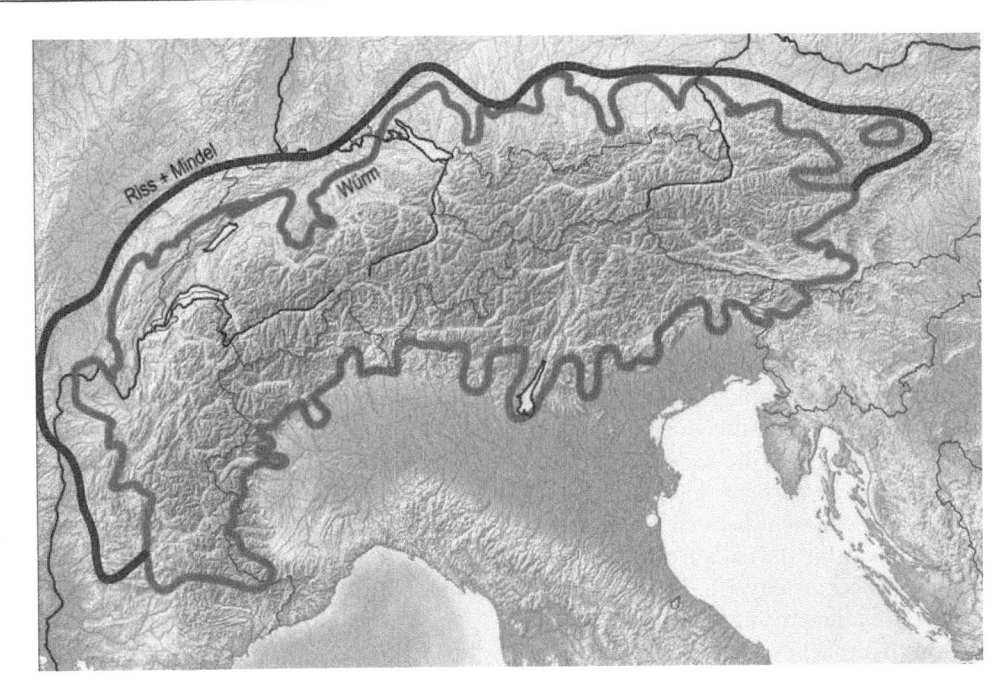

Approximate extent of alpine glaciations.

Earth's global temperature. Multimillion-year periods of prolonged reduced temperature, during which glaciation spreads, are sometimes called glacial epochs or ice ages.

The presence or absence of glaciers has a domino effect on Earth's geography and climate. Ocean depth is one factor that is highly dependent on the amount of water that is frozen in glacial zones. Estimates show that the world's glaciers in Antarctica and Greenland hold enough frozen water to raise the Earth's ocean levels by more than sixty meters (two hundred feet), thereby drastically altering the amount of land available for habitation.

Glaciers also can influence sea levels through their affect on isostacy, which is the gravitational equilibrium between the Earth's crust and the mantle below. The crust of the Earth floats along the mantle because the crust is less dense and the mantle is more fluid (caused by heat from the Earth's core). As a portion of the crust becomes heavier, that portion sinks into the mantle, causing a depression in the Earth's surface and causing water levels to rise in the surrounding area.

During the last ice age, which occurred between thirty thousand and one hundred thousand years ago, portions of North America and Eurasia were covered in thick glacial ice, creating a deep depression in the mantle. As most of this ice melted, the mantle and crust began to return to equilibrium, a phenomenon known as isostatic rebound. As the crust rose, sea levels receded in many parts of the Northern Hemisphere.

Global warming is a trend caused by human activity, a reduction in vegetation worldwide, and an increase in greenhouse gases. This warming phenomenon is noted by the level of glacial retreat, or the melting of glacial ice, which has been measured since the nineteenth century. One focus of the study of glaciers and glaciation is to develop a more complete understanding of climate cycles on Earth and the future of Earth's environmental evolution.

SHORT-TERM GLACIAL CYCLES

The Cenozoic period, which is the current geologic age, began approximately seventy million years ago. Evidence from geologic sources indicates that climate change and glaciation have occurred on a relatively regular cycle in this period of Earth's history.

Earth is in an ice age called the Pleistocene-Quaternary glaciation, which began more than 2.5 million years ago. During an ice age, short-term fluctuations in climatic variables lead to periods known as glacial stages, or simply glacials, characterized by relatively low temperatures and increased glacial buildup. These glacial periods alternate with interglacial periods, in which temperatures increase and glacial ice retreats as the climate becomes warmer. Geologic evidence indicates that in the Pleistocene-Quaternary glaciation, the Earth has experienced twenty or more alternating glacial and interglacial periods. The earth is now in an interglacial period.

Geologists believe that glacial and interglacial periods are partially related to changes in Earth's solar orbit. In the current geologic period, called the Holocene, the Earth orbits the sun in a circular pattern. However, the shape of the Earth's solar orbit gradually shifts over thousands of years, alternating between circular and elliptical patterns. The extent to which an object's orbit deviates from a circular path is known as its eccentricity. Changes in Earth's eccentricity can have a major impact on the amount of solar radiation that reaches the planet's surface at various times of the year.

In addition, Earth's axis of rotation is tilted with respect to the sun at an angle of about 23.45 degrees; this tilt is called its axial tilt or obliquity. Earth's obliquity oscillates between 21 and 24.5 degrees over a forty-two-thousand-year cycle. Finally, the Earth's axis of rotation itself rotates on a twenty-six-thousand-year cycle, a phenomenon known as precession. While Earth's North Pole is currently pointed toward the star known as Polaris, or the North Star, Earth's axis will gradually rotate toward the star known as Vega.

Serbian geophysicist and engineer Milutin Milanković (1879–1958) used the Earth's precession, obliquity, and eccentricity to calculate the cyclic pattern of the planet's orbital relationship with the sun; these cycles came to be called Milankovitch cycles. He then correlated these data with information on climate change to develop the Milankovitch hypothesis, which posits that repeating patterns of heating and cooling on Earth are related to the planet's angular orientation and relative distance from the sun.

The Milankovitch hypothesis predicts that Earth will experience major changes in climate on three separate cycles, every one hundred thousand years, forty-one thousand years, and twenty-one thousand years, corresponding closely to changes in the pattern of Earth's orbit. Geologic and climatological evidence indicates that Milankovitch cycles may be largely responsible for the alternating glacials and interglacials during the current ice age.

Long-Term Glacial Cycles

While Milankovitch cycles are helpful in explaining short-term variations in climate, geologic evidence indicates that the Earth has experienced periods of long-term glaciation, or glacial epochs, that cannot be explained by orbital variation. Some glacial periods last for hundreds of millions of years and involve glaciation far more extensive than that of the current Quaternary ice age.

One of the first glacial epochs to be studied occurred between 2.2 and 2.4 billion years ago, in the Paleoproterozoic era, and is thought to have lasted 200 million years or longer. Though this ancient ice age, called the Makganyene (or Huronian) glaciation, is poorly understood, global changes in tectonic movement and volcanism have been identified as potential causal factors. Geologists believe that the Makganyene glaciation might have marked the first time that most or all of the Earth was covered in glacial ice, a hypothesis that is now called the snowball Earth theory. Global glaciation results when contributing factors such as tectonic movement and atmospheric composition converge to create a positive feedback loop that allows glaciers to spread until they cover vast portions of the planet's surface.

The period from 850 to 630 million years ago, constituting a major portion of the Neoproterozoic period (1 billion to 540 million years ago), is sometimes called the Cryogenian period because of the two major glacial epochs that occurred during this span, which included the most extensive glaciations in the known history of the Earth. The Marinoan (635 million years ago) and Sturtian (710 million years ago) glaciations left widespread geologic evidence around the world, indicating that glacial ice was present at all latitudes and covered the Earth in thick layers.

During the Paleozoic era (540–340 million years ago), there existed two glacial epochs, both corresponding to major changes in the biosphere. The Andean-Saharan glaciation occurred between 460 and 430 million years ago, spanning the Ordovician and the Silurian periods, and corresponds with a mass extinction of life on Earth, after which terrestrial plants began to spread across the surface. The Karoo ice age (360–260 million years ago) spanned part of the Mississippian and Pennsylvanian periods and is partially explained by the spread of terrestrial plants, which removed greenhouse gases from the atmosphere and caused global cooling.

Geologists believe that glacial epochs are largely the result of tectonic shifting on the Earth's surface. Recent evidence suggests that glacial epochs occur when larger continents break into smaller continents, a process known as rifting. Increases in tectonic rifting might affect climate by altering levels of silicate weathering, which is the process by which atmospheric carbon dioxide is removed from the atmosphere through interaction with minerals dissolved from the Earth's crust into the oceans.

Silicate weathering increases with temperature. Thus, as continents move over the equator, where they are exposed to higher levels of solar radiation, silicate weathering increases, thereby leading to a reduction in atmospheric carbon dioxide and the beginning of a cooling cycle. Rifting and tectonic convergence are part of the theoretical supercontinent cycle, which suggests that the continents converge on a cycle of three hundred million to six hundred million years.

PLEISTOCENE GLACIATIONS

About 50 million years ago, Earth began to cool. Deep-ocean temperatures gradually dropped from 13°C to the present 1°C. Study of ocean sediment cores has identified some minor ice ages around 40 million years ago, another ice age formed the East Antarctica ice sheet 34 million years ago, and a third ice age around 13 million years ago left evidence in Alaska. Orbital factors probably influenced these advances, perhaps by affecting the carbon cycle, but the details remain obscure.

Starting about 3 million years ago, the climate developed two different states. Since then, it has oscillated between them, causing glaciations, with continental glaciers extending down to the 40° north parallel of latitude, and interglacials such as the current period, some with temperatures even higher than those of the present. Geologists identified glacial deposits, determined that some were on top of others, and gave names to each, but the names were not standardized internationally. The name of the most recent glaciation is Valdaian on the Russian Plains, Devensian in Britain, Weichselian in Scandinavia, Würm in the Alps, and Wisconsinan in North America. The deposits of one glacier are easily removed by subsequent glaciers, so there might be some for which no evidence remains on the continents.

Oxygen isotope ratios of marine sediments, which are not removed by later glaciations, show that there have actually been around fifty glaciations. Because these ratios indicate how much water is tied up in ice, this record is now seen as the best way of delineating when an advance ended and a retreat began. Each marine isotope stage (MIS) is given a number (odd ones for interglacials, even ones for glaciations) starting with the most recent. Interglacial MIS 103 occurred 2,580,000 years ago.

Once all these additional glaciations were known, analysis showed that astronomical cycles controlled their timing, as had been suggested by James Croll (1821-1890) in 1864 and Milutin Milanković (1879-1958) in 1941. These results are so robust that geologists now use these cycles to calibrate the more recent part of the geologic time scale.

That Milankovitch cycles govern the timing of glaciations is beyond dispute. How they do so, however, is not well understood. Feedback systems in the oceans, the atmosphere, the biosphere, or perhaps elsewhere are needed to amplify the tiny signal produced astronomically. Identifying and understanding these systems will help scientists understand current climate change.

EVIDENCE FOR GLACIATION

Ice-core samples help geologists learn about the climate of the distant past. Ice in glaciers is built of successive layers of snow that have become compacted to form ice. By taking vertical samples of glacial ice, geologists can examine the layers in the ice and can discern information about climate from the differences between and among layers.

Water vapor contains different molecular varieties of water, based on the types of isotopes contained within water molecules. Isotopes are variations of an element, such as hydrogen or oxygen, which have the same number of protons but differ in the number of neutrons. While these isotopes are of the same element, variations in the number of neutrons cause each isotope to behave differently in certain chemical reactions.

In general, the standard variety of water is H^{16}OH, which contains an isotope of oxygen known as oxygen-16 (^{16}O), considered the most common oxygen isotope. However, water also occurs in the formula $H^{18}OH$, which contains the oxygen-18 (^{18}O) isotope. In addition, some water molecules contain deuterium, which is an isotope of hydrogen, and may therefore occur in the chemical formula HOD. When water vapor rises before condensing because of heat, it tends to lose deuterium and oxygen-18 more readily than it will lose the oxygen-16 isotope. This means that colder temperatures favor the accumulation of ice with greater proportions of deuterium and oxygen-18. Therefore, by measuring the occurrence of various isotopes, geologists can measure changing temperatures in ancient ice samples. Ice cores taken from Antarctica preserve temperature-induced isotope variations from four hundred thousand years ago or earlier.

In addition to isotope chemistry, ice cores preserve a variety of other environmental elements representing the time when the ice was first deposited. Levels of dust and debris captured in glacial ice indicate that dry air and heavy winds dominated during glacial periods. In addition, geologists have found trapped gas inclusions within ice cores containing carbon dioxide and methane from the Quaternary ice age. Measurements indicate that these gases were more abundant during warmer periods and declined during glacial periods, providing evidence that greenhouse-gas levels are a major determinant in glaciation.

In some cases, geologists have found acids related to volcanic activity in glacial ice. Increased volcanism occasionally preceded periods of glacial activity because gases released by volcanoes can block solar radiation, leading to more pronounced differences between seasons.

—*Micah L. Issitt*
—*Otto H. Muller*

FURTHER READING

Dawson, Alastair G. *Ice Age Earth: Late Quaternary Geology and Climate.* Routledge, 1992.

Grotzinger, John P., and Thomas H. Jordan. *Understanding Earth.* 7th ed. Freeman, 2014.

Hambrey, Michael J., and Jürg Alean. *Glaciers.* 2nd ed. Cambridge University Press, 2004.

Monroe, James S., Reed Wicander, and Richard Hazlett. *Physical Geology: Exploring the Earth.* 6th ed. Thomson, 2007.

Rapp, Donald. *Ice Ages and Interglacials: Measurements, Interpretations, and Models.* Springer, 2009.

Ruddiman, William F. *Earth's Climate Past and Future.* 2nd ed. W. H. Freeman, 2008. This elementary college textbook has several sections concerning glaciations and their causes, with particular attention paid to how they can elucidate climate change. Illustrations, figures, tables, maps, bibliography, index.

Stanley, Steven M., and John A. Luczaj. *Earth System History.* Freeman, 2015.

ICE CORES

FIELDS OF STUDY

Engineering; Analytical Chemistry; Atmospheric Chemistry; Biochemistry; Environmental Chemistry; Geochemistry; Photochemistry; Mass Spectrometry; Spectroscopy; Meteorology; Climate Modeling; Statistics; Hydroclimatology; Hydrometeorology; Bioclimatology; Oceanography; Hydrology; Physical Geography; Ecology; Crystallography; Optics

SUMMARY

An ice core is a cylinder of ice typically measuring 10 centimeters in diameter and obtained by drilling vertically into a glacier. Glacial ice is produced by the natural, gradual transformation of snow into ice. Snow is made of fragile ice crystals surrounded by air. In the process that transforms snow into ice, this air is trapped in tiny bubbles that are identical in composition to the atmosphere that produced the snowfall. As snow falls year after year and is compacted into ice, each layer can be identified visually or electronically. In this way, glaciers preserve a record of atmospheric changes that have occurred over time, and ice cores can be drilled to retrieve that record.

PRINCIPAL TERMS

- **Dome C:** the Antarctic Plateau location of the Concordia Research Station

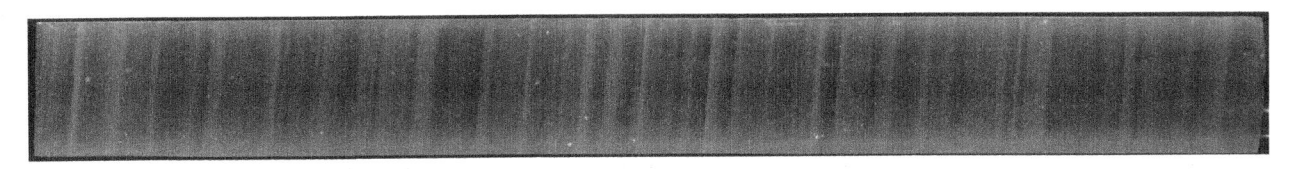

This photograph shows a section of the GISP2 ice core from 1837 to 1838 meters in which annual layers are visible. The appearance of layers results from differences in the size of snow crystals deposited in winter versus summer. This ice was formed ~16250 years ago during the final stages of the last ice age.

- **glacial ice:** ice created by the compression of snow into glaciers
- **paleoclimatology:** the study of past climates
- **Vostok:** a Russian Antarctic research station built in 1957 during the First Geophysical Year

Variation in the Atmosphere's Composition

Because they contain a chronologically ordered record of atmospheric composition, ice cores can be used to reconstruct the history of both natural and anthropogenic pollution. The chemical composition of Earth's atmosphere varies over time. It reacts to volcanic eruptions, for instance, which add carbon monoxide, carbon dioxide (CO_2), and sulfur dioxide to its normal content of nitrogen, oxygen, argon, CO_2, and methane. Thus, changes in the chemical composition of the air trapped in ice cores make it possible to date past volcanic eruptions.

In the same way, some human activities can be identified and dated. Analysis of ice cores from the Alps and Greenland has identified the pollution resulting from the smelting of lead ore by the Romans more than two thousand years ago. More recently, traces of chemicals linked to nuclear explosions or to the Chernobyl nuclear reactor meltdown have been detected in glacial ice. Today, at a time when issues of climate change and pollution are uppermost in the minds of many, glaciologists have sought to retrieve the longest possible ice cores in order to produce a substantial historical record to contextualize current atmospheric trends.

The Longest Ice Cores

The first giant ice core was drilled in Vostok, the Russian base in Antarctica. The region of Antarctica in which Vostok is located is extremely cold and dry. The advantage of low precipitation in glaciology is that, because each year produces less ice, a large number of years is recorded relative to core length. The disadvantage of such low precipitation is that, because each annual layer of ice is extremely thin, it can be very difficult to produce an accurate count. When all analyses were complete, Vostok glaciologists concluded that the 3,623-meter core drilled in 1996 represented the last 420,000 years of Earth's history.

The normal analysis of an ice core consists of analyzing the content of the ice for the presence of deuterium (an isotope of hydrogen) and the presence of oxygen 18 (O^{18}; an isotope of oxygen), because both are reliable indicators of Earth's temperature at the time of the snowfall. The temperature records obtained thanks to the Vostok ice core compare well with cores obtained in the center of Greenland. Since the Greenland ice sheet is smaller and thinner (1,600 meters deep on average) than the Antarctic ice sheet (2,400 meters deep on average), the paleoclimatic record derivable from Greenland ice cores is shorter. However, that record is consistent with the Vostok ice core record. Greenland also receives more precipitation than does Vostok, resulting in thicker annual layers of ice; a layer as old as 100,000 years can still be as much as 1 centimeter thick. Greenland ice cores therefore are outstanding at establishing a chronology for the last glacial period for which the estimated error does not exceed 2 percent for the first forty thousand years.

In an effort to increase the length of the paleoclimatic record, scientists from Europe have drilled in two other regions of the East Antarctic Ice Sheet under the European Project for Ice Coring in Antarctic (EPICA). A site on Dome C located at 75°06' south latitude and longitude 123°21' east was selected. There, where the ice sheet is thicker while both the ice and the air remain extremely cold, a permanent research base, Concordia, was built by France and Italy (partly financed by the European Union). Dome C is ideal for the thickness of its ice and its low precipitation; it initially provided a 3,140-meter ice core, revealing 740,000 years of climatic history. Since then, even deeper ice has been retrieved, extending knowledge of climatic history to over 800,000 years.

TESTING THEORIES OF CLIMATE CHANGE

In order to analyze the ice core samples that have been retrieved, scientists crush a sample of the ice in a vacuum, releasing the air trapped within the sample without allowing it to mix with the modern atmosphere. They can then analyze these air samples, compute the proportion of greenhouse gases (GHGs) such as CO_2 and methane in the air, and compare their proportions to the temperature record derived from the deuterium and ^{18}O analyses. The goal is to see if an increase of GHGs in ice always leads to an increase in the Earth's temperature or, conversely, if an increase in the Earth's temperature releases CO_2 and methane into the atmosphere.

CONTEXT

Scientists continue to search for deeper deposits of ice from which even longer historical sequences can be identified. The Dome C site holds the promise of ice cores that will extend the scientific record to more than one million years into the past. Meanwhile, research also continues on ways to recover more, and more precise, information from ice cores. Analysis of core samples containing volcanic dust taken at Dome C, for instance, has demonstrated that the magnetic polarity of the Earth reversed about 780,000 years ago. As ice cores get longer and scientists find ways to learn more from them, the historical record will become more detailed over longer periods of time.

—*Denyse Lemaire*
—*David Kasserman*

FURTHER READING

Alley, Richard B. *The Two-Mile Time Machine. Ice Cores, Abrupt Climate Change and Our Future.* Princeton University Press, 2014. Describes the analytical methods and interpretations of physical clues of past climates trapped in ice that is thousands of years old. Provides ice coring research information and discusses the work of geoscientists in relation to the cryosphere. The text is concise and clear, with superb artwork and photographs.

Trewby, Mary. *Antarctica: An Encyclopedia from Abbott Ice Shelf to Zooplankton.* Firefly, 2005. General encyclopedia of Antarctica. Explains all the features found on the continent using a clear and simple language.

Turney, Chris. *Ice, Mud, and Blood: Lessons from Climates Past.* New York: Macmillan, 2008. Describes the discoveries derived from ice cores and how these discoveries led to a better understanding of paleoclimates.

ICE SHELF COLLAPSES

FIELDS OF STUDY

Physical Geography; Oceanography; Fluid Dynamics; Thermodynamics; Heat Transfer; Physical Sciences; Process Modeling; Climate Modeling; Hydrology; Earth System Modeling; Ecology

SUMMARY

Ice shelves are the seaward extensions of continental ice sheets. They are found extending from 50 percent of the coast of Antarctica and account for 11 percent of the mass of ice of the Antarctic Ice Sheets. The two largest ice shelves are the Ross (472,960 square kilometers) and Ronne-Filchner (422,420 square kilometers) Shelves. Together, they have represented as much as 67 percent of the total area of Antarctic ice shelves, but in recent years large segments have broken away. Ice shelves are typically thinner at their seaward edge; the largest shelves reach a depth of 1,300 meters near the grounding line where the ice begins to float on open water, but thin to 200 meters at the leading edge.

PRINCIPAL TERMS

- **glacier:** a mass of ice that flows downhill, usually within the confines of a former stream valley
- **ice sheets:** masses of ice covering large areas of land
- **ice shelf:** a platform of freshwater ice floating over the ocean
- **West Antarctic ice sheet:** the smallest ice sheet in Antarctica, located west of the Transantarctic Mountains

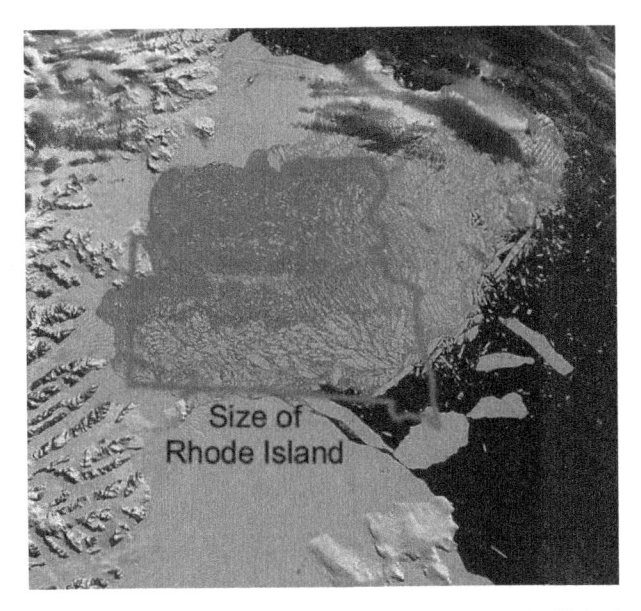

Glacier Larsen B is about the size of Rhode Island in the United States.

Formation of Ice Shelves

Ice shelves gain mass in three different ways. The primary source is the ice sheet that slowly moves off the land surface onto the water under the force of gravity, but accumulated snow on their upper surface and the freezing of seawater to the lower one also contribute. On the other hand, ice shelves lose mass in five ways: by calving of icebergs, by the melting of ice in contact with the sea, by snow blowing off the edge of ice shelves, by enhanced sublimation of ice under the very strong Antarctic winds, and by superficial melting. Since surface melting in Antarctica is negligible, and since wind ablation is limited in general, nearly all mass loss is the result of either iceberg calving at the seaward margin or bottom melting of ice in contact with the sea.

Ice Shelves of the West Antarctic Peninsula

With the widespread availability of video clips on the Internet, many people have observed the breaking of very large tabular icebergs off the Antarctic Peninsula. The calving of icebergs is a natural phenomenon that occurs regularly under normal climatic conditions. This calving should not be interpreted as a sure sign of global warming. In the extreme north of the Antarctic Peninsula, ice shelves are called Larsen A, B, C, and D, named in sequence from north to south. The Larsen A ice shelf began to retreat at a very rapid pace in January, 1995. The breakup of Larsen A was irrefutably a response to rising ocean temperature. South of Larsen A, Larsen B calved large tabular icebergs beginning in 1995, prior to a larger disintegration that occurred in the first three months of 2002. More recently, in 2008, the Wilkins Ice Shelf that had broken up in 1998 partially disintegrated. The culprit seemed to be the presence of relatively warmer ocean waters melting ice shelves from beneath and destabilizing the ice shelf at the grounding line.

Ice Shelves as Buttresses

Because ice shelves may be very thick, they are often anchored on the continental shelf, where they begin to float. The grounding of these ice shelves allows them to slow down the outward movement of ice sheets by acting as a buttress if they are thick enough. This effect is lost if the ice shelf disintegrates. In the case of the Larsen B Ice Shelf, its disintegration in 2002 led to an acceleration of the outward flow of ice from the glacier that feeds it.

Context

Although there is no doubt that ice shelves in the West Antarctic Peninsula have decreased in size over the last three decades, their complete melting would not raise sea level by more than a few millimeters. The great majority of floating ice is below sea level, so most of the volume of ice shelves is already within the ocean. Their destruction could, however, contribute significantly to rising sea level when the glaciers feeding these ice shelves speed up their seaward movement once the ice shelf stops being grounded. A 2004 study by researchers from the National Snow and Ice Data Center and NASA described accelerations of such glaciers reaching four to six times their normal speed.

Interestingly, the ice shelves of the southern part of the West Antarctic ice sheet and those of the East Antarctic ice sheet do not seem to be affected by global climate change. Their temperature remains below freezing, probably because they are surrounded by abundant sea ice that acts as a buffer, protecting them from the advection of warmer water.

—*Denyse Lemaire*
—*David Kasserman*

FURTHER READING

Alley, Richard B. *The Two-Mile Time Machine. Ice Cores, Abrupt Climate Change and Our Future* Princeton University Press, 2014. Describes the analytical methods and interpretations of physical clues of past climates trapped in ice that is thousands of years old.

Copeland, Sebastian. *The Global Warning.* Earth Aware Editions, 2007. This abundantly illustrated book about Antarctica tends to focus on the potential calamitous effects of global climate change using a clear and simple language.

Jouzel, Jean, Lorius, Claude and Raynard, Dominique *The White Planet. The Evolution and Future of Our Frozen World.* Princeton University Press, 2013. Examines the past, present and future roles of ice on Earth, concluding that climate change and global warming will ultimately have disastrous consequences for human societies on Earth regardless of technological advances.

Marshall, Shawn J. *The Cryosphere.* Princeton University Press, 2012. A complete introduction to the physical properties and thermodynamics of ice and snow as the basic matter of the cryosphere and its dynamics.

McGonigal, David. *Antarctica: Secrets of the Southern Continent.* Firefly Books, 2008. Describes the discoveries of the International Polar Year, 2007–2008, in various fields, including geology, geography, and climatology. Discusses potential effects of global warming.

Turney, Chris. *Ice, Mud, and Blood: Lessons from Climates Past.* Macmillan, 2008. Examines various discoveries derived from ice cores and how these discoveries led to a better understanding of paleoclimates.

Wadhams, Peter *A Farewell to Ice. A Report from the Arctic.* Oxford University Press, 2017. Discusses the role of Arctic and Antarctic ice shields, and how feedback systems function. Special attention is given to the potential for the release of quantities of methane from polar sea floors and Arctic permafrost as Earth's average temperature increases.

ICEBERGS AND ANTARCTICA

FIELDS OF STUDY

Physical Geography; Oceanography; Fluid Dynamics; Thermodynamics; Heat Transfer; Physical Sciences; Process Modeling; Climate Modeling; Hydrology; Earth System Modeling; Ecology

SUMMARY

About 98 percent of the surface of Antarctica, the coldest, driest, and southernmost continent on Earth, is covered in a thick, dense layer of ice. Some of this ice moves to the Southern Ocean surrounding the continent and breaks off to become icebergs. Antarctica is of profound interest to scientists because of its impact on the world's oceans and climate and because of its role as a rich source of data about Earth's history.

PRINCIPAL TERMS

- **ablation:** the removal of material from a glacier, ice shelf, or other mass of ice through evaporation, melting, or splitting
- **albedo:** a measure of the proportion of incoming light or radiation that is reflected from a surface, such as snow, ice, or water; also known as reflectivity
- **calving:** the breaking away of a smaller piece of ice from a larger one
- **glacier:** a river of freshwater ice that is massive enough to be put into motion by gravity; usually contains ice, air, rock, and some water
- **iceberg:** a large mass of freshwater ice that has broken from an ice shelf or a glacier; floats in a body of water
- **ice core:** a cylinder-shaped piece of ice that is collected by drilling into a glacier; can be used to analyze the history of Antarctica's climate
- **ice shelf:** a large, flat sheet of freshwater ice formed from a glacier or an ice sheet; floats in a body of water
- **ice stream:** a rapidly moving current of freshwater ice flowing from an ice sheet and moving more quickly than the ice that surrounds it; carries ice from the ice sheet

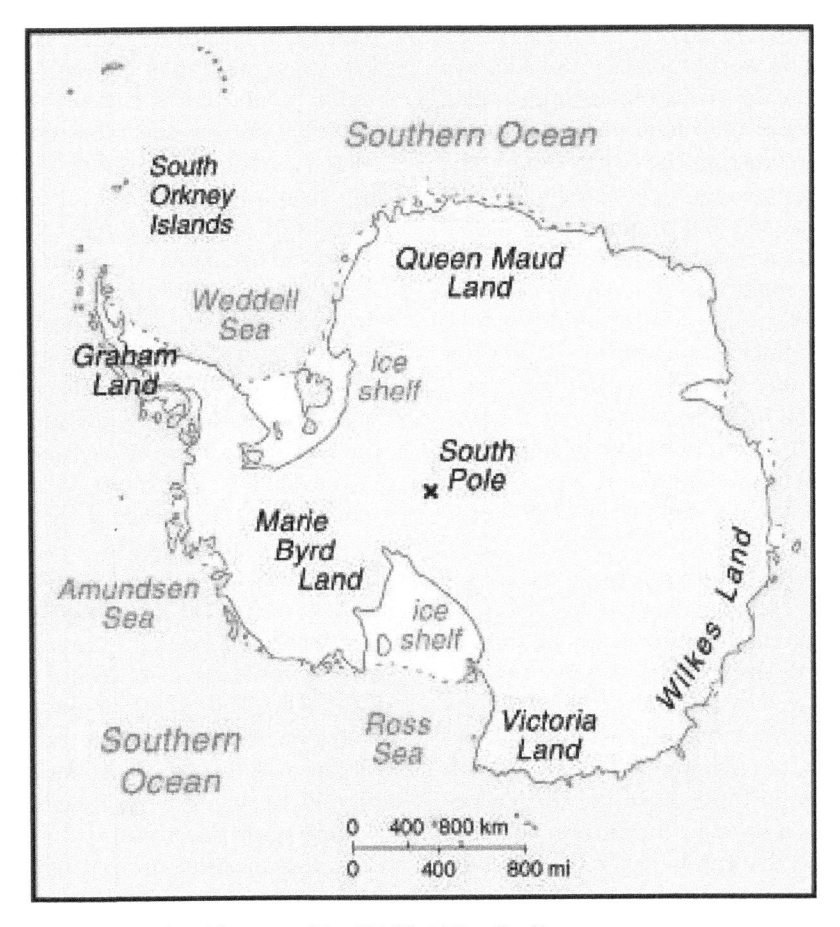

Map of Antarctica. (Courtesy of the CIA World Fact Book)

- **nipping:** process in which ice pushes forcibly against the edge of a ship
- **pack ice:** large, mobile masses of frozen, floating seawater that are not attached to a landform; also known as sea ice

ANTARCTICA, ICE SHEETS, AND ICE CORES

Antarctica is the southernmost of Earth's seven continents. Surrounded by the Southern Ocean, Antarctica is situated largely within the Antarctic Circle, the parallel of latitude that runs 66.5622 degrees south of the equator.

Antarctica is the only continent with no permanent human residents. In 1959, an international treaty was signed to establish Antarctica as a scientific preserve, to be used for peaceful purposes only. The treaty does not, however, recognize or dispute claims on any part of Antarctica as a national territory; it also does not allow the making of new territorial claims. About thirty countries operate scientific research stations on Antarctica, among them the United States, Australia, India, France, Germany, and Japan.

Antarctica spans about 14 million square kilometers (about 5.4 million square miles) of land area, slightly less than 1.5 times the area of the United States. Antarctica is the continent with the lowest surface temperatures, the greatest amount of wind, and the least precipitation. Much of Antarctica is considered a desert because the continent receives on average only about 5 centimeters, or 2 inches, of precipitation per year.

Almost the entire surface of Antarctica, about 98 percent, is covered in a thick and heavy layer of ice, which measures up to about 4.8 kilometers (3 miles) in depth in some places. This ice, which occupies about 29 million cubic kilometers of total volume (about 7 million cubic miles), represents approximately 90 percent of the world's total ice and is known as the Antarctic polar ice sheet. (A second, smaller, ice sheet covers most of the island of Greenland in the Arctic Circle surrounding the North Pole.) The Antarctic ice sheet contains enough frozen water that if it were to melt completely, it would cause a global rise in sea level of about 60 meters (200 feet).

Specific weather conditions are required for an ice sheet to form. Winter snowfall should not melt completely when summer temperatures rise. If snow persists throughout the year for thousands of years, each season's snowfall forms a new layer on the previous season's layer; these layers eventually freeze into ice. As new snow falls, its weight also compresses the existing ice, making the ice sheet even more dense and thick.

Because the ice sheet comprises thousands of layers of snow, it presents a detailed historical record of the environment and climate at the time of each snow deposit. By drilling into the sheet to collect samples known as ice cores, scientists can access that record. Ice cores have been used to determine, among other things, the temperature at which each snowfall occurred. A measure called the oxygen isotope ratio is used as a stand-in for air temperature. Other measurements that can be taken include dust levels, greenhouse gas levels (such as carbon dioxide and methane), hydrogen peroxide, and sulfate.

By analyzing these data, scientists have been able to map a picture of climate change throughout Earth's history, pinpointing when historic ice ages and interglacial, or warmer, periods occurred. Ice core data also have confirmed that the concentration of greenhouse gases in Earth's atmosphere has increased significantly since the late eighteenth century, the beginning decades of the Industrial Revolution.

ICE SHELVES, ICE CALVING, AND ICEBERGS

Because of the force exerted by its own weight, the massive Antarctic ice sheet is in a state of extremely slow but constant movement. Glaciers and ice streams, which are both flowing masses of ice, ooze from the inland portions of the ice sheet into the Southern Ocean.

In the Southern Ocean the ice extends over open water as a floating, frozen mass beginning at the coastline. These masses are known as ice shelves. Ice shelves are dynamic, not static, in nature. They can gain more ice from the ice sheet, but they also can lose ice to melting or calving.

Ice calving is a form of ablation, a term used to describe any of several processes through which some icy material is removed from a larger whole. In calving, a large block of ice separates from the margin of a mass of ice, usually a glacier or an ice shelf, and breaks away. Calving occurs when fractures in the ice grow and spread, causing the ice to become more brittle. A variety of factors contribute to the likelihood of a calving event, including the extent to which an ice shelf has melted below the water line, the speed at which glaciers and ice streams flow, and the temperature of the ice itself.

Calving occurs where ice meets or stands over water, as at the leading edge of a glacier or at the coastal margin where the ice sheet loses contact with the underlying ground and becomes free-floating. The resulting calved blocks become known as icebergs, which have an important place in the Antarctic ecosystem. Many seabirds live on their surfaces, and krill, fish, and phytoplankton form communities below the icebergs. (Ships traveling in the Southern Ocean and the North Atlantic are at risk of being nipped, or struck, by floating icebergs.)

One study has found that when icebergs melt far from the coastline, they introduce trapped nutrients, including iron and krill, to the water. This process speeds the growth of algae and other photosynthesizing phytoplankton, which take up carbon dioxide from the atmosphere. The effects of global warming have been causing ice shelves to disintegrate and to calve icebergs more frequently, and scientists predict that this trend will continue. If icebergs do contribute to the removal of carbon dioxide, they may add an important additional feedback mechanism to the complex, dynamic global climate system.

SEA ICE AND THE CLIMATE

The seawater surrounding Antarctica freezes at a slightly lower temperature than the freshwater that makes up its ice sheet, glaciers, ice shelves, and icebergs. Typically, seawater freezes at -1.9°C (28.58°F). Normally, water freezes at 0°C (32°F).

The ice that forms when seawater freezes is less dense than the unfrozen water. This ice floats on the surface, acting as a barrier that prevents heat and gases from traveling between the atmosphere and the ocean. Even when sea ice breaks into large pieces, known as pack ice, the ice tends to be driven together into a nearly continuous mass that provides excellent cover. This prevents the atmosphere above the Southern Ocean from cooling drastically, making sea ice a critical control factor in keeping global climate stable.

Sea ice is important for another reason. It has a bright, light surface that reflects back into space about 80 percent of the sunlight that strikes it. This process is known as the albedo effect. (A surface's albedo is a measure of how reflective it is.) The surface of the ocean itself is much darker, and it has an albedo that is closer to about 10 percent. As a result, when sea ice melts, a dramatically greater amount of sunlight and radiation is absorbed by the ocean, which heats up, creating a cycle that leads to even more melting.

Decades of scientific data have shown that sea ice in the Arctic north has become thinner and has shrunk more each summer since about 1980, as temperatures there slowly warm. Antarctic sea ice is reacting somewhat differently. Around the Antarctic Peninsula, a long, thin, stretch of land that extends outside the Antarctic Circle, higher temperatures have led to a reduction in the extent of sea ice cover. This is also the region where two major ice shelves, known as the Larsen B and the Ross Ice Shelves, disintegrated (calved into icebergs). In the rest of the continent, sea ice has actually increased slightly since the start of the twenty-first century, partly because the circumpolar current buffers warm water from the tropics from reaching it.

—*M. Lee*

FURTHER READING

Alley, Richard B. *The Two-Mile Time Machine. Ice Cores, Abrupt Climate Change and Our Future* Princeton University Press, 2014. Describes the analytical methods and interpretations of physical clues of past climates trapped in ice that is thousands of years old.

Bennett, Matthew M., and Neil F. Glasser. *Glacial Geology: Ice Sheets and Landforms.* John Wiley & Sons, 2009. A concise, beginner-friendly overview of the geologic concepts required to understand the icy landforms of Antarctica. Written for undergraduates but accessible to advanced high school students, each chapter includes text boxes, color photographs, illustrations, and suggestions for further reading.

Joughin, Ian, and Richard B. Alley. "Stability of the West Antarctic Ice Sheet in a Warming World." *Nature Geoscience* 4 (2011): 506–513. Analyzes satellite observations and historical data to estimate the magnitude of the contribution of west Antarctic ice sheet melting to past sea level rises, and to model its potential future impact.

Jouzel, Jean, Lorius, Claude and Raynard, Dominique *The White Planet. The Evolution and Future of Our Frozen World.* Princeton University Press, 2013. Examines the past, present and future roles of ice on Earth, concluding that climate change and global warming will ultimately have disastrous consequences for human societies on Earth regardless of technological advances.

McGonigal, David. *Antarctica: Secrets of the Southern Continent.* Francis Lincoln, 2009. An illustrated volume that covers Antarctic geology, geography, wildlife, historical expeditions, and scientific endeavors. Stunning photographs and clear, absorbing writing make this an excellent introduction to the continent for high school students and older.

Ravindra, Rasik, et al. "Antarctica." In *Encyclopedia of Snow, Ice, and Glaciers*, edited by Vijay P. Singh, Pratap Singh, and Umesh K. Haritashya. Springer, 2011. A brief but comprehensive overview of Antarctica's ice sheet, glaciers, icebergs, sea ice, and ice cores, complete with color photographs and figures.

Thomas, David N., and Gerhard Dieckmann, eds. *Sea Ice.* Blackwell, 2010. Examines the role of sea ice in the global ecosystem, particularly in relation to the impact climate change is having on sea ice in the Antarctic and other regions. Fully illustrated with color photographs. A technical collection best suited for college students with some grounding in climate science, oceanography, or geology.

Walker, Sally M. *Frozen Secrets: Antarctica Revealed.* Carolrhoda Books, 2010. This highly readable book, written for high school students, focuses on the work of Earth scientists. Explains the techniques used to analyze ice cores, bedrock, water, and even samples of air, and covers what is known about the continent's past and future. Includes a glossary.

ICE-OUT STUDIES

FIELDS OF STUDY
Hydrology; Environmental Studies, Ecosystem Management; Statistics; Physical Geography

SUMMARY
Across the northern United States, a popular tradition for many decades has been guessing the date of ice breakup ("ice-out") on local lakes. Not only is it a local news story and harbinger of spring, but ice-out has numerous practical ramifications as well. It marks the end of winter activities on the lake, such as ice fishing and travel across the ice, and the beginning of summer activities, such as open-water fishing. Some localities also use the ice-out as the basis of charitable fund-raising events.

PRINCIPAL TERMS
- **ice-out:** synonymous with "spring break-up," when winter ice begins to break up as winter ends
- **meltwater:** liquid water from the thawing of ice and snow

HISTORICAL RECORDS
In earlier days, ice-out was important for commercial activities such as log drives and steamship navigation. As a result, for many lakes ice-out data are available since the middle of the nineteenth century. Moosehead Lake, the largest lake in Maine, has continuous data since 1848; Sebago Lake in southern Maine has sporadic data since 1807. Areas outside New England also maintain ice-out statistics. Minnesota has at least four lakes with more than 100 years of data, and the longest record is 139 years. Lakes Mendota and Monona, which flank the city of Madison, Wisconsin, have records extending back more than 150 years.

Ice-out dates are quite variable from year to year and depend on air temperature, the flux of meltwater into the lake, and the discharge of the outlet stream. For example, the earliest and latest ice-out dates for Moosehead Lake differ by forty-five days. Nevertheless, scientists from the U.S. Geological Survey analyzed ice-out dates for New England lakes and found that the average date of ice-out was nine days earlier in 2000 than it was in 1850. Generally, ice-out data are tabulated in terms of day of the year (the Julian date) rather than the calendar date. For example, February 20 is Julian day 51. The data indicate generally steady ice-out dates until 1900, decreasing (earlier) dates until roughly 1950, a slight increase in ice-out dates until about 1970, and decreasing dates thereafter. Despite the wide variability from year to year, the overall pattern is consistent across numerous lakes.

SIGNIFICANCE FOR CLIMATE CHANGE
There are many complexities in the use of ice-out data, largely due to variations in what defined "ice-out." Sometimes a marker placed on the ice has been used to signal ice breakup. In some places the marker was even a junked car (an environmentally dubious practice), for which people could purchase the right to guess which day the car would fall through the ice, as a fund-raising event. In other places the criterion has been the ability to cross the lake by boat without being blocked by ice. Still other places have used visual estimates of ice cover. Such estimates can be quite subjective, however. It may take several days for ice to clear out, and ice may persist in restricted coves long after most of a lake is ice-free.

Unfortunately, it is not always clear what criterion was used for many early records. Also, ice-out data

Frozen lake on Mather Island, Prydz Bay, Antarctica. The area was visited during an expedition of the german research vessel RV POLARSTERN (ANT-XXIII/9).

are extremely "noisy," in that the range of the data is much larger than the long-term change in average date. For example, for Maine's Moosehead Lake, the best-fit trend line for the ice-out data shows a nine-day decrease in ice-out day between 1849 and 2005, but the variation between earliest and latest ice-out is forty-five days. Even from year to year, the variations can be quite large: In 2001 ice-out occurred on day 124, dropping to day 110 in 2002 and increasing to day 127 in 2003.

Ice-out data for any individual lake are usually so variable that there is a significant possibility that the data, just by chance, happen to show a decrease over time. For example, if one flips a coin and keeps score of the difference between heads and tails, even though the flips are completely random there may be long runs where only one side of the coin shows up. Similarly, in ice-out data there may be long runs in which the difference increases or decreases steadily. However, when data from many lakes, especially those separated by large distances, show the same pattern, the statistical significance of the data becomes much greater.

Taking all these considerations into account, ice-out data for lakes across the northern United States show a broad trend of earlier ice breakup. Climate change skeptics have not paid much attention to ice-out data. Their most common approach has been to use biased anecdotal data such as pointing to unusually late ice-out dates on a particular lake, rather than to examine long-term trends over a large number of lakes

—*Steven I. Dutch*

FURTHER READING

Hodgkins, Glenn A., and Ivan C. James. "Historical Ice-Out Dates for Twenty-nine Lakes in New England." U.S. Geological Survey Open-File Report 02–34. U.S. Geological Survey, 2002. Listings of ice-out dates for all years with data and discussion of methods and interpretation. The data listings enable users to perform their own analyses of the data.

Hodgkins, Glenn A., Ivan C. James, and T. G. Huntington. "Historical Changes in Lake Ice-Out Dates as Indicators of Climate Change in New England, 1850–2000." U.S. Geological Survey Fact Sheet FS 2005–3002. Available online at http://pubs.usgs.gov/fs/2005/3002/. A summary of ice-out trends in New England lakes, with graphs of data trends but no data.

U.S. Geological Survey. "Lake Ice-Out Data for New England." 2007. Available online at http://me.water.usgs.gov/iceout.html. Contains links to other articles and an interactive map that links to data for each lake. The data are in simple columnar format, making it easy to copy and paste into a spreadsheet.

Wisconsin State Climatology Office. "Wisconsin Lake Ice Climatologies: Duration of Lake Ice." 2004. Available online at http://www.aos.wisc.edu/~sco/lakes/WI-lake_ice-1.html. Links to data for twenty-seven Wisconsin lakes. The links show duration of ice cover rather than specific dates of freezing and breakup. Lakes Mendota and Monona, with more than 150 years of data, show clearly decreasing trends, as do numerous other lakes.

INDIAN OCEAN

FIELDS OF STUDY

Oceanography; Physical Geography; Ecosystem Management; Earth System Modeling; Ecology; Heat Transfer; Thermodynamics; Engineering; Environmental Sciences; Environmental Studies; Waste Management; Meteorology; Climate Modeling; Fluid Dynamics; Hydroclimatology; Hydrometeorology; Atmospheric Science; Bioclimatology; Climate Classification; Climate Zones; Hydrology; Petroleum Refining

SUMMARY

The Indian Ocean shares the broad ecological and oceanographic features of the other major oceans of the world, but it possesses many unique and interesting characteristics. In general terms, the Indian Ocean's size means that both its shores and its depths are so varied that it is difficult to describe it as a single geographical unit. Its westernmost tides arrive at the shores of a continent that bears little resemblance to the shores of its easternmost limits. Most of the truly

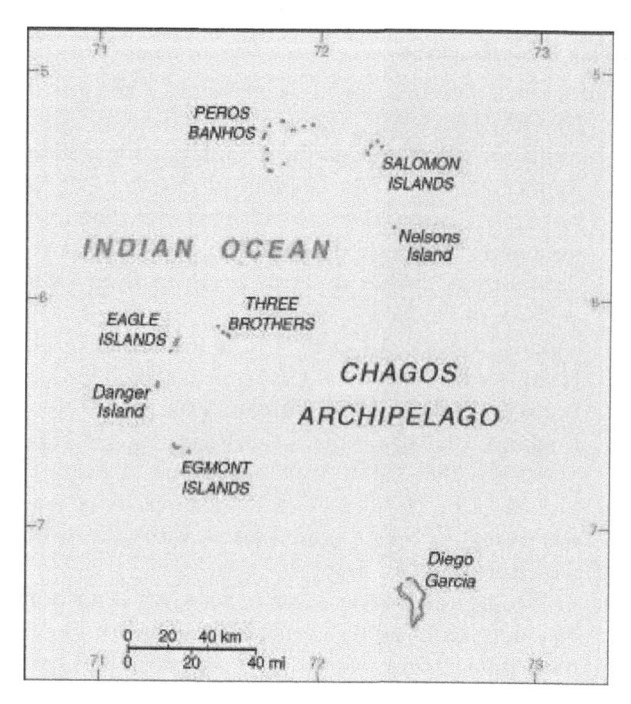

Map of the British Indian Ocean Territory. (Courtesy of the CIA World Fact Book)

unique characteristics of the Indian Ocean, however, lie beneath its surface in the form of extensive mountain ridges and, in one case at least, a very deep rift that represents an unparalleled underwater world of its own.

PRINCIPAL TERMS

- **Gondwanaland:** an ancient supercontinent that geologists theorize broke into at least two large segments; one segment became India and pushed northward to collide with the Eurasian landmass, while the other, Africa, moved westward
- **gyres:** circular patterns in the movement of surface currents that create nearly self-contained local subsections within the larger pattern of a typical ocean current
- **Java Trench:** one of the deepest areas of the Indian Ocean, located off the southern coast of Java in Indonesia; it is a form of geological canyon created by the upward thrust of mountain ridges from the ocean floor
- **monsoon:** a seasonal movement of winds into and out of the Indian Ocean region caused by variations of atmospheric pressure over the Indian Ocean and the interior land mass of Asia
- **Somali Current:** a seasonally reversing current that moves between the eastern coasts of Africa and the Arabian Peninsula

GEOLOGICAL ORIGINS

The geological origins of the Indian Ocean make it unique among the world's major oceans. Compared with the Pacific and Atlantic, the Indian Ocean is considerably smaller (covering an area of about 73 million square kilometers, in contrast to the Atlantic's nearly 84 million and the Pacific's nearly 166 million square kilometers). It is also of more recent geological origin than the Atlantic, Pacific, or Arctic Oceans. Geologists specializing in the evolutionary history of Earth and plate tectonics estimate that about 150 million years ago, a giant southern continent called Gondwanaland began to break apart. The movement of segments both westward (to what became the African continent) and northeastward (to what became India, a central section of which became known as Gondwana) took at least 100 million years. The collision of the Indian subcontinent with the Eurasian landmass about 50 million years ago brought about the violent upheaval of the Himalaya Mountains that continues in the present day. One of the effects of this phenomenon was to define new shorelines of the "youngest" of the world's major oceans.

The final product of these major geological upheavals was an oceanic body extending over the area between Australia in the east and Africa to the west. Its northernmost point corresponds to the Tropic of Cancer, where the Indian subcontinent joins the Eurasian landmass. From India's western coast to the southeastern tip of Arabia, the waters of the Indian Ocean form the Arabian Sea. On the opposite side of India, the Bay of Bengal and the Andaman Sea extend eastward to the coasts of Southeast Asia (Myanmar, Thailand, the Malay Peninsula, and the Indonesian Archipelago). If one includes the two smaller subsidiary seas, the Persian Gulf and the Red Sea, the Indian Ocean extends even farther north, to 30 degrees north latitude. To the south, the Indian Ocean technically goes as far as Antarctica. Two features of the Indian Ocean's floor define the point at which it separates the Atlantic and Pacific Oceans: the Atlantic Indian Basin and the South Indian Basin.

Associated Seas and Major Rivers

Although the formation of the Indian Ocean created several important seas and gulfs as distinct subsections of its total surface, there are fewer such bodies here than in the other oceans of the world. One should contrast general geographical denominations such as the Bay of Bengal (comprising most of the area east of India and touching the coasts of Southeast Asia) or the Arabian Sea (separating western India from the coasts of the Arabian Peninsula) with the geographical uniqueness of the Red Sea and the Persian Gulf. The latter two are, in fact, clearly separated from the main body of the Indian Ocean by the narrow Mandab Straits and the Straits of Hormuz, respectively. Both the Red Sea and the Persian Gulf have ecologies that are very different from that of the main body of the Indian Ocean. This is not the case for the two other semi-contained gulfs off the Arabian coast, the Gulf of Oman and the Gulf of Aden. The area known as the Great Australian Bight is simply the slightly curved central southern coast of Australia and is therefore even less circumscribed than the Bay of Bengal west of India. The Andaman Sea lies north of the Indonesian Archipelago and is enclosed geographically from the Bay of Bengal by a line of islands (actually an extension of the Indonesian islands) called the Nicobar and Andaman Islands.

Several major rivers pour large amounts of freshwater into the Indian Ocean. Such rivers are probably much older than the Indian Ocean itself, even though their pattern of flow was different in earlier geological ages. The Zambezi in East Africa, the Indus in northwest India, and the Ganges in northeast India were all probably flowing from their respective continental freshwater sources toward what eventually became the Indian Ocean's coastline. Each of these has had, over the long period since the formation of the ocean, a notable effect on the configuration of the coast where it empties into the ocean. Freshwater currents have, for example, cut actual canyons into the continental shelf area adjacent to the coast. In the case of the Ganges, an immense zone of sediment has built up, affecting both marine life and local currents in its delta area in the Bay of Bengal.

Ocean Depths and Submarine Geological Features

The average depth of the Indian Ocean is in the range of 3,636 to 3,940 meters. Several extremely deep but limited areas, notably the Java Trench to the south of Indonesia, are nearly twice as deep. The continental shelf along the coasts of the Indian Ocean is generally narrower than that of the other oceans, averaging 122 kilometers before deeper waters begin. The area west of the Indian coast, off the major city of Mumbai, is an exception. There, the continental shelf extends almost 325 kilometers into the ocean.

The floor of the Indian Ocean is crisscrossed by a number of underwater mountain ranges. Although notable ridges exist, its underwater topography is nowhere near as complex or spectacular as the eastern half of the Pacific, where extensive archipelagos with many small islands and some very large island formations (Japan, the Philippines, New Guinea, and New Zealand, for example) predominate. The most concentrated area of subsurface mountains in the Indian Ocean is centered near 30 degrees longitude, which is about halfway between the west coast of India and the Gulf of Aden, south of Arabia. A number of small but historically important islands mark high points along the mountain ridges between 30 and 60 degrees longitude. Mauritius and the Seychelles are examples of these. The huge island of Madagascar (588,000 square kilometers) is the most prominent surface example of the complex north-to-south submarine mountain systems located east of the African coast. The first of these is the Mauritius Ridge, marked on the surface at its southernmost point by Mascarene Island (due east from Madagascar) and by the Seychelles Islands to the north. The next range, the Carlsberg Ridge, is longer than the Mauritius Ridge, but none of its peaks emerge to form islands.

Finally, a long ridge extends due south off the southwest coast of India. This range is marked at the surface by the Laccadive Islands and the Maldives. Again, like Mascarene Island and the Seychelles near Madagascar, the Maldives are dwarfed by the single major island just to the southeast of the tip of India, Sri Lanka (formerly Ceylon). Sri Lanka, however, is the tailing part of the Indian subcontinental landmass rather than the tip of a submarine mountain range.

Beginning with the 90-degree east longitude line and moving toward the eastern shores of the Indian Ocean, the topography of the ocean floor is quite different from that of the western half. First, the very name of one subsurface range, the Ninety East Ridge, suggests a very regular pattern extending from north to south. The Ninety East Ridge, discovered only as recently as the 1960s, has gained the distinction of being the longest and straightest underwater mountain range in the world. Unlike the other ridges of the eastern basin of the Indian Ocean, most notably those south of the Indonesian coast, the Ninety East Ridge appears to be seismically inactive.

Between 90 degrees east longitude and the western shores of Australia one finds the third-deepest point in the Indian Ocean, the Wharton Basin, measuring nearly 6,364 meters deep. Farther south, off the southwestern tip of Australia, is the Diamantina Deep. Neither of these deep points is, however, associated with the rapid fall-off from mountain ridges to deep valleys that is characteristic of ocean trenches. Trenches are characteristic of the eastern rim of the Pacific Ocean, but only one such phenomenon occurs in the Indian Ocean. The Java Trench off the southern coast of Indonesia is more than 6,060 meters deep. Pioneer scientists examining flora and fauna in this area of what were then still undiscovered ocean trenches hypothesized that Indonesia was very close to the dividing line between tectonic plates. They observed that plants and animals to the east of Java appear to be biologically isolated in their evolution from species farther west.

The long, curved pattern of the ranges that constitute the Indonesian Archipelago actually extends far beyond the northern tip of Sumatra. Its peaks can be found in the chain of islands known as the Nicobar and Andaman Islands. The presence of these island chains west of the coasts of Thailand and Myanmar helps define the Andaman Sea area east of the main body of the Bay of Bengal.

TIDES AND CURRENTS

The immense size of the Indian Ocean means that tidal phenomena are variable, both in type and in the volume of tidal movement registered. The most common tides are semidiurnal (occurring twice daily). These are characteristic of the subequatorial eastern shores of Africa and, farther north and much farther east, in the Bay of Bengal. Australia's southwest coast, which is roughly opposite the subequatorial eastern coast of Africa, has an entirely different tidal pattern. Australia's Indian Ocean shores experience diurnal (once per day) tides that are extremely light by comparison to those of other coasts.

The Indian Ocean is the only ocean in the world with asymmetric reversing surface currents. Asymmetric conditions apply when currents in the northern half of the ocean are moving in a different direction from those of the southern half. The complexity of Indian Ocean currents close to the surface goes well beyond the relatively simple question of north-south asymmetry. One finds, for example, that wind conditions contribute to the creation of gyres, circular or spiral movements that break the broad pattern of the surface current into localized segments.

It is particularly important to note that broad patterns of currents in the Indian Ocean reverse according to the season. Currents, like so many other factors determining the overall ecology of the Indian Ocean, are affected in large part by the major monsoonal wind and weather conditions that are characteristic of this region of the world.

Probably the most famous current in the Indian Ocean is the so-called Somali Current, which moves, in certain seasons of the year, in a fairly rapid clockwise direction from the northeast coast of the Horn of Africa. In the summer season, this current goes as far as the coast of India. At its farthest point moving east in the summer months, it meets the southwesterly monsoon current off the Indian subcontinent. During the winter, the direction of the Somali Current reverses, a situation that created near-ideal seasonal sailing conditions for centuries for ships sailing between the Arabian peninsular zone, particularly the Persian Gulf, and the Horn of Africa.

CLIMATIC ZONES AND WIND PATTERNS

Because of the great north-to-south distance covered by the Indian Ocean and its associated seas, there are several quite distinct climatic zones according to geographical location. The most famous, and most important for sustaining seasonal agriculture in the entire Indian Ocean area, is the so-called monsoon zone, which runs from 10 degrees north of the equator to about 10 degrees south. The region between 10 and 30 degrees south of the equator is the zone of what has traditionally been called the trade winds. It is the predictability of steadily blowing southeast winds in this wide region (as distinct from the area of the Somali Current) that made maritime communication, and therefore trade, between the opposite shores of the Indian Ocean possible.

Farther south, near the global band of the Tropic of Capricorn (running through the island of Madagascar on the western shores and Australia on the east) lies the subtropical to temperate zone, between 30 and 45 degrees latitude. Some 20 degrees south of the Tropic of Capricorn, climatic conditions begin to show the temperate cooling influence of the extreme southern extent of Indian Ocean waters leading to the last climate zone. From 45 degrees latitude southward to the Antarctic ice cap, the beginnings of sharply cold Antarctic waters mark the end of the gradual transition separating some of the world's hottest climates in the Indian Ocean proper from the extreme cold of the southern zone of the globe. Here, three of the world's four oceans almost literally fuse in the Antarctic ice cap.

MONSOON WIND CONDITIONS

Generally stated, monsoon conditions are semiannual reversing wind patterns. Extensive areas of high pressure "empty" their air in the direction of equally vast low-pressure zones. When this happens, winds moving across water carry the moisture they pick up, which is typically precipitated as rain before they reach their low-pressure destinations. In the case of Indian Ocean monsoons, the widespread heating of the landmass in the Northern Hemisphere during summer creates conditions of low atmospheric pressure over Asia. This low-pressure zone becomes an attractive force for masses of air that are pressed downward by high-pressure conditions over Australia. The resultant winds that move in a northwesterly direction across Southeast Asia and the Indian subcontinent bring with them a much-needed monsoon season of heavy rains that lasts until the particular atmospheric conditions cease to apply. Generally, the monsoon rainy season is predictable, but the arrival of torrents of rain to quench the dry agricultural fields of south Asia and Southeast Asia is not guaranteed. When the typical monsoon wind pattern develops but insufficient moisture is collected to bring rains, areas that depend on monsoon waters can face serious drought for at least one year, as there is no chance of humid air movements over the landmass once the directional wind pattern created by Asia's summer heating ends.

NATURAL RESOURCES

The Indian Ocean contains many key minerals that are extracted to supplement the local economies of several of the countries along its shores. Manganese is found in several areas off South Africa and Australia. Other minerals include tin and chromite. Mineral wealth in the water is, however, overshadowed by the vast petroleum reserves concentrated in one of its seas: the Persian Gulf. These are estimated to be perhaps the largest oil reserves in the world. Other locations where petroleum wealth is important are on the island of Sumatra and in its offshore waters. Similar intermediate potential for petroleum production is found in the Red Sea and off the shores of western India.

The fishing industries that depend on the Indian Ocean are varied. Many depend on the phenomenon of upwellings, movements of water from lower depths that carry phytoplankton, a basic food source for many fish species, close to the surface. The most common commercialized fisheries seek large schools of sardines, mackerel, and anchovies. The principal single species fished in many areas of the Indian Ocean is shrimp.

SIGNIFICANCE

The Indian Ocean and its associated seas represent one of the most diversified ecological marine environments in the world. This applies both to its coastlines and to the world beneath its surface. The African landmass along the ocean's western shores reveals a variety of natural characteristics that hardly resemble what one finds along its eastern shores. Dry coastal regions characterize both the southern and northern reaches of the African landmass, with some of the driest deserts in the world extending from the Somalia Horn of Africa into the Red Sea and around the southern shores of Arabia into the Persian Gulf. Although these conditions continue all the way along the northwestern coastline to the Gujarati coast of India, the Indian Ocean ecosystem is clearly distinct once one passes the Indian subcontinent.

From the Bay of Bengal southward along the eastern shores of the Indian Ocean, the environment becomes increasingly tropical, passing through some of the most extensive tropical rainforests in the world, especially in the Indonesian Archipelago. Indeed, parts of the west coast of Australia, generally thought to be nearly a desert environment, exhibit a tropical ecology that contrasts notably with the dry climates of the African coastline of the Indian Ocean. With this physical ecological diversification comes an enormous variation in plant and animal life along the ocean's shorelines and in the relatively shallow waters of its continental shelf. Studying the diversity of the Indian Ocean's biological ecology and submarine geology is like observing multiple different worlds on one segment of the planet.

—*Byron D. Cannon*

FURTHER READING

American Museum of Natural History. *Ocean*. Dorling Kindersley, 2006.

Charabe, Yassine, and Salim al-Hatrushi. *Indian Ocean Tropical Cyclones and Climate Change*. Springer, 2010.

Clift, Peter D., and R. Alan Plumb. *The Asian Monsoon*. Cambridge University Press, 2008.

Lighthill, James, and Robert Pearce, eds. *Monsoon Dynamics*. Cambridge University Press, 2009. Print.

Mukhopadhyay, Ranadhir, Anil K. Ghosh, and Sridhar D. Iyer. *The Indian Ocean Nodule Field: Geology and Resource Potential*. Ed. M. Hale. Elsevier, 2012.

Neprochnov, Y. P., et al., eds. *Intraplate Deformation in the Central Indian Ocean Basin*. Geological Society of India, 1998.

Pearson, M. N. *Trade, Circulation, and Flow in the Indian Ocean World*. 2015.

Rao, P. V., ed. *The Indian Ocean: An Annotated Bibliography*. Kalinga, 1998.

Wiggert, Jerry D. *Indian Ocean Biogeochemical Processes and Ecological Variability*. American Geophysical Union, 2009.

INDICATORS OF CLIMATE CHANGE AND GLOBAL WARMING

Instrument records from land stations and ships indicate that the global annual mean surface air temperature rose during the twentieth century. The warming occurred more quickly in high latitudes than it did in the tropics. It was also faster over land than it was over the ocean and faster in the Northern Hemisphere than in the Southern Hemisphere. Winters warmed more than did summer, and nights warmed more than did days. Contemporary daily temperature ranges have narrowed, precisely because nights have warmed more than have days.

Extensive heat waves and intense floods have become more frequent in recent decades. Globally,

Global Warming Potentials of Major Greenhouse Gases

Greenhouse Gas	Global Warming Potential*
CO_2 (carbon dioxide)	1 (reference)
CH_4 (methane)	25
N_2O (nitrous oxide)	298
HFC-23 (hydrofluorocarbon-23)	14,800
SF_6 (sulfur hexafluoride)	22,800

*One-hundred-year time horizon.

the average number of tropical storms (about ninety per year) changed little during the twentieth century, although historical data are poor for some regions. In the North Atlantic, where the best records are available, there has been a clear increase in the number and intensity of tropical storms and major hurricanes. From 1997 to 2006, there were about fourteen tropical storms per year, including about eight hurricanes in the North Atlantic, compared to about ten storms and five hurricanes between 1850 and 1990.

On timescales of thousands of years or greater, Earth's climate has been both warmer and colder than it is today, although temperatures around the turn of the twenty-first century were the warmest in the past two thousand years. Based on ice-core proxy data, four major global glaciations occurred in past 450,000 years, about one every 100,000 years, correlating well with the cyclical variations in Earth's orbit known as the Milankovitch cycles. Various ice ages occurred, with the most recent one ending about 11,500 years ago. Before that, much of North America was covered in permanent ice. Over the course of Earth's history, its temperature has swung more than 10°C between cold and warm modes.

CLIMATE CHANGE SCENARIO

Future climate changes are predicted by climate models based on assumed greenhouse gas (GHG) emission scenarios. The scenarios range from high fossil fuel consumption, resulting in atmospheric carbon dioxide (CO_2) concentration of 800 parts per million, to low consumption, with CO_2 concentration reaching 550 parts per million. The reliability of these predictions depends on future global environmental, energy, and climate policy, as well as the accuracy of the models.

Most models project that climate change will accelerate during the twenty-first century and that the global average temperature will increase by between 1.8°C and 4.0°C by 2100. As in the past, warming will be more pronounced in the polar Northern Hemisphere during winter. Precipitation amounts are likely to increase in high latitudes and to decrease in most subtropical lands. Heat waves and heavy precipitation events will very likely increase in frequency. With warmer oceans, future tropical storms will become more intense, with greater peak wind speeds and heavier precipitation.

CONTEXT

Climate change may be attributed to natural processes or to human activity. Natural factors include Earth's internal processes, such as volcanic eruptions, as well as external parameters, such as solar luminosity and Earth's orbital pattern around the Sun. Anthropogenic activity includes GHG and aerosol emission and, to a lesser degree, changes in land use. Separating natural and anthropogenic causes of climate change is challenging, if it is possible at all. Since no controlled laboratory setting exists in which to conduct climate change experiments, climate scientists have developed computer models based on the laws governing climate systems. By altering model settings, one can simulate natural and anthropogenic effects on climate, separately or in combination, thereby tracing the causes of climate change. In general, on scales of a decade to a century, climate change is attributable to atmosphere-ocean interaction and to human activity. On scales of millennia to hundreds of thousands of years, the variations in Earth's orbit directly controls the planet's climate. This orbit is described by the Milankovitch cycles, which repeat every 20,000 to 100,000 years. Beyond the million-year timescale, tectonic drift is likely the main driver of climate change.

—*Zaitao Pan*

FURTHER READING

Alley, Richard B. *The Two-Mile Time Machine. Ice Cores, Abrupt Climate Change and Our Future*. Princeton University Press, 2014. Describes the analytical methods and interpretations of physical clues of past climates trapped in ice that is thousands of years old.

Maslin, Mark. *Global Warming: A Very Short Introduction*. Oxford University Press, 2009. Provides a concise general introduction to the major factors that define the climate of Earth.

"NASA, NOAA Analyses Reveal Record-Shattering Global Warm Temperatures in 2015." *Goddard Institute for Space Studies*. NASA, 20 Jan. 2016, www.nasa.gov/press-release/nasa-noaa-analyses-reveal-record-shattering-global-warm-temperatures-in-2015. Accessed 17 May 2018.

Serreze, Mark C., and Roger G, Barry. *The Arctic Climate System.* Cambridge University Press, 2014. Describes the Arctic environment and the climate-related changes to that environment that have occurred since 1980.

Mélieres, Marie-Antoinette, and Chloé Maréchal. *Climate Change: Past, Present, and Future.* John Wiley & Sons, 2015. This book first provides an introduction to the basics of climatology and then investigates the physical evidence of past and present climate change in the context of that basic framework.

INTERGLACIAL PERIOD

FIELDS OF STUDY

Atmospheric Science; Atmospheric Chemistry; Analytical Chemistry; Environmental Chemistry; Geochemistry; Physical Chemistry; Photochemistry; Thermodynamics; Fluid Dynamics; Environmental Sciences; Environmental Studies; Waste Management; Meteorology; Process Modeling; Hydroclimatology; Bioclimatology; Climate Zones; Oceanography; Hydrology; Earth System Modeling; Ecology

SUMMARY

How does the present climate compare to the climate during other interglacials? Are current temperatures, sea levels, and carbon dioxide (CO_2) concentrations unprecedented, or should they be considered within the expected range? When, if ever, is the current interglacial going to end? These are among the important questions that climate science seeks to answer. Ice cores, loess deposits, pollen analysis, and other data and techniques provide information that comes together to form a detailed picture that will help answer them.

PRINCIPAL TERMS

- **Holocene:** the current interglacial, which began 11,700 years ago
- **isotopes:** variants of an element that are chemically identical but have different atomic mass numbers and vary in radioactivity
- **marine isotope stage:** half of a glacial cycle, as identified in the oxygen isotope data from ocean cores; advances are given even numbers, and retreats are given odd numbers
- **Milankovitch cycle:** cyclical variance in Earth's orbital parameters, including axial inclination, climatic precession, and orbital eccentricity
- **MIS 11:** an interglacial that may be the best analogue for the Holocene; also called the Holsteinian or Termination V
- **MIS 5e:** the most recent interglacial before the Holocene, also known as the Eemian, LIG (Last InterGlacial), or Termination II
- **$^{18}O/^{16}O$ ratio:** ratio between two oxygen isotopes that is altered by global average temperatures associated with the advance and retreat of continental glaciers

BIOLOGICAL DATA

Over the last three million years, there have been forty to fifty glacial/interglacial cycles. Scientists have sought to characterize conditions during these periods using a variety of approaches. Plant remains, particularly pollen, have been analyzed to estimate temperature and humidity. The carbonate shells of planktonic marine organisms, preserved as fossils in the sediments beneath the sea, have been analyzed to infer sea surface temperatures. It is difficult to separate regional effects from global ones, and often data from different time periods is only available in different locations, so acquiring a global picture is not easy.

In general, interglacial climate is seen as being quite similar to the current climate. In fact, the term is often restricted to periods during which temperatures were at least as high as they have been during the Holocene, the name given to the current interglacial.

ICE CORE DATA

Cores have been drilled out of the ice in Greenland and Antarctica. Within the ice are bubbles of air that have been preserved since the ice was formed.

An ice core sample containing rock and other materials that have not seen sunlight for hundreds of thousands of years. Scientists at the North Greenland Eemian Ice Drilling site announced the historical significance of hitting bedrock at the depth of 8,324 feet.

The deepest core, from an area called Dome C in Antarctica, has samples of air from more than 800,000 years ago. These core samples provide detailed information on eight complete glacial cycles. Ratios of oxygen isotopes ^{18}O to ^{16}O are used to infer how much of the Earth's water was tied up in glacial ice, since glaciers sequester ^{16}O and thereby increase the proportion of ^{18}O in the ocean. The concentration of deuterium (^2H) in a sample, moreover, can be used to infer the temperature of the air when the snow formed. Age is determined by combining depth, snow accumulation rates, ice flow rates, compaction rates, and so on, in a complex but reproducible way and calibrating the results by using radiometric dating techniques on volcanic dust incorporated in the ice. Dust is examined in detail, and often its place of origin can be determined. While no two cycles are identical, they all share a number of traits.

Each cycle has a saw-tooth shape. A glacial advance ends abruptly, with rapid melting of the ice and a rapid rise in air temperatures. For example, in MIS 9, the temperature rose 13°C in eight thousand years. This warming can stop abruptly, with cooling starting almost immediately, or it can taper off, warming at a slower rate for perhaps a dozen millennia, before cooling begins. Of the eight previous terminations in the Dome C core, four started cooling immediately and four tapered off. Once cooling begins, it is far more gradual than is warming, continuing, with some reversals, for about 100,000 years. Ice is sequestered as cooling occurs, and CO_2 levels fall.

The air temperatures in Antarctica were higher than the average for the current millennium during each of the last four interglacials, but lower during those more than 500,000 years old. The Holocene (MIS 1) began 11,700 years ago. The three most recent interglacials before the Holocene (MIS 5, 7, and 9) had durations of about twelve thousand years or less, so simply by looking at the ice core data one might guess that the next glacial advance may be imminent. However, a theory proposed by Milutin Milanković (1879-1958) suggests things may not be this simple.

Periodic variations in some of the orbital parameters of the Earth are known to govern the timing of glacial cycles. One past interglacial, during which those variations were similar to today's, was MIS 11, which lasted for twenty-eight thousand years and had warming taper off after an early rapid rise. However, MIS 19, which also had similar orbital variations, saw cooling right after its initial warming and was of shorter duration. There seems to be a degree of randomness within the regularity of the Milankovitch cycles such that the timing of the changes produced is predictable, but the scale and intensity of the changes are dependent on a number of other factors.

Context

Earth is currently in an interglacial state and has been in that state for nearly twelve thousand years. Geology and oceanography have shown that over the last three million years Earth has switched between glacial and interglacial states some forty to fifty times. With no anthropogenic influence, one would conclude that continental glaciers will advance again, the only question being when. In the presence of anthropogenic influences, however, it is not clear that Earth will return to another glacial state.

Although there is some discussion over whether the orbital conditions for triggering a glacial advance have already occurred or are yet to occur, it is still prudent to examine the conditions on the planet today to evaluate the risk posed by another glacial advance. Of particular interest is why, during times of elevated CO_2 and elevated temperatures, former interglacials succumbed to the minor fluctuations of solar energy produced by those orbital conditions.

—*Otto H. Muller*

FURTHER READING

Luthi, Dieter, et al. "High-Resolution Carbon Dioxide Concentration Record 650,000–800,000 Years Before Present." *Nature* 453, no. 7193 (2008): 379–82. Includes a plot of CO_2 concentrations and temperatures over eight glacial cycles. Charts, tables, bibliography.

Rapp, Donald. *Ice Ages and Interglacials: Measurements, Interpretations, and Models.* Springer, 2009.

Ruddiman, William F. *Earth's Climate Past and Future.* 2nd ed. W. H. Freeman, 2008. This elementary college textbook has a thorough discussion of how scientists have learned about the climate of interglacials. Illustrations, figures, tables, maps, bibliography, index.

Ruddiman, William. *Plows, Plagues, and Petroleum: How Humans Took Control of Climate.* Princeton University Press, 2016. Written for the lay public, this book provides the background and thinking behind the theory that humans have influenced the climate for the last nine thousand years, primarily through agriculture. Illustrations, figures, tables, maps, bibliography, index.

K

KILIMANJARO'S ICE CAP

FIELDS OF STUDY
Physical Geography; Volcanology; Glaciology; Environmental Sciences; Environmental Studies; Meteorology

SUMMARY
At 5,895 meters, Kilimanjaro is Africa's highest mountain. It is of volcanic origin, and is located alongside the East African rift zone where Earth's crust is actively spreading. Its trio of volcanic peaks gradually became inactive, first Shira, then Mawenzi, and finally Kibo, the highest. Ice and snow gathered on Kilimanjaro's upper slopes during the late Pleistocene epoch, about eleven thousand years ago. Ice cores taken from the mountain's glaciers have revealed a history of expansions and contractions of the ice cap, with the contractions occurring during periods of drought eighty-three hundred, fifty-two hundred, and four thousand years ago. Periods of expansion resulted from climatic conditions that were warmer and wetter than those of today. Recently, Kilimanjaro's ice sheets have been confined to Kibo.

PRINCIPAL TERMS
- **ice cap:** semipermanent glacial crown of ice atop a mountain or other geologic formation
- **sublimation:** the conversion of a solid directly into a gas, without passing through an intermediate liquid state
- **volcano:** a usually mountainous rift in Earth's crust caused by magma erupting through fissures onto the planet's surface

THE SHINING MOUNTAIN
One-half million years ago, early human inhabitants of the Rift Valley first saw this majestic mountain, but its names derive from a much later period. African tribes called it the "white" or "shining" mountain, because of its ice cap. Its present name comes from the Swahili name for the mountain "*Kilima-Njaro*" meaning "Mountain of Greatness" because of its massive height and bulk. The first European to see the mountain was the German missionary Johann Ludwig Krapf on December 3, 1849. This not only sparked interest in searches for the source of the Nile but also in the scientific study of Kilimanjaro and its ice sheets. The existence of these features flew in the face of the "common wisdom" of the time, which held that snow and ice could not exist in such close proximity to the equator.

VANISHING ICE SHEETS
In the late nineteenth century, explorers created maps and drawings of the mountain and its environs. From these, scholars derived an estimate that, in 1880, ice covered about 20 square kilometers of Kilimanjaro's principal peak. In 1889, two Europeans were the first to reach Kilimanjaro's summit, and they brought back information about the extent and depth of the ice sheets. In the twentieth century, photographs from the mountain's base and, later, from airplanes, provided benchmarks by which the shrinkage of the ice cap could be measured. In 1912, a precise map was constructed based on photogrammetric evidence. At that time, the ice sheets had diminished to 12.1 square kilometers, though they still existed on all sides and descended to about 4,400 meters. From 1912 to the early twenty-first century, Kilimanjaro's icefields were periodically surveyed, and the data gathered indicated that the time of greatest contraction had been from about 1880 to 1950.

Ernest Hemingway, an American writer, made Kilimanjaro's ice cap world famous through his

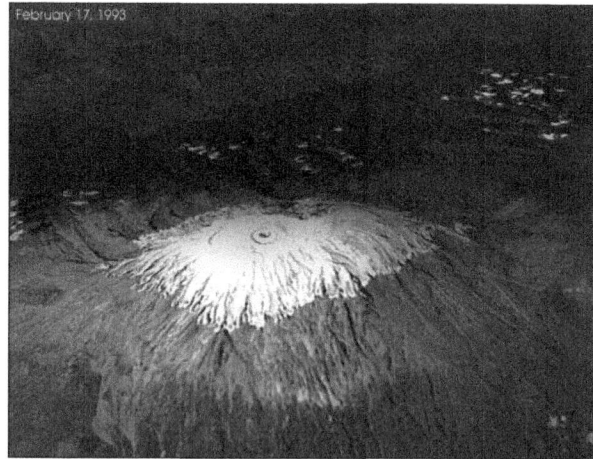

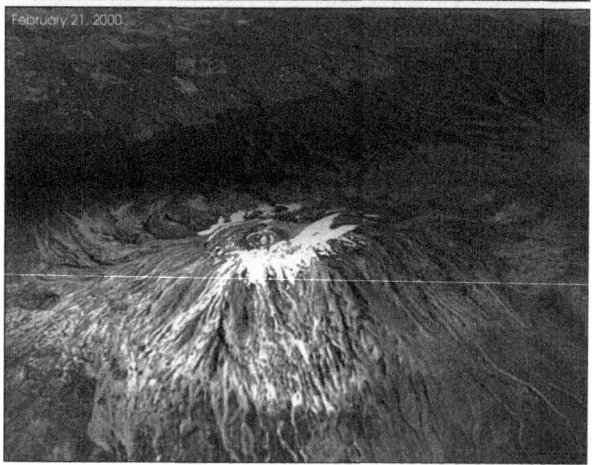

These two images taken by NASA show the changes in snow accumulations on the summit of Mount Kilimanjaro. The top image was photographed on February 17, 1993 and the lower one was photographed on February 21, 2000.

popular short story, "The Snows of Kilimanjaro," (1936) which was made into a successful Hollywood movie in 1952. By the early 1950s, the mountain's icefields had diminished to 6.7 square kilometers. Sufficient data had been collected to determine that the rarely observed trickles of meltwater were unable to account for the retreating glaciers. Some scientists began to attribute glacial contraction to the sublimation of summit ice into water vapor under the influence of the tropical sun in dry air at below-freezing temperatures. However, during the second half of the twentieth century, as evidence multiplied that midlatitude mountain glaciers were shrinking because of global warming, some scientists extended this explanation to such equatorial mountain glaciers as Kilimanjaro's.

Aerial photographs of Kibo's ice sheets had been supplemented by satellite pictures, and this new information showed that, although the ice cap's contraction had slowed since 1953, 75 to 80 percent of its area had vanished during the past century, along with deep reductions in the volume of its ice and snow. Some scientists predicted that the ice cap would be completely gone by 2015 or 2020. Since the "Snows of Kilimanjaro" were famous as well as photogenic, it was natural for advocates of greenhouse gas (GHG) reduction to use dramatic images of the mountain's disappearing ice cap to bolster their cause. Greenpeace advocates held a news conference from Kilimanjaro's summit, and former U.S. Vice-President Al Gore argued that the vanishing ice cap was evidence of global warming.

SIGNIFICANCE FOR CLIMATE CHANGE

Although many scientists agreed that temperate-zone mountain glaciers were retreating because of global warming, an increasing number of climatologists and glaciologists believed that the specific case of Kilimanjaro's ice loss was caused by other factors. Data collected from balloons, satellites, and an automatic weather station on one of Kibo's icefields revealed that reduced precipitation of ice and snow in desiccated air, along with extensive exposure to solar radiation in below-freezing temperatures, led to the accelerated sublimation of ice and snow. Fluctuating weather systems in the Indian Ocean, which influenced humidity and cloud cover over Kilimanjaro, may also have played a role.

CONTEXT

Most scientists familiar with the data held that global warming had little or no effect on the ice decline. These scientists chided global-warming enthusiasts for their misuse of Kilimanjaro's vanishing ice cap to support their views on climate change, but they also criticized global-warming deniers who overgeneralized the Kilimanjaro case to include midlatitude glaciers. Kilimanjaro can therefore serve as a cautionary tale of how politics and overheated rhetoric often lead passionate advocates to distort scientific data and images in unacceptable ways.

—*Robert J. Paradowski*

FURTHER READING

Bowen, Mark. *Thin Ice: Unlocking the Secrets of Climate in the World's Highest Mountains.* Macmillan, 2005. This book has been called one of the best yet published on climate change. Part 6 is dedicated to Kilimanjaro. Notes, references, index.

Darling, Seth B. and Sisterson, Douglas L. *How to Change Minds About Our Changing Climate.* The Experiment LLC, 2014. Addresses the issue of climate change denial in the face of the demonstrable effects of climate change.

Dessler, Andrew *Introduction to Modern Climate Change* 2nd ed. Cambridge University Press, 2016. Provides a good overview of the science and sociology of climate change.

Huggel, Christian, Carey, Mark, Clague, John J. and Kääb, Andreas, eds. *The High-Mountain Cryosphere. Environmental Changes and Human Risks.* Cambridge University Press, 2015. A extensive examination of the global environmental factors that dynamically affect the formation of mountain ice and snow caps and glaciers.

Mann, Michael E. *The Hockey Stick and the Climate Wars. Dispatches from the Front Lines.* Columbia University Press, 2014. This very readable book lays out the science behind global climate change and identifies the conflict between climate change proponents and deniers.

Michaels, Patrick. *Meltdown: The Predictable Distortion of Global Warming by Scientists, Politicians, and the Media.* Cato Institute, 2004. Although Michaels, a climatologist, accepts the reality of global warming, he argues that alarmists have exaggerated its future effects, and he uses the debate over Kilimanjaro's shrinking ice cap to show how science can be distorted for political ends.

Mote, Philip W., and Georg Kaser. "The Shrinking Glaciers of Kilimanjaro: Can Global Warming Be Blamed?" *American Scientist* 95 (July/August, 2007): 318–325. This article by a climatologist and glaciologist musters much scientific evidence to show that factors other than global warming are responsible for the retreat of Kibo's ice sheets. Illustrated with photographs, graphs, and a map. Bibliography.

Stanley, Steven. *Earth Systems History.* 3rd ed. W. H. Freeman, 2009. Lays out the history of Earth, including the factors affecting global climates and climate change, over the vastness of geological time. Selected chapters focus on climate change in terms of glaciation and deglaciation.

L

LA NIÑA

FIELDS OF STUDY
Oceanography; Physical Geography; Meteorology; Fluid Dynamics; Thermodynamics; Heat Transfer; Environmental Sciences; Environmental Studies; Climate Modeling; Process Modeling; Hydroclimatology; Earth System Modeling

SUMMARY
Two large weather anomalies alternate south of an equatorial band across the middle and eastern portions of the Pacific Ocean. One is El Niño, characterized by warmer-than-average surface waters; the other is La Niña, marked by cooler surface temperatures (the two are collectively designated as the El Niño-Southern Oscillation or ENSO).

PRINCIPAL TERMS
- **eastern trade winds:** winds blowing in a northeasterly or southeasterly direction near the equator as a result of the Coriolis Effect and the Hadley circulation, so named because they favored sailing ships traveling eastward on trade missions

THE CHRISTMAS CHILDREN
Two large weather anomalies alternate south of an equatorial band across the middle and eastern portions of the Pacific Ocean. One is El Niño, characterized by warmer-than-average surface waters; the other is La Niña, marked by cooler surface temperatures (the two are collectively designated as the El Niño-Southern Oscillation or ENSO). La Niña is defined by Pacific surface temperatures of 0.5°C or more below average for a period of at least five months. The cooling occurs as stronger-than-usual eastward trade winds blow across the Pacific, drawing cold water from the ocean depths to the surface.

The Spanish names of these weather events originate from the appearance of *El Niño* off the coast of Peru as Christmastime approaches; thus, *el niño* (boy child) refers to the coming of the infant Jesus. The opposite pattern is termed *La niña* (girl child), although at one time it was called *El Viejo* (the old man). ENSO does not occur on a scheduled basis but does happen regularly and with widespread, serious consequences.

SIGNIFICANCE FOR CLIMATE CHANGE
The effects of global warming on ENSO, the causes of which are only incompletely known, remain the subject of scientific debate, particularly because so great a number of variables affect ENSO. To obtain more data and understand long-term weather patterns, a monitoring system in the Pacific collects data from buoys, satellites, and computer models.

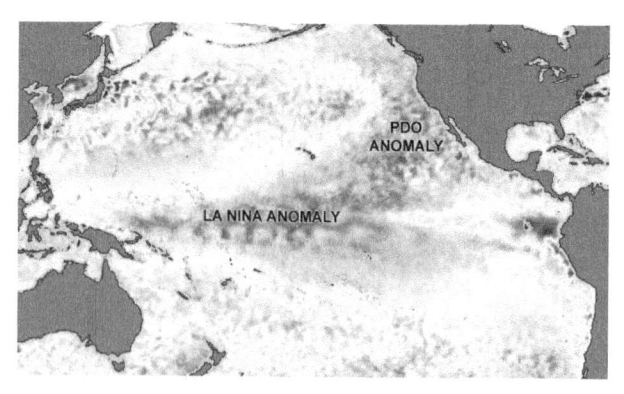

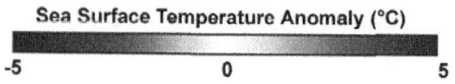

A cool-water anomaly known as La Niña occupied the tropical Pacific Ocean throughout 2007 and early 2008. The cool water anomaly in the center of the image shows the lingering effect of the year-old La Niña.

Hotter or colder surface Pacific waters alter overhead trade winds, thereby shifting the North American jet stream. Altering the track and strength of the jet stream produces exceptional rain or drought. The cold, heavy air from La Niña pushes the jet stream to the upper part of the United States.

On average, a La Niña event may last from nine to twelve months, appearing at the end of one year and extending into the next. Consequently, in the southeastern portions of the United States, winter temperatures are warmer and dryer than normal. In northwestern regions, they are cooler and wetter. Hurricanes in the Atlantic and tornadoes in the United States tend to increase in number and force during La Niña events. In South America, dryer-than-usual conditions prevail in southern regions.

El Niño events occur more frequently than do La Niña events, in a ratio of approximately two to one. However, La Niña events last longer. One such event lasted, with a brief interlude, from mid-1998 to early 2001. Another occurred during the latter half of 2007 through the first half of 2008, provoking epic rainfall in Australia and record snowfalls in parts of China.

—*Edward A. Riedinger*

FURTHER READING

Bell, Gerry. "Impacts of El Niño and La Niña on the Hurricane Season." *Climate.gov*. NOAA, 30 May 2014. Web. 24 Mar. 2015.

D'Aleo, J. S., and P. G. Grube. *El Niño and La Niña*. ORYX, 2002. Print.

Di Liberto, Tom. "ENSO + Climate Change = Headache." *Climate.gov*. NOAA, 11 Sept. 2014. Web. 24 Mar. 2015.

Fagan, Brian *Floods, Famines and Emperors. El Niño and the Fate of Civilizations*. Basic Books, 1999. A readable historical examination of how the El Niño–Southern Oscillation (ENSO) phenomenon have affected civilizations on a global scale throughout the past.

Glantz, M. H. *Currents of Change: Impacts of El Niño and La Niña on Climate and Society*. 2nd ed. Cambridge University Press, 2001. Print.

"La Niña." *National Geographic Education*. National Geographic Society, 1996–2015. Web. 24 Mar. 2015.

LAKES

FIELDS OF STUDY

Hydrology; Limnology; Physical Geography; Environmental Chemistry; Analytical Chemistry; Physical Chemistry; Photochemistry; Thermodynamics; Environmental Sciences; Environmental Studies; Waste Management; Meteorology; Climate Modeling; Earth System Modeling; Fluid Dynamics; Hydroclimatology; Hydrometeorology; Bioclimatology; Climate Zones; Ecosystem Management; Earth System Modeling; Ecology; Spectroscopy; Mass Spectrometry

SUMMARY

Lakes are geologically short-lived features, and sediments deposited in lakes (called lacustrine sediments) have constituted only a tiny fraction of the sedimentary rocks on Earth. Nevertheless, lake sediments are important sources of information about past climates. Several important economic resources, including oil shales, diatomaceous earth, salt and other evaporites, some limestones, and some coals, originate in lakes.

PRINCIPAL TERMS

- **allogenic sediment:** sediment that originates outside the place where it is finally deposited; sand, silt, and clay carried by a stream into a lake are examples
- **biogenic sediment:** sediment that originates from living organisms
- **clastic sediments:** sediments composed of durable minerals that resist weathering
- **clay:** a mineral group whose particles consist of structures arranged in sandwich-like layers, usually sheets of aluminum hydroxides and silica, along with some potassium, sodium, or calcium ions

- **clay minerals:** any mineral particle less than 2 micrometers in diameter
- **endogenic sediment:** sediment produced within the water column of the body in which it is deposited; for example, calcite precipitated in a lake in summer
- **mineral:** a solid with a constant chemical composition and a well-defined crystal structure
- **mineraloid:** a solid substance with a constant chemical composition but without a well-ordered crystal structure
- **plankton:** plant and animal organisms, most of which are microscopic, that live within the water column
- **seston:** a general term that encompasses all types of suspended lake sediment, including minerals, mineraloids, plankton, and organic detritus

GEOLOGICAL ORIGIN OF LAKES

Several geologic mechanisms can create the closed basins that are needed to impound water and produce lakes. The most important of these mechanisms include glaciers, landslides, volcanoes, rivers, subsidence, and tectonic processes.

Continental glaciers formed thousands of lakes by the damming of stream valleys with moraine materials. Glaciers also scoured depressions in softer bedrock, and these later filled with water to form lakes. Depressions called kettles formed when buried ice blocks melted. Mountain glaciers also continue to produce numerous small, high alpine lakes by plucking away bedrock. The bowl-shaped depressions that occur as a result of this plucking are called cirques; lakes that occupy cirques are called tarns. Sometimes a mountain glacier moves down a valley and carves a series of depressions along the valley that, from above, look like a row of beads along a string. When these depressions later fill with water, the lakes are called paternoster lakes, the name (Latin for "our father") having been given for their similarity to beads on a Christian rosary.

Landslides sometimes form natural dams across stream valleys. Large lakes then pond up behind the dam. Volcanoes may produce lava flows that dam stream valleys and produce lakes. A volcanic explosion crater may fill with water and so produce a lake. After an eruption, the area around the eruption vent may collapse to form a depression called a caldera. Some calderas, such as Crater Lake in Oregon, fill with water. Rivers can produce lakes along their valleys when the loop of a meandering channel finally is enclosed by sediment and leaves behind an oxbow lake, isolated from the main channel. Sediment may accumulate at the mouth of a stream, and the resulting delta may build, bridging across irregularities in the shoreline, to create a brackish coastal lake.

Natural subsidence creates closed basins in areas underlain by soluble limestones or evaporite deposits. As the underlying limestone is dissolved away, the ground above collapses into the cavity, forming a sinkhole that may later fill with water. Finally, large-scale downwarping of tectonic plates has produced some very large lakes. Large basins form when the crust warps or sinks downward in response to deep forces. The subsidence produces very large closed basins that can hold water. A few immense lakes owe their origins to tectonic downwarping.

SEDIMENTATION

With few exceptions, most lakes exist in relatively small depressions and serve as the catch basins for sediment from the entire watershed or drainage basin around them. The natural process of sedimentation ensures that most lakes fill with sediment before long periods of geologic time have passed. Lakes with areas of only a few square kilometers or less will fill within a few tens of thousands of years. Very large lakes and inland seas may endure for more than 10 million years. Human-made lakes and reservoirs have unusually high sediment-fill rates in comparison with most natural lakes. Human-made lakes may fill with sediment within a few decades to a few centuries.

Lake sediments come from four sources: allogenic clastic materials that are washed in from the surrounding watershed; endogenic chemical precipitates that are produced from dissolved substances in the lake waters; endogenic biogenic organic materials produced by plants and animals living in the lake; and aeolian or airborne substances, such as dust and pollen, transported to the lake in the atmosphere.

Allogenic clastic materials are mostly mineral in nature, produced when rocks and soils in the drainage basin are weathered by mechanical and chemical processes to yield small particles. These particles are moved downslope by gravity, wind, and running water to enter streams, which then transport them to the lake. Clastic materials also enter the lake via waves, which erode the materials from the shoreline, and via landslides that directly enter the lake. In winter, ice formed on the lake can expand and push its way a few centimeters to 1 meter or so onto the shore. There, the ice may pick up large particles, such as gravel and cobbles. When the spring thaw comes, waves can remove that ice, together with its enclosed particles, and float it out onto the lake. The process by which the large particles are transported out on the lake is called ice-rafting. As the ice melts, the large clastic particles drop to the bottom. They are termed dropstones when found in lake sediments. A landslide into a lake or a flood in a stream that feeds into the lake can produce water heavily laden with sediment. The sediment-laden water is denser than clean water and therefore can rush down and across the lake bottom at speeds sufficient to carry even coarse sand far out into the lake. These types of deposits are called turbidite deposits.

Endogenic chemical precipitates in freshwater lakes commonly consist of carbonate minerals (calcite, aragonite, or dolomite) and mineraloids that consist of oxides and hydroxides of iron, manganese, and aluminum. In some saline and brine lakes, the main sediments may be carbonates, together with sulfates such as gypsum (hydrated calcium sulfate), thenardite (sodium sulfate), or epsomite (hydrated magnesium sulfate), or with chlorides such as halite (sodium chloride) or more complex salts. Of the endogenic precipitates, calcite is the most abundant. Its precipitation represents a balance between the carbon dioxide content of the atmosphere and that of the carbon dioxide dissolved in the lake water.

Diatoms are distinctive microscopic algae that produce a frustule (a kind of shell) made of silica glass that is highly resistant to weathering. When seen under a high-powered microscope, diatom frustules appear to be artwork, looking like beautiful and highly ornate saucer- and pen-shaped works of glass. A tiny spot of lake sediment may contain millions of them.

A lake's sediment may contain from less than 1 percent to more than 90 percent organic materials, depending upon the type of lake. Most organic matter in lake sediments is produced within the lake by plankton and consists of compounds such as carbohydrates, proteins, oils, and waxes that are made up of carbon, hydrogen, nitrogen, and oxygen, with a little phosphorus. Plankton has an approximate bulk composition of 36 percent carbon, 7 percent hydrogen, 50 percent oxygen, 6 percent nitrogen, and 1 percent phosphorus (by weight). Plankton includes microscopic plants (phytoplankton) and microscopic animals (zooplankton) that live in the water column. Lakes that are very high in nutrients (eutrophic lakes) commonly have heavy blooms of algae, which contribute much organic matter to the bottom sediment. Terrestrial (land-derived) organic material such as leaves, bark, and twigs form a minor part of the organic matter found in most lakes. Terrestrial organic material is higher in carbon and lower in hydrogen, nitrogen, and phosphorus than is planktonic organic matter.

Airborne substances usually constitute only a tiny fraction of lake sediment. The most important of such material is pollen and spores. Pollen usually constitutes less than 1 percent of the total sediments, but that tiny amount is a very useful component for learning about the past climates that have existed on Earth. Pollen is among the most durable of all natural materials. It survives attack by air, water, and even strong acids and bases. Thus, it remains in the sediment through geologic time. As pollen accumulates in the bottom sediment, the lake serves as a kind of recorder for the vegetation that existed around it at a given time. By taking a long core of the bottom sediment from certain types of lakes and identifying the various pollen grains that it contains, a geologist may look at the pollen changes that have occurred through time and reconstruct the history of the climate and vegetation in an area.

Volcanic ash thrown into the atmosphere during eruptions enters lakes and forms a discrete layer on the lake bottom. When Mount St. Helens erupted in 1980, it deposited several centimeters of ash in lakes more than 160 kilometers east of the volcano. Geologists have used layers of ash in lakes to reconstruct the history of volcanic eruptions in some areas. Although dust storms contribute sediment to lakes, such storms are usually too infrequent in most areas to contribute significant amounts. In addition to wind-blown dust, a constant rain of tiny particles enters the atmosphere from space as micrometeorites, some of which reaches the surface and becomes a component of lake and seafloor sediments.

WATER CIRCULATION

Lake waters are driven into circulation by temperature-induced density changes and by wind. Most freshwater lakes in temperate climates circulate completely twice each year and so are termed dimictic lakes. Circulation exerts a profound influence on water chemistry of the lake and the amount and type of sediment present within the water column. During summer, lakes become thermally stratified into three zones. The upper layer of warm water (epilimnion) floats above the denser cold water and prevents wind-driven circulation from penetrating much below the epilimnion. The epilimnion is usually in circulation, is rich in oxygen (from algal photosynthesis and diffusion from the atmosphere), and is well lighted. This layer is where summer blooms of green and blue-green algae occur and calcite precipitation begins. The middle layer (thermocline) is a transition zone in which the water cools downward at a rate of greater than 1°C per meter. The bottom layer (hypolimnion) is cold, dark, stagnant, and usually poor in oxygen. There, bacteria decompose the bottom sediment and release phosphorus, manganese, iron, silica, and other constituents into the hypolimnion.

Sediment deposited in summer includes a large amount of organic matter, clastic materials washed in during summer rainstorms, and endogenic carbonate minerals produced within the lake. The most common carbonate mineral is calcite (calcium carbonate). The regular deposition of calcite in the summer is an example of cyclic sedimentation, a sedimentary event that occurs at regular time intervals. This event occurs yearly in the summer season and takes place in the upper 2 or 3 meters of water. On satellite photos, it is even possible to see these summer events as whitenings on large lakes, such as Lake Michigan.

As the sediment falls through the water column in summer, it passes through the thermocline, into the hypolimnion, and onto the lake bottom. As it sits on the bottom during the summer months, bacteria, particularly anaerobic bacteria (those that thrive in oxygen-poor environments), begin to decompose the organic matter. As this occurs, the dissolved carbon dioxide increases in the hypolimnion. If enough carbon dioxide is produced, the hypolimnion becomes slightly acidic, and calcite and other carbonates that fell to the bottom begin to dissolve. This is essentially the same process that occurs at the carbonate compensation depths of the oceans. The acidic conditions also release dissolved phosphorus, calcium, iron, and manganese into the hypolimnion, as well as some trace metals. Clastic minerals such as quartz, feldspar, and clay minerals are not affected in such brief seasonal processes, but some silica from biogenic material such as diatom frustules can dissolve and enrich the hypolimnion in silica. As summer progresses, the hypolimnion becomes more and more enriched in dissolved metals and nutrients.

Autumn circulation begins when the water temperature cools and the density of the epilimnion increases until it reaches the same temperature and density as the deep water. Thereafter, there is no stratification to prevent the wind from circulating the entire lake. When this happens, the cold, stagnant hypolimnion, now rich in dissolved substances, is swept into circulation with the rest of the lake water. The dissolved materials from the hypolimnion are mixed into a well-oxygenated water column. Iron and manganese that formerly were present in dissolved form now oxidize to form tiny solid particles of manganese oxides, iron oxides, and hydroxides. The sediment therefore becomes enriched in iron, manganese, or both during the autumn overturn, with the amount of enrichment depending upon the amount of dissolved iron and manganese that accumulated during summer in the hypolimnion. Dissolved silica is also swept from the hypolimnion into the entire water column. In the upper water column, where sunlight and dissolved silica become present in great abundance, diatom blooms occur. The diatoms convert the dissolved silica into solid opaline frustules.

As circulation proceeds, the currents may sweep over the lake bottom and actually resuspend 1 centimeter or more of sediment from the bottom and margins of the lake. The amount of resuspension that occurs each year in freshwater lakes is primarily the result of the shape of the lake basin. A lake that has a large surface area and is very shallow permits wind to keep the lake in constant circulation over long periods of the year.

As winter stratification develops, an ice cover forms over the lake and prevents any wind-induced circulation. Because the circulation is what keeps the lake sediment in suspension, most sediment quickly falls to the bottom, and sedimentation then is minimal through the rest of winter. If light can penetrate the ice and snow, some algae and diatoms can utilize this weak light, present in the layer of water just below the ice, to reproduce. Their settling remains contribute small amounts of organic matter and diatom frustules. At the lake bottom, the densest water (that at 4°C) accumulates. As in summer, some dissolved nutrients and metals can build up in this deep layer, but because the bacteria that are active in releasing these substances from the sediment are refrigerated, they work slowly, and not as much dissolved material builds up in the bottom waters.

When spring circulation begins, the ice at the surface melts, and the lake again goes into wind-driven circulation. Oxidation of iron and manganese occurs (as in autumn), although the amounts of dissolved materials available are likely to be less in spring. Once again, nutrients such as phosphorus and silica are circulated out of the dark bottom waters and become available to produce blooms of phytoplankton. Spring rains often hasten the melting, and runoff from rain and snowmelt in the drainage basin washes clastic materials into the lake. The period of spring thaw is likely to be the time of year when the maximum amount of new allogenic (externally derived) sediment enters the lake.

Spring diatom blooms continue until summer stratification prevents further replenishment of silica to the epilimnion. Thereafter, the diatoms are succeeded by summer blooms of green algae, closely followed by blooms of blue-green algae. Silica is usually the limiting nutrient for diatoms; phosphorus is the limiting nutrient for green and blue-green algae.

DIAGENESIS

After sediments are buried, changes occur. This process of change after burial is termed diagenesis. Physical changes include compaction and de-watering. Bacteria decompose much organic matter and produce gases such as methane, hydrogen sulfide, and carbon dioxide. The "rotten-egg" odor of black lake sediments, often noticed on boat anchors, is the odor of hydrogen sulfide. After long periods of time, minerals such as quartz or calcite slowly fill the pores remaining after compaction.

One of the first diagenetic minerals to form is pyrite (iron sulfide). Much pyrite occurs in microscopic spherical bodies that look like raspberries; these particles, called framboids (from the French *framboise*, meaning "raspberry"), are probably formed by bacteria in areas with low oxygen within a few weeks. In fact, the black color of some lake muds and oozes results as much from iron sulfides as from organic matter. Other diagenetic changes include the conversion of mineraloid particles containing phosphorus into phosphate minerals such as vivianite and apatite. Manganese oxides may be converted into manganese carbonates (rhodochrosite). Freshwater manganese oxide nodules may form in high-energy environments such as Grand Traverse Bay in Lake Michigan.

STUDY OF LAKES

Scientists who study lakes (limnologists) must study all the natural sciences, including physics, chemistry, biology, meteorology, and geology, because lakes are complex systems that include biological communities, changing water chemistry, geological processes, and interactions among water, sunlight, and the atmosphere.

Modern lake sediments are collected from the water column in sediment traps (cylinders and funnels into which the suspended sediment settles over periods of days or weeks) or by filtering large quantities of lake water. Living material is often sampled with a plankton net. Older sediments that have accumulated on the bottom are collected with dredges and by piston coring, which involves pushing a sharpened hollow tube (usually about 2.5 centimeters in diameter) downward into the sediment. Cores are valuable because they preserve the sediment in the order in which it was deposited, from oldest at the bottom to the most recent at the top. Once the sample is collected, it is often frozen and taken to the laboratory. There, pollen and organisms may be examined by microscopy, minerals may be determined by X-ray diffraction, and chemical analyses may be made.

Varves are thin laminae that are deposited by cyclic processes. In freshwater lakes, each varve represents one year's deposit; it consists of a couplet with a dark layer of organic matter deposited in winter and a light-colored layer of calcite deposited in summer. Varves are deposited in lakes where annual circulations cannot resuspend bottom sediment and therefore cannot mix it to destroy the annual lamination. Some lakes that are small and very deep may produce varved sediments; Elk Lake in Minnesota is an example. In other lakes, the accumulation of dissolved salts on the bottom eventually produces a dense layer (monimolimnion), which prevents disturbance of the bottom by circulation in the overlying fresher waters. Soap Lake in Washington State is an example. Because each varve couplet represents one year, a geologist may core the sediments from a varved lake and count the couplets to determine the age of the sediment in any part of the core. The pollen, the chemistry, the diatoms, and other constituents may then be carefully examined to deduce what the lake was like during a given time period. The study is much like solving a mystery from a variety of clues. Eventually, the history of climate changes of the area may be learned from the study of lake varves.

—*Edward B. Nuhfer*

FURTHER READING

Botts, Lee, et al. *The Great Lakes: An Environmental Atlas and Resource Book*. Govt. of Canada, 2002.

Dennis, Jerry. *The Living Great Lakes*. Thomas Dunne Books, 2003. Written for the layperson. Covers the geology, natural history, biology, and industry of the Great Lakes. Discusses the structure of the lakes from their formation to the current human impact of resource mining, construction of dams and canals, and introduction of invasive species.

Dodds, Walter K., and Matt R. Wiles. *Freshwater Ecology: Concepts and Environmental Applications of Limnology*. 2nd ed. Academic Press, 2010. Covers the physical and chemical properties of water, the hydrologic cycle, nutrient cycling in water, as well as biological aspects. Written by two of the leading scientists in freshwater ecology. An excellent resource for college students as each chapter has a short summary of main topics.

Grady, Wayne, et al. *Great Lakes: The Natural History of a Changing Region*. Greystone Books, 2011. Discusses the natural history, geology, ecology, and conservation of the Great Lakes. A chapter discusses the human impact of invasive species on the ecology of the lakes. Includes further readings, a list of common and scientific names, illustration credits, and indexing.

Håkanson, Lars, and M. Jansson. *Principles of Lake Sedimentology*. Blackburn Press, 2002. A reference for professionals in the field of lake sedimentology, though parts of it may be accessible to high school students. Focuses on lake sediments in detail. Provides methods of sampling and discusses the influence of lake type and shape on the sediments formed in the lake, the circulation of lake waters, the chemistry of sediments, and the pollution of lakes.

Kaye, Catheryn Berger, and Philippe Cousteau. *Going Blue*. Free Spirit Publishing, 2009. Discusses conservation issues related to oceans, lakes, rivers, and other bodies of water, as well as topics such as pollution, watershed management, coral bleaching, and ocean acidification. Also provides a guideline for action with multiple chapters discussing how students can get involved in water conservation. Lists many resources to help the reader find more information or get started on projects.

Kolumban, Hutter, Yongqi Wang, and Irina P. Chubarenko. *Physics of Lakes: Foundation of the Mathematical and Physical Background*. Springer, 2011. Assumes a certain level of mathematical knowledge on the part of the reader, but the mathematical concepts presented are not beyond ready comprehension, beginning with the fundamental equations of lake hydrodynamics and progressing to angular momentum and vorticity.

Micklin, P., and N. V. Aladin. "Reclaiming the Aral Sea." *Scientific American* 298, no. 4 (April, 2008): 64–71. Outlines the collapse of the Aral Sea due to wasteful irrigation of the desert; the geography of residual lakes; the 2005 dam construction and future of the Amu Dar'ya; and application of lessons learned from the Aral disaster to other regions with similar risks.

Wetzel, R. G., ed. *Limnology*. 3rd ed. Elsevier Science, 2001. A well-written textbook typical of those used by undergraduates and graduates in introductory limnology courses. Covers physical, biological, and chemical aspects of lakes. Assumes a knowledge of high school algebra, chemistry, physics, and biology.

LITTLE ICE AGE (LIA)

FIELDS OF STUDY

Physical Geography; Environmental Science; Environmental Studies; Climate Modeling; Meteorology; Atmospheric Science; Earth System Modeling; Astronomy; Observational Astronomy; Tree Ring Analysis

SUMMARY

Indirect and anecdotal evidence indicates that it was colder than normal during the seventeenth century in Europe and possibly worldwide as well. Without accurate weather records for this time period, the exact dates of the Little Ice Age are not known. The coldest period was the seventeenth century, but date estimates range from as early as about 1350 or 1400 to as late as 1850. Indeed, some writers use the term "Little Ice Age" to refer only to that coldest period, while others designate the entire period from 1350 to 1850 with the term. The coldest portion of the Little Ice Age occurred during the time of the Maunder Minimum, a period of virtually no sunspot activity. If the Sun's luminosity is slightly lower during periods of reduced sunspot activity, the Little Ice Age may have been caused by a temporary decrease in the Sun's luminosity.

PRINCIPAL TERMS

- **luminosity of Sun:** the total energy output of the Sun every second, measured in watts
- **Maunder Minimum:** a period from about 1645 to about 1715 when very few sunspots were observed
- **sunspot cycle:** also known as the solar activity cycle, an eleven-year cycle in the number of sunspots and amount of other solar magnetic activity
- **sunspot minimum/maximum:** the time when there is the minimum/maximum number of sunspots during the eleven-year sunspot cycle
- **sunspots:** dark spots on the surface of the Sun caused by solar magnetic activity

LITTLE ICE AGE

The Little Ice Age was not a true "ice age." During the various true ice ages, all of which occurred many millennia ago, glaciers invaded temperate midlatitude regions. The evidence for these ice ages is geologic, because they occurred before recorded history. The Little Ice Age occurred only a few centuries ago. It was generally colder than normal, but not nearly as cold as the ice ages. Historians know about the Little Ice Age from various anecdotal documents and indirect proxies. There are, however, no accurate weather records for this period, because the instrumentation needed to produce such records had for the most part not yet been invented. Therefore, scientific knowledge of the extent and severity of the Little Ice Age is not precise.

The exact dates of the start and end of the Little Ice Age are somewhat controversial. Authors generally agree that the Little Ice Age encompassed the seventeenth century, but there is some disagreement as to both how long afterward it lasted and how much sooner it started. This disagreement can be understood by examining reconstructed temperatures for the past millennium. The seventeenth century, which was about 0.5°C cooler than the 1961–1990 average temperature, was both the longest-lasting cool period and the coldest period of the past one thousand years. Thus, nearly all researchers agree that this century was part of the Little Ice Age.

The entire period from about 1300 to the late nineteenth century was cooler than normal for the millennium. However, most of this time was not as cold as the seventeenth century, and there were some relatively warm periods during these centuries. The first half of the nineteenth century was colder than normal, but the eighteenth century was about as warm as the average for the millennium. Thus, the Little Ice Age may have lasted as late as 1850, but it may have ended in the early 1700s.

The period from 1000 to shortly after 1200, the Medieval Warm Period, was nearly as warm as the end of the twentieth century. Average temperatures then dropped fairly quickly during the late thirteenth century, making it possible to date the beginning of the Little Ice Age as early as about 1350 to 1400. The middle of the fourteenth and fifteenth centuries were nearly as cold as the seventeenth century, but there were warmer periods about 1400 and during the first portion of the sixteenth century. Thus, the Little Ice Age may not have started until the seventeenth century.

The Frozen Thames, 1677.

The uncertainty of dating the beginning and end of the Little Ice Age results from the facts that the cool climate from approximately 1350 to 1850 was interspersed with relatively warm periods and that the longest-lasting and coolest period was the seventeenth century. The exact time period of the Little Ice Age is therefore fairly loosely defined.

EVIDENCE FOR THE LITTLE ICE AGE
Thermometers were not invented until the end of the sixteenth century, and widespread, systematic use of accurate thermometers did not occur until much later. There are therefore no accurate weather records to verify the Little Ice Age. Climate researchers must use other lines of evidence, including both various proxies and anecdotal evidence.

The most common proxy studies for climate involve tree rings. The thickness and density of the rings varies with various climatic conditions, including temperature and rainfall. In polar regions, studies of various properties of ice cores provide climate information. The properties include the rate at which ice accumulates, layers that have melted, and isotope ratios. Growth thicknesses and other properties of corals can also provide climate information.

For all of these proxy studies, climate researchers statistically analyze the relationship between the proxy and climate conditions during recent periods for which accurate weather records exist. The researchers then extrapolate the climate conditions back to the dates before accurate weather records.

The further back researchers extrapolate, the less accurate the proxy is in reconstructing climate conditions. Hence, climate estimates for the first half of the millennium are less accurate than are more recent estimates. Proxy studies are further complicated by the fact that multiple variables can affect the proxy. For example, tree rings are affected by both temperature and rainfall conditions.

In addition to proxy studies, there is anecdotal evidence for the Little Ice Age. Examples of this type of evidence include such things as diary entries and paintings. Diary entries might include reports of unusual freezings of various bodies of water, extreme snowfalls, and so forth. Paintings in eras when artists strove for realism can also depict frozen landscapes and bodies of water. If the paintings made during a particular time period show a large number of frozen landscapes of locations that seldom freeze now, researchers can conclude that the time period was cooler than normal. These lines of evidence are not scientific, but many reports of a particular time period being colder than normal strongly suggest that it actually was colder, even in the absence of scientifically reliable weather records. These lines of evidence apply primarily to Europe, so the Little Ice Age could have been either a strictly European phenomenon or a global phenomenon.

THE MAUNDER MINIMUM AND THE LITTLE ICE AGE
There is fairly strong evidence that variations in the Sun's luminosity related to sunspot activity caused the Little Ice Age. Sunspots are dark regions on the surface of the Sun caused by solar magnetic activity. Solar magnetic activity also causes bright areas, or faculae, on the Sun's surface. The Sun undergoes an eleven-year cycle regulating the amount of sunspots, faculae, and related solar magnetic activity it experiences. Satellite measurements over the most recent solar cycles show that the Sun's luminosity is a very small amount higher during sunspot maximum than during sunspot minimum. The net effect of the bright areas on the Sun is slightly larger than the net effect of the dark areas, so the Sun is brighter during sunspot maximum.

There are also less well established longer cycles in solar activity. Notably, the Maunder Minimum was a period from about 1645 (possibly as early as 1620) to 1715 when there were very few sunspots. This period corresponds to the coldest portion of the Little Ice Age. If the observation that the Sun emits less energy during periods of minimal sunspot activity holds, then the Sun's lower luminosity during the Maunder Minimum may have caused the coldest portions of the Little Ice Age.

Closer comparison of the sunspot activity and global temperatures over the past thousand years supports this hypothesis. The warm period from 1000 to 1200 corresponds to the Medieval Grand Maximum in sunspot activity: During sunspot maxima, there were many more sunspots than is usual during sunspot maxima. There were also other extended periods of very few sunspots similar to the Maunder Minimum, including the Spörer, Wolf, and Dalton minima. Like the Maunder Minimum, these other minima correspond to the cooler periods of the extended Little Ice Age. It is not proven that variations in the Sun's luminosity caused the Little Ice Age, but it seems to be the most likely explanation.

CONTEXT

Global warming, especially since the later half of the twentieth century, has become a serious worldwide concern. Most climate researchers attribute this global warming to anthropogenic causes, particularly increased emissions of carbon dioxide (CO_2) and other greenhouse gases (GHGs). The extremely hot surface temperatures on Venus clearly demonstrate that CO_2 can warm a planet.

If, however, the Little Ice Age resulted from solar luminosity variations related to long-term solar activity cycles, then there is the possibility that similar solar variations are contributing to current global warming. Some, but not all, late-twentieth-century sunspot maxima were higher than normal, suggesting the possibility that Earth is entering another sunspot grand maximum similar to the Medieval Grand Maximum. If this is the case, then global warming may have solar variations as well as increased GHGs as a component cause.

—*Paul A. Heckert*

FURTHER READING

Eddy, J. A. "The Maunder Minimum." *Science* 192 (1976): 1189–1192.

Foukal, P., C. Fröhlich, H. Spruit, and T. M. L. Wigley. "Variations in Solar Luminosity and Their Effect on Earth's Climate." *Nature* 443 (2006): 161–166.

Golub, Leon, and Jay M , Pasachoff. *Nearest Star: The Surprising Science of Our Sun*. Cambridge University Press, 2014.

Hoyt, Douglas V., and Kenneth H. Schatten. *The Role of the Sun in Climate Change*. New York: Oxford University Press, 1997.

Jones, P. D., T. J. Osborn, and K. R. Briffa. "The Evolution of Climate Over the Last Millennium." *Science* 292 (2001): 662–667.

Maunder, E. Walter. "A Prolonged Sunspot Minimum." *Knowledge* 17 (1894): 173–176.

Maunder, E. Walter. "The Prolonged Sunspot Minimum, 1645–1715." *Journal of the British Astronomical Society* 32 (1922): 140.

Soon, Willie, and Steven H Yaskell. *The Maunder Minimum and the Variable Sun-Earth Connection*. World Scientific, 2004.

LONG-TERM WEATHER PATTERNS

FIELDS OF STUDY

Atmospheric Science; Oceanography; Physical Geography; Thermodynamics; Environmental Sciences; Environmental Studies; Meteorology; Climate Modeling; Statistics; Process Modeling; Fluid Dynamics; Hydroclimatology; Hydrometeorology; Bioclimatology; Heat Transfer

SUMMARY

Much work in meteorology is dedicated to the accurate prediction of short-term weather patterns. However, an important part of meteorology is the pursuit of information on long-term trends and conditions. This field involves the study of consistent weather conditions that develop and continue through months and years. The field also entails an understanding of certain phenomena that contribute to these extended weather periods. Research on long-term weather patterns can help scientists understand past climate changes and predict future shifts, enabling society to better prepare for such weather changes.

PRINCIPAL TERMS

- **Arctic oscillation:** long-term weather pattern in which the different air pressures in the Arctic and middle latitude regions cause varying weather conditions
- **El Niño:** meteorological condition in which the waters of the tropical, eastern Pacific Ocean are warmed by the atmosphere
- **La Niña:** meteorological condition in which the waters of the eastern, tropical Pacific Ocean are cooled by a lack of radiation from the atmosphere
- **Madden-Julian oscillation:** intraseasonal tropical wave that travels around the globe, causing monsoons and other high-water storms and also suppresses them
- **trade winds:** winds at the level of the ocean surface that blow from the east to the west in the tropical Pacific

EL NIÑO

A carefully watched long-term weather pattern is the phenomenon known as El Niño, also known as El Niño Southern Oscillation (ENSO). El Niño is a cyclical event in which warm ocean water (which is heated at the equator) moves eastward from the western Pacific Ocean because of stagnation in the trade winds (which blow from east to west along the equator).

Normally, the trade winds send the solar-heated water westward, where the resulting humidity creates weather patterns that travel around the world. With El Niño, the warm water moves eastward and creates new weather systems and (because of El Niño's distinctive sea-level pressure signature) a shift in the manner in which the global atmosphere circulates.

El Niño has different effects on different regions of the world. For example, evidence suggests that El Niño contributes to the generation of powerful, tornado-producing storms in the prairie regions of the United States and Canada. For other regions, such as the Atlantic seaboard, El Niño can mean more precipitation in the form of snowstorms, nor'easters (large rotating Atlantic coastal storms capable of high winds and heavy precipitation that typically blow in from the north-east), and other systems. In South America, ENSO can cause storms that produce severe flooding.

In some regions, El Niño can lead to less precipitation. For example, in the Rockies and southwestern Canada, El Niño often means less snow than usual. Meanwhile, in Australia, El Niño patterns frequently cause droughts. ENSO is even known to reduce the number and severity of hurricanes, which form in the typically warm waters off western Africa.

El Niño can last between two and seven years. While it is a cyclical event, its appearance is typically erratic. For example, the longest-lasting El Niño event recorded during the modern era lasted from 1990 to 1995. However, since 1995, the El Niño cycle has been shorter, lasting one or two years on average.

LA NIÑA

When ENSO periods come to a close, the trade winds intensify because of the difference in surface-level air pressure. The winds blow the warmer water westward and from the west coasts of North and South America. As the warm water is carried west, cooler water from beneath the surface of the eastern Pacific is drawn to the surface. As is the case during El Niño, the cooling water again causes a change in atmospheric circulation, creating changes in the jet streams (a global band of strong air currents several miles above the Earth's surface). As a result, weather patterns change. This pattern is known as La Niña.

La Niña is considered the latter half of the ENSO cycle. Like El Niño, La Niña's effect on the weather varies, based on the region. For example, the warmer water in the western Pacific leads to wetter weather conditions in Indonesia and Australia. In the United States, colder weather frequently exists in the northwest region, while the southern and mid-Atlantic see warmer and drier conditions.

La Niña also results in changes in severe weather patterns. For example, the generation of major storms that produce hail and tornadoes is reduced by the cooler, drier air manifest during La Niña. However, La Niña disrupts the wind shear (the difference in wind speed and direction in two close areas of the atmosphere) that can hinder the development of tropical storms in the Atlantic. The result of this disruption is an increase in the number and severity of Atlantic hurricanes.

ARCTIC OSCILLATION
ENSO is not the only atmospheric cycle. Another long-term weather pattern, the Arctic Oscillation (AO) system, involves two areas of atmospheric pressure. These two patterns are located at the polar latitude (above the Arctic Circle) and at the middle latitude (between southern Florida and the Arctic Circle).

There are two phases in which the AO is manifest. These phases have alternated frequently during the last century, with each phase lasting between ten and forty years. During the negative phase, higher-than-normal pressure is in place in the Arctic regions, while lower pressure exists at the middle latitudes (such as most of North America and Europe). These conditions cause cold air to move into the middle latitudes, which leads to colder-than-average winters in these regions. During the positive phase, however, the patterns are reversed: Ocean storms and wetter weather are drawn into the northernmost regions of North America and Europe, while the middle latitudes remain drier and warmer.

A number of factors contribute to the change in phases within the AO cycle, including the pressures associated with rising and lowering sea levels, water temperatures, and even greenhouse gases. AO is closely related to two other regional oscillation patterns: the North Atlantic Oscillation and the northern annual mode, which affect winter weather patterns in the North American and Northern European and Asian regions, respectively.

MADDEN-JULIAN OSCILLATION
Still another type of long-term weather pattern is the Madden-Julian Oscillation (MJO). The MJO is intraseasonal (it can develop within thirty, sixty, or ninety days of a given season), occurring across the planet's tropical regions (particularly in the Indian and Pacific Oceans). The MJO is global as well, traveling in a wave in the atmosphere.

Depending on the phase of the MJO, this long-term weather pattern is responsible for the generation and suppression of heavy tropical rainfall (including the high precipitation that accompanies monsoons). The MJO's phase of high-volume precipitation first begins in the Indian Ocean and then proceeds east toward the western and eastern Pacific. As it approaches the cooler waters of the eastern Pacific, the MJO tends to become less laden with precipitation, as the heat from warmer water dwindles. The MJO does show some limited life in the tropical Atlantic but becomes noticeable again as it re-approaches the Indian Ocean.

The MJO is one of the more vexing long-term weather patterns to study and predict because of its relatively slow movement and development along the tropical path. The key to its ability to develop and strengthen storm systems is the warm waters over which it passes. Through the use of satellite sensors, scientists are therefore attempting to pinpoint the locations of these pockets of warm water, developing models that may one day enable them to better understand and predict MJO movements.

MODELS FOR FORECASTING LONG-TERM PATTERNS
Meteorologists and climatologists alike see a great deal of value in the analysis and prediction of long-term weather patterns. A season-by-season weather forecast can help scientists prepare for weather patterns. For example, tracking La Niña during the winter season can help researchers forecast enhanced or subdued storm systems in a given geographic area. For this reason, scientists are working to better track and predict the more erratic AO and MJO. Furthermore, any evidence of above- or below-average precipitation is helpful for the residents of these regions in preparing for the effects of these long-term weather patterns.

To track and predict long-term weather patterns, scientists utilize a number of approaches. First, they gather data on surface-level conditions (such as water temperatures and wind speeds) and weather patterns collected from satellite sensors, airborne remote sensors, and ground-based systems. Using these data, they generate mathematical equations that represent each of the physical processes at work. For example, scientists have developed a series of mathematical equations designed to assign values to circulation anomalies within the MJO; such equations help researchers better understand how such elements influence the MJO's atmospheric dynamics.

Once the data have been compiled and collated, they are used to form computer models that can predict future conditions created by a given long-term weather pattern. One example comes from Japan, whose Ministry of Science utilized the Earth simulator supercomputer to compile global weather data, including long-term weather patterns, ocean currents, and other factors, into a massive model. Scientists believe that, once this model is brought online, they will be able to predict droughts, severe storms, and other weather and climate conditions as far as thirty years into the future.

Other computer models focus on a particular long-term weather pattern and analyze its individual elements, such as regional cold temperature zones along an ENSO track or the relationship between atmospheric circulation and surface-level conditions. For example, the National Weather Service in the United States operates a climate prediction center, which compiles a wide range of sensor data on such elements as surface and water temperature, wind velocity, precipitation, and air currents (and on anomalies in these conditions based on seasonal averages). Based on this information the center generates seasonal short- and long-term forecasts, tracking weather systems based on the influences of ENSO, AO, and MJO and other patterns.

—*Michael P. Auerbach*

FURTHER READING

Ahrens, C. Donald, and Perry J. Samson. *Extreme Weather and Climate*. Brooks Cole, 2010. Provides a nontechnical review of the different types of severe weather, such as hurricanes, tornado-producing storms, and flooding. Discusses the long-term weather patterns that produce such weather.

Allen, Robert J., and Charles S. Zender. "Forcing of the Arctic Oscillation by Eurasian Snow Cover." *Journal of Climate* 24, no. 24 (2011): 6528–6539. Describes the factors that contribute to the Arctic Oscillation, including sea level, greenhouse gases, and warm seawater. Argues that snow cover, particularly in northern Europe and Asia, plays an important role in this process.

Bai, Xuezhi, et al. "Severe Ice Conditions in the Bohai Sea, China, and Mild Ice Conditions in the Great Lakes During the 2009/10 Winter: Links to El Niño and a Strong Negative Arctic Oscillation." *Journal of Applied Meteorology and Climatology* 50, no. 9 (2011): 1922–1935. The authors conducted a comparative study of the Great Lakes and China's Bohai Sea, focusing on the creation of large amounts of ice caused by negative phases in the Arctic Oscillation.

Chang, Chih-Pei, and Mong-Ming Lu. "Intraseasonal Predictability of Siberian High and East Asian Winter Monsoon and Its Interdecadal Variability." *Journal of Climate* 25, no. 5 (2012): 1773–1778. Discusses scientific efforts to predict severe weather in East Asia based on the positive and negative phases of Arctic oscillation.

Climate Central, Inc. *Global Weirdness: Severe Storms, Deadly Heat Waves, Relentless Drought, Rising Seas and the Weather of the Future*. Vintage Books, 2013.

D'Aleo, Joseph S., and Pamela G. Grube. *The Oryx Guide to El Niño and La Niña*. Greenwood Press, 2002. Examines the causes and effects of the various stages of ENSO. In addition to analysis of the history, economic impacts, and natural forces creating ENSO, reviews attempts to more effectively predict this phenomenon.

Mogil, H Michael. *Extreme Weather: Understanding the Science of Hurricanes, Tornadoes, Floods, Heat Waves, Snow Storms, Global Warming, and Other Atmospheric Disturbances*. Black Dog & Leventhal Publisher, 2010. A detailed analysis of the science of severe weather and climate change. Provides a review of the long-term weather patterns that produce this weather and suggestions about how society can prepare for future weather disasters.

M

MARITIME CLIMATE

FIELDS OF STUDY

Climate Classification; Climate Zones; Climatology; Climate Sciences; Physical Geography; Earth System Modeling; Climate Modeling

SUMMARY

Maritime climates are generally considered to be those that are moderated by the sea. However, a true maritime climate is in most cases a climate in which it is neither very warm nor very cold, with adequate rainfall throughout most of the year.

PRINCIPAL TERMS

- **climate:** long-term, average, regional or global weather patterns
- **moderation:** the effect of buffering against extremes of temperature by the presence of a large body of water

Maritime Climates

Maritime climates are generally considered to be those that are moderated by the sea. However, a true maritime climate is in most cases a climate in which it is neither very warm nor very cold, with adequate rainfall throughout most of the year A typical maritime climate occurs on the coast of Oregon and Washington in the United States, in many parts of New Zealand, in Tasmania, and in much of western Europe. Although surrounded by the sea, tropical islands are not normally considered to have a maritime climate but rather a tropical climate. True maritime climates usually have winter daytime temperatures of about 15°C and winter nighttime temperatures of about 5°C. In the summer, daytime temperatures average about 25°C and nighttime temperatures average about 15°C. Rainfall in a true maritime climate occurs throughout the year, with no pronounced wet or dry season, and averages 10 centimeters per month.

Despite the moderate nature of the true maritime climate, extremes do occur. In many maritime climates, daytime temperatures can reach above 35°C, and occasionally even above 40°C, while nighttime temperatures may reach 0°C, and occasionally as low

Seattle's skyline taken from the Puget Sound.

as -10°C. Sunshine is generally more than adequate for plant growth, and annual bright sunshine hours of two thousand to twenty-five hundred hours are the norm. Rainfall, although generally adequate for plant growth, can vary: Maritime climates can have periods of up to six weeks without any appreciable rainfall, and there is at least one known instance of no rain at all falling during almost every month on the calendar. In contrast, monthly rainfalls of over 20 centimeters in a month are reasonably common, but monthly rainfalls of over 50 centimeters are not impossible. Although a daily rainfall of more than 0.3 centimeter is uncommon, at times, the rainfall in a twenty-four-hour period may exceed 20 centimeters. In summary, a true maritime climate is generally an easy climate to live in for people, plants, and animals, but relatively extreme events do occur.

SIGNIFICANCE FOR CLIMATE CHANGE

Although the true maritime climate has milder winters, generally cooler summers, and greater temperature ranges from nighttime to daytime than do continental climates, there are many variations from place to place. For example, the climate of New York City, although situated on a coast, has a climate quite different from that of Vancouver, on the coast of British Columbia. Alterations in the true maritime climate brought about by global climate change are likely to be relatively small, and in most places plants and animals should be able to adapt to the change. However, if there is an overall increase in temperatures in the middle latitudes of both hemispheres, where most of the true maritime climates occur, summers are likely to become warmer, albeit not as warm as current continental climates, and winters are likely to have fewer days below freezing. Sunshine hours are unlikely to change, but rainfall extremes, especially on a daily basis, are likely to increase somewhat.

New Zealand is considered to have a typical maritime climate, although there are some areas in New Zealand that are quite dry, with an average annual rainfall of only 30 centimeters. Nevertheless, most of New Zealand, as well as much of western Europe, has a typical maritime climate. Questions arise as to the effects of global warming on the temperature and agricultural production of such a climate. Considering New Zealand as an example, vineyards over the past one hundred years have flourished in many parts of the country, but as yet, in the far south of New Zealand in the Southland district, there are no commercial vineyards. Instead, the district is dairy country, covered with very green pastures and grazing cattle. The Southland District Council and the Southland Regional Council might consider the probable effects on the region of changes predicted in the Intergovernmental Panel on Climate Change's 2007 report. If they did, they might advertise the prospects of the Southland district to become a major grape-growing region in the twenty-first century.

—*W. J. Maunder*

FURTHER READING

Aguado, Edward, and James E. Burt. *Understanding Weather and Climate.* 7th ed. Pearson Prentice Hall, 2015. The formation of midlatitude storms is surveyed. Continental climates and the structure of climate are discussed.

Barry, Roger G and Hall-McKim, Eileen A. *Essentials of the Earth's Climate System.* Cambridge University Press, 2014. Describes many aspects of the influence of ocean currents and oscillations as drivers of the state of Earth's climate.

Maslin, Mark. *Global Warming: A Very Short Introduction.* Oxford University Press, 2009. Provides a concise general introduction to the major factors that define the climate of Earth.

Saha, Pijushkanti *Modern Climatology.* Allied Publishers Pvt Ltd, 2012. An entry-level textbook that provides an concise, yet thorough, overview of the science of climatology and climate relationships.

Schimel, David *Climate and Ecosystems.* Princeton University Press, 2013. One of the "Princeton Primers on Climate" series, this book describes the interaction mechanisms of various ecosystems and climate.

MEAN SEA LEVEL

SUMMARY

Sea level is the height of the surface of the ocean at any given location. Sea level is highly variable and can undergo very rapid changes due to such events as tides, tsunamis, changes in barometric pressure, wind-generated waves, and even freshwater floods. While these events can produce changes in sea level of several meters, they are local in scale and of a very short duration, generally lasting only for hours. Mean sea level is the average, global height of the sea surface, independent of these local, short-term changes. Changes in mean sea level are on the order of a few millimeters per year.

Mean sea level at specific locations can be calculated using tide gauge records and subtracting the effects of annual changes in atmospheric pressure and long-term changes in tidal ranges, which are driven by astronomical factors. Changes in global mean sea level can be calculated using satellite-based radar altimetry, such as with the TOPEX/Poseidon satellite. The radar altimeter measures the height of the satellite above the ocean, based on the time it takes for a radio signal to travel from the satellite to the sea surface and back. Since the actual altitude of the satellite is known, any changes in the altimeter measurement reflect changes in the height of the sea surface itself.

SIGNIFICANCE FOR CLIMATE CHANGE

On short timescales (decades to centuries), mean sea level is a function of the amount of water stored as ice in glaciers and ice sheets. As global temperature rises, less water is stored as ice, contributing to a rise in mean sea level. A rise in mean sea level in response to global warming has important societal consequences. First, such a rise contributes to a loss of land, as coastal areas are slowly inundated by water. This is a concern for certain low-lying island nations such as the Maldives or Tuvalu. The Maldives is a nation made up of twelve hundred islands in the Indian Ocean, which has a maximum elevation of only 2.5 meters above current sea level. Thus, the Maldivian population of approximately 380,000 people is highly vulnerable to even a slow rise in mean sea level. For other nations, a rise in mean sea level is also a concern because of increased hazards from flooding during high tides—especially spring tides—and storms. A rise in mean sea level provides a higher baseline upon which tidal fluctuations build. According to the Fourth Assessment Report released by the Intergovernmental Panel on Climate Change, from 1993 to 2003, mean sea level rose approximately 3.1 millimeters per year.

Sea Level marker on the side of the road from Jerusalem to the Dead Sea.

—*Anna M. Cruse*

MEDIEVAL WARM PERIOD (MWP)

FIELDS OF STUDY

Climatology; Meteorology; Tree Ring Analysis; Analytical Chemistry; Thermodynamics; Environmental Sciences; Climate Modeling; Hydroclimatology; Atmospheric Science; Bioclimatology; History

SUMMARY

The Medieval Warm Period (MWP) is a term used to describe a period of several centuries that preceded the Little Ice Age (LIA). The MWP (also the Medieval Warm Epoch or the Little Climatic Optimum) was proposed in 1965 by Hubert Horace

Authentic Viking recreation, Newfoundland, Canada.

Lamb, who deduced that the North Sea region had experienced a warming period. His reasoning was based on historical records that appeared to be consistent with a milder climate and improved agricultural production.

PRINCIPAL TERMS

- **anthropogenic:** deriving from human sources or activities
- **dendroclimatology:** the study of tree rings as indicators of climatic conditions
- **multidecadal:** occurring over a period of multiple decades
- **proxies:** observable factors that retain the effects of past climates
- **speleotherm:** stalactites, stalagmites and other features that develop in caves

THE LAMB DEDUCTION

The Medieval Warm Period (MWP) is a term used to describe a period of several centuries that preceded the Little Ice Age (LIA). The MWP (also the Medieval Warm Epoch or the Little Climatic Optimum) was proposed in 1965 by Hubert Horace Lamb (1913–1997), a British meteorologist and groundbreaking climate historian, who believed it lasted from roughly 900 to 1300 CE. During the MWP, Lamb believed, the North Atlantic and northern and western Europe experienced warmer conditions on average. He presented evidence drawn primarily from historical documentary data such as the expansion of agriculture to higher-altitude fields in mountainous regions, shifts in the cultivation of certain crops (such as wine production in the British Isles), changes in tree lines, and reports of weather and weather-related events (such as floods and droughts) in historical writings. This period also coincided with the Viking settlement of Greenland and excursions to Labrador, as well as a population boom in Europe, which grew from roughly 35 million to 80 million people between 1000 and 1347 CE. All of these events seemed consistent with a milder climate and improved agricultural production.

In the decades following Lamb's assertion, temperature proxies were examined to better define the MWP, the transition to the LIA, and the extent to which they were regional or global phenomena. For example, measurements of oxygen isotope ratios in marine sediments from a Sargasso Sea core indicated an ocean temperature around 1100 that was about 1°C warmer than present levels. This was followed by a nearly 2°C decrease between 1100 and 1700.

Dendroclimatology, or the study of climate through tree-ring growth, conducted in the Sierra Nevada and the Great Basin of the western United States suggests periods of increased warmth and later periods of severe drought during the MWP. Similar tree-ring studies in northern Sweden and the Polar Ural Mountains provide evidence of increased temperatures from 971 to 1100 and from 1110 to 1350, respectively. Ice core and borehole studies in Greenland indicate a warm period peaking around 1000, followed by a 3°C decrease into the LIA. Studies of glaciers and their moraines suggest that the MWP generally coincided with glacial retreat and the shift to the LIA with glacial advance. Other types of proxy studies include examining lake sediments, speleotherm (stalactite/stalagmite) growth, and coral growth. Many, but not all, proxy studies of sufficient length do indicate a temperature peak during the MWP and a decrease into the LIA, but with considerable variability in the details.

SIGNIFICANCE FOR CLIMATE CHANGE

The MWP is a central issue in the debate over climate change, as it is the most recent period of climatic warmth assumed to be free of anthropogenic factors. Likewise, the transition to the LIA is the most recent significant temperature shift before the warming of the twentieth century. The initial estimates by Lamb suggest MWP temperatures were about 0.5°C warmer than those of the late twentieth century. This peak subsequently decreased, as more Northern Hemisphere data sets were combined. While many proxy data sets showed a temperature peak in the MWP, the heterogeneity of the peak locations, magnitudes, and durations meant that averaged temperature changes were reduced from those seen at individual sites.

Even with multidecadal filtering to extract long-term trends during the MWP from year-to-year fluctuations, interludes of cold have appeared in periods of relative warmth. The 2001 report of the Intergovernmental Panel on Climate Change (IPCC) largely dismissed the existence of a distinct, global MWP and LIA, epitomized by the famous hockey stick graph that showed a small, nearly linear decrease in average temperature from 1000 to the late nineteenth century with no distinct MWP to LIA transition. After some controversy, the next IPCC report in 2007 revived the MWP and LIA, using eight proxy-based temperature reconstructions to define a mild Northern Hemisphere temperature peak between 950 and 1100 that was 0.1°–0.2°C cooler than the mean global temperature between 1961 and 1990 and 0.3°–0.4°C warmer than the coolest LIA period.

Attempts to measure the MWP illustrate the challenges of using temperature proxies rather than instrumental temperature readings. Proxies are influenced by effects other than temperature—for example, growth patterns in tree-ring studies can reflect precipitation, diseases, and atmospheric carbon dioxide concentration. Studies that combine multiple proxy data sets must accurately calibrate their results against a common temperature standard, but different researchers use different approaches. Only a limited number of proxy sites extend back to the MWP, with few in the Southern Hemisphere and unbalanced coverage in the Northern. Overall, there is considerable potential error in each reconstruction, such that current views of the MWP are tentative and will likely change with future data.

Interest in the cause of the MWP has been muted by the debate over its nature. If the event was only regional, then changes in regional meteorological features such as the North Atlantic Oscillation might provide a sufficient explanation. If the MWP was more global, solar activity may account for it. Sunspot records do not exist for the MWP, but concentrations of the cosmogenic isotopes carbon 14 and beryllium 10 in tree rings and ice cores suggest that there was stronger than normal solar radiation around 1000 and again between 1100 and 1250. These isotope concentrations are also consistent with evidence of weaker radiation during much of the LIA.

—*Raymond P. LeBeau, Jr.*

FURTHER READING

Fagan, Brian. *The Great Warming: Climatic Change and the Rise and Fall of Civilizations.* New York: Bloomsbury Press, 2008.

Hogan, C. Michael. "Medieval Warm Period." *Encyclopedia of Earth.* Boston U, 1 Jan. 2013. Web. 23 Mar. 2015.

Hughes, Malcolm, and Henry F. Diaz, eds. *The Medieval Warm Period.* Berlin: Springer-Verlag, 1994.

Kaufman, Darrell S., et al. "Continental-Scale Temperature Variability during the Past Two Millennia." *Nature Geoscience* 6 (2013): 339–46. PDF file.

Lamb, Hubert Horace. *Climate, History and the Modern World.* Routledge, 2006.

Singer, S. Fred, and Dennis T. Avery. *Unstoppable Global Warming: Every Fifteen Hundred Years.* Rev. ed. Rowman & Littlefield, 2008.

MEDITERRANEAN CLIMATE

FIELDS OF STUDY

Climate Classification; Climate Zones; Environmental Sciences; Environmental Studies; Ecology; Physical Geography; Hydroclimatology; Meteorology; Hydrometeorology; Bioclimatology; Oceanography; Climate Modeling; Earth System Modeling

SUMMARY

A Mediterranean climate is characterized by wet winters and dry summers, with mild winter and hot summer temperatures. This type of climate covers just under 2 percent of Earth's land area. It is found only in the middle latitudes (around 35° to 45°

Sunset view on the Mediterranean Sea towards Cannes from Juan-Les-Pins near Nice, France.

north or south latitude) and near an ocean or the Mediterranean Sea. Most of the area included in this climatic zone is the Mediterranean Sea basin, for which it was named.

PRINCIPAL TERMS

- **desertification:** the gradual expansion of desert conditions into other climate regions
- **Iberian Peninsula:** the part of Europe extending from southern France, comprising Spain, Portugal and Malta

MEDITERRANEAN CLIMATES

Climate is the long-term average weather of a particular location. A Mediterranean climate is characterized by wet winters and dry summers, with mild winter and hot summer temperatures. This type of climate covers slightly less than 2 percent of Earth's land area. It is found only in the middle latitudes (around 35° to 45° north or south latitude) and near an ocean or the Mediterranean Sea. Most of the area included in this climatic zone is the Mediterranean Sea basin, for which it was named. Other Mediterranean climate regions include the coastal regions of central and southern California, central Chile, the west side of the tip of South Africa, and parts of southern Australia, particularly the southwest.

The summer dryness in Mediterranean climates is caused by stable atmospheric high-pressure systems preventing storm systems from entering the area. These semipermanent high-pressure systems move away from the area and toward the equator during the fall, allowing rain-producing low-pressure systems to move in. When high pressure is reestablished in the spring, low-pressure areas can no longer move through the area, and rain-free conditions are reestablished for a number of months.

Because this climatic zone is always near a large body of water and such water bodies moderate temperatures (water heats and cools more slowly than does land), Mediterranean winter temperatures rarely reach the freezing point. Summer temperatures are greatly affected by the relative warmth or coolness of the body of water adjoining the area. The Mediterranean Sea is a warm body of water, causing the Mediterranean basin to be relatively warm in the summer. Other areas, such as California, have relatively cool summer temperatures near the coast, because the Pacific Ocean, for example, is cooler than the Mediterranean Sea. The dryness of summer allows large temperature fluctuations during the day/night cycle. It also contributes to Mediterranean areas experiencing a relatively large number of wildfires, both because the air contains less moisture to resist fire and because the vegetation native to dry areas is itself drier and more prone to burn than that of wetter climates.

SIGNIFICANCE FOR CLIMATE CHANGE

Mediterranean climates depend for their stability upon a combination of specific atmospheric conditions and the presence of a large body of water. Thus, if the atmospheric conditions are altered by climate change, or if they cease to exist near large bodies of water, Mediterranean climates will disappear. On the other hand, new Mediterranean climates could be established near different water bodies under the right circumstances.

Areas such as California or Chile are on north-south coastlines, and it is possible that climate change could move the climatic zone farther north or south. If the change is not overly abrupt, Mediterranean flora and fauna could migrate along with the climate zone. However, in the Mediterranean basin, a substantial increase in global warming or another climatic change could destroy the ecosystem altogether, because there is no large body of water to the north to which the climate, flora, and fauna could shift. Within Europe, there could be some movement north, as seems to be taking place in the British Isles.

The little egret, a species of little heron commonly found in the Mediterranean Sea area, was unknown in England prior to 1996. As southern England has begun to develop a more Mediterranean climate, it was the second most populous heron in the United Kingdom by 2008. While birds have the ability to move from the European mainland to Great Britain, however, plants and land animals are unable to cross the English Channel by any natural means.

Climatic changes have been documented in several Mediterranean and nearby regions. These changes raise particular concerns regarding biodiversity, which is more affected by climate than by other environmental factors. The Iberian Peninsula, for example, has the greatest animal diversity of any region in Europe, and many Iberian species are endangered by global warming. Average summer temperatures in the region are generally rising, which is putting a substantial heat stress upon both plants and animals. Humans are not immune to such factors, and a study documenting a 200 percent increase, or doubling, in the number of extremely hot days on the peninsula concluded that frail people's lives were at risk from the increase.

Desertification is also taking place in the Mediterranean, as rainfall decreases and higher temperatures increase the rate of evaporation. A large number of droughts in the early twenty-first century have contributed to the encroachment of the desert from the south. Although between the mid-twentieth and early twenty-first century, there was generally a decrease in the amount of rainfall, in some specific locations there was an increase in precipitation.

It is estimated that up to 20 percent of plant species live in Mediterranean regions. In some areas, such as California, population growth has played a major factor in decreasing the number and diversity of plants, but changes in the precipitation patterns have also been a contributing factor, as has the resulting increase in wildfires. Even in areas where climate change has not substantially altered the total rainfall, the number of days on which it rains has changed. Many areas have fewer but harder rains than normal, resulting in less moisture being absorbed by the soil. An Intergovernmental Panel on Climate Change study has indicated that Mediterranean climates will very probably experience an increase in severe droughts, heat waves, and wildfires as a result of the changing climate.

—Donald A. Watt

FURTHER READING

Bolle, Hans-Jurgen, ed. *Mediterranean Climate: Variability and Trends*. Springer, 2003.

Giupponi, Carlo, and Mordechai Shechter, eds. *Climate Change and the Mediterranean: Socio-economic Perspectives of Impacts, Vulnerability, and Adaptation*. Edward Elgar, 2003.

Lionello, P., ed. *The Climate of the Mediterranean Region: From the Past to the Future*. Elsevier Science, 2012.

McDade, Lucinda. "Plant Communities and Climate in Southern California." *Rancho Santa Ana Botanic Garden*. Rancho Santa Ana Botanic Garden, n.d. Web. 23 Mar. 2015.

Moreno, José, and Walter C. Oechel, eds. *Anticipated Effects of a Changing Global Environment in Mediterranean-Type Ecosystems*. Springer-Verlag, 1995.

MEGACITIES' EFFECTS ON LOCAL CLIMATES

FIELDS OF STUDY

Climate Zones; Environmental Sciences; Physical Geography; Meteorology; Climate Modeling

SUMMARY

A megacity is a city with a population greater than 10 million. When cities such as Mexico City, Los Angeles, or Hong Kong get this large, it becomes difficult to determine their precise boundaries or true population. The U.S. Census Bureau and the United Nations often disagree in nonsystematic ways about the population of some of the world's largest cities, and the discrepancies between their estimates can represent several million people.

PRINCIPAL TERMS

- **megalopolis:** a megacity that sprawls over a large area, rather than being concentrated spatially in the manner of traditional cities

Tokyo Skyline.

- **urban heat island:** an urban region that is significantly warmer than the surrounding rural areas
- **urbanization:** the process of concentration of the human population in cities

MEGACITIES

The bureaucracy and infrastructure of megacities can be as complex as those of small nations, and resource allocation within them is particularly difficult. In 1950, New York City was arguably Earth's only megacity. In less than one human lifetime there were at least thirty megacities on the planet. One of the most challenging aspects of this dramatic increase in the number of megacities is that they are increasingly appearing in some of the poorest nations of the world. In 1950, only three of the world's most populous cities were in the developing world. In 2005, about 75 percent of the world's largest cities were in the developing world. The rise of these megacities represents a profound development in the history of humanity, and the challenges they present with respect to climate change may also be among the greatest opportunities to address that change.

REALITY ON THE GROUND

A megacity is a complex structure consisting of a sophisticated built environment that shelters and sustains millions of human agents. These agents, living in close proximity to one another, often represent extremes of human experience, as megacities juxtapose great wealth and poverty, as well as the diversity of human cultures. In their vast urban landscapes, millions of people live and die in sad and tragic conditions of poverty and low life expectancy. At the same time, others live lives of almost unfathomable wealth and freedom to travel about the globe. These urban environments simultaneously represent both a pinnacle of human achievement and a shameful failure to realize human potential.

CHANGE CHALLENGES AND OPPORTUNITIES

Cities almost always depend on a hinterland beyond their spatial extent to provide food, water, energy, and raw materials to sustain the lives of their citizens. They also increasingly depend on this hinterland to absorb their sewage, solid waste, and greenhouse gas (GHG) emissions. Historically this hinterland was predominantly nearby. Increasingly, however, hinterlands are farther and farther away from megacities, and, in regard to GHG emissions, the global atmosphere itself may be considered part of the hinterland. The hinterland of Los Angeles is global, as the city receives oil from the Middle East, water from the eastern Sierra Nevada, and food from Mexico, Europe, and Asia. Mexico City has built vast tunnels to divert sewage to distant hinterlands. Denver, Colorado, uses a network of tunnels to divert water from the western slope of the Rocky Mountains that would normally flow into Mexico's Sea of Cortez.

Almost all of these processes relate to climate change forcing factors in direct or indirect ways. Nonetheless, a fundamental and primary impact of megacites with respect to climate change is the energy used by these cities to provide electricity (often provided by coal-fired power plants) and the energy used to provide transportation (predominantly generated by fossil-fuel combustion). These urban areas are the most densely populated areas of the world. This density is an opportunity for numerous efficiencies with respect to energy consumption for electricity, transportation, and the myriad other related needs of urban residents that require electricity and transportation. Leveraging the energy efficiency opportunities that these densely populated areas represent will be of paramount importance with respect to humanity's collective response to the challenges of climate change.

—Paul C. Sutton

BIBLIOGRAPHY

Burdett, Ricky, and Deyan Sudjic. *The Endless City: The Urban Age Project by the London School of Economics and Deutsche Bank's Alfred Herrhausen Society.*

MERIDIONAL OVERTURNING CIRCULATION (MOC)

FIELDS OF STUDY

Oceanography; Physical Geography; Heat Transfer; Fluid Dynamics; Climate Modeling; Hydroclimatology; Earth System Modeling

SUMMARY

Meridional overturning circulation (MOC) is an oceanographic term for water flows in the plane defined by the vertical and meridional (or north-south) axes. It is calculated by averaging those north-south, up-down flows from east to west across the width of an ocean basin.

PRINCIPAL TERMS

- **advection:** horizontal movement of heat carried by the atmosphere, as opposed to the vertical movement associated with convection
- **meridional:** parallel with the north-south direction of the meridians of longitude
- **thermohaline:** indicating that the property depends on both temperature and salinity

The MOC

Most discussions of the MOC focus on the deep, overturning circulations that connect the ocean abyss to the surface. Deepwater formation (the "sinking branches" of the deep MOC) occurs in two broad regions of the global ocean: the high-latitude North Atlantic Ocean, predominantly in the Labrador Sea and the Nordic Sea, and the Southern Ocean, near Antarctica. Water's density increases when it is cold and salty. Thus, the densest surface waters occur in the Polar Regions. Temperatures there are low, and brine-rejection during ice-formation increases the salinity of surface ocean water.

Dense polar water sinks to form deepwater masses that spread out horizontally along the meridional axis to fill most of the global deep oceans. The return branch of this deep MOC is more diffusely distributed. Deep water becomes more buoyant, and it will return back toward the surface if it is heated or made less saline. Vertical mixing with less dense water higher in the water column will decrease the density of deep water, producing a return flow toward the surface. The deep MOC is sometimes called the "thermohaline circulation," although that term refers to such movement in all oceans, not just the Atlantic. The term "thermohaline circulation" is meant to evoke the idea that vertical motions are caused by changes in the temperature and salinity of seawater.

Shallow, wind-driven overturning circulations exist closer to the ocean's surface, the most prominent such feature being the subtropical cells in the Atlantic and Pacific Oceans. The winds blowing over the subtropical oceans force a convergence of surface waters, which pushes surface water downward in a process known as "Ekman pumping." This water travels at depth toward the equator, where the pattern of winds forces surface waters to diverge, bringing the water that was pumped downward in the subtropics back to the surface. Surface winds then force the surface waters back toward the subtropics, completing the subtropical cell.

Significance for Climate Change

The MOC plays a crucial role in maintaining Earth's climate. The majority of the solar heat energy captured by Earth obtains primarily in the tropics. Warm ocean currents move a large amount of tropical heat to higher latitudes in the subtropical cell and the deep MOC. This transfer of heat helps keep the high latitudes warm. The poleward heat flux is particularly strong in the North Atlantic Ocean. Heat brought poleward in the North Atlantic is then advected by large-scale winds eastward, where it warms Europe. Scientists have suggested that the North Atlantic

Thermohaline Circulation

This map shows the pattern of thermohaline circulation also known as "meridional overturning circulation." This collection of currents is responsible for the large-scale exchange of water masses in the ocean, including providing oxygen to the deep ocean.

MOC may slow down or shut down completely as a result of global warming; if this happens, the heat flux associated with the MOC would decrease, and Europe would become colder.

A climatic change that warms the high latitudes will introduce freshwater into the high-latitude North Atlantic Ocean as ice sheets melt and precipitation increases. This added freshwater would decrease the salinity of surface waters, reducing their tendency to sink. If polar water stops sinking, the warm surface current that brings lighter water to replace the sinking water would be disrupted, so the flow of heat to the North Atlantic and Europe would diminish and perhaps cease entirely.

Geologic evidence supports the argument that the introduction of freshwater into the North Atlantic can weaken heat transport by the Atlantic MOC. During the last ice age, ice sheets several kilometers thick covered a large portion of North America and northern Europe. Around 14,000 years ago, glaciers in the Northern Hemisphere retreated as a result of astronomically forced changes in Earth's orbit. Temperatures rose, and the ice sheets began to melt. Then, approximately 12,800 years ago, temperatures dropped rapidly back into the glacial range and ice sheets returned for another 1,300 years. This rapid drop in temperatures, known as the Younger Dryas, is believed to have been caused by the rapid input of freshwater into the North Atlantic from the melting North American ice sheet and the emptying of the very large glacial Lake Agassiz into the Atlantic Ocean, which dramatically decreased the Atlantic MOC. While there is no large ice sheet covering North America today, scientists are concerned that global warming could cause freshwater melt from Greenland to trigger analogous processes.

—*Alexander R. Stine*

Further Reading

Broecker, Wally *The Great Ocean Conveyor: Discovering the Trigger for Abrupt Climate Change.* Princeton University Press, 2010. Describes the global oceanic overturning circulation system and how its disruption by climate change may affect the climate of Earth.

Gornitz, Vivien *Rising Seas. Past, Present, Future.* Columbia University Press, 2013. Presents an overview of sea level rise due to climate change and describes how rising sea levels may affect the stability of the ocean heat conveyor system.

Jones, E. Peter and Anderson, Leif G. "Is the Global Conveyor Belt Threatened by Arctic Ocean Fresh Water Outflow?" Chapter in Dickson, Robert

R., Meincke, Jens and Rhines, Peter, eds. *Arctic-Subarctic Ocean Fluxes. Defining the Role of the Northern Seas in Climate.* Springer Science + Business Media B.V., 2008. Provides a detailed analysis of the interaction of Arctic waters with southerly ocean waters and the potential effects of those interactions on the global conveyor system.

Kuhlbrodt, T., et al. "On the Driving Processes of the Atlantic Meridional Overturning Circulation." *Reviews in Geophysics* 45 (April 24, 2007).

Richardson, P. L. "On the History of Meridional Overturning Circulation Schematic Diagrams." *Progress In Oceanography* 76 (2008): 466–486.

Schmittner, Andreas, Chiang, John C.H. And Hemming, Sidney R., eds. *Ocean Circulation. Mechanisms and Impacts.* John Wiley & Sons, 2013. Provides a thorough description of the effects on Earth's climate due to large "turnover" circulation currents, particularly the Atlantic conveyor system.

"The Ocean Conveyor." *Woods Hole Oceanographic Institution.* Woods Hole Oceanographic Inst., 2014. Web. 23 Mar. 2015.

METEOROLOGY AND CLIMATE POLICY

FIELDS OF STUDY

Atmosperic Sciences; Climate Modeling; Meteorology

SUMMARY

Meteorology is the scientific study of the atmosphere, weather, and climate. It combines most of the basic scientific disciplines, such as mathematics, physics, chemistry, statistics, and computer science, and applies them to Earth's atmosphere and its phenomena. Thus, meteorology is a branch of Earth science and of physical science. Meteorology provides the core knowledge about climate change and global warming. It can be used to analyze planetary climate patterns as well as continental, regional, and local patterns. Meteorologists measure weather and other specific atmospheric phenomena and abstract from those measurements to determine the climate, the long-term average conditions in a given location.

PRINCIPAL TERMS

- **climate:** long-term average weather conditions
- **mesoscale:** the atmospheric scale between microscale and synoptic scale, ranging from a few kilometers to hundreds of kilometers
- **microscale:** the smallest scale of atmospheric motion, ranging from meters to kilometers
- **planetary scale:** the largest scale of atmospheric motion, covering the entire globe
- **synoptic scale:** the typical scale of weather maps, showing such features as high- and low-pressure systems, fronts, and jet streams over an area spanning a continent
- **weather:** a particular atmospheric state at a given time and place

A meteorologist at work at the Storm Prediction Center in Norman, Oklahoma.

DYNAMIC METEOROLOGY

Dynamic meteorology is the core discipline of the atmospheric sciences. It employs dynamics, fluid mechanics, and classical mechanics, coupled with rigorous mathematics, to study atmospheric motion and evolution. Dynamic meteorology treats the atmosphere as a fluid continuum, applying Newtonian principles to atmospheric systems.

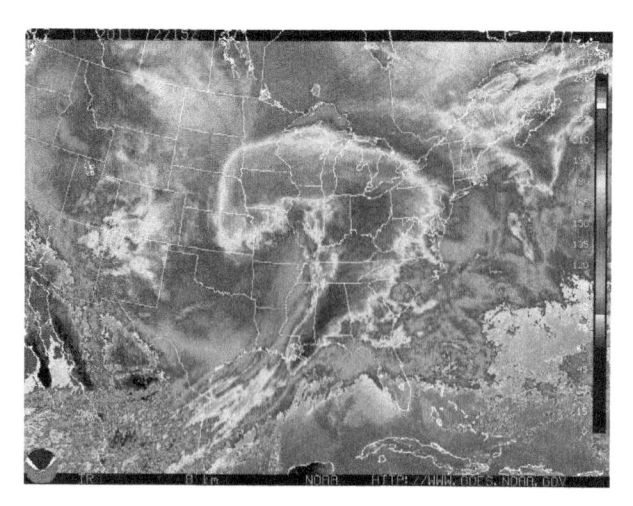

NOAA Satellite photo of Central US blizzard on February 2, 2011.

Modern numerical weather prediction is a result of this approach. Many methods in dynamic meteorology are also extended to study climate systems in the closely related discipline of climate dynamics. Climate dynamics may provide a good tool for studying climate change. Various global circulation models (GCMs) are examples of this application.

Physical Meteorology
In addition to its kinetic properties and dynamic evolution, the atmosphere possesses many other physical properties, such as its thermal content, its humidity, its electrical and optical properties, and so forth. To study these physical properties of the atmosphere, meteorologists incorporate the principles and approaches of physics. A wide range of subjects is helpful in studying such atmospheric phenomena, including cloud physics, thermodynamics, precipitation physics, boundary-layer meteorology, thermal convection, atmospheric electricity, and atmospheric optics.

Many atmospheric physical properties are directly related to global climate change. For example, clouds are an important factor in global warming. Clouds play a dual role in the global climate system. On one hand, they contribute significantly to Earth's albedo, reflecting a large amount of solar energy back into space and producing a global cooling effect. On the other hand, they absorb long-wave radiation from the Earth's surface and re-emit it, thereby heating the atmosphere and surface. Increased atmospheric humidity due to global warming will increase cloud cover and influence severe weather patterns as well. Physical meteorology can provide a detailed understanding of these aspects of climate change.

Applied Meteorology
As meteorologists' understanding of the complexity of Earth's climate system has increased, new applied meteorological specializations have emerged. Thus, the field now includes satellite meteorology, radar meteorology, statistical meteorology, agricultural micrometeorology, and climatology. These new subdisciplines are fundamentally interdisciplinary. They not only help translate meteorological concepts and research methods to other scientific disciplines but also strengthen meteorology by incorporating the technologies and methods of its sister sciences. For example, modern technologies, such as radar and satellites, add fresh content to meteorology and provide new observational tools for studying the atmosphere. These new areas are also important for global climate studies. For example, satellites can provide a global view of the global warming effect.

Atmospheric Chemistry
Traditional meteorology is mostly concerned with the physical aspects of the atmosphere. Climate researchers, however, have found that atmospheric chemistry is just as important for understanding climate change. Greenhouse gases are central to global climate change, and other aspects of atmospheric composition may play similarly important roles. Ozone depletion in the stratosphere is both an atmospheric dynamics and atmospheric chemistry problem that concerns climate change. In addition, Earth's carbon cycle is an important area for climate study.

With an increasing level of global industrialization and urbanization, environmental conservation and protection become more concerned issues. Air pollution and air quality are central to these environmental problems. Acid rain and environmental acidification are also concerns for environment protection and conservation. For all these problems, atmospheric chemistry can provide fundamental understanding.

Context
Meteorology is one of the primary sciences employed in the study of Earth's climate system. Earth's climate is a complex system; however, that includes five components: atmosphere, hydrosphere, lithosphere,

cryosphere, and biosphere. It is the interaction of all five of these components that determines Earth's climatic environment. Therefore, meteorology, although providing a core understanding of global climate and climate change, must be combined with knowledge from other scientific disciplines, such as oceanography, geology, hydrology, chemistry, biology, ecology, astronomy, and glaciology, to address global climate change.

—*Chungu Lu*

FURTHER READING

Ahrens, C Donald, and Robert Henson. *Meteorology Today: An Introduction to Weather, Climate, and the Environment.* Cengage, 2019. Discusses global climates, climate change, and classification. One of the most widely used introductory books on atmospheric science; covers a wide range of topics on weather and climate.

Arnold, Dennis G., ed. *The Ethics of Global Climate Change.* Cambridge University Press, 2011. Discusses the ethics of wealthy nations contributing disproportionately to global climate change that affects the poorest nations and the majority of the global population.

Gramelsberger, Gabriele and Feichter, Johann, eds. *Climate Change and Policy. The Calculability of Climate Change and the Challenge of Uncertainty.* Springer, 2011. A curated collection of articles addressing problems of uncertainty about climate modeling and climate change on an individual basis.

Lutgens, Frederick K., Edward J. Tarbuck, Tasa, Dennis G. and Herman, Redina. *The Atmosphere: An Introduction to Meteorology.* 14th ed. Pearson Education, 2018. An updated version of this textbook, in loose-leaf format, provides students with a general overview of the structure and properties of Earth's atmosphere.

Mbane Biouele, Cesar *Earth's Atmosphere Dynamic Balance Meteorology.* Scientific Research Publishing, 2014. Without becoming overly technical, this book discusses the structure of the atmosphere and many of the physical phenomena that occur there.

METHANE'S GLOBAL WARMING POTENTIAL

FIELDS OF STUDY

Atmospheric Chemistry; Biochemistry; Environmental Chemistry; Geochemistry; Photochemistry; Physical Chemistry; Thermodynamics; Chemical Engineering; Petroleum Refining; Environmental Sciences; Waste Management; Process Modeling; Fluid Dynamics; Atmospheric Science; Bioclimatology; Oceanography; Physical Geography

SUMMARY

Methane is a colorless, odorless gas with the molecular formula CH_4. It is the main chemical component of natural gas (accounting for 70–90 percent of such gas). Natural gas accounts for up to 20 percent of the U.S. energy supply. Methane was discovered by the Italian scientist Alessandro Volta (1745–1827), who collected it from marsh sediments and demonstrated that it was flammable. He called it "combustible air."

PRINCIPAL TERMS

- **alternative fuel:** clean or renewable fuel that can replace traditional fossil fuels
- **archaea:** a taxonomic group of prokaryotic, single-celled microorganisms similar to bacteria, but evolved differently
- **energy from waste:** technologies that are designed to produce energy and reduce or eliminate waste at the same time
- **fuel:** an energy source that is burned to release energy
- **fuel alternative:** replacement energy source that can be used instead of fuel
- **greenhouse gas (GHG):** a gas in the atmosphere that traps heat on Earth that would otherwise radiate into space

METHANE AS A GREENHOUSE GAS

As with all greenhouse gases (GHGs), methane in the atmosphere acts similarly to glass in a greenhouse. It allows light energy from the Sun to reach Earth's surface, but it traps heat energy radiated back from the surface in the form of infrared radiation. Since the beginning of the Industrial Revolution in the mid-eighteenth century, methane concentrations have more than doubled in the atmosphere, causing

343

Global Methane Emissions by Sector, 2000

Economic Sector	Percent of Total Emissions
Agricultural by-products	40.0
Fossil fuel retrieval, processing, and distribution	29.6
Waste disposal and treatment	18.1
Land use and biomass burning	6.6
Residential, commercial, and other sources	4.8
Power stations	0.9

Data from the Netherlands Environmental Assessment Agency.

nearly one-quarter of the planet's anthropogenic global warming. Continuous release of methane into the atmosphere causes rapid warming, because methane's contribution to the greenhouse effect is much more powerful than that of carbon dioxide (CO_2).

Global warming itself may trigger the release of methane trapped in tundra permafrost or ocean deposits, thereby accelerating climate change in a positive feedback loop. The release of large volumes of methane from such geological formations into the atmosphere has been suggested as a possible cause for global warming events in the past. Methane oxidizes to CO_2 and therefore remains in the atmosphere for a shorter time-period of nine to fifteen years, compared to CO_2, which may remain in the atmosphere for one hundred years.

Sources of Methane

According to the U.S. Environmental Protection Agency (EPA), about 60 percent of global methane emissions are a direct result of human-related activities. These activities include creating landfills, treating wastewater, animal husbandry (through enteric fermentation and manure production), cultivating rice fields, mining coal, and producing and processing natural gas. For instance, the livestock sector (including cattle, chickens, and pigs) generates 37 percent of all anthropogenic methane. Landfills are the second largest anthropogenic source of methane in the United States.

Natural sources of methane include wetlands, lake sediments, natural gas fields, termites, oceans, permafrost, and methane hydrates. Wetlands are responsible for up to 76 percent of global natural methane emissions. Surprisingly, according to EPA data, termites contribute about 11 percent of global natural methane emissions. In most of these processes, methane is produced by microorganisms called archaea as the integral part of their metabolism. Such microbes are called methanogens, and the route of methane generation is called methanogenesis.

Archaea live in oxygen-depleted habitats, because the presence of oxygen would kill them instantly. For their food source, methanogens use products of bacterial fermentation such as CO_2 and molecular hydrogen (H_2); different acids such as acetate, pyruvate, or formate; or even carbon monoxide. That is why methanogenic archaea usually exist in consortium with other microorganisms (bacteria). They also live in symbiotic relationships with other life forms, such as termites, cattle, sheep, deer, camels, and rice crops.

Methane as a Fuel

In the 1985 science-fiction film *Mad Max Beyond Thunderdome* starring Mel Gibson, a futuristic city was run on methane generated by pig manure. In reality, methane can be a very good alternative fuel. It has a number of advantages over other fuels produced by microorganisms. It is easy to make and can be generated locally, obviating the need for long-distance distribution. Extensive natural gas infrastructure is already in place to be utilized. Utilization of methane as a fuel is a very attractive way to reduce wastes such as manure, wastewater, or municipal and industrial

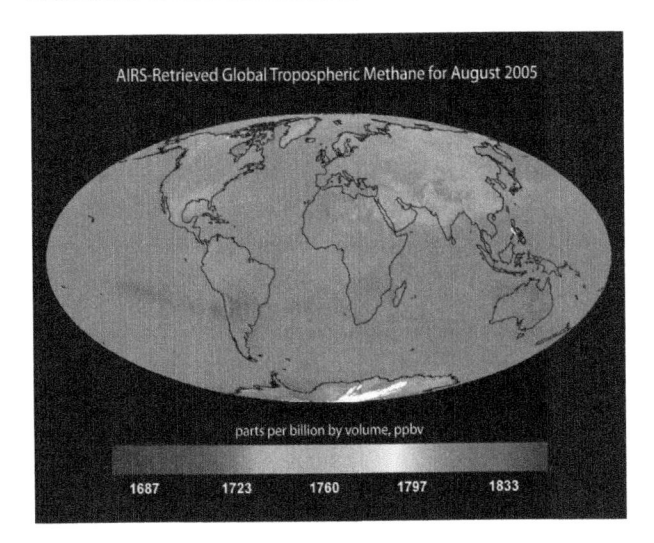

A satellite image of the global distribution of methane in Earth's troposphere.

wastes. In local farms, manure is fed into digesters (bioreactors), where microorganisms metabolize it into methane. Methane can be used to fuel electrical generators to produce electricity.

In China, millions of small farmers maintain simple, small, underground digesters near their houses. There are several landfill gas facilities in the United States that generate electricity using methane. San Francisco has extended its recycling program to include conversion of dog waste into methane to produce electricity and to heat homes. With a city dog population of 120,000, this initiative promises to generate significant amounts of fuel with a huge reduction of waste at the same time.

Methane was used as a fuel for vehicles for a number of years. Several Volvo car models with bi-fuel engines were made to run on methane, with gasoline as a back up. Methane is more environmentally friendly than are fossil fuels. Burning methane results in production of CO_2 and contributes to global warming, but with less impact on Earth's climate than methane itself would have in the atmosphere, for a net benefit. Even though the use of methane as an energy source releases CO_2, the process as a whole can be considered CO_2 neutral, in that the released CO_2 can be assimilated by archaea.

Methane Removal Processes

The natural mechanism of methane removal from the atmosphere involves its destruction by the hydroxyl radical (OH). Significant amounts of methane are also consumed by microorganisms called methanotrophs, which use the methane for energy and biosynthesis. These bacteria are prevalent in nature and potentially could be used for methane mitigation.

Context

Since methane is a powerful contributor to global warming, any efforts to reduce methane emissions will have a rapid impact on Earth's climate. One way to avoid methane release into the atmosphere is to turn it into a fuel. Supply of fossil fuels, particularly oil, is limited and does not satisfy world energy demands, which consistently increase. The extensive use of fossil fuels causes global warming. Methane utilization in place of fossil fuels as an energy source can provide significant environmental and economic benefits. In the future, landfills and wastewater treatment facilities can possibly be redesigned to optimize methane production. However, further research is needed to better understand archaean-bacterial methanogenic communities in landfills and wastewater treatment facilities in order to improve methane generation.

Some technical obstacles exist to efficiently converting landfill wastes that primarily contain plant lignocellulosic material. Lignocellulose is a combination of lignin, cellulose, and hemicellulose that strengthens plant cell walls. Microbial communities in landfills cannot utilize lignin. Creating efficient methane-producing facilities that are also capable of reducing waste is a feasible option for sustainable development to provide fuel to heat homes, run cars, generate electricity, and eliminate powerful GHG and health hazards.

—*Sergei Arlenovich Markov*

Further Reading

Archer, David. *Global Warming: Understanding the Forecast*. Wiley, 2012. Devotes a chapter to discussion of methane and its greenhouse effects.

"Climate Change Indicators: Greenhouse Gases." *EPA*, 6 Oct. 2016, www.epa.gov/climate-indicators/greenhouse-gases. Accessed 2 Feb. 2017.

Dlugokencky, E. J., et al. "Continuing Decline in the Growth Rate of the Atmospheric Methane Burden." *Nature*, 393, 1998, pp. 447–50.

Madigan, Michael T., et al. *Brock Biology of Microorganisms.* 12th ed. Pearson/Benjamin Cummings, 2009. Introductory microbiology textbook. Chapter 17 describes methane production by archaea.

National Academy of Sciences. *Methane Generation from Human, Animal, and Agricultural Wastes.* Lulu Com, 2013. Describes the role of disparate sources in methane production.

Nebel, Bernard J., and Richard T. Wright. *Environmental Science: Towards a Sustainable Future.* Prentice Hall, 2008. Several chapters describe methane as a GHG and an alternative fuel.

Reay, David *Greenhouse Gas Sinks.* CAB International, 2007. Covers the importance of carbon dioxide, methane and nitrous oxide as greenhouse gases, and describes the function of sinks for each gas.

Wadhams, Peter *A Farewell to Ice. A Report from the Arctic.* Oxford University Press, 2017. Discusses the role of Arctic and Antarctic ice shields, and how feedback systems function. Special attention is given to the potential for the release of quantities of methane from polar sea floors and Arctic permafrost as Earth's average temperature increases.

MODES OF CLIMATE VARIABILITY

FIELDS OF STUDY

Climatology; Climate Modeling; Statistics; Fluid Dynamics; Atmospheric Science; Environmental Sciences; Environmental Studies; Hydroclimatology; Bioclimatology; Climate Classification; Climate Zones; Oceanography; Physical Geography; Earth System Modeling

SUMMARY

A college town with 2,000 residents that hosts 10,000 students for eight months every year has an average population of 8,666, but it will almost never have that number present. In "academic year" mode, it has a population of 12,000, and in "vacation" mode, it has 2,000. The climate also has modes, and its behavior varies dramatically between them. Temperatures, rainfall, winds, and other climatic phenomena during El Niño are very different from those during La Niña. Recognizing these modes is essential to understanding how the climate operates.

PRINCIPAL TERMS

- **El Niño-Southern Oscillation (ENSO):** a coupled oceanic/atmospheric seesaw that occurs in the equatorial Pacific but often has global climatic consequences
- **modes:** phases of a climatic seesaw; for example, El Niño is the warm mode of the ENSO seesaw, whereas La Niña is the cold mode
- **North Atlantic Oscillation (NAO):** a seesaw in pressure between the Azores and southwestern Iceland, thought by some scientists to be an expression of the Northern Annular Mode
- **Northern annular mode (NAM) and Southern Annular Mode (SAM):** also called, respectively, the Arctic and Antarctic Oscillations, seesaws in pressure between the latitudes near 45° northern (or southern) latitude and the North (or South) Pole
- **Pacific Decadal Oscillation (PDO):** a temperature, pressure, and wind seesaw in the Pacific Ocean
- **Pacific-North American (PNA) pattern:** a seesaw between northern Pacific and North American pressures
- **regimes:** another word for "modes," fitting in with meteorological metaphors such as "fronts"
- **seesaw:** a change in opposite directions, such as high pressure in one region and low pressure in the other
- **teleconnection:** a connection between two widely separated regions of the planet that have highly correlated changes in some climatic parameter, usually resulting from a seesaw

SEESAWS

Modes of climate variability are often referred to as "seesaws" because what is missing from one region (such as warmth, atmospheric pressure, or precipitation) is found in excess in the other region. Such seesaws result from the fact that the atmosphere is a finite body of gas that obeys the laws of physics.

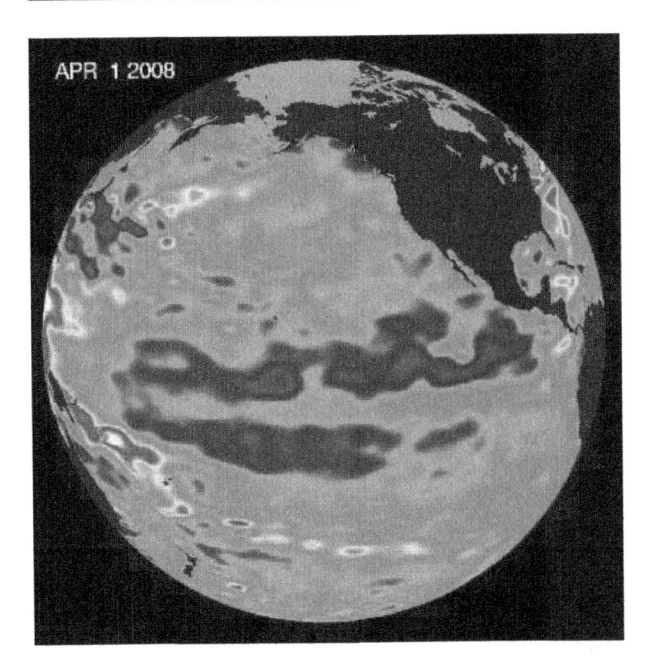

The dark area above reveals a cooling trend corresponding to La Niña and the cool phase of the Pacific Decadal Oscillation.

By 1924, a number of seesaws had been identified by Sir Gilbert Walker (1868–1958). The data set he presented in 1932 had 183 stations widely spaced across the globe, with multiyear records that permitted statistical analysis. He put his North Atlantic Oscillation (NAO) and North Pacific Oscillation (NPO) on a statistical footing, detailing the strength of correlations between what he called "action centers." He also established the existence of the Southern Oscillation, which he defined in terms of a pressure seesaw. By the 1960s others had shown that this coincided with a pattern of sea surface temperature fluctuations called El Niño, and so it is now known as the El Niño-Southern Oscillation (ENSO).

Additional workers found more seesaws, and often an index was determined by combining the values of some climatological variable at two or more locations in a simple algebraic way: The Southern Oscillation Index was obtained by subtracting the sea-level barometric pressure at Darwin, Australia, from that at Tahiti; the North Atlantic Oscillation Index was obtained by subtracting the sea-level pressure at Iceland from that at the Azores.

STATISTICAL IDENTIFICATION OF SEESAWS

Since Walker's work, the data series from many of his stations have been extended by more than seventy years, hundreds of new stations have been established, and satellite and other remote-sensing techniques have contributed immense amounts of climate information. New statistical techniques involving eigenvector analysis have been developed to analyze these data, particularly Principal Component (PC)/Empirical Orthogonal Function (EOF) analysis.

These techniques take data sets, which can be enormous, and rearrange them into separate, independent components that reveal how the data points are linked. As an example, consider the pressure data for points in the Northern Hemisphere, but outside the tropics (that is, at latitudes greater than 20° north). Eigenvector analysis finds that much of the variability in these data can be explained by two regional seesaws, EOF1, and EOF2. A map of which regions are controlled by each EOF shows that EOF1 corresponds to the NAM, and EOF2 corresponds to the PNA. Because this representation of PNA uses criteria that differ from its original index-based definition, it is often referred to as PNA.

CONTEXT

Teleconnections show that Earth's climate is not entirely random. Spatial patterns exist, and the polarity of the seesaws within these patterns alternates, often with far-reaching consequences. What is less well known is the temporal behavior of these patterns, what causes them to switch polarity, and how they interact.

ENSO is the shortest and best-known seesaw, having an average period of four years, but this period can vary from two to seven years. Efforts to explain why its period should change, or to predict how long a particular El Niño or La Niña will last, have so far been unsuccessful.

The PDO has effects that are geographically similar to those of the ENSO but has a period of twenty to fifty years. Cool before 1924, then warm until 1947, cool again until 1976, and warm again until at least 1998, it has had ambiguous behavior since then. In addition to not knowing why it reverses or when it might reverse again, climatologists do not know whether the PDO causes increased ENSO fluctuations or is caused by them.

Scientists' understanding of the modes of climate variability is incomplete, but most climate scientists agree that they play an important role over periods of years to decades. Our ability to interpret climate data correctly depends on being able to place them in context with respect to these modes. Predictions of climate change would improve if it were possible to predict when these modes will reverse and how strong they would be.

—*Otto H. Muller*

FURTHER READING

Ahrens, C Donald, and Robert Henson. *Meteorology Today: An Introduction to Weather, Climate, and the Environment.* Cengage, 2019. One of the most widely used and authoritative introductory textbooks for the study of meteorology and climatology. Explains complex concepts in a clear, precise manner and supports them with numerous images and diagrams. Discusses hurricanes and the mechanisms that generate them extensively.

Alley, Richard B. *The Two-Mile Time Machine. Ice Cores, Abrupt Climate Change and Our Future* Princeton University Press, 2014. Describes the analytical methods and interpretations of physical clues of past climates trapped in ice that is thousands of years old.

Anderson, David E., Goudie, Andrew S. and Parker, Adrian G. *Global Environments Through the Quaternary. Exploring Environmental Change.* 2nd ed., Oxford University Press, 2013. Provides concise descriptions of several ocean oscillations in relation to their cyclic effects on climatic conditions.

Rohli, Robert V., and Anthony J. Vega. *Climatology.* Jones & Bartlett, 2008. With excellent images and clear descriptions, this textbook offers an outstanding examination of the various modes of climate variability. The treatment of statistical methods, particularly those involving eigenvalues and principal components, provides a good feel for how they work without requiring the reader to become enmeshed in equations and linear algebra.

Sarachik, Edward S., and Mark A. Cane. *The El Niño-Southern Oscillation Phenomenon.* Cambridge University Press, 2018. The book offers a thorough description of the history and mechanics of the two components of ENSO, before delving into a technical analysis of the phenomenon.

MONSOONS

FIELDS OF STUDY

Atmospheric Science; Physical Geography; Oceanography; Thermodynamics; Fluid Dynamics; Environmental Sciences; Environmental Studies; Meteorology; Climate Modeling; Hydroclimatology; Hydrometeorology; Climate Zones; Hydrology; Ecosystem Management; Ecology; Earth System Modeling; Heat Transfer

SUMMARY

Monsoons are seasonal wind systems that reverse directions biannually and are crucial for the economic stability and agricultural productivity of affected geographic areas.

PRINCIPAL TERMS

- **austral:** referring to an object or occurrence that is of the Southern Hemisphere
- **boreal:** referring to an object or occurrence that is of the Northern Hemisphere
- **convection:** heat transfer by the circulating movement that occurs in fluid materials as warmer, less dense material rises above cooler, denser material
- **geostrophic:** descriptive of wind that occurs when the Coriolis force is in exact balance with the force of a horizontal pressure gradient and therefore blows in a straight line
- **Intertropical Convergence Zones (ITCZ):** low-pressure areas where southern and northern trade winds meet
- **orography:** study of mountains that incorporates assessment of how they influence and are affected by weather and other variables
- **oscillation:** variation of some physical property or condition between two opposing states, much like the rising and falling of a wave between its maximum and minimum heights

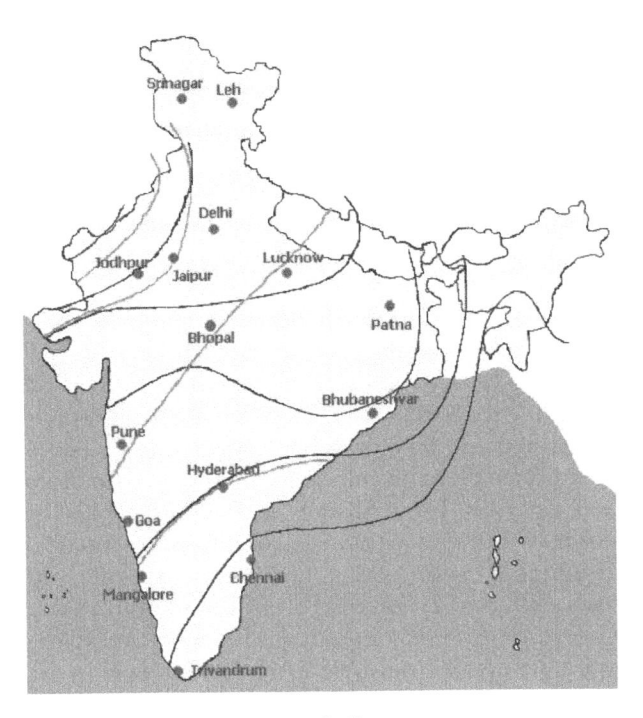

Map of monsoon's progress over India.

- **troposphere:** the level of the atmosphere closest to the ground, extending from the surface to an altitude of eleven kilometers
- **trough:** a long and relatively narrow area of low barometric pressure
- **vortex:** the central locus of a whirling liquid or gas, about which the fluid mass circulates

Origins of Monsoons

The term "monsoon" originated from the Arabic word *mausim*, meaning season, used by sailors to comment about changing winds above the Arabian Sea. Solar heat produces winds that shift to the north and south according to the sun's position each season. Scientists cite geological evidence that suggests conditions favorable for monsoons began millions of years ago when the Indian subcontinent collided with the Asian plate, eventually creating the Himalaya mountain range and the Tibetan Plateau. Warm and cool ocean water and landmasses affect atmospheric circulation, creating recurring wind systems that sweep over large regions. The majority of monsoons occur in tropical areas adjacent to the Indian Ocean, although seasonal winds also affect Africa, northeast Asia, Australia, and North and South America.

Monsoons are described by geographical terminology; the major monsoon systems are the South Asian or Indian monsoon, the East Asian monsoon, the Malaysian-Australian monsoon, the North American monsoon, and the West African monsoon. Monsoons are also characterized as either boreal or austral, according to their location in the Northern or Southern Hemisphere, respectively.

For thousands of years, monsoons have been incorporated into the literature, folklore, and religious rituals of numerous cultures. The monsoon motif is a universal symbol for rebirth and fertility as well as devastation and death. Monsoons are predictable to the extent that it is known that they probably will occur during specific seasons. However, the winds are often erratic, being delayed or appearing prematurely and precipitating extreme or minute amounts of rainfall, and sometimes bypassing regions entirely. Usually, between April and October, monsoon winds develop in the southwest, shifting direction to originate from the northeast between October and April. Monsoons are a prolonged series of winds and not restricted to a single storm.

Monsoons manipulate the planet's climate, often proving to be beneficial and occasionally detrimental. The winter monsoon blowing from land to sea is usually associated with dryness, while the summer monsoons storming from the sea onto land produce torrential, sustained rainfall that is vital for agricultural activity. Sufficient precipitation assures growth of ample crops to nourish domestic populations, to export for economic profits, and to provide employment for farm laborers. The absence of monsoon rains can result in famines and impoverishment.

Monsoon Dynamics

For several centuries, scientists have analyzed monsoons to determine the physical forces that generate the winds and regulate their behavior. Researchers have agreed on basic explanations regarding fundamental aspects of monsoons, such as their relationship to atmospheric and oceanic conditions and how they tend to vary instead of conforming to exacting standards. Scholars continue to seek answers to complex questions about monsoons, especially concerning fluctuations displayed in time periods ranging from seasons to decades. Such information might enable meteorologists to predict possible occurrences and outcomes that could impact both local and global populations and economies. Researchers want to understand more about the onset of monsoons and their active and break periods, thought to be caused by shifting troughs. The actual beginning of a monsoon is often disputed, with some scientists saying that increased humidity indicates the monsoon's start, while others say precipitation or formation of a vortex signals the beginning.

Monsoons are caused by sea and land breezes that create temperature and air-pressure differences between landmasses and bodies of water. Land absorbs heat to a different extent than water does, and the difference between land and water temperature is one factor in the instigation of monsoons. Areas in low latitudes near the equator undergo circulatory and precipitation changes because of temperature deviations on adjacent continents and seas. The amount of solar radiation emitted each season affects the temperature and air pressure over continents. During the Northern Hemisphere's summer, when the northern half of the planet tilts toward the sun, high-pressure systems move from the cooler ocean into the land's low-pressure area. As the landmass cools during winter, its high-pressure air mass moves from the land to the low-pressure air mass over the warmer ocean.

The jet stream moves south during winter and north in the summer, transporting air masses to and from monsoon regions. In winter, the Siberian high-pressure system causes air to circulate clockwise. Winds move from the northeast, down the Himalayas, and across land cooled during shortened days toward the sea, creating dry conditions and causing monsoon rains in Indonesia and Australia. In summer, the winds reverse direction because the land is heated by increased solar energy. Moving counterclockwise, the winds carry moisture from the sea in the southwest toward land. For example, the Somali jet stream near Africa moves across the equator to the Arabian Sea and alters the direction of ocean currents, which causes cold water to rise from the depths and surface temperatures to drop. Humid winds shift into India, rising when they reach the Himalayas. As the winds are lifted over the Tibetan Plateau, the air cools sufficiently to become saturated, triggering thunderstorms and convectional rainfall. The variation and strength of the Indian monsoon also seems to be related to the Southern Oscillation, which is characterized by a reversal of air pressure at opposite ends of the South Pacific Ocean at irregular intervals of three to seven years. The mechanics of this relationship are not yet well understood.

Scientists have designated three monsoon circulation patterns. The lateral component indicates a monsoon that circulates across the equator. Transverse circulation moves from North Africa and the Middle East into southern Asia. The Walker circulation, also a transverse pattern, moves across the Pacific Ocean. The size, shape, coastal positions, and elevation of landmasses influence these patterns, and the shifting of the low-pressure Intertropical Convergence Zone (ITCZ) between north and south can determine how monsoon systems move. These factors contribute to variations in monsoon intensity and duration. In areas such as Australia, monsoon winds do not rise over or descend from mountains, and high-pressure masses undergo geostrophic adjustment as they ascend over monsoon troughs. Orography influences the nature of monsoons globally. African and Australian monsoons tend to be weaker because they are not elevated as high as those lifted by Asian winds. Because the Rocky Mountains and the Sierra Madre lift air masses somewhat like the Himalaya, though not as high, some researchers claim that a monsoon circulation pattern occurs in North America, causing increased precipitation in Mexico and the southwestern United States every summer.

IMPACT ON CIVILIZATION
The effects of monsoons permeate the civilizations of regions where they occur. Approximately 60 percent of people worldwide are economically dependent on the climate affected by the Malaysian-Australian monsoon system. Humans, animals, and plants rely on monsoons to provide essential moisture. The absence of monsoon rains can cause droughts and famine, killing millions of people by starvation. Any fluctuation in the monsoon cycle, whether within one season, year, or decade, can be detrimental. At least two billion people rely on monsoons to irrigate rice and wheat crops. In India, grain production is essential to feed the growing population that expands annually at a rate greater than agricultural yields increase. At least one-fourth of the Indian economy is based in agriculture, and 60 percent of laborers are engaged in agricultural employment. Monsoon rain is critical to maintain this balance.

Ironically, monsoons also flood areas and drown people, cause landslides that destroy communities, and inundate crops. Millions of acres often remain underwater for months. Several thousand people die annually during monsoons, and thousands more become homeless. Many people are reported missing after a monsoon deluge. Floodwaters wash away unstable dams, buildings, and graves in cemeteries. Extreme humidity is stifling for most people. Humanitarian and relief agencies such as the International Red Cross and Red Crescent Society provide emergency food rations and shelter. Shipping along trade routes in monsoon regions is often both helped and hindered by winds.

Folklore, proverbs, and prayers provide insights into personal experiences with monsoons. Natives and visitors to monsoon regions have documented their encounters with monsoons. Some tourists purposefully travel to Asia during monsoons because hotels and businesses are not crowded, prices are lower, and reservations are easier to secure. However, some visitors note the inconvenience of always carrying an umbrella, traveling on monsoon-damaged roads, and encountering storm-related delays. Astronomers can observe celestial phenomena because of clear skies during the winter monsoons. Oceanographers also consider the monsoons useful because nutrient-rich waters rise to the surface, allowing scientists to study how plants, animals, the sea, and the atmosphere exchange carbon dioxide.

Modeling Monsoons

Because monsoons are essential for affected populations to thrive despite the wind's uncertain behavior, scientists have initiated programs of cooperative research regarding monsoons. They hope to collect sufficient data to develop computer models that can predict when monsoons will occur and how they will impact landmasses. Such forecasting efforts have often frustrated researchers because of the monsoons' capricious nature. Speculation is primarily hindered by intraseasonal oscillations between the active and passive precipitation phases of monsoons. The Center for Ocean-Land-Atmosphere Studies (COLA) examines the relationship of the ocean, the atmosphere, and heat sources. While empirically based forecasts have often proved more reliable than modeled simulations during the twentieth century, scientists seek to perfect their experimental methods.

Researchers recognize that regional topographic differences and varying hydrodynamic situations impede monsoon modeling attempts. Incorporating information about variables such as the temperature of the sea surface and snow cover, scientists try to understand how these factors alter monsoon behavior. They also want to comprehend how the monsoons affect ocean and atmosphere interactions in addition to how the water-air pressure system relationship influences monsoons. Studies have been designed to expand knowledge about the role of convection in monsoons. Researchers are also interested in studying the impact of monsoons on climates inside and outside the monsoons' immediate zone. Scientists disagree whether monsoons affect or are affected by El Niño; studies have been done to evaluate global precipitation data and drought conditions to determine any correlations.

Monsoon variability hinders forecasting and potential benefits to agriculture based on information concerning the onset of rainfall. For example, with advance knowledge, farmers could plant crops that require less water in case a weaker monsoon is forecast. Monsoons, however, do not always begin when expected, and variations can happen within the yearly cycle or fluctuate over several years. Using different global circulation models, researchers seek to comprehend the fundamental physical processes of monsoons, and then apply this knowledge to create specific models representing a seasonal cycle based on their observations, statistics, and hypotheses. Interpreting results produced by these models compared with satellite data, researchers become aware of how monsoons vary according to the landmasses they traverse and according to unique oceanic and atmospheric conditions.

Realizing these factors are linked, researchers create models to consider numerous variables concurrently but realize that more sophisticated modeling of factors such as sea-surface temperature, solar radiation, water vapor, cloud cover, soil moisture, and terrain is necessary to improve prediction methods. Such models must accurately simulate a monsoon's average seasonal rainfall and behavior, taking into account how it varies within one season and one year. The models must also consider anomalies with other documented monsoons in that area to isolate how external conditions such as altered topography may affect the monsoon's internal dynamics.

FUTURE RESEARCH

Monsoons fascinate the researchers who strive to acquire more complex understanding of the seasonal wind systems. Scientists hope to predict monsoons more precisely, because the winds impact global economies and populations. Meetings of international monsoon experts have been scheduled to share knowledge of monsoon cycles, coordinate research methods, and set future goals. Primarily, scientists want to explain why monsoons vary in behavior and how such variables as sea-surface temperature and location of land-based heat sources interact with and influence the monsoons' fluctuations. Researchers also want to explore the relationship between monsoons and the El Niño/Southern Oscillation (ENSO) phenomenon.

Scientists realize that to achieve accurate predictive techniques, full understanding of sea-surface temperature anomalies that affect monsoon circulation must be obtained through an analysis of oceanic processes that currently remain vague. Future models will assess these temperature patterns in different geographic regions. Models will also further evaluate water surface fluxes, land surface coverings of snow and vegetation, and global warming, all of which affect temperatures and alter monsoon cycles. Until such comprehension is attained through enhanced model simulations designed to analyze numerous dynamic variables simultaneously, predictions will be minimized, slowing endeavors to manage the winds, and monsoons will continue to affect life both advantageously and harmfully in tropical zones.

—*Elizabeth D. Schafer*

FURTHER READING

Ahrens, C. Donald. *Essentials of Meteorology: An Invitation to the Atmosphere.* Brooks, 2012.

Ahrens, C Donald, and Robert Henson. *Meteorology Today: An Introduction to Weather, Climate, and the Environment.* Cengage, 2019. Discusses global climates, climate change, and classification.

Chang, Chih-Pei, et al., eds. *The Global Monsoon System: Research and Forecast.* 2nd ed. Hackensack: World Scientific, 2011. Print.

Clift, P. D., R. Tada, and H. Zheng, eds. *Monsoon Evolution and Tectonic-Climate Linkage in Asia.* Bath: Geological Soc., 2010. Print.

Fein, Jay S., and Pamela L. Stephens, eds. *Monsoons.* Wiley, 1987.

Hodges, Kip. "Climate and the Evolution of Mountains." *Scientific American* Aug. 2006: 72–79. Print.

Lighthill, James, and Robert Pearce, eds. *Monsoon Dynamics.* Cambridge University Press, 2009.

Wang, Bin. *The Asian Monsoon.* Springer, 2007t.

N

NORTH ATLANTIC OSCILLATION (NAO)

FIELDS OF STUDY

Oceanography; Fluid Dynamics; Heat Transfer; Atmospheric Science; Physical Geography; Thermodynamics; Environmental Sciences; Climate Modeling; Hydroclimatology; Hydrology; Earth System Modeling

SUMMARY

The North Atlantic Oscillation (NAO) is the dominant pattern of atmospheric circulation in the North Atlantic region ranging from central North America to Europe and northern Asia. The NAO is usually developed in the winter and is caused by fluctuations in atmospheric pressure between a subpolar, low-pressure center near Iceland and a subtropical, high-pressure center near the Azores-Gibraltar region. The NAO is generally described by the North Atlantic Oscillation index, which is a weighted measurement of the difference between the subpolar low-pressure zone and the subtropical high-pressure center during the winter season of the North Atlantic region.

PRINCIPAL TERMS

- **anthropogenic:** deriving from human sources or activities
- **convection:** motion in a fluid that results in the vertical transport and mixing of the fluid's physical properties, such as heat
- **decadal:** describing a cyclical event that occurs over a period of at least ten years
- **trade winds:** winds blowing in a northeasterly or southeasterly direction near the equator as a result of the Coriolis Effect and the Hadley circulation

THE NAO INDEX

The NAO Index is a weighted measurement of the difference between the subpolar low-pressure zone and the subtropical high-pressure center during the winter season of the North Atlantic region. The positive NAO index phase corresponds with time periods when a stronger subtropical high-pressure center and

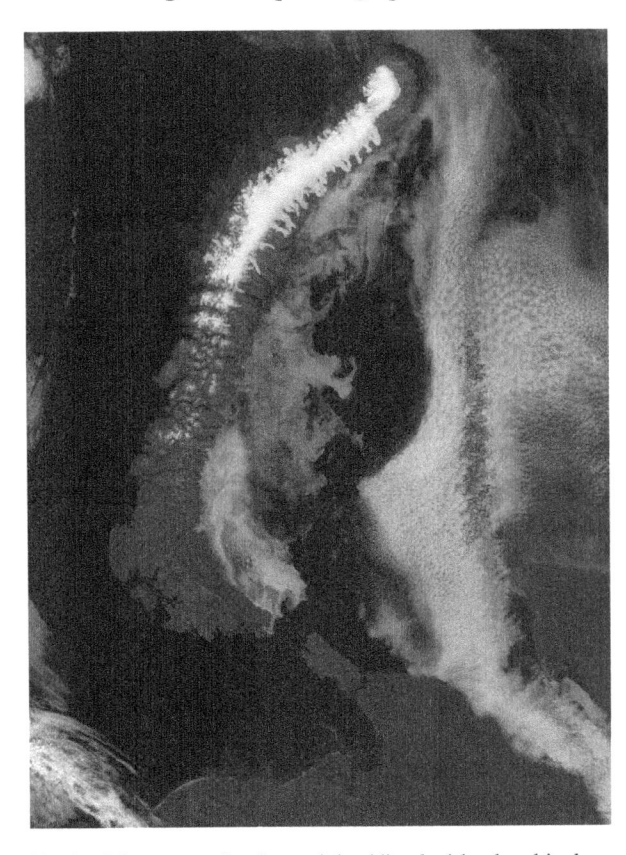

North of the western Russian mainland lies the island archipelago of Novaya Zemlya. The northern island is glacier covered and is the site of ongoing research into the effects of the North Atlantic Oscillation and climate change on the glaciers.

a deeper-than-normal subpolar low-pressure zone exist in the North Atlantic region, increasing the atmospheric pressure gradient in this region. During positive-NAO-index years, the western subtropical North Atlantic Ocean is warm. Strengthened westerly winds blow warmth and moisture into north-central Europe. The warm, moisture-bearing winds arriving from the subtropical Atlantic Ocean make Europe warmer and wetter. In the meantime, northern Canada and Greenland experience cold and dry winters. Cooler temperatures occur off the west coast of Africa. Strong trade winds send more dust out across the ocean toward the Caribbean Sea. The eastern United States undergoes a mild and wet winter season.

The negative NAO index phase corresponds with time periods when both the subtropical high-pressure center and the subpolar low-pressure zone are weakened, which would reduce the atmospheric pressure gradient in the North Atlantic region. As a result, fewer and weaker winter storms occur in this region. More moist air is brought to the Mediterranean, and cold air is brought to northern Europe. Northeastern Canada and Greenland experience mild and wet winters and the eastern United States undergoes a cold and dry winter season.

Significance for Climate Change

The NAO index varies from year to year and has evidenced a cyclicity of decadal scales over the past 150 years. The NAO index was persistently positive in the early 1900s, negative in the 1960s and 1970s, and considerably more positive during the 1980s and early 1990s. Since the heat capacity of the ocean is much greater than that of a continent, the NAO accounts for approximately one-third of the changes in average winter surface temperatures in the northern hemisphere. Variations in the NAO have significant impacts on many aspects of North Atlantic societies and the environment, such as agricultural harvests, water resources, fishery yields, industrial energy production, and ecosystems. Significant changes in the NAO may in turn influence climatic changes, including changes in sea surface temperature (SST), ocean circulation patterns, and Arctic sea-ice coverage.

Many mechanisms have been proposed to account for NAO index variability, including atmospheric response to changes in SST, variability of atmospheric convection in the tropics, internal and nonlinear dynamics of the extratropical atmosphere, and anthropogenic forcing caused by greenhouse gas (GHG) emissions and ozone depletion. Tropical heating has been proved to influence the atmospheric circulation over the North Atlantic region. Since tropical convection is sensitive to the underlying SST distribution, recent warming of the tropical oceans may lead to persistently positive values for the NAO index.

Some scientists think that changes in atmospheric circulation associated with the NAO index contributed to the winter warming of the Northern Hemisphere. Statistical evidence has demonstrated that the forcing of increased GHG concentration in the atmosphere may have affected the long-term variability of the NAO. Recent comparisons of NAO index records between the 1800s and the late twentieth century demonstrate that global warming may cause the increased variability of the NAO. Even though studies have linked climate change to the NAO, however, the mechanism of the NAO is still not fully understood. The NAO needs to be further investigated to advance understanding of the linkages between anthropogenic forcing and NAO variability.

—*Yongli Gao*

Further Reading

Anderson, David E., Goudie, Andrew S. and Parker, Adrian G. *Global Environments Through the Quaternary. Exploring Environmental Change*. 2nd ed., Oxford University Press, 2013. Provides concise descriptions of several ocean oscillations in relation to their cyclic effects on climatic conditions.

Appenzeller, C., T. F. Stocker, and M. Anklin. "North Atlantic Oscillation Dynamics Recorded in Greenland Ice Cores." *Science* 282 (1998): 446–449.

Goodkin, N. F., K. A. Hughen, S. C. Doney, and W. B. Curry. "Increased Multidecadal Variability of the North Atlantic Oscillation Since 1781." *Nature Geoscience* 1 (2008): 844–848.

Holland, D. M., et al. "Acceleration of Jakobshavn Isbrae Triggered by Warm Subsurface Ocean Waters." *Nature Geoscience* 1 (2008): 659–664.

Hurrell, J. W., et al., eds. *The North Atlantic Oscillation: Climate Significance and Environmental Impact*. Washington, D.C.: American Geophysical Union, 2003.

Paeth, H., et al. "The North Atlantic Oscillation as an Indicator for Greenhouse-Gas Induced Regional Climate Change." *Climate Dynamics* 15, no. 12 (1999): 953–960.

NORTH SEA

FIELDS OF STUDY

Oceanography; Physical Geography; Ecosystem Management; Earth System Modeling; Ecology; Heat Transfer; Thermodynamics; Engineering; Environmental Sciences; Environmental Studies; Waste Management; Meteorology; Climate Modeling; Fluid Dynamics; Hydroclimatology; Hydrometeorology; Atmospheric Science; Bioclimatology; Climate Classification; Climate Zones; Hydrology; Petroleum Refining

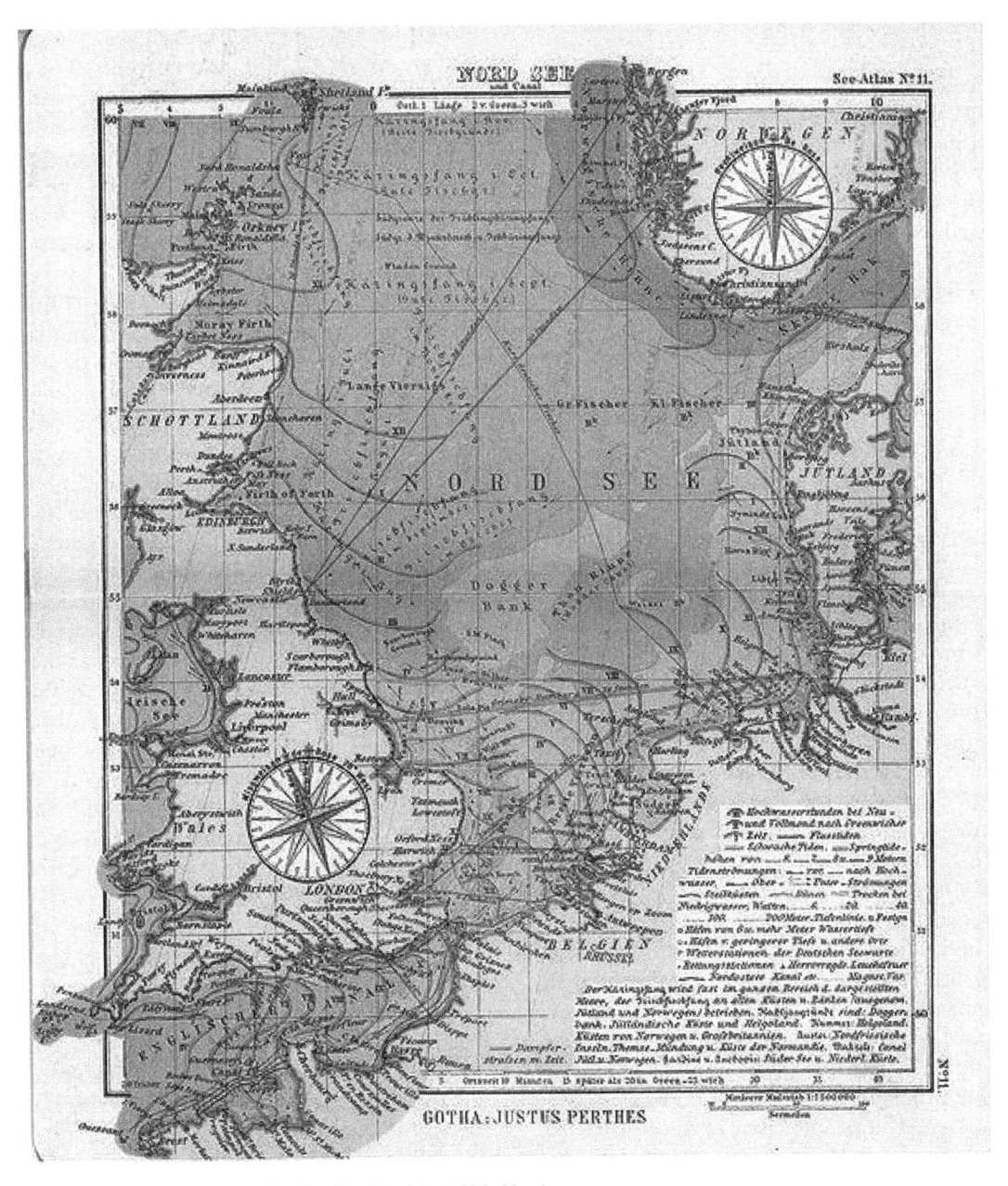

Good map of North Sea, Justhus Perthes See Atlas 1906 - Nordsee.

SUMMARY

The North Sea, located between the United Kingdom and continental Europe, is one of the most economically important bodies of water in the world. The impact of human activity on the North Sea is of major concern of environmentalists.

PRINCIPAL TERMS

- **Baltic Sea:** the body of water between Scandinavia and Eastern Europe
- **bank:** an elevated area of land beneath the surface of the ocean
- **fjord:** a steep-sided narrow inlet eroded into the face of seaside cliff, typical of Scandinavia but found throughout the world
- **Norwegian Sea:** the body of water north of the North Sea
- **strait:** a narrow waterway connecting two larger bodies of water
- **trench:** a long, narrow, depressed area in the ocean floor

PHYSICAL CHARACTERISTICS

The North Sea is an arm of the Atlantic Ocean located between the islands of Britain and the mainland of northwestern Europe. It is bordered by the island of Great Britain to the southwest and west, by the Orkney Islands and the Shetland Islands to the northwest, by Norway to the northeast, by Denmark to the east, by Germany and the Netherlands to the southeast, and by Belgium and France to the south.

To the north, the North Sea opens to the Norwegian Sea. To the south, it is connected to a narrow waterway known as the Strait of Dover, located between southeast England and northwest France. The Strait of Dover connects the North Sea to a wider waterway known as the English Channel, located between southern England and northern France. The English Channel opens to the Atlantic Ocean. To the east, the North Sea is connected to a strait known as the Skagerrak, located between Norway and Denmark. The Skagerrak is connected to a strait between Sweden and Denmark known as the Kattegat, which opens to the Baltic Sea.

The North Sea covers an area of about 570,000 square kilometers. It contains about 50,000 cubic kilometers of water. The North Sea is generally shallow, with an average depth of about 94 meters (about 308 feet). By comparison, the Atlantic Ocean has an average depth of about 3,930 meters (about 12,900 feet). The southern part of the North Sea is the most shallow, with the northern part growing deeper as it approaches the much deeper Norwegian Sea.

The floor of the North Sea is rough and irregular. In the southern part, where the water is often less than 40 meters (131 feet) deep, many areas of elevated underwater land known as banks are shifted and reworked by tides and currents. These moving banks often present a hazard to navigation. The Dogger Bank, a large bank located roughly in the center of the North Sea, is only about 15 to 30 meters (49 to 98 feet) below sea level.

Several areas of greater-than-average depth, known as trenches, are located in the North Sea. In the otherwise shallow waters of the south, a trench known as Silver Pit reaches a depth of about 97 meters (318 feet). Not far north of the Dogger Bank is a trench known as Devils Hole that reaches a depth of more than 450 meters (1,476 feet). The deepest part of the North Sea is the Norwegian Trench, a large trench that runs parallel to the southern coast of Norway. The Norwegian Trench is between 25 and 30 kilometers (15.5 and 18.5 miles) wide, with depths ranging from 300 to 700 meters (984 to 2,297 feet).

The coastline of the North Sea varies from rugged highlands in the north to smooth lowlands in the south. The coast of Norway is mountainous and broken, with thousands of rocks and small islands that are frequently indented by steep-sided inlets called fjords. The coasts of Scotland and northern England are high and rocky, but less broken. The coasts of middle England and the Netherlands are low and marshy with places that have been isolated from the sea by human-made dykes that regulate water levels. The coasts of southern England, France, and Belgium are low and sandy.

GEOLOGICAL HISTORY, HYDROLOGY, AND CLIMATE

The shape and size of the North Sea have varied greatly over time. At the end of the Pliocene Epoch, about 1.6 million years ago, the southern half of the North Sea was part of the mainland of Europe. At that time, the Rhine River, which now empties into the North Sea in the southern part of the Netherlands, ran to a point about 400 kilometers north of London.

The Thames River, which now empties into the North Sea east of London, continued eastward until it met the Rhine River.

During the Pleistocene epoch, from about 1.6 million to 10,000 years ago, vast ice sheets advanced and retreated several times and deposited a thick layer of clay on the bottom of the North Sea. At its greatest extent, the ice covered the entire North Sea. About 8,000 years ago, the ice retreated for the last time. A few hundred years later, the rising waters of the North Sea broke through the land bridge connecting England and France, forming the Strait of Dover. The modern coastlines of the North Sea were formed about 3,000 years ago.

The movements of the ice sheets were largely responsible for the rugged floor of the North Sea. The Dogger Bank and smaller banks in the southern part of the North Sea were created when the ice deposited large amounts of earth and stones in a particular area. Some of the trenches in the North Sea are believed to be located in areas where ancient rivers emptied into the North Sea when it was much smaller.

Water enters the North Sea from the Atlantic Ocean by way of the Strait of Dover and the Norwegian Sea. This water is relatively warm and salty. It is heated by the warm North Atlantic Current, which moves north along the western side of the British Isles and enters the Norwegian Sea. Colder, less salty water from the Baltic Sea enters by way of the Skagerrak, creating a counterclockwise current in the North Sea.

Freshwater enters the North Sea from the Thames, the Rhine, and other large rivers. The salt content of the North Sea varies from about thirty-four to about thirty-five parts per thousand. Higher concentrations are found off the coast of Great Britain, and lower concentrations are found off the coast of Norway.

The average temperature of the surface water in the North Sea in January varies from about 2°C (36°F) east of Denmark to about 8°C (46°F) between the Shetland Islands and Norway. In July, the average temperature varies from more than 15°C (59°F) along the coast from the Strait of Dover to Denmark, to about 12°C (54°F) between the Shetland Islands and the Orkney Islands.

The average air temperature varies from between 0°C and 4°C (32° and 39°F) in January to between 13°C and 18°C (55° and 64°F) in July. Winters are stormy, and gales are frequent. The average difference between low tide and high tide is between about 4 meters and 6 meters (13 feet and 20 feet) along the coast of the British Isles and the southern coast of the European mainland. Along the northern coast of the European mainland, the difference is usually less than 3 meters (10 feet).

HUMAN ACTIVITIES

Because of its long coastline and the many rivers that empty into it, the North Sea has long been an area of important human activity. The exchange of people, goods, and ideas made possible by the North Sea had a profound influence on the cultural development of northwestern Europe during the Renaissance.

In modern times, the North Sea is one of the busiest shipping areas in the world. Adding to its economic importance is the fact that it provides the only waterway between the Baltic Sea and the Atlantic Ocean. Because of the North Sea, the Netherlands and the United Kingdom are among the world's leading nations in the volume of cargo carried by sea. The Europoort complex at Rotterdam, in the Netherlands, handles more cargo than any other seaport in the world. Other major ports located on the North Sea include Antwerp, Belgium; Dunkirk, France; London, England; and the three ports of Hamburg, Bremerhaven, and Wilhelmshaven in Germany.

The accessibility of the North Sea also made it an early area of scientific research. The *Challenger* expedition, launched by the British in 1872, began a new era in oceanography. In 1902, the International Council for the Exploration of the Seas was founded in Denmark for the purpose of studying the North and Baltic Seas. It has since compiled the longest record of marine ecological conditions in the world. In more recent years, a large number of marine laboratories, research centers, and scientific vessels have been active in the area.

The North Sea has also long been an area of land reclamation and flood-control projects, particularly in the Netherlands. For centuries the Dutch have reclaimed land from the North Sea by building dikes. In the 1930s, a dike 30 kilometers (18.5 miles) long was built in the Netherlands, creating a large freshwater lake. Three-fifths of an area of land formerly under the North Sea was then reclaimed as farmland. After abnormally high tides flooded a large part of the Netherlands in 1953, the Dutch built a large flood-control system on rivers emptying into the North Sea.

The British built a similar system on a smaller scale on the Thames River in 1984.

Fishing has also been a major activity in the area for a long time. The constant mixing of shallow water in the North Sea provides a rich supply of nutrients to plankton, the microscopic organisms that support a wide variety of commercially important fish. The main species caught are cod, haddock, herring, and saithe. Lesser quantities of plaice, sole, and Norway pout are also caught. Sand eel, mackerel, and sprat are caught for the production of fish meal.

The major fishing nations in the region are Norway, Denmark, the United Kingdom, and the Netherlands. In 1983, the nations of Europe created the Common Fisheries Policy. This arrangement establishes the amount of each species that each nation may catch in the open waters of the North Sea. This policy does not apply to fish caught in coastal waters, which are considered to belong to a particular nation.

Natural Gas and Oil

New economic resources were discovered in the North Sea in the second half of the twentieth century. In 1959, the first known source of natural gas in the North Sea was identified. This source was an extension into the sea of a large natural gas field in the northeastern part of the Netherlands. By the end of the 1970s, a large number of natural gas production sites were developed in the North Sea. These sites are located primarily along a line running east and west for about 150 kilometers (93 miles) between the Netherlands and England. A smaller number of natural gas sources are located in the central and northern regions of the North Sea.

The first nation to obtain oil from the North Sea was Norway, which began operating its first offshore oil well in 1971. The United Kingdom began extracting oil from the North Sea in 1975. The oil wells are primarily located in the northern and central regions of the North Sea. By the 1980s, offshore oil wells were in operation from north of the Shetland Islands to an area about 650 kilometers (404 miles) to the south.

Oil is found in the North Sea in a basin of sedimentary rock thousands of meters thick. Almost all the world's supply of oil is found in similar basins. Such basins are believed to exist in areas where the outer layer of the Earth has been stretched and thinned. This causes a basin to be formed beneath the outer layer, which then collects sediments for millions of years. If the sediments are subjected to certain temperatures and pressures, organic matter within them is transformed into oil. Seismic measurements using controlled explosions have confirmed that the outer layer of land beneath the central portion of the North Sea is about half as thick as the same layer elsewhere. The stretching and thinning of this area took place millions of years ago, when Europe and the island of Great Britain were drifting apart.

Significance

The most critical issue facing the nations that border the North Sea is the impact of human activity in the area. Pollution from ships and from land-based operations is a major concern. Although international agreements on limiting pollution of the North Sea have been in place since 1969, enforcement remains difficult. Events such as the Ecofisk oil well blowout, which spilled more than 30 million liters of oil into the North Sea in 1977, point out the importance of controlling contamination.

An important problem is the disposal of offshore equipment that is no longer in use. As long ago as 1958, international agreements stated that all such equipment must be removed from the sea. However, because of the extremely high cost of removing large installations, new agreements were made in 1989 that allowed such equipment to be disposed of at sea. The agreement required that the water where the equipment is disposed be at least 100 meters deep and that at least 35 meters of water remain above the disposed debris. This policy met with worldwide controversy in 1995, when the Shell company decided to dispose of an unused oil well known as the Brent Spar in this way. Under intense pressure from environmentalists and organizations such as Greenpeace, Shell reversed this decision and made plans to remove the Brent Spar entirely.

Less dramatic, but possibly even more important, is the problem of oil pollution from muds used to lubricate drills in offshore oil wells. Studies have indicated that oil from this and other sources is far more widespread than previously thought. The most significant impact was on an organism known as the burrowing brittle-star. The digging of this animal brings oxygen into sediments, encouraging other organisms to grow there. The burrowing brittle-star is also an important food source for many fish. The number of burrowing brittle-stars per square meter fell from more than one hundred in unpolluted waters to zero within 1 or 2 kilometers (.5 to 1 mile) of oil wells. Such losses may have been a major factor in the decrease of the numbers of North Sea fish such as cod, whose population fell by two-thirds between 1980 and 1995.

—Rose Secrest

FURTHER READING

Flemming, Nicholas Coit, Council for British Archaeology, and English Heritage. *Submarine Prehistoric Archaeology of the North Sea: Research Priorities and Collaboration With Industry.* Council for British Archaeology, 2004.

Gaffney, Vincent, Kenneth Thomson, and Simon Fitch, eds. *Mapping Doggerland: The Mesolithic Landscapes of the Southern North Sea.* Archaeopress, 2007.

Glennie, Kw. *Petroleum Geology of the North Sea: Basic Concepts and Recent Advances.* 4th ed. John Wiley & Sons, 2009.

Harvie, Christopher. *Fool's Gold: The Story of North Sea Oil.* Hamish Hamilton, 1994.

Ilyina, Tatjana. *The Fate of Persistent Organic Pollutants in the North Sea: Multiple Years Model Simulations of g-HCH, a-HCH and PCB 153.* Springer-Verlag, 2007.

Ozsoy, Emin, and Alexander Mikaelyan, eds. *Sensitivity to Change: Black Sea, Baltic Sea, and North Sea.* Kluwer Academic Publishers, 1997.

Shennan, Ian, and Julian E. Andrews, eds. *Holocene Land-Ocean Interaction and Environmental Change Around the North Sea.* Special Publication No. 166. Geological Society of London, 2000.

Smith, Norman J. *North Sea Oil and Gas, British Industry and the Offshore Supplies Office.* Vol. 7. Edited by John Cubitt. Elsevier, 2011.

Smith, Norman J. *The Sea of Lost Opportunity: North Sea Oil and Gas, British Industry and the Offshore Supplies Office.* Elsevier, 2011.

Warren, E. A., and P. C. Smalley. *North Sea Formation Waters Atlas.* Geological Society, 1994.

Ziegler, Karen, Peter Turner, and Stephen R. Daines, eds. *Petroleum Geology of the Southern North Sea: Future Potential.* Geological Society, 1997.

O

OBSERVATIONAL DATA OF THE ATMOSPHERE AND OCEANS

FIELDS OF STUDY

Atmospheric Science; Oceanography; Environmental Sciences; Environmental Studies; Thermodynamics; Meteorology; Climate Modeling; Statistics; Process Modeling; Hydroclimatology; Hydrometeorology; Bioclimatology; Hydrology; Physical Geography; Earth System Modeling; Ecosystem Management; Ecology; Tree Ring Analysis; Spectroscopy; Mass Spectrometry

SUMMARY

Two important areas of concern for Earth scientists are the oceans and atmosphere. Technologies have improved significantly in this arena, with sensors and other equipment integrated onboard satellites, aircraft, and ships and on buoys placed in key areas. These systems provide extensive observational data on such issues as air quality, pressure, ocean currents, salinity, and temperature changes. Although areas of the oceans and atmosphere remain difficult to gauge, the continued evolution of technologies and research approaches may soon enable effective analyses of these regions.

PRINCIPAL TERMS

- **algorithm:** a set of instructions used to perform a task
- **downburst:** a convective windstorm associated with strong thunderstorm systems
- **El Niño/La Niña:** cyclical increases and decreases, respectively, in Pacific Ocean water temperature that foster shifts in weather patterns worldwide
- **polar low:** a severe, mesocyclonic winter storm that occurs in higher ocean latitudes
- **salinity:** concentration of salt in a given area of the ocean
- **tsunami:** a series of long, high sea waves caused by seismic activity or other disturbances

BASIC PRINCIPLES

Earth's oceans and atmosphere are two key areas to which scientists are increasingly paying attention as they examine climate change. For most of the twentieth century, however, scientific technology used in these arenas was limited in the amount of data it could collect. Many sensor systems, for example, could detect conditions at the ocean's surface but not in areas well beneath the surface. Similarly, meteorological sensors were for many years hampered by an inability to penetrate cloud cover when studying atmospheric phenomena. Since the latter half of the twentieth century, however, technologies have improved steadily, enabling researchers to examine a wide range of aspects related to trends in the oceans and atmosphere.

In addition to improvements to data-gathering systems found in ground observation facilities; aboard planes, ships, and buoys; and attached to weather balloons, satellite-based remote sensors have generated an even greater amount of data on much larger target areas. This technological evolution means that a greater amount of observational data can be collected, collated, and utilized to generate comprehensive models of the environment and of climate change. Such data also can help scientists to understand and even predict atmospheric and oceanic phenomena with greater precision than ever before.

COLLECTING AND ANALYZING DATA

In the pursuit of observational data, scientists utilize a wide range of technologies. Some of these systems (such as weather balloons) are sent directly into a layer of the atmosphere or are placed on a research ship. Others are placed on board satellites, enabling the observation of a much wider target range. In the oceans, for example, a growing network of surface and subsurface buoys using sophisticated sensors is generating data. One such data area is sea-level tracking. Here, buoys can be used to track tidal changes, currents, and even high-water events such as tsunamis (a series of long, high sea waves that are caused by seismic activity or other disturbances). Buoys designed by the Science Applications International Corporation are among the types of state-of-the-art buoy systems commonly used, many of which can be configured for multiple tasks, such as detecting tsunamis or monitoring temperature or chemical changes associated with climate change.

The National Oceanic and Atmospheric Administration (NOAA) in the United States utilizes an extensive network of such buoy technologies. NOAA operates nearly three dozen buoys along the Atlantic and Pacific coastlines and in the Gulf of Mexico and the Great Lakes. This network gathers observational data on sea-level changes brought on by the tides and other phenomena, and it detects changes in water temperatures, currents, and salt content.

Meanwhile, meteorologists utilize airborne and satellite-based sensors to collect observational data on atmospheric conditions and related phenomena. Using such technologies as passive and active radars, scientists can detect high concentrations of gases and particles in a given target area. For example, international scientific organizations (such as the World Meteorological Organization) are using data compiled from these technologies to study the emission and distribution of greenhouse gases from their sources.

The growing ability of research technologies to compile larger and more comprehensive observational data sets in turn fosters the need for programs and systems that can compile the data and help generate models. In some cases, models of oceanic and atmospheric trends and events are generated using complex mathematical algorithms (sets of instructions used to perform a task), assigning values to certain data sets. Algorithms also are utilized to analyze irregularities within data sets; by isolating these inconsistencies (and, potentially, errors), researchers can develop observational data sets with greater simplicity and reliability.

In addition to using mathematical models, researchers examining oceanic and atmospheric observational data can call upon computer modeling systems. This software can compile large amounts of observational data, generating frameworks on local, regional, and global scales as they occur through time. These models can be created using multiple grids, allowing the user to analyze data collected within a specific region and to compare conditions in one grid with conditions in another.

Studying Climate Change

For decades, scientists from a wide range of fields (including oceanography and meteorology) have looked critically at the effects of climate change. Remote sensors and other technologies have been used to analyze changes in atmospheric and surface temperatures, precipitation patterns, and other environmental changes associated with global warming and climate shift. These technological systems can gather large volumes of observational data and use it to generate models that analyze climate changes.

Scientists, for example, are monitoring the rate of temperature changes and salinity (the concentration of salt in a given area of the ocean) to monitor the rate of climate change occurring in the oceans. Combining the observational data acquired from sensor networks from all over the world, researchers can create a composite of the changes in temperature and salinity associated with global climate change.

Observational data on the relationship between salinity and temperature also are proving useful in predicting the length and potential effects of El Niño and La Niña (cyclical increases and decreases, respectively, in water temperature that foster shifts in weather patterns worldwide). As scientists argue that El Niño and La Niña patterns may become prolonged because of climate change, observational data on such phenomena are becoming an important component of this aspect of climatological research.

The use of observational data on different types of severe weather events also can assist researchers who are attempting to track and predict future climate shifts. For example, one study analyzed strong winter storms in the higher latitudes of the oceans. Such storms are powerful and frequently cause damage to offshore oil rigs and disrupt shipping lanes. However, these polar lows, as they are called, are relatively small in size and have seen little scientific study as distinct phenomena. Still, scientists believe that, as the climate continues to shift, polar lows may increase both in terms of size and volume. Such a trend could mean further disruptions to international trade routes and the energy industry and could mean harm to populated areas.

Meteorologists are studying observational data from polar lows to model their development and behavior. It is hoped that such data can help them predict future polar lows and assist interested parties in adapting to the likely increase in polar low occurrences.

In addition to the copious amounts of data gathered by individual studies of oceanic and atmospheric phenomena are the combined research efforts of regional and global networks and organizations. For example, the World Climate Research Programme, an international network of governmental and nongovernmental organizations, operates the Climate Variability and Predictability program. This network enables participants to collaborate, share observational data, and use this information to generate models on a broad spectrum of research areas, such as sea temperatures, atmospheric circulation, and ocean currents.

PREDICTING SEVERE PHENOMENA

In addition to providing localized data on climate change, observational data of the oceans and atmosphere help in the prediction of severe events and phenomena. In the oceans, for example, the presence of buoy nets in areas susceptible to volcanic and seismic activity (and hurricanes) has helped scientists understand the conditions that can create tsunamis and other high-water events.

NOAA, in collaboration with other organizations around the world, has formed an extensive network of observational data-collecting buoys. The National Data Buoy Center, which provides real-time data on conditions at a single buoy, provides useful data on sea-level changes and on underwater pressure (a key indicator of a tsunami). These data can help scientists track so-called killer waves and dangerous surf. In turn, emergency authorities can notify coastal residents of potential danger.

Similarly, observational data gathered from ground-based, airborne, and satellite-borne sensors are being used to more accurately track the development and movement of severe atmospheric storms.

In some cases, observational data can help scientists generate models that can cast a light on the nature of previously unpredictable atmospheric phenomena.

For example, meteorologists are learning more and more about tornadoes but have been somewhat hampered in the study of downbursts. Unlike a tornado, in which winds spin violently, a downburst is a convective windstorm, in which cool air may rush from the storm cloud downward toward the surface. Once the downburst reaches the ground it quickly spreads outward, sometimes at gusts of more than 160 kilometers (100 miles) per hour.

Predicting and gauging downburst activity has long been a challenge for researchers. In 2010, however, scientists developed a three-dimensional model of downbursts based on a tremendous amount of observational data gathered at storm sites. This computer model, which assigns numerical values to observational data, can help researchers simulate a wide range of storm events that may produce downbursts. While such research cannot prevent these dangerous storm phenomena, it can help residents prepare for such storms, potentially reducing casualties and property damage.

—*Michael P. Auerbach*

FURTHER READING

Feudale, Laura, and Jagadish Shukla. "Influence of Sea Surface Temperature on the European Heat Wave of 2003 Summer. Part I: An Observational Study." *Climate Dynamics* 36, nos. 9/10 (2011): 1691–1703. A review of observational data on ocean surface temperatures and their contributions to a dangerous heat wave in Europe in 2003.

Jury, Mark R. "An Intercomparison of Observational, Reanalysis, Satellite, and Coupled Model Data on Mean Rainfall in the Caribbean." *Journal of Hydrometeorology* 10, no. 2 (2009): 413–420. Discusses the use of a wide range of technologies to generate data on rainfall trends in the Caribbean in a twenty-one-year period. Different types of data and models were generated based on this data.

Melanotte-Rizzoli, P., ed. *Modern Approaches to Data Assimilation in Ocean Modeling*. Science, 1996. A compilation of articles on oceanographic data assimilation. Contributors focus on developing improved data assimilation techniques for ocean-oriented computer-model development.

Mozdzynski, George, ed. *Use of High Performance Computing in Meteorology.* World Scientific, 2007. Features discussions on the use of comprehensive observational data in atmospheric and oceanographic computer modeling.

Steere, Richard C., ed. *Buoy Technology: An Aspect of Observational Data Acquisition on Oceanography and Meteorology.* University of California Press, 1967. This book, based on presentations at the 1964 Buoy Technology Symposium, features an analysis of engineering developments in buoy technologies and how its evolution generates improved observational data.

Steinacker, Reinhold, Dieter Mayer. and Andrea Steiner. "Data Quality Control Based on Self-Consistency." *Monthly Weather Review* 139, no. 12 (2011): 3974–3991. Reviews methods for addressing different types of errors associated with the acquisition of observational data in meteorological studies. Introduces quality control approaches in this field.

OCEAN DYNAMICS

FIELDS OF STUDY

Oceanography; Physical Geography; Fluid Dynamics; Climate Modeling; Meteorology; Earth System Modeling; Thermodynamics; Process Modeling; Statistics; Mathematics

SUMMARY

Motions in the Earth's oceans include tides, which are caused by the gravity of the Sun and Moon; surface currents, which are driven by wind; and thermohaline circulation, which is driven by density differences in seawater. All of these motions move large amounts of water from place to place. In so doing, they also transport heat from equator to pole and from the surface to the deep ocean, and they also transport dissolved chemicals, including greenhouse gases. Although the saltiness of ocean water varies from place to place, the relative proportions of dissolved materials are extremely uniform. Ocean dynamics keep the oceans thoroughly mixed and profoundly affect Earth's temperature; hence, they are important in climate modeling. Although waves are the most obvious water motion to most people, the water in waves merely oscillates back and forth; that is, waves do not transport water long distances and thus are not discussed in this article.

PRINCIPAL TERMS

- **gyres:** large, rotating loops of ocean current found in all major oceans; they are driven westward by the trade winds near the equator and eastward by the westerlies at high latitudes
- **meridional:** referring to the motion of air or water in a generally north-south direction, that is, along meridians
- **thermohaline circulation:** a vertical circulation in the oceans that is mostly driven by water-density differences, which in turn are governed by temperature and salinity
- **trade winds:** twin wind belts on either side of the equator that generally blow westward
- **westerlies:** belts of wind in midlatitudes that generally blow from west to east

TIDES

Tides are the result of the gravitational attraction of the Moon and Sun. Although the Sun is far more massive than the Moon, its much greater distance means that its tidal effect is only about half that of the Moon. Nevertheless, if the Earth lacked a Moon, it would still have appreciable tides.

The continents prevent water from moving freely, so the actual movement of the tides is very complex. In most ocean basins the high and low tides revolve like the spokes of a wheel or the wave in a drinking glass oscillating in a circle. Tides move water through the connections between oceans and are important in keeping ocean water uniformly mixed. In small bodies of water it can take hours for the tides to progress from one end to the other, so tide predictions have to be based on local observations as well as the positions of the Sun and Moon.

Solar and lunar tides affect each other appreciably. Solar and lunar high and low tides can reinforce each other or partially cancel each other out. When the Earth, Sun, and Moon are in a straight line, at new or full moon, solar and lunar tides reinforce each other. The range between low and high tide is large, a condition called spring tide. When the Sun and Moon are 90° apart, as at first or last quarter moon, solar and lunar tides partially cancel each other out. The range between low and high tide is unusually small, a condition called neap tide.

OCEAN CURRENTS

Surface ocean currents are driven by the winds. Generally, water near the equator is pushed west by the trade winds until it strikes a continent. Most of it then is diverted poleward, where it encounters the prevailing westerlies and is pushed east. Once it reaches the eastern side of the ocean, most of the water is diverted toward the equator. Thus, in all the ocean basins there is a large loop, or gyre, rotating clockwise in the Northern Hemisphere and counterclockwise in the Southern.

Around the Antarctic is a unique geography, a belt of latitude consisting entirely of ocean. With no topography to hinder them, the westerly winds in the Southern Hemisphere, called the Roaring Forties, create some of the roughest seas in the world. They also create a globe-girdling current, the Circum-Antarctic Current, that continuously circles Antarctica and is the principal mechanism for transferring water from one ocean to another.

THERMOHALINE CIRCULATION

The densest seawater on Earth is found around the Antarctic. The water is dense because it is both cold and salty. The water is salty because freezing of sea ice leaves dissolved salt concentrated in the remaining liquid. This cold, dense water sinks to the bottom and flows northward as a dense layer called Antarctic bottom water. The largest amount of Antarctic bottom water flows beneath the Pacific until it reaches Alaska, where it rises and merges into the surface circulation. It then flows around the North Pacific gyre until it reaches the southwest Pacific, where it has warmed to over 30°C. Some warm water circulates through Indonesia to the Indian Ocean, and even though the amount of water involved is fairly small, the amount of heat transferred is large. Warm Indian Ocean water rounds Africa, cooling somewhat, then warms again in the South Atlantic. Some warm water crosses into the North Atlantic, travels up the eastern coast of North America as the Gulf Stream, then crosses

Scientists in the Argo program deploy a float to monitor ocean temperature and salinity. More than three thousand such floats are deployed worldwide.

to Europe. Finally, in the Arctic, the water cools and sinks. It then begins traveling south as North Atlantic deep water. Antarctic bottom water is also creeping northward in the Atlantic, colder and denser than North Atlantic deep water. Thus, in the North Atlantic, Gulf Stream water is moving northward on the surface, Antarctic bottom water is moving north along the bottom, and North Atlantic deep water is moving south just above the Antarctic bottom water. Because much of the water movement is northward or southward, thermohaline circulation (THC) is sometimes called meridional overturning circulation (MOC), although this is not entirely correct as the MOC is localized in the Atlantic Ocean but the THC is global in extent. The MOC is just part of the THC.

CONTEXT

Large-scale movements of ocean water affect local and global climate by transporting heat. They also affect climate change by transporting dissolved greenhouse gases into the deep ocean for storage and back to the surface for release. One aspect of thermohaline circulation has received particular attention. Many scientists are convinced that a sudden release of freshwater from glacial lakes in North America covered the North Atlantic with a layer of freshwater that prevented the exchange of heat between the ocean and the atmosphere and caused a sharp cooling event, called the Younger Dryas, about twelve thousand years ago. Some have suggested that melting of the Greenland ice cap might have a similar effect, so that warming of the climate might, paradoxically, produce a cooling episode.

—*Steven I. Dutch*

FURTHER READING

Broecker, Wally *The Great Ocean Conveyor. Discovering the Trigger for Abrupt Climate Change.* Princeton University Press, 2010. Describes the global oceanic overturning circulation system and how its disruption by climate change may affect the climate of Earth.

Denny, Mark *How the Ocean Works. An Introduction to Oceanography.* Princeton University Press, 2012. Summary of oceanography for introductory college students.

Garrison, Tom S. *Essentials of Oceanography.* 8th ed. Brooks/Cole, 2013. Summary of oceanography for introductory college students.

Pinet, Paul R. *Invitation to Oceanography.* Jones & Bartlett, 2009. Discusses geology, biology, chemistry, and physics of the oceans at an introductory college level. Includes a chapter on climate change.

Trujillo, Alan P., and Harold V. Thurman. *Essentials of Oceanography.* 11th ed. Prentice, 2014.

Stewart, Robert H. *Introduction to Physical Oceanography.* A & M University: Robert H. Stewart, 2008. This treatment of oceanography offers extensive discussion of the physics of water movement in the oceans. Intended for upper-level college students.

OCEAN LIFE AND RISING WATER TEMPERATURES

FIELDS OF STUDY

Oceanography; Environmental Sciences; Environmental Studies; Waste Management; Climate Modeling; Climatology; Bioclimatology; Atmospheric Science; Physical Geography; Earth System Modeling; Ecology; Toxicology

SUMMARY

Oceans cover 71 percent of Earth's surface and account for a little less than half of the photosynthetic conversion of carbon dioxide (CO_2) into the organic compounds that make up the bodies of living organisms. Earth's oceans have long been thought of as being impervious to anthropogenic degradation. Nonetheless, human activity has had an adverse effect on marine life, from phytoplankton to top marine predators such as sharks. Effects attributable to elevated land temperatures include displacement of currents and upwelling zones and increased runoff in major river systems. Effects attributable to elevated sea surface temperatures include melting sea ice in the polar regions.

PRINCIPAL TERMS

- **dead zones:** areas of deepwater oxygen depletion due to surface algal blooms or disruption of thermohaline circulation
- **El Niño–Southern Oscillation (ENSO):** periodic fluctuation of temperatures and currents in the Pacific Ocean on a four-, ten-, and ninety-year cycle
- **primary production:** production of fixed carbon through photosynthesis
- **thermohaline circulation:** the rising and sinking of water caused by differences in water density due to differences in temperature and salinity

HUMAN IMPACT ON OCEAN LIFE

Until quite recently, scientists and the general public considered Earth's oceans to be impervious to anthropogenic degradation. Oceans cover 71 percent of Earth's surface and account for a little less than half of its primary production, that is, the photosynthetic conversion of carbon dioxide (CO_2) into the organic compounds that make up the bodies of living organisms.

The ocean is far from being a uniform habitat; however, with the exception of some near-shore environments, ecological niches cover wide areas and intergrade, meaning that species can readily adapt by shifting their ranges. In consequence, environmental pressures producing elevated extinction rates on land have a less dramatic effect in the open ocean.

Nonetheless, human activity has had an adverse effect on marine life, from phytoplankton to top marine predators such as sharks. While overfishing,

Coral reefs. The picture was taken in Papua New Guinea.

pollution, and damming of rivers that serve as spawning grounds for marine fish have all taken their toll, these are only indirectly related to global warming.

Present effects attributable to elevated land temperatures include displacement of currents and upwelling zones and increased runoff in major river systems. Effects attributable to elevated sea surface temperatures include melting sea ice in the Arctic and Antarctic, reducing habitat for polar bears, penguins, and the many humbler species of plants and animals that thrive at the margins of the polar ice caps. In the tropics, higher sea surface temperatures alter coral metabolism, causing corals to bleach when they lose symbiotic algae (zooxanthellae) critical to their growth. Degradation of coral reefs profoundly affects the many organisms restricted to this habitat.

As atmospheric CO_2 has continued to increase, altered seawater chemistry has become a growing concern. An estimated one-quarter of the CO_2 generated by burning fossil fuels does not remain in the atmosphere but rather is dissolved in the oceans, increasing seawater acidity. According to the U.S. National Oceanic and Atmospheric Administration (NOAA) Pacific Marine Environmental Laboratory (PMEL), since the start of the Industrial Revolution, ocean acidity has increased by approximately 30 percent. One of the effects of ocean acidification is the stunting of shell production. Shells and similar structures found in marine organisms are formed from calcium carbonate, a process known as calcification; as the acidity of the ocean increases, the calcium

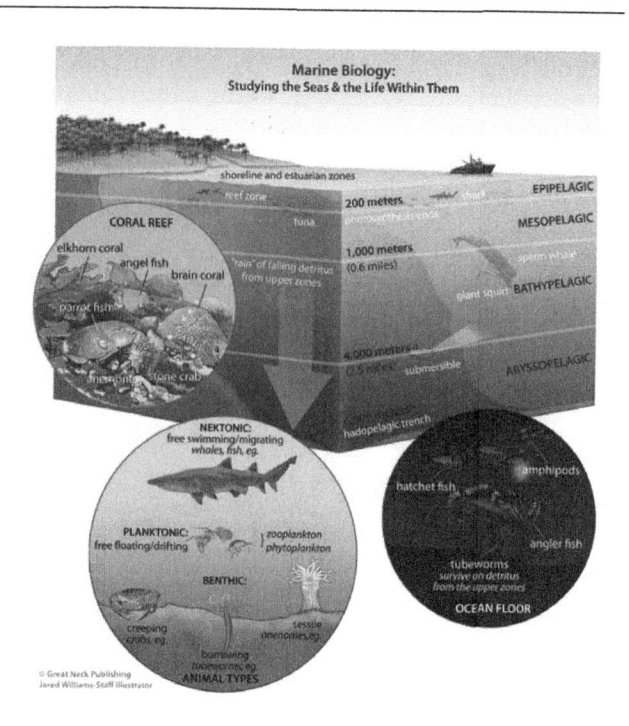

Marine Biology: Marine biology is the study of the interactions of living things in and near the ocean ecosystem.

carbonate in these structures becomes more vulnerable to dissolution. Coelenterates such as corals, whose skeletons are made up of aragonite (one form of calcium carbonate), are more susceptible than are mollusks, which have shells of calcite (another form of calcium carbonate), but both remain vulnerable. The threat posed to marine calcifying organisms by ocean acidification was largely theoretical until 2012, when an article published in the journal *Nature Geoscience* reported the "extensive dissolution" of the shells of pteropods found in the Antarctic Ocean.

CLIMATE CHANGE AND MARINE LIFE IN THE GEOLOGIC RECORD

Scientists recognize at least five major global mass extinction events, of which the Permian extinction, which took place 251 million years ago at the Permian-Triassic boundary, was the most devastating. At that time, approximately 95 percent of marine species and 70 percent of land species became extinct in three distinct pulses over a period of about eighty thousand years.

While several theories have been proposed to account for the Permian extinction, among the best supported of these theories is one that attributes it to drastic climate change triggered by massive volcanic eruptions in Siberia. According to this theory, each eruption caused brief cooling episodes due to volcanic dust and sulfuric acid aerosols in the atmosphere; once the dust and aerosols dispersed, this cooling was followed by a period of warming caused by the huge amounts of carbon dioxide, sulfur dioxide, and other greenhouse gases (GHGs) that were released along with the lava. Over the course of a million years, repeated eruptions, dwarfing anything humans have experienced in their brief tenure on Earth, eventually overwhelmed the planet's capacity to self-correct.

Another theory postulates that the ocean depths became increasingly oxygen depleted, favoring the growth of bacteria that produce hydrogen sulfide. High pressures and cold temperatures in the abyss allowed the hydrogen sulfide to build up, only to be released in a gigantic "burp" of highly toxic fumes. Still another theory points to the storage of large quantities of methane in the form of clathrates in deep-sea sediments, suggesting that this methane was abruptly released when warming raised the temperature of the deeper regions of the ocean by 5 degrees Celsius. In addition to being toxic and a powerful GHG, methane is explosive at concentrations as low as 5 percent. Both of these events—deoxygenation of the oceans and the release of trapped methane—could have similarly been triggered by the Siberian eruptions and the significant global warming that resulted.

Whatever the cause, the Permian extinction, which devastated every group of plants and animals, was extremely abrupt by geological standards. Sedimentary rocks dating from after the cataclysm are nearly bare of fossils for the first ten million years of the Triassic period.

The Present and Near Future

Unless the most carefully researched models are far off the mark, nothing resembling the devastating geochemical upheavals of the Permian-Triassic period looms in the foreseeable future, even if present levels of fossil fuel consumption persist. These models presuppose that volcanic activity will continue at levels typical of the Holocene and that no asteroids are headed in Earth's direction.

Possible changes due to increasing ocean acidity are being closely monitored, with the discovery of shell dissolution in Antarctic pteropods presenting new cause for concern. Acidity alone is not expected to reach lethal levels in the near future, but temperatures are rising rapidly, and marine organisms in the Antarctic cannot adjust their ranges southward. A wide variety of fish and birds depend on this snail for continued survival.

Polar Regions

In both the Arctic and the Antarctic, chilled surface seawater sinks, allowing nutrient-rich waters to well up from below and support high phytoplankton productivity. The polar seas teem with life. The lower surfaces of ice sheets also support dense growth of attached algae. Global warming near the North Pole causes the most productive zone to retreat northward and contract in extent. This restricts the number of both herbivores and carnivores the system can support. Most polar animals are unable to extend their ranges into temperate seas, because their unique adaptations to frigid temperatures make them poor competitors and susceptible to disease in warmer climates. The situation in the Southern Hemisphere is even more acute, as species migrating southward encounter the continental margin.

The plight of polar bears has received considerable attention. These huge carnivores prey almost entirely on seals that they hunt on sea ice. As the seals are declining in numbers and retreating farther from shore in response to the shrinking of the ice cap, bears are starving and failing to reproduce. Whale populations that had begun to recover from overexploitation by the whaling industry are also declining again as a result of low food supplies. Antarctic penguins also face declining food supplies and an influx of predators, including sharks, which are extending their ranges southward.

Coral Reefs

Reef-building corals, and the numerous species that depend on them, have a narrow temperature range for optimum growth. They are also vulnerable to changes in sea level due to either global warming or global cooling. During the last Pleistocene glaciation, the resulting drop in sea level exposed much of Australia's Great Barrier Reef, restricting this unique ecosystem to isolated pockets. A rapid rise in sea level would damage existing reefs by reducing light levels below those needed by symbiotic algae.

367

A 2°C rise in surface temperature is sufficient to cause bleaching in corals as the individual polyps eject symbiotic algae. Bleaching initially causes growth to cease and eventually kills the coral colony. In recent years, there have been massive die-offs of corals. Notably, a severe bleaching event in 2016 killed an average of 67 percent of shallow-water corals in the northern part of the Great Barrier Reef over an eight- to nine-month period. A previous die-off of corals in the South Pacific was associated with a severe El Niño–Southern Oscillation (ENSO) event to which global warming may have contributed. Near-shore pollution also devastates coral reefs in populated areas.

Dead Zones

In a number of parts of the world, extensive areas of ocean have become depleted in oxygen, turning once-productive fisheries into wastelands. Most of these dead zones are associated with rivers that drain populated areas; one of the largest lies offshore of the mouth of the Mississippi River. This dead zone owes its existence to influxes of nutrient-laden freshwater to the Gulf of Mexico. These nutrients stimulate massive algal blooms. There is an indirect connection to global warming, in that warming generally causes increased precipitation and therefore increased runoff. Dead zones off the west coasts of the United States and South Africa result from disruption of cold currents and associated upwelling zones and thus are believed to be directly related to global warming.

Productivity

While global warming due to elevated CO_2 levels can cause local drops in productivity due to drought on land and disruption of thermohaline cycles in the ocean, the long-term predicted effect of such warming on a global scale is a net increase in photosynthesis, with an upper limit that far exceeds any projections based on realistic economic indicators. In the long term, if Earth begins producing more food, both the numbers and the diversity of herbivores and predators can be expected to increase.

In the short term, however, such changes lead to the proliferation of weedy species with high reproductive rates and broad ecological ranges, loss of diversity, and generally unstable conditions. Species with specialized ecological requirements become extinct, and natural ecosystems increasingly resemble intentional agriculture or aquaculture. A glimpse of the future may be gleaned from the formerly rich fisheries off the West Coast of North America. These have been in decline for several decades, mainly because of pollution and overfishing. Warmer waters coupled with a persistent dead zone off the coast of California and Oregon have further reduced stocks of commercial and sport fishes, but they have favored proliferation of the Humboldt giant squid, an aggressive predator adapted to warm temperatures and low oxygen levels.

Rising temperatures can be expected to reduce areas of high planktonic productivity near the poles while expanding them near the equator and at continental margins, threatening polar species with starvation and extinction while increasing in numbers in warmer climates without a corresponding increase in diversity.

There may be reef-building organisms ready to replace corals should the seas become inhospitable. During the very warm late Cretaceous, rudists, a group of bivalve mollusks related to clams, were the main reef builders. Several types of algae also have limestone skeletons. If any of these groups were to replace corals, the structural integrity of reefs would be preserved, but the beauty and diversity of the ecosystem would be sadly compromised on any conceivable human time scale.

Context

The main threats to the abundance and diversity of marine life derive from Earth's human population explosion and its concomitant overexploitation and pollution of coastal waters. The exploding population is also a major factor in global warming. In terms of direct threats posed to marine life from rising temperatures, dwindling sea ice in the Arctic and especially the Antarctic is probably the most clear-cut. While anthropogenic climate change is undoubtedly a factor in coral-reef destruction and the decline of fisheries, it is likely not the only one. As scientists learn more about long-term cycles involving ENSO and analogous oscillating pressure and current systems in other oceans, a better understanding of the relationship of current extreme events to long-term trends should emerge.

—*Martha A. Sherwood*

Further Reading

Bednaršek, Nina, et al. "Extensive Dissolution of Live Pteropods in the Southern Ocean." *Nature Geoscience*, vol. 5, no. 12, 2012, pp. 881–85.

Benton, Michael J., and Richard J. Twitchett. "How to Kill (Almost) All Life: The End-Permian Extinction Event." *Trends in Ecology and Evolution*, vol. 18, no. 7, 2003, pp. 358–65.

Fleming, Nic. "When a Volcanic Apocalypse Nearly Killed Life on Earth." *BBC Earth*, BBC, 18 Dec. 2015, www.bbc.com/earth/story/20151218-when-a-volcanic-apocalypse-nearly-killed-life-on-earth. Accessed 21 Feb. 2017.

"Life and Death after Great Barrier Reef Bleaching." *ARC Centre of Excellence for Coral Reef Studies*, 29 Nov. 2016, www.coralcoe.org.au/media-releases/life-and-death-after-great-barrier-reef-bleaching. Accessed 22 Feb. 2017.

"What Is Ocean Acidification?" *Pacific Marine Environmental Library*, National Oceanic and Atmospheric Administration, www.pmel.noaa.gov/co2/story/What+is+Ocean+Acidification percent 3F. Accessed 21 Feb. 2017.

OCEAN STRUCTURE AND ATMOSPHERE COUPLING

FIELDS OF STUDY

Oceanography; Heat Transfer; Fluid Dynamics; Atmosphere Science; Thermodynamics; Environmental Sciences; Environmental Studies; Hydroclimatology; Hydrometeorology; Hydrology; Physical Geography; Earth System Modeling

SUMMARY

The ocean has a complex structure, both at its surface and in the vertical dimension descending to the ocean floor. This internal structure results in layering with respect to temperature, salinity, density, and the way in which the ocean responds to the passage of light and sound waves. Interaction between the oceans and the atmosphere is affected by the entire structure of the oceans and is perhaps the single most important driver of weather and climate on Earth.

PRINCIPAL TERMS

- **convective overturn:** the renewal of the bottom waters caused by the sinking of surface waters that have become denser, usually because of changes in temperature or salinity
- **doldrums:** the equatorial zone where winds are calm and variable and there is heavy thunderstorm rainfall
- **halocline:** a zone within a body of water, characterized by a rapid rate of change in salinity
- **horse latitudes:** the belts of latitude approximately 30 degrees north and 30 degrees south of the equator, where the air pressure is high, the winds are very light and the weather is hot and dry
- **pycnocline:** a zone within a body of water, characterized by a rapid rate of change in density
- **salinity:** the quantity of dissolved salts in seawater, usually expressed as parts per thousand
- **saltwater wedge:** a wedge-shaped intrusion of seawater from the ocean into the bottom of a river; the thin end of the wedge points upstream
- **thermocline:** a zone within a body of water, characterized by a rapid change in temperature

TEMPERATURE LAYERING

One highly significant aspect of ocean structure is the layering of water based on temperature and salinity differences. In order to understand the reasons for the thermal layering of the ocean, one must bear in mind that the primary source of heating for the ocean is sunlight. About 60 percent of this entering radiation is absorbed within the first meter of seawater, and about 80 percent is absorbed within the first 10 meters. As a result, the warmest waters in the ocean are found at its surface.

That does not mean, however, that surface temperatures in the ocean are the same everywhere. Because more heat is received at the equator than at the poles, ocean surface temperatures are closely related to latitude. As a result, they are distributed in bands of equal temperature extending east and west, parallel to the equator. Temperatures are highest along the equator because of the near-vertical angles

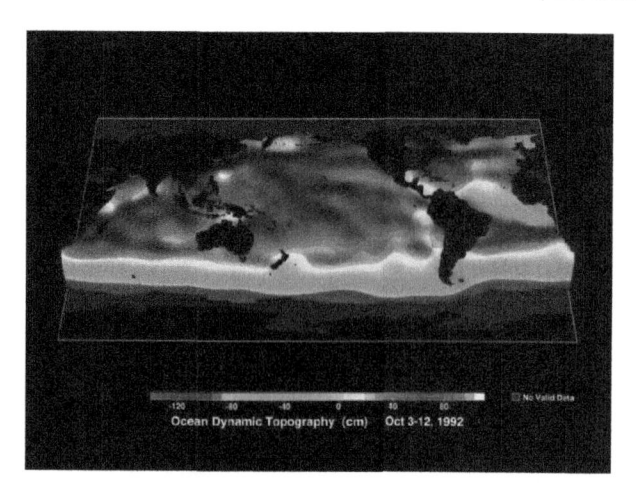

TOPEX/Poseidon was the first space mission that allowed scientists to map ocean topography with sufficient accuracy to study the large-scale current systems of the world's oceans.

at which the sun's rays are received here. As latitude increases toward the poles, ocean temperatures gradually cool as a result of the decreasing angle of incidence of the incoming solar radiation.

Measurements of ocean surface temperature range from a high of 33°C in the Persian Gulf, a partly landlocked, shallow sea in a desert climate, to a low of -2°C in close proximity to ice in polar regions. There, the presence of salt in the water lowers the water's freezing point below the normal 0°C level. Because salinity of this cold water is so high, it sinks to the ocean floor and travels along it for substantial distances. Ocean surface temperatures may also vary with time of year, with warmer waters moving northward into the Northern Hemisphere in the summertime and southward into the Southern Hemisphere in the wintertime. These differences are most noticeable in midlatitude waters. In equatorial regions, water and air temperatures change little seasonally, and in polar regions, water tends to be cold all year long because of the presence of ice.

Vertically downward from the equator, toward the ocean floor, water temperatures become colder. This results from the facts that solar heating affects the surface waters only and that cold water is denser than warm water. When waters at the surface of the ocean in the polar regions are chilled by extremely low winter temperatures, they become denser than the underlying waters and sink to the bottom. They then move slowly toward the equator along the sea floor, lowering the temperature of the entire ocean.

As a result, deep ocean waters have much lower temperatures than might be expected by examination of the surface waters alone. Although the average ocean surface temperature is 17.5°C, the average temperature of the entire ocean is a frigid 3.5°C.

Oceanographers recognize the following layers within the ocean, based on its temperature stratification: First, there is an upper, wind-mixed layer, consisting of warm surface water up to 500 meters thick. This layer may not be present in polar regions. Next is an intermediate layer, below the surface layer, where the temperature decreases rapidly with depth; this transitional layer can be 500 to 1,000 meters thick and is known as the main thermocline. Finally, there is a cold, deep layer extending to the ocean floor. In polar regions, this layer may reach the surface, and its water is relatively homogeneous, with temperature slowly decreasing with depth.

Because the upper surface layer is influenced by atmospheric conditions, such as weather and climate, it may contain weak thermoclines as a result of the daily cycle of heating and cooling or because of seasonal variations. These are temporary, however, and may be destroyed by severe storm activity. Nevertheless, the majority of ocean water lies below the main thermocline and is uniformly cold, the only exception being hot springs on the ocean floor that introduce water at temperatures of 300°C or higher. Plumes of warmer water emanating from these hot springs have been detected within the ocean.

SALINITY

A second phenomenon responsible for layering within the ocean is variation in the water's salinity. For the ocean as a whole, the salinity is 35 parts per thousand. Considerable variation in the salinity of the surface waters from place to place results from processes that either add or subtract salt or water. For example, salinities of 40 parts per thousand or higher are found in nearly landlocked seas located in desert climates, such as the Red Sea or the Persian Gulf, because high rates of evaporation remove the water but leave the salt behind. High salinity values are also found at the surface of the open ocean at the same latitudes where there are deserts on land (the so-called horse latitudes). There, salinities of 36 to 37 parts per thousand are common.

At the equator, however, much lower salinity values are encountered, despite the high temperatures and nearly vertical rays of the sun. The reason is that the equatorial zone lies in the so-called doldrums, a region of heavy rainfall. The ocean's surface waters are therefore diluted, which keeps the salinity relatively low. Similarly low salinities are also found in coastal areas, where rivers bring in large quantities of freshwater, and in higher latitudes, where rainfall is abundant because of numerous storms.

Despite the variation in salinity in the ocean's surface water, the deep waters are well mixed, with nearly uniform salinities ranging from 34.6 to 34.9 parts per thousand. Consequently, in some parts of the ocean, surface layers of low-salinity water overlie the uniformly saline deep waters and, in other parts of the ocean, surface layers of high-salinity water overlie the uniformly saline deep layer. Between these layers are zones of rapidly changing salinity known as haloclines. An important exception to this picture is a few deep pools of dense brine such as are found at the bottom of the Red Sea, where salinities of 270 parts per thousand have been recorded.

Haloclines are very common in coastal areas. Off the mouth of the Amazon River, for example, a plume of low-salinity river water extends out to sea as far as 320 kilometers, separated from the normally saline water below by a prominent halocline. In many tidal rivers and estuaries, a layer of heavier seawater will extend many kilometers inland beneath the freshwater discharge as a conspicuous saltwater wedge.

Density Stratification

A prominent density stratification within the ocean results from the variation in ocean temperatures and salinity just described. As noted, two factors make water heavier: increased salinity, which adds more dissolved mineral matter, and decreased temperature, which results in the water molecules being more closely packed together. Therefore, the least-dense surface waters are found in the equatorial and tropical regions, where ocean temperatures are at their highest. Toward the higher latitudes, the density of surface ocean waters increases because of the falling temperatures. In these areas, low-density surface water is found only where large quantities of freshwater are introduced, by river runoff, by high amounts of precipitation, or by melting ice.

Vertical density changes are even more pronounced. As water temperatures decrease with depth, water densities increase accordingly. This increase in density, however, is not uniform throughout the ocean. At the poles, the surface waters are almost as cold as the coldest bottom waters, so there is only a slight increase in density as the ocean floor is approached. By contrast, the warm surface waters in the equatorial and tropical regions are underlain by markedly colder water. As a result, a warm upper layer of low-density water is underlain by an intermediate layer in which the density increases rapidly with depth. (This middle layer is known as the pycnocline.) Below it is a deep zone of nearly uniform high-density water.

Convective overturning takes place when this normal density stratification is upset. In a stable density-stratified system, the less dense surface water floats on top of the heavier, deeper water. Occasionally, however, unstable conditions will arise in which heavier water forms above lighter water. Then, convective overturning takes place as the mass of heavier water sinks to its appropriate place in the density-stratified water column. This overturning may occur gradually or quite abruptly. In lakes and ponds, it occurs annually in regions where winter temperatures are cold enough. In the ocean, convective overturning is primarily associated with the polar regions, where extremely low winter temperatures result in the sinking of vast quantities of cold water. In addition, convective overturning has been observed in the Mediterranean during the wintertime, when chilled surface waters sink to replenish deeper water.

Light Penetration and Sound Waves

Oceanographers also recognize stratification in the ocean based on the depths to which light penetrates, and they divide the ocean into two zones. The upper zone, which is known as the photic zone, consists of the near-surface waters that have sufficient sunlight for photosynthetic growth. Below this zone is the aphotic zone, where there is insufficient light for photosynthetic growth. The lower limit of the photic zone is generally taken as the depth at which only 1 percent of the surface intensity of sunlight still penetrates. In the extremely clear waters of the open ocean, this depth may be 200 meters or more.

Stratification in the ocean based on the behavior of sound waves has also been observed. Because sound waves travel nearly five times faster under water than in the air, their transmission in the ocean has been extensively studied, beginning with the development of sonar, the echo sounder. So-called scattering layers have been recognized as regions that reflect sound, usually because of the presence of living organisms that migrate vertically, as layers within the water column, depending on light intensity. The sofar (sound fixing and ranging) channels are density layers within the ocean where sound waves can become trapped and can travel for thousands of kilometers with extremely small energy losses. These channels have the potential to be used for long-distance communications. Shadow zones are also caused by density layers within the ocean. These layers trap the sound waves and prevent them from reaching the surface. One advantage of shadow zones is that submarines can travel in them undetected.

OCEAN-ATMOSPHERE COUPLING

Ocean-atmosphere coupling describes the interdependency between the temperatures and circulation of water in the ocean and those of air in the atmosphere. Changes in the surface temperature of ocean water produce changes in the atmosphere above the water, which alters wind patterns and leads to further changes in surface ocean temperature. If these changes are significantly large or long-lived, changes in atmospheric patterns capable of producing changes in global weather patterns can result. The most prominent example of this is the El Niño/La Niña weather cycle, which gained notoriety when it was recognized as the cause of numerous climate disruptions in North America during the late twentieth and early twenty-first centuries.

The specific heat of water, the amount of energy required to alter water's temperature, is extremely high. As a result, Earth's water systems are a significant stabilizing influence on global surface temperatures. The ocean-atmosphere coupling is among the most significant points of distribution of water in the global hydrologic cycle. Water evaporates from the oceans into the atmosphere, which carries water vapor over land, where it precipitates, providing freshwater to terrestrial ecosystems. This cycle, and particularly the interface between air and sea, both directly affect and are affected by global climate patterns, particularly those involving temperature.

STUDY OF THE OCEAN

The measurement of water temperatures at the ocean's surface is quite simple. A thermometer placed in a bucket of water that has been scooped out of the ocean at the bow of a boat will suffice, provided the necessary precautions have been taken to prevent temperature changes caused by conduction and evaporation. For oceanwide or global studies, satellites provide near-simultaneous readings of ocean surface temperatures within an accuracy of 1°C. These satellites utilize infrared and other sensors, which are capable of measuring the amount of heat radiation emitted by the ocean's surface to within 0.2°C.

The significance of ocean-atmosphere coupling to global warming increased as cyclical meteorological phenomena, such as the El Niño/La Niña cycle in the southern Pacific and the North Atlantic Oscillation patterns began accelerating in the early 1980s. The study of ocean-atmosphere coupling thus intensified during the late twentieth and early twenty-first century, as scientists sought to determine the causes and extent of global warming, as well as its future duration and potential for escalation. Many of these studies sought to determine whether global warming patterns have human causes, while others attempted to forecast changes in global weather patterns to determine the potential for future extreme weather events such as severe droughts, flooding, and drastic changes in temperature. Long-term studies of patterns of climate variability involving ocean-atmosphere coupling have failed to yield definitive answers to these inquiries, leading many scientists to conclude that these patterns must be examined for longer periods in order to determine their implications for global warming and its causes.

Measuring the temperature of the deeper subsurface waters posed a problem, however, because a standard thermometer lowered over the side of a ship will "forget" a deep reading on its way back to the surface. As a result, the so-called reversing thermometer was developed in 1874. This thermometer has an S-bend in its glass tube. When the thermometer is inverted at the desired depth, the mercury column breaks at the S-bend, thus recording the temperature at that depth. Today, electronic instruments record subsurface water temperatures continuously. These devices can be either dropped from a plane or ship or moored to the ocean floor.

The measurement of the salinity of seawater is not as easy as one might think. An obvious way to determine water salinity would be to determine the amount of dried salts remaining after a weighed sample of seawater has been evaporated, but in actual practice, that is a messy and time-consuming procedure, hardly suitable for use on a rolling ship. A variety of other techniques have been used over the years, based on such water characteristics as buoyancy, density, or chloride content. By far the most popular relies on seawater's electrical conductivity. In this method, an electrical current is passed through the seawater sample; the higher the salt content of the water, the lower the electrical resistance of the solution and the faster this current is observed to pass. Using this method, oceanographers have been able to determine the salinity of seawater samples to the nearest 0.003 part per thousand. This is an important advantage, in view of the fact that the salinity differences between deep seawater masses are very minute.

Measurement of the densities of surface water samples can easily be accomplished by determining the water sample's buoyancy or weight, but the real difficulty comes with attempts to determine a subsurface water sample's density. If this water sample is brought to the surface, its temperature, and therefore its density, will change. Although sophisticated techniques are available for the determination of density at depth, in actual practice the density is not measured at all; instead, it is computed from the sample's known temperature, salinity, and depth. The density of the water is almost wholly dependent on these three factors.

Various methods are available for measuring the depth of light penetration in seawater. A crude estimate can be made using the Secchi disk, which was first introduced in 1865. This circular, white disk is slowly lowered into the water, and the depth at which the disk disappears from sight is noted visually. More sophisticated measurements can be made using photoelectric meters. Another good indicator of the maximum depth of light penetration in the sea is the lowest level at which photosynthetic growth can take place. For sound studies within the ocean, various methods are used to create the initial sound, including the use of explosives in seismic profiling. The returning echo is detected by means of a receiver known as a hydrophone.

—Donald W. Lovejoy
—Michael H. Burchett

Further Reading

Broecker, Wally. *The Great Ocean Conveyor*. Princeton, N.J.: Princeton University Press, 2010. Discusses ocean currents, focusing specifically on the great conveyor belt. Written by the great ocean conveyor's researcher. Explains the conception of this theory and the resulting impact on oceanography. Written in a manner easy to follow with some background in science, yet still relevant to graduate students and scientists.

Charlier, R. H., and C. W. Finkl. *Ocean Energy: Tide and Tidal Power*. Springer-Verlag, 2009. Discusses fundamentals of oceanic energy harvesting. Describes historical and current technologies, with examples drawn from around the world. Also discusses social, economic, and environmental impacts. Contains dense, verbose, and technical writing and is therefore best suited for graduate students and researching oceanographers or engineers with some prior knowledge of ocean dynamics.

Emelyanov, Emelyan M. *The Barrier Zones in the Ocean*. Berlin: Springer-Verlag, 2005. Suitable as a textbook for advanced oceanography and ocean geochemistry courses, and for research reference. Discusses the properties of forty different ocean barrier zones with regard to salinity, hydrodynamics, temperature, and light, as well as processes that affect sedimentation in the open ocean and in bodies of water such as the Baltic and Mediterranean Seas.

Schwartz, M. *Encyclopedia of Coastal Science*. Springer, 2005. Contains many articles specific to ocean and beach dynamics. Discusses coastal habitat management topics, hydrology, geology, and topography. Articles may run multiple pages and have diagrams. Each article has bibliographical information and cross referencing.

Segar, Douglas A., and Elaine Stamman, Segar. *Introduction to Ocean Sciences*. Douglas A. Segar, 2012. Comprehensive coverage of all aspects of the oceans, their chemical makeup, and circulation. Readable and well illustrated. Suitable for high school students and above.

Seibold, Eugen, and Wolfgang H. Berger. *The Sea Floor: An Introduction to Marine Geology*. 3rd ed. Springer-Verlag, 2010. Offers an introduction to many topics in marine geology that covers geological structures from the continental shelf to deep-ocean trenches. Discusses processes such as seafloor spreading, the sediment cycle, currents, and pelagic rain.

Soloviev, Alexander, and Roger Lukas. *The Near-Surface Layer of the Ocean: Structure, Dynamics and Applications.* Springer, 2006. Uses the results of major air-sea interaction experiments to present the physics and thermodynamics of this oceanic system, providing a detailed treatment of the surface microlayer, upper-ocean turbulence, thermohaline and coherent structures, and the high-speed wind regime.

Talley, Lynne D., et al. *Descriptive Physical Oceanography: An Introduction.* Academic Press, 2012. Designed to introduce oceanography majors to the field of physical oceanography. Has useful sections on the temperature, salinity, density, light, and sound structure of the ocean. Includes a discussion of the various instruments and methods used for measuring these properties.

OZONE DEPLETION AND OZONE HOLES

FIELDS OF STUDY

Atmospheric Science; Atmospheric Chemistry; Analytical Chemistry; Environmental Chemistry; Photochemistry; Thermodynamics; Chemical Engineering; Petroleum Refining; Waste Management; Meteorology; Climate Modeling; Statistics; Process Modeling; Fluid Dynamics; Chemical Kinetics; Earth System Modeling; Toxicology; Spectroscopy

SUMMARY

One of the naturally-occurring gases in the atmosphere, ozone absorbs ultraviolet radiation from the sun. Chlorofluorocarbons, once used as primary refrigerants, aerosol propellants, and in the making of certain plastics, react in the atmosphere and release chlorine atoms, which then react with and destroy molecules of ozone. Depletion of stratospheric ozone allows incoming ultraviolet radiation to reach the surface, resulting in severe damage to all living organisms if ozone depletion continues. The term "ozone hole" refers to the seasonal decrease in stratospheric ozone concentration occurring over Antarctica. Ozone-hole formation directly related to the use of CFCs is evidence that human activities can significantly alter the composition of the atmosphere.

This image provided by NASA was compiled by the Ozone Monitoring Instrument on NASA's Aura satellite from September 21-30, 2006. The average area of the ozone hole was the largest ever observed, at 10.6 million square miles according to government scientists.

PRINCIPAL TERMS

- **catalyst:** a substance that increases the rate of a chemical reaction without itself being altered in the process
- **chlorofluorocarbon (CFC):** a group of chemical compounds containing carbon, fluorine, and chlorine, used in air conditioners, refrigerators, fire extinguishers, spray cans, and other applications
- **Dobson spectrophotometer:** a ground-based instrument for measuring the total column abundance of ozone at a particular geographic location
- **food chain:** the arrangement of the organisms of an ecological community according to the order

Sept. 22, 2012

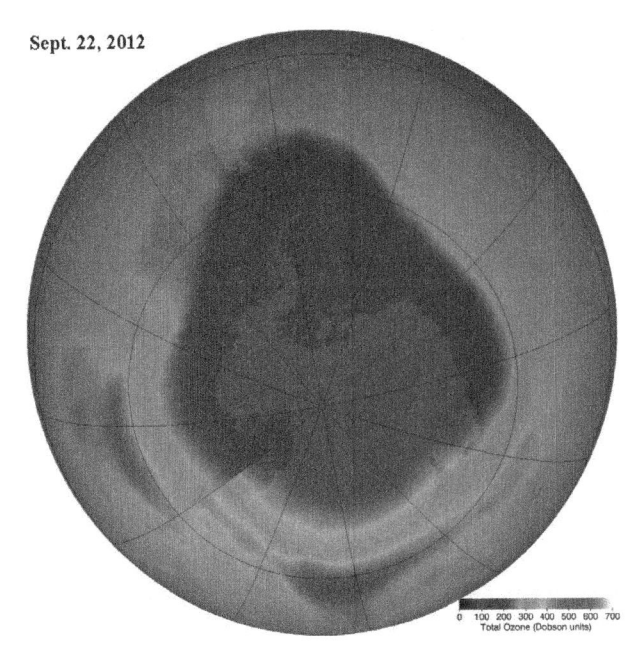

Antarctic Ozone Hole as of September 22, 2012.

of predation in which each consumes the next, usually lower, member as a food source
- **ozone layer:** a region in the lower stratosphere, centered about 25 kilometers above the surface of Earth, which contains the highest concentration of ozone found in the atmosphere
- **ozone:** the molecular form of oxygen containing three atoms of oxygen per molecule as O_3, as compared to elemental oxygen having the molecular formula O_2
- **phytoplankton:** free-floating microscopic aquatic plants that use sunlight to convert carbon dioxide and water into food for themselves and for other organisms in the food chain
- **polar stratospheric clouds:** clouds of ice crystals formed at extremely low temperatures in the polar stratosphere
- **polar vortex:** a closed atmospheric circulation pattern around the South Pole that exists during the winter and early spring; atmospheric mixing between the polar vortex and regions outside the vortex is slow
- **stratosphere:** the region of the atmosphere between 10 and 50 kilometers above the surface of Earth
- **total column abundance of ozone:** the total number of molecules of ozone above a 1-centimeter-square area of Earth's surface

- **Total Ozone Mapping Spectrometer (TOMS):** a space-based instrument for measuring the total column abundance of ozone globally
- **ultraviolet solar radiation:** electromagnetic radiation having wavelengths between 4 and 400 nanometers

CONCENTRATION OF OZONE IN ATMOSPHERE

Ozone, although only a minor component of the atmosphere, plays a vital role in the survival of life on Earth. Ozone molecules in the stratosphere absorb incoming high-energy ultraviolet (UV) light from the sun. Absorption of UV light in the stratospheric ozone layer, a region that contains the maximum concentration of atmospheric ozone, though only about 12 parts per million, prevents most UV light from reaching the surface of the planet. If none of the sun's ultraviolet radiation were blocked by the ozone layer, it would be difficult, if not impossible, for most forms of life, including humans, to survive on land.

The concentration of ozone in the atmosphere is highly variable, changing with altitude, geographic location, time of day, time of year, and prevailing local atmospheric conditions. Long-term fluctuations in ozone concentration are also seen, some of which are related to the solar sunspot cycle. While long-term average ozone concentrations are relatively stable, short-term fluctuations of as much as 10 percent in total column abundance of ozone as a result of the natural variability in ozone concentration are often observed.

Beginning in the early 1970s, a new and unexpected decrease in stratospheric ozone concentration was first observed. The decrease was localized near Antarctica, and appears in early spring (which begins in September in the Southern Hemisphere). The initial decrease in ozone was small, but by 1980, decreases in total column abundance of ozone of as much as 30 percent were being recorded, well outside the range of variation expected as a result of random fluctuations. This seasonal depletion of stratospheric ozone in a circular region above Antarctica, which by 1990 had reached 50 percent of the total column abundance of ozone, was soon given the label "ozone hole."

ROLE OF CFCs

While it was initially unclear whether the formation of the Antarctic ozone hole stemmed from natural causes or from anthropogenic effects on the environment, extensive field studies combined with the results of laboratory experiments and computer modeling of the atmosphere quickly led to a consistent and detailed explanation for ozone-hole formation. The formation of the ozone hole has two principal causes: chemical reactions that occur generally throughout the stratosphere, and special conditions that exist in the Antarctic region.

Under normal conditions, the concentration of ozone in the stratosphere is determined by an equilibrium balance between reactions that remove ozone and those that produce ozone. The removal reactions are mainly catalytic chain reactions, in which trace atmospheric chemical species destroy ozone molecules without themselves being consumed. In such processes, it is possible for one chain component to remove many ozone molecules before being itself removed. A single chlorine atom, for example, is estimated to remove as many as 100,000 ozone molecules through chemical chain reactions before it is itself removed by forming a nonreactive species. The trace species involved in ozone removal include hydrogen oxides and nitrogen oxides, formed primarily by naturally occurring processes, and chlorine and bromine atoms and their corresponding oxides.

A major source of chlorine in the stratosphere is the decomposition of a class of compounds called chlorofluorocarbons (CFCs). Such compounds are used as refrigerants in refrigeration and air conditioning applications, and were commonly used as aerosol propellants and solvents, freely released into the atmosphere. Their use and handling is now strictly regulated. Chlorofluorocarbons are extremely stable in the lower atmosphere, with lifetimes of several decades, due to their extreme lack of chemical reactivity. The main fate of chlorofluorocarbons in the atmosphere, however, is slow migration into the stratosphere, where they absorb ultraviolet light and fragment to release chlorine atoms. The chlorine atoms produced from the breakdown of chlorofluorocarbons in the stratosphere provide an additional catalytic process by which stratospheric ozone is destroyed. A similar set of reactions involving a class of bromine-containing compounds called halons, used in some types of fire extinguishers, leads to additional ozone destruction by similar photochemical processes. By 1986, the average global loss of stratospheric ozone caused by the release of chlorofluorocarbons, halons, and related compounds into the environment was estimated to be 2 percent.

Antarctic Conditions

While the decomposition and subsequent reaction of chlorofluorocarbons, halons, and other synthetic compounds can explain the slow general decline in ozone concentration observed in the stratosphere, additional processes are needed to account for the more massive seasonal ozone depletion observed above Antarctica. These processes involve a set of special conditions that in combination are unique to the stratosphere above Antarctica.

During daylight hours, a portion of the chlorine present in the stratosphere is tied up in the form of reservoir species, compounds such as hydrogen chloride and chlorine nitrate that do not react with ozone. This slows the rate of removal of ozone by chlorine. Processes that directly or indirectly involve absorption of sunlight transform reservoir species and release ozone-destroying chlorine atoms. During the Antarctic winter, when sunlight is entirely absent, stratospheric chlorine is rapidly converted into reservoir species.

In the absence of additional chemical processes, the onset of spring in Antarctica and the return of sunlight convert a portion of the reservoir compounds into reactive chlorine species and reestablish the balance between ozone-producing and ozone-destroying processes. However, the extremely low temperatures occurring in the stratosphere above Antarctica during the winter months lead to the formation of polar stratospheric clouds, which, because of the extremely low concentration of water vapor in the stratosphere, do not form during other seasons or outside the polar regions of the globe.

The ice crystals that compose the clouds act as catalysts that convert reservoir species into diatomic chlorine and other gaseous chlorine compounds that, in the presence of sunlight, re-form ozone-destroying species. At the same time, nitrogen oxides in the collection of reservoir species are converted into nitric acid, which remains attached to the ice crystals. As these ice crystals are slowly removed from the stratosphere by gravity, the potential for conversion of active forms of chlorine into reservoir species is greatly reduced. Because of this, when spring arrives, large amounts of ozone-destroying chlorine species are produced by the action of sunlight, and only a small fraction of this reactive chlorine is converted into reservoir species. The increased rate of ozone removal caused by the abundance of reactive chlorine present in the stratosphere leads to ozone depletion and formation of the ozone hole.

An additional process important in formation of the ozone hole is the unique air-circulation pattern in the stratosphere above Antarctica. During the winter and early spring, a vortex of winds circulates about the South Pole. This polar vortex minimizes movement of ozone and reservoir-forming compounds from other regions of the stratosphere. As this polar vortex breaks up in midspring, ozone concentrations in the Antarctic stratosphere return to normal levels, and the ozone hole gradually disappears.

Atmospheric Ozone Study and Interpretation

Researchers utilize a great diversity of devices and techniques in the study and interpretation of atmospheric ozone. One popular technique is the use of simulation models. A good model is one that simulates the interrelationships and interactions of the various parts of the known system. The weakness of models is that, often, not enough is known to give an accurate picture of the total system or to make accurate predictions. Most modeling is done on computers. Scientists estimate how fast chemicals such as CFCs and nitrous oxide will be produced in the future and build a computer model of the way these chemicals react with ozone and with one another. From this model, it is possible to estimate future ozone levels at different altitudes and at different future dates.

Similar processes appear to be at work in the Arctic stratosphere, leading to ozone depletion, as in the Antarctic. However, the National Oceanic and Atmospheric Administration (NOAA) Aeronomy Laboratory in Boulder, Colorado, reported a discrepancy between observed ozone depletion and predicted levels, based on models that account accurately for Antarctic ozone depletion. This report suggests that some other mechanism is at work in the Arctic. Thus, while good models can be very useful in studying new data, observed discrepancies highlight the need for better system models and modeling algorithms.

There are two models favored by most scientists in this area. Some scientists put forth a chemical model that says the depletion is caused by chemical events promoted by the presence of chlorofluorocarbons created by industrial processes. Acceptance of this model was promoted by the discovery of fluorine in the stratosphere. Fluorine does not naturally occur there, but it is related to and can be formed photochemically from CFCs. The other model assumes that the ozone hole was formed by dynamic air movement and mixing. This model best fits data gathered by ozone-sensing balloons that sample altitudes up to 30 kilometers and then radio the data back to Earth. Ozone depletion is confined to the atmosphere at altitudes between 12 and 20 kilometers. While the total ozone depletion is 35 percent, different strata have shown various amounts of depletion from 70 to 90 percent. Surprisingly, about half the ozone was gone in twenty-five days. This finding does not fit the chemical model very well.

Besides ozone-sensing balloons, satellite survey data provide more direct measurements obtained over longer periods of observation time. The National Aeronautics and Space Administration (NASA) obtains measurements with its Nimbus 7 satellite. Ozone measurements made by this satellite helped to develop flight plans for the specialized aircraft NASA also deploys in ozone studies. NASA's ER-2 aircraft is a modified U-2 reconnaissance plane that carries instruments up to 20 kilometers in altitude for seven-hour flights to 80 degrees north latitude. A DC-8, operating during the same period, is able to survey the polar vortex, owing to its greater range. In addition, scientists utilize many meteorological techniques and instruments, including chemical analysis of gases by means of infrared spectroscopy, mass spectroscopy and gas spectroscopy combined, gas chromatography, and oceanographic analysis of planktonic life in the southern Atlantic, Pacific, and Indian oceans. As new research methods and techniques become available, they are also applied to this essential study.

Public Health Concerns

Stratospheric ozone provides global protection from the lethal effects of ultraviolet radiation from the sun. This ability to absorb ultraviolet radiation protects all life-forms on Earth's surface from excessive ultraviolet radiation, which destroys plant and animal cells. Currently, between 10 and 30 percent of the sun's ultraviolet B (UV-B) radiation reaches the ground surface. If ozone levels were to drop by 10 percent, the amount of UV-B radiation reaching Earth would increase by 20 percent.

Present-day UV-B levels are responsible for the fading of paints and the yellowing of window glazing and for car finishes becoming chalky. These kinds of degradation will accelerate as the ozone layer is depleted. There could also be increased smog, urban air pollution, and a worsening of the problem of acid rain in cities. In humans, UV-B causes sunburn, snow blindness, skin cancer, and cataracts, and promotes aging and wrinkling of the skin. Skin cancer is the most common form of cancer, with more than 400,000 new cases reported every year in the United States alone. The National Academy of Sciences has estimated that each 1 percent decline in ozone would increase the incidence of skin cancer by 2 percent. Therefore, a 3 percent depletion in ozone would be expected to produce some 24,000 more cases of skin cancer in the United States every year.

Ecological Concerns

Many other forms of life, from bacteria to forests and food crops, are adversely affected by excessive radiation. Ultraviolet radiation affects plant growth by slowing photosynthesis and by delaying germination in many plants, including trees and food crops. Scientists have a great concern for the organisms that live in the ocean and the effect ozone depletion may have on them. Phytoplankton, zooplankton, and krill (a shrimplike crustacean) could be greatly depleted if there were a drastic increase in ultraviolet A and B. The result would be a tremendous drop in the population of these free-floating organisms, which are extremely important because they are the foundation species of the global food chain. Phytoplankton use the energy of sunlight to convert inorganic compounds into organic plant matter. This process provides food for the next step in the food chain, the herbivorous zooplankton and krill. They, in turn, become the food for the next higher level of animals in the food chain. Initial studies of this food chain in the Antarctic suggest that elevated levels of ultraviolet radiation impair photosynthetic activity. Recent studies show that a fifteen-day exposure to UV-B levels 20 percent higher than normal can kill off all anchovy larvae down to a depth of 10 meters. There is also concern that ozone depletion may alter the food chain and even cause changes in the organisms' genetic makeup. An increase in the ultraviolet radiation is likely to lower fish catches and upset marine ecology, which has already suffered damage from human-made pollution. On a worldwide basis, fish presently provide 18 percent of all the animal protein consumed.

International Response

The United Nations Environmental Program (UNEP) is working with governments, international organizations, and industry to develop a framework within which the international community can make decisions to minimize atmospheric changes and the effects they could have on Earth. In 1977, UNEP convened a meeting of experts to draft the World Plan of Action on the Ozone Layer. The plan called for a program of research on the ozone layer and on what would happen if the layer were compromised. In addition, UNEP created a group of experts and government representatives who framed the Convention for the Protection of the Ozone Layer. This convention was adopted in Vienna in March, 1985, by twenty-one nations and the European Economic Community and has subsequently been signed by many more nations. The convention pledges that the nations that sign are to protect human health and the environment from the effects of ozone depletion. Action has already been taken under the Convention to protect the ozone layer. Several countries have restricted the use of CFCs or the amounts produced. The United States banned the use of CFCs in all nonessential applications in 1978. Some countries, such as Belgium and the Nordic countries, have in effect banned CFC production altogether. The group has also worked with governments on a Protocol to the Convention that required signatory nations to limit their production of CFCs. It is the hope and aim of these nations that such international cooperation will lead to a better global environment.

In 2017 it was announced that the ozone hole was the smallest it had been since 1988. However, this was mainly due to unusually warm temperatures, not to human efforts to reduce ozone-depleting chemicals. Scientists cautioned against seeing the smaller hole as being a sign of recovery.

—*George K. Attwood*
—*Jeffrey A. Joens*

Further Reading

Bielle, David. "Ozone Hole Closing Up, Thanks to Global Action." *Scientific American*. Scientific American, 15 Sept. 2014. Web. 23 Mar. 2015.

Douglass, Anne, Natalya Kramarova, and Susan Strahan. *Inside the Ozone Hole*. NASA's Goddard Space Flight Center, 2015. PDF file.

Lang, Kenneth. *Sun, Earth and Sky*. 2nd ed. Springer, 2006. Discusses the sun, its energy, solar radiation, and magnetism. The final two chapters cover the relationship between the sun and earth. Covers geomagnetic storms, aurora, the greenhouse effect, and the ozone layer. Also contains a glossary, quotation references, a further reading list, websites, and both author and subject indexing.

Marshal, John, and R. Alan Plumb. *Atmosphere, Ocean and Climate Dynamics: An Introductory Text*. Elsevier Academic Press, 2008. Provides an excellent introduction to atmospheres and oceans. Discusses topics such as the greenhouse effect, convection and atmospheric structure, oceanic and atmospheric circulation, and climate change. Suited for advanced undergraduates and graduate students with some background in advanced mathematics.

Mersmann, Katy, and Theo Stein. "Warm Air Helped Make 2017 Ozone Hole Smallest Since 1988." *NASA*, 2 Nov. 2017, www.nasa.gov/feature/goddard/2017/warm-air-helped-make-2017-ozone-hole-smallest-since-1988. Accessed 14 Dec. 2017.

Seinfeld, J. H., and S. N. Pandis. *Atmospheric Chemistry and Physics: From Air Pollution to Climate Change*. 2nd ed. Wiley, 2006.

Somerville, Richard C. J. *The Forgiving Air: Understanding Environmental Change*. 2nd ed. American Meteorological Society, 2008. Presents a thorough investigation of the relationship between human activities and changes in the atmosphere and global climate. Written by a scientist involved in atmospheric research, but aimed at a nonscientific audience. Chapter 2 offers a discussion of the ozone hole.

Sportisse, Bruno. *Fundamentals in Air Pollution: From Processes to Modelling*. Springer, 2009. Discusses issues arising from air pollution such as emissions, the greenhouse effect, acid rain, urban heat islands, and the ozone hole. Topics are well organized and clearly explained, making this text accessible to the layperson, although it was written for the undergraduate.

Wallace, John M., and Peter V. Hobbs. *Atmospheric Science*. 2nd ed. Academic Press, 2006. Represents a complete study of the atmosphere that covers fundamental physics and chemistry topics as well as specific topics in atmospheric science such as radiative transfer, weather forecasting, and global warming. Contains significant detail and technical writing, but is still accessible to the undergraduate studying meteorology or thermodynamics.

World Meteorological Organization. *Global Atmosphere Watch: Antarctic Ozone Bulletin* 1 (August 28, 2008).

P

PACIFIC OCEAN

FIELDS OF STUDY

Oceanography; Physical Geography; Ecosystem Management; Earth System Modeling; Ecology; Heat Transfer; Thermodynamics; Engineering; Environmental Sciences; Environmental Studies; Waste Management; Meteorology; Climate Modeling; Fluid Dynamics; Hydroclimatology; Hydrometeorology; Atmospheric Science; Bioclimatology; Climate Classification; Climate Zones; Hydrology; Petroleum Refining

SUMMARY

The Pacific Ocean is the world's largest ocean and has an overall area of 182 million square kilometers. It is twice the size of the next largest ocean, the Atlantic. The Pacific Ocean covers approximately one-third of Earth's surface and is larger than Earth's entire landmass area.

PRINCIPAL TERMS

- **continental drift:** the gradual movement of continental landmasses within Earth's crust, driven by magmatic convection in the underlying mantle layer
- **continental shelf:** part of a continental landmass, usually gently sloping, that extends beneath the ocean from the water's edge to the continental slope
- **continental slope:** the defining edge of a continental shelf where it drops off sharply toward the ocean's floor
- **delta:** a triangular area with its longest side abutting the sea where a river deposits silt, sand, and clay as it flows into an ocean, lake, or other body of water
- **equator:** an imaginary line, equidistant from the North and South Poles, around the middle of the planet where day and night are of equal length
- **guyot:** an undersea mountain, or seamount, that has formed by volcanic activity, from which the peak has been eroded through wave action
- **lagoon:** a body of saltwater separated from the ocean by a bank of sand
- **tectonic plate:** a segment of Earth's crust that is put into movement by magmatic convection in the underlying mantle layer
- **tidal range:** the difference in water depth between high and low tides
- **trench:** an extraordinarily deep region of the sea floor where two tectonic plates meet and the one making up the deep sea floor is subducted beneath the less dense, and therefore lighter, continental mass of the adjacent tectonic plate

LOCATION OF THE PACIFIC OCEAN

The Pacific Ocean, with a mean depth of 4,255 meters and a maximum depth of 10,970 meters, extends from its southern extreme at Antarctica some 15,550 kilometers north to the Bering Strait, which separates North America from Asia. Its east-west area extends almost 19,440 kilometers from the western coast of Colombia in South America to Asia's Malay Peninsula.

The Pacific and Arctic Oceans meet at the Bering Strait in the north. In the far south at Drake Passage, south of Cape Horn, the Atlantic and Pacific Oceans come together. Exactly where the Pacific meets the Indian Ocean is more difficult to define because the two bodies of water intermingle along a string of islands that extends to the east from Sumatra through Java to Timor and across the Timor Sea to Australia's Cape Londonderry. South of Australia, the Pacific

The Pacific coast.

runs across the Bass Strait to the Indian Ocean and continues from Tasmania to Antarctica.

The eastern Pacific generally follows the Cordilleran mountain system, which runs the entire length of North and South America from the Bering Strait to Drake Passage and includes both the Rocky Mountains and the Andes. The east coast of the Pacific is, comparatively, quite regular except for the Gulf of California and the fjord regions in its northern and southern extremes. The continental shelf of the eastern Pacific is narrow, and the continental slope at times quite steep, as is characteristic of subduction zones.

The Pacific's western extreme in Asia is, by contrast, quite irregular. As on the eastern coast of the Pacific, the western coast is bordered by mountain systems that run roughly parallel to the coast. The western Pacific has many dependent seas, notable among them the Bering Sea, the East China Sea, the Sea of Japan, the Sea of Okotsk, the South China Sea, and the Yellow Sea. The eastern extremes of these seas are characterized by peninsulas jutting out toward the south, by island arcs, or by both.

Unlike the Atlantic, much of whose water flows into it from the rivers that feed it, only about one-seventh of the Pacific's water comes from direct river flow. Most of its water comes from the dependent seas that are nourished by such East Asian rivers as the Amur, the Xi, the Mekong, the Yangtze, and the Yellow Rivers.

Origins and Divisions

The Pacific Ocean is the present-day remnant of the world ocean called Panthalassa and has been evolving to its present state for some 200 million years, since the single landmass called Pangaea began to split apart through tectonic action. Through continental drift that has occurred over the intervening millions of years, the landmass divided into the several smaller fragments that are the present-day continents. As the spaces between the continental segments became larger, they filled with the water that surrounded them and, through eons, formed the various oceans and seas that encompass the surface of the planet.

The Pacific Ocean is so large that it is difficult to discuss it as a single ocean. Various parts of it are substantially different from other parts. For purposes of discussion, the Pacific is usually thought of as comprising three distinct regions.

The western region begins in Alaska's western Aleutian Trench, extends through the Kuril and Japanese Trenches, runs south to the Tonga and Kermadec Trenches, then continues on to an area northeast of New Zealand's North Island. This region is characterized by large strings of islands, the largest constituting Japan, New Guinea, New Zealand, and the Philippines. Some of these strings, notably Japan, New Guinea, and New Zealand, were, through time, sheered off from the continent by tectonic movements. The entire western Pacific is also dotted with volcanic islands, which are the peaks of high underwater mountains. These were formed as molten rock from successive volcanic eruptions cooled and solidified.

The ocean floor of the central area of the Pacific is the largest underwater expanse on Earth. It is the most geologically stable area on the Pacific floor. It contains sprawling underwater plains, most at a depth of about 4,500 meters, that are essentially flat, although they contain some irregularities and some geological formations, called guyots, which resemble the mesas found in above-water plains.

The eastern part of the Pacific, which abuts the United States' west coast, has a narrow continental shelf and a steep continental slope so that close to shore there are deep areas, such as the Monterey Canyon, which is more than 3 kilometers below the water's surface. The two most significant trenches in this part of ocean are the Middle America Trench in the North Pacific and the Peru-Chile Trench in the South Pacific.

Volcanic activity has historically been more pronounced along the western edge of the Pacific than

381

along its eastern edge, although the East Pacific Rise, a chain of underwater mountains that runs from Southern California almost to the tip of South America, is also quite volcanically active. In the central and western Pacific, however, there is a profusion of islands that are the tops of underwater volcanic mountains.

Pacific Currents

The Pacific, like the Atlantic, has two large systems of currents called gyres. The Northern Gyre moves in a westerly direction from near the equator and is a surface ocean current that carries warm surface water into the Kuroshio Current and gives Japan moderately warm temperatures. The Kuroshio Current then moves northeast away from the Kamchatka Current, which brings cold water from the area around the Bering Strait. Part of it divides, moving north into the North Pacific Drift. The other part heads southeast.

Off the west coast of North America, part of the North American Drift moves north and becomes the Alaska Current. It brings to the Alaskan coastline waters warm enough to keep its shoreline and harbors from freezing in winter. Another part of the North American Drift veers south and forms the California Current, whose waters are cooled as they flow south to join the North Equatorial Current, which again flows west, completing its clockwise motion.

The South Equatorial Current also moves west toward Australia and New Guinea, but it then veers south to begin its counterclockwise course. The East Australian Current skirts Australia's east coast and passes between Australia and New Zealand, south of which it feeds into the West Wind Drift, which moves east toward the coast of South America. Just north of South America's tip, it feeds into the Peru or Humboldt Current, which runs north along South America until it connects with the South Equatorial Current, which moves west to complete the South Pacific Gyre.

In both the northern and southern areas of the Pacific, winds from the west move the water in an easterly direction. Closer to the equator, however, trade winds move the water toward the west. This balance permitted early sailors to head east by riding the westerly winds and to return to the east by riding the trade winds. There are virtually no winds on the equator itself. Sailors can founder there, in the so-called doldrums, for days with none of the propulsive forces that nature provides both north and south of the Earth's dividing line.

The currents of the Pacific are important both for transportation and for bringing warm and cold waters to parts of the ocean, thereby giving them and the land adjoining them more moderate temperatures than might be expected at their latitudes. Occasionally, nature runs afoul of this balance as the southern trade winds diminish and fail to push the cool Peru Current north. Instead, the Peru Current is replaced by the Pacific Equatorial Countercurrent. The Peruvians first dubbed this phenomenon El Niño, which means "the Christ child," because it usually comes at the Christmas season. In some years, El Niño is followed by its opposite sibling La Niña, resulting in two successive years of hardship. As waters reach unaccustomed temperatures, much sea life fails to appear where it would otherwise be, bringing catastrophe to people dependent upon fish and seaweed for their diets or for income.

Formation of Pacific Islands

The Pacific Ocean has more islands than any other ocean. In the west, most of these islands are volcanic, although some, such as Japan, New Guinea, and New Zealand, were formed when the corresponding chunks of land became separated from the continental landmass in prehistoric times through tectonic action. The volcanic islands in the western Pacific often rise just barely above sea level. Any change in the ocean's level or tidal range can inundate them. Global warming threatens to melt ice in the polar regions, which will in turn cause sea levels to rise and, in some cases, cause islands to disappear. Indeed, some small, politically independent island nations face total inundation. It is estimated that by the year 2050, the oceans of the world might rise as much as 50 centimeters, which could wipe out currently habitable regions occupied by millions of people in the South Pacific.

Another major structure in the South Pacific is the atoll, a tropical island on which a massive coral reef, often ring-like, generally rests on a volcanic base. Kwajalein is the largest atoll in the world, with a circumference of nearly 324 kilometers. Other coral formations, such as ridges or reefs, can grow to enormous lengths, such as the Great Barrier Reef east of Australia. Coral formations are composed of living organisms that require light and a stable environment in order to survive. They are, therefore, always close to the surface of the ocean and often protrude above the surface as tidal sea levels change. Most of them teem with aquatic life. In some cases, when they form around low-lying islands, the island disappears as the ocean surrounding it rises. The coral formation, however, often remains, filled with entrapped sand and rising above the water's surface as an atoll.

Such island chains as the Hawaiian, Pitcairn-Tuamotu, and Tubai Islands are the result of great plumes of extraordinarily hot magma that have risen through the Earth's mantle over millions of years and erupted through fissures in the oceanic plate above. The resulting formation that occurs when the molten rock cools is an island composed largely of basaltic rock. Over great time spans, the island is slowly carried away to the northwest as the Pacific Ocean tectonic plate continues its normal motion, clearing the way for another such island to be formed above the plume at the same relative location in the oceanic plate. Hundreds of islands, lined up almost like pearls in a necklace, have resulted from this ongoing volcanic activity, though many have not survived to breach the ocean surface due to erosion by waves and currents in the ocean water.

Pacific Rim

The areas around the Pacific have many volcanic mountains, large ranges of which line the coasts and still larger ranges of which are submerged. The highest mountains on Earth are the Himalayas, with Mount Everest towering to 8,795 meters; many of the mountains submerged in the Pacific, however, are higher. From the ocean's floor at the Mariana Trench, undersea mountains, or seamounts, rise more than 10,600 meters as they approach the ocean's surface.

Many of the mountains on the Pacific Rim arose as, over time, the oceanic crust is subducted beneath the continental crust. Intense heat is generated by friction in the subduction zone beneath a continental mass, while the pressure exerted by the inexorable motion of the crustal plates causes fractures to form in the overlying rock, permitting magma to erupt through the continental crust as volcanoes. This process occurs all around the so-called Ring of Fire, from the North Island of New Zealand, through the Philippines, Malaysia, Japan and Siberia, to the Aleutian Islands and on down the west coast of North and South America to Peru and Chile.

The coast along the Pacific Rim is narrow, giving way to the mountain ranges that run close to the shoreline. An exception to this is in eastern China where the Yellow and the Yangtze Rivers have, for many centuries, carried silt toward the ocean, creating fertile coastal plains.

Geological activity beneath the Pacific results in thousands of small earthquakes every year along the Pacific Rim as oceanic plates move relative to each other. These earthquakes, most of which are too insignificant to be felt by humans, result in little or no damage. They relieve pressure that builds up between tectonic plates as they collide. When such pressure builds up without relief over a period of many years, however, a major earthquake, usually causing considerable property damage and loss of life, may occur. Severe earthquakes in the sea floor may generate extraordinarily large waves called tsunamis. Tsunamis, known to have achieved heights of more than 30 meters, can engulf large stretches of shoreline, as happened in Alaska in 1964. The worst tsunamis are usually caused by sudden displacements of crustal material in the subduction process, in turn causing a massive displacement of seawater that translates to the surface as an energy wave. Such waves travel through the ocean at speeds of up to 800 km per hour, and can cross the entire Pacific Ocean in a matter of hours. The tsunami that devastated the coast of Japan in 2011, producing worldwide economic consequences, was produced by a subduction zone earthquake.

Food Resources

The feeding hierarchy that exists among the creatures of the sea is usually referred to as a food chain. One can envision this hierarchy as a triangle, at whose broad base is phytoplankton, microscopic plant organisms that require sunlight for their survival and are plentiful near the surfaces of oceans. Zooplankton, microscopic animal organisms, feed on the phytoplankton. They, in turn, constitute the diet of numerous species of small fish called anchoveta. These fish in turn are food for such larger fish as tuna and dolphin, as well as such birds as cormorants and pelicans, which dive into the water to catch fish. Animal waste, decaying plants, and dead fish and birds sink to the ocean floor, where scavengers consume them. Bacteria at those levels cause such droppings to decay, in the process releasing nutrients into the seawater that provide the phytoplankton with nourishment, thereby completing the food chain.

The food chain triangle becomes smaller as it approaches the top. Small fish survive by eating plankton. Larger fish eat the smaller fish, then sharks and whales eat the larger fish. A food chain that begins with millions of plankton may end up producing hundreds of thousands of anchoveta; thousands of cod, hake, and mackerel; and one-half dozen sharks for every whale that is part of this intricate hierarchy.

Although fish are the major food source harvested from the world's oceans, the Pacific also yields a great deal of kelp and seaweed, which, particularly in Asian countries, constitute a significant part of human diets. Seaweed is often harvested and laid in strips beside the ocean, where the sun dries it, thereby preserving it and making it easy to transport and to store. Even though they are not consumed as readily in the eastern parts of the Pacific Rim as in Asia, kelp and other sea plants are used in many food and pharmaceutical preparations. They have valuable pharmacological properties that are in great demand by drug manufacturers throughout the world.

Mineral Resources

The mineral wealth beneath the Pacific Ocean has barely been tapped. Among the minerals available near the shore are chromite, gold, iron, monazite, phosphorus, tin, titanium, and zircon. The greatest exploitation of the ocean's treasure trove has been by the petroleum industry. Offshore drilling takes place worldwide for the recovery of crude oil and natural gas. In addition, large quantities of sand and gravel are harvested every year for use in construction and manufacturing.

The deep sea has remained a mysterious place. The historic descent in 1960 of the bathyscaphe *Trieste* into the Mariana Trench, the deepest area of ocean anywhere in the world, unlocked many mysteries. Subsequent exploration of the deepest areas of the ocean has challenged many long-held beliefs and revolutionized deep-sea research. It was long thought impossible for life to exist under the enormous pressure of the ocean below 760 meters, but the *Trieste* discovered life at the very bottom of the Mariana Trench, much to the surprise of oceanographers. In early 2012, James Cameron undertook a solo dive mission to the floor of the Mariana Trench in a specially-built one-man submarine, returning with a priceless video record of that environment.

The deep ocean is richest in its deposits of cobalt, copper, manganese, and nickel. These deposits remain largely unharvested because of the difficulty of getting to them, but such problems will undoubtedly be overcome as deep-sea mining technology is developed.

—*R. Baird Shuman*

Further Reading

Allen, Gerald R., and D. Ross Robertson. *Fishes of the Tropical Eastern Pacific*. University of Hawaii Press, 1994.

American Museum of Natural History. *Ocean*. Dorling Kindersley Limited, 2006.

Benson, Keith Rodney, and Philip F. Rhebock. *Oceanographic History: The Pacific and Beyond*. University of Washington Press, 2002.

Broad, William J. *The Universe Below: Discovering the Secrets of the Deep Sea*. Simon and Schuster, 1997.

Clarke, Allan J. *An Introduction to the Dynamics of El Niño and the Southern Oscillation*. Academic Press/Elsevier, 2008.

Glavin, Terry, and David Suzuki Foundation. "The Last Great Sea a Voyage through the Human and Natural History of the North Pacific Ocean." *The Last Great Sea a Voyage through the Human and Natural History of the North Pacific Ocean*, www.deslibris.ca/ID/416854.

Henderson, Bonnie. *Strand: An Odyssey of Pacific Ocean Debris*. Oregon State University Press, 2008.

Kowahata, Hodaka, and Yoshio Awaya. *Global Climate Change and Response of the Carbon Cycle in the Equatorial Pacific and Indian Oceans and Adjacent Landmasses*. Elsevier, 2006.

Nunn, Patrick D. *Climate, Environment and Society in the Pacific During the Last Millenium*. Elsevier, 2007.

Ogawa, Yujiro, Ryo Anma, and Yildirim Dilek. *Accretionary Prisms and Convergent Margin Tectonics in the Northwest Pacific Basin*. Springer Science+Business Media, 2011.

Severin, Tim. *The China Voyage: Across the Pacific by Bamboo Raft*. Addison-Wesley, 1994.

Taylor, Brian, and James Natland, eds. *Active Margins and Marginal Basins of the Western Pacific*. American Geophysical Union, 1995.

PALEOCLIMATES AND PALEOCLIMATE CHANGE

FIELDS OF STUDY

Analytical Chemistry; Atmospheric Chemistry; Biochemistry; Environmental Chemistry; Physical Chemistry; Photochemistry; Geochemistry; Thermodynamics; Climate Modeling; Bioclimatology; Hydroclimatology; Climate Classification; Oceanography; Physical Geography; Earth System Modeling; Ecology; Crystallography; Spectroscopy; Mass Spectrometry; Optics

SUMMARY

Earth's climate has not been uniform in geologic or even historic time. Climatic changes form the backdrop for much of human prehistory and are viewed by some as a driving force in the rise and fall of civilizations. Looking further back into the geologic record, the abrupt transitions between eras indicate periods of rapid climatic change. Despite decades of intensive research, there remain many uncertainties and unanswered questions about the causes and impact of climatic change, particularly in remote geologic time.

PRINCIPAL TERMS

- **astronomical forcing:** climatic change triggered by changes in solar luminosity, variation in the Earth's orbit, and bolide impact
- **Milankovitch cycles:** variations in the eccentricity of the Earth's orbit, the tilt of the Earth's axis, and the precession of equinoxes, resulting in climatic variation on a scale of tens of thousands of years
- **positive climate feedback loops:** self-reinforcing climatic processes, such as increased snow cover increasing planetary albedo, promoting additional cooling and therefore more snow cover
- **proxies:** measurable parameters, correlated with climate, that are preserved in the geologic record

THE BEGINNING OF PALEOCLIMATOLOGY

Paleoclimatology as a science owes its beginnings to the work of Louis Agassiz (1807–1873), who in 1840 tuned his attention from the study of fossilized fish to observations on glaciers in Switzerland and concluded that the presence of characteristic glacial geologic formations in widely separated localities indicated that an ice sheet had once covered northern Europe. At about the same time, systematic study of fossilized plants in Britain, Germany, and Pennsylvania suggested that the climate at a more remote period had been tropical. paleoclimatology as a scientific discipline remained largely a tool to study ancient ecosystems until the last quarter of the twentieth century, when the focus turned to understanding mass extinctions in the remote past and historical climatic extremes in order to predict and manage global warming in the present.

Sediment coring from a raft on a lake in East Greenland; performed to recover Quaternary sediments for the reconstruction of paleoclimate and the glaciation history of Greenland.

RESEARCH TECHNIQUES

Systematic instrumental records of temperature, precipitation, and wind speed in parts of Europe and North America exist for the last 150 years, with spotty records for the eighteenth century. Data such as timing of the grape harvest in Southern Europe and the maximum extent of the annual flooding of the Nile extend the range of human observations several centuries backward in time, and historic and even legendary sources chronicle catastrophic events throughout human history.

Most paleoclimatology relies on the study of proxies, of which tree rings (dendrochronology) are a good example. If a variation in ring width in historic times correlates with fluctuations in temperature or precipitation, a similar pattern in the more remote past indicates similar fluctuations. Within the roughly four-thousand-year lifespan of the oldest living trees, dendrochronology provides very high resolution of climatic fluctuations. It has helped establish that the El Niño-Southern Oscillation (ENSO) cycle of wet and dry years has occurred for many centuries in coastal Peru and the American Southwest. Some corals and mollusks also develop annual rings that are indicative of fluctuation in water temperature at a given locality.

The type of vegetation and species present in a region are both good climatic indicators. Leaves, woody material, and particularly pollen occur in abundance in bogs, lake sediments, and sedimentary rocks. Plant species and genera have changed very little over the last few million years, making it a safe assumption that the environment in which a fossil was deposited closely resembled the environment where the same species occurs today. The genus *Metasequoia* (dawn redwood) today thrives in wet temperate climates. Its occurrence in a fifty-million-year-old fossil assemblage of the Eocene Age on Ellesmere Island in the Canadian Arctic is one of many pieces of evidence for an unusually warm period in geologic time.

Pollen analysis is a powerful tool, because pollen is extraordinarily resistant to decay, many pollen grains can be identified to genus, and analysis of relative abundance provides a fairly complete picture of the vegetation that produced it. Pollen in bog cores provided the first evidence for the Younger Dryas, an episode of drastic cooling in Europe between 12,900 and 11,500 years ago. The prevalence of pollen of an arctic plant, *Dryas octopetala*, indicated an abrupt return to arctic-tundra conditions within a ten-year period.

Further back in the geologic record, general morphology of plant fossils can be used as a climate proxy. Forest vegetation indicates relatively high precipitation; broad-leaved evergreens, a tropical or subtropical climate; and slow-growing woody plants with small vessel elements, a semiarid ecosystem.

The abundance and species composition of oceanic plankton are highly sensitive to water temperatures. Foraminifera (protozoa) and diatoms (algae) produce distinctive resistant outer coatings. The ratio of stable isotopes of carbon and oxygen in carbonate-containing marine sediments reflects their concentration in seawater. Plants selectively use carbon 12 in photosynthesis, so oceanic sediments become enriched in carbon 13 during warm, moist periods. Evaporated water is enriched in oxygen 16 relative to oxygen 18, leaving seawater with a higher proportion of oxygen 18 during periods of glaciation.

There are many geologic indicators of past climatic conditions. Cool, arid conditions produce deposits of loess (windblown dust). Volcanic ash distributed over a wide area may signal the onset of a cooling period. Low sea levels indicate glaciation, and high sea levels indicate warming episodes. organic matter accumulates on land during cool wet periods. Continental glaciers produce characteristic scouring of rocks and glacial moraines. Ice-rafted debris found far from the continental margin, especially at low latitudes, indicates extensive glaciation.

The effects of geological and astronomic forcing mechanisms can be modeled based on a knowledge of plate tectonics and changes in the Earth's orbit and solar luminosity, and the results can be compared with proxy measurements. The relative proportions of sea and land mass and the latitudinal distribution of continents have changed substantially over nearly four billion years of life on Earth, but continental drift occurs too slowly to account for dramatic climate changes in the Phanerozoic. plate tectonics do, however, spawn massive episodes of volcanism, such as those associated with the Permian-Triassic and Miocene-Pliocene transitions. Milankovitch cycles, which predict changes in the absolute amount of solar radiation reaching the Earth, show some positive correlation with Pleistocene glaciation pulses. Cores taken from glaciers in Greenland and Antarctica provide evidence of climate over the last 400,000 years in the form of rates of precipitation, amounts of atmospheric dust, and concentrations of carbon dioxide (CO_2) in trapped air.

CLIMATIC CHANGE IN THE GEOLOGIC RECORD

Scientists who developed the geologic timescale used today based their division into eras and periods on marked discontinuities in both the inorganic constituents and fossils in sedimentary rocks. The discovery of radiometric dating based on decay of uranium isotopes at the end of the nineteenth century allowed an absolute time scale to be superimposed on the stratigraphic sequence. From a climatic perspective, the periods represent spans of millions of years, during which global climate was relatively uniform, divided by much briefer spans that were characterized by climatic extremes, elevated extinction rates, and the subsequent emergence of numerous new taxa.

The geologic record is punctuated by five recognized episodes of catastrophic extinction, when 60–80 percent of the then extant species disappeared in less than a million years. The best-known of these is the end-Cretaceous event marking the end of the dinosaurs 65 million years ago. Most biologists accept that the precipitating factor was an asteroid collision. The end-Ordovician extinctions, 440 million years ago, coincided with a drastic lowering of the sea level, possibly representing extensive glaciation. The causes of the late Devonian and Triassic-Jurassic extinctions are uncertain. From the point of view of present policy makers attempting to learn from the lessons of geologic history, the Permian-Triassic extinctions, 250 million years ago, are probably the most instructive, both because they were the most profound and because the postulated causes most closely mirror global anthropogenic changes that are rapidly creating what Richard Leakey and others have termed "the sixth extinction."

A somewhat controversial interpretation of late Precambrian geologic formations postulates a "snowball Earth" between 790 and 630 million years ago. This period encompassed three or possibly four major glacial episodes, during which the Earth's oceans may have completely frozen, halting the evolution of multicellular life. At the time, most of the Earth's land mass centered over the South Pole. There is also evidence of major glaciation during the Huronian glaciation, 2 billion years ago. The postulated cause is depletion of methane due to release of oxygen by photosynthesis.

For most of the Phanerozoic, temperatures on Earth, as inferred from sea levels, isotope ratios, and vegetation at high latitudes have been significantly higher than they are at present. The late Cretaceous and the Eocene were particularly warm periods, characterized by temperate deciduous forests in Antarctica and the Canadian Arctic, and tropical jungles in Central Europe and the Pacific Northwest. The past 50 million years have witnessed gradual cooling and increasing aridity. Pronounced aridity and a drop in sea level in the Pliocene caused both the Mediterranean and Black Seas to become landlocked and shrink to a fraction of their present size.

CLIMATIC CHANGE IN HUMAN PREHISTORY

The tenure of modern humans on Earth encompasses the last Pleistocene glaciation and the ten thousand years of the Holocene, during which the Earth's climate has fluctuated, with a temperature maximum roughly six thousand years ago and a temperature minimum, the Little Ice Age, from the fourteenth to early nineteenth centuries. There are many studies correlating prehistoric cultural changes with climatic changes. For agricultural societies, the droughts associated with colder periods are more devastating than lower temperatures themselves. The historic and geologic records contain no compelling evidence of rapid rises in temperature such as the Earth is currently experiencing—global warming in geologic time appears to be a gradual process to which life adapts itself. Cooling, on the other hand, can be extremely rapid and catastrophic. Massive volcanic eruptions cause global cooling by ejecting fine ash and sulfates into the atmosphere. Historic cold episodes beginning in 1470 BCE, 535 CE, 1315, and 1815 are dwarfed by the eruption of Mount Toba on the island of Sumatra seventy-four thousand years ago, which is believed by some to have nearly wiped out the human race.

CONTEXT

Probably the most important lesson to be learned from paleoclimatology in respect to global warming is that of the disruption of North Atlantic currents and resulting deep freeze in Europe in the 8.2ka event. Very rapid melting of the Greenland ice cap and release of freshwater into the Atlantic could well mimic the effects of the rapid draining of glacial Lake Agassiz through the St. Lawrence at the end of the Wisconsin Glaciation. The other lesson to be learned, although it does not so readily translate into international policy making, is that a massive volcanic eruption could, within a period of a few months, cause a substantial drop in global temperatures and global agricultural productivity, a much grimmer scenario in today's overpopulated and environmentally degraded world than in 1315 or 1815.

—Martha A. Sherwood

FURTHER READING

Alley, Richard B. *The Two-Mile Time Machine. Ice Cores, Abrupt Climate Change and Our Future* Princeton

University Press, 2014. Describes the analytical methods and interpretations of physical clues of past climates trapped in ice that is thousands of years old.

Alverson, Keith D., Raymond S. Bradley, and Thomas Pedersen, eds. *Paleoclimate, Global Change, and the Future.* Springer Verlag, 2003. Collection of scholarly papers comparing natural paleoclimate changes with modern anthropogenic trends; discusses ancient civilizations.

Bradley, Raymond S. *Paleoclimatology: Reconstructing Climates of the Quaternary.* Academic Press, 2015. A definitive introduction to the science of paleoclimatology, and the use of climate proxies.

Cronin, Thomas M. *Principles of Paleoclimatology.* Columbia University Press, 1999. A standard textbook explaining in detail the methodology of paleoclimatological research. Includes an extensive discussion of the implications of late twentieth century discoveries for global energy policies.

Mackay, Anson, Battarbee, Rick, Birks, John and Oldfield, Frank, eds. *Global Change in the Holocene.* Routledge, 2014. Chapters by several contributors describe Holocene climate reconstruction from the resources of many different climate proxies.

Pap, Judit M., and Peter Fox, eds. *Solar Variability and Its Effects on Climate.* American Geophysical Union, 2004. In addition to Milankovitch cycles, discusses sunspots and changes in solar output in geologic time.

Saltzman, Barry. *Dynamical Paleoclimatology: Generalized Theory of Global Climate Change.* New York: Academic Press, 2002. Treats interactions between abiotic, biotic, and anthropogenic variables; discusses controversies about the magnitude of human climatological impacts.

PHYSICS OF WEATHER

FIELDS OF STUDY

Thermodynamics; Atmospheric Science; Atmospheric Physics; Environmental Sciences; Environmental Studies; Meteorology; Climate Modeling; Statistics; Process Modeling; Mathematics; Fluid Dynamics; Oceanography; Physical Geography

SUMMARY

The thermal structure of Earth's atmosphere determines the development of large-scale weather systems, the dispersion of atmospheric pollutants, and the potential for severe weather. A clear understanding of the thermodynamic processes of the atmosphere is important in many applications, including weather prediction, the completion of environmental impact studies, and the modification of select weather phenomena.

PRINCIPAL TERMS

- **available potential energy:** that part of the total potential energy of the atmosphere that is available for conversion to the kinetic energy of atmospheric winds
- **dry adiabatic lapse rate:** the rate at which the temperature of a dry air parcel decreases with altitude as it rises through the atmosphere
- **equivalent potential temperature:** the potential temperature an air parcel would have if its water vapor were condensed and the latent heat added to the parcel
- **hydrostatic balance:** the balance between the downward force of gravity and the upward force resulting from the decline in air pressure with altitude
- **latent heat:** heat released during the condensation of water vapor to form liquid water or the vapor deposition of water vapor to form ice
- **potential temperature:** the temperature an air parcel would have if it were adiabatically expanded or compressed to a standard pressure of 100 kilopascals
- **total potential energy:** the sum of the internal energy and the gravitational potential energy of an air column
- **wet adiabatic lapse rate:** the rate of temperature decrease with altitude of a rising air parcel in which condensation occurs

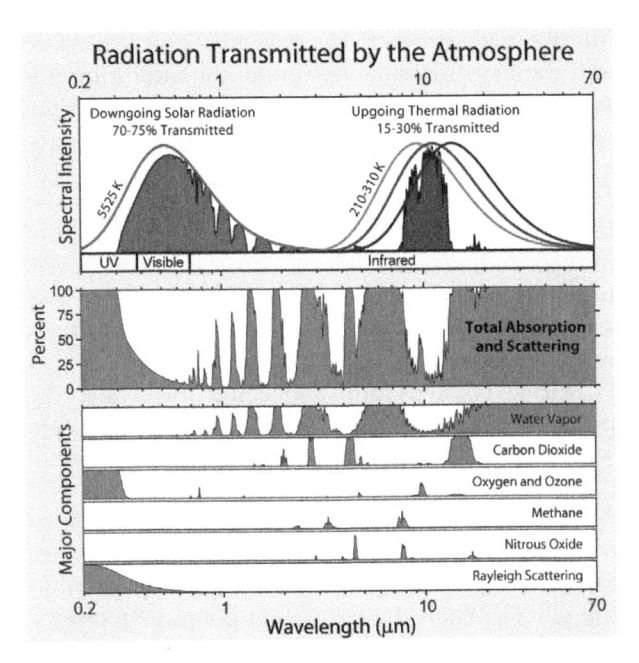

This figure shows the absorption bands in the Earth's atmosphere (middle panel) and the effect that this has on both solar radiation and upgoing thermal radiation (top panel). Individual absorption spectrum for major greenhouse gases plus Rayleigh scatter.

Hurricane Abby approaching the coast of British Honduras. The complete eyewall cloud is visible. Location: Near British Honduras (Belize).

Overview

Earth's atmosphere is a giant heat engine that converts radiant energy from the sun into the kinetic energy of atmospheric winds. Rising air motions associated with atmospheric circulations are responsible for the formation of important weather phenomena such as clouds, precipitation, and severe storms. Depending upon the vertical variation of air temperature, rising air motions may be either enhanced or suppressed by buoyant forces, so the thermal structure of the atmosphere is a key factor in the development of weather.

For the larger-scale motions of the atmosphere, the acceleration of air in the vertical direction is always small. Therefore, an approximate balance exists between the downward force of gravity and the upward force because of the decline in air pressure with height. Motions that satisfy this approximate balance are said to be hydrostatic. The rate at which pressure decreases with height depends on the air density, and, since air behaves as an ideal gas, the density at any given pressure level is determined by the air temperature. Assuming a hydrostatic balance, the pressure at any level of the atmosphere may then be determined from a knowledge of the surface pressure and the vertical distribution of temperature.

Vertical air motions and air temperature are strongly interdependent. This may be understood by considering the buoyant force that acts on a rising air parcel—a small, isolated mass of air that neither mixes nor exchanges heat with the surrounding air. As the parcel rises through the atmosphere, it expands in response to the decrease in atmospheric pressure. During expansion, the parcel works to push back the air in the surrounding atmosphere. If the expansion can be considered adiabatic—that is, if the parcel does not lose or gain heat—the work is done at the expense of the energy associated with the random motions of the parcel's air molecules, the internal energy. Therefore, the temperature of the parcel will decrease as it rises through the atmosphere. In a hydrostatic atmosphere, the first law of thermodynamics may be used to show that the temperature of the parcel will decrease with increasing altitude at a rate of 9.8 Kelvins per kilometer. This rate of temperature decrease is called the dry adiabatic lapse rate.

The buoyant force experienced by the air parcel will depend on how the density of the parcel compares with that of its environment. If the temperature of the air parcel is less than that of the surrounding air, it will experience a downward buoyant force, which acts to return it to the level at which it originated. Similarly, if the air parcel has a temperature that is greater than its surroundings, it will experience a buoyant force, which acts to accelerate its upward motion.

Since the temperature of a parcel changes with its motion, it is useful to define a related thermodynamic variable that is conserved during adiabatic air movements. Potential temperature is defined as the temperature an air parcel would have if it were adiabatically expanded or compressed to a standard pressure of 100 kilopascals. Air temperature and pressure uniquely determine the potential temperature. While the potential temperature of a rising air parcel is constant, that of the surrounding atmosphere may differ from one level to another. The variation of the potential temperature of the atmosphere with height determines whether vertical air motions will be enhanced or suppressed by buoyant forces. If the potential temperature increases with height, the atmosphere is said to be stable, since rising or sinking air parcels will experience buoyant forces that act to return them to their original level. Similarly, an unstable atmosphere is one in which the potential temperature decreases with height.

The presence of water vapor has only a slight effect on the thermodynamic properties of air unless condensation takes place. If the temperature of an air parcel becomes cool enough, water vapor in the parcel will condense to form droplets of liquid water. The potential energy lost by the condensing water molecules is transferred to the surrounding air molecules as latent heat, so the temperature of a rising air parcel in which condensation takes place will decrease less rapidly than the temperature of a parcel in which it does not. The so-called wet adiabatic lapse rate is not constant but depends on temperature and pressure. The altitude at which a rising parcel of surface air begins to experience condensation is called the lifting condensation level. This level is often a good estimate of the altitude at which the bases of convective clouds are found.

The rate at which air temperature decreases with increasing altitude is called the "lapse rate." If the lapse rate of the atmospheric environment is intermediate between the wet adiabatic lapse rate and the dry adiabatic lapse rate, the atmosphere is said to be conditionally unstable. In such an environment, a rising parcel of surface air will experience a downward buoyant force unless it is lifted high enough that condensation occurs and the release of latent heat causes its temperature to exceed that of its surroundings. The altitude above which the parcel becomes positively buoyant and continues to rise without an external source of lifting is called the level of free convection.

During condensation and cloud formation, the potential temperature of an air parcel is not conserved. Yet, it is possible to define an equivalent potential temperature that is conserved.

Equivalent potential temperature is the potential temperature an air parcel would have if all of its water vapor were condensed and the latent heat added to the parcel. When the equivalent potential temperature of the atmospheric environment decreases with height, the atmosphere is said to be convectively unstable. Under these conditions, if a layer of air is lifted through a large enough distance, the bottom of the layer will experience a greater heating because of the release of latent heat than will the top. Thus, the layer will become unstable even if the original lapse rate within the layer was stable. Most outbreaks of severe weather occur when the atmosphere is convectively unstable.

Once clouds form, some liquid water droplets and/or ice crystals within the cloud may grow large enough to fall to Earth as precipitation. At middle latitudes, most clouds contain some ice, and the growth of ice particles dominates the production of precipitation. This results from the fact that the amount of water vapor that is needed within a cloud for the growth of an ice crystal is significantly less than that needed for the growth of a liquid cloud droplet of the same size. Ice crystals, therefore, will grow at the expense of liquid droplets within the same cloud. The droplets evaporate and supply water vapor to the growing ice particles. Once the ice crystal has grown to a large enough size by vapor deposition, it will fall through the cloud and grow further by the collection of other cloud particles until it reaches precipitation size.

Atmospheric conditions that control local weather phenomena, such as clouds and precipitation, are determined by the large-scale motions of Earth's atmosphere, which are, in turn, driven by radiant energy from the sun. Near Earth's equator, the total solar energy incident at the top of the atmosphere exceeds the energy emitted by the atmosphere to space in the form of infrared radiation. By contrast, the polar regions of Earth's atmosphere lose more energy to space than they receive from the sun. The net effect is to increase the internal energy of the atmosphere near the equator and to deplete that energy near the poles. The winds of Earth's atmosphere play a dominant role in transporting excess energy poleward to balance the energy deficit at high latitudes.

In a hydrostatic atmosphere, the gravitational potential energy of an air column bears a constant ratio, about 0.4, to the internal energy. If internal energy is either added by heating or converted to kinetic energy, the gravitational potential energy changes to maintain the same ratio.

Therefore, it is convenient to treat gravitational potential energy and internal energy together as the total potential energy. The amount of potential energy available to drive atmospheric circulations depends on the distribution of potential temperature. In the case of a stable and horizontally uniform atmosphere, no potential energy would be available for conversion to kinetic energy. Thus, the north-south temperature gradient created by the latitudinal variation in solar heating makes energy available to atmospheric circulation systems.

The portion of the potential energy that is available for conversion to kinetic energy during an adiabatic redistribution of air to create a horizontally uniform and stable atmosphere is called the available potential energy. Such a redistribution of air requires the sinking of cold air and the rising of warm air so that gravitational potential energy and internal energy are released simultaneously. At middle latitudes, these rising and sinking motions occur in association with large-scale circulations around low-pressure systems that develop along fronts. As warm air collides with and is lifted over cold air along the frontal boundary, potential energy is converted into kinetic energy from the developing circulation. Most significant weather events at middle latitudes are associated with this type of circulation system.

APPLICATIONS

The concepts of atmospheric thermodynamics are of considerable practical importance in many areas of atmospheric science, including weather analysis and forecasting, the assessment of environmental impact, and weather modification.

Much of Earth's weather results from the interaction of adjacent air masses having different characteristics of temperature and moisture content. It is therefore important to be able to follow the motions of air masses by identifying characteristic properties that are conserved during their motion. Since most air motions are approximately adiabatic, air parcels will tend to move along surfaces of constant potential temperature. If the mixing ratio—the ratio of the mass of water vapor to the mass of dry air in an air parcel—is plotted on a map representing a surface of constant potential temperature, the motions of air masses become apparent, since mixing ratio is a conserved quantity outside regions where condensation occurs. Such an analysis is called an isentropic analysis.

Severe weather in the United States is most commonly associated with the southeasterly flow of warm moist air from the Gulf of Mexico near the surface of Earth and the westerly flow of cool dry air aloft. Under these conditions, the atmosphere is convectively unstable, and the change in wind direction with height favors the development of long-lived thunderstorms. If the afternoon heating of Earth's surface is sufficient, the resulting convective motions may trigger the formation of severe weather.

One measure of atmospheric stability that is used in severe weather forecasting is the lifted index. To calculate this index, consider an imaginary parcel of air with a temperature equal to the forecast high temperature for the day and a moisture content equal to that averaged over the lower atmosphere. If one imagines such a parcel to rise to the middle of the atmosphere, about 50 kilopascals, its temperature there can be calculated on theoretical grounds. As it is lifted, the parcel's temperature decreases at the dry adiabatic lapse rate until it reaches its lifting condensation level. From that point upward, its temperature decreases more slowly at the wet adiabatic lapse rate. At its destination, the calculated temperature of the parcel is subtracted from the measured temperature of its environment to obtain the lifted index. A large negative lifted index indicates a high probability of severe weather. Lifted indexes more negative than about -2 are correlated with tornado-producing storms.

Atmospheric stability also has a profound influence on the rate at which atmospheric pollutants are diluted by mixing with air. Turbulent eddies, which are responsible for most of the mixing, may be enhanced by an unstable environment or suppressed by a stable environment. At night, the radiational cooling of Earth's surface and adjacent air typically produces a stable atmosphere in which the dilution of atmospheric pollutants is very slow. Pollutants may be transported large distances from their source by the wind before their concentrations are appreciably reduced. During the daylight hours, the lower atmosphere is generally unstable, and convective motions driven by solar heating mix pollutants upward efficiently.

Often, the layer mixed by convection is capped by a temperature inversion, a stable layer of air in which temperature increases with height. The depth of the atmospheric layer through which pollutants are mixed is an important factor in determining their concentrations.

Therefore, accurate modeling of air quality requires a knowledge of the level at which the upper-level inversion occurs. Normally, the height of the inversion is determined from temperature measurements taken at several altitudes by an instrumented balloon called a radiosonde.

Strong upper-air inversions are usually formed by sinking air motions, subsidence, associated with systems of high pressure. The convective mixing in the lower atmosphere prevents these sinking motions from extending all the way to Earth's surface, so only air parcels above the mixed layer move downward, and the adiabatic temperature increase the sinking motions experience results in the formation of a so-called subsidence inversion. Most severe air pollution episodes are caused by subsidence inversions created by slow-moving high-pressure systems.

Since weather is such an important factor in many economic endeavors, including agriculture, the prospect of weather modification has been the focus of much attention. Most attempts at weather modification are based on the fact that ice crystals, if they exist within a cloud at the correct concentration, will grow at the expense of water droplets in the same cloud and may eventually produce precipitation. If a cloud whose upper parts are at a temperature below 0 degrees Celsius has too little ice to produce precipitation efficiently, it may be possible to enhance the precipitation by introducing ice nuclei into the cloud particles, which promote the growth of ice crystals. The addition of ice nuclei into a cloud is referred to as cloud seeding.

Unfortunately, most studies on the enhancement of precipitation by cloud seeding have not been carried out with enough care to allow proper statistical evaluation of the results. With further study, however, this technique may eventually prove to be a useful tool for weather modification.

Context

By the early eighteenth century, it was already widely accepted that atmospheric winds must somehow be the result of differences in solar heating between the equator and the poles.

Nevertheless, the exact relationship between surface winds and the latitudinal variation in air temperature was not generally understood. In 1735, George Hadley (1685-1768) proposed a model of the general circulation of Earth's atmosphere, in which atmospheric winds were driven by a single thermal circulation cell between the equator and the poles. The trade winds—the easterly winds in the tropics—were explained in terms of the deflection of winds in the circulation cell by the rotation of Earth. This model more firmly established the connection between solar heating and large-scale air motions. William Ferrel modified Hadley's model in 1856 to include three main circulation cells in order to account for observed winds at higher latitudes.

In 1903, when Max Margules (1856-1920) first introduced the idea of total potential energy, an understanding of the energetics of middle-latitude cyclones or low-pressure systems began to emerge. Nevertheless, it was not until 1918 that Vilhelm (1862-1951) and Jacob (1897-1975) Bjerknes developed the polar front theory of cyclones and convincingly showed that those storms receive their energy from the interaction of air masses of different temperatures. In 1926, Sir Harold Jeffreys (1891-1989) showed that middle-latitude cyclones play an important role in maintaining the general circulation. The concept of available potential temperature and its usefulness in the study of the general circulation were presented by Edward Lorenz (1917-2008) in 1955.

The forecasting of weather by solving the fundamental equations describing the physics of the atmosphere has been realized only in recent years. The difficulty lies in part in the fact that the equations are very complicated and general solutions have not been forthcoming. In 1904, Vilhelm Bjerknes first suggested that forecasts might be made by solving the equations numerically. Lewis Fry Richardson (1881–1953) showed, in 1922, that equations describing the state of the atmosphere could be reduced to many approximate algebraic equations using appropriate techniques. Since digital computers were not yet available, however, the computations needed to obtain solutions were too lengthy to allow the technique to be useful for weather forecasting. In 1937, Carl-Gustaf Arvid Rossby (1898–1957) developed a simple equation for forecasting the movement of large-scale troughs and ridges of pressure in the middle atmosphere by using the concept of vorticity—the spin of air around a vertical axis. Rossby's equation assumed that the atmosphere was barotropic, that is, that the temperature of the atmosphere was uniform in the horizontal.

Thus, the equation was unable to predict changes in the strength of atmospheric disturbances that result from the release of available potential energy. Nevertheless, the first successful numerical forecasts performed in 1950 were based upon Rossby's work. By 1953, forecast models capable of predicting the development of middle-latitude cyclones were being integrated numerically. As more powerful generations of digital computers become available, numerical forecast models continue to include more detailed descriptions of the physics of weather.

—R. D. Russell

Further Reading

Aguado, Edward, and James E. Burt. *Understanding Weather and Climate.* 7th ed. Pearson Prentice Hall, 2015. Discusses meteorology and climatology concepts with reference to common, everyday events. Presents conclusions from the IPCC as well as many other scientific studies on climate change. Examines weather events, structure and dynamics of atmosphere, and the past, present, and future climate on earth.

Frederick, John E. *Principles of Atmospheric Science.* Jones & Bartlett, 2008. Introductory text describing the various fields of atmospheric sciences, including atmospheric chemistry, atmospheric physics, and climatology. Describes techniques and research methods utilized in modern climate and atmospheric research.

Friedman, Robert Marc. *Appropriating the Weather: Vilhelm Bjerknes and the Construction of a Modern Meteorology.* Cornell University Press, 1989. The author traces the development of meteorological thought that led to a modern understanding of middle-latitude cyclones. The relationship between the development of the commercial applications of meteorology and that of the science itself is explored. The career of the Norwegian meteorologist Vilhelm Bjerknes is central to the discussion.

Lutgens, Frederick K., Edward J. Tarbuck, Tasa, Dennis G. and Herman, Redina. *The Atmosphere: An Introduction to Meteorology.* 14th ed. Pearson Education, 2018. An updated version of this textbook, in loose-leaf format, provides students with a general overview of the structure and properties of Earth's atmosphere.

Marshal, John, and R. Alan Plumb. *Atmosphere, Ocean and Climate Dynamics: An Introductory Text.* Elsevier Academic Press, 2008. An excellent introduction to atmospheres and oceans. Discusses topics such as the greenhouse effect, convection and atmospheric structure, oceanic and atmospheric circulation, and climate change. Suited for advanced undergraduates and graduate students with some background in advanced math.

PLANKTON AND THE MITIGATION OF THE GREENHOUSE EFFECT

FIELDS OF STUDY

Biochemistry; Environmental Chemistry; Environmental Sciences; Environmental Studies; Waste Management; Bioclimatology; Oceanography; Physical Geography; Ecosystem Management; Earth System Modeling; Ecology; Pathology; Toxicology

SUMMARY

Plankton are comprised of a large variety of different species, classed as phytoplankton or as zooplankton, ranging in size from single-celled organisms to tiny gatherings of algae. Oceanic nutrient levels are the major determinant of phytoplankton productivity,

Planctonic polychaete worm from genus Tomopteris.

while zooplankton productivity is dependent upon the availability of phytoplankton. Together, phytoplankton and zooplankton directly or indirectly feed all the rest of the animals in the ocean, including human fisheries.

PRINCIPAL TERMS

- **bloom:** the sudden or rapid development of a large population of plankton and algae
- **carbon cycle:** the cyclic mechanism whereby carbon, as carbon dioxide, is removed from the atmosphere, converted to glucose and other compounds in photosynthesis, or to carbonate solids, and subsequently returned to the atmosphere when those materials break down or decompose
- **carbon sink:** any physical process, such as forest or plant growth, that consumes carbon dioxide in the atmosphere
- **euphotic zone:** the upper layer of the oceans, where light-dependent life forms can exist
- **photosynthesis:** the process occurring in green plants for the production of glucose from carbon dioxide and water in the presence of chlorophyll and sunlight

PLANKTON

Plankton (from the Greek planktos, or wanderer) comprise a vast assortment of life-forms with limited or no swimming power that largely drift in the ocean. They are divided into the plantlike phytoplankton and the animal-like zooplankton.

Phytoplankton comprise at least four thousand species of plants that, as on land, use sunlight in the process of photosynthesis to generate sugars and other high-energy organic compounds. They live in the ocean's euphotic zone, the sunlit upper layer, which is only tens of meters deep. Phytoplankton range in size from microscopic to a millimeter in length and include single-celled organisms and tiny clumps of algae.

Because the ocean covers 71 percent of the Earth's surface, phytoplankton are a major driver of Earth's carbon cycle. Their photosynthesis extracts carbon dioxide (CO_2) from surrounding water and replaces it with oxygen. The resulting changes in oceanic gas levels change atmospheric gas levels as well. Some phytoplankton (such as coccolithophores) accrete calcium carbonate ($CaCO_3$) for shielding. If the carbon contained in their bodies reaches the ocean depths, it may be sequestered there for years, or even for millions of years.

Oceanic nutrient levels are major determinants of phytoplankton productivity. The most fertile areas are river estuaries, shallow waters, and upwelling areas in the deep ocean. Phytoplankton are eaten by zooplankton. Zooplankton range mostly from microscopic to the size of small snails, although jellyfish, some more than 2 meters in diameter, also swim weakly and can be classed as zooplankton.

Zooplankton slow the carbon cycle, because they consume phytoplankton and emit CO_2. Phytoplankton "blooms" of increased productivity are quickly followed by surging zooplankton populations. Conversely, zooplankton contribute to carbon cycling by accreting carbonate shells, dropping fecal pellets, and falling to the sea bottom upon death.

Together, phytoplankton and zooplankton directly or indirectly feed all the rest of the animals in the ocean, including human fisheries. Oceanic acidification may affect both planktonic food production and planktonic capture of CO_2. Increasing CO_2 levels in the atmosphere translate into higher levels in oceanic surface waters and favor increased photosynthesis. However, increased CO_2 levels also cause greater ocean acidity. Increased acidity in the waters may strongly hinder carbonate shell-building among plankton and other marine life-forms. If so, it would slow the marine carbon-capture process and contribute to even greater atmospheric CO_2 levels.

SIGNIFICANCE FOR CLIMATE CHANGE

In 1988, oceanographer John Martin declared, "Give me half a tanker of iron, and I can start a new ice age." Martin was referring to phytoplankton growth in the "bluewater desert" areas of the deep ocean. The creatures' growth rate is often limited by the availability of trace amounts of iron. It has been hypothesized that ice ages may result from large amounts of wind-blown dust enriching the oceans with iron. Such enrichment could have caused phytoplankton blooms that reduced atmospheric CO_2 levels, thus reducing the greenhouse effect and cooling the planet.

Similarly, Martin suggested that artificial iron fertilization in the oceans might reduce global warming. (A major campaign of oceanic fertilization would be a species of geoengineering.) Several limited experiments of a few hundred square kilometers and a few days duration have confirmed major iron fertilization in the Pacific Ocean and in the southern ocean around Antarctica.

Phytoplankton emit the sulfur-bearing gases dimethyl sulfide (DMS, CH_3SCH_3) and carbonyl sulfide (COS). Their breakdown product, sulfur dioxide (SO_2), produces airborne particles (aerosols) that reflect visible light but allow infrared light (heat waves) to pass through, thus causing more cooling. Ocean fertilization could be self-funding, because a part of the increased planktonic production could be harvested via increased fisheries production.

The major objections to oceanic fertilization are as follows: First, as noted, zooplankton populations grow to feed on phytoplankton blooms, and they quickly return much CO_2 to the atmosphere. Second, even if it is not eaten, much planktonic biomass decays and gives up captured carbon before it can sink to the bottom, so it is not an efficient carbon sink. Third, if fertilization were widely implemented, organic material reaching the deep ocean would vastly increase. In decaying, it could harm marine life by lowering oxygen levels in the deep waters. Fourth, plankton emit some greenhouse gases (GHGs), such as oxides of nitrogen, that might cancel some aerosol cooling. Finally, aerosols released by fertilization in sufficient amounts to cause noticeable cooling would also significantly decrease the sunlight available for photosynthesis. For these reasons, opponents of oceanic fertilization argue that it might result in some mitigation of global warming, but it could never represent a comprehensive solution to the problem.

—*Roger V. Carlson*

Further Reading

Adhiya, Jagat, and Sallie W. Chisholm. "Is Ocean Fertilization a Good Carbon Sequestration Option?" MIT Press, 2001. This white paper presents the case against oceanic fertilization.

Field, Christopher B., and Michael R. Rapauch, eds. *The Global Carbon Cycle: Integrating Humans, Climate, and the Natural World.* Island Press, 2004. This compendium of articles from the Scientific Committee on Problems of the Environment (SCOPE) discusses the major atmospheric sources and sinks. The final chapter details possible advantages and limitations of ocean fertilization.

Mitra, Abhijit, Kakoli Bannerjee, and Avijit Gangopadhyay. *Introduction to Marine Plankton.* Daya, 2008. Although the authors emphasize plankton in the waters near India, they provide an excellent overall description of plankton, its role in the carbon cycle, fertilization effects, identification, and even culturing. However, the book does not comment directly on oceanic fertilization or acidification.

"The Ocean in a High-CO_2 World." Special Section in *Journal of Geophysical Research* 110, no. C9 (2005). Includes papers on the carbon cycle, as well as articles making arguments for and against oceanic fertilization.

Royal Society. *Ocean Acidification Due to Increasing Atmospheric Carbon Dioxide.* Author, 2005. The Royal Society in the United Kingdom assembled a group of distinguished scientists to summarize the issues of oceanic acidification in this study.

PLEISTOCENE CLIMATE

FIELDS OF STUDY

Analytical Chemistry; Atmospheric Chemistry; Biochemistry; Environmental Chemistry; Physical Chemistry; Photochemistry; Geochemistry; Thermodynamics; Climate Modeling; Bioclimatology; Hydroclimatology; Climate Classification; Oceanography; Physical Geography; Earth System Modeling; Ecology; Crystallography; Spectroscopy; Mass Spectrometry; Optics

Woolly mammoths were driven to extinction by climate change and human impacts. The image depicts a late Pleistocene landscape in northern Spain with woolly mammoths (*Mammuthus primigeniu*), equids, a woolly rhinoceros (*Coelodonta antiquitati*), and Europe and European cave lions (*Panthera leo spelaea*) with a reindeer carcass.

SUMMARY

Characterized by repeated cycles of glacial cold and interglacial warmth, Pleistocene climate was always changing. Scientists believe that the timing of these changes was controlled by periodic variations in the orbit of the Earth. They are still trying to figure out how the small changes produced by these variations were amplified into dramatic climatic shifts. Of particular interest is the question of if and when another glacial advance will occur.

PRINCIPAL TERMS

- **climatic precession:** cycle of variations in the Earth-Sun distance at summer solstice
- **loess:** deposits of very fine grained, wind-blown material often associated with glacial deposits
- **marine isotope stage:** half of a glacial cycle, as identified in the oxygen isotope data from ocean cores
- **orbital eccentricity:** cyclically variant deformities in Earth's orbit
- **Pleistocene epoch:** an epoch within the geologic timescale, ending at 11,700 years ago
- **Pliocene epoch:** the epoch preceding the Pleistocene and beginning around 5.332 million years ago

CLIMATE DURING THE PLEISTOCENE

Geologists divide up the 4.6 billion years of Earth's history, usually based on the planet's fossil record. The planet's history is divided into eras, which are divided into periods, which are divided into epochs, which are divided into ages. Eras are neatly separated by major mass extinctions, and most of the other divisions are also reasonably clear. However, the division between two period/epoch pairs has proven to be more difficult: The base of the Quaternary period and the base of the Pleistocene epoch are either at the top or the bottom of the Gelasian age.

The ages within both the Pliocene and the Pleistocene epochs are determined by the marine fossils from those epochs, which are not difficult to date. A problem arises, however, from an attempt to put the beginning of the ice ages at the right age boundary. In 1948, at a meeting of the International Union of Geological Sciences, the top of the Gelasian age (1.81 million years ago) was chosen as the advent of the ice ages. Since then, a tremendous amount of climate-related research has suggested that the bottom of the Gelasian, 2.59 million years ago, might have been a better choice. As often happens, both dates are used. The discrepancy is not significant in the context of climate issues. Because the climate began to behave differently a little before 2.59 million years ago and it did not change much 1.81 million years ago, a description of Pleistocene climate will also apply to Gelasian climate, whether one considers the Gelasian age to be part of the Pleistocene or part of the Pliocene epoch.

Around 2.6 million years ago, the climate began to oscillate between glacial and interglacial states. When only land-based data on glacial advances were available, only a few advances were described, but as the ability to interpret isotope data from ocean cores developed, it was determined that there have been at least fifty advance/retreat cycles in the last 2.6 million years. The shift from continental rocks to ocean sediment cores as the source of definitive data was important in providing this additional data.

The surfaces of the continents, particularly when being scraped repeatedly by large glaciers, do not keep very good archives. In some places, glacial deposits lie on top of other glacial deposits, and those earlier deposits often have complete soil profiles. Geologists thus knew that there had been more than one glacial period, but they had little detailed knowledge about the succession of such periods. Each new advance usually removed any former glacial deposits, obscuring the record. In the deep oceans, however, remains of dead organisms accumulate, and there is little to remove them. As isotope geochemistry developed, it became apparent that the ratio of oxygen-18 (^{18}O) to oxygen-16 (^{16}O) preserved in these remains changed systematically in all of the world's oceans during the last three million years or so. Evaporation removes more ^{16}O, and during glaciations will sequester much of this isotope in the ice sheets, thereby enriching the oceans in ^{18}O. Fluctuations in the isotope ratio can indicate whether the Earth was in a glacial (high $^{18}O/^{16}O$ level) or interglacial (low $^{18}O/^{16}O$ level) state.

As cores of sediments were retrieved from the ocean floor, changes in the $^{18}O/^{16}O$ ratio as a function of time were observed and analyzed. Shifts were numbered, starting with 1 for the present interglacial and using odd numbers for interglacials and even numbers for glacials. The time period represented by each number is called a marine isotope stage (MIS). Glacial cycles already known from continental evidence correlated with some of the MIS cycles, but there were many more MIS cycles, they were global in nature, and their dates were known more accurately. Terms such as "Illinoian" and "Moscovian Dnieper" were replaced by designations such as "MIS 6."

Using Fourier analysis and other methods, scientists found that some periods of time were very strongly represented in the marine isotope record. These peaks in information seemed to correspond to natural cycles. (Imagine recording the sounds from a neighborhood continuously for a decade and then analyzing them. One might expect to see information peaks at intervals corresponding to the cycles governing human activity: every twenty-four hours, every week, and possibly every year.) The peaks in the sediment cores occurred every forty-one thousand years, about every twenty-two thousand years, and every hundred thousand years.

Causes

A glacier will grow if snow accumulates over periods of years, as snow that fell during the previous winter does not melt completely during the summer. Thus, summer temperatures are the limiting factor in glaciation. If summers are not warm enough to melt all of winter's ice accumulation, the glaciers will advance. Winter temperatures do not matter to this process, so long as they remain below freezing.

Temperatures on Earth are determined primarily by the Sun, and its influence is modulated by periodic changes in the tilt of the planet's rotational axis, the precession of that axis, and the eccentricity of Earth's orbit. These changes were known by the middle of the nineteenth century, and James Croll suggested in 1864 that they might be responsible for the advance and retreat of continental glaciers. Data, computations, and dating techniques were not yet sufficient to support the theory, however, and interest in it waned until Milutin Milanković resurrected the idea in 1913. He worked on the problem for decades, publishing his completed book in 1941; it was translated into English in 1969.

Milanković was able to calculate how much solar energy would reach a latitude of 65° north during the summer. He obtained this result by combining the effects of a 100,000-year cycle in the eccentricity of the Earth's orbit, a 41,000-year cycle in the inclination of the Earth's axis, and a roughly 22,000-year cycle in climatic precession. Thus, the peaks evident in ocean sediment cores matched precisely Milanković's cycles.

When the ocean core analyses were complete, there was no question that glacial advances and retreats followed the timing of the Milankovitch cycles. As a result, serious concerns were raised for a time about a coming glacial advance. Some papers written in the 1970s—as well as meetings, symposia, and water cooler discussions—concerned the global cooling that Milanković's work seemed to predict. In part, this was due to a decade-long cooling period in the Northern Hemisphere, but it also stemmed from the recognition that, according to Milanković, the current interglacial should end soon.

Trying to guess when it will end, scientists examined the record. Between 3 million and 0.8 million years ago, the cycles were dominated by the 41,000-year cycle. Afterward, however, a period began that is often called the Mid-Pleistocene Transition, and the cycles became dominated by the 100,000-year cycle. Although many theories have been put forward to explain this transition, it is still not well understood. In addition, it is unknown whether the current interglacial period has been extended by the effects of agriculture and other human activities over the last eight thousand years or whether it would have been longer than average regardless of human behavior.

Pleistocene Weather

Estimates made from isostatic rebound studies suggest that the continental glaciers at their height during Pleistocene glaciations were 1–2 kilometers thick. The ice moved from north to south, reaching a latitude of 40° north, near the northern boundary of the state of Pennsylvania. Even when the glacier stopped advancing, the ice within it continued to move from north to south, grinding away at the material beneath it. At its southern edge, melting converted the ice to liquid water.

Where the glaciers terminated, the ice rose steeply, with large accumulations of ice adjacent to large lakes and hills. In the summer, this ice must have caused wind and precipitation patterns quite different from those experienced during interglacial periods. All year long, the presence of the glacier is likely to have altered global wind patterns. It was white, cold, and added considerably to the elevation. With so much water tied up in ice, sea level fell by about 100 meters, exposing huge expanses of the continental shelves to weathering and erosion.

CONTEXT

If Earth's climate continues to behave cyclically, as it has for the past 2.6 million years, then another glacial advance will occur in the future, perhaps within a millenium or two. Alternatively, anthropogenic inputs of greenhouse gases may have warmed the planet so much that another glacial advance will only occur after humans stop burning fossil fuels and the resulting emissions have left the atmosphere. A third possibility is that anthropogenic outputs have altered the land so much that continental glaciations are no longer possible.

Clearly, these alternatives represent very different climate change scenarios. Although scientists are confident that Milankovitch cycles were the pacemakers for Pleistocene climate change, there is little agreement on the mechanisms connecting these pacemakers to the various drivers and amplifiers of the climate system. Nor is there agreement as to why the dominant cycle shifted from one of 41,000 years to one of 100,000 years in the middle of the Pleistocene epoch. As these and other issues become better understood, it may become possible to predict the future with greater confidence.

—*Otto H. Muller*

FURTHER READING

Alley, Richard B. *The Two-Mile Time Machine. Ice Cores, Abrupt Climate Change and Our Future* Princeton University Press, 2014. Describes the analytical methods and interpretations of physical clues of past climates trapped in ice that is thousands of years old.

Bradley, Raymond S. *Paleoclimatology: Reconstructing Climates of the Quaternary*. Academic Press, 2015. A definitive introduction to the science of paleoclimatology, and the use of climate proxies.

Jouzel, Jean, Lorius, Claude and Raynard, Dominique *The White Planet. The Evolution and Future of Our Frozen World*. Princeton University Press, 2013. Examines the past, present and future roles of ice on Earth, concluding that climate change and global warming will ultimately have disastrous consequences for human societies on Earth regardless of technological advances.

Macdougall, J. D. *Frozen Earth: The Once and Future Story of Ice Ages*. University of California Press, 2004. Compelling tale for the general reader about how the Pleistocene ice ages were discovered, studied, and understood. Provides biographical sketches of many of the major contributors. Photos, diagrams, maps, annotated bibliography.

Mackay, Anson, Battarbee, Rick, Birks, John and Oldfield, Frank, eds. *Global Change in the Holocene*. Routledge, 2014. Chapters by several contributors describe Holocene climate reconstruction from the resources of many different climate proxies.

Ruddiman, W. F. *Plows, Plagues, and Petroleum: How Humans Took Control of Climate*. Princeton University Press, 2005. Written for the lay public, this book provides the background and thinking behind the theory that humans have influenced the climate for the last nine thousand years, primarily through agriculture. Illustrations, figures, tables, maps, bibliography, index.

POLAR BEARS AND CLIMATE CHANGE

FIELDS OF STUDY

Bioclimatology; Environmental Sciences; Environmental Studies; Ecosystem Management; Physical Geography; Waste Management; Meteorology; Oceanography

SUMMARY

The polar bear has become an icon of global warming. In 1973, the five "polar bear nations" (Canada, Denmark, Norway, the former Soviet Union and the United States) agreed to limit polar bear hunting to

Polar bears.

the Inuit and other native peoples. The bear population recovered, only to face a new threat: global climate change.

PRINCIPAL TERMS

- **anthropogenic climate change:** changes in overall long-term weather patterns due to human activity
- **climate:** long-term average weather conditions
- **greenhouse effect:** an atmospheric phenomenon in which certain molecular components of the atmosphere capture infrared radiation from a planet's surface and prevent it from radiating back out into space
- **greenhouse gases (GHGs):** atmospheric gases whose molecules absorb infrared radiation from a planet's surface and then emit a significant portion of it back into the atmosphere, most commonly carbon dioxide, methane, water, and ozone.

THE LIVES OF POLAR BEARS

Polar bears are the largest and most dangerous carnivorous species of North America. Males and females live solitary lives, except for a receptive female encounters an acceptable male. At all other times, polar bears avoid each other as much as possible, with occasional meetings sometimes ending in deadly fights. After mating, the male and female polar bears go their separate ways, and as winter begins the female makes a den, where she will give birth while hibernating. The cubs will suckle from their mother while she continues to hibernate, depleting her stores of body fat and protein. A normal litter consists of two cubs, with three being born occasionally. When the mother emerges from hibernation, she is typically weakened and emaciated, and must immediately find food in sufficient quantities to meet the nutritional needs of herself and her cubs. She will shepherd her cubs for two years or more before sending them away to live on their own. It is imperative that she avoid any meeting with a male during the first year of this cycle, as the male will do his best to kill the cubs. After the first year, however, the cubs can be left to fend for themselves briefly as the female goes through another mating cycle.

Changes in sea ice, permafrost, and prey species in the Arctic and subarctic have been linked to reduced body condition and smaller litter size in polar bears. In western Hudson Bay, long-term studies conducted by wildlife biologist Ian Stirling of the Canadian Wildlife Service and his colleagues documented climate warming and a significant reduction in the amount, location, and persistence of sea ice adjacent to the shore. The polar bear is heavily dependent on the ringed seal as a food source. Its usual hunting technique requires sea ice. A bear will locate the breathing hole of a seal and then ambush the animal when it comes up for air or will prey upon young seals in their dens. As sea ice diminishes, access to seals diminishes accordingly and the bears are less able to catch their prey.

Hungry polar bears waiting on shore for sea ice to form have become a problem, invading northern villages and encountering native hunters. At one time, this increase in sightings was interpreted as evidence of increasing numbers and used to justify higher quotas for the taking of polar bears. The modern consensus is that polar bears are increasingly attracted to human settlements by trash and other potential food sources due to a decline in their natural hunting conditions. The interaction between bears and humans further threatens the bear population, bringing disease and harmful behavioral changes as well as the killing of bears that pose threats to humans.

SIGNIFICANCE FOR CLIMATE CHANGE

A series of studies coordinated by the United States Geological Survey and released on September 7, 2007, predicted that two-thirds of polar bears would perish by the year 2050. All polar bears in Alaska would be gone. The studies—conducted by American and Canadian scientists—used conservative assumptions to project loss of sea ice due to global warming.

At the time of the studies, there were an estimated twenty thousand to twenty-five thousand polar bears worldwide.

Because the polar bear, *Ursus maritimus*, is thought to have evolved from the brown bear, *Ursus arctos*, 200,000 years ago, some have suggested that the polar bear might reverse course and adapt to and live in the terrestrial ecological niche now occupied by the brown bear. Indeed, in recent years a number of northern bears have been identified that appear to have characteristics of both brown and polar bears, thought to have resulted from the mating of the two species. However, brown bears are primarily vegetarian, whereas polar bears are primarily carnivorous. Brown bear claws are adapted for digging, polar bear claws for grabbing and holding prey. Polar bears are slow to reproduce, and it is unlikely that they could change both their behavior and their physical characteristics quickly enough to cope with disappearing sea ice. Sea ice is central to the life of a polar bear. It provides habitat to prey animals and hunting, mating, and denning sites for the bears themselves. In western Hudson Bay, for example, polar bears gather on the shore each year to wait for sea ice to form.

While polar bears and their response to global warming have been studied longer and in greater depth than has any other Arctic species, they are far from the only animal affected by a changing climate. Gray whales migrating south to their breeding grounds off the coast of Mexico have recently appeared to be emaciated, suggesting a shortage of their normal diet of tube worms and amphipods in their summer feeding grounds of the Bering and Chukchi Seas.

Other marine animals that are affected by the loss of sea ice and climate warming in the Arctic include ringed seals, bearded seals, ribbon seals, sea lions, walruses, narwhals, fish, and many sea birds. Biologists have only begun to study how these creatures will respond to the profound changes in their habitat. Polar bears—known to eat dead whales washed up on the shore as well as other carrion—might benefit temporarily from the die-off of marine animals, but will suffer eventually as their prey species are reduced.

On May 14, 2008, the polar bear was listed as threatened under the provisions of the Endangered Species Act. This designation is one step down from that of endangered species and offers less protection from human activities. However, Dirk Kempthorne, then secretary of the interior under President George W. Bush, made clear that the Department of the Interior would not allow the polar bear's plight to be used to justify limits on greenhouse gases (GHGs) saying: "When the Endangered Species Act was adopted in 1973, I don't think terms like 'climate change' were part of our vernacular." The act, he said, "is not the instrument that's going to be effective" in dealing with climate change.

Images of polar bears have been used by conservation groups to advocate reductions in GHGs in hopes of slowing and perhaps reversing global warming and also in fund-raising by these groups. Global warming skeptics complain that such images appeal to emotion and not to reason and demand further proof that a warming climate will result in great reduction and even extinction of polar bear populations by mid-century.

In the 2010s it became increasingly clear that the threat posed to polar bears by climate change was indeed backed up by science. In early 2017, the U.S. Fish and Wildlife Service lent its support to the argument that declining sea ice was the primary threat to polar bears, and suggested decisive action to halt the warming of the Arctic was necessary to prevent the species' extinction. Experts agreed that limiting greenhouse gas emissions would improve the survival chances of polar bears, but noted that the apparent climate change skepticism of the administration of President Donald Trump meant immediate action was unlikely.

—*Thomas Coffield*

FURTHER READING

Fears, Darryl. "Without Action on Climate Change, Say Goodbye to Polar Bears." *The Washington Post*, 9 Jan. 2017, www.washingtonpost.com/news/energy-environment/wp/2017/01/09/without-action-on-climate-change-say-goodbye-to-polar-bears/?utm_term=.dbce906341f1. Accessed 8 Nov. 2017.

Kazlowski, Steve. *The Last Polar Bear: Facing the Truth of a Warming World*. Mountaineers Books, 2008. The author has spent thousands of hours observing and photographing polar bears in the wild.

"Polar Bears and Climate Change." *World Wildlife Fund,* 2017, www.worldwildlife.org/pages/polar-bears-and-climate-change. Accessed 8 Nov. 2017.

Stirling, Ian. *Polar Bears.* University of Michigan Press, 1999. Written for the general reader by the world's most experienced polar bear researcher. Richly illustrated with photographs by Dan Guravich.

Stirling, Ian, and Claire L. Parkinson. "Possible Effects of Climate Warming on Selected Populations of Polar Bears (*Ursus Maritimus*) in the Canadian Arctic." *Arctic* 59, no.3 (2006): 261–275. Research paper postulating that climate warming is responsible for Canadian polar bears' declining populations, reduced body condition, and reduced survival among young.

POLAR CLIMATE

FIELDS OF STUDY

Climate Classification; Climate Zones; Environmental Sciences; Environmental Studies; Ecology; Physical Geography; Hydroclimatology; Meteorology; Hydrometeorology; Bioclimatology; Oceanography; Climate Modeling; Earth System Modeling

SUMMARY

Polar climate is characterized by year-round cold conditions, tundra characterized by low-growing vegetation and the absence of trees, permafrost characterized as a permanently frozen soil layer under the topsoil, and ultimately by an ice cap in the farthest northern reaches. These conditions do not apply to the southern polar regions, where the ice-covered continent of Antarctica is completely surrounded by ocean waters.

PRINCIPAL TERMS

- **permafrost:** soils that remain frozen throughout seasonal changes for a period of two or more years
- **topsoil:** the uppermost layer of soil on Earth's surface, being the interface between the solid surface and the atmosphere, and the locus of essentially all plant growth
- **tundra:** the treeless region between the northern tree line, characterized by low-growing vegetation and permafrost subsoil

POLAR CLIMATES

A polar climate is characterized by year-round cold conditions. In the northern polar region, the tundra is characterized by low-growing vegetation and the absence of trees, with permafrost characterized as a permanently frozen soil layer under the topsoil, and ultimately by an ice cap in the farthest northern reaches. The tundra is bordered on the south by forests and on the north by the Arctic lands and has a layer of permafrost. The Arctic is characterized by short, cool summers and long, cold winters. Average Arctic high temperatures range from 0°C in Russia to 12.2°C in Alaska, and average lows range from -16.3°C in Canada to -28.1°C at the highest point of Greenland's ice sheet. These conditions do not apply to the southern polar regions. In Earth's southern polar region is Antarctica, 95 percent of which is covered in ice. Average summer and winter temperatures on the Antarctic continent range from 0°C to -70°C, respectively.

Permafrost is a permanently frozen soil layer under the topsoil. The top layer, or active layer, melts and refreezes each year, while the lower level stays frozen year-round. The active layer provides water and a place for plants to grow. The temperature in the lower levels is at or below 0°C, and usually

Mt. Herschel (3335m asl) from Cape Hallet with Seabee Hook penguin colony in Foreground. Antarctica.

ranges from -6° to -9°C. One-fourth of the land in the Northern Hemisphere is in permafrost zones, and any permafrost areas in the Southern Hemisphere are in Antarctica.

Significance for Climate Change

Although the ice sheet is not expected to melt completely in the near future, the Antarctic is warming ten times faster than the average anywhere else in the world. Since the mid-1940s, the average temperature is -16°C warmer year-round and -13° to -14°C warmer in early winter. The far western edges and ground are becoming warmer, and the ice sheet edge is rapidly eroding. The amount of sea ice was stable from 1840 to 1950, but has decreased since then by roughly 20 percent.

The icebergs are shrinking, glaciers are receding, and the immense Larsen ice shelf began disintegrating in 1995. Nearly 2,590 square kilometers of the ice shelf collapsed between 1998 and 2000. The breakups of ice chunks lead to the faster flow of ice from the glaciers upstream, which in turn can increase the sea level over time. With the edges of the ice sheet melting, the wildlife populations are fluctuating. Adelie penguins are dying off, because they cannot adjust to the changes in their environment. Meanwhile, other species are migrating south and establishing colonies. Scientists have observed the southern migration of elephant and fur seals, as well as other species of penguins. Additionally, low grasses, mosses, and tiny shrubs are thickening in several areas, and the sea-floor areas that had previously been shrouded by overlying ice shelves are now receiving a much higher influx of solar energy that is driving the development of new productive habitat.

Climate changes are more dramatic in the Arctic. The Arctic is ringed by the lands of Canada, Denmark (through Greenland), Finland, Iceland, Norway, Russia, Sweden, and the United States. Sea ice covers the entire Arctic Ocean and nearby waters every winter. This sea ice provides a home for polar bears and transportation routes for people, and shields coastal towns and underwater creatures. Normally, the ice melts on land in the summer and breaks up some in the Arctic Ocean, but there was 80 percent less ice in 2005 than in the 1970s, with more rapid melting since 2003. According to the U.S. National Oceanic and Atmospheric Administration, in the summer of 2016 Arctic sea ice reached its second minimum extent on record. The minimum extent recorded was in 2012, and the 2016 minimum was 29 percent smaller than the average minimum between 1981 and 2010. Scientific data suggest that the Arctic Ocean may be ice-free in the summer as early as the 2030s, a condition that has not existed in at least 800,000 years. A rise in ice sheet melting can lead to a rise of sea level, particularly as land-bound glaciers are enabled to advance more freely into ocean waters without sea ice formations to impede their movement, and even a 1–2 millimeter rise increases flooding at the coastlines.

Summertime temperatures in parts of the Arctic are warmer than those recorded in the last four hundred years, and winter temperatures have increased by 2° to 4°C since 1976. The Arctic tends to cool Earth, because it reflects more incoming solar radiation than it absorbs and increase Earth's albedo, with ice and snow being the most reflective, followed by tundra vegetation. The warmer temperatures have led to unpredictable weather patterns that threaten the delicate natural balance of the environment and endangered the plants and animals, as well as the humans who depend on them. The ice sheets near the coastline are smaller or nonexistent, which greatly affects subsistence whalers and seal hunters and subjects the coastline to damaging waves. Several coastal Alaskan villages have had to relocate inland and others are threatened because of coastal erosion.

When the lower layers of permafrost thaw, they can create an underground lake which may drain off, leaving a cavity. The surface will then slump inward, creating a sinkhole into which trees, roads, and buildings slowly fall. Such melting has already happened in Siberia and Fairbanks, Alaska, where the permafrost is at the warmest levels since the last ice age. The sinkholes are patchy so far and usually occur where digging and construction opened up the landscape. Thawing permafrost can also cause the sea levels to rise slowly and, because it traps large amounts of carbon, melting permafrost releases the carbon back into the atmosphere as methane, a greenhouse gas.

—*Virginia L. Salmon*

Further Reading

"Arctic Report Card: Update for 2016." *Arctic Program,* National Oceanic and Atmospheric

Administration, http://www.arctic.noaa.gov/Report-Card/Report-Card-2016. Accessed 31 Jan. 2017.

Bonan, Gordon *Ecological Climatology: Concepts and Applications.* 3rd ed., Cambridge University Press, 2015. Provides a concise definitive description of Earth's recognized climate zones within the broader context of climatology.

Maslin, Mark. *Global Warming: A Very Short Introduction.* Oxford University Press, 2009. Provides a concise general introduction to the major factors that define the climate of Earth.

Saha, Pijushkanti *Modern Climatology.* Allied Publishers Pvt Ltd, 2012. An entry-level textbook that provides an concise, yet thorough, overview of the science of climatology and climate relationships.

Schimel, David *Climate and Ecosystems.* Princeton University Press, 2013. One of the "Princeton Primers on Climate" series, this book describes the interaction mechanisms of various ecosystems and climate.

POLAR STRATOSPHERIC CLOUDS (PSCS)

FIELDS OF STUDY

Atmospheric Chemistry; Atmospheric Science; Photochemistry; Thermodynamics; Meteorology; Spectroscopy

SUMMARY

Polar stratospheric clouds (referred to as "PSCs" by scientists and called nacreous clouds by the general public) form in the lower stratosphere between heights of 15 kilometers and 25 kilometers in the lee of high-latitude mountain ranges in winter. Ice crystals and particles composing the visible cloud are of uniform size, about 10 micrometers in diameter. After sunset, light from the Sun below the horizon continues to illuminate particles within PSCs, producing iridescent bright red and blue coloration zones in the clouds.

PRINCIPAL TERMS

- **catalytic surface:** an interface or surface that promotes a chemical reaction but is not consumed in that reaction
- **nacreous:** having the iridescent appearance of nacre, or 'mother of pearl'
- **NOX:** an acronym denoting oxides of nitrogen
- **ozone layer:** the portion of the Earth's stratosphere (from 10–50 kilometers altitude) where ozone has formed and absorbs dangerous ultraviolet radiation from the Sun
- **photochemical:** chemical reactions or processes that are driven by interaction of the reacting materials with light
- **ultraviolet:** electromagnetic radiation having wavelengths shorter than those of the visible spectrum, associated with the transport of energy capable of promoting chemical reactions

Colorful Clouds

Polar stratospheric clouds are typically referred to as "PSCs" by meteorologists and atmospheric scientists. They are commonly called nacreous clouds by the general public, and form in the lower stratosphere between heights of 15 kilometers and 25 kilometers in the lee of high-latitude mountain ranges in winter. Ice crystals and particles composing the visible cloud are of uniform size, about 10 micrometers in diameter. After sunset, light from the Sun below the horizon continues to illuminate particles within PSCs, producing iridescent bright red and blue coloration zones in the clouds. Colorful PSC displays are most frequently photographed before dawn and after sunset when they appear in Oslo, Norway, but photographs and videotapes have also been taken in other countries where suitably high mountain conditions can be found, such as Scotland, England, Scandinavia, Iceland, Alaska, and Antarctica.

Atmospheric scientists have divided PSCs into two "Types" as defined by chemical and thermal differences. Type I PSCs have temperatures of about -78°C. Type I(a) PSCs contain crystalline compounds of water and nitric acid; Type I(b) PSCs contain droplets of nitric (HNO_3) and sulfuric (H_2SO_4) acids; Type I(c) PSCs contain nonspherical particles of metastable nitric acid in water; and Type II PSCs are composed of water ice crystals at temperatures of -85°C.

403

Nacreous Clouds over the NASA Radome, McMurdo Station, Antarctica.

Significance for Climate Change

Discovery of the formation of the ozone hole in the stratosphere above the Antarctic continent each spring focused scientific attention on chemical reactions depleting ozone. Scientists deduced that ultraviolet light from the rising Sun promoted chemical reactions leading to ozone destruction, and investigations linked chemical reactions going on within the PSCs to regional ozone depletion above the Antarctic.

Particles within Antarctic PSCs act as catalytic surfaces for reactions involving the destruction of relatively stable atmospheric chlorine compounds (including artificial chlorofluorocarbons sometimes called Freons). Through a series of reactions, these chlorine compounds are transformed into highly reactive molecular chlorine gas (Cl_2) and hypochlorous acid (HCl) molecules, which rapidly form chlorine radicals, which in turn destroy ozone molecules. During winter and early spring, there is a strong easterly flow of air in the Antarctic stratosphere, the Antarctic Polar Vortex. Little air from lower latitudes mixes with the ozone-depleted air within the vortex, allowing greater depletion of ozone to occur, creating the ozone hole.

PSCs also catalyze other important chemical reactions, including denoxification, the conversion of oxides of nitrogen to nitric acid, which is found in all three Type I PSCs. Because decreasing the level of nitric oxide (NO_2) allows high levels of ozone-removing hypochlorite (ClO) to remain in the stratosphere, this denoxification promotes development of the ozone hole. Denoxification leaves the waste product nitric acid behind in the cloud particles, which eventually exits the stratosphere, a process called denitrification.

—*Anita Baker-Blocker*

Further Reading

Ahrens, C Donald, and Robert Henson. *Meteorology Today: An Introduction to Weather, Climate, and the Environment.* Cengage, 2019. Discusses global climates, climate change, and classification.

Andronache, Constantin, ed. *Mixed-Phase Clouds. Observations and Modeling.* Elsevier, 2018. An advanced text in which the basic principles of cloud formation are discussed and analyzed.

Lohman, Ulrike, Lüönd, Felix and Mahrt, Fabian *An Introduction to Clouds From the Microscale to Climate.* Cambridge University Press, 2016. A thorough introduction to the formation and function of clouds and their role in determining the climate of Earth.

Lutgens, Frederick K., Edward J. Tarbuck, Tasa, Dennis G. and Herman, Redina. *The Atmosphere: An Introduction to Meteorology.* 14th ed. Pearson Education, 2018. An updated version of this textbook, in loose-leaf format, provides students with a general overview of the structure and properties of Earth's atmosphere.

Randall, David A. *Atmosphere, Clouds, and Climate.* Princeton University Press, 2012. An overview of the major atmospheric processes. Covers the function of the atmosphere in the regulation of energy flows and the transport of energy through weather systems, including thunderstorm and monsoons. Examines obstacles in predicting the weather and climate change.

Tanvir, Islam, Hu, Yongxiang, Kokhanovsky, Alexander and Wang, Jun, eds. *Remote Sensing of Aerosols, Clouds and Precipitation* Elsevier, 2017. Describes the methods, theoretical foundations and applications of satellite-based monitoring of aerosols, clouds and precipitation events occurring in Earth's atmosphere.

POLAR VORTEX

FIELDS OF STUDY

Atmospheric Science; Fluid Dynamics; Thermodynamics; Meteorology; Heat Transfer; Observational Astronomy

SUMMARY

When liquids or gases occupy a region of space while circulating, or spinning, scientists call that pattern of motion a vortex. A vortex can occur in a bathtub, in a kitchen sink, or in the atmosphere. When it takes place about the poles of a planet, it is called a polar vortex.

PRINCIPAL TERMS

- **greenhouse gases (GHGs):** atmospheric gases whose molecules absorb infrared radiation from a planet's surface and then emit a significant portion of it back into the atmosphere, most commonly carbon dioxide, methane, water, and ozone.
- **stratosphere:** an atmospheric region extending from about 17 to 48 kilometers above the Earth's surface
- **troposphere:** an atmospheric region extending from the Earth's surface to about 17 kilometers high over equatorial regions and to about 8 kilometers high over polar regions

CHARACTERISTICS OF POLAR VORTICES

Polar vortices are so large that they surround the polar heights and often straddle the troposphere and the stratosphere. Their strength varies with the seasons; they are strongest in winter, reaching speeds of 100 meters per second, and they are weakest, or even nonexistent, in summer. Polar vortices have also been observed on several other planets, including Jupiter, Mars, Saturn, and Venus.

When the temperature drops below a certain critical value of about −80°C, special clouds called polar stratospheric clouds (PSCs) form inside the vortex. Unlike ordinary clouds near the surface of the Earth,

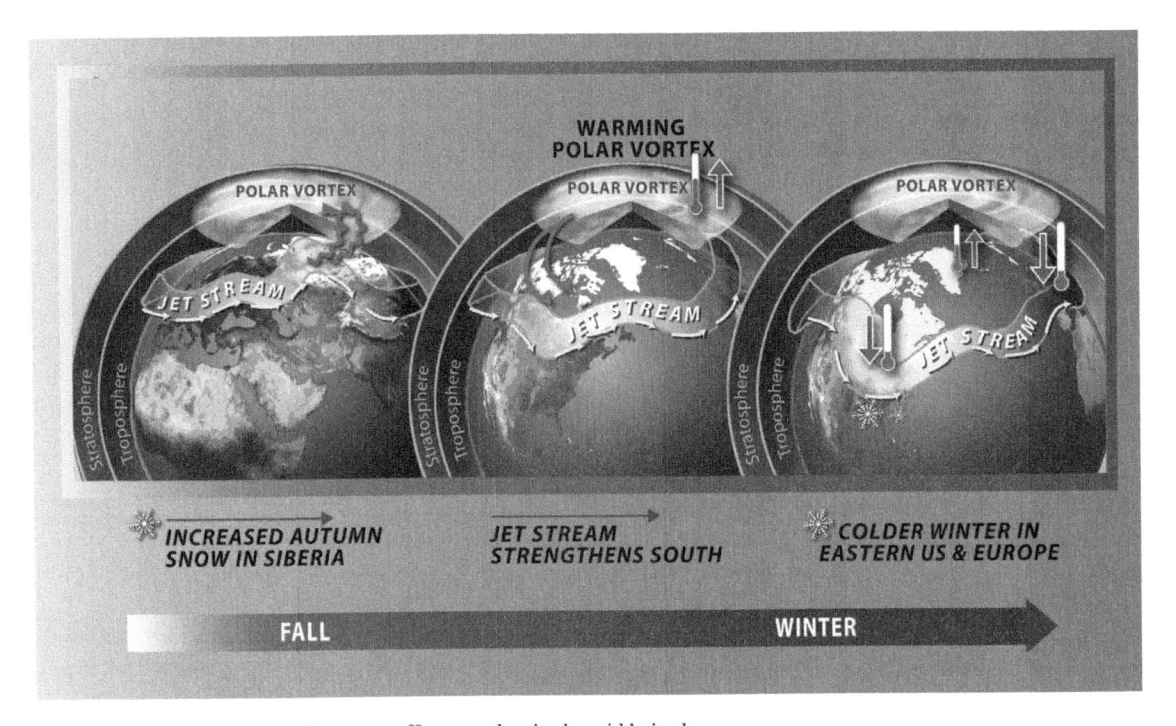

Shows how variations in the polar vortex affects weather in the mid-latitudes.

PSCs contain water-ice droplets mixed with a variety of other materials, and the chemical interactions among them are extremely sensitive to changes in temperature.

Significance for Climate Change

Factories and homes produce chlorine-containing chemicals that interact in the atmosphere with sunlight, subsequently forming chlorine compounds such as hydrochloric acid (HCl) and chlorine nitrate ($ClONO_2$). Prior to being banned throughout the world the most notorious of these chemicals were the chlorofluorocarbons, or CFCs. Ozone, which absorbs harmful ultraviolet radiation from the Sun, is found in a thin layer of the upper atmosphere, which has come to be called the "ozone layer." Chlorine compounds in the atmosphere combine with each other, with water, and with other chemicals to create products that attack ozone. The key chemical reactions that create these products only occur on the surface of PSCs. There is evidence that they cannot take place elsewhere in Earth's atmosphere.

Greenhouse gases (GHGs), which include the CFCs that are still present in the atmosphere, enhance the depletion of the ozone layer. While they cause atmospheric warming near Earth's surface, GHGs actually cool the stratosphere, where ozone resides; since the chemistry of the ozone layer is very sensitive to temperature, even very small decreases in the temperature of the stratosphere increase the loss of ozone.

Scientists have observed a statistical correlation between the cycles of the polar vortex and the weather. Severe cold weather in the Northern Hemisphere correlates well with a weak polar vortex. Similarly, when that vortex is strong, weather in the Northern Hemisphere turns warm. Moreover, a weak polar vortex is susceptible to interruption, and southward movement from the North Pole can compound the effects of typical surface cold fronts and result in bitterly cold weather far from the Arctic.

—*Josué Njock Libii*

Further Reading

Goddard Space Flight Center. *Arctic Ozone Watch*. NASA Ozone Watch, 25 Sept. 2013. Web. 23 Mar. 2015.

Kennedy, Caitlyn. "Wobbly Polar Vortex Triggers Extreme Cold Air Outbreak." *Climate.gov*. NOAA, 8 Jan. 2014. Web. 23 Mar. 2015.

Müller, Rolf, ed. *Stratospheric Ozone Depletion and Climate Change*. RSC, 2012.

Polvani, L. M., A. H. Sobel, and D. W. Vaugh, eds. "Stratospheric Polar Vortices." *The Stratosphere: Dynamics, Transport, and Chemistry*. Washington: Amer. Geophysical Union, 2010. Print.

"What Is a Polar Vortex?" *SciJinks*. NASA and NOAA, 18 Mar. 2015. Web. 23 Mar. 2015.

PRECIPITATION

FIELDS OF STUDY

Atmospheric Science; Physical Geography; Ecosystem Management; Ecology; Physical Chemistry; Thermodynamics; Heat Transfer; Environmental Sciences; Environmental Studies; Earth System Modeling; Waste Management; Meteorology; Climate Modeling; Process Modeling; Fluid Dynamics; Hydroclimatology; Hydrometeorology; Bioclimatology; Oceanography; Hydrology

SUMMARY

Precipitation consists of particles of liquid or frozen water that fall from clouds toward the ground surface. Thus, precipitation links the atmosphere with the other reservoirs of the global hydrologic cycle, replenishing oceanic and terrestrial reservoirs. In addition, precipitation is the ultimate source of freshwater for irrigation, industrial consumption, and supplies of drinking water.

PRINCIPAL TERMS

- **acid precipitation:** rain or snow that is more acidic than normal, usually because of the presence of sulfuric and nitric acid
- **Bergeron process:** precipitation formation in cold clouds whereby ice crystals grow at the expense of supercooled water droplets
- **cold cloud:** a visible suspension of tiny ice crystals, supercooled water droplets, or both at tem-

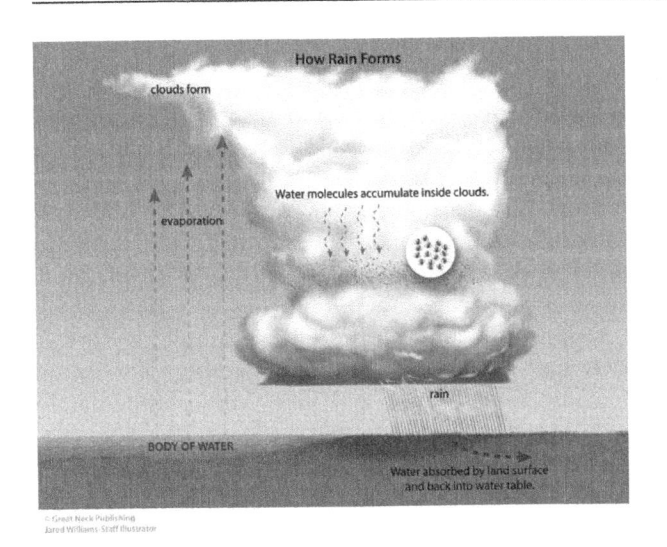

Rain: Rain is a form of precipitation. The water cycle has three main steps: evaporation, condensation, and precipitation.

peratures below the normal freezing point of water
- **collision-coalescence process:** precipitation formation in warm clouds whereby larger droplets grow through the merging of smaller droplets
- **rain gauge:** an instrument for measuring rainfall, usually consisting of a cylindrical container open to the sky
- **supercooled water droplets:** droplets of liquid water at temperatures below the normal freezing point of water
- **warm cloud:** a visible suspension of tiny water droplets at temperatures above freezing

CLOUD PARTICLES

Precipitation consists of liquid or frozen particles of water that fall from clouds and normally reach the ground surface. Under certain conditions, however, a type of precipitation known as virga forms and falls normally, but evaporates before reaching the ground. The most familiar types of precipitation are rain and snow. Perhaps surprisingly, most clouds, even those associated with large storm systems, do not produce precipitation. A typical cloud particle is about one-millionth the size of a raindrop. Special circumstances are required for the extremely small water droplets or ice crystals that compose a cloud to grow into raindrops or snowflakes.

Cloud particle diameters are typically in the range of 2 to 50 micrometers, with one micrometer being one-millionth of a meter. They are so small that they remain suspended within the atmosphere unless they vaporize or somehow undergo considerable growth. Upward-directed air currents, or updrafts, are usually strong enough to prevent cloud particles from leaving the base of a cloud. Even if cloud droplets or ice crystals descend from a cloud, their fall rates are so slow that they quickly vaporize in the relatively dry air under the cloud. In order to precipitate, therefore, cloud particles must grow sufficiently massive that they counter updrafts and survive thousands of meters of descent to the ground surface. Cloud physicists have identified two processes whereby cloud particles grow large enough to precipitate: the Bergeron process and the collision-coalescence process.

THE BERGERON AND COLLISION-COALESCENCE PROCESSES

Most precipitation originates via the Bergeron process, named for the Scandinavian meteorologist Tor Bergeron (1891–1977), who, in about 1930, first described the process. It occurs within cold clouds at a temperature below freezing (0°C). Cold clouds are composed of ice crystals or supercooled water droplets or a mixture of the two. Supercooled water droplets are tiny drops that remain liquid at temperatures below their normal freezing point. Bergeron discovered that precipitation is most likely to fall from cold clouds composed of a mixture in which supercooled water droplets at least initially greatly outnumber ice crystals. In such a circumstance, ice crystals grow rapidly while supercooled water droplets vaporize. As ice crystals grow, their fall rates within the cloud increase. They collide and merge with smaller ice crystals and supercooled water droplets in their paths and thereby grow still larger. Eventually, the ice crystals become so heavy that they fall out of the cloud base. If the air temperature is below freezing during most of the descent, the crystals reach the surface as snowflakes. If, however, the air below the cloud is above freezing, the snowflakes melt and fall as raindrops.

Growth of ice crystals at the expense of supercooled water droplets in the Bergeron process is linked to the difference in the rate of escape of water molecules from an ice crystal versus a water droplet. Water molecules are considerably more active in the liquid phase than in the solid phase. Hence, water molecules escape from water droplets more readily than they do from ice crystals. Within a cold cloud, air that is saturated for water droplets is actually supersaturated for ice crystals. Consequently, water molecules diffuse from the water droplets and deposit on the ice crystals. That is, the water droplets vaporize and release water molecules to the air as the ice crystals grow by accruing those water molecules from the air.

The collision-coalescence process occurs in warm clouds (clouds at temperatures above 0°C). Such clouds are composed entirely of liquid water droplets. Precipitation may develop if the range of cloud droplet sizes is broad. Larger cloud droplets have greater fall velocities than do smaller droplets, as they are less affected by upwelling air currents. As a result, larger droplets collide and coalesce with smaller droplets in their paths. Collision and coalescence are repeated a multitude of times until the droplets become so large and heavy that they fall from the base of the cloud as raindrops. Since the force of upwelling air currents varies, the forming droplets may be pushed back up to higher altitudes time and time again, becoming larger each time as their fall path through the cloud is extended.

Once a raindrop or snowflake leaves a cloud, it enters drier air, a hostile environment in which some of the precipitation vaporizes. In general, the longer the journey to the ground surface and the drier the air beneath the cloud, the greater the amount of rain or snow that returns to the atmosphere as vapor. It is understandable, then, why highlands receive more precipitation than do lowlands, which are hundreds to thousands of meters farther from the base of the clouds.

Types of Precipitation

Precipitation occurs in a variety of liquid and frozen forms. Besides the familiar rain and snow, precipitation also occurs as drizzle, freezing rain, ice pellets, and hail. Drizzle consists of small water drops less than 0.5 millimeter in diameter that drift very slowly downward to the ground. The relatively small size of drizzle drops stems from their origin in low stratus clouds or fog. Such clouds are so shallow that droplets originating within them have a limited opportunity to grow by coalescence.

Rain falls most often from thick nimbostratus and cumulonimbus (thunderstorm) clouds. The bulk of rain originates as snowflakes or hailstones, which melt on the way down as they enter air that is warmer than 0°C. Because rain originates in thicker clouds, raindrops travel farther than does drizzle, and they undergo more growth by coalescence. Most commonly, raindrop diameters range from 0.5 to 5 millimeters; beyond this range, drops are unstable and break apart into smaller drops. Freezing rain (or freezing drizzle) develops when rain falls from a relatively mild air layer onto the ground-level objects that are at temperatures below freezing. The drops become supercooled, and then freeze immediately on contact with subfreezing surfaces. Freezing rain forms a layer of ice that sometimes grows thick and heavy enough to bring down tree limbs, power lines, and grid towers; disrupt traffic; and make walking or transportation hazardous.

Snow is an assemblage of ice crystals in the form of flakes. Although it is said that no two snowflakes are identical, all snowflakes have hexagonal (six-sided) symmetry. Snowflake form varies with air temperature and water vapor concentration and may consist of flat plates, stars, columns, or needles. Snowflake size also depends in part on the availability of water vapor during the crystal growth process. At very low temperatures, the water vapor concentration is low so that snowflakes are relatively small. Snowflake size also depends on collision efficiency as the flakes drift toward the ground. At temperatures near freezing, snowflakes are wet and readily adhere to each other after colliding, so flake diameters may eventually exceed 5 centimeters. Snow grains and snow pellets are closely related to snowflakes. Snow grains originate in much the same way as drizzle, except that they are frozen. Their diameters are generally less than 1 millimeter. Snow pellets are soft conical or spherical white particles of ice with diameters of 1 to 5 millimeters. They are formed when supercooled cloud droplets collide and freeze together, and they may accompany a fall of snow.

Ice pellets, often called sleet, are frozen raindrops. They develop in much the same way as does freezing rain except that the surface layer of subfreezing air is so deep that raindrops freeze before striking the ground. Sleet can be distinguished readily from freezing rain because sleet bounces when striking a hard surface, whereas freezing rain does not.

Hail consists of rounded or irregular pellets of ice, often characterized by an internal structure of concentric layers resembling the interior of an onion. Hail develops within severe thunderstorms as vigorous updrafts propel ice pellets upward into the higher reaches of the cloud. It is not unusual for clouds in severe thunderstorms reach altitudes of more than 10 kilometers. Along the way, ice pellets grow via coalescence with supercooled water droplets and eventually become too heavy to be supported by updrafts. The ice pellets then descend through the cloud, exit the cloud base, and enter air that is typically above freezing. As ice pellets begin to melt, those that are large enough may survive the journey to the ground as hailstones. Most hail consists of harmless granules of ice less than 1 centimeter in diameter, but violent thunderstorms may spawn destructive hailstones the size of golf balls or larger. Hail is usually a spring and summer phenomenon that can be particularly devastating to crops as it shreds leaves and flowers, breaks fruit loose from the branches, and can even render damages to agricultural equipment.

Changes in Precipitation Chemistry

Over the past few decades, considerable concern has been directed at the environmental impact of changes in the chemistry of precipitation. Water vapor in the atmosphere is molecularly pure water. The global hydrologic cycle purifies water through what is essentially the process of distillation, and the droplets of liquid water that form from the vapor are also composed of molecularly pure water, which has a pH value of 7. But as raindrops and snowflakes fall from clouds to the ground, they dissolve and interact with pollutants in the air. In this way, the chemistry of precipitation is altered. Rain is normally slightly acidic because it dissolves atmospheric carbon dioxide, producing a very weak solution of carbonic acid having a pH only slightly less than 7. Where air is polluted with oxides of sulfur and oxides of nitrogen, however, these gases interact with moisture in the atmosphere to produce droplets of sulfuric acid and nitric acid solutions. These acidic droplets greatly increase the acidity of precipitation. Precipitation that falls through such polluted air may become orders of magnitude more acidic than normal, and was once measured in Scotland, in 1974, as having a pH of only 2.4, more than ten thousand times more acidic than normal, unpolluted rainfall.

Field studies have confirmed a trend toward increasingly acidic precipitation in the form of rains and snows over the eastern one-third of the United States. Much of this upswing in acidity can be attributed to acid rain precursors emitted during fuel combustion. Coal-burning for electric power generation is the principal source of sulfur oxides, while high-temperature industrial processes and motor vehicle engines produce nitrogen oxides. Where acid rains fall on soils or bedrock that cannot neutralize the acidity, lakes and streams become more acidic. Excessively acidic lake or stream water disrupts the reproductive cycles of fish and has numerous other negative environmental effects. Acid rains leach metals (such as aluminum) from the soil, washing them into lakes and streams, where they may harm fish and aquatic plants.

Study of Precipitation

Precipitation is collected and measured with essentially the same device that has been used since the fifteenth century: a container open to the sky. The standard U.S. National Weather Service rain gauge consists of a cone-shaped funnel that directs rainwater into a long, narrow cylinder that sits inside a larger cylinder. The narrow cylinder magnifies the scale of accumulating rainwater so that rainfall can be resolved into increments of 0.01 inch. (Rainfall of less than 0.005 inch is recorded as a "trace.") Rainwater that accumulates in the inner cylinder is measured against a graduated scale. Rainfall is measured at some fixed time once every twenty-four hours, and the gauge is then emptied.

With regard to snow, scientists are interested in measuring snowfall during each twenty-four-hour period between observations, as well as the meltwater equivalent of that snowfall, and the depth of snow on the ground at each observation time. New snowfall is usually collected on a simple board that is placed on top of the old snow cover. When new snow falls, the depth is measured to the board; the board is then swept clean and moved to a new location. The meltwater equivalent of new snowfall can be determined by melting the snow collected in a rain gauge (from which the funnel has been removed). Snow depth is usually measured with a special yardstick or meterstick. In mountainous terrain where snowfall is substantial, it may be necessary to use a coring device to determine snow depth (and meltwater equivalent). Snow depth is determined at several representative locations and then averaged.

The average density of fresh-fallen snow is 0.1 gram per cubic centimeter. As a general rule, 10 centimeters of fresh snow melts down to 1 centimeter of rainwater. This ratio varies considerably depending on the temperature at which the snow falls. "Wet" snow falling at surface air temperatures at or above 0°C has a much greater water content than does "dry" snow falling at surface air temperatures well below freezing. The ratio of snowfall to meltwater may vary from 3:1 for very wet snow to 30:1 for dry, fluffy snow.

Monitoring the timing and rate of rainfall is often desirable, especially in areas prone to flooding. Hence, some rain gauges provide a cumulative record of rainfall. In a weighing-bucket rain gauge, the weight of accumulating rainwater (determined by a spring balance) is calibrated as water depth. Cumulative rainfall is recorded continuously by a device that either marks a chart on a clock-driven drum or sends an electrical pulse to a computer or magnetic tape. During subfreezing weather, antifreeze in the collection bucket melts snow as it falls into the gauge so that a cumulative meltwater record is produced.

Both rainfall and snowfall are notoriously variable from one place to another, especially when produced by showers or thunderstorms. The emplacement of a precipitation gauge is particularly important in order to ensure accurate and representative readings. A level site must be selected that is sheltered from strong winds and is well away from buildings and vegetation that might shield the instrument. In general, obstacles should be no closer than about four times their height.

Significance

Without precipitation, Earth would have no freshwater and thus no life. When water vaporizes from oceans, lakes, and other reservoirs on the ground surface, all dissolved and suspended substances are left behind. Hence, water is purified (distilled) as it cycles into the atmosphere and eventually returns to the surface as freshwater precipitation. In this way, the global hydrologic cycle supplies the planet with an essentially fixed quantity of freshwater.

As the human population continues its rapid growth, however, demands on the globe's fixed supply of freshwater are also increasing. In some areas, such as the semiarid American Southwest, water demand for agriculture and municipalities has spurred attempts to enhance precipitation locally through cloud seeding. Usually, cold clouds that contain too few ice crystals are seeded by aircraft with either silver iodide crystals (a substance with molecular properties similar to ice) or dry-ice pellets (solid carbon dioxide at a temperature of -78°C) in an effort to stimulate the Bergeron precipitation process.

Cloud seeding, although founded on an understanding of how precipitation forms, is not always successful and at best may enhance precipitation by perhaps 20 percent. The question remains as to whether the rain or snow that follows cloud seeding would have fallen anyway. Even if successful, cloud seeding may merely bring about a geographical redistribution of precipitation so that an increase in precipitation in one area is accompanied by a compensating reduction in a neighboring area. Cloud seeding that benefits agriculture in eastern Colorado, for example, might also deprive farmers of rain in the downwind states of Kansas and Nebraska. The uncertainties of cloud seeding underscore the need for conservation of the planet's freshwater resource. Conservation should entail not only strategies directed at wise use of freshwater but also measures to manage water quality. Abatement of water pollution not only reduces hazards to human health and aquatic systems but also increases the available supply of freshwater.

—*Joseph M. Moran*

Further Reading

"Acid Rain Program." *EPA*. Environmental Protection Agency, 26 Jan. 2015. Web. 4 Mar. 2015.

Ahrens, C Donald, and Robert Henson. *Meteorology Today: An Introduction to Weather, Climate, and the Environment*. Cengage, 2019. Discusses global climates, climate change, and classification.

Christopherson, Robert W., and Ginger H Birkeland. *Geosystems: An Introduction to Physical Geography*. Pearson, 2018.

Oliver, John E., ed. *The Encyclopedia of Climatology*. Springer, 2004.

Pruppacher, Hans R., and James D. Klett, eds. *Microphysics of Clouds and Precipitation*. 2nd ed. Springer, 2010.

Randall, David. *Atmosphere, Clouds and Climate*. Princeton University Press, 2012.

Schneider, Bonnie. *Extreme Weather: A Guide to Surviving Flash Floods, Tornadoes, Hurricanes, Heat Waves, Snowstorms, Tsunamis and Other Natural Disasters.* Palgrave Macmillan, 2012.

Schneider, Stephen Henry, and Michael D. Mastrandrea. *Encyclopedia of Climate and Weather.* 2nd ed. Oxford University Press, 2011.

Smith, Jerry E. *Weather Warfare: The Military's Plan to Draft Mother Nature.* Adventures Unlimited Press, 2006.

Straka, Jerry M. *Cloud and Precipitation Microphysics: Principles and Parameterizations.* Cambridge University Press, 2009.

Strangeways, Ian. *Precipitation: Theory, Measurement and Distribution.* Cambridge University Press, 2010.

Vasquez, Tim. *Weather Analysis and Forecasting Handbook.* Weather Graphics Technologies, 2011.

Wang, Pao K. *Physics and Dynamics of Clouds and Precipitation.* Cambridge University Press, 2013.

PROBABILITY AND STATISTICS

FIELDS OF STUDY

Mathematics; Physical Chemistry; Chemical Kinetics; Thermodynamics; Environmental Studies; Waste Management; Climate Modeling; Statistics; Process Modeling; Fluid Dynamics; Earth System Modeling

SUMMARY

Probability and statistics are two related fields covering the science of collecting, measuring, and analyzing information in the form of numbers. Both probability and statistics are branches of applied mathematics. Probability focuses on using numeric data to predict a future outcome. Statistics incorporates theory into the gathering of numerical data and the drawing of accurate conclusions. Because nearly all fields in applied science rely on the analysis of numbers in some way, probability and statistics are one of the most diverse areas in terms of subjects and career paths. Statisticians also practice in areas of the academic world outside of science and throughout industry.

PRINCIPAL TERMS

- **confidence:** a value assigned to a statistical result that indicates how close the calculated result is expected to be to the true value of the property
- **probability:** in its simplest definition, the likelihood of a specific event happening, although determining a probability can be a very complex problem as the number of independent variables increases
- **statistics:** the mathematical treatment of a set of collected values in order to determine essential ratios and trends

A pair of dice, symbolic of probability.

DEFINITION AND BASIC PRINCIPLES

Probability and statistics are two interconnected fields within applied mathematics. In both fields, principles of scientific theory are applied to the analysis of groups of data in the form of numbers. The main objective of probability and statistics is to ask and answer questions about data with as much accuracy as possible.

Defining "probability" can be a challenge, as multiple schools of thought exist. In one view, held by a group of scholars known as frequentists, probability is defined as the likelihood that a statement about a set of data will be true in the long run. Frequentists focus on the big picture, specifically at the collective outcome of multiple experiments conducted over time, rather than on specific data items or outcomes. In contrast, scholars known as Bayesians prefer to start with a probability-based assumption about a

411

set of data, then test to see how close the actual data come to the initial assumption. On both sides of the debate, probabilists are seeking to understand patterns in data to predict how a population might behave in the future.

Statistics is a field with a broader scope than probability, but in some ways, it is easier to define. The academic discipline of statistics is based on the study of groups of numbers in three stages: collection, measurement, and analysis. At the collection stage, statistics involves issues such as the design of experiments and surveys. Statisticians must answer questions such as whether to examine an entire population or to work from a sample. Once the data are collected, statisticians must determine the level of measurement to be used and the types of questions that can be answered with validity based on the numbers.

No matter how rigorous an individual study might be, statistical findings are often met with doubt by scholars and the general public. A quote repeated often (and mistakenly attributed to former British Prime Minister Benjamin Disraeli) is "There are three kinds of lies: lies, damned lies, and statistics."

BACKGROUND AND HISTORY
Probability has been a subject of interest since dice and card games were first played for money. Gambling inspired the first scholarly discussions of probability in the sixteenth and early seventeenth centuries. The Italian mathematician Gerolamo Cardano (1501–1576) wrote *Libellus de ratiociniis in ludo aleae* (*The Value of All Chances in Games of Fortune*, 1714) in about 1565, although the work was not published until 1663. In the mid-1600s, French mathematicians Blaise Pascal (1623–1662) and Pierre de Fermat (1601–1665) discussed principles of probability in a series of letters about the gambling habits of a mutual friend.

The earliest history of a statistical study is less clear, but it is generally thought to involve demographics. British scholar John Graunt (1620–1674) studied causes of mortality among residents of London and published his findings in 1662. Graunt found that statistical data could be biased by social factors, such as relatives' reluctance to report deaths due to syphilis. In 1710, John Arbuthnot (1667–1735) analyzed the male-female ratio of babies born in Britain since 1629. His findings that there were more males than females were used to support his argument in favor of the existence of a divine being.

A third branch of statistics—the design of experiments and the problem of observational error—has its roots in the eighteenth century work of German astronomer Tobias Mayer (1723–1762). However, a paper by British theologian Thomas Bayes (1702–1761) published in 1764 after his death is considered a turning point in the history of probability and statistics. Bayes dealt with the question of how much confidence could be placed in the predictions of a mathematical model based on probability. The convergence between probability and statistics has increased over time. The development of modern computers has led to major advances in both fields.

HOW IT WORKS
In terms of scope, probability and statistics are some of the widest, most diverse fields in applied mathematics. If a research project involves items that must be counted or measured in some way, statistics will be part of the analysis. It is common to associate statistics with research in the sciences. Similarly, probability is used by anyone relying on numbers to make an educated guess about events in the future. This is the fundamental principle of climate modeling and climate prediction, as "climate" is defined by the occurrence of weather events such as rainfall, temperature, wind velocity, etc., over a long period of time. Statistical analysis of those events is essential to determining and identifying climatic trends.

The word "statistic" can refer to almost any number connected to data. When statistics are discussed as a discipline, though, there is a multistep process that most projects follow: definition and design, collection, description and measurement, and analysis and interpretation.

Definition and Design
Much of the scholarship in the field of statistics focuses on data definition and the design of surveys and experiments. Statistical projects must begin with a question, such as "Does grade-point average in high school have an effect on income level after graduation?" In the data definition phase, the statistician chooses the items to be studied and measured, such as grades and annual earnings. The next step is to define other factors, such as the number of people to be studied, the areas in which they live, the years

in which they graduated from high school, and the number of years in which income will be tracked. Good experimental design ensures that the rest of the project will gather enough data, and the right data, to answer the question.

Collection
Once the data factors have been defined, statisticians must collect them from the population being studied. Experimental design also plays a role in this step. Statistical data collection must be thorough and must follow the rules of the study. For example, if a survey is mailed to one thousand high school graduates and only three respond, more data must be collected before the survey's findings can be considered valid. Statisticians also must ensure that collected data are accurate by finding a way to check the reliability of answers.

Description and Measurement
Collected data must be stored, arranged, and presented in a way that can be used by statisticians to form statistics and draw conclusions. Grade-point averages, for example, might be compared easily if all the survey participants attended the same school. If different schools or grading systems were used, the statistician must develop rules about how to convert the averages into a form that would allow them to be compared. Once these conversions are made, the statistician would decide whether to present the data in a table, chart, or other form.

Analysis and Interpretation
In terms of statistical theory, the most complex step is the analysis and interpretation of data. When data have been collected, described, and measured, conclusions can be drawn in a number of ways—none of which is right in every case. It is this step in which statisticians must ask themselves a few questions: What is the relationship between the variables? Does a change in one automatically lead to a change in the other? Is there a third variable, or a lurking variable, not covered in the study that makes both data points change at the same time? Is further research needed?

One of the most common methods used for statistical analysis is known as modeling. A model allows the statistician to build a mathematical form, such as a formula, based on ideas. The data collected by the study can then be compared with the model. The results of the comparison tell a story that supports the study's conclusions. Some models have been found to be so innovative that they have earned their creators awards such as the Nobel Prize. However, even the best models can have flaws or can fail to explain actual data. Statistician George Box once said, "All models are wrong, but some models are useful."

Prediction
Probability deals with the application of statistical methods in a way that predicts future behavior. The goal of many statistical studies is to establish rules that can be used to make decisions. In the example, the study might find that students who achieve a grade-point average of 3 or higher earn twice as much, as a group, as their fellow students whose averages were 2.9 or lower. As an academic discipline, probability offers several tools, based on theory, that allow a statistician to ask questions such as: How likely is a student with a grade-point average of 3.5 to earn more than $40,000 per year?

For both statistics and probability, one of the primary objectives is to ask mathematical questions about a smaller population to find answers that apply to a larger population.

APPLICATIONS AND PRODUCTS
It would be nearly impossible to find a product or service that did not rely on probability or statistics in some way. In the case of a cup of coffee, for example, agricultural statistics guided where the coffee beans were grown and when they were harvested. Industrial statistics controlled the process by which the beans were roasted, packaged, and shipped. Statistics even influenced the strength to which the coffee was brewed—whether in a restaurant, coffeehouse, or a home kitchen. Probability played a role in each step as well. Forecasts in weather and crop yields, pricing on coffee bean futures contracts, and anticipated caffeine levels each had an effect on a single brewed cup.

One way to understand the applications and products of probability and statistics is to look at some general categories by function. These categories cover professional fields that draw some of the highest concentrations of professionals with a statistical background.

Process Automation

One of the broadest and most common ways in which statistical methods are applied is in process automation. Quality control is a leading example. When statistical methods are applied to quality control, measures of quality such as how closely products match manufacturing specifications can be translated into numbers and be tracked. This approach allows manufacturers to ensure that their products meet quality standards such as durability and reliability. It also verifies that products meet physical standards, including size and weight. Quality control can be used to evaluate service providers as well as manufacturers. If measures of quality such as customer satisfaction can be converted into numerical data, statistical methods can be applied to measure and increase it. One well-known quality control application, Six Sigma, was developed by the Motorola Corporation in the early 1980s to reduce product defects. Six Sigma relies on both probability and statistical processes to meet specific quality targets.

Another field in which process automation is supported by statistical analysis is transport and logistics. Transport makes up a significant amount of the cost of manufacturing a product. It also plays a major role in reliability and customer satisfaction. A manufacturer must ensure that its products will make the journey from one place to another in a timely way without being lost or damaged. To keep costs down, the method of transport must be as inexpensive as possible. Statistical methods allow manufacturers to calculate and choose the best transportation options (such as truck versus rail) and packaging options (such as cardboard versus plastic packaging) for their products. When fuel costs rise, the optimization of logistics becomes especially important for manufacturers. Probability gives manufacturers tools such as futures and options on fuel costs.

Biostatistics

Statistics are used in the biological sciences in a variety of ways. Epidemiology, or the study of disease within a population, uses statistical techniques to measure public health problems. Statistics allow epidemiologists to measure and document the presence of a specific illness or condition in a population and to see how its concentration changes over time. With this information, epidemiologists can use probability to predict the future behavior of the health problem and recommend possible solutions.

Other fields in the biological sciences that rely on statistics include genetics and pharmaceutical research. Statistical analysis has played a key role in the Human Genome Project. The amount of data generated by the effort of mapping human genes could be analyzed only with complex statistical processes, some of which are still being fine-tuned to meet the project's unique needs. Probability analyses allow geneticists to predict the influence of a gene on a trait in a living organism.

Pharmaceutical researchers use statistics to build clinical trials of new drugs and to analyze their effects. The use of statistical processes has become so widespread in pharmaceutical research that it is extremely difficult to obtain approvals from the U.S. Food and Drug Administration (FDA) without it. The FDA publishes extensive documentation guiding researchers through the process of complying with the agency's statistical standards for clinical trials. These standards set restrictions on trial factors ranging from the definition of a control group (the group against which the drug's effects are to be measured) to whether the drug effectively treats the targeted condition.

Spatial Statistics

Understanding areas of space requires the analysis of large amounts of data. Spatial statistics are used in fields such as climatology, agricultural science, and geology. Statistical methods provide climatologists with specialized tools to model the effects of factors such as changes in air temperature or atmospheric pollution. Meteorology also depends on statistical analysis because its data are time based and often involve documenting repeated events over time, such as daily precipitation over the course of several years. One of the best-known applications of probability to a field of science is weather forecasting.

In agricultural science, researchers use statistics and probability to evaluate crop yields and to predict success rates in future seasons. Statistics are also used to measure environmental impact, such as the depletion of nutrients from the soil after a certain kind of crop is grown. These findings guide recommendations, based on probabilistic techniques, about what kinds of crops to grow in seasons to come. Animal science relies on statistical analysis at every stage, from genetic decisions about the breeding of livestock to the environmental and health impacts of raising animals under certain kinds of farming conditions.

Geologists use statistics and probability in a wide range of ways. One way that draws a significant amount of interest and funding from industry is the discovery of natural resources within the Earth. These efforts include the mining of metals and the extraction and refining of products such as oil and gasoline. Mining and petroleum operations are leading employers of statisticians and probabilists. Statistical processes are critical in finding new geologic sites and in measuring the amount and quality of materials to be extracted. To be done profitably, mining and petroleum extraction are functions that must be carried out on a large scale, assisted by sizable amounts of capital and specialized equipment. These functions would not be possible without sophisticated statistical analysis. Statistics and probability are also used to measure environmental impact and to craft effective responses to disasters such as oil spills.

Risk Assessment
As a science, statistics and probability have their roots in risk assessment—specifically, the risk of losing money while gambling. Risk assessment remains one of the areas in which statistical analysis plays a chief role. A field in which statistical analysis and risk assessment are combined at an advanced level is security. Strategists are beginning to apply the tools of statistics and game theory to understanding problems such as terrorism. Although terrorism was once regarded as an area for study in the social sciences, people are developing ways to control and respond to terrorist events based on statistics. Probability helps strategists take lessons from terrorism in one political context or world region and apply them to another situation that, on the surface, might look very different.

Actuarial science is one of the largest and most thoroughly developed fields within risk assessment. In actuarial science, statistics and probability are used to help insurance and financial companies answer questions such as how to price an automobile insurance policy. Actuaries look at data such as birth rates, mortality, marriage, employment, income, and loss rates. They use this data to guide insurance companies and other providers of financial products in setting product prices and capital reserves.

Quantitative finance uses statistical models to predict the financial returns, or gains, of certain types of securities such as stocks and bonds. These models can help build more complex types of securities known as derivatives. Although derivatives are often not well understood outside of the finance industry, they have a powerful effect on the economy and can influence everyday situations such as a bank's ability to loan money to a person buying a new home.

Survey Design and Execution
Surveys use information gathered from populations to make statements about the population as a whole. Statistical methods ensure that the surveys gather enough high-quality data to support accurate conclusions. A survey of an entire population is known as a census. One prominent example is the United States Census, conducted every ten years to count the country's population and gather basic information about each person. Other surveys use data from selected individuals through a process known as sampling. In nearly all surveys, participants receive questions and provide information in the form of answers, which are turned into mathematical data and analyzed statistically.

Aside from government censuses, some of the most common applications of survey design are customer relationship management (CRM) and consumer product development. Through the gathering of survey data from customers, companies can use customer relationship management to increase the effectiveness of their services and identify frequent problems to be fixed. In creating and introducing new products, most companies rely on data gathered from consumers through Internet and telephone surveys and on the results of focus groups.

SOCIAL CONTEXT AND FUTURE PROSPECTS
There is an increasing need for professionals with a knowledge of probability and statistics. The growth of the Internet has led to a rapid rise in the amount of information available, both in professional and in private contexts. This growth has created new opportunities for the sharing of knowledge. However, much of the information being shared has not been filtered or evaluated. Statistics, in particular, can be fabricated easily and often are disseminated without

a full understanding of the context in which they were generated.

Statisticians are needed to design experiments and studies, to collect information on the basis of sound research principles, and to responsibly analyze and interpret the results of their work. Aside from a strong knowledge of this process, statisticians must be effective communicators. They must consider the ways in which their findings might be used and shared, especially by people without a mathematical background. Their results must be presented in a way that will ensure clear understanding of purpose and scope.

Fields likely to see a higher rate of growth are those involving statistical modeling, such as climate modeling. An increase in computer software and other tools to support modeling has fueled higher demand for professionals with a familiarity in this area. Statistical modeling is useful in many contexts related to probability.

—*Julia A. Rosenthal, MS*

FURTHER READING

Black, Ken. *Business Statistics: For Contemporary Decision Making.* 8th ed. Wiley, 2014.

Boslaugh, Sarah. *Statistics in a Nutshell.* 2nd ed. O'Reilly, 2013.

Fung, Kaiser. *Numbers Rule Your World: The Hidden Influence of Probabilities and Statistics on Everything You Do.* McGraw, 2010.

Mlodinow, Leonard. *The Drunkard's Walk: How Randomness Rules Our Lives.* Pantheon, 2008.

Nisbet, Robert, John Elder, and Gary Miner. *Handbook of Statistical Analysis and Data Mining Applications.* Elsevier, 2009.

Ott, R. Lyman, and Michael T. Longnecker. *An Introduction to Statistical Methods and Data Analysis.* Brooks, 2010.

Sharpe, Norean R., Richard De Veaux, and Paul F. Velleman. *Business Statistics: A First Course.* 2nd ed. Pearson, 2013.

Takahashi, Shin. *The Manga Guide to Statistics.* No Starch, 2009.

R

RAIN FORESTS AND THE ATMOSPHERE

FIELDS OF STUDY

Ecosystem Management; Ecology; Environmental Sciences; Environmental Studies; Waste Management; Climate Modeling; Hydroclimatology; Atmospheric Science; Hydrology; Physical Geography; Earth System Modeling

SUMMARY

Rain forests are ecosystems noted for their high biodiversity and high rate of photosynthesis. The rapid deforestation of such areas is of great concern to environmentalists both because it may lead to the extinction of numerous species and because it may reduce the amount of photosynthesis occurring on the Earth. Because photosynthesis releases large amounts of oxygen into the air, a curtailment of the process by rain-forest deforestation may have negative effects on the global atmosphere. All life on Earth depends on the "sea" of air surrounding them. The atmosphere includes abundant, permanent gases such as nitrogen (78 percent) and oxygen (21 percent) as well as smaller, variable amounts of other gases such as water vapor and carbon dioxide. Organisms absorb and use this air as a source of raw materials and release into it by-products of their life activities.

PRINCIPAL TERMS

- **photodissociation:** a chemical process in which a molecule spontaneously separates into two or more stable components by absorbing light energy
- **photolysis:** a chemical process in which single atoms are separated fro a molecule by the absorption of light energy
- **photosynthesis:** the chemical process occurring in green plants by which atmospheric carbon dioxide is combined with water in the presence of chlorophyll and sunlight to produce the simple sugar glucose and molecular oxygen
- **respiration:** effectively the reverse of photosynthesis, in which glucose is broken down and reconverted to carbon dioxide and water
- **stoma:** a pore within the surface of leaves, through which various gases such as carbon dioxide and oxygen enter and exit the leaf

Cellular Respiration

Cellular respiration is the most universal of the life processes. A series of chemical reactions beginning with glucose and occurring in cytoplasmic organelles called mitochondria, cellular respiration produces a chemical compound called adenosine triphosphate (ATP). This essential substance furnishes the energy cells need to move, to divide, and to synthesize the multitude of chemical compounds that are required in the biochemical processes of living systems. to perform all the activities necessary to sustain life. Cellular respiration occurs in plants as well as animals, and it occurs during both the day and the night. In order for the last of the series of chemical reactions in the process to be completed, oxygen from the surrounding air (or water, in the case of aquatic plants) must be absorbed. The carbon dioxide that is formed in the respiration process of metabolism is released into the air.

For cellular respiration to occur, a supply of glucose (a simple carbohydrate compound) is required. Photosynthesis, an elaborate series of chemical reactions occurring in chloroplasts, produces glucose and oxygen from carbon dioxide and water in the presence of chlorophyll and sunlight. Energy present in light must be trapped by the chlorophyll within the chloroplasts to drive photosynthesis. Therefore,

417

photosynthesis occurs only in green plants and related organisms, such as algae, and only during the daylight hours. Carbon dioxide, required as a raw material, is absorbed from the air, while the resulting oxygen is released into the atmosphere. The exchange of gases typically involves tiny openings in leaves, called stomata. In animals, the process is reversed, as glucose is broken down by the action of numerous enzymes and reverts to carbon dioxide and water. The carbon dioxide is exchanged through the lungs and exhaled back into the atmosphere. Animal respiration is rather more complicated than this, however, as oxygen must also be breathed in for use in other biochemical processes that recover the energy stored in the various compounds, including glucose.

OXYGEN CYCLE

Oxygen is required for the survival of the majority of microorganisms and all plants and animals. From the surrounding air, organisms obtain the oxygen used in cell respiration. Plants absorb oxygen through the epidermal coverings of their roots and stems and through the stomatal openings of their leaves.

The huge amounts of oxygen removed from the air during respiration must be replaced in order to maintain a constant reservoir of oxygen in the atmosphere. There are two significant sources of oxygen. One involves water molecules of the atmosphere that undergo a process called photodissociation: Oxygen remains after the lighter hydrogen atoms are released from the molecule and escape into outer space.

The other source is photosynthesis. Chlorophyll-containing organisms release oxygen as they use light as the energy source to split water molecules in a process called photolysis. The hydrogen is transported to the terminal phase of photosynthesis called the *Calvin cycle*, where it is used as the hydrogen source necessary to produce and release molecules of the carbohydrate glucose. In the meantime, the oxygen from the split water is released into the surrounding air.

Early in Earth's history, before certain organisms evolved the cellular machinery necessary for photosynthesis, the amount of atmospheric oxygen was very low. As the number and sizes of photosynthetic organisms gradually increased, so did the levels of oxygen in the air. A plateau was reached several million years ago as the rate of oxygen release and absorption reached an equilibrium.

OZONE

Another form of oxygen is ozone. Unlike ordinary atmospheric oxygen, in which each molecule contains two atoms, ozone molecules have three oxygen atoms each. Most ozone is found in the stratosphere at elevations between 10 and 50 kilometers (6 and 31 miles). This layer of ozone helps to protect life on earth from the harmful effects of ultraviolet radiation. Scientists, especially ecologists, are concerned as the amount of ozone has been reduced drastically over the last few decades. Already, an increase in the incidence of skin cancer in humans and a decrease in the efficiency of photosynthesis has been documented. Another concern related to ozone is that of an increase in ozone levels nearer to the ground, where living things are harmed as a result. The formation of ozone from ordinary oxygen within the atmosphere is greatly accelerated by the presence of gaseous pollutants released from industrial processes.

CARBON CYCLE

All forms of life are composed of organic (carbon-containing) molecules. Carbohydrates include glucose as well as lipids (fats, oils, steroids, and waxes), proteins, and nucleic acids. The ability of carbon to serve as the backbone of these molecules results from the uniquely facile ability of carbon atoms to form chemical bonds with other carbon atoms and also with oxygen, hydrogen, and nitrogen atoms.

Like oxygen, carbon cycles in a predictable manner between living things and the atmosphere. In photosynthesis, carbon is "fixed" as carbon dioxide in the air (or dissolved in water) is absorbed and converted into carbohydrates. Carbon cycles to animals as they feed on plants and algae. As both green and nongreen organisms respire, some of their carbohydrates are oxidized, releasing carbon dioxide into the air. Each organism must eventually die, after which decay processes return the remainder of the carbon to the atmosphere.

GREENHOUSE EFFECT

Levels of atmospheric carbon dioxide have fluctuated gradually during past millennia, as revealed by the analysis of the gas trapped in air bubbles of ice from deep within glaciers, polar ice shields and subterranean ice. In general, levels were lower during glacial periods and higher during warmer ones. After the nineteenth century, levels rose slowly until about

1950 and then much more rapidly afterward. The apparent cause has been the burning of increased amounts of fossil fuels associated with the Industrial Revolution and growing energy demands in its wake. The global warming that is now being experienced is believed by most scientists to be caused, at least in part, by increased carbon dioxide levels. The "greenhouse effect" is the term given to the insulating effects of the atmosphere with increased amounts of carbon dioxide. A portion of the solar energy captured by Earth's surface that would be radiated as heat back into outer space is retained within the atmosphere, thus maintaining Earth's average temperature at a higher level than it would be if all of the insolation energy were to be radiated away as heat. Without the atmosphere and the greenhouse effect, the heat energy radiated back to space would necessarily have been equal in order to satisfy thermodynamic laws of energy. However, as the atmosphere developed a "bottleneck" was introduced into this exchange such that the energy going out must still be equal to the energy coming in, but with a heat reservoir to act as a buffer. This heat reservoir is the difference that allows the average temperature of Earth to be about +15°C rather than the -18°C that it would be without the atmosphere. This is an equilibrium level exists because of the essentially fixed amount of carbon dioxide and other greenhouse gases present in the atmosphere. Increasing or decreasing those amounts would demand that a new equilibrium balance be achieved according to the new levels of greenhouse gases, either raising or lowering Earth's average temperature accordingly.

FOREST ECOSYSTEMS
The biotic (living) portions of all ecosystems include three ecological or functional categories: producers (plants and algae), consumers (animals), and decomposers (bacteria and fungi). The everyday activities of all organisms involve the constant exchange of oxygen and carbon dioxide between the organisms of all categories and the surrounding atmosphere.

Because they release huge quantities of oxygen during the day, producers deserve special attention. In both fresh and salt water, algae are the principal producers. On land, this role is played by a variety of grasses, other small plants, and trees. Forest ecosystems, dominated by trees but also harboring many other plant species, are major systems that produce a disproportionate amount of the oxygen released into the atmosphere by terrestrial ecosystems.

Forests occupy all continents except for Antarctica. A common classification of forests recognizes these principal categories: coniferous (northern evergreen), temperate deciduous, and tropical evergreen, with many subcategories for each. The designation "rain forest" refers to the subcategories of these types that receive an amount of rainfall well above the average. Included are tropical rain forests (the more widespread type) and temperate rain forests. Because of the ample moisture they receive, both types contain lush vegetation that produces and releases oxygen into the atmosphere on a larger scale than do other forests.

TROPICAL AND TEMPERATE RAIN FORESTS
Tropical rain forests exist at relatively low elevations in a band about Earth that parallels the equator. The Amazon basin of South America contains the largest continuous tropical rain forest. Other large expanses are located in western and central Africa and the region from Southeast Asia to Australia. Smaller areas of tropical rain forests occur in Central America and on certain islands of the Caribbean Sea, the Pacific Ocean, and the Indian Ocean. Seasonal changes within tropical rain forests are minimal. Temperatures, with a mean near 25°C, seldom vary more than 4°C. Rainfall each year measures at least 400 centimeters.

Tropical rain forests have the highest biodiversity of any terrestrial ecosystem. Included is a large number of species of flowering plants, insects, and animals. The plants are arranged into layers, or strata. In fact, all forests are stratified but not to the same degree as tropical rain forests. A mature tropical rain forest typically has five layers. Beginning with the uppermost, they are an emergent layer (the tallest trees that project above the next layer); a canopy of tall trees; understory trees; shrubs, tall herbs, and ferns; and low plants on the forest floor.

Several special life-forms are characteristic of the plants of tropical rain forests. Epiphytes are plants such as orchids that are perched high in the branches of trees. Vines called lianas wrap themselves around trees. Most tall trees have trunks that are flared at their bases to form buttresses that help support them in the thin soil.

This brief description of tropical rain forests helps to explain their role in world photosynthesis and the related release of oxygen into the atmosphere. As a result of the many layers of forest vegetation, the energy from sunlight as it passes downward is efficiently utilized. Furthermore, the huge amounts of oxygen released are available for use not only by the forests themselves but also, because of global air movement, by other ecosystems throughout the world. Because of this, tropical rain forests are often referred to as "the Earth's lungs."

Temperate rain forests are much less extensive than tropical rain forests. Temperate rain forests occur primarily along the Pacific Coast in a narrow band from southern Alaska, through British Columbia and on to central California. Growing in this region is a coniferous forest but one with higher temperatures and greater rainfall than those to the north and inland. This rainfall of 65 to 400 centimeters per year is much less than that of a tropical rain forest but is supplemented in the summer by frequent heavy fogs. As a result, evaporation rates are greatly reduced. Because of generally favorable climatic conditions, temperate rain forests, like tropical ones, support a lush vegetation. The rate of photosynthesis and release of oxygen is higher than in most other world ecosystems.

Ecologists and conservationists are greatly concerned about the massive destruction of rain forests. Rain forests are being cut and burned at a rapid rate to plant crops, to graze animals, and to provide timber. The ultimate effect of deforestation of these special ecosystems is yet to be seen.

—*Thomas E. Hemmerly*

FURTHER READING

Ashton, Mark S., Tyrrell, Mary S., Spalding, Deborah and Gentry, Bradford, eds. *Managing Forest Carbon in a Changing Climate.* Springer Science + Business Media, 2012. Provides a close examination of the role of forests and forest management in the regulation of atmospheric carbon dioxide.

Humphreys, David. *Logjam: Deforestation and the Crisis of Global Government.* Earthscan, 2006. Highly critical of the failure of international organizations to take on the task of preventing deforestation.

RECENT CLIMATE CHANGE RESEARCH

FIELDS OF STUDY

Atmospheric Chemistry; Analytical Chemistry; Spectroscopy; Mass Spectrometry; Geochemistry; Environmental Sciences; Environmental Studies; Climate Modeling; Atmospheric Science; Oceanography; Physical Geography; Tree Ring Analysis; Crystallography; Molecular Biology; Computer Science

SUMMARY

Scientists are studying the effects of greenhouse gases emitted through industrialization and transportation, particularly in regard to vehicles. Greenhouse gases are believed to be contributing to an increase in global temperatures and triggering climate change. The technologies and research methods employed toward this end are steadily improving. Climatologists can now analyze prehistoric evidence of periods in which climate change occurred, can analyze current trends, and can create models that can predict future conditions.

PRINCIPAL TERMS

- **aerosol:** a gaseous suspension of fine liquid and solid particles
- **geographic information system:** a network of satellite mapping technologies that can capture detailed images of the land surface
- **ice core:** long, cylindrical sample of ice bored from glaciers that provides evidence of ancient climate conditions
- **paleoclimatology:** study of climate conditions in Earth's ancient and prehistoric past
- **sedimentary rock:** rock that has broken from igneous, metamorphic, or other sedimentary rocks to form new deposits

PALEOCLIMATOLOGY

One way climatologists and other scientists analyze climate change on Earth is to compare current conditions with conditions that existed during ancient and prehistoric eras. For example, in 1835, prominent scientist Louis Agassiz (1807–1873), after having listened to a number of theories offered by his peers that the world's glaciers were likely retreating, developed a hypothesis that there was once an Ice Age, during which the massive glaciers that cover the North Pole actually covered most of North America and Europe.

As Agassiz's theory gained popularity in the nineteenth century, another scientist, Svante Arrhenius (1859–1927), offered another groundbreaking idea. In 1895, Arrhenius theorized that the Ice Age was caused by a drop in carbon dioxide levels in the atmosphere, which in turn triggered a dramatic drop in global temperatures. Arrhenius also offered a warning that few people took seriously until almost one century later: that the emissions caused by industrialization could eventually trigger another shift in climate.

The theories offered by Agassiz and Arrhenius are examples of paleoclimatology. In this scientific field, researchers analyze evidence of past events of climate change. This evidence is found in a number of areas. For example, paleoclimatologists may examine the rings inside trees, which can reveal periods of prolonged drought. Other studies entail the analysis of layers of sedimentary rock (rock that broke from igneous, metamorphic, or other sedimentary rocks to form new deposits). Scientists also study sediment at the bottom of the ocean or beneath lakes and swamps. Such sediment can provide clues about the climates in which they were formed millions of years ago. They also can reveal much about the origins of the sediment and, therefore, how far it was carried by ancient glaciers or volcanoes.

One of the most useful types of paleoclimatological evidence is the ice core. Ice core samples are long, cylindrical samples that are removed from glaciers by boring downward from a certain area. The core samples that are removed contain gas bubbles, pollen, sediment, and other compounds and elements from hundreds, thousands, and even millions of years ago. Ice core samples help scientists obtain a

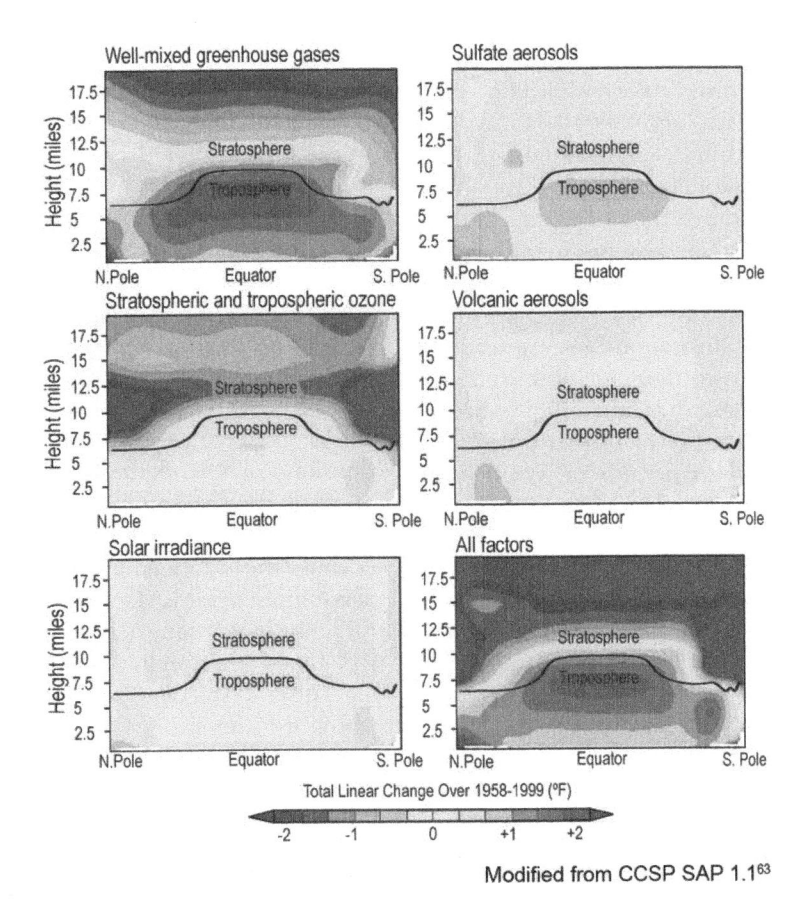

Modified from CCSP SAP 1.1[63]

This set of climate simulations provides evidence of human-induced climate change.

simple, vertical time line of different periods of climate change in Earth's ancient history.

COMPARATIVE STUDIES

To analyze the climate changes in Earth's history (and to analyze more recent changes), scientists conduct comparative studies. In some cases, comparative studies might entail the analysis of climate conditions in a single region through millennia.

Climatologists may examine sedimentary basins for evidence of climate temperatures during a particular era and compare the findings with recent and current conditions. This type of study can help scientists develop models that catalog climatological changes in time and help predict conditions.

Other types of studies involve the comparative analysis of paleoclimatological evidence from several areas around the globe. One such study entailed the compilation of data in the subpolar northern Atlantic regions and the warmer Pacific waters. Using computer models, scientists constructed a profile of the temperature changes that occurred during one of the warmer periods in Earth's history: the middle Pliocene epoch (about 3.5 million years ago). Based on this evidence, paleoclimatologists and other scientists have generated forecast models for climate changes.

A third type of comparative climate change study entails the analysis of similar systems in different parts of the world. Scientists might focus on rainforests in South America and Southeast Asia, collecting data that can provide an illustration of the sensitivities of such forested areas to dramatic and gradual temperature changes. Similarly, climatologists have conducted comparative studies of the sensitivity of rivers and coastal areas with periods of significant temperature shifts.

METHODS OF STUDYING CLIMATE CHANGE

Scientists utilize a number of technologies to study climate change. These technologies have evolved steadily, particularly in the late twentieth century and early twenty-first century. Among these technologies are remote sensors, which are systems that target a region from a distance.

Remote sensors are used to detect temperature and thermal pockets near the surface or in the water, to detect dense particle clouds in the atmosphere, and to detect other environmental conditions. Some examples of remote sensors are active and passive radars, thermal imagers, and spectral scanners. Many later developments in remote sensing have enabled technologies to penetrate cloud cover, study the ocean floor, and scan targets at all times of day and in most weather conditions.

In addition to providing clearer images of a target and gaining access to previously daunting targets, advanced remote sensors are providing larger and larger amounts of data. Scientists compiling the voluminous data are aided by the ongoing evolution of computer modeling systems. This evolution in computer modeling is particularly evident in the amount of data that can be gathered, compiled, assessed, and incorporated into computer models. Computer modeling can in turn provide detailed conceptualizations of past examples of climate change and can predict regional and global climate change trends.

SATELLITE-BASED CLIMATE STUDIES

A significant development in the field of climate change study is the use of satellite-based sensors. Unlike aerial, ground-based, and shipboard remote sensors, satellite-based radars, thermal scanners, and infrared and other sensors can perform scans of considerably larger target areas.

Satellite-based sensors have been in operation since the 1960s, but in recent decades, satellite systems have undergone significant upgrades. The latest in satellite sensors can detect and quantify carbon monoxide levels, atmospheric temperature shifts, water vaporization rates, and other detailed climatological areas.

For example, studying the aerosols (gaseous suspensions of fine liquid and solid particles) in the atmosphere is an important aspect of climate change research. Satellites can be highly useful in this regard because they can scan broad areas of the atmosphere at multiple angles. However, some sensors have had difficulty in scanning aerosol depth. In 2007, scientists developed a corrective measure. By creating a new algorithm to calculate aerosol depth and by then applying it to remote sensors that are designed to scan a wide range of land areas, scientists have scanned aerosol depths with greater precision than before.

Another major innovation for the study of climate change is the Geographic information system (GIS), a network of satellites developed to generate detailed maps of the ground surface. However, as the study of climate change has grown as a scientific discipline, more scientists are turning to GIS to map such trends as coastal erosion and the retreat of vegetation.

Climate Change Networks

Developments in climate change research include the formation of climate change networks. These groups comprise scientists, advanced students, government officials, business leaders, and others who are interested in playing a role in understanding and preventing (where possible) climate change.

Climate change networks commonly employ the use of the Internet, allowing participants to share global data, take part in seminars and webinars, and collaborate on research projects. The U.S. Forest Service, for example, runs the Climate Change Resource Center (CCRC). The CCRC, part of a larger network of U.S. and global climate-change research groups, also produces a wide range of scholarly papers on such subjects as vegetation distribution, air pollution, and innovations in sensor technologies.

Climate change networks also are found at major universities and at nonprofit organizations. The University of New Hampshire's Climate Change Research Center, for example, offers undergraduate and graduate students a number of grant programs for climate change studies. The nonprofit Electric Power Research Institute conducts research on climate policy costs, energy market viability, and other topics. This group also has an information-sharing forum for its members.

Climate change networks represent the continuing evolution of climate change research, as they integrate the latest in technology and research practices into the global information system. These innovations help researchers, government decision makers, business leaders, and private citizens alike understand in greater detail (and with greater speed) any changes in the environment and how those changes contribute to regional and global climate change.

—*Michael P. Auerbach*

Further Reading

Aghedo, A. M., et al. "The Impact of Orbital Sampling, Monthly Averaging, and Vertical Resolution on Climate Chemistry Model Evaluation with Satellite Observations." *Atmospheric Chemistry and Physics* 11, no. 13 (2011): 6493–6514. Describes the combined use of satellite-based sensors and computer climate-change models for a series of assessments performed by the Intergovernmental Panel on Climate Change in 2011.

Alley, Richard B. *The Two-Mile Time Machine. Ice Cores, Abrupt Climate Change and Our Future.* Princeton University Press, 2014.

Dassenakis, Manos, et al. "Remote Sensing in Coastal Water Monitoring: Applications in the Eastern Mediterranean Sea (IUPAC Technical Report)." *Pure and Applied Chemistry* 84, no. 2 (2012): 335–375. Discusses the wide range of applications of satellite-based remote sensing to the study of climate change in coastal areas. Demonstrates how such systems can help scientists study large areas.

Organization for Economic Cooperation and Development. *Space Technologies and Climate Change: Implications for Water Management, Marine Resources, and Maritime Transport.* Author, 2008. An overview of the various types of technologies used in the study of climate change as it pertains to water resource management. Focuses on space-based systems.

Seidel, Klaus, and Jaroslav Martinec. *Remote Sensing in Snow Hydrology: Runoff Modelling, Effect of Climate Change.* Springer, 2010. Describes the use of remote sensors to gauge snow- and ice-melting trends and how the resulting melt runoffs affect the nearby and global environments. The book reviews one hundred regions worldwide.

Ward, Peter D. *Under a Green Sky: Global Warming, the Mass Extinctions of the Past, and What They Can Tell Us About Our Future.* Harper Perennial, 2008. A discussion of the Permian extinction, which occurred more than 252 million years ago and killed nearly 97 percent of all living organisms on Earth. Theorizes that a dramatic rise in global temperatures, triggered by an increase in carbon dioxide in the atmosphere, was likely the biggest contributor to this mass extinction.

Zhong, B., S. Liang, and B. Holben. "Validating a New Algorithm for Estimating Aerosol Optical Depths Over Land from MODIS Imagery." *International Journal of Remote Sensing* 28, no. 18 (2007): 4207–4214. Describes new algorithms for the estimation of aerosol depth in the atmosphere using remote sensors. Aerosols provide vital clues about elements that lead to climate change.

REMOTE SENSING OF THE ATMOSPHERE AND OCEANS

FIELDS OF STUDY

Atmospheric Science; Oceanography; Physical Geography; Earth System Modeling; Heat Transfer; Computer Science; Software Engineering; Thermodynamics; Engineering; Environmental Sciences; Physical Sciences; Electronics; Spectroscopy; Meteorology; Climate Modeling; Process Modeling; Mathematics; Fluid Dynamics; Hydroclimatology

SUMMARY

Scientists utilize a variety of remote sensory technologies, including passive imaging systems, radar, and lidar (light direction and ranging), to study gases and aerosols in the five layers of Earth's atmosphere. Remote sensory systems are also essential pieces of technology in oceanic studies Remote sensors analyze and forecast meteorological phenomena and conditions and are effective in detecting temperature changes, currents, sea levels, and even topography in the oceans. Remote sensory technologies are ground-based, airborne, and space-based. Such technologies continue to evolve, providing clearer and more detailed images (and greater volumes of data). The field of remote sensing has particular relevance in assessing global warming caused by anthropogenic greenhouse gas emissions. In light of concerns about global climate change and about environmental protection, wide-ranging remote sensors are integral for oceanic research.

PRINCIPAL TERMS

- **active sensor:** type of remote sensor that emits radiation at a target to study its composition and condition
- **gravest empirical mode:** concept in which the relationship between oceanic subsurface density and surface elevation is examined
- **lidar:** type of remote sensor that operates similarly to radar but uses lasers instead of radio waves
- **lower atmosphere:** region of the atmosphere comprising the troposphere and the tropopause, reaching an altitude as high as 19 kilometers, or 12 miles
- **middle atmosphere:** region of the atmosphere comprising the mesosphere, mesopause, stratosphere, and stratopause
- **Northern Ocean:** Arctic portion of Earth's oceanic system
- **passive sensor:** type of remote sensor that detects naturally emitted energy, such as reflected sunlight, from target sources
- **pulse Doppler:** type of radar system that emits waves of electromagnetic energy at an atmospheric target; provides a detailed profile of motion, precipitation, and other conditions and objects
- **scatterometer:** active radar that emits high-frequency microwave pulses at a target
- **Southern Ocean:** Antarctic portion of Earth's oceanic system
- **synthetic aperture radar:** type of radar that emits a high volume of radio waves at a target as it passes overhead, creating a multidimensional image of the target
- **thermosphere:** region of the atmosphere marked by thin gases and ultraviolet radiation; also known as the upper atmosphere

THE ATMOSPHERE

One of the keys to life on Earth is the planet's atmosphere. The atmosphere is the origin of weather patterns and acts to regulate surface temperatures. Although the atmosphere consists almost entirely of the three primary gases oxygen, nitrogen, and argon, it also contains aerosols (mists), particles, and many other gases.

Five basic layers make up the atmosphere, each having distinct thermal patterns, rates of motion, chemical compositions, and density. The first layer is the troposphere, which begins at the planet's surface and extends skyward between 6.5 kilometers, or 4 miles (at the poles), and 20 km, or 12.5 mi (at the equator), to the tropopause, a transitional boundary between the troposphere and the layer above. These two regions make up the lower atmosphere. Earth's weather patterns occur in this layer.

The second layer is the stratosphere, extending from the tropopause to about 50 km (31 mi) above the ground surface. In the stratosphere, heat produced by the development of ozone increases with height. This phenomenon is contrary to what occurs in the troposphere, wherein heat decreases with increased height. Although little water vapor is present in the stratosphere, about 19 percent of the gases of Earth's atmosphere are found in the troposphere. Separating the stratosphere from the level above it is another transitional boundary called the stratopause.

The third level is the mesosphere, extending from the stratopause outward about 90 km (56 mi) above the surface. The gases in this layer are thinner, and temperatures are caused by ultraviolet radiation from the sun and decrease with an increase in height. It is in the mesosphere where most space debris, such as meteors, burn up. The section separating the mesosphere from the layer above is called the mesopause. Combined, the mesopause, mesosphere, stratopause, and stratosphere are called the middle atmosphere.

The fourth level, located about 600 km (375 mi) above the Earth, is the thermosphere. In this layer, comprising the upper atmosphere, gases are thin and increasingly heated with height, a condition created by high-energy ultraviolet and X-ray solar radiation. Finally, the fifth level, the exosphere, which is located about 1,000 km (620 mi) above the surface, is the region in which satellites orbit the planet.

Types of Remote Sensor Systems

Ground-based technologies and systems are usually housed in observatories or in mobile units such as ships and weather balloons. Ground-based remote sensors are used to analyze precipitation patterns, turbulence, temperature, and other atmospheric conditions and trends.

Aircraft-borne remote sensors often prove highly effective. Scientists may fly high-altitude aircraft into clouds to take readings and photograph images that are not available to sensors on the ground. One such study occurred in 2000, when a high-altitude aircraft operated by the National Aeronautics and Space Administration was flown into cloud fields over Oklahoma. Using a system known as a multi-angle imaging spectroradiometer, researchers generated simulations of weather patterns at high altitudes.

The evolution of satellite technologies has added another dimension to the study of the atmosphere. Satellites now play an integral role in the examination of the many atmospheric layers, providing greater range and depth to atmospheric science. For example, the European scientific satellite *MetOp* includes in its cache of onboard equipment a thermal, infrared, spectral-imaging system. This technology enables scientists to develop a comprehensive and detailed profile of the changes in atmospheric composition caused by pollution.

Remote sensing equipment is also used to forecast changes in atmospheric conditions. Weather forecasting continues to evolve, enabling meteorologists and other scientists to analyze complex weather patterns and predict atmospheric phenomena, based on the data compiled. Doppler radar, lidar (LIight Direction And Ranging), thermal imaging, and other active and passive remote sensors are employed for this purpose, developing models that predict how weather systems (such as storms and even fog) form, how they will be constituted, and the tracks they will follow.

Radar and Lidar

Radar directs radio waves at a target and interprets the returning waves. Scientists use radar to study wind speeds, precipitation, and other elements of weather systems. Radar has evolved considerably since the 1980s to include systems that can produce highly detailed images of atmospheric conditions and events. One such radar system is pulse Doppler radar, which focuses electromagnetic radiation at a target at various frequencies. Arrays of Doppler radars can provide a comprehensive profile of a given target, generating three-dimensional models and enabling researchers to more accurately predict phenomena. Another increasingly popular application in the study of the atmosphere is a variation of radar that uses lasers rather than radio waves and electromagnetic radiation. Lidar, as it is known, focuses laser beams primarily at lower-level atmospheric targets and gathers data on precipitation, particulates, clouds, and other features. Lidar can in many cases provide data on specific targets with greater detail than can traditional radar.

Remote Sensing, Absorption, and Scattering

Different types of remote sensors rely on spectral images of targets as illuminated by the sun. Depending largely on the size and composition of the target in question and on the wavelength at which the light is irradiated, light is either absorbed or scattered by the target. Oxygen and nitrogen molecules in the atmosphere, for example, scatter solar radiation along a short wavelength of light only. This selective scattering ability is known as Rayleigh scattering. The colors that are emitted through this scattering process are blue and violet, which is why the sky appears blue on clear days. However, through Mie scattering, some molecules are large enough to scatter virtually any light wavelength equally. Cloud droplets, for example, are large enough to demonstrate Mie scattering properties: With every wavelength of solar radiation scattered, clouds appear white (a combination of all colors on the spectrum).

Scattering has evolved into an important research method for studying the atmosphere. Many different types of molecules, particulates, gases, and clouds are found in this broad region, and each type of molecule is composed in such a way that it absorbs some light and scatters others. Lidar has proved to be an effective tool in this arena. Using ground-based lidar, scientists emit a series of beams into a cloud or target area in the atmosphere. The backscattering (Rayleigh or Mie) that occurs enables researchers to profile the composition of the cloud. Some airborne remote sensors employ lidar, pointing the beam downward into a specific area. Satellites, which can profile a much larger area by virtue of their location, also sometimes include lidar.

An example of the successful study of scattering as it pertains to specific phenomena is a 2011 survey conducted near Bozeman, Montana. Scientists attempting to assess the composition and origin of a layer of aerosol in the overhanging atmosphere used lidar, in addition to other remote sensor systems designed to detect scattered light signatures, to generate backscatter. Based on the data retrieved from a variety of angles, scientists determined that the aerosol cloud was made up of smoke from forest fires that were burning in California more than 1,600 km (1,000 mi) to the southwest.

Active and Passive Remote Sensors

In a highly complex environment like Earth's atmosphere, containing many different types of particles, aerosols, and gases, a variety of remote sensing technologies must be used. Passive sensory systems focus on targets that reflect natural energy or emit their own energy. Often, passive sensors are effective during the day, when sunlight reflects off target molecules and other objects. However, remote sensors that detect infrared signatures such as thermal energy emitted naturally from a source are also used during the day or night. Active sensors, however, emit radiation at a target to analyze that target. Such radiation includes microwaves, light waves, or radio waves. Active sensors are useful for a number of reasons. First, when analyzing a target that does not emit an energy signature, active sensors are as effective in daylight as they are during periods without illuminating sunlight. Second, these sensors can be adjusted to detect energy emissions other than solar radiation. Third, by adjusting the wavelengths of the emissions focused on a target, different properties of a target can be studied from multiple angles.

Many active and passive remote sensors focus on microwave emissions. Microwaves exist along wavelengths of between 1 centimeter (0.4 inch) and 1 meter (3 feet). Because of these relatively long wavelengths, microwaves do not scatter as sunlight does. Additionally, many microwaves, either emitted by an active sensor system or captured by a passive one, that are radiated at longer wavelengths penetrate thick clouds, haze, and even steady rainfall. Thus, microwave-oriented passive and active sensors can be utilized in most types of weather. Because of this flexibility, meteorologists and other atmospheric scientists utilize such sensors not only to study phenomena but also to create models for weather prediction.

Computer Modeling

An important complement to remote atmospheric sensing is computer modeling. Models are highly useful for their ability to compile large amounts of data from studies that encompass broad target areas or complex systems. This application is particularly useful for conducting large-scale observations, such as global studies on air pollution and national weather forecasting. One successful application of computer modeling to remote sensing is the ongoing research on transpiration, the process by which water evaporates and enters the atmosphere. Since the early 1990s scientists have used a number of ground-based, airborne, and satellite-based remote sensory systems to detect temperature, humidity, surface radiation, and other global atmospheric elements to connect the conditions of many different target areas and create a composite of transpiration systems. Such studies require the consolidation of enormous amounts of remote sensor data and the creation of a number of computer models to better understand trends and to more accurately predict future conditions.

Studying the Oceans

Oceans cover about 65.7 percent of Earth, spanning about 334 square kilometers (129 million square miles), with a volume of 1,370 million cubic km (329 million cubic mi). About 97 percent of Earth's water is found in oceans, which reach an average depth of about 3.8 km (2.4 mi). Oceans play a critical role in many natural processes, including generating precipitation and weather patterns, transferring sediment, supporting countless marine ecosystems, and delivering chemical elements and compounds such as carbon and methane that have important roles in climatology.

An extensive series of processes and systems functions under the ocean surface. Key indicators of shifts in oceanic systems and processes include coastline erosion, temperature changes, changes in ocean currents, and the presence of materials such as ice and algae blooms. Researchers also monitor the discharge of human-made compounds, such as oil and waste materials, into the oceans. In addition to studying isolated geographic areas, scientists increasingly focus on the system of oceans as a whole. Such pursuits require the application of satellite-based and airborne research technologies, which can compile data and images of broad oceanic regions.

Remote Sensors and the Oceans

In the study of oceans, remote sensors are usually placed aboard aircraft or satellites. Remote sensors may detect thermal conditions, wave height, current speeds, water vaporization, and debris movements (including the movements of icebergs and sediment). There exists a wide range of remote sensor technologies, most of which rely on the object of study emitting some form of detectable energy such as microwaves or reflected solar radiation.

Data and Image Types

Remote sensors may be used to glean information about the spread of pollutants and erosion. Such technologies are used to study a key indicator of ocean pollution and erosion: the coastal plume, which appears along many coasts. In a coastal plume, denser water pushes lighter, unpolluted water upward and, along with it, plankton and other materials. To detect coastal plumes, thermal radiometers, which use infrared to capture the heat signatures of the dense substances caught in the plume, are frequently used.

Another highly useful tool for studying oceanic systems is a scatterometer. Scatterometers are active radar systems that usually are mounted on a satellite. The device transmits high-frequency microwave pulses at the ocean surface. The pulses that bounce back to the satellite are then measured. The different return waves provide information about the winds at the ocean surface, creating a detailed composite of wave sizes and other aspects of the ocean's surface. Knowledge of these surface winds, combined with data from other instrumentation such as the aforementioned thermal radiometers, radar, lidar, and other sensor systems, helps scientists understand the speed at which coastal plumes, algae blooms, and oceanic storm fronts move and provides information on surface water temperatures and wave sizes.

Microwave sensors also have a number of other highly useful applications to oceanic studies, as every object on the surface emits some type of low-level microwave energy. Sea ice is one type of surface object that emits a detectable microwave signature that can be scanned from orbit, even through cloud cover and at any time of day. Passive microwave sensors re found on many different satellites and are used to track sea ice, revealing clues about surface temperatures, wind conditions, and currents. Passive microwave sensors cannot assess the temperature of a piece of ice. Still, they can be used to detect the physical and chemical properties of a block of sea ice, including its crystalline structure.

Synthetic Aperture Radar (SAR)

Like other radar systems, SAR emits radio waves at a target, reading the echo to create an image of the target. However, SAR, which is usually attached to an aircraft or a satellite, continuously sends such waves at the target as it moves overhead. The constant echoes create a more comprehensive, multidimensional image of the target. SARs are useful in the study of coastal plumes, as they generate images of the plumes themselves and track their growth and movements. SAR systems are useful in gathering comprehensive data about oceanic conditions. Wind velocities and wave height are essential factors for monitoring changes in ocean currents, sedimentation transfer, and evaporation. Many SAR systems can even penetrate the ocean surface and provide detailed images of the topography of the ocean floor. Such detailed information can help scientists generate computer models and more accurately predict changes.

INNOVATIVE REMOTE SENSORS

In addition to using well-established remote sensors, scientists are exploring the use of innovative sensor technologies. When added to a series of other sensors, these sensors provide even greater clarity to the oceanic target at hand. For example, polarimetric passive radiometers can detect electromagnetic waves emanating from the source target. Used in concert with radiometers and other remote sensors, polarimetric passive radiometers can provide readings of ocean salinity, adding a new dimension to the study of oceanic conditions.

Earth's oceans are dynamic and are changing in a wide range of areas almost constantly. Knowing this, scientists must consider certain environmental conditions when studying oceanic phenomena and characteristics. For example, in the study of circulation, scientists must account for rising and falling water levels. Ideally, such studies are conducted when the ocean is at its lowest tidal level so that elements such as high waves can be avoided as such conditions can disguise the results of examinations.

An innovative tool in the calculation of oceanic elevation is the interferometric radar altimeter. This type of remote sensor observes two sets of waves from a target such as the distance between high waves. An examination of this wave interference provides data on ocean's elevation and provides greater resolution to scientific measurements.

Remote sensors have proved effective at analyzing shallow-depth and surface conditions. A vexing pursuit, however, in the study of oceans (particularly in Arctic and Antarctic regions) is deep currents. Remote sensors have had problems penetrating the dense underwater environment, creating only low-resolution images that are largely unusable.

Scientists have applied another innovative remote sensory approach to this problem. In many areas of the Northern and Southern oceans, subsurface density can be used to provide clues about deep currents. In one study, researchers analyzed the relationship between subsurface density and surface elevation (a theoretical concept known as the gravest empirical mode [GEM]). The study's authors argued that these sizable areas of density can be used to calculate salinity and temperatures, two key elements that play a role in thermohaline circulation currents. Because of these findings, researchers may begin to use satellite-based and aircraft-mounted remote altimeters to isolate these columns of dense surface and subsurface water. From the data collected through this GEM-altimeter approach, researchers may find another useful tool in studying the oceans' dynamic systems.

New techniques in the study of sea ice through the use of microwave sensors are also being developed. Passive microwave sensors do not detect the temperature of a piece of ice. However, some researchers have proposed utilizing the weather to assist in this capacity. This approach entails analysis of an ice block's microwave emissions and its brightness (as created by local weather conditions). Using a two-step mathematical algorithm, scientists are generating a catalog of global sea ice that may be used to track temperature changes on the ocean's surface.

FUTURE IMPLICATIONS

Remote sensors continue to evolve as a technology. They are creating greater image clarity and providing more comprehensive data on atmospheric targets and trends, and in many ways the evolution of remote sensors has helped scientists better understand some of the most challenging aspects of the oceanic system.

Satellite-based systems provide data on atmospheric trends and systems on a global scale. Researchers have long used contact sensors such as those mounted on buoys or dragged behind moving ships to study ocean elevations, wave height, floor topography, and temperature changes. Coupled with the ability of scientists to view and share remote sensor data through the Internet, often in real time, this technology is likely to continue its contributions to meteorology, oceanography, and earth science. Remote sensors used today add many more dimensions to these studies.

This evolution is important, particularly in light of demands to better predict severe weather and to accurately monitor greenhouse gas emissions. An apparent increase in severe weather (such as the 2011 outbreak of tornadoes across the United States) has prompted scientists to explore the factors that contribute to severe storms. Remote sensors that can detect wind velocities, transpiration, and other key elements are vital. The ever-evolving technologies can be used in locations in which previous systems would acquire unreliable data only; also, remote sensors can now be used to speed up thorough studies that used to take much longer. The remote and harsh environments of the Arctic and Antarctica (the Northern and Southern ocean regions) provide excellent examples of this. Today, remote sensors can detect thermal changes, the formation of icebergs, density shifts, currents, and other important phenomena from orbit. Such developments in research of the Northern and Southern oceans are critical, as scientists who are concerned with global climate change tend to focus on ocean elevation and debris in and around these geographic areas.

Furthermore, heightened public attention to the effects of human-made greenhouse gas emissions (such as industrial pollution and automobile exhaust) continues to create the need to examine the types and volumes of particulates in the atmosphere. In this arena too, state-of-the-art remote sensors are making such studies possible. The remote sensors found particularly on satellites are easily accessed by participating scientists around the world. There exists a wide range of internationally managed satellites with onboard remote sensors that are trained on Earth's oceans and atmosphere. Through such easy access and global communications networks, scientists can share data and develop large-scale international studies. With the continued evolution of remote sensors and innovative remote-sensing techniques, scientists will likely continue to generate voluminous and detailed data and images of Earth's oceanic and atmospheric systems.

—*Michael P. Auerbach*

FURTHER READING

Clerbaux, Cathy, Solene Turquety, and Pierre Coheur. "Infrared Remote Sensing of Atmospheric Composition and Air Quality: Toward Operational Applications." *Comptes Rendus Geoscience* 342, nos. 4/5 (2010): 349–356. Discusses the uses of remote sensors for studying pollution's effects on the atmosphere. Reviews satellite-based technologies such as thermal infrared sensors to study the troposphere on a global scale.

Guzzi, Rodolfo, ed. *Exploring the Atmosphere by Remote Sensing Techniques.* Springer, 2010. In this collection of lectures, the editor presents various remote sensing practices used in the study of gases, aerosols, and clouds in the atmosphere.

Hoff, Raymond, et al. "Applications of the Three-Dimensional Air Quality System to Western U.S. Air Quality: IDEA, Smog Blog, Smog Stories, AirQuest, and the Remote Sensing Information Gateway." *Journal of the Air and Waste Management Association* 59, no. 8 (2009): 980–989. Describes an extensive network of remote sensors, based on ground and onboard satellites used to monitor aerosols released into the atmosphere in the United States. The system, which creates three-dimensional images of the atmosphere, is a joint partnership of local, state, and federal governments and private citizens.

Kokhanovsky, Alexander A., ed. *Light Scattering and Remote Sensing of Atmosphere and Surface.* Vol. 6 in *Light Scattering Reviews.* Springer, 2011. A compilation of papers on remote sensing as it relates to light scattering and the images created by precipitation as influenced by light scattering phenomena.

Marzano, Frank S., and Guido Visconti, eds. *Remote Sensing of Atmosphere and Ocean from Space: Models, Instruments, and Techniques.* Springer, 2011. A collection of lectures that focus on space-borne remote-sensing techniques, including microwave, infrared, and passive sensors and weather forecasting practices.

Mityagina, M. I., O. Y. Larova, and S. S. Karimova. "Multi-sensor Survey of Seasonal Variability in Coastal Eddy and Internal Wave Signatures in the North-Eastern Black Sea." *International Journal of Remote Sensing* 31, nos. 17/18 (2010): 4779–4990. Describes a multisystem approach to the application of remote sensors in the study of coastal areas and waves in the Black Sea. Among the systems employed in this study are synthetic aperture radar, radiometers, infrared sensors, and other active and passive radar systems.

"Radiometric Normalization of Sensor Scan Angle Effects in Optical Remote Sensing Imagery." *International Journal of Remote Sensing* 28, no. 19 (2007): 4453–4469. Describes some of the optical effects, caused by the content of the atmosphere, which can be scanned using satellite-based and airborne remote sensors.

Rees, W G. *Physical Principles of Remote Sensing.* Cambridge University Press, 2013. Provides general information about the principles of remote sensing technology and their application to earth sciences (including oceanic studies). The primary types of sensors described are those mounted on satellites.

S

SATELLITE METEOROLOGY

FIELDS OF STUDY

Atmospheric Chemistry; Biochemistry; Environmental Chemistry; Geochemistry; Inorganic Chemistry; Physical Chemistry; Phorochemistry; Thermodynamics; Chemical Engineering; Environmental Sciences; Environmental Studies; Physical Sciences; Meteorology; Process Modeling; Fluid Dynamics; Chemical Kinetics; Hydroclimatology; Atmospheric Science; Oceanography; Hydrology; Physical Geography; Ecosystem Management; Ecology; Spectroscopy

SUMMARY

Satellite meteorology, the study of atmospheric phenomena using satellite data, is an indispensable tool for forecasting weather and studying climate on a global scale.

KEY TERMS

- *active sensor:* a sensor, such as a radar instrument, that illuminates a target with artificial radiation, which is reflected back to the sensor
- *albedo:* the percentage of incoming radiation that is diffusely reflected by a planetary surface
- *El Niño:* a periodic anomalous warming of the Pacific waters off the coast of South America; part of a large-scale oceanic and atmospheric fluctuation that has global repercussions
- *geosynchronous (geostationary):* describing a satellite that orbits about Earth's equator at an altitude and speed such that it remains above the same point on the surface of the planet
- *near-polar orbit:* an orbit of Earth that lies in a plane that passes close to both the North and South Poles
- *passive sensors:* sensors that detect reflected or emitted electromagnetic radiation that has issued from another source
- *radiometer:* an instrument that quantitatively measures reflected or emitted electromagnetic radiation within a particular wavelength interval
- *spatial resolution:* the extent to which a sensor is able to differentiate between closely spaced features
- *sun-synchronous orbit:* for an Earth satellite, a near-polar orbit at an altitude such that the satellite always passes over any given point on Earth at the same local time
- *synthetic aperture radar (SAR):* a space-borne radar imaging system that uses the motion of the spacecraft in orbit to simulate a very long antenna

TYPES OF SATELLITES AND INSTRUMENTS

Satellite meteorology is the study of atmospheric phenomena, notably weather and weather conditions, using information gathered by instruments aboard artificial satellites. These satellites, including the International Space Station, are equipped with instruments that monitor cloud cover, snow, ice, temperatures, and other parameters, to give scientists a continuous and up-to-date view of meteorological conditions and activity over a large area. The use of satellite data is an important tool not only for forecasting weather and tracking storms but also for observing climate change over time, monitoring ozone levels in the stratosphere, and studying numerous other aspects of global weather and climate on an ongoing basis.

The satellites from which meteorological measurements are made can be categorized by their orbits. Some weather satellites have a geosynchronous, or geostationary, orbit, meaning that they travel around the globe at an altitude and speed that keep them above the same point over the equator. A near-polar, sun-synchronous orbit, by contrast, is a north-south orbit that passes close to the poles such that a satellite

A hurricane, with the eye in its center, as seen in a satellite image.

on that orbital path passes over any given location on Earth at the same local time. Geosynchronous satellites have better temporal resolution: They provide updated information for an area every thirty minutes, while near-polar, sun-synchronous satellites may take anywhere from a few hours to several days to transmit updates. However, near-polar satellites have the higher spatial resolution; that is, they are better at providing images in which closely spaced features can be identified. Geosynchronous satellites provide images with comparatively poor spatial resolution because they must orbit at a greater altitude (at least 35,000 kilometers above the surface). Examples of geosynchronous meteorological satellites include the U.S. Geostationary Operational Environmental Satellite (GOES) series, the European Space Agency's METEOSAT series, Russia's Geostationary Operational Meteorological Satellite (GOMS) series, and Japan's Geostationary Meteorological Satellite (GMS), or Himawan, series. Near-polar, sun-synchronous meteorological satellites include the National Oceanic and Atmospheric Administration (NOAA) series of satellites and Russia's Meteor series.

The various orbiting platforms carry different sets of instruments. The weather satellites launched in the 1960's and early 1970's included television camera systems as part of their instrument packages. Later satellites have relied instead on instruments such as specialized radiometers (instruments that measure the amounts of electromagnetic radiation within a specific wavelength range) and radar systems. Radiometers measure such parameters as surface, cloud, and atmospheric temperatures; atmospheric water vapor and cloud distribution; and scattered solar radiation. Radar-system measurements include satellite altitude and ocean-surface roughness. Television cameras and radiometers are examples of passive sensors, which record radiation reflected or emitted from clouds, landforms, or other objects below. Radar systems are active sensors, sending out signals and recording them as they are reflected back. The data collected by a satellite's sensors are transmitted via radio to ground stations. If a near-polar orbiter is not within transmitting distance of a ground station, its onboard data-collection system will store the information until the satellite passes within range.

Weather Satellites

The Advanced Television Infrared Observation Satellite, or TIROS, and Next-Generation (ATN) near-polar orbiting satellites (NOAA 8 through 14), have carried an assortment of sophisticated instruments. Two of these satellites, NOAA 12 and NOAA 14, remained in operation into the late 1990's. All of the ATN series satellites have included an advanced, very-high-resolution radiometer (AVHRR), which detects specified wavelength intervals within the visible, near-infrared, and infrared wavelength ranges to generate information on sea-surface and cloud-top temperatures and ice and snow conditions; a TIROS operational vertical sounder (TOVS), which measures emissions within the visible, infrared, and microwave spectral bands to provide vertical profiles of the atmosphere's temperature, water vapor, and total ozone content from the ground surface to an altitude of 32 kilometers; and a solar proton monitor, which detects fluctuations in the sun's energy output, particularly those related to sunspot (solar storm) activity. With the exception of NOAA 8 and NOAA 12, all these satellites have included the Earth radiation budget experiment (ERBE), which uses long-wave and short-wave radiometers to provide data pertaining to Earth's albedo. All but NOAA 8, NOAA 10, and NOAA 12 have carried the solar backscatter ultraviolet (SBUV) radiometer, which measures the vertical structure of ozone in the atmosphere by monitoring the ultraviolet radiation that the atmosphere scatters back into space. Of the ATN series of satellites, NOAA 12 is the only one that has not also carried the Search and Rescue Satellite Aided Tracking (SARSAT) system, which detects distress signals from downed aircraft and emergency beacons from ocean vessels, then relays the signals to special ground stations.

The satellites of the U.S. GOES series that remained in operation in the late 1990's (GOES 8 and GOES 9) are geosynchronous orbiters with their own distinctive instrumentation. Instrumentation aboard GOES 8 (also called GOES-EAST) and GOES 9 (or GOES-WEST) includes a sounder, which uses visible and infrared data to create vertical profiles of atmospheric temperature, moisture, carbon dioxide, and ozone; a five-band multispectral radiometer, which scans visible and infrared wavelengths to obtain sea-surface temperature readings, detect airborne dust and volcanic ash, and provide day and night images of cloud conditions, fog, fires, and volcanoes; and a search-and-rescue support system similar to the ones flown on the NOAA near-polar orbiters. The GOES satellites are also equipped with a space environment monitor (SEM), which uses a solar X-ray sensor, a magnetometer, an energetic particle sensor, and a high-energy proton alpha detector to monitor solar activity and the intensity of Earth's magnetic field. The GOES-EAST satellite is positioned above the equator at an approximate longitude of 75 degrees west, while the GOES-WEST satellite orbits above the equator at approximately 135 degrees west longitude. These locations are ideal for monitoring the climatic conditions across North America. From them, GOES-EAST can provide images of storms approaching the eastern seaboard across the Atlantic Ocean during hurricane season (June through November), and GOES-WEST can monitor the weather systems that move in from across the Pacific Ocean, which affect the western seaboard during most of the year. However, when instrument malfunctions impair a satellite's ability to provide data, the National Weather Service can use small rocket engines aboard the satellites to reposition a more functional platform to provide the desired coverage until a replacement satellite can be launched. After the imaging system on GOES 6 failed in 1989, for example, GOES 7 was relocated several times to compensate.

RADAR AND SATELLITE DATA

While operational weather satellites have generally carried radiometers, radar instruments have been part of the instrument package aboard experimental satellites and space shuttle flights. One of the best-known orbiters using active sensors is Seasat, a short-lived experimental craft launched in 1978 to monitor the oceans. During its three months of operation, this Earth-resources satellite provided a wealth of data for meteorological study. Its radar altimeter determined the height of the sea surface, from which data scientists derived measurements of winds, waves, and ocean currents. Its radar scatterometer yielded information on wave direction and size that, in turn, provided insights into wind speed and direction. The most sophisticated of Seasat's active sensors was its synthetic aperture radar (SAR) system, a radar imaging system that created a "synthetic aperture" of view by using the motion of the platform to simulate a very long antenna. Images of the ocean's surface were obtained from SAR data.

Satellite data have become indispensable to the meteorologist. From orbit, information is readily available for any location on the planet, regardless of its remoteness, inaccessibility, climatic inhospitality, or political affiliation. Satellites yield regular, repeated, and up-to-date coverage of areas at minimal cost. They make it possible to view large weather systems in their entirety, and facilitate meteorological observations on a regional or global basis. Satellites also provide a single data source for multiple locations, alleviating the problem of individual variance of calibration and accuracy that would be associated with separate ground-based observations for each location. However, it is important to note that ground-based stations can make more accurate and detailed observations of a small area, and such details may be lost in a view from space. Important though it is to modern meteorology, the use of satellite data augments, rather than replaces, other methods of study.

WEATHER FORECASTING

Satellite meteorology provides a rapid and relatively inexpensive means of obtaining current and abundant information on temperature, pressure, moisture, and other atmospheric, terrestrial, and oceanic conditions that affect weather and climate. These data, collected in digital form, are readily processed and integrated with other information. Through these ongoing observations from orbit, scientists gain insights into the short-term and long-term implications of major atmospheric phenomena.

Weather forecasting is the best-known application of satellite meteorology. Anyone who has watched a televised weather report is familiar with geostationary satellite images, a series of which are usually presented in quick succession to show the recent movement of major weather systems. Meteorologists use computers to process the vast amounts of data provided by satellites and other information sources, including ground-based stations, aircraft, ships, and buoys. Data processing yields such forecasting aids as atmospheric temperature and water-vapor profiles, enhanced and false-color images, and satellite-image "movies." Computer models of atmospheric behavior also assist the meteorologist in short-range and long-range forecasting.

Satellite images of cloud cover alone yield a wealth of information for the forecaster. By comparing imagery from visible and infrared spectral regions, meteorologists can identify cloud types, structure, and degree of organization, then make assumptions and deductions concerning associated weather conditions. For example, the tall cumulus clouds that produce thunderstorms appear bright in the visible range, as they are deep and thus readily reflect sunlight. These clouds show up in infrared images as areas of coldness, an indicator of the altitude to which the clouds have climbed. Clouds that appear bright in visible-range imagery but that register as warm (low-altitude) in infrared scans may be fog or low-lying clouds. Wispy, high-altitude cirrus clouds, which are not precipitation-bearing, appear cold in infrared images but may not show up at all in visible-range scans.

Weather satellites have proved particularly useful in the science of hurricane and typhoon prediction. These large, violent, rotating tropical storms originate as relatively small low-pressure cells over oceans, where coverage by conventional weather-monitoring methods is sparse. Before the advent of satellites, ships and aircraft were the sole source of information on weather at sea, and hurricanes and typhoons often escaped detection until they were dangerously close to populated coastal areas. Using images and data obtained from orbiting satellites, meteorologists can track and study these storms continuously from their inception through their development and final dissipation. With accurate storm tracking and ample advance warning, inhabitants at risk can evacuate areas threatened by wind and high water, thereby minimizing loss of life.

Climate Studies

Meteorological satellites also provide scientists with a view of how human activity affects climate on a local, regional, and even global basis. Terrestrial surface-temperature measurements clearly show urban "heat islands," where cities consistently radiate more heat energy than the surrounding countryside. In images obtained from orbit, thunderstorms can be seen developing along the boundaries of areas of dense air pollution: The haze layer inhibits heating of the ground surface, leading to the unstable atmospheric conditions that produce rainfall. Satellite imagery has revealed that in sub-Saharan Africa, where the overgrazing of livestock owned by nomads has contributed substantially to the spread of desert areas, the resulting increase in albedo has led to a reduction in rainfall and a subsequent reinforcement of drought conditions. Studies of deforestation in tropical areas have incorporated satellite data in their efforts to determine whether replacing forests with agricultural land affects rainfall by reducing evaporation or altering albedo. Satellite data have also played a major role in the ongoing debates regarding how human activity has affected global temperature trends and the ozone layer.

Satellite meteorology is useful for monitoring the climatological effects of natural occurrences as well. The 1991 eruption of Mount Pinatubo in the Philippines marked the first time that scientists were able to quantify the effects of a major volcanic eruption on global climate. Satellites equipped with ERBE instruments tracked the dissemination of the ash and sulfuric acid particles resulting from the violent eruption. The larger ash particles more readily drift down into the lower portion of the atmosphere, where they are typically removed in precipitation. The smaller sulfuric acid particles, however, can remain suspended in the stratosphere for several years, eventually reaching lower altitudes where they combine with water to produce acidic precipitation. The ERBE instruments measured the amount of sunlight reflected by clouds, land surfaces, and particles suspended in the atmosphere and detected the contribution of suspended particles, clouds, and trace gases such as carbon dioxide to the amount of heat that the atmosphere retained. The eruption was found to have brought about a uniform cooling of Earth, temporarily slowing the ongoing global-warming trend that has been observed since the 1980's.

Another natural phenomenon, El Niño, has also been the subject of satellite-based study. This periodic anomalous warming of the Pacific waters off the coast of South America is part of a large-scale oceanic and atmospheric fluctuation known as the Southern Oscillation, in which atmospheric pressure conditions alternately decline and rise over the eastern Pacific Ocean and Australia and the Indian Ocean. These widespread pressure changes influence rainfall patterns around the world. Satellite measurement of sea-surface temperatures facilitates the early detection of El Niño conditions, and satellite observations on a global basis help scientists to discern the climatic patterns that make up this complex phenomenon.

Meteorological satellites have also been used to gather data pertaining to "solar weather." Using orbiting sensors that detect energetic particles from the sun, as well as more direct imaging methods, scientists can monitor and predict sunspots and other solar activity. The ability to predict the increases in solar emissions that are associated with sunspots allows scientists to anticipate the resulting ionospheric conditions on Earth, the effects of which include magnetic storms and disruption of radio transmissions.

—*Karen N. Kähler*

Bibliography

Ahrens, C. Donald. *Essentials of Meteorology: An Invitation to the Atmosphere*. Belmont, Calif.: Brooks/Cole Cengage Learning, 2012. Discusses various topics in weather and the atmosphere. Chapters cover topics such as tornadoes and thunderstorms, acid deposition and other air pollution topics, humidity and cloud formation, and temperature.

Bader, M. J., et al., eds. *Images in Weather Forecasting: A Practical Guide for Interpreting Satellite and Radar Imagery*. New York: Cambridge University Press, 1995. A collection of essays written by leading meteorologists to describe the equipment and techniques used in weather forecasting. Chapters focus on remote sensing, radar meteorology, and satellite meteorology. Color illustrations.

Barrett, E. C., and L. F. Curtis. *Introduction to Environmental Remote Sensing*. 3d ed. New York: John Wiley & Sons, 1992. Discusses weather analysis, forecasting satellite systems, and data applications. Deals with the atmosphere energy and radiation budgets and other aspects of global climatology. Each chapter includes references.

Burroughs, William James. *Watching the World's Weather*. Cambridge, England: Cambridge University Press, 1991. Focuses on the importance of satellite meteorology to an understanding of weather and climate on a global scale. Deals not only with satellites and instrumentation but also with the essentials of meteorology. Includes satellite images, a glossary, a list of acronyms, and an annotated bibliography.

Campbell, Bruce A. *Radar Remote Sensing of Planetary Surfaces*. New York: Cambridge University Press, 2002. Covers methodology and theory of radar remote sensing. Provides examples from Earth, the moon, and other planets. Includes references and indexing.

Campbell, James B., and Randolph H. Wynne. *Introduction to Remote Sensing*. 5th ed. New York: Guilford Press, 2011. Provides an interdisciplinary introduction to the topic, with background information on the electromagnetic spectrum. Covers many aspects of digital imagery, from aerial photography and coverage to image enhancement and interpretation.

Collier, Christopher G. *Applications of Weather Radar Systems: A Guide to Uses of Radar Data in Meteorology and Hydrology*. 2d ed. New York: Wiley, 1996. Offers a detailed look into scientific advancements regarding the tools used in meteorology and hydrology. Focuses on radar meteorology, precipitation measurement, hydrometeorology, and weather forecasting.

Fishman, Jack, and Robert Kalish. *The Weather Revolution*. New York: Plenum Press, 1994. Provides a nontechnical explanation of the basics of meteorology and outlines the evolution of weather forecasting, before and since the advent of weather satellites. The chapter on the development of satellite meteorology discusses programs from TIROS through GOES.

Gurney, R. J., J. L. Foster, and C. L. Parkinson, eds. *Atlas of Satellite Observations Related to Global Change.* Cambridge, England: Cambridge University Press, 1993. Suitable for college-level readers. Includes articles on the stratosphere, the troposphere, Earth's radiation balance, ocean-atmosphere coupling, and snow and ice cover. Illustrated with satellite imagery. An appendix describes selected satellites and sensors.

Hill, Janice. *Weather from Above.* Washington, D.C.: Smithsonian Institution Press, 1991. An overview of U.S. weather satellite programs, intended for a nonscientific audience, that includes a glossary of acronyms, a chronological list of meteorological satellites, and suggestions for further reading. Illustrations include photographs of weather satellites, many from the collection of the National Air and Space Museum.

Kelkar, R. R. *Satellite Meteorology.* Hyderabad: BS Publications, 2007. Traces the history of satellite meteorology as the youngest and fastest-growing branch of the science, to its present state, and speculates on the future. Suitable as a textbook or as a reference work for several related disciplines.

Lillesand, Thomas M., Ralph W. Kiefer, and Jonathan Chipman. *Remote Sensing and Image Interpretation.* 6th ed. Hoboken, N.J.: John Wiley & Sons, 2008. The chapter on Earth resource satellites includes a description of the U.S. NOAA and GOES series of satellites and the Air Force's Defense Meteorological Satellite Program. Earlier chapters provide detailed information on various scanning instruments.

Lubin, Dan, and Robert Massom. *Polar Remote Sensing: Atmosphere and Oceans.* Chichester, England: Praxis Publishing, 2006. A comprehensive multidisciplinary work covering the polar environment, with satellite remote sensing applications to atmospheric chemistry, meteorology, climate study, and physical oceanography.

Lutgens, Frederick K., Edward J. Tarbuck, and Dennis Tasa. *The Atmosphere: An Introduction to Meteorology.* 11th ed. Upper Saddle River, N.J.: Prentice Hall, 2010. Offers an excellent introduction and description of the atmosphere, meteorology, and weather patterns. Suitable for the reader new to the study of these subjects. Color illustrations and maps.

Menzel, W. Paul. *Applications with Meteorological Satellites.* Geneva: World Meteorological Organization, 2001. Presents a NOAA report that discusses the history and evolution of satellite meteorology, data processing, radiation principles, accounting for clouds, measuring surface temperatures, and other techniques used in taking atmospheric measurements. A technical report that provides detailed information best suited for graduate students and professional in the fields of meteorology and satellite imagery.

Monmonier, Mark S. *Air Apparent: How Meteorologists Learned to Map, Predict, and Dramatize Weather.* Chicago: University of Chicago Press, 1999. A college-level text that looks at the satellites and radar systems used to collect meteorological data, as well as the techniques used to interpret that information. Color illustrations, index, and bibliographical references.

Qu, John J., Robert E. Murphy, Wei Gao, Vincent V. Salomonson, and Menas Kofatos, eds. *Earth Science Satellite Remote Sensing. Science and Instruments.* Vol. 1. Beijing: Tsinghua University Press, 2006. A specialist reference work describing the scientific principles of remote sensing by satellite and the instruments used in the process.

Stevens, William Kenneth. *The Change in the Weather: People, Weather, and the Science of Climate.* New York: Random House, 2001. Describes various natural and human-induced causes of changes in the climate. Includes a twenty-page bibliography and an index.

Vallis, Geoffrey K. *Atmospheric and Oceanic Fluid Dynamics: Fundamentals and Large-scale Circulation.* New York: Cambridge University Press, 2006. Begins with an overview of the physics of fluid dynamics to provide foundational material on stratification, vorticity, oceanic and atmospheric models. Discusses topics such as turbulence, baroclinic instabilities, wave-mean flow interactions, and large-scale atmospheric and oceanic circulation. Best suited for graduate students studying meteorology or oceanography.

SEA ICE AND THE GLOBAL CLIMATE

FIELDS OF STUDY

Oceanography; Thermodynamics; Fluid Dynamics; Climatology; Climate Modeling; Hydroclimatology; Bioclimatology; Atmospheric Science; Physical Geography; Earth System Modeling

SUMMARY

Sea ice is formed when the surface layer of the ocean reaches the freezing point. For salt water, this occurs at a temperature of –1.8°C, slightly colder than the freezing point of freshwater. Because water is more buoyant in its solid state, the newly formed ice floats on top of the ocean, forming a barrier between the cool air and the warm water.

PRINCIPAL TERMS

- **buoyancy:** the ability of a material to float within a fluid medium according to their relative density
- **greenhouse gases (GHGs):** an atmospheric gas whose molecules absorb infrared radiation from a planet's surface and then emit a significant portion of it back into the atmosphere, most commonly carbon dioxide, methane, water, and ozone.

THE NATURE OF SEA ICE

There are two types of sea ice. Fast ice (also known as land fast ice) is seawater that has frozen along the shore line, or, in shallow areas, has frozen to the ocean floor. These formations are, by their very nature, stationary. Drift ice is frozen salt water that floats and is affected by winds and currents. First-year ice sheets, or floes, are typically thin and subject to shattering and refreezing, forming rafts and ridges.

A cluster of drift ice is called pack ice, and this is what covers both the Arctic and the Antarctic regions of the planet. The amount of pack ice in the polar regions varies with the season. During winter, sea ice usually covers between 14 and 16 million square kilometers in the Arctic and between 17 and 20 million square kilometers in the Antarctic. In the Antarctic, the sea ice melts during the summer and is therefore considered seasonal ice, but in the Arctic it remains year around.

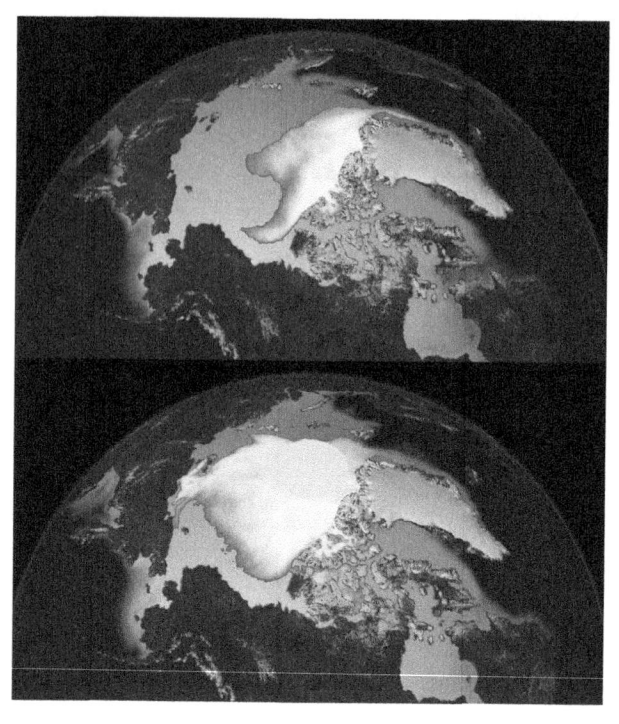

The image shows sea ice coverage in 1980 (bottom) and 2012 (top). Multi-year ice is shown in bright white, while average sea ice cover is shown in light blue to milky white. The data shows the ice cover for the period of November 1, 1980, through January 31 in their respective years.

Sea ice poses a hazard for shipping and has been noted in sailors' logs going back to the fourth century BCE. Sea ice is commonly confused with icebergs. Although both represent significant shipping hazards, icebergs are made of freshwater, usually from precipitation or snow melt, and break off from glaciers. Sea ice, being composed of salt water, is inherently denser than the freshwater ice of icebergs, making floating pieces of identical size somewhat more massive and capable of inflicting greater damage when impacted.

SIGNIFICANCE FOR CLIMATE CHANGE

Sea ice, especially Arctic sea ice, plays a significant role in the overall global climate by regulating the exchange of heat, moisture, and salinity in the ocean. As seawater freezes, its salt content is reduced in a process called brine rejection. The salt rejected from the frozen water increases the salinity of the surrounding ocean and can affect ocean currents. Sea ice insulates the warm ocean water and keeps it from losing heat to the much colder arctic or subarctic air. It also, however, reflects more sunlight than the surrounding ocean, preventing the sunlight from warming the ocean. When more sunlight reaches the water, the heat is absorbed, raising the temperature. Fissures or cracks in the ice release the trapped heat from the warmer ocean water, which can affect precipitation and cloud cover.

Greenhouse gases (GHGs) emitted through human activities and the resulting increase in global mean temperatures are the most likely underlying cause of sea ice decline, but the direct cause is a complicated combination of factors resulting from the warming and from climate variability.

Passive microwave satellite data show that, as of 2006, the Arctic sea ice decreased by 3.6 percent during each of the previous three decades. In 2007, Arctic sea ice was at a record low. Despite a recovery to near average levels during the following winter, 2008 summer sea ice coverage was nearly as low as the previous year. In the summer of 2016 the second lowest minimum extent Arctic sea ice was recorded. The previous minimum extent was recorded in 2012, and the 2016 minimum was 29 percent smaller than the average minimum between 1981 and 2010. Scientific data suggest that the Arctic Ocean may be ice-free in the summer as early as the 2030s, a condition that has not existed in at least 800,000 years. As a result, the remaining ice pack currently is thinner than in previous years and is less able to withstand the varying temperatures brought about by global warming.

As the global climate changes, the quantity of polar sea ice is diminishing, affecting regional ecosystems. Animals such as polar bears, walruses, and seals rely on ice floes for breeding, shelter, and especially hunting. As the summer ice diminishes, the summer hunting season grows shorter, which is a crucial consideration for animals that hibernate and then typically hunt for food that is dependent on those areas of sea ice. The large mammals have to travel farther to find food, and more time passes between kills. A 1980 study found that the average female polar bear in the Hudson Bay region of Canada weighed 300 kilograms. By 2004, that average weight had dropped to 230 kilograms. This decline has led to an overall 15 percent drop in polar bear birth rate. Scientists at the United States National Snow and Ice Data Center have predicted that, if current trends persist, by the year 2030, the Arctic Ocean will be without ice for the first time in about one million years.

P. S. Ramsey

FURTHER READING

"Arctic Report Card: Update for 2016." *Arctic Program*, National Oceanic and Atmospheric Administration, http://www.arctic.noaa.gov/Report-Card/Report-Card-2016. Accessed 31 Jan. 2017.

"Polar Bears and Climate Change." *World Wildlife Fund*, 2017, www.worldwildlife.org/pages/polar-bears-and-climate-change. Accessed 8 Nov. 2017.

Schimel, David *Climate and Ecosystems*. Princeton University Press, 2013. One of the "Princeton Primers on Climate" series, this book describes the interaction mechanisms of various ecosystems and climate.

Gosnell, Mariana. *Ice: The Nature, the History, and the Uses of an Astonishing Substance*. University of Chicago Press, 2007. Provides an overview of the various types of ice found all over the planet.

Melnikov, I. A. *The Arctic Sea Ice Ecosystem*. Gordon and Breach, 1997. Examines in detail ice and its impact on Earth and its inhabitants.

Mulvaney, Kieran. *At the Ends of the Earth: A History of the Polar Regions*. Island Press/Shearwater Books, 2001. A history of human interaction with the polar regions and how those regions are affected by the changing climate.

Thomas, David N. *Frozen Oceans: The Floating World of Pack Ice*. Natural History Museum, 2004. Accessible, illustrated overview of pack ice.

Thomas, David N., and Gerhard S. Dieckmann, eds. *Sea Ice: An Introduction to its Physics, Chemistry, Biology, and Geology*. Blackwell Science, 2003. Collection of research papers covering all aspects of sea ice.

SEA LEVELS

FIELDS OF STUDY

Oceanography; Physical Geography; Statistics; Hydrology; Engineering; Environmental Sciences; Environmental Studies; Waste Management; Ecosystem Management; Earth System Modeling; Ecology

SUMMARY

Sea level is the height of the surface of the ocean relative to the land at any given location. Sea level is highly variable and can undergo very rapid changes due to such events as tides, tsunamis, changes in barometric pressure, wind-generated waves, and even freshwater floods. While these events can produce changes in sea level of several meters, they are local in scale and of a very short duration, generally lasting only for hours. Mean sea level is the average, global height of the sea surface, independent of these local, short-term changes. Mean sea level provides a frame of reference for land elevations and ocean depths. Changes in mean sea level are on the order of a few millimeters per year.

PRINCIPAL TERMS

- **continental shelf:** the extension of a continental mass beneath the sea; a flat or gently sloping platform usually 10 to 100 kilometers wide, extending to a depth of 100 to 150 meters
- **eustatic sea-level change:** a change in sea level worldwide, observed on all coastlines on all continents
- **glaciation:** commonly known as an "ice age," the cyclic widespread growth and advance of ice sheets over the polar and high-latitude to midlatitude regions of the world
- **greenhouse effect:** the warming of the atmosphere caused by absorption and re-emission of infrared energy by carbon dioxide and other gases in the atmosphere
- **isostasy:** the passive, vertical rise or fall of the crust caused, respectively, by the removal or addition of a load on the crust
- **local sea-level change:** a change in sea level only in one area of the world, usually by land rising or sinking in that specific area
- **mean sea level:** the average height of the sea surface over a multiyear time span, taking into account storms, tides, and seasons
- **regression:** the retreat of the sea from the land, allowing land erosion processes to occur on material previously below the sea surface
- **tectonics:** the process of origin, movement, and deformation of large-scale structures of the crust
- **transgression:** the advance of the sea over the land, allowing marine sediments to be deposited on what had previously been dry land

CHANGING SEA LEVEL

Sea level is a major aspect of the modern world. The distinction between land and sea depends on the level of the sea. A rise in sea level means the flooding of adjacent low-lying land; a drop in sea level means exposure of some of the sea floor. Sea level is most accurately defined as mean (average) sea level, which is the average height of the sea surface measured over an extended period of time for all conditions of tides, seasons, and storms. Mean sea level at specific locations can be calculated using tide gauge records and subtracting the effects of annual changes in atmospheric pressure and long-term changes in tidal ranges, which are driven by astronomical factors. Changes in global mean sea level can be calculated using satellite-based radar altimetry, such as with the TOPEX/Poseidon satellite. The radar altimeter measures the height of the satellite above the ocean, based on the time it takes for a radio signal to travel from the satellite to the sea surface and back. Since the actual altitude of the satellite is known, any changes in the altimeter measurement reflect changes in the height of the sea surface itself.

When sea level rises and floods the land, it is called a marine transgression. The flooded land becomes part of the ocean environment, and marine sediments are deposited on what was once dry land. Marine regression, by contrast, occurs when sea level drops and the shallow sea floor is exposed. The exposed sea floor becomes part of the land environment and is subject to the same type of erosion processes that function on land. The rocks of the continents show that over Earth's history, sea level has transgressed onto, and regressed from, the continents many times and to many different levels.

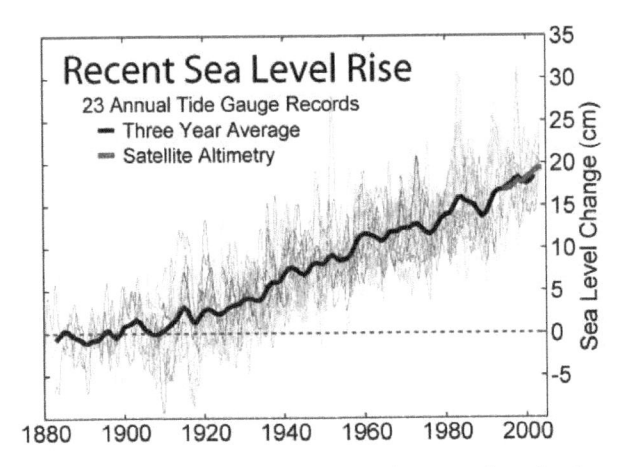

This figure shows the change in annually averaged sea level at twenty-three geologically stable tide gauge sites with long-term records as selected by Douglas (1997).

Sea level can be changed in six basic ways: by ocean surface oscillation; by changing the force of gravity; by moving the land up or down; by changing the characteristics of the ocean water; by changing the amount of water in the oceans; and by changing the volume of the ocean basins. The change in sea level brought about by these six ways results in two major types of sea-level change: local and eustatic. Local sea-level change means that only a specific area of coastline is involved and that coastlines relatively far away or on other continents are not changed. Oscillation of the water surface, changing gravity, and land moving up and down also produce local sea-level change. Eustatic sea-level change means that coastlines around the planet all experience a sea-level change of the same magnitude at the same time. Changing water characteristics, changing the amount of water on the planet, or changing the volume of the ocean basins produces eustatic sea-level change.

LOCAL SEA-LEVEL CHANGE

Sea-level change produced by oscillation of the ocean surface refers to waves, storm surges, and tides. Each wave that breaks onto the shore can be considered a short-term, low-magnitude sea-level change. In extreme cases of large waves, erosion and coastline modification can occur during each sea-level "microevent." Storm surges are changes in sea level brought about by the movement of surface waters under the influence of strong winds and low atmospheric pressures, such as occurs with hurricanes. This is the result of higher air pressure around the storm system pushing down on the surrounding water while lower air pressure inside the system allows the interior water to rise. As for the direct effect of wind on water levels, high onshore winds can pile water up on coastal areas by 3 or 4 meters, while offshore winds can lower sea level in the same way but by lesser amounts.

Tides are produced in the oceans by the force of lunar and solar gravity, and by the rotation of the Earth-Moon system about its common center of mass. Coastline configuration, latitude, time of the lunar month, and many other factors control the timing and magnitude of ocean tides. Tide characteristics are extremely variable from place to place, but in a given area, the changes in sea level are quite predictable, with magnitudes from a fraction of a meter to more than 10 meters possible. Because ocean surface oscillations are common, low in magnitude, and regular, they are often not recognized as sea-level changes. They are different for every coastline, producing local sea-level change.

The force of gravity is not perfectly uniform over Earth. It depends on the amount of mass beneath the surface and may vary by a small amount from location to location. This variation is most marked between the continental landmasses, where there is a great deal more matter to exert gravitational force, and the ocean floors, where there is much less crustal mass. The ocean, as a fluid, responds to the force of gravity: If gravity is slightly weaker in one area than in another, the sea will rise slightly higher; if it is slightly stronger, the sea will sink slightly lower. This sea-level variation occurs only in time frames of millions of years, but it is used to explain local sea levels that are different in certain areas from predicted values.

In many areas, the land itself is moving up or down. This movement has three sources: tectonics, isostasy, and subsidence. Tectonics is the movement and distortion of Earth's crust by convective forces generated within the magma of the mantle, influenced by the thermal activity in the planetary core. Tectonic action is responsible for the formation of volcanoes, mountains, and earthquakes. In an area where tectonic processes are active, the land may be forced up or down, causing it to rise or sink with respect to sea level. In tectonically active areas sea-level changes of many meters up or down have been historically documented. Isostasy is the vertical movement of the crust downward when a load is applied

Estimated Global Mean Sea-Level Rise, 1870-2008

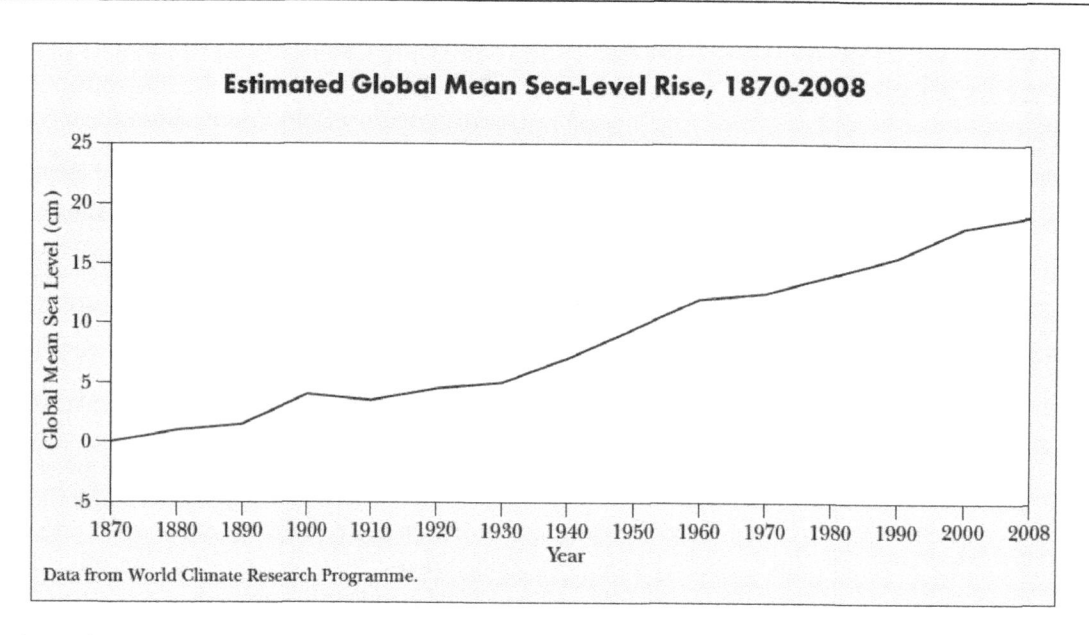

Data from World Climate Research Programme.

or its rebound upward when the load is removed. When isostasy occurs in a coastal area, the land will rise or sink, with a subsequent change in sea level. Glaciers are examples of how a load can be applied to the crust, causing isostatic subsidence. When the glacier melts, the load is removed, and the crust can isostatically rebound. This is the process responsible for the gradual rise of the land west of the Niagara Escarpment in southwestern Ontario, Canada, and of the rising east coast of Britain such that thousand-year-old coastal installations now stand as much as 400 meters above sea level. Coral islands and volcanic islands are other examples of a load being placed on the crust, which then isostatically sinks. Land can sink beneath sea level by a process called subsidence. Subsidence is often caused by compaction of the land material, which commonly happens when oil or water is withdrawn from the ground. The loss of the fluid allows the rock to compact, and the overlying land sinks. All three of these methods of moving land up or down occur in localized areas and so result in local sea-level change.

EUSTATIC SEA-LEVEL CHANGE

Changing sea level eustatically, or worldwide, requires changes that affect the oceans as opposed to the land. The ocean basins essentially have a fixed volume. If the nature of the water in it is changed, if the amount of water is changed, or if the shape of the container (ocean basin) is changed, sea level will change. The two characteristics of water that control sea level are its temperature and its salinity. For each degree Celsius that the oceans warm, thermal expansion will raise sea level 2 meters due to the lesser density (mass per unit volume) of the water.

The amount of water available to the oceans is changed in three ways: by the growth and melting of glaciers, by steam released from volcanoes, and by water lost to the formation of hydrated crystals in sediments. Glaciation is one of the most important sea-level controls. As ice sheets grow, they are fed by evaporated seawater falling as snow; this seawater is then "trapped" as ice on the continents, and sea level falls by the corresponding amount. This is a negligibly small amount per snowflake, but over thousands and millions of years it has produced glacial coverage up to 4 kilometers in thickness in some places such as Greenland and Antarctica. If the ice melts and the water flows back to the ocean, sea level rises. During the last 2 million years, glaciations have come and gone at least four times, and sea level has risen and fallen over a range of approximately 125 meters. This range is enough to almost totally expose continental shelves during maximum ice advance on the continents. Only eighteen thousand years ago, sea level was 125 meters below its present level, yet by three thousand years ago, it was essentially at today's level. If the remaining ice on Antarctica were to melt entirely, sea level would rise an additional 60 or more meters, drowning coastal cities worldwide.

Trend of Sea Level Change (1993-2008)

This chart of sea-level changes from 1993 to 2008 shows that such changes are very different in different regions of the globe.

The oceans are thought to have originated by the outgassing and release of water as steam from early volcanic activity. Volcanoes are still actively adding water to the oceans. When sediments are deposited in the ocean, they normally contain some entrapped water. Tectonic activity can return these water-bearing sediments back into the crust by the process of subduction. There appears to be a rough balance or equilibrium between water escaping the crust from volcanoes and water returning to the crust through sediment deposition, with no overall sea-level change caused by these processes in the modern world.

Sea level is also affected by changes in the volume of the ocean basins. Continents are continually losing sediments to the sea through processes of erosion. This addition of sediment is slowly filling in the ocean basins, pushing sea level up. At the same time, tectonic activity is distorting the edges of continents, often folding them, which increases the volume of the ocean basins. Tectonic activity in the ocean floor can force the sea floor upward, limiting ocean basin volume. Tectonic activities tend to occur episodically, in fits and starts, rather than as a smooth, continuous process. When tectonics is active, it may increase or decrease ocean volume, causing a drop or rise in sea level. Sediments deposited on the ocean floor may be plastered back onto the continents by tectonic activity, first decreasing and then increasing ocean basin volume. Sea-level changes, in response to sediment balance and tectonic location and magnitude, occur slowly over millions of years.

STUDY OF SEA LEVEL

When sea level changes, it leaves an imprint on the land. Transgressing ocean waters erode the land surface with waves and then, as waters deepen again, deposit marine sediments. When regression occurs, the shallowing water leads to wave erosion of the sea floor. Further regression exposes the sea floor to the air and to erosion by wind, rain, and running water from precipitation. Casual examination of the continents reveals clear evidence of past marine transgression and regression. Limestones, marine shales, and marine sandstones are common around the world on the dry land of the continents. They range in age from billions of years old to only a few hundred years old. The sea has transgressed and regressed many times in Earth's history. The timing and exact nature of the sea-level change may be more difficult to determine. In the Bahamas, fossil corals exist several meters above sea level. By sampling the corals and dating them by means of trace amounts of radioactive elements in the samples, scientists can determine their age. In this case, their age is about 125,000 years. Therefore, more than 100,000 years ago, the land to which the fossil coral is attached was under water. Did the land rise, or did the water drop? Measurement of tectonic activity shows that the islands of the Bahamas are not rising or sinking. Examination of the history of glaciation shows that 125,000 years ago, ice sheets had melted back a little more than they are presently. From this information, scientists concluded that the coral grew when sea level was eustatically higher than it is today.

Further evidence of sea-level change can be obtained by examining the tectonic, deposition, and erosion history of the planet. Examination of major episodes of tectonic activity, sediment deposition, and erosion can allow the construction of graphs showing how sea level rose and fell, in general terms, over much of Earth's history. Most of the sea-level changes preserved in the rock record were tectonically generated. Glaciation has occurred only a few times in the past, and the sea-level changes caused by glaciation have left a unique and distinctive record in rock structures.

If sea level never changed, the geology of all surface rocks would be very simple. There would be no exposed marine rocks at all. If there were no tectonics to help drive sea-level change, the continents would eventually erode to sea level. Earth would eventually become a flat, relatively featureless place.

Significance for Climate Change

The study of changing sea levels has become critical. Worldwide industrialization has greatly escalated climate change and global warming. As such, polar ice caps and other glaciers have been melting at an alarming rate, causing global sea levels to rise. Studies show that they are rising faster than they have for 2,800 years. James Hansen, a former NASA scientist and climate change expert, published research in March 2016 that predicted that sea levels across the world would rise by several meters within the next century or so if emissions do not change. General estimates vary, but scientists say that sea levels could rise anywhere from 1.3 to 3.9 feet by 2100. Coastal cities across the globe would be under threat small island countries, such as those in Micronesia, could disappear completely under the ocean.

On short timescales (decades to centuries), mean sea level is a function of the amount of water stored as ice in glaciers and ice sheets. As global temperature rises, less water is stored as ice, contributing to a rise in mean sea level. A rise in mean sea level in response to global warming has important societal consequences. First, such a rise contributes to a loss of land, as coastal areas are slowly inundated by water. This is a concern for certain low-lying island nations such as the Maldives or Tuvalu. The Maldives is a nation made up of twelve hundred islands in the Indian Ocean, which has a maximum elevation of only 2.5 meters above current sea level. Thus, the Maldivian population is highly vulnerable to even a slow rise in mean sea level. For other nations, a rise in mean sea level is also a concern because of increased hazards from flooding during high tides and storms. A rise in mean sea level provides a higher baseline upon which tidal fluctuations build. According to the Fourth Assessment Report released by the Intergovernmental Panel on Climate Change, from 1993 to 2003, mean sea level rose approximately 3.1 millimeters per year.

—*Anna M. Cruse*
—*John E. Mylroie*

Further Reading

Coe, Angela L., ed. *The Sedimentary Record of Sea-Level Change.* Cambridge University Press, 2003.

Davis, Richard A., Jr. *Sea-Level Change in the Gulf of Mexico.* Everbest Printing, 2011.

De Angelis, Hernan, and Pedro Skvarca. "Glacier Surge After Ice Shelf Collapse." *Science* 299 (2003): 1560–62.

Kelley, Joseph T., Orrin H. Pilkey, and J. A. G. Cooper. *America's Most Vulnerable Coastal Communities.* Geological Society of America, 2009.

Kennet, James P. *Marine Geology.* Prentice Hall, 1982.

Komar, Paul D. *Beach Process and Sedimentation.* Prentice Hall, 1998.

Milman, Oliver. "Climate Guru James Hansen Warns of Much Worse than Expected Sea Level Rise." *Guardian.* Guardian News and Media, 22 Mar. 2016. Web. 15 Apr. 2016.

Montgomery, C. W. *Environmental Geology*. 9th ed. McGraw-Hill, 2010.

Pilkey, Orrin H., and Rob Young. *The Rising Sea*. Island Press, 2009.

Sinha, P. C., ed. *Sea Level Rise*. Anmol Publications, 1998.

Stowe, Keith. *Exploring Ocean Science*. 2nd ed. John Wiley & Sons, 1996.

SEA SEDIMENTS AND CLIMATE CHANGE

FIELDS OF STUDY

Oceanography; Physical Geography; Earth System Modeling; Heat Transfer; Atmospheric Chemistry; Environmental Chemistry; Geochemistry; Physical Chemistry; Photochemistry; Thermodynamics; Environmental Sciences; Environmental Studies; Climate Modeling; Fluid Dynamics

SUMMARY

Sea sediments collect on the sea bottom, constituting the upper layer of the ocean floor. They consist of materials weathered from the continents, the remains of planktonic organisms, and minerals that are deposited by hydrothermal vents. Marine sediments are generally classified according to their source as terrigenous, metalliferous, or biogenic.

PRINCIPAL TERMS

- **biogenic:** originating from or formed by living organisms
- **iceberg:** any large block of freshwater ice that has separated from a glacier and floats freely on ocean waters
- **metalliferous:** metal-containing, formed from metal salts and metal-bearing minerals
- **Mid-Ocean Ridge:** a ridge of volcanic sea-floor mountains that approximately bisects the Atlantic Ocean, formed by the movement of the east and west Atlantic tectonic plates away from their shared boundary, driven by the upwelling of molten magma
- **terrigenous:** originating from terrestrial, or land-based, sources

SEDIMENT SOURCES AND TYPES

Terrigenous sediments are derived from the weathering of continental material, and are transported to the oceans by rivers, wind, and even glaciers.

Soft sediment deformation in exposed Dead Sea sediment, Israel.

Terrigenous sediments are found most often near the continental margins, where rivers deposit much of their load. Very fine grained terrigenous sediments (clays) are transported to the central abyssal plains by wind and are the dominant sediment type under surface waters characterized by low levels of plankton productivity. Local accumulations of terrigenous sediments deposited by calved icebergs after they melt are known as "ice-rafted debris." These terrigenous sediments are typically larger than are sediments transported by rivers or winds and accumulate near the polar regions.

Biogenic sediments, which dominate most regions of the seafloor, are the skeletal remains of microscopic plankton. Biogenic sediments are classified as either calcareous or siliceous depending on their chemical composition. Coccolithophores, foramaniferans, and pteropods produce calcareous skeletons composed of carbonate minerals, while diatoms and radiolarians produce siliceous skeletons composed of opaline silica.

Metalliferous sediments are found near the crests of mid-ocean ridges, where mixing of hot hydrothermal vent fluids with deep ocean waters causes the precipitation of metal (iron, zinc, copper) sulfide and oxide minerals. Sea sediments accumulate on the seafloor at very different rates. Accumulation of sediments near the continental margins can be as rapid as 1–10 centimeters every thousand years, while wind-blown terrigenous dust accumulates on the abyssal plains at rates of 1–2 millimeters per thousand years.

Significance for Climate Change

The distribution of marine sediments is a function of the location of the sediment source, the chemical reactions that occur in the water column and on the seafloor as the sediments accumulate, and the input rate of terrigenous sediments. Biogenic sediments are produced in surface waters in regions that contain sufficient concentrations of nutrients such as nitrogen, phosphorus, and silica. Production of siliceous plankton dominates in surface waters of the southern ocean, the Arctic Ocean, along continental margins where upwelling occurs, and along the Equatorial Divergence. Production of calcareous plankton dominates in the remaining regions.

When both calcareous and siliceous plankton settle through the water column, they are subject to chemical dissolution. Calcareous sediments are preserved where the ocean floor is generally less than approximately 4,000 meters deep and large inputs of terrigenous material are absent. Below 4,000 meters, the carbonate minerals that make up calcareous sediments are dissolved before they can be preserved, because there are low concentrations of dissolved carbonate ions in deep waters.

While the ocean is everywhere undersaturated with respect to silica, the rate of dissolution decreases with temperature, and thus with depth, in the ocean. Siliceous sediments are therefore preserved in deep regions of the ocean, under areas of high radiolarian or diatom productivity. The preservation of large amounts of terrigenous sediments in the deep abyssal plains, far removed from the continents, reflects the low levels of productivity found in the central gyres: There are no biogenic sediments produced in these waters to dilute the amount of terrigenous sediments delivered by the wind.

Sea Sediments Reflect Climate Changes

Changes in sea sediments preserved over geologic time reflect climate change in a variety of ways. For example, along a transect from a mid-ocean ridge toward a continental margin, the expected distribution of sediments will be as follows: metalliferous sediments near the ridge crest, calcareous sediments, siliceous sediments as the ocean floor deepens below 4,000 meters, and then terrigenous clay. During times of colder climate, when sea level drops in response to glacier formation, a larger region of the ocean floor will be shallow enough to preserve calcareous sediments. When the climate is colder, the atmospheric concentration of carbon dioxide (CO_2) decreases. This, in turn, leads to an increase in the concentration of carbonate ions in the deep ocean and a deepening of the depth at which calcareous sediments can be preserved. Thus, a wider swath of seafloor on either side of a mid-ocean ridge will preserve calcareous sediments.

Because siliceous oozes accumulate under highly productive surface waters in generally cool regions, colder climates can favor the expansion of areas on the seafloor where these sediments are found. Generally drier conditions during glacial times, moreover, contribute to an increase in the flux of terrigenous sediments to the deep abyssal plains. Warmer, wetter climates can be recorded in sea sediments by an increase in the deposition of terrigenous sediments near the continental margins. Finally, global temperature changes are recorded in the isotopic composition of oxygen preserved in calcareous or siliceous sediments, and of carbon found in calcareous sediments. Changes in the pH of the ocean—which reflect atmospheric concentrations of CO_2 and temperature—are recorded in the isotopic composition of boron that is preserved in some biogenic sediments.

—Anna M. Cruse

Further Reading

Bearman, Gerry, et al. *Marine Biogeochemical Cycles.* Elsevier, 2008. Introductory oceanography textbook that covers the processes by which marine sediments are formed, the chemical reactions by which they are modified, and the use of marine sediments in paleoceanographic studies to reconstruct past environments and climates.

Burroughs, William James. *Climate Change in Prehistory: The End of the Reign of Chaos.* Cambridge University Press, 2008. Covers a broad spectrum of data concerning human adaptation to climate change in prehistory, but has chapters specific to the evidence for climate change that is found in ocean sediments.

Chester, Roy. *Marine Geochemistry*. Springer, 2013. Collection of papers discussing the physical and chemical properties of marine sediments and the chemical reactions that occur to those sediments as they undergo diagenesis and are converted into sedimentary rocks.

Denny, Mark *How the Ocean Works. An Introduction to Oceanography*. Princeton University Press, 2012. This book provides an in-depth examination of the effect of the rotational motion of Earth and the Coriolis Effect.

Garrison, Tom S. *Oceanography: An Invitation to Marine Science*. Brooks/Cole, Cengage Learning, 2010. Discusses sediments, including calcium carbonate precipitates. Includes abundant diagrams to aid readers from the layperson to advanced undergraduates.

Pinet, Paul R. *Invitation to Oceanography*. Jones & Bartlett, 2009. Discusses geology, biology, chemistry, and physics of the oceans at an introductory college level. Includes a chapter on climate change.

Talley, Lynne D., et al. *Descriptive Physical Oceanography: An Introduction*. Academic Press, 2012. Provides an overview of the science of physical oceanography with emphasis on the chemical processes occurring in ocean water that determine such phenomena as the carbon compensation depth. College-level.

SEA SURFACE TEMPERATURES (SST)

SUMMARY

Sea surface temperature (SST) is a bulk measurement of the temperature of the surface of the ocean. Average SST ranges from approximately –3° Celsius in the Arctic Ocean to 28° Celsius in equatorial waters. SST is one of the parameters recorded by ships as part of their standard atmospheric and oceanic observation practices, which are governed by international codes first established in the late nineteenth century.

The earliest sea surface temperature measurements were taken using a simple bucket and thermometer. Such measurements reported the average value for the upper 1–2 meters of the ocean. Since the 1940's, it has been standard practice to record the temperature of water coming through a ship's intake ports. This measurement, however, can record a temperature of any depth up to 20 meters, depending on the ship's buoyancy, and can be affected by heat from the ship's engines. Today, temperature data are still recorded in this manner on research vessels, weather ships, and many commercial and military vessels, with the result that many of the data are concentrated along shipping routes.

SST data are transmitted to the World Meteorological Commission and then transferred to weather services around the globe. SST readings are also recorded using moored and drifting buoys. With buoy readings, temperature probes are placed

Deviations in Mean Global Sea Surface Temperature

The following table lists deviations in average global sea surface temperature from the baseline temperature average set during the period between 1951 and 1980.

Year	Deviation (in 0.1° Celsius)
1880	–2
1890	–2
1900	–1
1910	–3
1920	–2
1930	–1
1940	+1
1950	0
1960	+1
1970	0
1980	+1
1990	+3
2000	+4

Data from Goddard Institute for Space Studies, National Aeronautics and Space Administration.

at a standard depth of 1 meter below the surface. Data from buoys are often transmitted by satellite to a data center, such as the National Data Buoy Center. As part of the World Ocean Circulation Experiment

(WOCE), over forty-eight hundred drifting buoys were released to record SST among other data. Since the 1980's, SSTs have increasingly been obtained using satellite-borne radiometers. These instruments indirectly measure the temperature of the ocean's "skin" (the top 10 microns) based on the radiation intensity of select wavelengths, typically in the infrared spectrum.

Significance for Climate Change

Sea surface temperature is a critical parameter used in general circulation models (GCMs), as well as meteorological models, to predict future climate and weather, respectively. While air surface temperatures have increased since the 1900's, global SST records are more complicated. In fact, an examination of historical records of SSTs showed a warming in the Atlantic Ocean by up to 0.4° Celsius, while equatorial Pacific SSTs cooled by approximately 0.2° Celsius.

Because of the slow response time of the oceans, as compared to the atmosphere, most scientists consider that, in most cases, observed seasonal changes in SST reflect changes in atmospheric conditions. However, there are exceptions to this assumption, most notably El Niño-Southern Oscillation (ENSO) events. ENSO events are coupled atmosphere-ocean phenomena that occur when warmer SSTs feed back into changes in atmospheric circulation patterns. El Niño phenomena are marked by warmer SSTs in the eastern equatorial Pacific Ocean, centered near the coasts of Peru and Ecuador, and they lead to a shift in the strength and direction of atmospheric circulation across the equatorial Pacific. This shift is known as the Southern Oscillation. ENSO events appear to have a periodicity of approximately three to eight years, in a highly irregular pattern.

On seasonal timescales, SSTs are important factors in the development of storms and hurricanes (tropical cyclones). A threshold SST of approximately 27°-29° Celsius is required for a hurricane to develop. The reasons for this threshold remain unknown, but, generally, higher SSTs appear to be correlated with the development of stronger hurricanes. According to observations correlated by the Intergovernmental Panel on Climate Change (IPCC), there has been an increase in intense hurricane activity in the North Atlantic since 1970. The available data do not show a clear trend in the annual number of tropical cyclones, however, and data integrity prior to 1970 remains questionable. Climate models used by the IPCC, which are based on projections of global temperature to the year 2100, indicate that an increase in hurricane activity is likely.

SST measurements are crucial to the coupled atmosphere-ocean models used to predict weather and climate, but obtaining long-term and reliable data sets that cover a wide geographic region remains a challenge. Because satellite-based SST measurements and ship- or buoy-based measurements are taken at different depths in the water, these data sets cannot be directly compared. This complication arises because satellite-based measurements are strongly affected by daytime heating of the thin layer they are able to measure, as well as by surface evaporation and reflected radiation. Also, because radiometers often cannot obtain readings through cloud cover, satellite SST data contain a fair-weather bias: SST readings are not obtained from these instruments on cloudy days, so short-term variations related to these meteorological conditions are not recorded. Ships, meanwhile, retain their own problems, because any craft powered by a motor will necessarily alter the temperature of the water through which it travels.

—*Anna M. Cruse*

Further Reading

Burroughs, William James. *Climate Change: A Multidisciplinary Approach.* 2d ed. New York: Cambridge University Press, 2007. Describes the climate system and the physical behavior of the various parts of the system. Considers ocean circulation and SST variations in the context of atmospheric circulation and presents information on how this knowledge is used in the predictive models that inform policy makers and political leaders. Figures, references, index.

Intergovernmental Panel on Climate Change. *Climate Change, 2007—The Physical Science Basis: Contribution of Working Group I to the Fourth Assessment Report of the Intergovernmental Panel on Climate Change.* Edited by Susan Solomon et al. New York: Cambridge University Press, 2007. Comprehensive treatment of the causes of climate change, written for a wide audience. Figures, illustrations, glossary, index, references.

Open University. *Ocean Circulation.* 2d ed. Boston: Elsevier Butterworth-Heinemann, 2005. This introductory oceanography textbook covers atmospheric and oceanic circulation, with references to sea surface temperature throughout. Also of note is the discussion of ENSO. Knowledge of algebra and trigonometry is assumed. Illustrations, figures, tables, maps, references, index.

SEASONAL CHANGES

FIELDS OF STUDY

Environmental Sciences; Atmospheric Science; Climatology; Hydroclimatology; Bioclimatology; Oceanography; Heat Transfer; Tree Ring Analysis

SUMMARY

Weather and climate are variable phenomena in all regions of the world. Thus, irrespective of whether climate is undergoing a linear change to a new climate regime, climate is in some sense always changing. To determine whether climate change is of the type generally referred to by such phrases as "global warming," one would look at a relatively significant change in the overall weather pattern over an extended period of time. Minor variations over a short period of time are quite common and are simply part of the total climate package of any particular area.

PRINCIPAL TERMS

- **climate:** long-term, average, regional or global weather patterns
- **weather:** the set of atmospheric conditions obtaining at a given time and place

Changing Weather

Weather and climate are variable phenomena in all regions of the world. Thus, irrespective of whether climate is undergoing a linear change to a new climate regime, climate is in some sense always changing. To determine whether climate change is of the type generally referred to by such phrases as "global warming," one would look at a relatively significant change in the overall weather pattern For example, an arid region in Peru may well have several years without any appreciable rain, followed by one or two weeks of relatively wet conditions. This might in turn be followed by another long period with no appreciable rain. Such climatic changes are quite normal. To determine whether climate

Rainfall Patterns in Tauranga, New Zealand, 1898-2008

Weather averages may hide relatively intense variation in seasonal precipitation from year to year. For example, Tauranga, New Zealand, experienced rainfalls between 1898 and 2008 that were as much as 56 percent above or 43 percent below the average for that time period.

Year	Deviation from Average Rainfall (%)
1906	−28
1915	−41
1916	+47
1917	+43
1919	−27
1920	+36
1935	+27
1938	+38
1956	+35
1962	+56
1971	+33
1973	−25
1979	+31
1982	−36
1986	−25
1993	−33
1997	−26
1999	−24
2002	−43
2005	+28

change is of the type generally referred to by such phrases as "global warming," one would look at a relatively significant change in the overall weather pattern. For example, long periods of dry years may become less lengthy or lengthier, and periods of appreciable rain may also become more or less lengthy.

Another example would be snowfalls in the winter in much of northeastern North America, where snowfall varies considerably from one year to the next, and

447

one decade to the next. Thus, the 1940s was a period of relatively low snowfalls in some places, while the 1950s was a period of relatively high snowfalls in the same places. Such variations are quite common and are simply part of the total climate package of any particular area.

SIGNIFICANCE FOR CLIMATE CHANGE
Seasonal climates in all regions of the world have varied over all timescales, and the last one hundred to two hundred years is, in most cases and most areas, not significantly different than any other period of similar length. During the last millennium, for example, there have been long periods with warm and dry conditions (especially during the Viking explorations of northern Europe in the eleventh century), as well as long periods of cold conditions (especially during the Maunder Minimum from 1645 to 1715, which was a period when very low sunspot activity coincided closely with the coldest part of the Little Ice Age).

Global warming may cause the seasons to become generally warmer in most cases, but they can also become cooler in some places. There may also be greater variation or greater change in daytime or nighttime temperatures than in overall average temperatures. In addition, the average temperatures during winter months may change in a given area in a direction opposite that of the change in the average temperatures of the summer months. Rainfall patterns and frequencies of heavy rainfalls and droughts may be affected, as well as the frequency, path, and intensities of tropical storms such as hurricanes, tropical cyclones, and typhoons.

The climate predictions from the 2008 report of the Intergovernmental Panel on Climate Change (IPCC) suggest an overall warming in the winter in North America by 2°–3°C by the year 2050 and drier conditions during the summer in the Mediterranean. However, given the complexities of global, regional, and local climate systems, it is not at all easy to translate such broad-scale predictions into more focused predictions. Thus, even if the report is accurate, it is extremely difficult to know how those changes will affect winters in the American Midwest or tourism and agricultural water supplies in southern Europe, for example.

Thus, it could very well be that in the 2040s, Chicago could experience fewer days with snow but a greater intensity of snowfall. A 3°–4°C increase in summertime temperatures may well translate into a nighttime temperature increase of 1°–2°C, a daytime increase of 4°C, and an overall increase of 10 percent in the number of days with a maximum temperature of 38°C or more. While the climate models from the IPCC reports appear on the surface to be relatively specific for an area such as the western United States, the specific translation of such forecasts into, say, the Rocky Mountain area poses considerable difficulties, especially if one is interested in what will happen in the wintertime or the summertime in a place such as Denver during a specific decade.

—*W. J. Maunder*

FURTHER READING
Lutgens, Frederick K., Edward J. Tarbuck, Tasa, Dennis G. and Herman, Redina. *The Atmosphere: An Introduction to Meteorology.* 14th ed. Pearson Education, 2018. An updated version of this textbook, in loose-leaf format, provides students with a general overview of the structure and properties of Earth's atmosphere.

SEAWATER COMPOSITION

FIELDS OF STUDY
Analytical Chemistry; Physical Chemistry; Oceanography; Geochemistry; Thermodynamics; Environmental Studies; Waste Management; Physical Geography; Ecology

SUMMARY
The properties of seawater are determined primarily by the properties of pure water, and secondarily by its nature as a solution. Because water is a liquid and has great capacity as a solvent, seawater is well mixed

and salty due to the ions that have been dissolved from the rocks of the continental crust. Seawater is a source of mineral wealth for humankind.

PRINCIPAL TERMS

- **element:** one of a number of substances composed entirely of atoms that cannot be broken into smaller particles by chemical means
- **free oxygen:** the element oxygen by itself, not combined chemically with a different element
- **hydrosphere:** the areas of Earth that are covered by water, including the oceans, seas, lakes, and rivers
- **mineral:** an inorganic substance occurring naturally and having definite physical properties and a characteristic chemical composition that can be expressed by a chemical formula
- **nodule:** a lump of mineral rock typically found on the ocean floor
- **primary crystalline rock:** the original or first solidified molten rock of Earth
- **salinity:** a measure of the quantity of dissolved solids in ocean water, typically given in parts per thousand by weight
- **weathering:** the breaking down of rocks by chemical, physical, and biological means

THE FORMATION OF EARTH

The hydrosphere consists of the water areas of Earth, which include ponds, lakes, rivers, groundwater, and the oceans. The oceans form the largest portion of the hydrosphere, covering 71 percent of Earth's surface. The composition of Earth's water derives from the circumstances surrounding the formation of the solar system and the planet Earth.

In the beginning, it is hypothesized that a mass of gases and space dust came together and gravitational eddies eventually formed separate clouds of aggregate materials, which presumably later consolidated into planets. When the young planetary mass started to cool, it formed a crust, which being thin, allowed heat, molten material, and gaseous material to escape from the interior through numerous cracks. The gaseous material formed the first atmosphere, probably made up primarily of hydrogen and helium molecules. These molecules, being fairly light in weight and highly energized by atmospheric temperatures and incoming solar radiation, were probably lost to space. Eventually, they were replaced by other gases derived from the interior by the ongoing volcanic activity as the planet's surface continued to cool. This subsequent atmosphere was composed mainly of carbon dioxide and water vapor, with higher percentages of other gases released from entrapment in rock structures than are found today. The water in the oceans and all other water on the planet are believed to have been released from within the mineral structures of the material from which the planet formed.

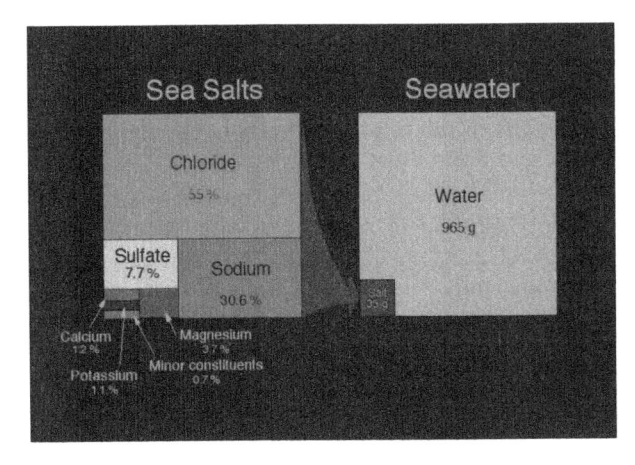

Proportion of salt to sea water (right) and chemical composition of sea salt.

Clouds formed from the condensing water vapor in the ancient atmosphere, shielding Earth's surface and allowing less than 60 percent of the sun's energy to penetrate. As the surface continued to cool, the water vapor condensed into liquid and began to fall and accumulate in depressions. With widespread volcanic activity continuing, water vapor was continuously being released, along with smaller amounts of carbon dioxide, chlorine, nitrogen, and hydrogen gas, which then underwent chemical reactions driven by sunlight to produce methane and ammonia. As the surface continued to cool, vast amounts of condensation eventually formed the oceans. Geologic evidence shows that the oceans have existed for at least 3 billion years. This evidence comes in the form of algal fossils presumed to have grown in a marine environment.

Earth's crust was formed by primary crystalline rocks, the first molten material that solidified into rock as the planet cooled. These rocks were weathered and eroded into the particles that became deposited and slowly dissolved in water, carried into the ocean basins, and accumulated. Also carried into the ocean

449

were great loads of particulate material weathered from the primary crystalline rocks and deposited as sediments. Thus the components of the primary rocks were freed by chemical weathering and dissolved in ocean water or chemically bonded with sediments and carried into the ocean by rivers. The chemical weathering took place because of the different compounds in the atmosphere, some of which combined with water vapor to form acidic rainwater ranging in strength from only slightly acidic solutions of carbonic acid to solutions of much stronger nitric and sulfuric acids. Each of these various acidic solutions had the ability to leach out various metallic ions, such as sodium, potassium, magnesium, iron, and others, from the materials that they contact. Seawater gained its characteristic saltiness through these processes.

Comparing the discharge of gases by hot springs in the United States to the average rate throughout the 3 billion years of the oceans' existence shows that enough water vapor is produced to fill the oceans to one hundred times their present volume. Water therefore must be recycled and not all newly formed; the excess amount does not represent new water released from the original crystallization of magma. Only 1 percent of this water requires such an origin to account for the present volume of the oceans.

Water on land comes daily from the sea. The seas hold about 4.4 billion cubic meters of saltwater. Of this amount, about 12 million cubic meters enters the atmosphere each year through evaporation and is returned by rainfall and the flow of rivers, and about 3 million cubic meters descends each year over continents, replenishing ponds, lakes, and rivers.

Gases and Solids in Seawater

Ocean water is a mixture of gases and solids dissolved in pure water (96 percent pure water and 4 percent dissolved elements, by weight). Nearly every natural element has been found or is expected to be found in seawater, although some occur only in very small amounts. The most abundant mineral found in ocean water is sodium chloride (familiar as common table salt), which makes up 85 percent of the dissolved minerals. It is interesting to note that the composition of human blood is very similar to that of seawater; in it are all the elements of the sea, dispensed in different proportions.

The seven most abundant minerals in seawater include sodium chloride, 27.2 parts per thousand; magnesium chloride, 3.8 parts per thousand; magnesium sulfate, 1.7 parts per thousand; calcium sulfate, 1.3 parts per thousand; potassium sulfate, 0.9 part per thousand; calcium carbonate, 0.1 part per thousand; and magnesium bromide, 0.1 part per thousand. It is important to note that these "minerals" do not occur in the form suggested by their molecular formulas, but as the component ions dissolved in water. Six ionic species actually make up 99.3 percent of the total mass of the dissolved material: chlorine, 55.2 percent; sodium, 30.4 percent; sulfate, 7.7 percent; magnesium, 3.7 percent; calcium, 1.2 percent; potassium, 1.1 percent; and others, 0.7 percent.

The salinity of seawater is fairly uniform across the oceans at different latitudes and at various depths, mainly because of winds, waves, and currents. Local variations in mineral content are attributed to freshwater streams entering oceans, glacial melt, and human activity, but the variations overall are small. The salinity averages about 35 parts per thousand. This has been verified by the *Glomar Challenger* expedition, using Nansen bottles to take samples of seawater at different depths around the world and salinometers to measure salinity, as well as many other types of tests. There is more variation of salinity at the surface than at depths because of freshwater (rivers, glacial melt) entering the ocean at a given location, plus the biological activity and climate at different latitudes.

In areas where freshwater is entering the ocean, the salinity will decrease. Biological activity changes the salinity according to which species and how many marine plants and animals reside in the area. In climates that are hot and dry, the rate of evaporation is high and rainfall is low, so the ratio of dissolved salts to water is higher, making the water more saline (saltier). Similarly, in the polar regions in winter, the water freezes but the minerals do not, which changes the water-to-mineral ratio and thus increases salinity. In most climates, rainfall is greater than evaporation, thus diluting seawater and decreasing salinity.

The most abundant dissolved gases in the ocean are nitrogen, carbon dioxide, and oxygen. The amount varies with depth, and as oxygen and carbon dioxide are vital to life, most living plants and animals are found in the top 100 meters of the ocean, or the sunlight penetration layer. The amount of dissolved gases will also vary with temperature. Warm water holds less dissolved gas than cold water because cold water is heavier and sinks, carrying oxygen-rich water to the ocean depths, which, in turn, allows fish and marine life to live in the deepest parts of the ocean. At the surface, gases (oxygen and carbon dioxide) are exchanged between the ocean and the atmosphere as well as between the plants and animals that live in the top layers of the ocean, where the most abundant life is found. Marine plants take in carbon dioxide along with water and then, with the aid of sunlight, produce the sugar glucose and oxygen in the process called photosynthesis. Marine animals and photosynthetic plants take in oxygen and exhale or release carbon dioxide in the process called respiration.

Most precipitation that falls finds its way back to the ocean largely by rivers, bearing salts from soils and rock in solution. Many things affect the salinity of the ocean, including the exchange of water between the ocean and the atmosphere, which is determined by climate, and the absorption of salts by plants and animals. Considering the history of ocean salinity, one might ask if the oceans have possessed a relatively uniform salinity throughout their history or if they are becoming more saline. By far the most important component of salinity is the chloride ion, which is produced in the same manner as water vapor. The ratio of chloride ions to water vapor has not fluctuated throughout geologic time, so scientists conclude, on the basis of present evidence, that ocean water salinity has been relatively constant over the lifetime of the oceans.

It is now evident that the oceans became salty early in their history because of weathering by acidic rainfall and the erosion of primary crystalline rock. Also, through continual volcanic activity, water vapor and gases were amply supplied to the atmosphere to aid in this weathering and eroding process. These particles were carried to the ocean and dissolved in seawater, making the water salty. This process has been going on since the formation of the oceans, so it is safe to assume that the ocean has been salty since it was formed approximately 3 billion years ago.

Comparing the salinity of oceans both past and present by studying vents and hot springs, the salinity appears to maintain a certain balance. Salinity ranges between 33 and 38 parts per thousand, by weight. This variation is caused by atmospheric effects at different latitudes, by freshwater entering into the oceans from rivers and glacial melt, and by biological activity in the ocean itself as plants and animals remove minerals needed for their growth and development (photosynthesis, respiration, shell building, and so on) from the seawater and put into it other forms (waste products, skeletal structures, shells, casts, and so on). In local areas, such as bays, coves, and estuaries, the salinity of seawater is further affected by human activities such as industry and agriculture, and by the pollution from these activities.

Study of Seawater Composition

The investigation of seawater composition is accomplished primarily by collecting samples of seawater at different depths around the world and then measuring the salinity of the samples. A variety of instruments are used to collect samples. Nansen bottles are special metal cylinders fastened at a measuring point on a strong wire, which is then lowered into the sea to the desired depth. A messenger weight is dropped down the wire; when it strikes the bottle, it releases a catch. The bottle then turns upside down, and its valves close, trapping the water at that depth inside. The Nansen bottle commonly used does not seal completely, however; a better apparatus is the Fjorlie sampler or Niskin bottle. The Fjorlie sampler or Niskin bottle is attached to a line at both ends with spring-closing hinged ends. A messenger weight closes the bottle with a good seal.

Corrosion of metal-lined samplers may cause changes in the water composition in an hour or so. Copper, zinc, lead, and iron in metal linings often contaminate seawater samples. Plastics have solved this problem for both collection and storage of seawater. It is also necessary to filter out any organic matter that could alter the seawater composition. For maximum accuracy of testing, samples should be tested as soon as possible and not stored.

A salinometer is used to electronically measure salinity. It is easy to use and gives immediate readings. Because the ions dissolved in seawater affect its properties as a conductor of electricity, the more ionic mineral matter in the sample, the better it conducts. The results are then compared to a table of standard measurements to obtain the sample's salinity.

—*Joyce Gawell*

Further Reading

Emerson, Steven, and John Hedges. *Chemical Oceanography and Marine Carbon Cycle.* Cambridge

University Press, 2008. Provides a good overview of geochemistry topics in oceanography. Discusses chemical composition, thermodynamics, carbonate chemistry, the carbon cycle, and calcium carbonate sedimentation. Appendices follow the chapters they pertain to. Contains excellent indexing.

Garrison, Tom S. *Essentials of Oceanography.* 6th ed. Brooks/Cole Cengage Learning, 2012. Written to provide a basic understanding of the formation and function of the world's ocean environments for students undertaking an introductory course in oceanography.

Grasshoff, K., et al. *Methods of Seawater Analysis.* Wiley-VCH, 2002. A widely used college text that is completely revised and extended, devoting more attention to the techniques and instrumentation used in determining the chemical makeup of seawater. Illustrations, bibliography, and index.

Harrison, Roy M. *Principles of Environmental Chemistry.* Royal Society of Chemistry, 2007. Discusses chemistry of the atmosphere, freshwater, oceans, and soils. Also discusses biogeochemical cycling and environmental organic chemistry. Suited for undergraduates and graduate students with a chemistry background.

Marshall, John, and R. Alan Plumb. *Atmosphere, Ocean and Climate Dynamics: An Introductory Text.* Burlington, Mass.: Academic Press/Elsevier, 2008. An introductory university-level textbook written to provide a grasp of the interactions of oceans and atmosphere both past and present. Assumes a certain amount of mathematical background on the part of the reader.

Segar, Douglas A., and Elaine Stamman, Segar. *Introduction to Ocean Sciences.* Douglas A. Segar, 2012. Provides comprehensive coverage of all aspects of the oceans and salinity. Readable and well illustrated. Suitable for high school students and above.

Steele, John H., Steve A. Thorpe, and Karl K. Turekian, eds. *Marine Chemistry and Geochemistry.* Academic Press, 2010. Pulls articles in from the Encyclopedia of Ocean Sciences to provide a topic-specific compilation of oceanography articles. Discusses the chemistry of seawater, radioactive isotopes in the oceans, pollution, and marine deposits.

Tarbuck, Edward J., et al. *Earth: An Introduction to Physical Geology.* Pearson, 2018. Provides a clear picture of earth systems and processes. Suitable for the high school or college reader. Includes an accompanying computer disc in addition to its illustrations and graphics. Bibliography and index.

Trujillo, Alan P., and Harold V. Thurman. *Essentials of Oceanography.* 10th ed. Prentice Hall, 2010. Designed to give the student a general overview of the oceans in the first few chapters. Follows up with a well-designed, in-depth study of the ocean—involving chemistry of the ocean, currents, air-sea interactions, the water cycle, and marine biology—and a well-developed section on the practical problems resulting from human interaction with the ocean, such as pollution and economic exploitation.

Usdowski, Eberhard, and Martin Dietzel. *Atlas and Data of Solid-Solution Equilibria of Marine Evaporites.* Springer Verlag, 2013. Offers the reader an illustrated guide to seawater composition, including phase diagrams of seawater processes. Accompanied by a CD-ROM that reinforces the concepts discussed in the chapters.

Williams, Richard G., and Michael J. Follows. *Ocean Dynamics and the Carbon Cycle: Principles and Mechanisms.* Cambridge University Press, 2011. Provides the fundamentals of oceanography. Discusses both biological and chemical aspects of ocean dynamics. Deals with carbonate chemistry and the carbon cycle in oceans.

SEVERE STORMS

FIELDS OF STUDY

Atmospheric Science; Meteorology; Climate Modeling; Thermodynamics; Heat Transfer; Fluid Dynamics; Environmental Sciences; Environmental Studies; Hydroclimatology; Hydrometeorology; Hydrology; Oceanography; Physical Geography; Earth System Modeling; Ecology

SUMMARY

A severe storm is a violent weather phenomenon that has a specific structure, often associated with heavy precipitation and air circulating in a cyclonic or anticyclonic manner. Winds affected by a storm are often of high velocity, a factor used to differentiate among storm stages.

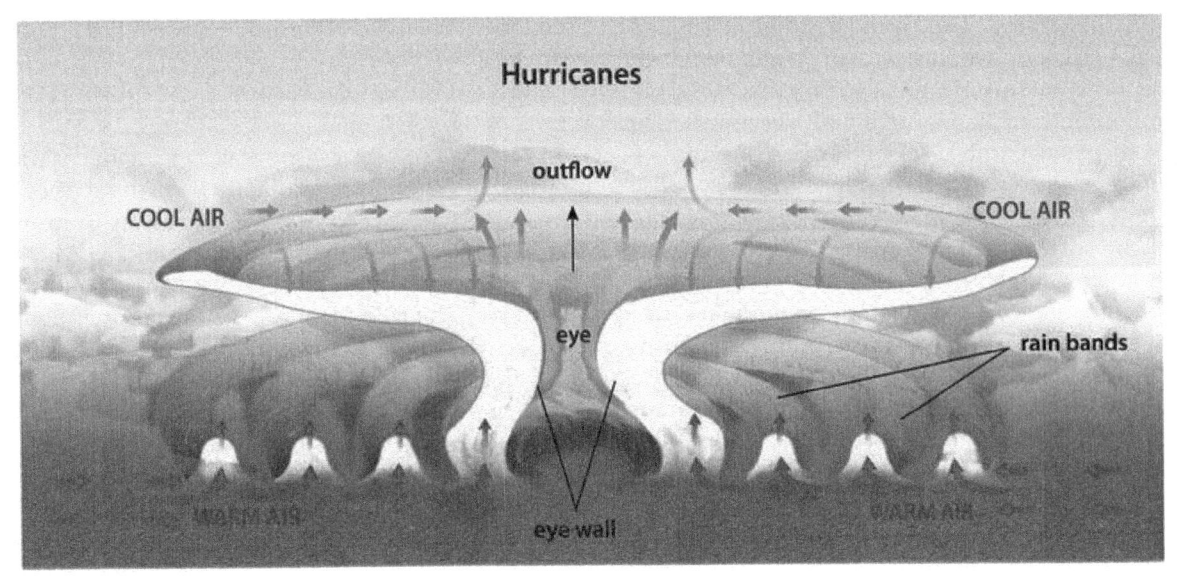

Principles of Climatology

Severe storms

Hurricanes: The illustration shows the features of a hurricane.

PRINCIPAL TERMS

- **blizzard:** a winter storm characterized by cold wind having a minimum velocity of 56 kilometers per hour, large amounts of blowing snow, and low levels of visibility
- **Doppler radar:** a radar method that uses the Doppler shift to measure the speed of targets moving either toward or away from the radar
- **Doppler shift:** a phenomenon in which the wavelength of electromagnetic radiation (or other type of wave) is lengthened by reflection from a surface moving away from the source or shortened by reflection from a surface moving toward the source
- **downburst:** a severe localized downward outflowing of air and associated wind shear below a thunderstorm
- **El Niño:** meteorological condition in which the waters of the eastern, tropical Pacific Ocean are warmed by the atmosphere
- **glaze:** a coating of ice formed on exposed objects by the freezing of a film of supercooled water deposited by rain, drizzle, or fog
- **hurricane:** a severe tropical cyclone, typically between 500 and 600 kilometers in diameter, with winds in excess of 65 knots (74 miles or 120 kilometers) per hour
- **ice storm:** a storm characterized by a fall of freezing rain, with the formation of glaze on objects below
- **La Niña:** meteorological condition in which the waters of the eastern, tropical Pacific Ocean are cooled by a lack of radiation from the atmosphere
- **rawinsonde:** a radiosonde with a radar target attached so that it can be tracked for collection of wind information
- **tornado:** a violent rotating column of air that forms and extends downward from a cumulonimbus cloud and has the appearance of a funnel, rope, or column that touches the ground
- **vortex:** the center of a whirling or rotating fluid, typically with low pressure in the center due to loss of the fluid mass in that locus by centrifugal force

WINTER STORMS

A storm is a disturbed state of the atmosphere that has an impact on the ground surface with powerful and potentially destructive weather. Some types of storms are confined to certain seasons or locations, and others can occur anywhere. Among the storms occurring in the atmosphere are winter storms, thunderstorms, tornadoes, and tropical cyclones. Winter storms in particular can be quite severe, destructive, and life-threatening. Two types of winter storm can be particularly catastrophic: ice storms and blizzards.

The term "ice storm" is given to rain that falls from the atmosphere in liquid form and freezes when it comes in contact with a surface having a temperature of 0°C or lower. This form of precipitation is also known as freezing rain. In order for freezing rain to occur, the surface temperature of the ground must be below 0°C, and above-freezing temperatures must be present aloft. Snowflakes fall through the layer of warmer air and are melted. The resulting raindrops then cool as they pass through the colder air near the surface, losing most or all of their latent heat content such that they freeze quickly on impact with any cold surface. This kind of precipitation produces layers of solid ice, called glaze, that coat streets, trees, automobiles, and power lines, sometimes to thicknesses of more than two centimeters. The added weight of the ice has been sufficient to collapse vulnerable structures and can cause major power disruptions when electrical grid towers fall under the additional load. One of the best-known examples of such a storm is a system that led to the accumulation of nearly 7 cm (3 in) of solid ice in northern New York and New England and parts of Canada in 1998. The damage from that storm included broken and downed trees and telephone wires, and collapsed electrical grid towers, causing power outages for millions of people. The storm caused more than $3 billion in damage and was responsible for the death of nearly forty people.

A blizzard is characterized by a strong wind with a velocity of 56 kilometers per hour or higher, temperatures lower than −7°C and enough snow to restrict visibility to less than 150 meters. The more common type of condition associated with blizzards is heavy snowfall of 30 centimeters (1 foot) or more. However, ground blizzards, in contrast, do not produce as much snow; instead, they have strong winds that kick up snow that is already on the ground. Whether the snow falls from the sky or is blown from the ground, one of the most common features of a blizzard, in addition to high winds, is extremely poor visibility of 0.5 km (0.25 mi) or less. The severe winds and poor visibility characteristic of a blizzard make these storms extremely dangerous. The high winds coupled with the low temperature produce extreme windchill factors, while the amount of falling or blowing snow driven by the wind can produce total whiteout conditions. Blizzards are associated with midlatitude cyclones. The type of cyclone most likely to produce blizzard conditions is one with a surface low pressure connected with a low pressure in the upper air at the level of the jet stream.

The area of heaviest snowfall in a cyclone is within about two hundred kilometers north of the low-pressure center. The heavy snow results from moist air from the south turning counterclockwise around the low-pressure center. Farther to the south, usually along the cold front, sleet and freezing rain may occur. Sleet, like freezing rain, occurs when above-freezing temperatures are present aloft; only the raindrops freeze before reaching the surface. The strongest winds are behind the cold front, and blizzard conditions are most likely to occur there.

Another type of severe weather system, the nor'easter, continues to garner study, especially because it affects more people and has a wider geographic effect, particularly in the northeastern United States, than both tornadoes and hurricanes. A nor'easter is a strong storm that involves the interplay of cold air from Canada with the warm air of the Atlantic Ocean. The two fronts create a slow-moving, counterclockwise storm that features high winds and heavy precipitation that typically blows in with the most impact from the north-east due to the counterclockwise rotation of the storm system. Nor'easters are frequently known as winter storms, although they occur year-round. One of the most famous examples of a nor'easter is the so-called "perfect storm" of October, 1991, which caused damage from as far south as Florida and as far north as Maine. Formed by the convergence of three large systems, their cumulative power made this storm so powerful that a hurricane actually formed inside the larger nor'easter. This hurricane was never named, becoming one of only eight unnamed cyclones since the naming practice was introduced in the 1950s; a more pressing need was to track the major storm and the devastation it caused.

Thunderstorms

A thunderstorm is a violent disturbance in the atmosphere and typically occurs along a weather front. These storms are especially frequent when a cold front moves into a mass of warm, moist air. Associated features of thunderstorms include lightning and thunder, occasionally hail, and frequently heavy precipitation, although in dry climates precipitation at the surface may not occur. When temperatures in the atmosphere decrease rapidly with height, the atmosphere is unstable. Moreover, if the air is moist, a considerable amount of energy is stored inactively in the water vapor, and this energy will be released when the vapor changes to liquid water or ice. When this moist, unstable air is given an initial lift by unequal heating of the ground surface, a mountain range, or an advancing front, a rising air current is set in motion. As long as the rising air is less dense than the surrounding air, it will continue to rise. As the water vapor is condensed, air density is decreased, and a towering thunderhead cloud forms.

Conditions favorable for the formation of thunderstorms most often occur in warm, moist tropical air. For this reason, thunderstorms form most frequently in states bordering the Gulf of Mexico and are also frequent around the Great Lakes region in North America. Thunderstorms are divided into two basic types, termed local and organized thunderstorms. Local storms are isolated, scattered, and usually short lived. Normally they occur on warm summer afternoons near the time of the peak daily temperature. Organized thunderstorms are long lived and occur over larger areas than local storms. They form in rows called squall lines, along cold fronts, occasionally along warm fronts, and adjacent to mountain peaks.

The initial stage of thunderstorm formation is the cumulus stage, in which the cloud is dominated by updrafts. Precipitation does not occur in this stage, but as the cloud gets larger, the updrafts get stronger and more widespread. In the top of the cloud, where liquid water and ice crystals are abundant and where buoyancy is less, a downdraft is initiated. As soon as the downdraft starts, the second stage in the life cycle is reached, called the mature stage.

The most violent weather in the thunderstorm occurs during the mature stage. A strong updraft and downdraft both exist in the cloud formation, and heavy rain is produced in the downdraft side. Also associated with the downdraft are strong wind gusts at the surface. Depending on the strength of upper winds and downdraft currents, surface winds may range from cool, gentle breezes to strong blasts of air. Gradually, the downdraft spreads throughout the cloud, and the dissipating stage is reached. In this stage, the storm is characterized by weak downdrafts and light rain.

Hazards associated with thunderstorms other than strong winds and heavy rain include lightning and hail. Lightning occurs because of the separation of positive and negative charges within clouds, between clouds, and between the clouds and the ground. When the electrical potential of a charge is sufficient to overcome the insulating effect of the air, a lightning stroke results. Thunder is caused by a rapid expansion of air as a lightning bolt—which is several times hotter than the sun's surface—passes through it. Hail is formed as an ice crystal is buffeted about within a cloud and successive layers of ice are added to the developing hailstone as supercooled water is encountered within the cloud. The eventual size of a hailstone depends on the length of time it undergoes these conditions during its passage through the cloud.

Tornadoes and Hurricanes

Tornadoes are also associated with severe thunderstorms. They are small, powerful storms usually less than half a kilometer in diameter, though at times they may extend for a kilometer or more. A tornado may have the shape of a funnel, rope, or cylinder extending from the base of a thunderstorm cloud to the surface of the ground. The tone of the tornado depends on its background, the debris, and condensed moisture within it. A blue sky behind a tornado makes it appear dark, whereas intense rain behind it makes it look white. The funnel will appear dark when it is filled with debris and dust picked up from the surface.

The motion of a tornado is highly variable, averaging 65 kilometers per hour, but some tornadoes have been observed to travel as fast as 110 kilometers per hour. Inside, the storm winds, which almost always turn in a counterclockwise direction, may whirl around the center in excess of 500 kilometers per hour. Within a tornado are smaller, more intense vortices called suction vortices. There may be anywhere from one to three such vortices in a tornado. When there is more than one suction vortex, they rotate around a common center and may account for total destruction in one area, while an area only meters away sustains little or no damage.

Tornadoes occur in many parts of the world, but the topography and pressure patterns of the central United States are especially suited to tornado formation, such that the region extending northeast from Texas through southwestern Ontario is often referred to as "Tornado Alley." In this region, the Rocky Mountains block winds from the west, while the cold, dry jet stream flows over the mountains to meet warm, moist air flowing northward from the Gulf of Mexico due to high pressure in the Atlantic Ocean at 30 degrees north latitude. When unstable air rises, the extreme vertical contrast between the two air masses provides conditions for an explosive upward movement of air. Another aspect of the tornado is the low pressure in the center. Pressure here often approaches 800 millibars, whereas average sea-level pressure is 1,013 millibars.

Tornado occurrence follows the seasonal migration of the jet stream. In winter, the jet stream is closer to the Gulf of Mexico, making the Gulf states more prone to tornadoes. In spring, the Southern Plains states are most likely to experience tornadoes; in late spring and summer, the Northern Plains and the eastern United States are most prone. May is the peak month of occurrence. Because of its size, Texas has more tornadoes than any other state, but when storms are averaged over area, Oklahoma ranks first.

Hurricanes are tropical cyclones that occur in the Atlantic Ocean and Gulf of Mexico. Their counterparts in the Pacific Ocean are typically referred to as cyclones and typhoons. A hurricane goes through a four-stage development. Stage one is the tropical disturbance, in which a low-pressure center has some clouds and precipitation but no enclosed isobars (lines connecting points of equal pressure) and only light winds. The second stage is the tropical depression, with lower pressure in the center and at least one enclosed isobar but with winds less than 60 kilometers per hour. Third is the tropical storm stage, with winds between 60 and 120 kilometers per hour around a low-pressure center with several enclosed isobars. The storm is given a name in this stage. Fourth is the tropical cyclone or hurricane stage, with pronounced rotation around a central core or eye. Winds are sustained at speeds in excess of 120 kilometers per hour, circling as bands in toward the center. Hurricanes form over warm oceans and derive most of their energy from water with temperatures greater than 26.5 degrees Celsius. These storms form only in late summer and fall. Because the motion of a hurricane is affected by the Coriolis force, hurricanes are not seen to form closer than 5 degrees of latitude to the equator, where the Coriolis force is not sufficient to induce rotation. Hurricanes have no fronts and are smaller than a midlatitude cyclone, while central pressures are lower and winds are stronger.

STUDY OF STORMS

One of the first major investigations into the nature of thunderstorms was undertaken from 1946 to 1947 during what was called the Thunderstorm Project. The project consisted of flying instrument-bearing aircraft through thunderstorms to obtain various kinds of data. The study also was augmented by collection of data from various instruments such as radar, radiosondes, and ground-based instruments. Much was learned about thunderstorm structure, internal activity, and life cycle from this investigation. It was largely through the Thunderstorm Project that the cumulus, mature, and dissipating stages in a thunderstorm's life cycle were identified. Aircraft are also used to fly through hurricanes for the purpose of collecting storm data.

One device used in data collection in the upper atmosphere is the rawinsonde, which is a package of weather instruments attached to a balloon that is sent aloft. The rawinsonde has instruments to record temperature and humidity, a reflector for collecting wind data by ground-based radar, and a transmitter to send data to recorders in the weather station. Rawinsonde stations are spaced several hundred kilometers apart and are released only twice daily at 0000Z and 1200Z (the time in Greenwich, England). Thus, the rawinsonde is limited, because data collected in this manner are insufficient to make precise forecasts of severe thunderstorms or tornadoes hours in advance.

Another valuable meteorological tool is radar, which can be used to detect and observe storms hundreds of kilometers away from a station. A radar set sends out pulses of radio waves through an antenna that rebound from objects such as raindrops, cloud drops, ice crystals, and hail. A radio receiver intercepts the returning pulses between transmitted pulses.

Doppler radar is a particularly useful forecasting tool, especially for severe thunderstorms and tornadoes. Doppler radar sends out continuous radio waves instead of pulses and uses the Doppler effect on reflected radio signals to measure the speed of objects moving toward or away from the radar antenna. Targets such as cloud drops, raindrops, and other liquid or solid particles reflect radio waves, so the direction of air movement can be discerned. Thus, air that is whirling within a storm can be detected by the appearance of shortened wavelength on one side of a central point and lengthened wavelength on the other side. Doppler radar can identify and locate a tornado within a thunderstorm as early as fifteen to twenty minutes before the funnel touches the ground. A network of Doppler radars is in place across North America, providing meteorologists with a great deal of information regarding storms and weather patterns.

Satellites are another useful tool for detecting and observing middle-latitude cyclones, thunderstorms, and hurricanes. Several meteorological satellites have been placed in orbit over the years, beginning with the Television Infrared Observation Satellites (TIROS). Satellites such as the Applications Technology Satellites (ATS) and Synchronous Meteorological Satellites (SMS) are in geosynchronous orbit, which means they make one revolution of Earth in a twenty-four-hour period so that they remain over the same location on the surface. These satellites sense conditions on and above Earth and send coded information back to weather stations. The information is transformed into the weather images that are frequently shown on television. Images and measurements are taken in both the visible and infrared wavelength ranges. Both wavelengths can produce visible images, but infrared sensors detect the temperature differences of objects. These images provide information on midlatitude cyclones, thunderstorms, and hurricanes.

SIGNIFICANCE

Storms occur somewhere in North America nearly every day. They can be quite destructive and are often life-threatening, although they are at the same time beneficial as essential components of the water cycle. One dangerous aspect of storms is lightning; about fifty-five people are killed in the United States each year from lightning strikes.

Thunderstorms, on an annual basis, can cause more cumulative damage than other storms simply because there are so many of them. In addition to lightning, thunderstorms can produce heavy downpours of rain, which can cause flash flooding. Hailstorms are also associated with thunderstorms. Hail often destroys crops and can cause damage to buildings, automobiles, and other structures. Strong gusts of wind on the downdraft side of a thunderstorm often do appreciable damage to vegetation, especially trees, and to buildings. Downbursts—air that rushes out of a thunderstorm downdraft and spreads out laterally near the surface—are particularly hazardous to aircraft taking off or landing and have caused several plane crashes.

Tornadoes are generally associated with severe thunderstorms, although they also occur in conjunction with hurricanes. They are difficult to forecast and often appear quickly, without warning. More tornadoes occur annually across the contiguous United States than in any other country, averaging between seven hundred and one thousand each year. Deaths from tornadoes average more than one hundred per year, and property damage is typically valued in the millions of dollars. The conditions for the formation of storms that generate tornadoes migrate with the movement of jet stream, and the month of May has more tornadoes than any other, although June has more days on which tornadoes occur.

Hurricanes are normally limited to late summer and fall, although global climate changes and rising ocean temperatures have brought about a significant extension of the hurricane season, which now extends into late as December. Hurricanes are not as intense as are tornadoes, but, being thousands of times larger and of much longer duration, they cause more damage. They form most frequently in the Gulf of Mexico and the Atlantic Ocean. Ocean-side hurricanes typically move along the eastern seaboard of the United States and the maritime provinces of Canada as they make their way eastward before dying out over the colder waters of the North Atlantic. But in 2017, for the first time on record, a hurricane has lived long enough to cross the Atlantic Ocean from the east coast of the United States and make landfall on the west coasts of the British Isles. Similar storms also form in the Pacific Ocean, where they are normally called typhoons. The most damaging aspect of hurricanes or typhoons is the storm surge that they bring with them. Surges have been as high as 8.5 meters and have carried floods inland for several kilometers. The damage from Hurricane Camille in 1969, which had a storm surge of more than 7.5 meters, totaled some $1.5 billion. In 2005, Hurricane Katrina struck the Louisiana coast, flooding the city of New Orleans; Katrina boasted sustained winds of almost 280 kilometers (175 miles) per hour, pushing a storm surge of up to 8.5 meters (28 feet). More than 1,500 people were killed by the storm, and property damage amounted to more than $151 billion and counting as the recovery process continues than a decade later. High-velocity winds also cause considerable damage, which is more widespread than surge damage.

CLIMATE CHANGE

The high number of devastating tornadoes, nor'easters, and hurricanes since the early 1990s may be attributed to a number of atypical (although not unnatural) factors. For example, two critical phenomena, El Niño and La Niña, are well-known contributors to weather patterns.

El Niño, a warming trend in the eastern tropical Pacific, is known to contribute to the creation of storms with heavy precipitation. La Niña, in contrast, is a period marked by cooler water temperatures that brings colder, drier air along the jet stream (the band of air currents that proceeds from the west to the east), and causes periods of cooler air in the United States.

El Niño and La Niña are cyclical events caused by the interaction of the atmosphere and the surface of the ocean in the tropical Pacific. However, a growing school of scientific thought, concerned with global warming and climate change, argues that the longtime emission of greenhouse gases into the atmosphere has caused the atmosphere to increase in temperature. Such changes are theorized to foster El Niño and La Niña conditions more frequently than in previous centuries. Such shifts could lead to more severe droughts and to severe hurricanes, tornado-producing storms, flooding, and blizzards. Many scientists attribute the occurrence of high-profile and devastating storms in recent decades to this trend. Although data are not complete (tornadoes, for example, are too difficult to predict and model), researchers continue to seek direct connections between global warming and severe weather.

—*Ralph D. Cross*
—*Michael P. Auerbach*

FURTHER READING

Ahrens, C Donald, and Robert Henson. *Meteorology Today: An Introduction to Weather, Climate, and the Environment.* Cengage, 2019. One of the most widely used and authoritative introductory textbooks for the study of meteorology and climatology. Explains complex concepts in a clear, precise manner and supports them with numerous images and diagrams. Discusses hurricanes and the mechanisms that generate them extensively.

Birch, Eugenie Ladner, and Susan M Wachter. *Rebuilding Urban Places after Disaster: Lessons from Hurricane Katrina.* University of Pennsylvania Press, 2006.

Bluestein, Howard B. *Tornado Alley: Monster Storms of the Great Plains.* New York: Oxford UP, 2006. Print.

Dunlop, Storm. *The Weather Identification Handbook.* Lyons, 2003.

Grazulis, Thomas P. *The Tornado: Nature's Ultimate Windstorm.* University of Oklahoma Press, 2003.

Holton, James R., and Gregory J. Hakim. *An Introduction to Dynamic Meteorology.* 5th ed. Academic, 2013.

Lighthill, James, and Robert Pearce, eds. *Monsoon Dynamics.* Cambridge University Press, 2009.

Mogil, H. Michael. *Extreme Weather: Understanding the Science of Hurricanes, Tornadoes, Floods, Heat Waves, Snow Storms, Global Warming, and Other Atmospheric Disturbances.* Black Dog & Leventhal, 2007. Provides a comprehensive review of extreme weather events. Also offers thoughts on storm and disaster preparedness in the face of such weather.

Mooney, Chris. *Storm World: Hurricanes, Politics, and the Battle over Global Warming.* Harcourt, 2007.

Repetto, Robert, and Robert Easton. "Climate Change and Damage from Extreme Weather Events." *Environment* 52, no. 2 (2010): 22–33. In this article, the authors describe severe weather events as they relate to global warming and climate change. Repetto and Easton attempt to draw links between global warming and the increase in weather disasters in recent decades.

Samaras, Tim, Stefan Bechtel, and Greg Forbes. *Tornado Hunter: Getting Inside the Most Violent Storms on Earth.* Washington: National Geographic Society, 2009.

Schneider, Bonnie. *Extreme Weather: A Guide to Surviving Flash Floods, Tornadoes, Hurricanes, Heat Waves, Snowstorms, Tsunamis and Other Natural Disasters.* Palgrave, 2012.

Schneider, Stephen Henry, and Michael D. Mastrandrea, eds. *Encyclopedia of Climate and Weather.* 2nd ed. 3 vols. Oxford UP, 2011.

SOLAR CYCLE

FIELDS OF STUDY

Astronomy; Observational Astronomy; Heat Transfer; Spectroscopy; Electromagnetism; Mass Spectrometry; Photochemistry; Particle Physics; Thermodynamics; Process Modeling; Climate Modeling; Mathematics; Atmospheric Science; Physical Geography; Earth System Modeling

SUMMARY

Temporary dark markings on the surface of the Sun, known as sunspots, have been observed for hundreds of years. As seen from Earth, though, most sunspots are too small to be observed with the naked eye. It was not until the invention of the telescope that sunspots could be observed with regularity. Sunspots were a regular feature on the face of the Sun in the early seventeenth century; however, sunspots all but disappeared later in that century for a period of time lasting into the eighteenth century. This period with almost no sunspots is called the Maunder Minimum.

PRINCIPAL TERMS

- **greenhouse gases (GHGs):** atmospheric gases whose molecules absorb infrared radiation from a planet's surface and then emit a significant portion of it back into the atmosphere, most commonly carbon dioxide, methane, water, and ozone.
- **solar flare:** a sudden eruption of material from the surface of the Sun, typically associated with sunspot activity, often followed by disturbances in Earth's magnetic field
- **solar irradiance:** the amount of light and energy emitted from the Sun at any particular time
- **solar maximum:** the period of time at which the number of sunspots is highest

SUNSPOT ACTIVITY

By the beginning of the nineteenth century, astronomers had recognized that the number of sunspots increased and decreased according to a pattern. A peak number of sunspots, called the solar maximum, occurs about every eleven years. The cycle of sunspot activity does not have an exact period, however, with some cycles lasting closer to ten years and others

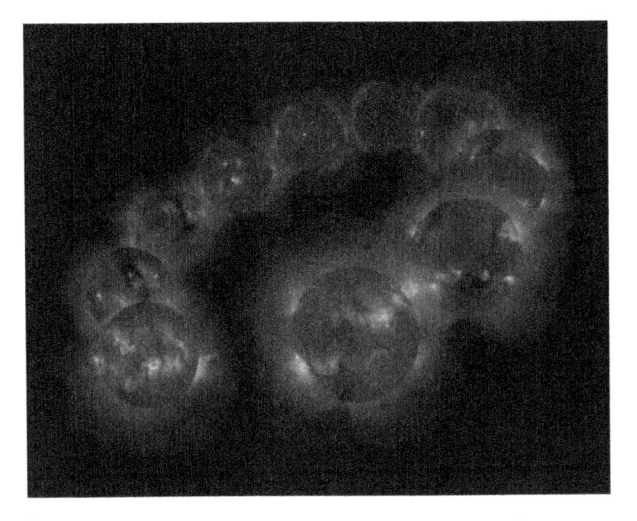

A solar cycle: a montage of ten years' worth of Yohkoh SXT images, demonstrating the variation in solar activity during a sunspot cycle, from after August 30, 1991, to September 6, 2001.

lasting close to twelve years. Furthermore, each sunspot cycle does not result in the same average number of sunspots at its maximum. Every sunspot cycle is a little different from the other cycles.

Sunspots are visible signs of magnetic activity on the Sun. Sunspots form in regions of intense magnetic field activity. These active regions are also the locations of huge explosive releases of magnetic energy called solar flares. The more active the Sun's magnetic field becomes, the more sunspots will appear on its surface. Astronomers have also found that, during each sunspot cycle, the Sun's magnetic field is reversed from that of the cycle before. Thus, the magnetic field follows a twenty-two-year cycle, composed of two eleven-year sunspot cycles.

SIGNIFICANCE FOR CLIMATE CHANGE

Since the advent of space-based observations of the Sun, astronomers have found that the Sun's energy output varies slightly with the sunspot cycle. The more active the Sun is, the more active the sunspot cycle, and the greater the solar irradiance. Since solar energy is the driving force behind Earth's climate, variations in the Sun's energy output could potentially result in climate change. The variation in solar energy is very small, typically less than 0.1 percent. Small changes of that magnitude, averaged

459

out over a solar cycle, should not have a long-term effect on the climate. However, not all sunspot cycles are equal in activity. Some are more active than average and some are less active than average. Historical data indicate that sunspot cycles sometimes exhibit trends in activity over long periods of time. An extended period of unusually active or unusually inactive sunspot cycles could have a cumulative effect on the climate.

During the Maunder Minimum in the seventeenth century, a significant shift toward a cooler climate was observed in many locations. This period of cooling is called the Little Ice Age. (The term "Little Ice Age" is used differently by different writers. Many use it to refer to the climate cooling from about 1300 to 1850, while others use it for the latter half of that interval, when cooling was greatest, beginning around 1550 or 1600.) There is no proof that there was a causal relationship between the two events, but most solar astronomers believe that the Maunder Minimum and the Little Ice Age are connected. Furthermore, there has been a general trend of increasing activity since the Maunder Minimum. Detailed measurements of solar irradiation go back only about a half century. During that time, there has been an observed correlation between solar energy output and the average number of sunspots. Thus, it makes sense that during the Maunder Minimum, the Sun was less active and was providing less solar energy to Earth than normal, resulting in a general cooling of the climate.

If the correlation between sunspot numbers and solar irradiation is consistent, the general increase in average solar activity since the Maunder Minimum would be expected to correspond with a period of increased solar heating of Earth. Some researchers believe that much of the global warming observed during this time period may be the result of natural solar energy increases. Other researchers, though, dispute the degree of change in solar energy output, suggesting instead that the increase in solar energy reaching Earth may only partially explain the observed temperature increases on Earth during the same period of time. If it is true that current global warming trends are the result of increased solar activity rather than human activity, it is unclear whether any human actions could reverse those trends, since they may not be a function of greenhouse gas (GHG) levels. On the other hand, a correlation between solar activity and global warming does not in and of itself establish that atmospheric GHG concentrations are unrelated to warming. Both factors could be contributors to climate change.

The Sun emits more charged particles when it is more active. These charged particles interfere with galactic cosmic rays, changing the amount of carbon-14 (^{14}C) produced on Earth. Increased activity results in less ^{14}C. These changes can be monitored by measuring the ^{14}C content of artifacts of known age. Such studies seem to support the hypothesis that increases in solar activity correlate with warmer periods in Earth's climate history and periods of reduced solar activity correlate with cooler periods in history.

—*Raymond D. Benge, Jr.*

FURTHER READING

Hoyt, Douglas V., and Kenneth H. Schatten. *The Role of the Sun in Climate Change.* Oxford, 1997. Provides an extremely well written and easy-to-follow explanation of how solar variations can affect climate. Extensive bibliography.

Leroux, Marcel. *Global Warming: Myth or Reality?: The Erring Ways of Climatology.* Praxis Publishing Ltd, 2010. Describes both natural and anthropogenic climate change, with many references. Some discussion is given to the role of solar variations in producing climate change.

Odenwald, Sten F. *The Twenty-Third Cycle: Learning to Live with a Stormy Star.* Columbia University Press, 2001. Well-referenced layperson's guide to solar activity over a sunspot cycle and the geomagnetic effects of solar storms that occur during high sunspot activity.

Soon, Willie, and Steven H Yaskell. *The Maunder Minimum and the Variable Sun-Earth Connection.* World Scientific, 2004. The bulk of the book is about the Sun's Maunder Minimum, with discussions of climatic changes occurring at that time. Good bibliography.

STORM SURGES

FIELDS OF STUDY

Oceanography; Physical Geography; Heat Transfer; Fluid Dynamics; Engineering; Earth System Modeling; Ecology; Environmental Sciences; Environmental Studies; Meteorology; Waste Management; Physical Sciences; Climate Modeling; Process Modeling; Statistics; Hydroclimatology; Hydrometeorology; Atmospheric Science; Bioclimatology; Hydrology; Ecosystem Management

SUMMARY

A storm surge, or tidal surge, is water pushed toward the shore by the winds of a storm. The combination of high winds, low pressure, and wave action is what makes a storm surge. The rush of water, combined with the regular action of the tide, causes water to rise significantly higher. This influx of water can cause significant flooding, especially in low-lying coastal areas.

PRINCIPAL TERMS

- **anticyclonic:** the clockwise rotation of an air mass in the Northern hemisphere, counterclockwise in the Southern hemisphere
- **Continental Shelf:** an area of relatively shallow waters surrounding the continental masses, the edges of which mark the beginning of the deep ocean waters
- **cyclonic:** the counterclockwise rotation of an air mass in the Northern hemisphere, clockwise in the Southern hemisphere
- **hurricane:** in the northern hemisphere, a cyclonic storm originating in the tropics
- **tsunami:** the influx of water on a coastline, associated with a sudden, massive displacement of water due to undersea seismic events and landslides

The Making of Storm Surges

Storm surges occur when a tropical storm or hurricane makes landfall. The pressure in the center, or eye, of a storm is low enough that surface water is drawn upward in a dome, just as beverages may be drawn up through a drinking straw. As the storm reaches land, this dome of water piles up against the shoreline. In the northern hemisphere, the area of

Storm surge from Hurricane Irene in Greenwich, Connecticut.

landfall within the right front quadrant of a cyclonic storm's forward motion is in particular danger. In the southern hemisphere, such storms are anticyclonic, placing the area of landfall in the left front quadrant in the greatest danger.

Storm surges are sometimes confused with tsunamis, though the two phenomena are not related in any way. A tsunami, from the Japanese word for "harbor wave," is generated by an undersea earthquake, volcanic eruption or a landslide that creates a set of waves that can grow in intensity as they reach shallow coastal regions. Storm surges only occur in the presence of tropical storms, hurricanes, or cyclones.

The amount of surge is related to the size of the storm, as well as the slope of the Continental Shelf. A shallow slope allows more water to travel inland, while a steeper slope limits the surge but poses a greater danger of generating a breaking wave. The effects of a storm surge are multiplied in confined areas such as harbors due to a funneling effect and can also be felt in inland rivers and lakes.

The National Hurricane Center (NHC) uses the Sea, Lake, and Overland Surges from Hurricanes (SLOSH) model to determine which areas should be evacuated in the event of a storm surge. Storm factors taken into consideration by this model include forward speed, pressure, size, track, and wind speed. These factors are weighed against the timing of the

461

tide and the topological configuration of the projected landfall.

SIGNIFICANCE FOR CLIMATE CHANGE

The inland reach of a storm surge is dependent on the slope of the Continental Shelf and the height of the land above sea level. In the United States, most of the population on the Gulf of Mexico and the Atlantic coastline is a little more than three meters above sea level, making those areas particularly vulnerable. If the sea level were to rise as a result of climate change, the population's vulnerability would increase significantly.

Waves and the action of the current during a storm surge can cause additional damage. Water weighs 1,000 kilograms per cubic meter, and the erosion of beaches and coastal highways is a particular problem. Moreover, the influx of salt water on freshwater bodies disrupts local ecosystems. Salt water that travels far inland may not dissipate for weeks.

The storm surge associated with Hurricane Katrina in August, 2005, was one of the most massive in U.S. history, spanning an area between Grand Isle, Louisiana, and Mobile Bay, Alabama. The storm surge drove water from the Gulf of Mexico up into Lake Pontchartrain and the Mississippi River, which in turn breached the levees in the city of New Orleans and the surrounding areas. Lake Pontchartrain, normally about 0.3 meter above sea level, peaked at 2.62 meters above sea level. According to the U.S. Geological Survey, more than 560 square kilometers of land were eroded into the ocean. The Chandeleur Islands, which formed the easternmost point of Louisiana, were completely destroyed.

The human and financial impact of the Katrina storm surge on the northern Gulf Coast is well documented, but there was a significant ecological price as well. Retreating flood waters carried raw sewage, pesticides, toxic chemicals, and other waste products into the surrounding wetlands. Overall, sixty national wildlife refuges were damaged as a result of Hurricanes Rita and Katrina. A contributing factor to the damage from the 2005 storm season was the ongoing loss of wetlands and barrier islands, which had previously protected the low-lying areas from the worst of the earlier hurricanes.

—*P. S. Ramsey*

FURTHER READING

Boon, John D. *Secrets of the Tide: Tide and Tidal Current Analysis and Applications, Storm Surges and Sea Level Trends*. Woodhead Publishing, 2011.

Larson, Eric. *Isaac's Storm: A Man, a Time, and the Deadliest Hurricane in History*. Vintage, 2000.

Mooney, Chris. *Storm World: Hurricanes, Politics, and the Battle over Global Warming*. Harcourt, 2007.

Pugh, David T. *Changing Sea Levels: Effects of Tides, Weather and Climate*. Cambridge University Press, 2008.

Sargent, William. *Storm Surge: A Coastal Village Battles the Rising Atlantic*. University Press of New England, 1995.

SURFACE OCEAN CURRENTS

FIELDS OF STUDY

Oceanography; Fluid Dynamics; Thermodynamics; Heat Transfer; Climate Modeling; Process Modeling; Hydroclimatology; Hydrometeorology; Atmospheric Science; Hydrology; Physical Geography; Earth System Modeling

SUMMARY

Ocean currents represent a dynamic system that, along with atmospheric circulation, helps to distribute heat evenly across the planet. Responding to the seasons, ocean currents play important roles in climate, marine life, and ocean transportation.

PRINCIPAL TERMS

- **core ring or core eddy:** a mass of water that is spun off of an ocean current by that current's meandering motion
- **Coriolis Effect:** the apparent deflection of any moving body or object from its linear course, caused by Earth's rotation
- **current:** a sustained movement of seawater in the horizontal plane, usually wind-driven

The Great Ocean Conveyor Belt

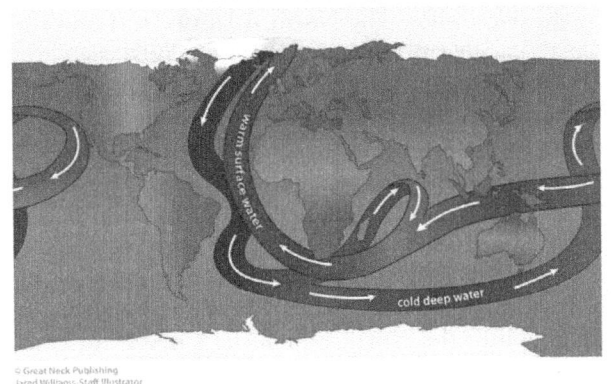

Great Ocean Conveyor Belt: Ocean currents follow patterns much like a conveyor belt. The warm surface water follows the red conveyor belt and the cold deep water follows the blue.

- **drift:** a movement similar to a current but more widespread, less distinct, slower, more shallow, and less easily delineated
- **gyre or gyral:** the very large, semi-closed surface circulation patterns of ocean currents in each of the major ocean basins
- **heat budget:** the balance between the incoming solar radiation and the outgoing terrestrial reradiation
- **planetary winds:** the large, relatively constant prevailing wind systems that result from Earth's absorption of solar energy and that are affected by Earth's rotation
- **thermohaline circulation:** any circulation of ocean waters that is caused by variations in the density of seawater resulting from differences in the temperature or salinity of the water

OVERVIEW

The "heat budget" of Earth results in a temperature range that makes life on the planet possible. Ocean currents play a vital role in the heat budget. These currents are major determinants of climates and strongly influence the distribution of marine life. Ocean currents must be studied in relation to other aspects of the environment with which they interact. The currents are, for example, closely associated with atmospheric circulation, because the planetary winds are the prime movers of the currents. The friction of the wind blowing over the ocean surface began the slow, shallow movement of surface waters that eventually became a global circulation of immense volumes of seawater. There are also deeper ocean currents that are much slower moving and are difficult to monitor, whose significance, therefore, is less well understood.

In part, the deep currents derive from the physical fact that water is continuous in structure, such that water moved from one location must be replaced by water from a different location. Such is the nature of fluids. The deep currents, however, are primarily caused by thermohaline circulation, driven by slight differences in seawater densities resulting from differences in temperature and salinities. The shallow, wind-driven currents affecting the surface waters, although they may be hundreds of meters deep, are effectively independent of such deep ocean currents.

The most significant features of ocean currents are their geographic locations and their directions of flow. It is helpful to recognize overall patterns. There are large-circulation gyres in each of the major ocean basins, discernible as an apparent overall circular movement of surface waters. These gyres, or gyrals, move clockwise in the Northern Hemisphere and counterclockwise in the Southern Hemisphere. The North Central Atlantic Gyre, for example, located east of the United States, is one of the best known and most studied. The Florida Current (part of the Gulf Stream system) is on the west side of the gyre and is a warm current flowing generally northward. The Canaries Current, on the east side of the gyre, is a cold current that flows generally southward. The North Atlantic Drift and the North Equatorial Current form the eastward and westward components of the gyre, respectively.

The result of circulation in the gyre is that warm water from the equatorial region is transported poleward to heat-deficient areas. Simultaneously, the Canaries Current transports colder water back toward the equator. The ocean currents thus help to distribute surface and atmospheric heat more evenly throughout the world. In the Atlantic Ocean south of the equator, a large gyre moves counterclockwise. The warm Brazil Current on the west side of the gyre flows southward, transporting heat away from the equator. The Benguela Current on the east side of the gyre moves colder water toward the equator.

In the Pacific Ocean, similar patterns of clockwise and counterclockwise gyres are apparent. North of the equator, the Japan Current (also known as the Kuroshio) transports warm water toward the pole, and the California Current moves colder water toward the equator. In the Pacific Ocean south of the equator, the cold, nutrient-rich Humboldt Current flows northward off the west coast of South America and is renowned historically as one of the most fertile commercial marine fishery areas in the world. The Indian Ocean possesses similar gyres, although the attenuated portion north of the equator presents some special features.

Wind and Solar Energy

The forces that drive the oceanic current circulations are the planetary winds. The planetary winds are in turn driven by solar energy. The ocean currents, therefore, are sun-driven as sunshine energizes the Gulf Stream and the other currents. Some general principles about Earth's heat budget can be stated. The sun heats Earth, its atmosphere, oceans, and land, but each portion heats differently. The atmosphere, the most fluid and most responsive of the three, has developed large bands of alternating pressure belts and wind belts, such as the Northeast Trade Winds and the Prevailing Westerlies. The Northeast Trade Winds lie between 5 and 25 degrees north latitude. The winds flow predominantly from the northeast and form one of the most constant of the wind belts.

The friction of the wind moving over the ocean surface causes the surface waters to move with the wind, but because of the Coriolis Effect, caused by Earth's rotation, the movement of the water current in the Northern Hemisphere tends to be about 45 degrees to the right of the winds that cause the current. The resultant current, the North Equatorial Current, is fragmented into different oceans because of the intervening continents. Largely as a result of the Coriolis Effect, the current deflects to its right and, in the Atlantic Ocean, eventually becomes the Gulf Stream. In the Pacific Ocean, the comparable current is the Japan Current. One can thus see the origins of the clockwise gyrals in the Northern Hemisphere. Another wind belt in the Northern Hemisphere, the prevailing westerlies, is located between 35 and 55 degrees north latitude. These winds are not as constant as the northeast trade winds, and they flow prevalently from the west. The correlation of the latitudes of the westerlies and the west-to-east-moving currents of the gyres is apparent. The currents slow down and become more widespread, shallower, and less distinguishable but are urged on toward the east by the Westerlies. The North Atlantic Drift and the North Pacific Current result from this relationship.

Northern Hemisphere

Again analyzing the North Central Atlantic Gyre, the blocking position of the Iberian Peninsula causes the North Atlantic Drift to split, part moving southward toward the equator as the cold Canaries Current and part moving poleward into the Arctic Ocean as the warm Norwegian Current. The Canaries Current merges into the North Equatorial Current to complete the gyre. The temperature characterizations of currents and drifts as "warm" or "cold" are relative. There are no absolute temperature divisions. Some warm currents are actually lower in temperature than some cold currents. For example, the Norwegian Current is considered a warm current only because it is warmer than the Arctic water into which it is entering. Only a few degrees above freezing in winter, the Norwegian Current nevertheless transfers significant amounts of heat into these high latitudes and moderates the winter temperatures in Western and Northern Europe. A compensating movement of cold water out of the Arctic is accomplished by the southward flowing Labrador Current between Greenland and North America.

The Gulf Stream is the world's greatest ocean current. There is, however, some confusion about what constitutes the Gulf Stream. The Gulf Stream is generally taken to include the entire warm-water transport system from Florida to the point at which the warm water is lost by diffusion into the Arctic Ocean. It would thus include both the North Atlantic Drift and the Norwegian Current. Technically, the Gulf Stream is a smaller segment of that transport system as the portion off the northeast coast of the United States. The Gulf Stream system thus includes the Florida Current, the Gulf Stream, the North Atlantic Drift, and the Norwegian Current.

Southern Hemisphere

Ocean current patterns in the Southern Hemisphere are almost a mirror image of those in the Northern Hemisphere, adjusted for differences in the configuration of the continents. The southeast trade winds drive the South Equatorial Current, while the Coriolis Effect causes it to deflect to the left. Resultant gyres are counterclockwise, but again, the poleward-moving currents transfer the heat away from the equator, and the equatorward-moving currents return colder water.

In general, cold currents are richer in nutrients, have a higher oxygen content, and support a greater amount of life than warm currents. Most products of the world's commercial fisheries are yielded by cold waters. In contrast, cold currents offshore are associated with desert climates onshore. The atmospheric circulations that drive the ocean currents also create conditions that are not conducive to precipitation in the latitudes of these cold currents. Examples are the Sahara Desert adjacent to the Canaries Current, the Atacama Desert adjacent to the Humboldt Current, the Sonoran Desert adjacent to the California Current, and the Kalahari Desert adjacent to the Benguela Current.

There are other currents that are sporadic in occurrence, such as the warm El Niño current that periodically develops off the west coast of northwestern South America for reasons that are not well understood. Numerous small local currents are also caused by tides, storms, and local weather conditions.

Study of Ocean Currents

The study of ocean currents has acquired new significance as scientists have discerned the role of the currents in climate and marine life. Information comes from many sources. One of the earliest attempts to identify and chart an ocean current was made by Benjamin Franklin (1706-1790) when he was postmaster general of colonial America. His map of the Gulf Stream was published in 1770 and has proved to be remarkably accurate when one considers his sources of information. Franklin noted that vessels sailing westward from England to America in the midlatitudes of the Atlantic Ocean were taking longer than ships moving eastward and longer than ships moving westward but in lower latitudes. He correctly concluded that the vessels were moving against a slow, eastward-moving current.

Since that time, vast amounts of data have been acquired to detect, measure, and chart the currents. One of the early methods still employed is the use of drift bottles. Sealed bottles are introduced into the sea at various locations and dates and are allowed to float with the currents. Finders are requested to note the date and location of the bottle-find and to return the data to the address in the bottle. Ocean current data have also been obtained serendipitously through the loss of floating goods in mid-ocean shipping accidents. The points of landfall of such materials, as well as of materials washed out to sea by some natural disaster such as a tsunami, provide important information about the surface movement of ocean waters.

Various types of more technologically advanced current meters are also used. Some are moored to the sea bottom and can transmit results by radio. It is difficult for a ship at sea to measure currents because the ship itself is drifting with the current. Currents are generally very slow and difficult to measure. A few currents may be measured at 6 to 8 kilometers per hour, but much more common are those less than 1 kilometer per hour. The average surface velocity of the North Atlantic Drift is about 1.3 kilometers per hour. The currents also vary in width and depth. The Florida Current off Miami is about 32 kilometers wide, 300 meters deep, and moving at about 5 to 8 kilometers per hour. It transports more than 4 billion tons of water per minute. The volume of flow is more than one hundred times that of the Mississippi River. As the flow proceeds north and then east as the North Atlantic Drift, it spreads, thins, slows, and splits into individual meandering flows that are difficult to follow. Spin-off eddies, or "core rings," occur that can persist for months.

One useful method of tracking ocean currents is to be able to identify water of slight temperature variations and salinity differences. When the flow movement is so slow as to be practically undetectable with current meters, the slight temperature and salinity differences can be used as "tracers" to identify current movements. This method is also used in identifying the even slower-moving deep-ocean currents. More recently, advanced satellite imagery and high-altitude aerial photography have become extremely important in monitoring ocean currents. Using sensors that detect radiation at selected bands of the electromagnetic spectrum, satellites collect data on broad patterns of seawater temperatures and thus help scientists to understand the movements and extent of the currents. This type of sea monitoring is also useful in detecting any changes that might occur in the oceans in the future.

Significance

Ocean currents play a vital role in the environment. Along with atmospheric circulation, ocean currents serve to distribute the heat absorbed from the sun to different parts of the world. Immense volumes of relatively warm seawater are slowly moving poleward, transporting heat from the heat-surplus equatorial regions to the heat-deficient regions nearer the poles. Cold-water currents in turn move colder water back toward the equator. Although neither solar energy nor rainfall is evenly distributed over the planet, the mixing actions of the ocean currents function to keep the global environment in a steady state. These moderating effects of the ocean currents affect the climates of coastal areas in the middle and high latitudes, especially in Europe. The densely populated nations of northwestern Europe experience much milder winters than would otherwise be expected for such high latitudes. Northwestern North America similarly benefits.

Life in the sea is also aided by this current-driven ocean water mixing. In addition to heat, ocean currents distribute oxygen and nutrients, the result being the formation of certain areas in the oceans where very favorable life-supporting conditions occur. These fertile areas of mixing are concentrated sources of commercial marine fishery products. Where mixing is limited, nutrient-poor regions arise in the ocean such as the Sargasso Sea, located in the center of the North Central Atlantic Gyre. Global warming may alter ocean currents, creating a further need to study these currents and their effects on climate and marine life.

—*John H. Corbet*

Further Reading

Broecker, Wally. *The Great Ocean Conveyor*. Princeton, N.J.: Princeton University Press, 2010. Discusses ocean currents, focusing specifically on the great conveyor belt. Written by the great ocean conveyor's discoverer. Explains the conception of this theory and the resulting impact on oceanography. Written in a manner easy to follow with some background in science, yet still relevant to graduate students and scientists.

Clarke, Allan J. *An Introduction to the Dynamics of El Niño and the Southern Oscillation*. Burlington, Mass.: Academic Press, 2008. Presents the physics of ENSO, including currents, temperature, winds, and waves. Discusses ENSO forecasting models. Provides good coverage of the influence ENSO has on marine life, from plankton to green turtles. Each chapter has references, and there are a number of appendices and indexing. Best suited for environmental scientists, meteorologists, and similar academics studying ENSO.

Schwartz, M. *Encyclopedia of Coastal Science*. Springer, 2005. Contains many articles specific to ocean and beach dynamics. Discusses coastal habitat management topics, hydrology, geology, and topography. Articles may run multiple pages and have diagrams. Each article has bibliographical information and cross-referencing.

Sverdrup, Keith A., and E. Virginia Armbrust. An Introduction to the World's Oceans. McGraw-Hill, 2009. Includes a clearly written section on the planetary winds and their effects on ocean currents. Maps the general patterns of ocean-current circulation and uses diagrams to explain the Coriolis Effect. Lists suggested readings.

Talley, Lynne D., et al. *Descriptive Physical Oceanography: An Introduction*. Academic Press, 2012. Covers only the physical aspects of oceanography in a less comprehensive and more technical manner than general-introduction oceanography texts. Presents a detailed description of ocean circulation, both surface currents and deep currents. Includes details on the types of current meters and the methods of current measurements. Includes an extensive bibliography.

Thurman, Harold V., and Alan P Trujillo. *Introductory Oceanography*. Prentice Hall, 2008. Provides an introduction to oceanography that is comprehensive but not too technical for the general reader. Very well illustrated and includes some high-quality color maps and diagrams. Each chapter includes questions and exercises and also lists references and suggested readings. Addresses the circulations of the ocean currents according to the major ocean basins, with maps and diagrams for each basin.

Trujillo, Alan P., and Harold V. Thurman. *Essentials of Oceanography*. 10th ed. Prentice Hall, 2010. Discusses the topic of deep ocean currents in the broader context of the whole ocean and oceanography using a systems approach that is amenable to use by all students.

Ulanski, Stan. *The Gulf Stream: Tiny Plankton, Giant Bluefin, and the Amazing Story of the Powerful River in the Atlantic.* University of North Carolina Press,

2008. Discusses the hydrodynamics of the Gulf Stream, followed by the biology, and lastly the impact of this current on humans through history. An interesting compilation of the known information on the Gulf Stream current written for the general public.

Voituriez, Bruno. *The Gulf Stream.* UNESCO, 2006. Offers an analytical narrative that examines complex scientific information to describe the causes and dynamics of the Gulf Stream through the history of its discovery and exploration.

THERMOHALINE CIRCULATION (THC) AND THE THERMOCLINE

FIELDS OF STUDY
Oceanography; Fluid Dynamics; Thermodynamics; Process Modeling; Mathematics; Physical Geography; Earth System Modeling; Heat Transfer

SUMMARY
Thermohaline circulation is the result of density differences between masses of ocean water. The density of ocean water is a function of both temperature and salinity. Water increases in density in the polar regions, as surface water cools and its salinity is increased through sea-ice formation. A column of water becomes unstable when surface water has a higher density than deeper water, causing the high-density surface water to sink until it reaches a depth at which it is neutrally buoyant. At that depth, the water will flow horizontally throughout the oceans. The thermocline is the middle layer between the warmer, less dense upper ocean water and the cold, deep ocean water. It is found only between 60° north and 60° south latitude and is replaced by a halocline in the polar regions.

PRINCIPAL TERMS

- **density:** strictly defined as mass per unit volume, this is a temperature-dependent property as volume changes with temperature, but mass does not
- **gyre:** the major rotating current system at the surface of an ocean, generally produced by a combination of wind-generated currents and geostrophic currents
- **halocline:** a layer of rapidly changing salinity in north and south polar seas, due to the exclusion of salt from sea ice formation
- **thermohaline:** indicates that both salinity and temperature are determining factors

An atmospheric thermocline on the South Side of Chicago.

THERMOHALINE CIRCULATION
Thermohaline circulation (THC) is the large-scale circulation of water in the ocean basins, driven by the dense deep water of the polar regions. The density of ocean water is a function of both temperature and salinity. Water increases in density in the polar regions, as surface water cools and its salinity is increased through sea-ice formation. Salt excluded from sea ice as it forms increases the salt content of polar liquid water. A column of water becomes unstable when surface water has a higher density than deeper water, causing the high-density surface water to sink until it reaches a depth at which it is neutrally buoyant. At that depth, the water will flow horizontally throughout the oceans.

High-density surface water that forms in this way in the Labrador and Greenland Seas sinks to become North Atlantic deep water (NADW), while the high-density water that forms in the Weddell and Ross Seas sinks to become Antarctic bottom water (AABW). Other water masses that sink to intermediate depths form in the Mediterranean Sea (Mediterranean

Thermohaline Circulation

Heat Transfer from Ocean to Atmosphere

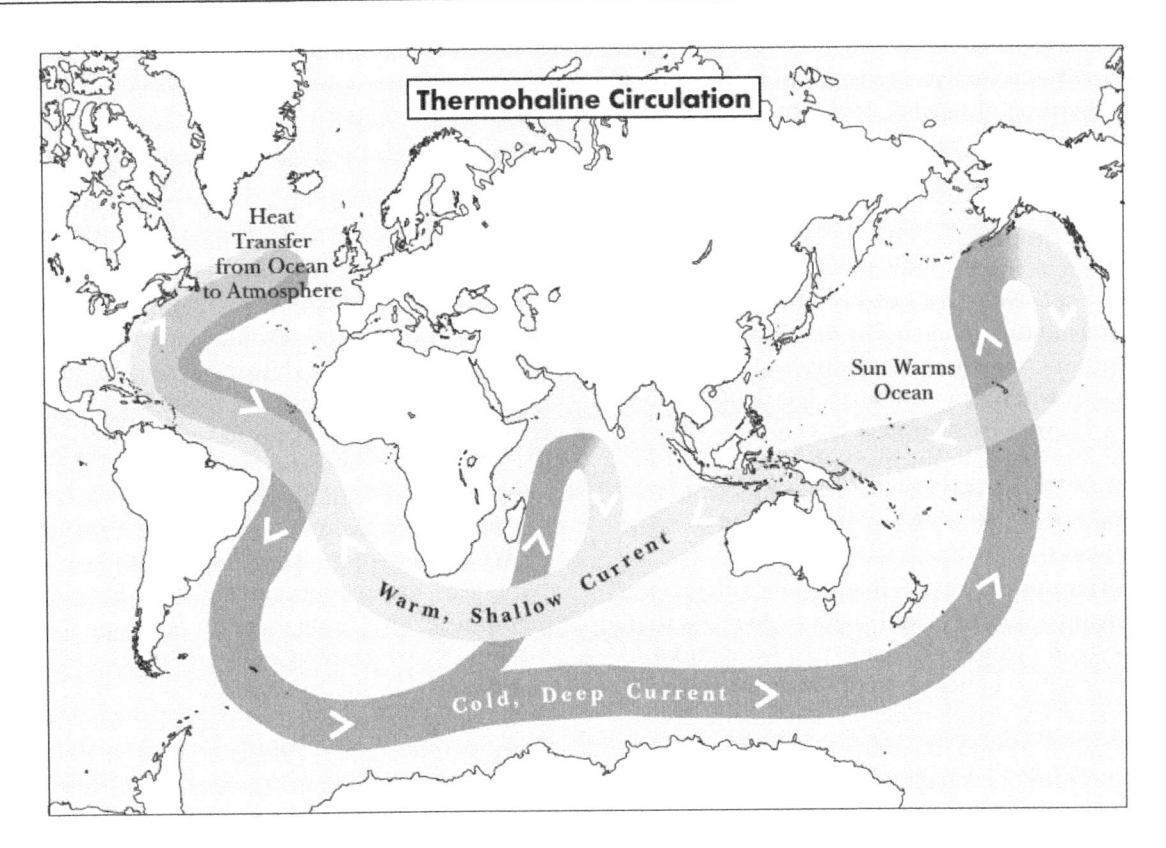

Sun Warms Ocean

Warm, Shallow Current

Cold, Deep Current

intermediate water) and along the edges of the Antarctic Circumpolar Current (Antarctic intermediate water); yet other water masses downwell in the centers of the subtropical gyres, somewhat like water swirling down a drain. NADW flows southward in the Atlantic Ocean, until it mixes with north-flowing AABW in the extreme southern Atlantic Ocean. Once mixed, these two masses form Pacific-Indian common water (PICO).

AABW flows north through the South Atlantic Ocean and can be detected as far north as the equator. After forming in the southern polar regions, deep water circulates around Antarctica and then flows northward into the Indian and Pacific Oceans. Although the mechanisms are not yet fully explained, it is thought that deep water slowly becomes less dense and ultimately returns to the surface as diffuse flow over a large area of the North Pacific Ocean. The net transport of warm surface water to the polar regions to replace surface water that sinks to become deep water is brought about by wind-driven surface currents.

Thermohaline circulation (THC) is sometimes used synonymously with meridional overturning circulation (MOC), although this is not technically correct. THC refers to a global circulation pattern linking surface- and deep-ocean circulation, while MOC refers only to deep circulation in the Atlantic Ocean.

THE THERMOCLINE

The global ocean consists of three layers of varying density. The density of ocean water is a function of changes in salinity and temperature. A shallow, mixed layer extends from the surface to a depth of approximately 30 meters. At the bottom of the ocean is a cold layer that extends from approximately 500 meters deep to the ocean floor. The thermocline is the middle layer between these two. Within the thermocline, the temperature of the ocean changes rapidly from approximately 20°C in the mixed layer to 2°C in deep waters.

The thermocline is found only in waters between 60° north and 60° south latitude; it is replaced by a halocline (a layer of rapidly changing salinity) in the polar regions, which is a significant factor in the operation of THC. The precise depth of the thermocline varies with seasonal changes in solar heating and wind strength. It is shallowest in the summer, deepest in the early spring or late fall, and sometimes absent during the winter. The influence of global warming on the thermocline can affect ocean biomes. The thermocline is shallowest and strongest (that is, it has the greatest temperature change over the shortest depth) when surface waters are warmest and they experience relatively weak winds and wave action, typically during the summer. The density of water decreases with increasing temperature, so a column of water is stable when a zone of warm, low-density water (the mixed layer) exists over a layer of cold, high-density water. The thermocline, when it is present, thus represents a barrier that prevents water from the mixed upper layer from mixing into the colder water of the deep ocean.

The surface mixed layer is the region of the ocean where phytoplankton live and produce biomass that forms the base of the food chain. As phytoplankton grow and produce biomass, they take up carbon and nutrients that are dissolved in surface waters. These nutrients are returned to surface waters during the winter, when the thermocline breaks down and disappears. Mixing of surface and deep waters also transports dissolved oxygen to the nearly-anaerobic deep waters. Thus, when the thermocline is present and prevents mixing between the mixed layer and deep waters, surface nutrients cannot be renewed, and biomass production by phytoplankton eventually stops. Also, since oxygen is not returned to deep waters, they can become uninhabitable by organisms that require dissolved oxygen, such as fish and corals. A warmer climate could cause the development of a permanent, stable thermocline, which, in turn, could lead to decreased biomass production in the surface waters and to the development of large areas in the deep ocean that are uninhabitable by marine life because of extremely low dissolved oxygen concentrations.

SIGNIFICANCE FOR CLIMATE CHANGE

THC can be thought of as a large conveyer belt that transports surface water and heat from the equator to the North and South Poles, with a return flow of cold, deep water to the equator along the ocean bottom. THC is driven by water sinking to the ocean depths in a few distinct locations. After sinking, this water returns to the surface via diffuse flow over a broad geographic region in the North Pacific Ocean. THC is rather slow, as a single cycle takes approximately one thousand years.

Because THC redistributes large amounts of heat on the Earth, it is an important facet of global climate and climate change. In particular, global warming could lead to an increase in the freshwater flux into the North Atlantic by melting glaciers or increasing precipitation and river flow. An increase in freshwater flux to the polar regions would lower the density of polar surface water, stabilizing water columns and preventing the surface water from sinking into the ocean depths. An influx of surface freshwater in the Greenland Sea would also serve to block the northward flow of the Gulf Stream, a surface current that brings heat and ocean water to the polar regions.

It has been hypothesized that a shutdown of THC would lead to localized cooling of the Northern Hemisphere. In 2005, scientists from Britain's National Oceanography Centre presented data to suggest that THC in the Atlantic had slowed during the late twentieth century. These data were subsequently challenged by other scientists, who presented other data sets that showed no such slowing. Despite the controversy regarding whether slowing is occurring, computer models consistently indicate that a shutdown of THC could lead to decreased warming or even cooling in the Northern Hemisphere. Geologic evidence suggests that the Younger Dryas, a time of global cooling that lasted from 12,800 to 11,500 years ago, may have been caused by THC collapse due to a large influx of freshwater when the Glacial Lake Agassiz emptied into the North Atlantic. In addition to climate changes, a shutdown of THC would have other important consequences, including the formation of anoxic conditions in large portions of the world oceans, reduction or collapse of phytoplankton productivity, and more frequent and severe El Niño events.

—*Anna M. Cruse*

FURTHER READING

Broecker, Wally *The Great Ocean Conveyor: Discovering the Trigger for Abrupt Climate Change*. Princeton University Press, 2010. Describes the global oceanic overturning circulation system and how its disruption by climate change may affect the climate of Earth.

Cessi, Paola. "Voyager: What Would Cause Thermohaline Circulation in the Oceans to Stop?" *Scripps Institution of Oceanography.* Scripps Inst. of Oceanography, UC San Diego, 19 Dec. 2013. Web. 23 Mar. 2015.

Gornitz, Vivien *Rising Seas. Past, Present, Future.* Presents an overview of sea level rise due to climate change and describes how rising sea levels may affect the stability of the ocean heat conveyor system.

Jones, E. Peter and Anderson, Leif G. "Is the Global Conveyor Belt Threatened by Arctic Ocean Fresh Water Outflow?" Chapter in Dickson, Robert R., Meincke, Jens and Rhines, Peter, eds. *Arctic-Subarctic Ocean Fluxes. Defining the Role of the Northern Seas in Climate.* Springer Science + Business Media B.V., 2008. Provides a detailed analysis of the interaction of Arctic waters with southerly ocean waters and the potential effects of those interactions on the global conveyor system.

Schmittner, Andreas, Chiang, John C.H. And Hemming, Sidney R., eds. *Ocean Circulation. Mechanisms and Impacts.* John Wiley & Sons, 2013. Provides a thorough description of the effects on Earth's climate due to large "turnover" circulation currents, particularly the Atlantic conveyor system.

TORNADOES

FIELDS OF STUDY

Atmospheric Science; Thermodynamics; Environmental Sciences; Environmental Studies; Meteorology; Climate Modeling; Statistics; Process Modeling; Fluid Dynamics; Physical Geography' Hydrology; Hydroclimatology; Earth System Modeling; Computer Science; Software Engineering; Electromagnetism

SUMMARY

Tornadoes are relatively small, violent, rotating storms that may produce devastating wind velocities of more than 400 kilometers per hour. The force of the cyclonic wind in a strong tornado can demolish well-built structures, and people in a tornado's path are at severe risk of physical harm and death.

PRINCIPAL TERMS

- **cold front:** the transition zone or zone of contact between two air masses when cold air moves into a region occupied by warmer air
- **Coriolis Effect:** a phenomenon in which, because of the planet's rotation, an apparent force is exerted on objects in motion, causing them to deflect from their intended path to the right in the Northern Hemisphere or to the left in the Southern Hemisphere

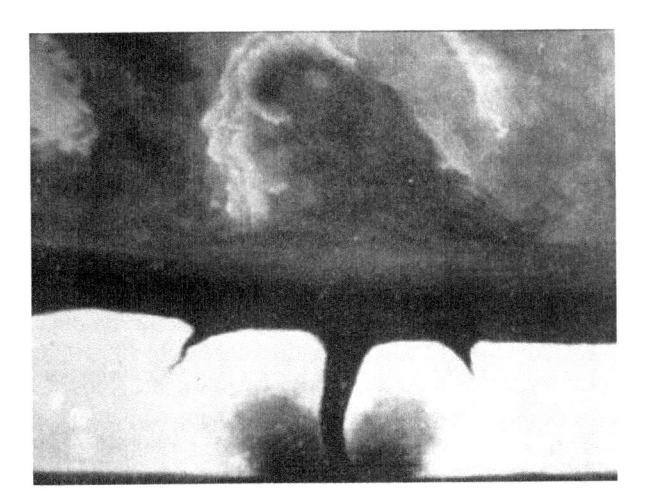

Southwest of Howard, South Dakota, on August 28, 1884, a tornado stirs the ground in this oldest known photograph of a tornado.

- **cumulonimbus cloud:** also called "thunderstorm cloud"; a very dense, tall, billowing cloud form that develops an anvil-shaped head due to high-altitude wind shear, and normally accompanied by lightning and heavy precipitation
- **dust devil:** a rotating column of air rising above a hot ground surface, made visible by the dust it contains; it is much smaller than a tornado, having winds of less than 60 kilometers per hour, and causing little or no damage

- **hurricane:** a huge, tropical low-pressure storm system with sustained winds in excess of 118 kilometers per hour, formed over warm ocean surface water and powered by thermodynamic heat transfer from the water
- **squall line:** any line of vigorous thunderstorms created by a cold downdraft that spreads out ahead of a fast-moving cold front
- **unstable air:** a condition that occurs when the air above rising air is unusually cool so that the rising air becomes relatively warmer and accelerates upward
- **vortex:** the central, low-pressure axis of any rotating fluid, as occurs in whirlpools and tornadoes
- **waterspout:** a tornado that exists over water; less violent and smaller waterspouts form in fair weather just as dust devils do over dry land

TORNADO FORMATION AND CLASSIFICATION

Tornadoes are relatively small, localized low-pressure areas associated with powerful thunderstorms under cumulonimbus clouds. For its size, the tornado is the most violent of the whirlwinds. The "typical" tornado is 250 meters in diameter, with whirling winds of about 240 kilometers per hour. The twisting funnel cloud typically travels at about 65 kilometers per hour over the surface and lasts about ten minutes, moving, in North America, along a northeasterly track. Very large, devastating tornadoes are relatively rare but have almost unbelievable destructive power. The Tri-State Tornado of March 18, 1925, touched down near Ellington, Missouri, at 1:00 PM. and ripped a trail of havoc for 352 kilometers across southern Illinois, finally breaking up at 4:30 PM. near Princeton, Indiana. The storm killed 695 people and injured 2,027. Damage, calculated in 1970 dollars, was $43 million. Compared with other tornadoes, it raced along the ground, averaging well over 100 kilometers per hour.

Tornadoes are most consistently associated with fast-moving cold fronts that sweep across the midsection of the United States, drawing warm, moist, tropical air from the Gulf of Mexico. The cold front is usually associated with a strong low-pressure storm system that rotates counterclockwise as it swirls across land in the prevailing westerly wind pattern. The counterclockwise rotation of the low-pressure system brings cold air in behind, which wedges underneath the warm Gulf air that is drawn in ahead of the storm center. When heavy, cold air wedges under the less dense warm, tropical air, the warm air is forced to rise. If the air is unstable, the cloud will accelerate upward, making a towering thundercloud. The upward surge stops only when the cloud has penetrated the excessively cold upper air. If the cloud tops can penetrate the tropopause at an altitude above 11 kilometers, severe storms, including supercells that spawn tornadoes, are possible. These high cloud tops indicate unstable air at abnormal heights, and occasionally the unstable conditions can drive storm cloud tops to 20 kilometers or higher. A high-altitude zone of strong wind from the west (the "jet stream") tends to increase the chance for violence when associated with a storm system.

Tornadoes are classified by their maximum wind velocity, which occurs on the skin of the spinning funnel. In 1971, T. T. Fujita of the University of Chicago developed the Fujita (F) Scale, which rated tornadoes from F0 to F5, or weakest to strongest. The ratings were based on the amount of potential damage created by the speed of a three-second wind gust. In 2007, the Enhanced Fujita (EF) Scale was adopted by the United States, with Canada using the enhanced version beginning in 2013. The updated scale was based on the original Fujita scale with six categories (0–5) of damage. However, the revision better reflected recent tornado damage surveys and aligned more closely with wind-associated damage by adding additional types and kinds of buildings, levels of construction quality, and various kinds of trees and vegetation. Tornadoes consisting of wind gusts between 65 and 85 miles per hour (mph) are ranked EF-0 (with "EF" representing "enhanced Fujita"); 86–110 mph gusts are ranked EF-1; 111–135 mph gusts are ranked EF-2; 136–165 gusts are ranked EF-3; 166–200 mph gusts are ranked EF-4; and storms with wind gusts over 200 mph are ranked EF-5.

Multiple tornadoes can occur when the weather conditions are ideal for severe weather. The worst tornado outbreak on record as of 2014 was from April 25 through April 28, 2011. A total of 355 tornadoes, with 211 on April 27 alone, created catastrophic and deadly paths through the northeastern, southern, and midwestern United States. Four of the tornadoes were rated as EF5, and over 340 people were killed with April 27 logging the most single-day, tornado-related deaths in the United States since the Tri-State Tornado that killed over seven hundred people. As of

2014, the worst twenty-four hours on record was April 3 to April 4, 1974, referred to the Jumbo Tornado Outbreak, when a remarkable 148 tornadoes struck eleven states, centering on Kentucky. Canada also reported an abnormally high number of tornadoes during that episode. The swarm of funnels claimed 300 lives and left 5,500 injured. The National Oceanic and Atmospheric Administration (NOAA) reported in 2014 that the United States experienced approximately 1,200 tornadoes per year with the average number of tornado deaths in the United States between the years 2012 and 2014 at 56.

Occurrence and Damage

Tornadoes have been reported in each of the fifty states, but they are rare in Alaska, Rhode Island, and Vermont where the average number of tornadoes in those states from 1980 to 2010 was closer to 0 than it was to 1. From 1950 to 2013, only four tornadoes were reported in Alaska and there were no tornado-related deaths. The state of Texas, on the other hand, had over 8,200 tornadoes in the same time period with over 560 fatalities. The area from the city of Fort Worth in northern Texas and then north through Oklahoma, Kansas, Nebraska and on through southwestern Ontario, Canada, has been dubbed Tornado Alley because the storms develop there so consistently. Another particularly vulnerable area in the United States that is prone to tornadoes is Florida, which has a higher than average amount of daily thunderstorms as well as intense tropical storms tend to produce tornadoes when the storms move ashore. Although rare, tornadoes are also known to develop in the midwestern Canadian provinces of Manitoba, Saskatchewan, and Alberta, but are generally restricted to the most southern parts of those provinces. The United States accounts for 75 percent of all tornadoes globally.

Conditions that favor the formation of tornadoes include broad flatlands with no obstructions to the flow of surface wind; an elevation near sea level to allow the full height of the atmosphere for the development of towering clouds; a position on a large continent where very cold air from the north can be swept into a low-pressure storm system that has access to hot, humid tropical air to the south; a southward bulge of strong jet stream currents aloft; and springtime weather patterns that provide intense low-pressure systems that can penetrate rather close to the Gulf of Mexico coast of the United States. March through July is the peak season, with May the most tornado-prone month. Winter tornadoes are mostly confined to the Gulf of Mexico coast, and the frequency moves north and swings toward Kansas as springtime progresses. By July and August, the area of tornado danger spreads through the northeastern states and into southern Canada. Oklahoma is in or near the worst areas most of the year. Most of the storms occur during the late afternoon, at the climax of daily heating, although a really violent frontal advance occasionally will generate night and early-morning tornadoes.

Some of the damage caused by tornadoes results from the rapid passage of low pressure over an area. Most houses are built to withstand downward pressure from much water, snow, or wind against the structure, especially weight on the roof. When a tornado passes over a house, however, the low pressure above, countered with high pressure inside that cannot leak out quickly enough plus wind pressure under the eaves, causes the house to appear to "explode" from within. A rapid pressure drop of 10 percent would give the pressure inside a house a lifting force of nearly 1 tonne per square meter. The roof is lifted slightly off the supporting walls, which, in turn, fall outward. The roof then drops back onto the interior of the house or blows away.

Whirlwinds can develop multiple vortices. Around the core of the funnel cloud, it is possible for high-speed "suction spots" to develop that might have wind velocities 100 kilometers per hour faster than the average velocity of the whirling funnel. Tornado paths observed in open ground often reveal a swirling pattern of streaks or scratches in the soil. On a larger scale, these suction traces match the paths of greatest destruction along the tornado's path. Even in the much smaller dust devils, multiple vortices have been observed. Two or more small dust columns may rotate around the perimeter of the central column of a dust devil. Hurricane Celia, which struck near Corpus Christi, Texas, in 1971, approached landfall with only 145-kilometer-per-hour winds, but the core (eye) of the storm broke into multiple vortices at Corpus Christi, and the damage was more typical of winds of 250 kilometers per hour. The hurricane was expected to cause minor damage, but the center behaved more like a cluster of F2-scale tornadoes and caused extensive damage, even to well-built structures.

STUDY OF TORNADOES

Although meteorologists are quite successful at predicting the general region where tornadoes are likely to occur, they have not been so successful at measuring the wind velocities and air pressure in tornadoes. The storms are so small and so violent that it is nearly impossible to get instruments into their direct path and have the instruments able to survive the passage of the funnel. The few weather stations that have been hit by tornadoes were destroyed or lost power to record the data. Barographs have recorded a drop in pressure of about 1 centimeter of mercury in about thirty seconds during the passage of tornadic storms.

Many of the best measurements of tornadoes are from indirect methods, such as calculating the wind velocity required to cause the kinds of destruction observed. Close estimations of wind velocity can be calculated from steel towers that have been toppled or from railroad boxcars that have been tipped over. Reinforced concrete grain silos have been ripped apart, and walls have collapsed in enough storms to get a large collection of approximate values. Surprisingly, a fairly reliable set of data involves the penetration of straw or splinters into wood surfaces. These projectiles are so small that they are quickly accelerated up to the velocity of the wind, and they are so common that most storms will hurl vast numbers of them at a variety of fixed targets. Experiments reported in 1976 provide many examples of the penetration of straw and toothpicks into all varieties of wood, both wet and dry. The data indicate that a velocity of 30 meters per second (108 kilometers per hour) is sufficient to drive a toothpick into soft pine. Broom straws need about twice as much velocity as do toothpicks to penetrate wood.

Meteorologists continue to improve their ability to forecast, locate, and track tornadoes. Space satellites, a worldwide network of manned weather stations, and sophisticated computer systems enable meteorologists to "see" weather as it develops. Balloons carry instrument packages aloft twice a day from about 90 of the 250 weather stations of the continental United States. The balloons give wind, humidity, temperature, and pressure data for the entire lower atmosphere where storms develop. Forecasters can determine which levels of the atmosphere are unstable and where the moist air is likely to be forced aloft with conditions that can generate violent storms. When dangerous storms begin to develop, Doppler radar is available in most parts of the nation. Doppler radar detects the wind component parallel to the radar beam, and then examines the pattern of the wind field to find locations of potential tornadoes. The radar helps to spot hail formation while other instruments monitor the location and frequency of lightning. All these clues to violent storms focus attention on the parts of storms that might spawn tornadoes. Critical information can be relayed to news services, which, in turn, can warn citizens who are in danger.

SIGNIFICANCE

There is no end to the documented stories of strange phenomena caused by tornadoes. Many instances are recorded where heavy boards have punched through steel plate or metal pipes. Automobiles have been pounded and rolled into battered wads of twisted steel by the storms, and some vehicles have been thrown into upper stories of buildings. Wire fences have been ripped up and wound into prickly balls up to 16 meters in diameter. Strange objects, such as animals, trash barrels, photographs, or blankets, may be picked up in one area by a tornado and deposited, sometimes completely unharmed, many kilometers away. A survivor of a tornado near Scottsbluff, Nebraska, reported seeing a head-sized boulder whirling around his car after the funnel engulfed the auto and its two occupants. After witnessing the flying boulder, the man was hurled from his car and nearly killed, regaining consciousness in a hospital. His passenger, also ejected from the car, died. The auto was destroyed and deposited in a nearby field.

The National Weather Service usually can predict severe weather regions several hours in advance, but the exact location of a tornado must wait for a visual sighting or the occurrence of a tornado signature on Doppler radar. When threatening storms develop, it is wise to monitor local weather broadcasts and to keep a lookout for the characteristic funnel cloud. Often the twister is causing major damage long before the dust swirl on the ground ever connects with the descending visible funnel. Usually a tornado will travel toward the northeast along the ground at about 60 kilometers per hour. The storms move erratically but do not alter course very much, and if the funnel appears to be heading toward the observer, it might be possible to leave the area of greatest danger by moving away in a direction perpendicular to the path of the storm.

Attempting to escape a tornado is often more dangerous than taking some protective measures, because drivers trying to flee tornadoes in cars are often involved in serious accidents resulting from panic. Conditions of traffic, congestion, and available time should be considered before attempting to run from a twister. Those in buildings should try to get to lower floors or in narrow, confined corridors. Above all, windows should be avoided: Many serious injuries during tornadoes result from flying pieces of glass. Taking shelter under heavy tables or inside a sturdy tub can prevent some injury from falling beams or masonry. Tornadoes tend to stay on or even above the ground surface, so a depression, pit, culvert, gutter, or ditch may provide safety. A deep storm cellar with a latched door provides excellent protection. A mobile home is perhaps the worst place to seek shelter from a tornado, due to their light and consequently weak construction. Automobiles also are not safe; cars are easily overturned and are often beaten into shapeless masses by tornadoes. Flying debris causes most of the injuries in tornadoes, and shelter should include protection from tumbling containers, planks, sections of fence, branches, splinters of glass, and other items that could be ripped loose by the storm.

—*Dell R. Foutz*

FURTHER READING

Ahrens, C. Donald and Henson, R. *Meteorology Today*. 11th ed. Cengage Learning, 2016. One of the most widely used and authoritative introductory textbooks for the study of meteorology and climatology. Explains complex concepts in a clear, precise manner and supports them with numerous images and diagrams. Discusses hurricanes and the mechanisms that generate them extensively.

Dunlop, Storm. *Meteorology Manual: The Practical Guide to the Weather*. Haynes, 2014.

Grazulis, Thomas P. *The Tornado: Nature's Ultimate Windstorm*. University of Oklahoma Press, 2003.

Lutgens, Frederick K., Edward J. Tarbuck, Tasa, Dennis G. and Herman, Redina. *The Atmosphere: An Introduction to Meteorology*. 14th ed. Pearson Education, 2018. An updated version of this textbook, in loose-leaf format, provides students with a general overview of the structure and properties of Earth's atmosphere.

Lusted, Marcia Amidon. *Extreme Weather Events*. Greenhaven Publishing, 2018.

Miller, Ron. *Chasing the Storm: Tornadoes, Meteorology, and Weather Watching*. Twenty-First Century, 2014.

"Monthly and Annual U.S. Tornado Summaries." *Storm Prediction Center*. NOAA, 6 Jan 2015. Web. 23 Feb. 2015.

Prothero, Donald R. *Catastrophes!: Earthquakes, Tsunamis, Tornadoes, and Other Earth-Shattering Disasters*. Johns Hopkins University Press, 2011.

Samaras, Tim, Stefan Bechtel, and Greg Forbes. *Tornado Hunter: Getting Inside the Most Violent Storms on Earth*. National Geographic Society, 2009.

Schneider, Bonnie. *Extreme Weather: A Guide to Surviving Flash Floods, Tornadoes, Hurricanes, Heat Waves, Snowstorms, Tsunamis and Other Natural Disasters*. Macmillan, 2012.

TREE RINGS AND CLIMATOLOGICAL INFORMATION

FIELDS OF STUDY

Tree Ring Analysis; Bioclimatology; Atmospheric Science; Atmospheric Chemistry; Environmental Sciences; Environmental Studies; Meteorology; Climate Modeling; Hydroclimatology; Hydrology; Earth System Modeling; Ecosystem Management; Ecology

SUMMARY

Tree rings, also known as growth rings or annual rings, are visible in the woody stems of trees. A woody stem is composed of cells collectively known as secondary xylem. These cells are responsible for moving water and nutrients upward in the plant and provide support for the above ground portion of the tree.

PRINCIPAL TERMS

- **dendrochronology:** identifying the chronological dates, by year, of past events by coordinating with the growth rings of living trees or trees of known age

- **xylem:** the supporting and water-conducting tissue of trees and other vascular plants

FORMATION OF TREE RINGS

Tree rings are formed by the production of cells of different widths. When growing conditions are more optimal, as is typical in the earlier part of the temperate growing season, the diameters or width of the cells are large and the cells appear light to the naked eye. This growth is typically called early wood. During less optimal growing conditions or later in the season, the diameters of the conducting cells are smaller and more fibers for support are produced. Both fibers and the conducting cells produced in the later portion of the growing season (or under less desirable environmental conditions) are smaller in size and have thicker walls than do the cells in the early wood. Because the cells are smaller and the walls are thicker, the wood appears darker to the eye. This darker, denser wood is often called late wood. Bands of lighter, early wood alternating with bands of darker, denser wood creates the appearance of rings in a cross-section of a tree.

In temperate areas with distinctive seasons, trees typically produce one ring for each growing season. Scientists are able to determine the age of a tree by examining the number of rings in the trunk of the tree. Since the width of the ring correlates to the quality of the growing season, one can use dendrochronology to infer past climatological information by measuring the width of the growth rings. The information gathered represents localized conditions for that species in that particular environment. In tropical areas, the growth in trees is more continuous and the use of rings to determine age and climatic conditions is less reliable.

Numerous factors contribute to the overall quality of a growing season. Environmental factors include moisture, temperature, nutrient availability, sunlight, carbon dioxide (CO_2) availability, wind, and pollution. Researchers have shown that the width of a ring in most trees is most closely associated with temperature and moisture availability. The age of the tree, competition with other organisms, herbivory, and disease can also influence the growth of a tree.

SIGNIFICANCE FOR CLIMATE CHANGE

Changes in climate affect tree growth. Increasing levels of CO_2 have been shown to enhance the rates

Tree rings, Hillsborough forest.

of photosynthesis and improve water efficiency in some tree species. This enhancement, however, has been observed in laboratory and modified field environments, for relatively short periods of time (up to three years), on seedlings, under weed-free and insect-free conditions. It is uncertain whether the same result would occur under field conditions with older trees. Enhanced growth would potentially result in the accumulation of more wood through the production of wider rings.

With increasing CO_2 concentrations and global warming, temperatures are increasing. It is likely that this temperature increase will affect the composition of tree species in some ecosystems. Ranges for some tree species are currently observed to be extending and decreasing for those species adapted to cooler environments. Temperature also affects conditions such as fire, drought, wind, and ice, thus influencing the growth of trees.

In addition to atmospheric warming, warming of the soil will affect the growth of trees. Research has shown that root growth and root turnover is enhanced with elevated CO_2 and temperatures. Much of the carbon in a tree is stored within its roots. Enhanced root growth might provide additional nutrients for the plant. In addition, as temperature increases, leaf litter decomposition also increases, which in turn increases the cycling of nutrients to the organisms growing in the soil.

Besides increases in CO_2 and temperature, other climatic factors are increasing that could have a significant impact on tree growth. Two such factors are ozone and sulfur. Both have been studied extensively, with seemingly conflicting results. Trees have been shown to be susceptible to increasing levels of ozone and acid rain (the latter produced from sulfur in the atmosphere). This has been well documented in the laboratory, in modified field experiments, in individual field observations, and in highly polluted localized areas. A difference in susceptibility among tree species has also been documented. In mixed forests, some tree species are damaged by acid rain, ozone, and other pollutants, while other species in the forest remain seemingly unaffected.

It has been more difficult for researchers to link pollutants with the decline of entire forests. The most dramatic and best-documented example of the impact of acid rain on a forest is the devastation of the Black Forest in Germany. Clearly the potential for damage by pollutants exists and is a reality in some areas, but additional field research and careful monitoring are needed to understand forest ecosystems and the levels at which pollutants are significantly impacting the ecosystem.

—*Joyce M. Hardin*

Further Reading

Bell, Nigel, and Michael Treshow. *Air Pollution and Plant Life.* Wiley, 2003. Details the impact of a wide variety of pollutants on plant life. Illustrations, figures, tables, bibliography, index, maps.

Karnosky, David, and Meeting of the International Union of Forestry Research Organizations. *Air Pollution, Global Change and Forests in the New Millennium.* Elsevier, 2006. Overview of the effects of climatic factors that emphasizes the impact of ozone, acid rain, and CO_2 on the growth and health of forests. Illustrations, figures, tables, bibliography, index, maps.

Raven, Peter Hamilton, et al. *Biology of Plants.* W.H. Freeman and Company Publishers, 2013. Text includes general information on plant growth, tree rings, and the impact of climate change. Illustrations, figures, tables, bibliography, index, maps.

TROPICAL CLIMATE

Fields of Study

Climate Classification; Climate Zones; Environmental Sciences; Environmental Studies; Ecology; Physical Geography; Hydroclimatology; Meteorology; Hydrometeorology; Bioclimatology; Oceanography; Climate Modeling; Earth System Modeling

Summary

The tropics are the equatorial region between the Tropic of Cancer (23.5° north latitude) and the Tropic of Capricorn (23.5° south latitude). Because of an atmospheric circulation pattern known as the Hadley circulation, the tropics tend to be warm and wet. This region contains a variety of ecosystems, including deserts, rain forests, savannas, as well as several major islands.

Principal Terms

- **ecosystem:** an ecological environment, and the flora and fauna that inhabit it, taken as a whole
- **hurricane:** in the northern hemisphere, a cyclonic storm originating in the tropics
- **savannah:** flat grasslands found in the tropics and sub-tropics

Tropical Variety

Although each of the tropical ecosystems resembles the others in having generally hot weather (with the exception of a few mountains, such as Kilimanjaro in Tanzania that lies only a few degrees south of the equator), they differ widely in average annual rainfall. In parts of Asia, Latin America, and Africa, rain forests are common, with average rainfall of more than 250 centimeters per year. The Sahara Desert also falls within the tropics, and it has one of the driest climates in the world. Some savanna regions, such as those in Kenya, receive moderate amounts of rainfall. Indeed, some savannas are at a high enough altitude that they have a fairly wide daily temperature range during some parts of the year. The tropics also include several important coral reefs, such as those found off the northern coast of Australia.

The Sugarloaf Mountain in Rio de Janeiro, Brazil, as seen from the Praia Vermelha (Red Beach).

Significance for Climate Change

One aspect of concern related to global climate change is that many of the tropical nations are among the poorest on Earth. These developing nations will be less able to adapt to climate change than the industrial nations of Europe and North America. All of the scenarios generated by the researchers connected with the Intergovernmental Panel on Climate Change (IPCC) indicate that global warming will have less of an impact on the temperature of the tropics than on the temperatures of regions at higher latitudes. Nonetheless, even minor temperature changes coupled with other factors can produce dramatic results.

In 1912, the average snow cover on Mount Kilimanjaro was 12 meters thick, yet by 2007 the snow cover had decreased to an average of only 1.5 meters thick. If the climate warms, in some places rainfall is expected to increase a great deal, while in other regions, such as northern Australia, drought conditions will intensify. In tropical rain forests in parts of Africa and South America, the intensity of rainstorms is predicted to increase, causing further erosion in cleared regions of the forests. Because many tropical ecosystems are quite fragile, even small changes in temperature, precipitation, and wind patterns will produce large adverse impacts on the affected nations. The major impacts in the tropics will come less from direct increases in temperature and more from changes in precipitation patterns, wind currents, and indirect impacts combining these two with temperature increases.

The impact of climate change on the tropics will differ from region to region, but it will become increasingly severe as global warming increases. For example, the IPCC projects that a 3°C increase in global temperature will lead to a major loss of tropical rain forests, and, with this increase, as many as one-third of all species will become extinct. A 4°C increase in temperature could lead to as many as 70 percent of all species worldwide, many in the tropics, becoming extinct. A temperature increase of this magnitude is also likely to lead increased ocean acidification and the bleaching and ultimate death of most coral reefs, many of which are found in the tropics.

Increased rainfall and the resulting runoff in tropical countries could lead to extensive flooding, a problem exacerbated in low-lying nations such as Bangladesh by coastal flooding. Coastal flooding in southern India is projected to threaten the water supplies of many communities by the latter part of the twenty-first century. Because most hurricanes are found in the tropics, tropical peoples have long been concerned by the number and magnitude of hurricanes. Several scientists predict that increasingly warmer oceans would be likely to spawn more hurricanes and cyclones of greater magnitude than in the past.

A combination of climate change and concomitant socioeconomic changes is likely to produce several negative social and economic effects for societies in the tropics. In Nigeria, for example, many people may be forced to migrate from the interior to already-crowded coastal cities. Increasing temperature combined with precipitation changes (either too little or too much) are likely to have a negative impact on the food supplies of tropical regions. In some cases, diseases such as malaria are likely to become more common because of increased standing water.

The various scenarios for climate change accepted by most scientists project an average increase in temperature of between 3°C and 7°C by the end of the twenty-first century. Although the temperature is projected to increase less in the tropics than in the temperate zones, the impact is likely to be severe. Many tropical nations lack the resources to adapt to climate change, so there is likely to be widespread suffering, in addition to extensive ecosystem damage.

—John M. Theilmann

FURTHER READING

Arnold, Dennis G., ed. *The Ethics of Global Climate Change.* Cambridge University Press, 2011. Discusses the ethics of wealthy nations contributing disproportionately to global climate change that affects the poorest nations and the majority of the global population.

Baer, Hans A. *Global Capitalism and Climate Change. The Need for an Alternative World System.* Alta Mira Press, 2012. Examines the manner in which the current global economic system directs the course of global climate change.

Dessler, Andrew *Introduction to Modern Climate Change* 2nd ed. Cambridge University Press, 2016. Provides a good overview of the science and sociology of climate change.

Intergovernmental Panel on Climate Change. *Climate Change, 2007—Impacts, Adaptation, and Vulnerability: Contribution of Working Group II to the Fourth Assessment Report of the Intergovernmental Panel on Climate Change.* Ed. by Martin Parry et al. New York: Cambridge U P, 2007. Print.

Mann, Michael E., and Lee R. Kump. *Dire Predictions.* D.K., 2008. Print.

"Tropics Feel the Heat of Climate Change." *Scientific American.* Scientific American, 31 Jan. 2014. Web. 20 Mar. 2015.

TROPICAL WEATHER

FIELDS OF STUDY

Meteorology; Climate Classification; Climate Zones; Physical Geography; Environmental Sciences; Environmental Studies; Heat Transfer; Climate Modeling; Fluid Dynamics; Hydroclimatology; Hydrometeorology; Atmospheric Science; Bioclimatology; Oceanography; Ecosystem Management; Ecology

SUMMARY

The tropical zone is the area surrounding the equator that is exposed to direct solar rays during part of the year. Tropical areas receive higher annual precipitation and experience low variation in temperature between seasons. Localized low-pressure systems in the ocean can develop into weather disturbances when the temperature of the ocean increases. As temperature and pressure continue to increase, weather disturbances develop into tropical depressions and tropical storms.

PRINCIPAL TERMS

- **circulation cell:** zones of concentrated air circulation caused by the rotation of the Earth and the cyclic distribution of heat through the atmosphere
- **cold front:** area in which a dominant stream of colder air pushes under a pocket of warmer air, causing the warm air to rise and leading to precipitation at the leading edge of the front
- **easterly waves:** localized zones of low pressure oriented parallel to the Earth's rotational axis and moving from east to west across the ocean; they form an important generative component of tropical weather patterns
- **Intertropical Convergence Zone:** low-pressure area created by the interaction of the tropical circulation cells on either side of the equator and by differential heating of the equatorial atmosphere
- **monsoon:** period of increased precipitation caused by the differential distribution of heat and pressure between the lithosphere and hydrosphere
- **storm surge:** rising water levels beneath and surrounding the warm core of a tropical storm system related to the effects of low- and high-pressure interactions within the storm
- **tropical cyclone:** tropical storm marked by clear rotation around a central, warm column of air and wind speeds above 119 kilometers (74 miles) per hour
- **tropical zone:** area between the Tropic of Cancer and the Tropic of Capricorn that falls directly under the sun during a part of the year
- **troposphere:** lowest level of Earth's atmosphere lying between 0 and 10 km (0 and 6 mi) above sea level within which weather patterns develop
- **warm front:** area in which a dominant warm current of air rises over a pocket of cold air to cause condensation and precipitation at the trailing edge of the front

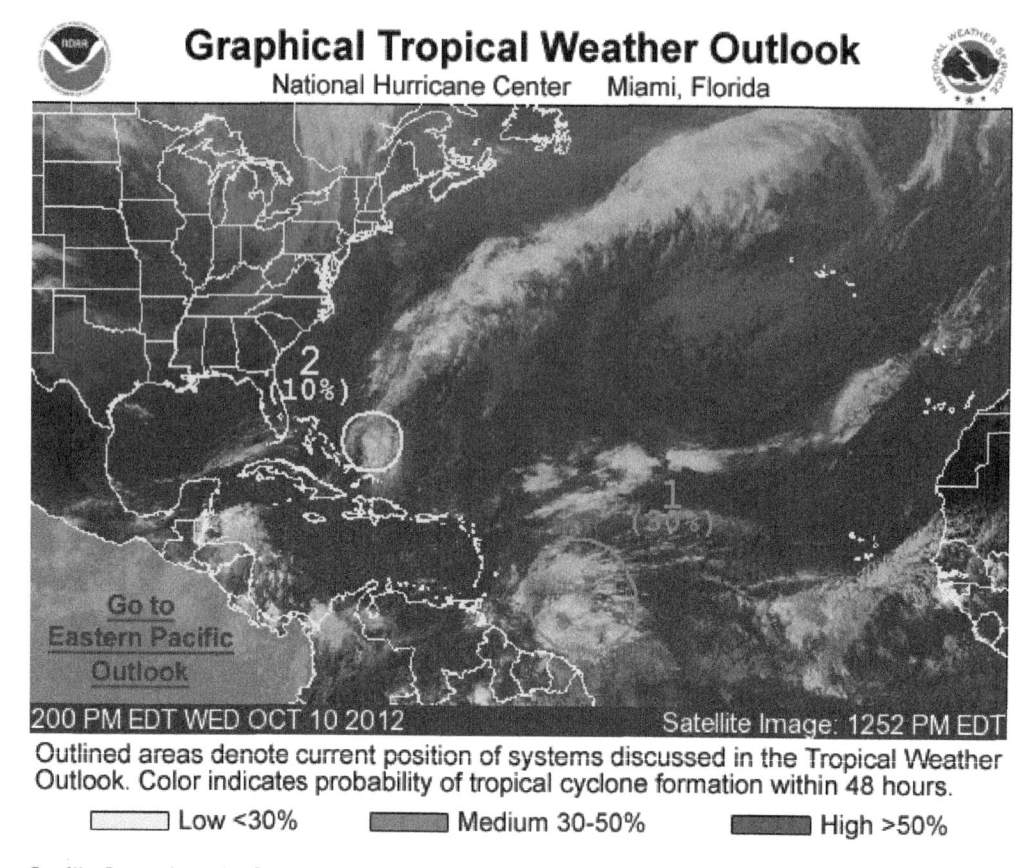

Graphical Tropical Weather Outlook
National Hurricane Center   Miami, Florida

Outlined areas denote current position of systems discussed in the Tropical Weather Outlook. Color indicates probability of tropical cyclone formation within 48 hours.

☐ Low <30%   ▨ Medium 30-50%   ■ High >50%

Satellite Image from the Caribbean Sea and Atlantic Ocean.

Origin of Tropical Weather

The tropical zone is the portion of the Earth that surrounds the Earth's equator. The zone ranges from approximately 23.5 degrees north (the Tropic of Cancer) to 23.5 degrees south (the Tropic of Capricorn).

Because of the relationship between the rotation of the Earth and its orbit around the sun, the portion of Earth that receives direct insolation varies between the Tropics of Cancer and Capricorn during the year. The tropical zone can be described as the portion of Earth that lies directly under the sun for some portion of the year.

In general, the tropics receive high annual precipitation and relatively stable temperatures that vary little between seasons. Rather than having warm and cold seasons, tropical areas have wet and dry seasons because precipitation in the tropics varies more markedly throughout the year than does temperature.

Weather patterns within the tropics are caused by the differential distribution of solar energy. Solar radiation strikes the Earth along the tropical zone, exciting the atoms of atmospheric gases, which causes them to vibrate and collide with one another. In the lowest level of the atmosphere, called the troposphere, excited pockets of atmospheric gases rise because they lose density and pressure as they spread. The gases continue to rise until they reach the limit of the troposphere, the tropopause, where, because the atmosphere is highly stratified, warm air is forced toward the poles.

As warm air moves toward the Tropics of Capricorn and Cancer it gradually cools and condenses and falls to the surface. By approximately 30 degrees north and south, tropical air currents moving close to the surface converge with cooler air moving toward the equator from the poles. Some of the tropical air is pushed back toward the equator, thus constituting a circulation cell, which is a pattern of rising and falling air currents that cycles air between the surface and the tropopause.

Two tropical or Hadley cells move air from the equator to about 30 degrees north and south and back. A portion of the tropical air from the Hadley cells continues moving toward the poles, cycling through two additional circulation cells and thereby distributing the solar energy contained within tropical air over the surface of the Earth.

Low-Pressure Systems and Storm Generation

Directly over the equator, where the northern and southern tropical cells meet, is the area called the Intertropical Convergence Zone (ITCZ). This zone is known historically as "the doldrums," for the lack of winds that were needed to propel sailing ships. A ship that got "stuck in the doldrums" might drift there for days until it might happen to get to a more favorable location of wind or water currents.

The ITCZ does not remain directly over the equator because the position of the sun relative to the Earth changes during Earth's orbit. The ITCZ therefore migrates with the seasons, moving through an area that takes the zone between 5 and 10 degrees north or south of the equator. The ITCZ migrates north in the summer months and south in the winter months, corresponding to temperature variations, increases in precipitation, and increased storm activity.

The ITCZ receives the highest level of solar heating during the year and therefore develops into a low-pressure area as the air within the ITCZ loses density in response to solar excitation. Low-pressure areas are characterized by cloud cover and strong winds. Because the Earth rotates counterclockwise, the winds spreading from the ITCZ are deflected relative to the poles; this a phenomenon known as the Coriolis Effect, which causes the ITCZ winds to move to the northeast and southeast from the equator. These winds, called the trade winds, carry air currents toward the ends of the tropical zone, where they gradually cool, thereby increasing in both density and pressure. This creates two high-pressure areas, called the subtropical high-pressure zones, which shift (along with the migration of the ITCZ) between 10 and 30 degrees north and south of the equator.

Weather fronts develop at the point where low- and high-pressure systems meet. A cold front is an area in which a dominant stream of cold air pushes under a bed of warmer air at its leading edge. This causes warm air to rise and condense rapidly; as the water vapor condenses, precipitation develops along the leading edge of the cold front.

A warm front is an area in which dominant streams of warm air encounter pockets of cold air. The warm air rises over the cold air, thereby leading to the condensation of water vapor and precipitation following in the wake of the front. In general, rising air produces rainfall while falling air creates dry conditions.

Tropical Monsoons

Monsoons are seasonal variations in precipitation that accompany the movement of the ITCZ. Monsoons result from the differential heating of the ocean and nearby landmasses and from changes in the direction of wind patterns. Monsoon seasons are the periods of highest annual precipitation during the year. Monsoons affect portions of Africa, Asia, North America, and South America.

Through most of the year, wind currents typically move from the continents to the ocean. In the summer months, the solid materials of the continental crust heat faster and reach higher average temperatures than the liquid water of the ocean, which disseminates and thereby regulates heat more efficiently. As this occurs, a low-pressure area develops over the land as hot air disperses into the atmosphere. This causes a reversal in the direction of prevailing winds, as cooler air pockets over the ocean are drawn over the land. The mixture of warm and cool air rises and leads to condensation and precipitation moving inward over the continent. This continues until heat levels reach equilibrium between the two areas and the monsoons begin to abate.

Monsoons also occur in the winter because the land loses heat more rapidly than does the water, thereby producing a high-pressure system over the land. When this reaches a certain level, warmer air pockets over the ocean are drawn over the land and begin to rise, again blending with cooler air to produce precipitation. These two systems are responsible for the seasonal cycle of monsoons alternating with dry seasons in the tropics.

Tropical Storms

Tropical storms develop within the trade winds out of localized disturbances called easterly waves, which are troughs of low pressure moving from east to west across the ocean. Easterly waves develop when a portion of the ocean is heated such that water vapor begins to rise from the surface through evaporation, thereby reducing pressure along a narrow band. As this zone of low pressure moves, the area behind it forms into a zone of divergence, where the winds flow out of the system.

To the front of the wave, a zone of convergence forms, where high- and low-pressure winds meet and spiral upward. This spiraling, rising column of current, which carries significant moisture from evaporated water vapor, is the origin of tropical storms.

The defining characteristic of a tropical storm is the column of warmer air at the center of the storm known as the core, which creates the differential pressure and density that drive the continued movement of the system. For a tropical storm to develop, the water underlying the storm system must be 26°C (79°F) or warmer, with high local humidity levels. When these factors merge with local weather disturbances, a tropical storm can develop.

Phases of Tropical Storms

Tropical storms begin as smaller, localized tropical disturbances. In most cases, tropical disturbances result in isolated thunderstorms that dissipate as the warm currents at the core are drawn into the upper troposphere. If the localized pressure difference grows, the storm becomes a tropical depression, which has a more defined core and wind speeds between 37 and 63 kilometers (23 and 39 miles) per hour. At this point, the tropical storm is generally given an identification number by meteorological organizations and is tracked for further developments.

When wind speeds increase to between 64 and 118 km (40 to 73 mi) per hour, the storm is reclassified as a tropical storm or tropical cyclone. At this point, the storm system is designated with a unique name and may pose a serious threat to coastal settlements. Cyclones with wind speeds exceeding 119 km (74 mi) per hour are called mature tropical cyclones and are characterized by a well-defined core, which is sometimes called the eye of the storm. Mature tropical cyclones that develop in the eastern part of the Pacific basin or in the Atlantic basin are called hurricanes, while those that develop in the western portion of the Pacific basin are called typhoons.

Once a cyclone forms, it will move in response to changes in local wind speed and pressure. The path of a cyclone is highly unpredictable, which increases the danger they pose to terrestrial ecosystems and human settlements. Cyclones can change speed and direction rapidly and without significant warning, making it difficult to predict the path that the cyclone will take. In general, most of the storms that originate above the equator tend to move northwest while those below the equator move southwest.

Tropical storms follow seasonal patterns, which are related to the north and south variation of the ITCZ. The peak for cyclones in the Southern Hemisphere falls between January and March, while the peak for cyclones in the Northern Hemisphere falls between June and November. The northwest Pacific basin is the most active area for tropical cyclones, with storms forming during any month of the year. Most cyclones are active for less than one week, though particularly strong storms have lasted for nearly one month before dissipating. Tropical storms dissipate when the latent heat energy in the ocean and atmosphere is reduced to such a level that the storm can no longer sustain its motion.

The National Weather Service in the United States names each tropical storm using a system of alternating male and female names chosen to reflect the alphabetical position of the storm for that year. The first named storm in any year will therefore be given a name beginning with the letter *A*. If the last storm of the preceding year was given a male name, then the new storm will be given a female name and vice versa. The list of potential names is recycled through a six-year period. When a hurricane causes significant damage to human settlements, the name of the hurricane may be retired from the list of potential storm names. Hurricane Allison (2001) and Hurricane Agnes (1972) are two hurricanes whose names have been retired from usage because of the level of damage wrought by those storms.

In some hurricanes wind speeds can exceed 180 km (112 mi) per hour, which is sufficient to destroy buildings and fell trees. When a cyclone is active, water levels in the area directly beneath and around the eye of the storm rise in response to the low pressure of the core, creating what is called a storm surge. Many hurricane- or typhoon-related deaths result from flood waters that accompany the storm surge. Hurricane Katrina, which struck the Atlantic and Gulf coasts of the United States in 2005, caused more than $80 billion in damage and led to nearly two thousand deaths.

In addition, approximately 25 percent of hurricanes that impact the continental surface have associated tornadoes, which are vortexes of wind currents that can move over dry land. Associated tornadoes can spread wind damage for miles surrounding the impact zone.

—*Micah L. Issitt*

FURTHER READING

Aguado, Edward, and James E. Burt. *Understanding Weather and Climate*. 7th ed. Pearson Prentice Hall, 2015. The formation of midlatitude storms is surveyed. Continental climates and the structure of climate are discussed.

Ahrens, C Donald, and Robert Henson. *Meteorology Today: An Introduction to Weather, Climate, and the Environment*. Cengage, 2019. Discusses global climates, climate change, and classification.

Ahrens, C. Donald. *Essentials of Meteorology: An Invitation to the Atmosphere*. 6th ed. Brooks/Cole, 2011. Covers the major principles in meteorological science, including methods used to study and predict changes in weather patterns. Discusses a variety of factors influencing the development of tropical weather patterns.

Denny, Mark *Making Sense of Weather and Climate. The Science Behind the Forecasts.* Columbia University Press, 2017. An informative yet entertaining book that looks at many readily-visible aspects of climate effects such as fog from a question and answer perspective.

Lutgens, Frederick K., Edward J. Tarbuck, Tasa, Dennis G. and Herman, Redina. *The Atmosphere: An Introduction to Meteorology*. 14th ed. Pearson Education, 2018. An updated version of this textbook, in loose-leaf format, provides students with a general overview of the structure and properties of Earth's atmosphere.

Maslin, Mark. *Global Warming: A Very Short Introduction.* Oxford University Press, 2009. Provides a concise general introduction to the major factors that define the climate of Earth.

TROPOSPHERE

FIELDS OF STUDY

Atmospheric Science; Atmospheric Chemistry; Physical Geography; Thermodynamics; Heat Transfer; Environmental Sciences; Environmental Studies; Fluid Dynamics; Hydroclimatology; Meteorology; Hydrometeorology; Bioclimatology; Oceanography; Earth System Modeling; Ecology

SUMMARY

The troposphere is the lowest layer of Earth's atmosphere. It typically extends from the surface to an altitude of about 10–16 kilometers, varying by latitude. In the tropics, the troposphere extends relatively high, whereas in the polar region, its upper boundary is significantly lower.

PRINCIPAL TERMS

- **greenhouse gases (GHGs):** atmospheric gases whose molecules absorb infrared radiation from a planet's surface and then emit a significant portion of it back into the atmosphere, most commonly carbon dioxide, methane, water and ozone.
- **longwave radiation:** electromagnetic radiation in the infrared range, having wavelengths longer than those of the visible spectrum, associated with heat and heating
- **radiative cooling:** the process of emitting energy as longwave radiation to balance the solar energy absorbed as shortwave radiation
- **shortwave radiation:** electromagnetic radiation in the ultraviolet range, having wavelengths shorter than those of the visible spectrum, associated with the transport of energy capable of promoting chemical reactions

PROPERTIES OF THE TROPOSPHERE

Because gravity weakens as the distance from a body increases, the atmosphere's density and pressure decrease exponentially with altitude. Most of the air molecules in the atmosphere are located in

Earth Structure

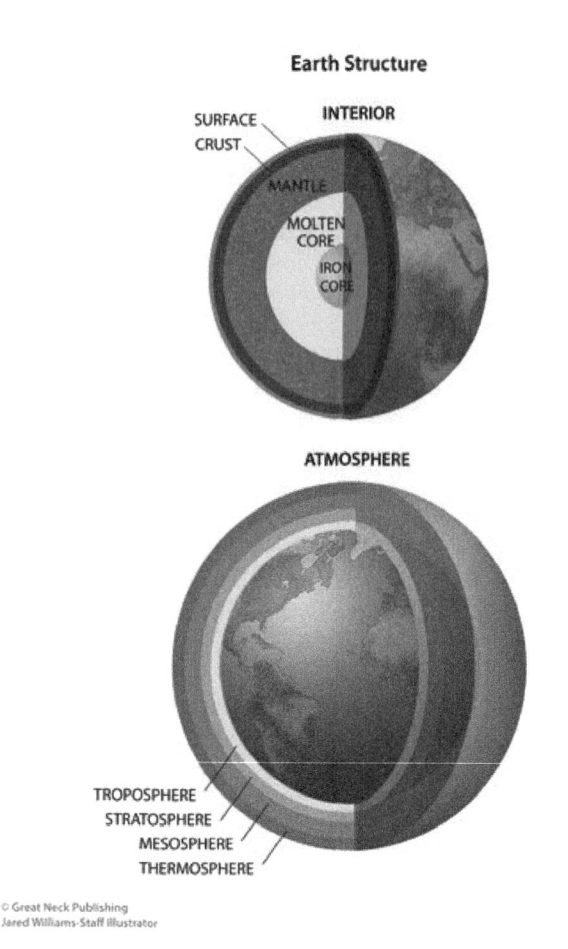

The interior structure of the earth includes the mantle, molten core, and solid iron core. The exterior of the earth, the atmosphere, includes the thermosphere, mesosphere, stratosphere, and troposphere.

the atmospheric layers closest to Earth's surface. As altitude increases, the number of atmospheric molecules per unit of volume decreases, so the air becomes thinner.

By contrast, air-temperature changes as a function of altitude are much more complex. The temperature of Earth's atmosphere decreases with altitude to the tropospheric boundary, 10–16 kilometers high. From that point to an altitude of about 20 kilometers, there is a "pause" in the temperature change, as temperature remains relatively constant. Above 20 kilometers, the temperature begins to increase. From about 45 to 50 kilometers up and from about 80 to 90 kilometers up, there exist two other zones of relatively constant temperature. Between them, from 50 to 80 kilometers up, the temperature decreases with height. Above 90 kilometers, the temperature increases with height throughout the rest of Earth's atmosphere. These zones of different types of temperature change define the structural layers of Earth's atmosphere: the troposphere, tropopause, stratosphere, stratopause, mesosphere, mesopause, and thermosphere.

Because more than 90 percent of atmospheric molecules are located in the troposphere, most weather and convection happen within this layer of the atmosphere. During the daytime, Earth's surface warms relatively quickly as shortwave solar energy is absorbed and subsequently radiated as longwave infrared radiation associated with heat. This radiative heating is necessary for convection to be generated. Warm, moist air rises, condensing when the air temperature drops at mid-to-high levels of the troposphere. This pattern of convection and condensation is the core process generating weather and storms.

During the nighttime, the air near Earth's surface cools faster than that above it. Very low in the troposphere, radiative cooling can cause a temperature inversion, in which a warm layer of air is situated above the cool air near the surface and below the cold air of the upper troposphere. This temperature structure acts as a lid in the mid-troposphere, limiting convection. This phenomenon is the reason storms do not typically occur in the late night or early morning.

Significance for Climate Change

The importance of the troposphere to global warming and climate change is twofold. First, most greenhouse gas (GHG) emissions, both natural and anthropogenic, are released into the troposphere, close to Earth's surface. Since gravity keeps more than 90 percent of air molecules within the troposphere, most of the emitted GHGs will stay there as well. Second, because the troposphere is in direct contact with Earth's surface, tropospheric reflections of global warming will be the most sensible and evident atmospheric reflections.

Many scientists predict that global warming will result in a significant increase of tropospheric temperatures. However, some studies based on the analyses of satellite measurements and radiosonde (weather balloon) data have shown no evidence of tropospheric warming. These studies have been critiqued in other studies, which have argued that results indicating no warming or even cooling in the troposphere were based on incomplete analyses, because they failed to take into account the cooling effect exerted by the stratosphere on the troposphere. However, some scientists believe that the overall temperature of the troposphere could remain unchanged or even decrease even while the Earth's surface temperature rises. They argue that an increase in atmospheric GHG concentrations may result in an increase in cloud cover. These clouds would increase Earth's albedo, reflecting solar radiation back into space and acting as a cooling influence.

Most weather occurs inside the troposphere. Thus, any changes affecting the troposphere will inevitably affect Earth's weather patterns. As the example of tropospheric temperature studies illustrates, however, it is difficult to predict the precise nature of these changes and their effects. If the troposphere does warm, it will extend higher than it does currently. If this happens, scientists predict that there will be much stronger convective storms, because an extended troposphere will allow deeper convection and thicker clouds. This is perhaps one reason that the Intergovernmental Panel on Climate Change (IPCC) predicted more severe storms would accompany future climate warming. Another possible effect of tropospheric warmth is an extension of tropical regions poleward. Such an extension would fundamentally alter the environment of many regions on Earth, with significant socioeconomic consequences.

—*Chungu Lu*

FURTHER READING

Ahrens, C Donald, and Robert Henson. *Meteorology Today: An Introduction to Weather, Climate, and the Environment*. Cengage, 2019. Discusses global climates, climate change, and classification. Widely used introductory textbook on atmospheric science; covers a wide range of topics on weather and climate.

Clerbaux, Cathy, Solene Turquety, and Pierre Coheur. "Infrared Remote Sensing of Atmospheric Composition and Air Quality: Toward Operational Applications." *Comptes Rendus Geoscience* 342, nos. 4/5 (2010): 349–356. Discusses the uses of remote sensors for studying pollution's effects on the atmosphere. Reviews satellite-based technologies such as thermal infrared sensors to study the troposphere on a global scale.

Intergovernmental Panel on Climate Change. *Climate Change, 2007—Synthesis Report: Contribution of Working Groups I, II, and III to the Fourth Assessment Report of the Intergovernmental Panel on Climate Change*. Edited by the Core Writing Team, Rajendra K. Pachauri, and Andy Reisinger. Geneva, Switzerland: Author, 2008. Comprehensive overview of global climate change published by a network of the world's leading climate change scientists under the auspices of the World Meteorological Organization and the United Nations Environment Programme.

Lutgens, Frederick K., Edward J. Tarbuck, Tasa, Dennis G. and Herman, Redina. *The Atmosphere: An Introduction to Meteorology*. 14th ed. Pearson Education, 2018. An updated version of this textbook, in loose-leaf format, provides students with a general overview of the structure and properties of Earth's atmosphere.

Mbane Biouele, Cesar *Earth's Atmosphere Dynamic Balance Meteorology*. Scientific Research Publishing, 2014. Without becoming overly technical, this book discusses the structure of the atmosphere and many of the physical phenomena that occur there.

U

URBAN HEAT ISLAND (UHI)

FIELDS OF STUDY

Heat Transfer; Atmospheric Science; Environmental Sciences; Environmental Studies; Thermodynamics; Engineering; Waste Management; Meteorology; Climatology; Climate Modeling; Fluid Dynamics; Hydroclimatology; Bioclimatology; Hydrometeorology; Climate Zones; Climate Classification; Physical Geography; Ecology; Ecosystem Management; Computer Science

SUMMARY

An urban heat island is a metropolitan area that is significantly warmer in surface and air temperature than its suburban and rural surroundings. The effect was first documented in 1881 in regard to the relative temperatures in London, England, and the surrounding countryside. City surfaces absorb and retain significantly more heat more effectively throughout the day due to multiple reflections from structures and the mass of artificial materials. The longer retention times result in cities having relatively warmer temperatures than their surroundings.

PRINCIPAL TERMS

- **greenhouse effect:** an atmospheric phenomenon in which certain molecular components of the atmosphere capture infrared radiation from a planet's surface and prevent it from radiating back out into space
- **greenhouse gases (GHGs):** atmospheric gases whose molecules absorb infrared radiation from a planet's surface and then emit a significant portion of it back into the atmosphere, most commonly carbon dioxide, methane, water and ozone.
- **greenspace:** areas within urban developments characterized by the presence of grass, trees and other plants, maintained using ecologically-friendly or "green" methods
- **urbanization:** the replacement of open countryside by housing and other urban developments

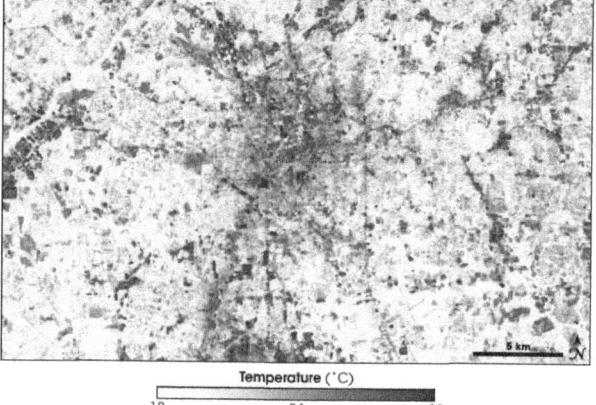

Satellite images of Atlanta, Georgia, in the visible and infrared spectra demonstrate the urban heat island effect of the city.

HOWARD'S OBSERVATION

An urban heat island is a metropolitan area that is significantly warmer in surface and air temperature than its suburban and rural surroundings. English amateur meteorologist Luke Howard (1772–1864), in his book *The Climate of London* (1818), described temperature differences between London and its surrounding countryside. He noted that it was warmer at night in London than in the country and speculated that the cause might be the burning of fuels. Research over the subsequent two centuries has charted a variety of differences between cities and their surroundings, as distinctive urban landscapes, domestic and industrial structures, and the behavior of urban dwellers affect the ways in which solar and other heat enters and exits the area. Urbanization causes changes to the preexisting natural landscape, as original materials (soil, vegetation, rock, water, and so on) are gradually replaced or modified with materials (concrete, tile, and many others) that, in conjunction with the activities and life patterns of a city's inhabitants, result in greater heat retention. This heat retention can in turn affect local weather patterns.

Significance for Climate Change

Scientists have identified many interconnected causes for the urban heat island effect. During the day, the sun warms buildings and roadways. Pollution from automotive traffic and industrial processes contributes to the formation of clouds and smog, which help trap heat, and tall buildings limit the ability of winds to disperse such formations (the "canyon effect"). The predominance of land structures and the paucity of bodies of water lessen the influence of evaporation, which would use some of the heat energy for the formation of water vapor. Instead, that energy raises the ambient temperature. The concentration of human bodies in cities also contributes to their heat level. The increase in city heat in turn encourages activities, such as the use of air conditioning, that further contribute to localized heat increases, since air conditioners make interiors cooler by making the exterior city warmer. Moreover, increased power plant emissions resulting from the additional consumption of electricity contribute to global warming.

To the extent that urban areas encourage structures or behaviors that stimulate additional activities, increasing emission of greenhouse gases (GHGs), urban heat islands may be viewed as contributing indirectly to global warming. (If cities are hotter than rural areas, then those who move to cities will be more likely to use energy to cool themselves than they otherwise would be.) Since 2004, according to a 2014 article in *Scientific American*, cities have more searing hot days each year. In fact, twelve U.S. cities averaged at least twenty more days a year above 90 degrees than nearby rural areas. In two-thirds of sixty cities analyzed, urbanization and climate change appear to be combining to increase summer heat faster than climate change alone is raising regional temperatures. By 2030, about 87 percent of North American's population is expected to live in urban areas, according to 2010 data from the World Health Organization. As large metropolitan areas fuse into megalopolises, or megacities, the problems only become exacerbated. Accordingly, students of the urban heat island effect have seen the need for various degrees of urban redesign as a key element in attempts to mitigate the noxious consequences of the effect. Chief among these is increase in, and optimal distribution of, urban vegetation and green space, including the planting and sustaining of suitable trees.

The urban heat island effect must be taken into account in any attempt to read the historical record of Earth's temperature. To compare today's temperature with that of a century ago requires comparability, and that can be affected by urban growth. Places where temperatures were measured in the past often have since experienced significant urban growth, with the associated increased temperature measurements. This growth makes direct comparisons of present and past measurements difficult. In light of this problem, climatologists must adjust the data on the basis of their best guess about distorting factors.

—*Rebecca S. Carrasco*

Further Reading

Gartland, Lisa. *Heat Islands: Understanding and Mitigating Heat in Urban Areas.* Earthscan/James and James, 2008.

Heikkinen, Niina. "How People Make Summer Hotter." *Scientific American*, 25 Nov. 2014. Web. 24 Mar. 2015.

Kenward, Alyson, Dan Yawitz, Todd Sanford, and Regina Wang. "Heat Islands Cook U.S. Cities Faster Than Ever." *Scientific American*, 22 Aug. 2014. Web. 24 Mar. 2015.

Maslin, Mark. *Global Warming: A Very Short Introduction.* Oxford University Press, 2009.

Spencer, Roy W. *Climate Confusion: How Global Warming Hysteria Leads to Bad Science, Pandering Politicians, and Misguided Policies That Hurt the Poor.* Encounter Books, 2008.

URBAN HEAT ISLAND (UHI) AND MEGACITIES

FIELDS OF STUDY

Heat Transfer; Atmospheric Science; Environmental Sciences; Environmental Studies; Thermodynamics; Engineering; Waste Management; Meteorology; Climatology; Climate Modeling; Fluid Dynamics; Hydroclimatology; Bioclimatology; Hydrometeorology; Climate Zones; Climate Classification; Physical Geography; Ecology; Ecosystem Management; Computer Science

SUMMARY

An urban heat island is a metropolitan area that is significantly warmer in surface and air temperature than its suburban and rural surroundings. The effect was first documented in 1881 in regard to the relative temperatures in London, England, and the surrounding countryside. City surfaces absorb and retain significantly more heat more effectively throughout the day due to multiple reflections from structures and the mass of artificial materials. The longer retention times result in cities having relatively warmer temperatures than their surroundings. A megacity is a city with a population greater than 10 million. When cities get this large, it becomes difficult to determine their precise boundaries or true population. The rise of megacities represents a profound development in the history of humanity, and the challenges they present with respect to climate change may also be among the greatest opportunities to address that change.

PRINCIPAL TERMS

- **greenhouse effect:** an atmospheric phenomenon in which certain molecular components of the atmosphere capture infrared radiation from a planet's surface and prevent it from radiating back out into space
- **greenhouse gases (GHGs):** atmospheric gases whose molecules absorb infrared radiation from a planet's surface and then emit a significant portion of it back into the atmosphere, most commonly carbon dioxide, methane, water and ozone.
- **greenspace:** areas within urban developments characterized by the presence of grass, trees and other plants, maintained using ecologically-friendly or "green" methods
- *megalopolis:* a megacity that sprawls over a large area, rather than being concentrated spatially in the manner of traditional cities
- **urbanization:** the replacement of open countryside by housing and other urban developments

HOWARD'S OBSERVATION

An urban heat island is a metropolitan area that is significantly warmer in surface and air temperature than its suburban and rural surroundings. English amateur meteorologist Luke Howard (1772–1864), in his book *The Climate of London* (1818), described temperature differences between London and its surrounding countryside. He noted that it was warmer at night in London than in the country and speculated that the cause might be the burning of fuels. Research over the subsequent two centuries has charted a variety of differences between cities and their surroundings, as distinctive urban landscapes, domestic and industrial structures, and the behavior of urban dwellers affect the ways in which solar and other heat enters and exits the area. Urbanization causes changes to the preexisting natural landscape, as original materials (soil, vegetation, rock, water, and so on) are gradually replaced or modified with materials (concrete, tile, and many others) that, in conjunction with the activities and life patterns of a city's inhabitants, result in greater heat retention. This heat retention can in turn affect local weather patterns.

REALITY ON THE GROUND

A megacity is a complex structure consisting of a sophisticated built environment that shelters and sustains millions of human agents. These agents, living in close proximity to one another, often represent extremes of human experience, as megacities juxtapose great wealth and poverty, as well as the diversity of human cultures. In their vast urban landscapes, millions of people live and die in sad and tragic conditions of poverty and low life expectancy. At the same time, others live lives of almost unfathomable wealth and freedom to travel about the globe. These urban environments simultaneously represent both a pinnacle of human achievement and a shameful failure to realize human potential.

Change Challenges and Opportunities

Cities almost always depend on a hinterland beyond their spatial extent to provide food, water, energy, and raw materials to sustain the lives of their citizens. They also increasingly depend on this hinterland to absorb their sewage, solid waste, and greenhouse gas (GHG) emissions. Historically this hinterland was predominantly nearby. Increasingly, however, hinterlands are farther and farther away from megacities, and, in regard to GHG emissions, the global atmosphere itself may be considered part of the hinterland. The hinterland of Los Angeles is global, as the city receives oil from the Middle East, water from the eastern Sierra Nevada, and food from Mexico, Europe, and Asia. Mexico City has built vast tunnels to divert sewage to distant hinterlands. Denver, Colorado, uses a network of tunnels to divert water from the western slope of the Rocky Mountains that would normally flow into Mexico's Sea of Cortez.

Almost all of these processes relate to climate change forcing factors in direct or indirect ways. Nonetheless, a fundamental and primary impact of megacites with respect to climate change is the energy used by these cities to provide electricity (often provided by coal-fired power plants) and the energy used to provide transportation (predominantly generated by fossil-fuel combustion). These urban areas are the most densely populated areas of the world. This density is an opportunity for numerous efficiencies with respect to energy consumption for electricity, transportation, and the myriad other related needs of urban residents that require electricity and transportation. Leveraging the energy efficiency opportunities that these densely populated areas represent will be of paramount importance with respect to humanity's collective response to the challenges of climate change.

Significance for Climate Change

Scientists have identified many interconnected causes for the urban heat island effect. During the day, the sun warms buildings and roadways. Pollution from automotive traffic and industrial processes contributes to the formation of clouds and smog, which help trap heat, and tall buildings limit the ability of winds to disperse such formations (the "canyon effect"). The predominance of land structures and the paucity of bodies of water lessen the influence of evaporation, which would use some of the heat energy for the formation of water vapor. Instead, that energy raises the ambient temperature. The concentration of human bodies in cities also contributes to their heat level. The increase in city heat in turn encourages activities, such as the use of air conditioning, that further contribute to localized heat increases, since air conditioners make interiors cooler by making the exterior city warmer. Moreover, increased power plant emissions resulting from the additional consumption of electricity contribute to global warming.

To the extent that urban areas encourage structures or behaviors that stimulate additional activities, increasing emission of greenhouse gases (GHGs), urban heat islands may be viewed as contributing indirectly to global warming. (If cities are hotter than rural areas, then those who move to cities will be more likely to use energy to cool themselves than they otherwise would be.) Since 2004, according to a 2014 article in *Scientific American*, cities have more searing hot days each year. In fact, twelve U.S. cities averaged at least twenty more days a year above 90 degrees than nearby rural areas. In two-thirds of sixty cities analyzed, urbanization and climate change appear to be combining to increase summer heat faster than climate change alone is raising regional temperatures. By 2030, about 87 percent of North American's population is expected to live in urban areas, according to 2010 data from the World Health Organization. As large metropolitan areas fuse into megalopolises, or megacities, the problems only become exacerbated. Accordingly, students of the urban heat island effect have seen the need for various degrees of urban redesign as a key element in attempts to mitigate the noxious consequences of the effect. Chief among these is increase in, and optimal distribution of, urban vegetation and green space, including the planting and sustaining of suitable trees.

The urban heat island effect must be taken into account in any attempt to read the historical record of Earth's temperature. To compare today's temperature with that of a century ago requires comparability, and that can be affected by urban growth. Places where temperatures were measured in the past often have since experienced significant urban growth, with the associated increased temperature measurements. This growth makes direct comparisons of present and past measurements difficult. In light of this problem, climatologists must adjust the data on the basis of their best guess about distorting factors.

—Rebecca S. Carrasco
—Paul C. Sutton

FURTHER READING

Gartland, Lisa. *Heat Islands: Understanding and Mitigating Heat in Urban Areas.* Earthscan/James and James, 2008.

Heikkinen, Niina. "How People Make Summer Hotter." *Scientific American,* 25 Nov. 2014. Web. 24 Mar. 2015.

Kenward, Alyson, Dan Yawitz, Todd Sanford, and Regina Wang. "Heat Islands Cook U.S. Cities Faster Than Ever." *Scientific American,* 22 Aug. 2014. Web. 24 Mar. 2015.

Maslin, Mark. *Global Warming: A Very Short Introduction.* Oxford University Press, 2009.

Spencer, Roy W. *Climate Confusion: How Global Warming Hysteria Leads to Bad Science, Pandering Politicians, and Misguided Policies That Hurt the Poor.* Encounter Books, 2008.

V

VOLCANOES AND CLIMATE

FIELDS OF STUDY

Physical Geography; Earth System Modeling; Environmental Sciences; Environmental Studies; Meteorology; Climate Modeling; Fluid Dynamics; Atmospheric Science; Oceanography; Heat Transfer; Tree Ring Analysis; Spectroscopy

SUMMARY

Volcanic activity has played a major geological and environmental role in the evolution of the Earth since the planet's formation. Studies of volcanic activity confirm the existence of a significant relationship between the effects of an eruption and long-term

Volcanoes:
Inner Earth Breaking Through

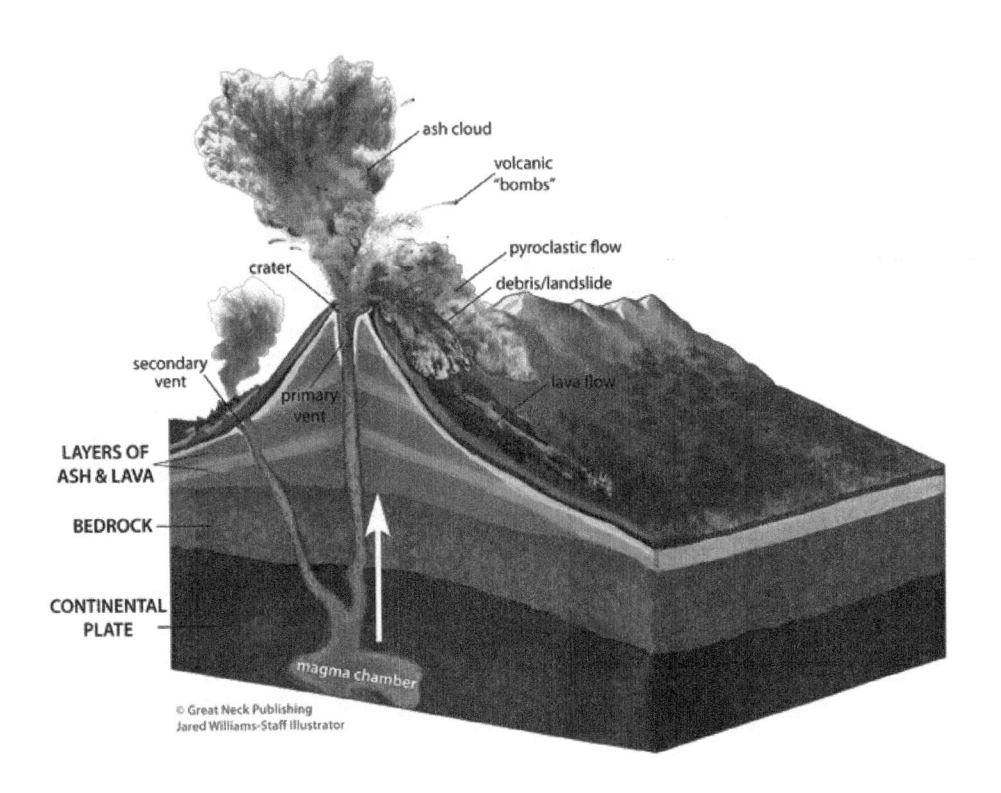

Volcanoes: Volcanoes erupt and explode spewing liquid rock or ash. Igneous rocks are formed from lava that cools above ground and from magma which cools slowly below the ground.

Composition of Air and of Volcanic Gases

Gas	Percent by Volume Air	Volcanic* Gas
N_2 (nitrogen)	77	5.45
O_2 (oxygen)	21	
H_2O (water vapor)	0.1 to 2.8	70.8
Ar (argon)	0.93	0.18
CO_2 (carbon dioxide)	0.033	14.07
Ne (neon)	0.0018	
CH_4 (methane)	0.00015	
NH_3 (ammonia)	0.000001	
SO_2 (sulfur dioxide)		6.4
SO_3 (sulfur trioxide)		1.92
CO (carbon monoxide)		0.4
H (hydrogen)		0.33

*Kilauea volcano, Hawaii.

climatic conditions. During such eruptions, huge amounts of gases and dust are expelled into the atmosphere. Water vapor accounts for over 90 percent of the total expelled gas, with the remainder being a mixture of carbon dioxide, sulfur dioxide, hydrogen sulfide, hydrogen, and fluorine. On a global scale, volcanic ash has its greatest effect on the upper atmosphere, where it prevents transmitted sunlight from reaching Earth's surface and contributes to a reduction in global temperatures, while at the same time acting to retain heat within the atmosphere.

PRINCIPAL TERMS

- **aerosol:** an aggregate of dispersed gas particles suspended in the atmosphere for varying periods of time because of their small size
- **atmosphere:** the thin layer of nitrogen, oxygen, and other gases surrounding Earth, whose density decreases rapidly with height
- **climate:** the sum total of the weather elements that characterize the average condition of the atmosphere over a long period of time for any one region
- **greenhouse effect:** the retention of solar heat in the lower atmosphere caused by the absorption and reradiation of infrared energy from the surface by various gases, creating an insulating effect similar to a greenhouse
- **ozone layer:** a region of the stratosphere, around 60 kilometers in altitude, containing ozone that absorbs ultraviolet radiation from the sun
- **stratosphere:** the atmospheric layer above the troposphere, characterized by little or no temperature change with altitude
- **sulfur dioxide:** SO_2, a colorless, nonflammable, suffocating gas formed when sulfur is oxidized
- **tropopause:** the transition zone at the top of the troposphere between the troposphere and the stratosphere
- **troposphere:** the lowest layer of atmosphere where temperature generally declines with altitude, containing about 95 percent of the mass of the atmosphere and the site of most atmospheric turbulence and weather features

ATMOSPHERIC ENVIRONMENT

Scientists have speculated for centuries on the impact of volcanic eruptions on the climate, and many in the past have noted the possible link between an eruption and climatic conditions. Benjamin Franklin (1706–1790) suggested that the blue haze that lingered over the city of Paris during his stay there was caused by a volcanic event. He believed the "fog," as he termed it, caused a decline in the temperature by absorbing portions of sunlight that otherwise would have reached the ground surface.

Technological advances such as satellite imagery have made it possible for scientists to pinpoint more precisely which types of matter injected into the atmosphere by a volcano have the greatest effect. Initially, researchers believed it was the amount of gas and ash hurled into the atmosphere that determined the impact. However, scientists have come to learn that it is the types of gases that are the critical determining factor. The molten rock of a volcano contains a variety of gases that are released into the atmosphere before, during, and after an eruption. They range from water vapor to dense clouds of sulfur. Other gases include carbon dioxide, sulfur dioxide, hydrogen sulfide, hydrogen, fluorine, chlorine, carbon monoxide, and hydrochloric acid, along with smaller amounts of other chemical compounds.

The aerosols emitted from volcanoes that could have influences on weather are very complex. Aerosols are not simply the dust of crushed rock; they are composed of various chemicals and also maintain various shapes and sizes. It is known that volcanoes are a source of sulfate aerosols. These are sulfur-based particles that have highly reflective surfaces that have the capability of scattering sunlight back into space. Additionally, they can act as condensation nuclei for the formation of clouds and affect the structure of clouds, adding to their reflectivity and their ability to block sunlight. The volcanic dust released into the lower atmosphere can create a temporary cooling effect by producing an ash cloud that blocks the sunlight. However, the ash particles are quickly washed out by the abundant supply of water and rain contained in the lower atmosphere. The majority of ash clouds stop rising at the troposphere, where they encounter increased air temperatures. Dust particles thrown into the stratosphere can linger for several weeks or months before they settle back to the surface.

Exposure to the acid gases such as sulfur dioxide, hydrogen sulfide, and hydrochloric acid can be life threatening, while exposure to fluorine is particularly dangerous. One of the most serious hazards posed by a volcano is the large amount of carbon dioxide it can emit during an eruption. Since it is heavier than air, carbon dioxide tends to settle in lower elevations.

As a result, numerous people residing near erupting volcanoes have been asphyxiated. When a volcano erupts, its immediate effect on the local area is evident, as it spews a great amount of ash, volcanic gases, and heat into the atmosphere. Violent eruptions usually are accompanied by thunderstorms, lightning, and torrential rains. In some cases, the intense heat can generate powerful whirlwinds, strong enough to topple nearby trees and destroy wildlife. Volcanic plumes also contain droplets of water in which acid gases have dissolved. These droplets eventually fall to the ground in the form of acid rain, which can have a corrosive effect on a variety of metal objects and other materials. Exposure to significant amounts of volcanic gases also can have a lethal effect on most varieties of vegetation.

For many years scientists believed that volcanic ash was the most important factor in climatic changes. They since have discovered that of all the ingredients tossed into the air, it is the volcanic gases that exert the greatest influence on climatic conditions. Specifically, the key factor is the conversion of the gases into sulfuric acid aerosols that are capable of remaining in the atmosphere for months. The droplets, though tiny in size, are able to reflect significant amounts of sunlight and have been detected at altitudes as high as 25 kilometers. They not only are able to cool the troposphere by reflecting solar rays back into space, but also can warm the stratosphere by absorbing infrared radiation. The reflective effect is particularly strong when it occurs in normally cloudless areas. It has been estimated that the aerosols can increase the planet's reflectivity, or albedo, by nearly 20 percent. Although aerosols eventually grow large enough to fall back to the surface from their own weight, the process can take years because of the rarefied and dry conditions in the stratosphere, which slows their growth. Aerosol cloud eruptions occur on an average of once every ten to twenty years, with major eruptions taking place at a rate of nearly one every one hundred years.

Only those volcanic eruptions that emit significant amounts of sulfur compounds have an effect on the global climate. Smaller eruptions can create atmospheric effects similar to the larger explosions if they release large concentrations of sulfur-rich elements directly into the stratosphere. Also, eruptions occurring at high latitudes have less impact on climatic conditions than those at lower latitudes, where air currents are greater. The aerosols travel more quickly around the globe when moving in an easterly or westerly direction. If they are traveling in a north-south direction, the movement is much slower, with many of the aerosols becoming confined for years in a zone surrounding the polar cap.

It has been suggested that volcanoes are the second-largest contributor to atmospheric chlorine, next to its mixing into the air from the oceans. During noneruptive stages as well as during violent eruptions, volcanoes emit chlorine gas. Volcanoes also inject this chlorine high into the stratosphere, where it has the potential to interact with natural ozone as well as to add more aerosols into the atmosphere. It has been found, however, that volcanic gases can be infused high into the stratosphere without violent eruptions.

MEASUREMENT METHODS

There are two primary indices that are used to measure the probable impact of volcanic activity. The first is the Dust Veil Index (DVI), which is based on an estimate of the volume of material injected into the atmosphere, surface temperatures, and the amount of sunlight reaching the surface. This index is derived from observations made at midlatitudes and thus is not entirely representative of the global scenario. The second index is the Volcanic Explosivity Index (VEI), which is based on the magnitude, intensity, dispersion, and destructiveness of an eruption. Other measurements of impact include tree-ring records, ice-core readings, and solar radiation measurements.

In addition to the impact indices, four primary methods are used to determine the volume and composition of gases that are emitted during large volcanic eruptions. The first is an examination of the composition and amount of aerosol layers in polar ice cores that are identified as being related to volcanic eruptions. Second is a comparison of gases from eruptions and glass inclusions in crystals that had formed in the volcanic rock prior to the event. The third is measurements of eruptions from satellite imagery, and the fourth is measurements of volcanic aerosols from the surface. Scientists have had some success with using a laser instrument that sends out a pulsing light beam, which reflects back when it detects aerosols. This method enables researchers to construct a profile indicating the density and height of the aerosols.

Ice core readings are especially beneficial in providing a clear record of older eruptions. Atmospheric aerosols from nearly every historic event have shown up in deep drillings of the polar ice. Since they tend to accumulate in layers, the ice cores offer clear annual records of climate and weather. As the aerosols fall to the ground surface over the poles, they begin to soil the surface snow with acid fallout in a process called sedimentation, by which thin layers of debris are formed. Over time, the snow becomes compacted into glacial ice and can be detected through electrical conductivity measurements. Some ice core discoveries have provided records going back more than 100,000 years.

Satellite measurements have enabled scientists from the U.S. Geological Survey to determine the amounts and compositions of gases emitted by several active volcanoes in the United States. Satellite sensors detected up to 1 million metric tons of sulfur dioxide tossed into the stratosphere during the main eruption of Mount St. Helens in 1980. Of particular benefit are the satellite observations conducted by the National Aeronautical and Space Administration's (NASA) Total Ozone Mapping Spectrometer (TOMS) instrument, which have helped to measure sulfur dioxide levels in the atmosphere following major events. TOMS was instrumental in tracking the band of sulfur dioxide across the Pacific produced by the eruption of Mount Pinatubo in the Philippines in 1991. Altogether, TOMS has made more than one hundred observations of volcanic events, including a major eruption of Chile's Cerro Hudson volcano in 1991. These measurements allow scientists to compare volcanic emissions of sulfur dioxide with injections of the gas from industrial plants and other human-based activities. Through comparative studies of volcanic activity, researchers are able to examine the effects of past and future eruptions with the aim of determining whether human or natural activities ultimately pose the greater threat to the environment.

HISTORICAL RECORD

Though less documented than modern eruptions, the larger historic events offer substantial proof of volcanic effects on climate for the simple reason they were bigger and left a more easily detectable trail of evidence. One reason for the occurrence of mass extinctions is a dramatic change in the Earth's surface conditions. Most species live in what can be called a "habitable zone" based upon a suitable range in temperature, the availability of sufficient water, and the right amount of sunlight. When a particular species is exposed to conditions outside of this zone, it either adapts to the changing conditions or dies. As compared to life in the oceans, life on land seems to be more fragile and more susceptible to change. Recent theories for the cause of mass extinctions have concentrated on cosmic impacts or extensive volcanic activity. In each case, the apparent mechanism can be related to a blockage of sunlight by the huge amounts of dust and debris that an impact or volcano would eject into the upper atmosphere.

By drilling into the sea floor south of Haiti and uncovering evidence of ash, researchers concluded that massive volcanic eruptions in the Caribbean Basin more than 55 million years ago created a sudden temperature inversion in the ocean waters that led to one of the most dramatic climatic changes in history. Scientists discovered distinctly colored volcanic ash layers that were far different from the sediments located above and below them. The period when the ash layers were deposited corresponds with a time of rapid warming globally. The presence of the ash indicates that a gigantic eruption took place just as the warming began. Scientists believe that the dust and gases from the eruptions initially cooled the atmosphere, increasing the density of seawater to the point where it sank into the deep ocean. The descending water, in turn, warmed the ocean floor and melted deposits of methane sediments, which then bubbled up into the air, creating a greenhouse effect that warmed the world. Evidence also indicates that the process resulted in the extinction of nearly one-half of all deep-sea animals, victims of asphyxiation because of the lower solubility of oxygen in the suddenly warmer waters. Conversely, the evolution of new plant and animal species, including many primates and carnivores, was accelerated. Scientists already were aware that volcanic eruptions had occurred in the North Atlantic Ocean nearly 61 million years ago and believe that the Caribbean Basin event somehow may have acted in connection with these earlier eruptions.

In another major undersea discovery, scientists suspect that the islands of Tonga and Epi, located about 1,930 kilometers east of Australia, were the products of a massive eruption that took place around the year 1453. During the course of their research, scientists found that the entire stretch of sea floor separating the two islands was a crater more than 11 kilometers in width. They also uncovered charred vegetation that was carbon-dated between 1420 and 1475. To further narrow down the date of the eruption to 1453, they analyzed ice cores from Greenland and Antarctica; tree-ring records from California, Europe, and China; and reports of worldwide crop conditions during the period.

In 1815, the volcano Tambora erupted in Indonesia, precipitating one of the clearest examples of an eruption-induced global cooling event. The volcano emitted a massive column of solid material into the upper atmosphere. The aerosol veil extended to both hemispheres with effects that lasted well into the following year. The aftereffects were such that the year 1816 came to be known as "the year without a summer." In some regions of New England, up to 15 centimeters of snow fell in the month of June. There were numerous other reports of abnormally cool weather, including record low temperatures that forced people to wear coats and gloves in July. The average temperature in the Northern Hemisphere was reduced by as much as 0.5°C, and parts of the United States and Canada experienced unusual summer frosts and crop failure. In Europe, the unusually cold readings resulted in widespread famine, though at the time the connection with the volcanic veil went unrecognized. Researchers identified the Tambora eruption from evidence uncovered in ice cores in Antarctica and Greenland. The event coincides with the fact that the decade between 1810 and 1820 is considered perhaps the coldest on record.

The famous 1883 eruption of Krakatau (or Krakatoa) in Indonesia marked perhaps the first time researchers became fully involved on a worldwide basis with a volcanically induced atmospheric event. Much of the global interest could be attributed to the advances in telegraphic communication, which enabled scientists to share their observations of the spectacular sunsets and other visible phenomena arising from the eruption. Measurements indicated that Krakatau generated a cloud of approximately 21 cubic kilometers of matter. Witnesses in the area recalled dramatic displays of lightning in the cloud veil and a strong odor of sulfur in the air. Researchers believe that the ash cast into the upper atmosphere by the eruption and the ensuing dust veil led to worldwide decreases in incoming solar radiation. Mean annual global temperatures fell close to 0.5°C in 1884, with the cooling period extending through the remainder of the 1880s. In 1884, there was a marked increase in the number of storms in the United States. Record snowfalls, an unusually high number of tornadoes, torrential rains, and severe flooding caused widespread damage. The abnormal atmospheric conditions attributed to Krakatau included brilliant sunsets and a blue or green tinge to the sun, depending on which part of the globe the observation was made. In 1888, the Royal Society of London published a volume that documented the eruption and formally established the connection between major volcanic activity and subsequent changes in worldwide atmospheric conditions.

By historic standards, the eruption of Mount Pinatubo may appear insignificant, but it stands as a watershed event in the ability of scientists to monitor the interaction of a volcanic cloud and the upper atmosphere. The eruption is believed to have sent nearly 20 billion kilograms of sulfur dioxide about 20 to 27 kilometers into the atmosphere, resulting in 30 billion kilograms of sulfuric acid aerosols. For several months the TOMS instrument tracked the sulfur cloud created by Pinatubo with images verifying that its particles circled the globe in about three weeks, forming an almost continuous band.

The Pinatubo eruption had a gigantic atmospheric impact. The year following its eruption turned out to be one of the coldest on record. Temperature measurements in the lower and middle atmospheres indicated a change of nearly 0.5°C between 1991 and 1992. By 1994, readings revealed that the volcano's effect had waned and that global temperatures had returned to previous levels. Scientific studies of more recent volcanic eruptions, such as those of Mount St. Helens (1980), El Chichon (1982), and Mount Pinatubo (1991), have confirmed how much of an effect a single volcanic eruption can have on global weather conditions. One can only imagine the devastating effects that another giant volcanic eruption such as the one that created the Yellowstone Basin in Wyoming (48 kilometers in diameter) would have on the world.

Significance

Much has been learned and much remains a mystery concerning the chemical and physical processes that occur between a volcanic eruption and climatic change. It is an area of intense study because of the belief among scientists that the balance of the global climate is dependent on the phenomenon of volcanism. The atmosphere essentially was developed through intermittent volcanic emissions of carbon dioxide and water vapor, along with nitrogen and possibly methane. There is no reason to believe that the relationship between the two forces has changed in any dramatic way.

Volcanic eruptions can be an agent for global cooling or global warming, depending on their interactions with other environmental elements. The historical record indicates that when the atmospheric balance is threatened by natural or human forces, the consequences can be severe. To most people, fractions of degrees may not appear significant, but in the grand scheme of the environment, they can produce dramatic effects. During the ice ages, the global temperature was only about 5°C cooler than it was at the close of the twentieth century. Scientists believe that a global rise in temperature of as little as 3°C could bring about dramatic changes, including accelerated glacial melting, rising sea levels, more frequent and more severe storms, and droughts. The temperature increase during the twentieth century is considered by many as evidence that the human production of greenhouse gases such as carbon dioxide is affecting the climate. However, it also is believed that multiple eruptions of large volcanoes over a long period of time can raise the carbon dioxide levels enough to cause substantial global warming. To add to the equation, studies also indicate a possible association between other volcanic vapors and the depletion of the ozone layer. A few researchers have even advanced the idea that there is a link between volcanism and El Niño, the periodic warm ocean conditions that appear along the tropical west coast of South America. A number of eruptions have preceded El Niño in years past, leading to speculation that volcanic eruptions and gases may trigger or strengthen the phenomenon.

Major events such as the Caribbean Basin volcanic eruption pose a special problem for climatic equilibrium. The sudden warming of the deep ocean resulting from a series of large volcanic eruptions is a scenario that scientists believe could occur again, causing a major disruption of atmospheric circulation. In attempting to trace the connections between volcanic activities and climate, scientists have begun to think in global terms and to look upon their task as an interdisciplinary effort. In so doing, they have been able to make significant strides in developing the depth of understanding necessary to form reasonably accurate forecasts of future catastrophic events.

—William Hoffman
—1. Marian Mustoe
—Paul P. Sipiera

FURTHER READING

Bourseiller, Philippe, and Jacques Duriex. *Volcanoes*. Harry Abrams, 2008.

Christopherson, Robert W., and Ginger H Birkeland. *Geosystems: An Introduction to Physical Geography*. Pearson, 2018. A highly readable and well-illustrated book that delivers a thorough overview of the various systems that interact to make up the physical environment. Closely discusses the effects of volcanism on climate.

de Boer, Jella Zeilinga, and Donald Theodore Sanders. *Volcanoes in Human History: The Far-Reaching Effects of Major Eruptions*. Princeton University Press, 2002. Provides an overview of volcanism followed by specific volcanoes and eruptions. Includes the interaction of human history and volcanic activity. Covers the Hawaiian Islands, Crete, Vesuvius, Iceland, Tambora, Krakatau, Pelée, Tristan da Cunha, and Mount St. Helens.

Decker, Robert W., and Barbara Decker. *Volcanoes*. 4th ed. Freeman, 2005. An introductory work on the study of volcanoes that contains a concise chapter, with clear illustrations, on the connection between volcanic eruptions and climate.

Fisher, Richard V., Grant Heiken, and Jeffrey B. Hulen. *Volcanoes: Crucibles of Change*. Princeton University Press, 1997. Covers a long list of volcanoes, from Mount Vesuvius to Mount St. Helens. A segment on the effects of volcanic gases helps explain the complex interactions among gaseous, liquid, and solid volcanic matter, the surrounding atmosphere, and solar radiation.

Francis, Peter, and Clive Oppenheimer. *Volcanoes*. 2nd ed. Oxford University Press, 2004. Covers volcanic activity from lava flows to pyroclastic currents. Provides many examples from Hawaii, Italy, Ethiopia, Japan, Mount St. Helens, and Krakatau.

Written for the layperson, but some knowledge of geology is helpful.

Kuhn, Gerald G. "The Impact of Volcanic Eruptions on Worldwide Weather." *Twenty-first Century Science and Technology* (Winter, 1997–1998): 48–58. A well-documented article that provides a comprehensive review of the evidence supporting a link between volcanic activity and climatic changes. Gives special attention to a wide range of weather-related abnormalities associated with the aftereffects of specific events.

Marti, Joan, and Gerald Ernst. *Volcanoes and the Environment*. Cambridge University Press, 2005. Presents a discussion of the various ways in which volcanism affects terrestrial, aquatic, and atmospheric environments, from both historical and present-day perspectives. Provides information on volcanoes from the physical mechanisms and behaviors to hazards and effects on humans and nature. Written for students and geologists studying environmental science and geology.

Oppenheimer, Clive. *Eruptions That Shook the World*. Cambridge University Press, 2011. Describes the mechanics of volcanoes and different types of eruptions. Also describes the effects of eruptions on the atmosphere, humans, and other organisms. Examines specific volcanoes and provides an appendix of major eruptions.

Savino, John, and Marie Jones. *Supervolcano*. New Page Books, 2007. Discusses the classification of supervolcanoes and their eruptions. Describes the effects of volcanic eruptions on climate and the ecosystem. Includes a glossary, bibliography, further resources, and indexing.

Siebert, Lee, Tom Simkin, and Paul Kimberly. *Volcanoes of the World*. 3rd ed. University of California Press, 2010. Reviewed as "the most comprehensive source on dynamic volcanism," presenting chronological, statistical, environmental, and historical information about volcanism over the past ten thousand years.

Winchester, Simon. *Krakatoa: The Day the World Exploded, August 27, 1883*. Harper-Collins Publishers, 2003. An excellent and entertaining description and placement of the violent eruption of Krakatoa within the context and theory of plate tectonics as a planetary process. Well researched. Includes historical observation of atmospheric and climatic events.

WEATHER FORECASTING: NUMERICAL WEATHER PREDICTION

FIELDS OF STUDY

Climatology; Meteorology; Climate Modeling; Computer Science; Software Engineering; Hydrometeorology; Hydroclimatology; Oceanography; Atmospheric Science; Environmental Sciences; Environmental Studies; Thermodynamics; Fluid Flow; Heat Transfer; Physical Sciences; Physical Geography; Ecosystem Management; Ecology; Astronomy

SUMMARY

Most industrialized nations use numerical weather prediction (NWP) techniques to formulate weather forecasts. NWP is based on physical laws that are incorporated into a mathematical model, with solutions determined using computer algorithms. Large amounts of data are used to initialize the numerical models and verification of short-range NWP models shows a considerable improvement over climatology. NWP models require large mainframe supercomputers to formulate each forecast.

PRINCIPAL TERMS

- **climatology:** the scientific study of climate that depends on the statistical database of weather observed over a period of twenty or more years for a specific location
- **ensemble weather forecasting:** repeated use of a single model, run many times using slightly different initial data; the results of the model runs are pooled to create a single "ensemble" weather forecast
- **forecast verification:** comparison of predicted weather to observed weather conditions to assess forecasting accuracy and reliability
- **global atmospheric model:** computational model of global weather patterns based on a spherical coordinate system representing the entire planet
- **hemispheric model:** a numerical model that extends over the whole Northern or Southern Hemisphere, or just one half of the planet
- **long-range prediction:** a weather forecast for a specific region for a period greater than one week in advance, often supplemented with climatological information
- **mesoscale model:** a weather forecast for an area of up to several hundred square kilometers in extent on a time scale of between one and twelve hours
- **nondeterminism:** chaotic, random events that cannot be predicted but that have a significant influence on the development of weather systems
- **nowcasting:** a very short-term weather forecast usually for the prediction of rapidly changing, severe weather events within a time of no more than a few hours

HISTORICAL DEVELOPMENT

The atmosphere is a very complex, chaotic natural system. Before the science of meteorology was developed, keen observers of the natural environment developed forecasting rules epitomized by such sayings as "Red sky at morning/ Sailors take warning/ Red sky at night/ Sailors delight." In the early nineteenth century, a generally acceptable systematic classification of clouds was introduced by Luke Howard (1772–1864), known as the father of British meteorology, and scientists began making routine daily weather observations at major cities and universities. A systematic study of the atmosphere as a chemical and physical entity began.

In 1904, Vilhelm Bjerknes (1862–1951) brought the scientific community to understand that atmospheric motions were largely governed by the first law of thermodynamics, Newton's second law of motion, conservation of mass (the "continuity equation"), the equation of state, and conservation of water in all its

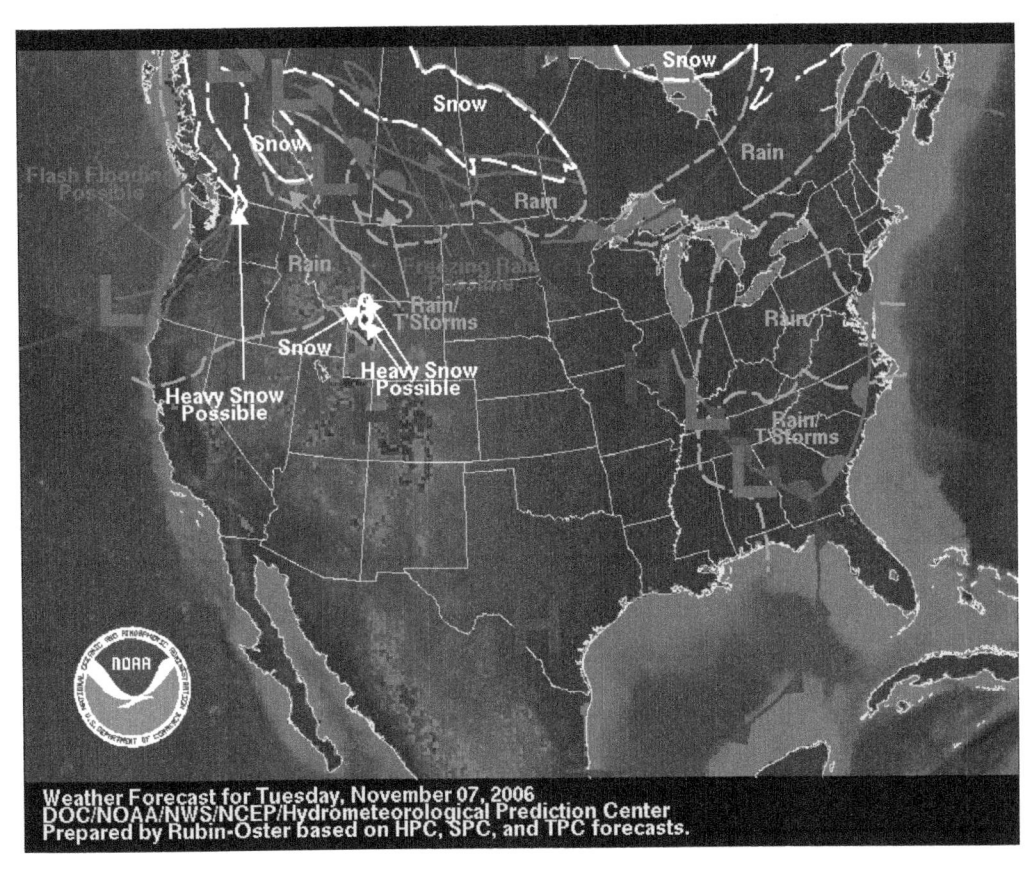

NOAA continental US weather forecast map for Election Day 2012-18-8.

forms. (These physical laws form what is now known as the "governing equations" for numerical weather prediction.) Bjerknes, writing that the fundamental governing equations constituted a determinate, nonlinear system, realized that the system had no analytic solution. He also recognized that available data to determine initial conditions were inadequate. In 1906, Bjerknes devised graphical methods to use in atmospheric physics. During the next decade, he adopted an approach to apply physics in a qualitative, as opposed to a numerical, technique to weather forecasting.

In 1916, Bjerknes began working at the Bergen Museum in Norway, a move that would be decisive in establishing meteorology as an applied science. In 1918, Bjerknes's son Jakob (1897–1975) noticed distinctive features on weather maps that led him to publish an essay entitled "On the Structure of Moving Cyclones." This was followed by the development of the concept of "fronts" as a forecasting tool at the Bergen School in 1919. Following World War I, Norwegian and Swedish meteorologists educated at the Bergen School began using the theory of fronts operationally. Their ability to correctly predict severe weather events led other Europeans to adopt this frontal forecasting method. Enthusiasm for this empirical method blossomed and remains strong among the general public today.

Meanwhile, attempts to develop numerical weather prediction techniques languished. A method for numerically integrating the governing equations was published by L. F. Richardson (1881–1953) in 1921. It contained several problems and errors. Following World War II, the development of mainframe computers allowed development of objective forecasting models incorporating many meteorological variables. By 1950, the first scientific results of a computer model, based on numerical integration of barotropic vorticity equations, were published. In the following decades, computer

499

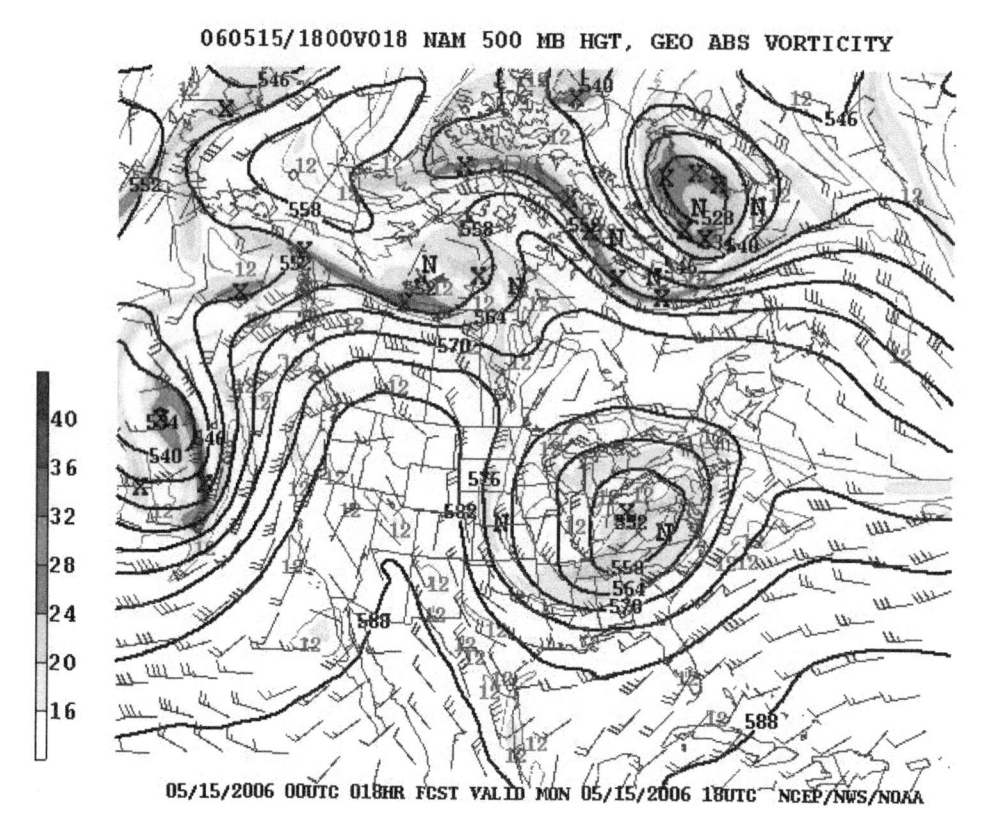

500 mb geopotential height forecast by the United States numerical weather prediction model NAM. Also an example of an Omega Block.

models grew increasingly sophisticated through the incorporation of larger and larger numbers of operations, iterations, and data input. Computer speeds increased dramatically, allowing use of sophisticated equations modeling atmospheric behavior. Larger models with relatively fine grids were developed.

USE OF SUPERCOMPUTERS

Modern weather forecasting attempts to model processes in the atmosphere by representing the appropriate classical laws of physics in mathematical terms. Models use calculations of assemblages of approximations of physical equations (algorithms). Large (mainframe) supercomputers generate numerical forecasts using these algorithms and an array of observations. Most models have at least fifteen vertical levels and a grid with a mesh size less than 200 kilometers. The exact formulation of any particular model depends upon the amount of data being input, how long in advance, how detailed, and which variables are being forecast.

Historically, weather observations for numerical forecasts were taken mainly at 0000 Coordinated Universal Time (UTC), and 1200 UTC using ground stations and radiosondes (weather balloons). Now these observations are supplemented by asynchronous observations from infrared and microwave detectors on satellites, radar, pilot reports, and automatic weather observation stations (including ocean buoys). These new data sources have substantially enlarged the amount of input data. However, because these data are fed into the models as they run, the distinction between input data and model predictions has become blurred. Thus, a weather map showing a numerically generated forecast (as seen on the Internet or in a television presentation) may contain information from many data sources of varying accuracy taken at various times and locations.

Data from all available observations are interpolated to fit the model grid size. Every forecast incorporates inherent errors from the observations, data interpolations, model approximations, the instabilities of the mathematics, and the limitations of the computers used. Models can be separated by their time frame into nowcasts, short-term forecasts, medium-range forecasts, long-term forecasts, and outlooks. In addition, specialized forecast models for hurricanes and other specific types of severe weather have been developed. Because of hazards associated with severe, rapidly developing weather threats, there is increasing interest in developing specialized nowcast and short-range mesoscale (local) forecast models.

Many countries staff their own meteorological offices, using satellite images and local weather observations to develop computer models uniquely suited to their needs. Countries with specific needs and adequate resources develop models to suit their own unique needs. Japan, with its high population concentrations exposed to tropical storm threats, has directed its meteorological efforts toward mesoscale forecasting. The entire country was covered by a network of weather radars by 1971, and Japan supplements its ground-based observations with geostationary weather satellites. This maximizes the Japanese ability to predict heavy rains from tropical typhoons and to issue flash flood forecasts.

In 1999, the United States National Weather Service (NWS) completed a twelve-year modernization program that included 311 automatic weather observing systems and 120 new Nexrad Doppler radars. Observed weather and forecasts were displayed on state-of-the-art computer systems at NWS forecasting offices.

In the mid-1970s, the European Centre for Medium Range Weather Forecasting (ECMWF) and the U.S. National Weather Service began utilizing high-speed supercomputers to generate and solve numerical weather models. Both of these entities have developed global atmospheric models for medium-range forecasts, producing a three- to six-day forecast. The numerical methods used in these global models involve a spectral transform method. One reason for using this method is that models frequently experience numerical problems for computations in polar latitudes. The ECMWF in 1999 was using a model with grid points spaced every 60 kilometers around the globe at thirty-one levels in the vertical. Its initial conditions used observations over the previous twenty-four hours and added in several early model runs forecasting about twenty minutes ahead to augment the observations. Wind, temperature, and humidity were then forecast at 4,154,868 points throughout the atmosphere. Although these models do very well compared with earlier, more primitive models, there is considerable room for improvement in forecasting ability, especially in the tropical latitudes. Medium-range forecast models continue to be improved periodically.

FUTURE OF NUMERICAL WEATHER PREDICTION

The greatest advancements in numerical weather prediction may arise through continued improvement in quality and quantity of weather data used as input in the models, through better methods of data interpolation, through improved algorithms of atmospheric behavior, and in the use of increasingly sophisticated supercomputers capable of more and faster iterations. Some modelers believe that departing from a latitude-longitude grid would prevent clustering of grid points near the poles, which now induces some problems in numerical computations.

Nowcasting, especially for rapidly developing severe weather conditions including severe thunderstorms, flash floods, and tornadoes, has received a great deal of attention by scientists and governments. Within the United States, it is hoped that the inclusion of the Nexrad Doppler radar observations into operational models will lead to better nowcasting.

Forecasts demonstrate both accuracy and skill. A forecast may be entirely accurate simply because of the normal weather pattern in a location, and not because of the skill of the forecaster. Conversely, a forecaster or a forecasting method demonstrates skill by accurately predicting weather that is out of the expected normal pattern for that location. Forecasts have been improving in a slow, steady fashion since the introduction of supercomputers. However, even for short-range time periods, there is still room for improvement. One method for improving any model's forecasting ability is to make ensemble forecasts by running the same model many times, each time with slightly different initial conditions, and then pooling the results statistically. This technique yields better results than any single model run, based on comparing verifications of the individual and ensemble results for a large number of forecasts. A variation of this technique is the superensemble forecast in which forecasts from several different computer models runs or even ensemble runs are pooled. Superensemble forecasts have been found to show greater accuracy than ensemble forecasts. One way in which superensemble forecasts are thought to achieve this is in the minimization of "forecast bias" in which any one model develops a spurious trend as the increasing numbers of iterations magnify small errors.

Worldwide, better weather satellite imagery for input to more detailed hemispheric models could lead to enhanced quality of numerical weather predictions. Greater availability of weather observations for the world's oceans would also be helpful. While hemispheric models are good at forecasting at midlatitudes, less accuracy and skill are shown in tropical forecasting, as determined by model verification. Improved models incorporating more detailed initial data could provide better forecasts in the tropics.

One of the greatest challenges confronting numerical weather prediction lies in the development of reliable long-range predictive models. Long-range predictions often fail because of nondeterministic or random events. The El Niño/Southern Oscillation is an example. El Niño events are now viewed as virtually nondeterministic events; when a strong El Niño event is occurring, the output from some long-range models may be subjectively altered by meteorologists to reflect past historical events. The goal of numerical weather prediction is to provide objective forecasts using a model that verifies data more frequently than do subjective forecasts, even those formulated by highly trained meteorologists. To forecast El Niño events better, it would be advantageous to develop a predictive model that could successfully incorporate Pacific Ocean surface temperatures as initial conditions. With longer and better data records, and further study of why existing long-range predictions have failed, better models may be developed.

—Dennis G. Baker
—Anita Baker-Blocker

Further Reading

Browning, K. A., and R. J. Gurney. *Global Energy and Water Cycles.* Cambridge University Press, 2007. This excellent book features reviews by Europeans and Americans on how weather forecasting models are developed and used.

Hamilton, Kevin, and Wataru Ohfuchi, eds. *High Resolution Numerical Modeling of the Atmosphere and Ocean.* Springer Science+Business Media, 2008. Presents a technical account of the efforts and progress in accurately modeling the climate and weather patterns of the world by computational means. Includes several papers presented by invited speakers to the first international convention focused on this endeavor.

Hodgson, Michael. *Basic Essentials: Weather Forecasting.* 3rd ed. Globe Pequot, 2007. Describes how people are trained to make forecasts using human observations of the weather around them. Before attempting to use computer forecasts, people should have a basic understanding of the physical principles operating in the atmosphere.

Monmonier, Mark S. *Air Apparent: How Meteorologists Learned to Map, Predict, and Dramatize Weather.* University of Chicago Press, 2016. Using weather maps, this book shows the growth of technology used in weather forecasting.

Santurette, Patrick, and Christo Georgicv. *Weather Analysis and Forecasting.* Academic Press, 2005. Discusses the use of satellite imagery to detect and analyze atmospheric moisture. Identifies factors that indicate severe weather occurrence and discusses the use of water vapor imagery to improve forecasting.

Vasquez, Tim. *Weather Analysis and Forecasting Handbook.* Weather Graphics Technologies, 2011. Discusses the technology, techniques, and physics principles used in modern weather forecasting. Describes thermal structure and dynamics of weather systems. Also covers model use in weather forecasting and weather system visualization. Easily accessible yet still technical, useful to anyone studying weather forecasting and meteorology.

Zdunkowski, Wilfred, and Andreas Bott. *Thermodynamics of the Atmosphere. A Course in Theoretical Meteorology.* Cambridge University Press, 2004. Aimed at graduate students and researchers in meteorology and related sciences. Focuses on the thermodynamics at work in the atmosphere. Agreement with thermodynamic principles is fundamental to the development of numerical methods of weather prediction.

WEATHER MODIFICATION

FIELDS OF STUDY

Atmospheric Science; Atmospheric Chemistry; Environmental Chemistry; Photochemistry; Thermodynamics; Chemical Engineering; Environmental Sciences; Environmental Studies; Climate Modeling; Meteorology; Process Modeling; Fluid Dynamics; Hydroclimatology; Hydrometeorology; Ecosystem management; Earth System Modeling; Heat Transfer

SUMMARY

Human activities can cause intentional or accidental changes in local weather situations. Many intentional weather modification experiments have focused on creating conditions to benefit agriculture.

PRINCIPAL TERMS

- **cloud seeding:** the injection of nucleating particles into clouds to enhance precipitation formation
- **dynamic mode theory:** a theory proposing that enhancement of vertical movement in clouds increases precipitation
- **fog dissipation:** removal of fog by artificial means
- **hail suppression:** a technique aimed at lessening crop damage from hailstorms by converting water droplets to snow to prevent hail formation or, alternatively, by reducing hailstone size
- **hygroscopic particulates:** minute particles that readily take up and retain moisture
- **static mode theory:** a theory assuming that natural clouds are deficient in ice nuclei, whereby clouds must be within a particular temperature range and contain a certain amount of supercooled water for cloud seeding to be successful
- **supercooled:** a liquid cooled below its normal freezing point without crystallizing or becoming solid, typically referring to water

DELIBERATE WEATHER MODIFICATION

Inadvertent weather modification, including increases or decreases in precipitation downwind from large industrial sites and the formation of fog, creates problems in some locales. Scientific attempts to deliberately modify weather activity and conditions have been pursued since World War II. The most popular techniques involve cloud seeding, the injection of cloud-nucleating particles into likely clouds to alter the physics and chemistry of condensation. Proponents of this technique claim that it may enhance precipitation amounts by 5 to 20 percent. However, some scientists believe that deliberate efforts to enhance precipitation often yield questionable results, even in favorable situations. In 1977 the United Nations passed a resolution prohibiting the use of weather modification for hostile purposes because of the threat to civilians. The United States signed the resolution but has continued defense research on operational weather modification in battlefield situations, as summarized in the U.S. Air Force position paper *Weather as a Force Multiplier: Owning the Weather in 2025*.

Studies have field-tested various methods of weather modification; results have varied widely. Weather modification has been attempted in many countries around the world, by government agencies, agricultural cooperatives, private companies, and research consortia. In agricultural areas farmers are convinced that hail suppression and precipitation augmentation have been achieved by weather modification. In some of these same locales, meteorologists have been unable to determine if weather modification has produced any change from what would have occurred without intervention. Attempts to duplicate weather modification efforts that have apparently been successful in one locale have often been met with questionable results. Meteorologists occasionally disagree among themselves as to whether a specific attempt at weather modification has succeeded. Reexamination of data from American studies undertaken in the past has led many scientists to conclude that the efficacy of cloud seeding has been overstated.

It should be clearly understood that it is impossible to change the climate of an entire region at will for a desired outcome through weather modification. It is also impossible to end a drought by seeding clouds. This is a result of long-term dynamic conditions and interactions that are essentially global in nature. Cloud seeding for agricultural purposes assumes that some enhancement of regional rainfall amounts over the course of the growing season will increase crop yields. Weather modification for hail suppression assumes that reduction in regional crop losses over the growing season is an attainable goal.

INADVERTENT WEATHER MODIFICATION

Pulp and paper mills produce huge quantities of large-and giant-diameter cloud condensation nuclei (CCN) in the effluent from their exhaust stacks. Downwind of these mills, precipitation appears to be enhanced about 30 percent above what was observed prior to construction of the mills. It is also thought that the heat and moisture emitted by these mills may play an active role in precipitation enhancement. One specific study of a kraft paper mill near Nelspruit in the eastern Transvaal region of South Africa has indicated that storms modified by the mill emissions lasted longer, grew taller, and rained harder than other nearby storms occurring on the same day. Radar measurements supported the theory that hygroscopic particulates released by this mill accelerated or amplified growth of unusually large-diameter raindrops.

An egregious example of inadvertent weather modification is the formation of ice fog over Arctic cities in Siberia, Alaska, and Canada. During winter, cities such as Irkutsk, Russia, and Fairbanks, Alaska, experience drastic reductions in visibility as particles released by combustion act as nuclei for the formation of minute ice crystals. No techniques are available to modify ice fogs.

During an investigation of the meteorological effects of urban St. Louis, Missouri, conducted during the 1970s, it was found that urban summer precipitation was enhanced by 25 percent relative to the surrounding area. Most of the increased precipitation occurred in the late afternoon and evening as a result of convective activity. The frequency of summer thunderstorms was enhanced by 45 percent, and the frequency of summer hailstorms was higher by 31 percent over the city and adjacent eastern and northeastern suburbs. During the late 1960s, studies demonstrated that widespread burning of sugar cane fields in tropical areas released large numbers of cloud condensation nuclei. Downwind, rainfall decreases of about 25 percent were noted.

CLOUD SEEDING

For millennia, people attempted to influence the weather by using prayers and incantations. Sometimes rain followed, and sometimes no rain fell for extended periods. Scientists began attempting various techniques to modify weather during World War II. In 1946 Vincent Schaefer (1906–1993) of the General Electric Research Laboratory observed that dry ice put into a freezer with supercooled water droplets caused ice crystals to form. On November 13, 1946, Schaefer demonstrated that dry ice pellets dropped from an aircraft into stratus clouds caused liquid water droplets to change to ice crystals and fall as snow. Bernard Vonnegut (1914–1997), a coworker, determined that silver iodide (AgI) particles also caused ice crystals to form. Project Cirrus involved apparently successful scientific attempts to seed clouds with ground-based AgI generators in New Mexico. These researchers then tried seeding a hurricane on October 10, 1947. The hurricane changed direction, making landfall in Georgia, resulting in a number of lawsuits against General Electric.

Early cloud-seeding experiments were empirical. AgI was dropped from aircraft, shot into clouds by rockets, or dispersed from ground-based generators. Researchers could selectively seed a pattern such as an "L" into a supercooled stratus cloud and see a visible "L" appear, thus "proving" that they could achieve results. When any rain occurred near a seeded area, it was attributed to the intervention. The apparent success of cloud seeding using AgI caused the technique to be modified and adopted in France, Canada, Argentina, Israel, and the Soviet Union. Wine-growing regions such as the south of France and Mendoza, Argentina, installed ground-based AgI generators. The former Soviet Union opted for rocket-borne AgI, which was launched in agricultural areas during thunderstorms in an effort to suppress hail.

In 1962 the U.S. Navy and Weather Bureau began an ambitious cooperative plan to modify hurricanes called Project Stormfury. Only a few hurricanes were seeded in attempts to reduce the intensity of the storms. Proponents of Stormfury suggested that seeding of Hurricane Debbie in 1969 caused a reduction of 30 percent in wind speed on one day. The following day, no seeding was done, followed by another seeding attempt. The second seeding was thought to have caused a 15 percent reduction in wind speeds. Proponents believed that 10 to 15 percent reductions in wind speeds might result in a 20 to 60 percent reduction in storm damage if similar results could be achieved by seeding other hurricanes. Stormfury was terminated in the late 1970s, with no definitive results.

During winters between 1960 and 1970, the Climax I and Climax II randomized cloud seeding studies were conducted in the Colorado Rockies. Although it was initially thought that precipitation enhancements on the order of 10 percent may have resulted, more recent examination of the results appears to indicate that cloud seeding had no statistically discernible effect on precipitation. During the Vietnam War, the U.S. military attempted to increase precipitation along the Ho Chi Minh Trail in an effort to impede enemy forces. In the United States during the 1970s, some entrepreneurs deployed ground-based AgI generators in selected agricultural regions, billing farmers for their services. Aircraft delivery of AgI became increasingly popular. By the late 1990s, a number of private companies were delivering airborne cloud seeding services in various areas worldwide.

Cloud physicists have explored why cloud seeding might be effective. The evidence suggests that seeding increases the size of droplets or ice crystals, allowing them to fall as precipitation. Two concepts have emerged: a static mode theory, which assumes that natural clouds are deficient in ice nuclei, and a dynamic mode theory, which assumes that enhancement of vertical movement in clouds increases precipitation. The static mode assumes that a "window of opportunity" exists for seeding cold continental clouds during which clouds must be within a particular temperature range and contain a certain amount of supercooled liquid water.

Fog Dissipation and Hail Suppression

During World War II, when improvements in visibility were crucial for military operations, efforts were made to dissipate fog. Fog may be dissipated by reducing the number of droplets, decreasing the radius of droplets, or both. Decreasing droplet radius by a factor of three through evaporation can provide a nine-fold increase in visibility. Possible methods of fog removal include using dry ice pellets or hygroscopic materials, heating the air, and mixing the foggy air with drier air. Airports that are plagued by supercooled fog in winter, such as Denver and Salt Lake City, can dissipate the fog by dropping dry ice pellets. Dry ice causes some liquid water droplets to freeze and grow, evaporating the remaining liquid droplets and allowing the larger frozen ice crystals to fall. One way of clearing fog at military airports when there is a shallow radiation fog close to the ground is to use helicopters to provide mixing. Entrenched jet engines can also be used to heat the air over runways by directing their hot exhaust gases. This is an expensive technique that has been used operationally in France, but it is also one that poses significant risk to smaller aircraft during landings and take-offs.

Farmers and vintners worldwide fear damaging hailstorms that can devastate crops. There are three approaches to suppressing hail damage: converting all liquid water droplets to snow to prevent hail formation, seeding to promote growth of many small hailstones instead of larger damaging hail, and introducing large condensation nuclei to reduce the average hailstone size. Most weather modification proponents believe that seeding with lead iodide or AgI to cause many small hailstones to form can substantially reduce hailstone size. Because small hailstones are less damaging than large ones, this technique could potentially lessen (but not eliminate) crop losses. It has been claimed that rocket-borne lead iodide seeding in Bulgaria reduced crop losses from hail by 50 to 60 percent. Similar seeding operations in the former Soviet Union were said to have reduced crop damage by 50 to 95 percent. A randomized study in North Dakota over four summers claimed that seeding helped reduce hail severity.

—*Anita Baker-Blocker*

Further Reading

Aguado, Edward, and James E. Burt. *Understanding Weather and Climate.* 7th ed. Pearson Prentice Hall, 2015. Discusses meteorology and climatology concepts with reference to common, everyday events. Presents conclusions from the IPCC as well as many other scientific studies on climate change. Examines weather events, structure and dynamics of atmosphere, and the past, present, and future climate on earth.

Congressional Research Service. *Weather Modification: Programs, Problems, Policy and Potential.* University Press of the Pacific, 2004. A thorough, scholarly report, originally prepared in 1978, that reviews the history, technology, activities, and various special aspects of weather modification of interest to the agricultural, scientific, commercial, and governmental fields.

Cotton, William R., and Roger A. Pielke. *Human Impacts on Weather and Climate*. 2nd ed. University Press, 2007. A comprehensive overview of weather modification, written as a text for both undergraduate and graduate study in atmospheric and environmental science. Looks at weather modification with regard to both intentional and unintentional effects.

Hoffman, Matthew J. *Ozone Depletion and Climate Change: Constructing A Global Response*. State University of New York, 2005. Discusses the challenges of global policies to mitigate climate change. Examines models of climate change due to ozone depletion. Focuses on universal participation and governance of these issues.

House, Tamzy J. *Weather as a Force Multiplier: Owning the Weather in 2025*. Biblioscholar, 2012. Discusses military weather modification.

Lutgens, Frederick K., Edward J. Tarbuck, Tasa, Dennis G. and Herman, Redina. *The Atmosphere: An Introduction to Meteorology*. 14th ed. Pearson Education, 2018. An updated version of this textbook, in loose-leaf format, provides students with a general overview of the structure and properties of Earth's atmosphere.

Marshal, John, and R. Alan Plumb. *Atmosphere, Ocean and Climate Dynamics: An Introductory Text*. Elsevier Academic Press, 2008. An excellent introduction to atmospheres and oceans. Discusses topics such as the greenhouse effect, convection and atmospheric structure, oceanic and atmospheric circulation, and climate change. Suited for advanced undergraduates and graduate students with some background in advanced math.

Smith, Jerry E. *Weather Warfare: The Military's Plan to Draft Mother Nature*. Kempton, Ill.: Adventures Unlimited Press, 2006. Covers processes that cause weather events and natural disasters, and our ability to influence these processes. Discusses cloud seeding, electromagnetic wave production, weather modification legislation, contrails, and stratospheric engineering. Written for the general population and lacking in "hard science," but presents a wide range of topics that provoke further research.

Strahler, Alan H., et al. *Physical Geography: Science and Systems of the Human Environment: Canadian Version*. John Wiley & Sons, 2008. A thorough, well-illustrated book containing considerable information about atmospheric processes and issues. Suitable for college students.

WEATHER VS. CLIMATE

FIELDS OF STUDY

Climatology; Meteorology; Environmental Sciences; Climate Modeling; Fluid Dynamics; Hydroclimatology; Hydrometeorology; Atmospheric Science; Bioclimatology; Climate Classification; Climate Zones; Oceanography; Hydrology; Physical Geography; Earth System Modeling; Ecology; Heat Transfer

SUMMARY

The weather is one of the most obvious and tangible aspects of Earth's atmospheric environment. One can see, feel, hear, smell, and even taste weather. It consists of the measurable meteorological conditions of the atmosphere at any given time. In contrast, climate is a concept that is constructed by the averages of the component elements of the weather: temperature, wind, pressure, precipitation, clouds, and visibility. Complicating the understanding of climate is the fact that the Earth's atmosphere reflects many kinds of climates over many different areas at any given time.

PRINCIPAL TERMS

- **archaeological record:** the human-made objects and residues of the past that are used by archaeologists to reconstruct history
- **desert climate:** a hot climate characterized by extremely low and sometimes deficit amounts of precipitation
- **insolation:** the amount of solar energy received at a given point on the Earth
- **orographic barrier:** a mountain or a hilly area that acts as a control on the climate on either side

Top Weather, Water, and Climate Events of the Twentieth Century

	Year
Top U.S. Events	
Galveston Hurricane	1900
Tri-state Tornado	1925
Great Okeechobee Hurricane/Flood	1928
Dust Bowl	1930's
Florida Keys Hurricane	1935
New England Hurricane	1938
Storm of the Century	1950
Hurricane Camille	1969
Super Tornado Outbreak	1974
New England Blizzard	1978
El Niño episode	1982-1983
Hurricane Andrew	1992
Great Midwest Flood	1993
Superstorm	1993
El Niño episode	1997-1998
Oklahoma/Kansas Tornado Outbreak	1999
Top Global Events	
India Droughts	1900, 1907, 1965-1967
China Droughts	1907, 1928-1930, 1936, 1941-1942
Sahel Droughts, Africa	1910-1914, 1940-1944, 1970-1985
China Typhoons	1912, 1922
Soviet Union Drought	1921-1922
Yangtze River Flood, China	1931
Great Smog of London	1952
Europe Storm Surge	1953
Great Iran Flood	1954
Typhoon Vera, Japan	1958
Bangladesh Cyclone	1970
North Vietnam Flood	1971
Iran Blizzard	1972
El Niño	1982-1983
Typhoon Thelma, Philippines	1991
Bangladesh Cyclone	1991
Hurricane Mitch, Honduras/Nicaragua	1998

Note: Chosen by NOAA scientists from among the world's most notable tornadoes, floods, hurricanes, climate events, and other weather phenomena, taking into account an event's magnitude, meteorological uniqueness, economic impact, and death toll.

Source: National Oceanic and Atmospheric Administration.

- **semiarid climate:** a climate of low annual precipitation (250–500 millimeters)
- **tropical climate:** a climate generally found in equatorial or tropical regions that exhibit high temperatures and precipitation year-round
- **west coast marine climate:** a mild, annually moist climate found along the western coasts of continents

WEATHER

The term meteorology, the study of the weather, originates from the work *Aristotelous peri genese s kai phthoras* (n.d.; *Meteorologica*, 1812), by the Greek philosopher Aristotle (384–322 BCE). Much of what Aristotle wrote was based around his assumptions about how the weather worked. He provided a model

The troposphere is the part of Earth's atmosphere where weather happens.

that was accepted without question for nearly two thousand years. However, with the advent of weather instruments (1600–1850s) such as the thermometer, barometer, and hydrometer, a science of meteorology slowly emerged. The invention of the telegraph in 1844 allowed for the rapid distribution of weather observations accumulated from a large geographical area. It was the telegraph that assisted meteorologists in constructing the first weather maps and charts.

MODERN OBSERVATIONS

Today, meteorologists depend on a wide range of instruments and communication systems to measure and distribute information about the weather. Weather observations at the surface of the Earth include measurements of wind direction and velocity, air temperature, visibility, types of observed clouds, atmospheric pressure, precipitation amount and type, and humidity. Usually, weather stations record surface conditions. However, some stations take

The small white balls on the ground are hailstones.

observations of the upper atmosphere with the aid of a balloon (radiosonde). Many stations use Doppler radar. Hourly weather observations are logged at these stations. These observations are then sent to forecasters, who use them to develop long-term forecasts.

Climate is defined as the average amount of precipitation and average temperature at a given location within a given season. A location's climate is determined by its latitude, altitude, and locational characteristics. The latitude of a station determines the amount of insolation the station will receive at any given time during the year. Higher latitudes experience less insolation than do stations located within the tropics. Altitude is the elevation of a station. Given the environmental lapse rate, or the general decline in temperature with altitude (4°C per 1,000 meters), stations found at higher elevations exhibit generally cooler climates. A station's climate is also controlled by its location with respect to land and water.

Seattle, Washington, although somewhat high in latitude, is also climatically influenced by its orientation with respect to Puget Sound. Its maritime location keeps Seattle's climate from exhibiting extreme temperatures. In contrast, locations such as Wenatchee, Washington, less than 160 kilometers to the east but on the leeward side of the Cascades, is located in a more continental position. Temperatures there are considerably colder in the winter and warmer in the summer. Additionally, the Cascades act as an orographic barrier to precipitation, making Wenatchee's climate a desert.

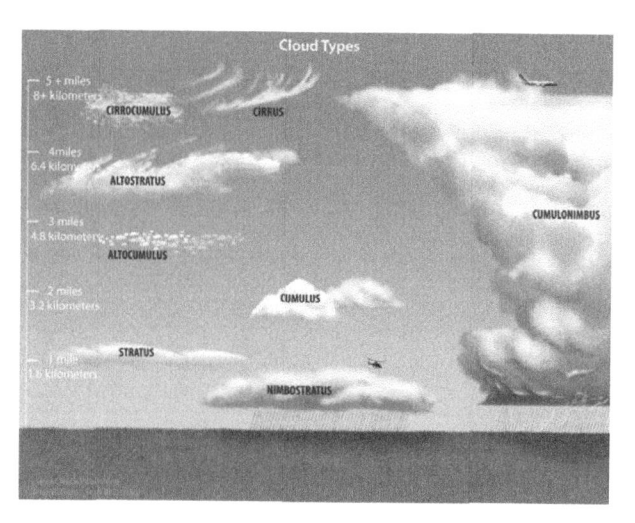

This illustration shows the different types of clouds.

CLIMATE

To understand climate, scientists must have access to weather data over long periods of time. Relatively accurate weather observations have only been available for a minute period of time. However, past climatic conditions can also be studied by indirectly interpreting the signature left by the movement of animals and plants, as well as the archaeological record of human activity. For example, seven-thousand-year-old cave drawings in the Sahara Desert suggest that in the past this desert region retained a climate that was conducive to habitation. From 400 to 1200 CE, historic records suggest, the Vikings exploited the warm, ice-free conditions on their coastlines and ventured out in ships in exploratory enterprises, settling Greenland and Iceland during this time. However, historic records are subject to problems of precision and human interpretation. It would not be until the late 1800s that weather records would be scientifically collected in the United States. It is the reliability of these meteorological data that dictates how accurate scientists' understanding of climate can become.

CLIMATE CLASSIFICATION

Climates on Earth can be classified and defined by their characteristics and their environmental qualities. However, classification systems are limited by the number of variables they utilize. One way to classify climate is to look at the distribution of different types of animals or plants that live within that climate.

Characteristics of Weather Fronts

	Cold Front	Warm Front	Stationary Front	Occluded Front
Cause of the Front	Cold air mass pushing up warm air mass	Warm air mass slides over cold air mass	Warm air mass and cold air mass do not move	Cool or cold air mass changes places on the ground with warm air mass
Type of Precipitation	Rain, snow, sleet, hail	Rain, snow, sleet	none	Rain or snow

For example, the Köppen classification system uses temperature and moisture levels related to plants to define the patterns of climate types. On the other hand, humans are found in every climatic type on Earth.

PRESENT CLIMATE DISTRIBUTIONS

The general categories of climates include: climates of the tropics, equatorial climates, savanna climates in areas such as the African region of the Sudan and Sahel and tropical semiarid climates. Additionally, desert climates can be found anywhere the geographical conditions are right for moisture to be depleted from the air. Monsoonal climates are found in India, Northern Australia, and regions within Southeast Asia. Mediterranean climates are located around the Mediterranean Sea, but they can also be found in Central Chile, California, and South Africa. Humid subtropical climates are found on the eastern margins of continents, in regions such as the southeastern United States, Uruguay, Brazil, Argentina, southern Japan, the Natal of South Africa, Taiwan, and parts of Eastern China.

West coast marine cool temperature climates are found along the west coasts of continents and in the middle latitudes. Seattle is a good example of this climate, which is also found on the Atlantic coasts of France and the United Kingdom. The warm summer continental climates are found in areas such as the central United States, northeastern China, and Korea. Cool summer continental climates are found in the northeastern United States, northern Japan, and Eastern Europe. Steppe climates are found in continental areas of North America, Argentina, and Patagonia. Sub-Arctic climates are known as the Taiga, a Russian term for the coniferous forests found in the areas of Siberia and Canada. Tundra is a cold climate just below the Arctic region. ice cap climates are found at the North and South Poles. Vegetation does not grow, and precipitation is sparse.

CONTEXT

In contrast to weather, climate is an abstract concept made up of the measured averages and extremes of weather. Climate, then, cannot be experienced firsthand. Over the course of a year, one can experience the change of seasons, but only on the basis of what one might expect the weather for those seasons to bring. Complicating this further is the variation in the types of climate that exist and the limitations to the models by which they are defined. The evidence for changes in climate is clearly found in the fossil record, as well as in geologic structures showing the advance and retreat of ice over the land. Additionally, the archaeological record indicates climate change by shifts in cultural activity. Core samples from glaciers can be analyzed to reveal oxygen content and thermal conditions of past atmospheres. Ice from these past periods continues its hold on parts of the globe.

—*M. Marian Mustoe*

FURTHER READING

Aguado, Edward, and James E. Burt. *Understanding Weather and Climate.* 7th ed. Pearson Prentice Hall, 2015. The formation of midlatitude storms is surveyed. Continental climates and the structure of climate are discussed.

Ahrens, C Donald, and Robert Henson. *Meteorology Today: An Introduction to Weather, Climate, and the Environment*. Cengage, 2019. Discusses global climates, climate change, and classification.

Denny, Mark *Making Sense of Weather and Climate. The Science Behind the Forecasts*. Columbia University Press, 2017. An informative yet entertaining book that looks at many readily-visible aspects of climate effects such as fog from a question and answer perspective.

Lutgens, Frederick K., Edward J. Tarbuck, Tasa, Dennis G. and Herman, Redina. *The Atmosphere: An Introduction to Meteorology*. 14th ed. Pearson Education, 2018. An updated version of this textbook, in loose-leaf format, provides students with a general overview of the structure and properties of Earth's atmosphere.

Maslin, Mark. *Global Warming: A Very Short Introduction*. Oxford University Press, 2009. Provides a concise general introduction to the major factors that define the climate of Earth.

Teague, Kevin Anthony and Gallicchio, Nicole *The Evolution of Meteorology. A Look Into the Past, Present and Future of Weather Forecasting*. Wiley-Blackwell, 2017. A comprehensive presentation of weather forecasting that includes descriptions of barometers and other devices in context.

WETLANDS AND SEA-LEVEL RISE

FIELDS OF STUDY

Physical Geography; Environmental Sciences; Environmental Studies; Waste Management; Climate Modeling; Fluid Dynamics; Hydroclimatology; Bioclimatology; Climate Classification; Climate Zones; Oceanography; Hydrology; Ecosystem Management; Ecology

SUMMARY

Wetlands are found from the equatorial region to the arctic regions of Canada. Even in the deserts, wetlands are found near oases. Wetlands are important to society as they provide sanctuary for fish and wildlife, water resources for streams and aquifers, and, in most cases, serve to mitigate the severity of flooding. Wetlands also provide support for fishing, hunting, and other recreational and educational activities such as bird watching, photography, painting, tourism, and research. They provide humans and wildlife with food sources and function as sources of energy and sinks for greenhouse gases (GHGs), sequestering carbon through the formation of peat, a precursor to the formation of coal.

PRINCIPAL TERMS

- **ecotone:** a transition area between open water and dry land ecosystems
- **greenhouse gases (GHGs):** an atmospheric gas whose molecules absorb infrared radiation from a planet's surface and then emit a significant portion of it back into the atmosphere, most commonly carbon dioxide, methane, water and ozone
- **hydrophyte:** a plant that is adapted to aquatic and semi-aquatic environments
- **riparian:** associated with the ecosystem existing along the shores and banks of natural freshwater bodies such as rivers, streams and lakes

THE NATURE OF WETLANDS

As the name implies, wetlands are a watery environment consisting of standing and slow-moving shallow water. As such, wetlands can be harbors of disease-causing insects and may limit some human activities, such as construction and farming. Though wetlands typically mitigate the violence and magnitude of flooding, some wetlands can also contribute to the duration of flooding in some cases, as they reduce the rate at which water removes from the surface. Wetlands are significant sources of methane gas, as the decomposition of plants accounts for about 20 percent of the methane found in wetlands. Wetlands are sometimes referred to as ecotones, transition zones between open bodies of water and land, and sometimes are described as the kidneys of the landscape or as biological supermarkets because of the roles they play.

Wetland.

Because of the varied nature of wetlands, it is often difficult to say what a wetland is and is not. Wetlands have different meanings to different people, depending on their background, exposure, knowledge, and political stand. There is no universal definition of what constitutes a wetland. The main reason wetlands are difficult to define is that they are found in widely varied locations, have different climates, soils, landscapes, water quantity and quality, flora, fauna, and other characteristics, including human disturbances such as dikes, draining, pollution, and so on. The definition of wetland typically depends on the specific purpose for defining it, such as research studies, general habitat classification, natural resource inventories, or environmental regulations.

All wetland definitions fall into two categories: regulatory definitions and nonregulatory definitions. Because there are myriad, variable definitions and changes in definitions, there is some confusion among legislative and regulatory bodies. This confusion has caused several isolated wetlands to be destroyed, as these wetlands were not connected to navigable waters and may not have seemed to qualify for protection.

Three criteria are necessary to define an area as a wetland: hydrology, hydric soil, and hydrophytes (vegetation adapted to wet conditions). Wetlands are areas that are typically wet most of the time or where the groundwater is very close to the surface. Hydrology is the main factor for identifying wetlands, because not all wetlands exhibit hydric soil conditions. Water quantity and quality determine soil characteristics and which plants and wildlife communities can inhabit the area. The soil, plants, and wildlife affect water quantity and quality.

Wetlands can be broadly divided into two major types: coastal wetlands and inland wetlands. Coastal wetlands are found along the oceans and seas, whereas inland wetlands are found along rivers and streams, near lakes, in low-lying land depressions, or where the groundwater meets the surface. Because of these conditions, the major types of wetlands that have been recognized are bogs, swamps, marshes, fens, potholes, and player lakes.

Significance for Climate Change

Any climate change could modify, create, or eliminate several types of wetlands. With an increase in global warming and sea-level rise, most of the coastal wetlands may be altered, destroyed, or eliminated. However, with the demise of coastal wetlands, new wetlands would be created in places such as Canada and Siberia's arctic regions as glaciers melt. Change in the climate would affect the effectiveness of wetlands to remove nutrients from flow-through systems such as lakes or riparian wetlands. Aquatic life forms would be lost as a result of prolonged nutrient loading to such wetlands. Wetland values would be altered, as the hydrology regime was altered in those wetlands. Climate change that would involve a sea-level rise would be detrimental to coastal regions and human life.

Many of the world's largest cities, located near coastal areas, have wetlands buffering the effect of ocean waves or flooding. Any meaningful sea-level rise would cause serious problems to these wetlands, changing the water chemistry and thus the flora and fauna. It would also affect those coastal cities by increasing the risk, incidence, and severity of flooding. Global warming would lead to destruction of some coastal wetlands and the creation of new wetlands, mainly in the Northern Hemisphere. Tourism may suffer in places where tourism helps the economy, such as the Florida Everglades, with shifts in locations of wetlands. Southeast Asian wetlands may be decimated with a rise in sea level resulting from climate change.

The use of wetlands to reduce GHGs may not be beneficial across the board, as it has been found that developmental strategy may be better for developing nations than emission reduction of GHGs. Specific regions would need to assess their wetlands according to their sustainability and economic persuasions,

511

noting that wetlands play significant roles in relation to carbon and GHGs such as methane. They can be carbon storage sinks, sequestering several metric tons of organic carbon and helping offset carbon emissions elsewhere.

—Solomon A. Isiorho

FURTHER READING

Berry, Sandra L., et al. *Green Carbon the Role of Natural Forests in Carbon Storage.* ANU E Press, 2010.

Bullock, A., and A. Acreman. "The Role of Wetlands in the Hydrological Cycle." *Hydrology and Earth System Sciences* 7, no. 3 (2003): 358–389.

McFadden, L., R. Nicholas, and E. Penning-Rowsell, eds. *Managing Coastal Vulnerability.* Elsevier Press, 2007. Good reading for those interested in coastal vulnerability areas of the world, including wetlands.

Mitsch, William J., et al. *Wetlands.* John Wiley & Sons, 2015. A more technical reading for those interested in the science of wetlands.

Tiner, Ralph W., and CRC Press. *Wetland Indicators: A Guide to Wetland Identification, Delineation, Classification and Mapping.* CRC Press/Taylor & Francis Group, 2017. A good source of information for identification of wetlands.

WIND

FIELDS OF STUDY

Atmospheric Science; Fluid Dynamics; Thermodynamics; Heat Transfer; Environmental Sciences; Environmental Studies; Meteorology; Climate Modeling; Process Modeling; Climate Zones; Oceanography; Physical Geography; Earth System Modeling

SUMMARY

Wind is the horizontal movement of air resulting from differences in atmospheric pressure and air densities. Pressure differences may develop on a local or global scale in response to differences in the distribution of solar energy, which affect the density of air masses and, therefore, the pressure they exert relative to each other.

PRINCIPAL TERMS

- **constant pressure chart:** a chart that shows the altitude of a constant pressure, such as 500 millibars
- **convergence:** the movement of different air masses flowing toward a common point
- **divergence:** a net outflow of air in different directions from a specified region
- **geostrophic wind:** an upper-level wind that flows in a straight path in response to a balance between pressure gradient and Coriolis acceleration
- **hurricane-force wind:** a wind with a speed of 64 knots (118 kilometers per hour) or higher
- **isobar:** a line on a meteorological chart delineating points of equal pressure,
- **local winds:** winds that, over a small area, differ from the general pressure pattern owing to local thermal or orographic effects
- **pressure gradient:** the rate of change of pressure with distance at a given time
- **rawinsonde:** a radiosonde tracked by radar in order to collect wind data in addition to temperature, pressure, and humidity

FACTORS AFFECTING WIND FLOW

Wind, as defined by meteorologists, is the horizontal movement of air. Differences in heating and internal motion in the atmosphere create differences in atmospheric pressure; when a change in pressure over distance is established, air accelerates down this pressure gradient from higher to lower pressure. The acceleration of this moving air depends on the amount of pressure change over a given distance.

Moving air associated with pressure change over a distance will either spread out over the surface (diverge) or will flow inward (converge). High-pressure areas are regions of divergence, and low-pressure areas are regions of convergence. The force associated with the air moving from high to low pressure is called the pressure gradient force. Pressure gradient force sets the wind into motion. If it were the only force affecting the wind, then winds would blow

A granite outcrop eroded by windblown sand, Llano de Caldera, Atacama Province, Chile.

directly from high to low pressure. However, other forces affect wind direction and velocity.

A second major factor affecting wind flow is Coriolis acceleration, which results from the Earth turning on its axis. An object moving over the surface of the Earth, except at the equator, moves in a curved path when observed from the rotating Earth. In the Northern Hemisphere, there is an acceleration to the right of the path of motion; in the Southern Hemisphere, the acceleration is to the left. Thus, in the Northern Hemisphere, a wind blowing from north to south becomes a northeast wind, and a wind blowing from south to north becomes southwesterly. The reverse occurs in the Southern Hemisphere.

A third force affecting wind flow is centripetal acceleration. Air currents seldom move on a straight path for long but rather tend to develop a curved pattern as they flows parallel to curved isobars. When this type of flow pattern evolves, centripetal acceleration is directed into the center of the cell or curve, the force acting perpendicularly to the direction of flow. This acceleration is directed outward from both high- and low-pressure cells in the equally opposite sense. Therefore, airflow affected by a high pressure; centripetal acceleration is in the opposite direction as that around a low pressure. Thus air movement about a low-pressure center is cyclonic, or counterclockwise, in the Northern Hemisphere, and anticyclonic, or clockwise, about a high-pressure center. Centripetal acceleration plays a more significant and immediate role in smaller circulations such as hurricanes and tornadoes than in larger, midlatitude cyclones.

A fourth factor affecting wind velocity and direction is frictional drag, which works in a direction opposite to wind motion; therefore, friction tends to slow wind velocity. A decrease in wind velocity, however, is accompanied by a decrease in Coriolis acceleration, which causes a slight change in wind direction back toward the direction of the pressure gradient. The effect is inherent in fluid dynamics.

A fluid, whether gas or liquid, flowing without restraint of any kind exhibits the property of laminar flow in which every particle of which the fluid is composed moves in unison. The presence of any kind of containing surface exerts a restraining force upon the flowing particles nearest to it as they move, thus disrupting their unity of movement. This, coupled with interactions between the particles themselves, results in the condition of turbulent flow, in which particles at different distances from the surface move at different rates. Air moving over the surface of the planet is constrained at the surface by the surface itself, while the density of the fluid decreases with altitude. Horizontally, there are no surface constraints to affect the flow of air, other than minor differences in density and pressure (the pressure gradient force). Frictional drag is thus at a maximum over land where an uneven surface consisting of trees, buildings, and hills provides barriers to the even flow of wind. Also, friction affects the flow of wind only in the first or second kilometers of the atmosphere. Wind direction and velocity in the lowest kilometer of the atmosphere are based on the sum of pressure gradient acceleration, Coriolis acceleration, centripetal acceleration, and frictional drag.

Above 1 kilometer, winds blow in response to pressure gradient, Coriolis, and centripetal acceleration. Frictional deceleration is negligible or completely absent. Consider the situation in which pressure is distributed in a linear fashion so that lines connecting points of equal pressure are straight. In this situation, pressure gradient acceleration and Coriolis acceleration are the only forces acting on the wind. Here, pressure gradient force is balanced by Coriolis acceleration so that the wind flows in a direction parallel to the isobars. Such winds are called geostrophic winds. Around circular highs and lows above the friction level, pressure gradient acceleration is balanced by both Coriolis and centrifugal acceleration. Thus, winds blow parallel to isobars in a clockwise direction around highs and in a counterclockwise direction around lows. In the Southern Hemisphere, the reverse is true. The winds thus described are called gradient winds.

MONSOON AND PRESSURE CHANGES
The monsoon is a seasonal wind system that changes direction from winter to summer and exists over eastern Asia and the adjacent oceans, and less significantly at other locations in the world. In winter, over the large landmass of Asia, air is cooled, and a cold, dense high-pressure center forms with a clockwise circulation of winds about it. Generally, these winds flow from land to sea during winter. In summer, a thermal low forms over India, and the airflow pattern reverses, with cyclonic flow bringing air on shore from the ocean. Reinforcing the thermal low is the migration of the Intertropical Convergence Zone northward over India. Moreover, the jet stream breaks down during summer, which reinforces the monsoon flow. With this annual wind-flow reversal, the climate of Asia is greatly affected. The offshore winds in winter bring dry weather to much of eastern Asia. Conversely, the onshore winds of summer bring copious amounts of precipitation to India and adjacent areas of southeast Asia.

Daily changes in temperature at many places around the world result in daily pressure changes, which cause distinctive wind patterns. One such system is that of the land and sea breezes. This system develops along coastal areas and along the shorelines of large lakes and inland seas. During the day, as the land heats rapidly, the air above heats up, expands, and becomes less dense, forming a thermal low. The warm, buoyant air rises, and cooler air from the water surface flows in to replace the rising air. In this fashion, a sea breeze develops during the day and usually reaches a peak in mid-afternoon, when the daily high temperature is attained. At night, conditions are reversed. The land cools more rapidly than water. In this way, the pressure relationship between land and water is reversed day to night. At night, pressure is higher over land and lower over water, so air flows from land to water, producing a land breeze.

The land breeze is usually not as well developed as is the sea breeze because the temperature contrast between land and water is not as great at night as it is during the day. Another wind system that has a day-to-night change in wind direction is that of the mountain and valley breezes. During the day, the mountain slopes warm the air, and it expands. The warm, less dense air rises and is called a valley breeze after its place of origin. At night, the slopes cool, and the air's density increases. The cool, dense air flows downslope in response to gravity and is called a mountain breeze.

LOCAL WINDS
Several local winds occur in response to topographic peculiarities and are difficult to explain on the basis of pressure patterns as they might appear on a weather map. The "Chinook" in the Rockies and the "foehn" in the Alps of Europe result from a combination of topographic effects and large-scale atmospheric systems. In response to these systems, winds flow down the lee side and are heated adiabatically by compression. The warming brought by these winds is often rapid.

The "sirocco" ("khamsin" in Egypt and "sharov" in Israel) and the "haboob" are hot, dusty winds that occur on flat terrain. The sirocco precedes a low-pressure system moving across the Sahara Desert. As it crosses the Mediterranean Sea, it picks up moisture and becomes a hot, humid wind by the time it reaches the coast of Europe. The haboob is created by air spilling out of the base of a thunderstorm and attains high speeds and picks up small soil particles, creating a sand storm extending upward as high as 1 kilometer or more.

A katabatic wind is a cold wind flowing downslope from an ice field or glacier. Wind velocities range from as little as 10 knots up to hurricane speeds. One such wind is the "bora," which originates in Russia and blows out across the Adriatic coast of Yugoslavia with speeds sometimes in excess of 100 knots (182.5 kilometers per hour). In France, a wind known as the "mistral" blows out of the French Alps and through the Rhone Valley to chill the Riviera along the Mediterranean Sea.

STUDY OF WIND
A number of instruments are used to collect data about wind direction and velocity at the surface or in

the upper troposphere. The wind vane is commonly used for determining surface wind direction. Most wind vanes are simple, relatively long planar structures that will self-align to the direction of movement of the wind, such as an arrow with a tail. The arrow, or other type of vane, is attached to a vertical pole about which it can move freely and always points in the direction from which the wind is blowing.

The anemometer is an instrument used to record wind velocity. It normally consists of three hemispherical cups attached to crossbars, which are in turn attached to a vertical shaft about which it can spin freely. The cups are pushed by the wind preferentially on their open side, causing the shaft to turn, and the wind speed is recorded by a counting device at the base of the shaft. An instrument used for recording both wind direction and velocity is the aerovane. It consists of a three-bladed propeller mounted on the end of a streamlined rod, with a vertical fin at the opposite end. The propeller rotates at a rate proportional to the wind speed. The fin and aerodynamic shape keep the propeller blades facing into the wind, so wind direction is easily determined. When a recorder, often remote, is connected, a continuous record of both wind velocity and direction can be obtained.

A series of instruments also has been developed to determine wind directions and velocities at higher levels. One is the pilot balloon, a small balloon released at the surface that rises at a known rate. The balloon is tracked using a small telescope called a theodolite, and periodic measurements of the balloon's horizontal and vertical angles are taken, giving the speed and direction of the winds carrying it. The pilot balloon principle can also be applied to a radiosonde ascent. Measurements of the vertical and horizontal angles tracking the radiosonde's ascent, taken periodically along with its distance from the observing station, can supply information on wind direction and speed.

A rawinsonde can be tracked using radar, so wind speed and direction can be obtained. Radar can also be used in conjunction with rockets to collect wind data at a distance above 30 kilometers. One type of rocket ejects a parachute carrying an instrument package, which can be tracked by radar. Another type of rocket ejects metallic strips at predetermined levels that can also be tracked by radar. Doppler radar can be used to determine direction and speed of wind by making use of the Doppler effect on reflected signals. Doppler radar measures speeds of objects moving toward or away from the antenna. When a signal is sent out and reflected from a raindrop or ice crystal, the returning signal will have a higher frequency if the particle is moving toward the radar and a lower frequency if it is moving away. One drawback of Doppler radar is that velocities of objects at right angles to the unit cannot be determined, so to achieve a three-dimensional effect, two or more units must be used.

Wind directions and speeds are plotted on charts using a symbol called the wind arrow. The shaft of an arrow shows wind direction, while barbs on the end of the arrow indicate speed. A barb represents 10 knots (18.25 kilometers per hour), one-half barb 5 knots (9.125 kilometers per hour), and a flag (a triangle-shaped symbol) represents 50 knots (91.25 kilometers per hour). These symbols may be used singularly or in combination to show any wind speed. On a surface chart, the wind arrows point out from a station in the direction from which the wind is coming. In the upper air above the friction level, the wind arrow points in the direction to which the wind is moving.

Winds above the friction level are plotted on constant pressure charts. A constant pressure chart is drawn using contour lines to show the elevation above Earth's surface of a constant pressure level, such as the 500-millibar level. When pressure is particularly high in an area in relation to surrounding areas, the height of a constant-pressure surface is higher than surrounding areas, and heights of a low-pressure region are lower than surrounding regions. The average elevation of the 500-millibar level is 5.5 kilometers but can vary from less than 5 to more than 6 kilometers.

Various constant pressure charts are used, ranging from just above the surface, such as the 850-millibar level, up to the tropopause at roughly the 200-millibar level. With the use of these constant pressure charts, meteorologists can gain a sense of the three-dimensional wind-flow profile from the surface up to the tropopause.

SIGNIFICANCE

Winds, in conjunction with temperature and humidity, can greatly affect human comfort and safety. The effects of wind influence the exchange of heat between the human body and the atmosphere. The body, particularly the surface of the skin, is continually exchanging heat with the environment. On a cold day when a wind is blowing, air molecules impact the skin, then move away, taking body heat with them. Clothing provides insulation, creating a shallow layer of warm air molecules, which form a shield that protects the skin from heat loss. The "windchill factor" relates the rate of heat loss due to wind action to the temperature having the equivalent rate of heat loss in the absence of wind.

Humankind's use of wind power may stem from the use of winds to propel sailboats or ships. Sails have been used as power sources on ships and boats for thousands of years, and were the chief source of power for water transportation until the use of steam in the latter part of the nineteenth century. The next step in the use of wind for power was through the use of windmills. The first known windmill appeared in Europe in 1105, and by the following century, thousands of windmills were in use in Europe. Then, as the burning of coal as a source of power became less expensive, it and other energy sources replaced windmills. In the early part of the twentieth century, windmills became popular as an inexpensive means of pumping water for agricultural uses on farms and ranches.

Today, wind power is again being considered as a partial solution to growing energy needs. The advantages of using wind power are that windmills are nonpolluting, and they are not limited to daylight hours as are solar cells. The liabilities of using wind power are several in number: Windmills can only be used in windy areas where wind flow is steady and neither too weak nor too strong; a weak wind will not turn the blades, and a strong wind might damage the machine. Windmills detract from the aesthetics of the landscape, and their cost factor can be quite high.

—*Ralph D. Cross*

FURTHER READING

Ackerman, Steven, and John Knox. *Meteorology: Understanding the Atmosphere.* 3d ed. Jones and Bartlett, 2012.

Aguado, Edward, and James E. Burt. *Understanding Weather and Climate.* 7th ed. Pearson Prentice Hall, 2015. The formation of midlatitude storms is surveyed. Continental climates and the structure of climate are discussed.

Ahrens, C Donald, and Robert Henson. *Meteorology Today: An Introduction to Weather, Climate, and the Environment.* Cengage, 2019. Discusses global climates, climate change, and classification.

Christopherson, Robert W., and Ginger H Birkeland. *Geosystems: An Introduction to Physical Geography.* Pearson, 2018.

Clarke, Allan J. *An Introduction to the Dynamics of El Niño and the Southern Oscillation.* Academic Press, 2008.

De Villiers, Marq. *Windswept: The Story of Wind and Weather.* Walker Publishing Company, 2006.

Frederick, John E. *Principles of Atmospheric Science.* Jones and Bartlett, 2008. A complete introduction to atmospheric science. Presents the fundamental scientific principles and concepts related to the Earth's climate system.

Gombosi, Tamas I. *Physics of the Space Environment.* Cambridge University Press, 2004.

Lutgens, Frederick K., Edward J. Tarbuck, Tasa, Dennis G. and Herman, Redina. *The Atmosphere: An Introduction to Meteorology.* 14th ed. Pearson Education, 2018. An updated version of this textbook, in loose-leaf format, provides students with a general overview of the structure and properties of Earth's atmosphere.

Saha, Kshudiram. *The Earth's Atmosphere: Its Physics and Dynamics.* Springer-Verlag, 2008.

Watts, Alan. *Instant Wind Forecasting.* Adlard Coles Nautical, 2010.

WORLD OCEAN CIRCULATION EXPERIMENT

FIELDS OF STUDY

Oceanography; Fluid Dynamics; Heat Transfer; Environmental Sciences; Environmental Studies; Waste Management; Physical Geography; Physical Sciences; Meteorology; Climate Modeling; Statistics; Process Modeling; Hydroclimatology; Hydrometeorology; Bioclimatology; Atmospheric Science; Climate Zones; Earth System Modeling

SUMMARY

The World Ocean Circulation Experiment is an ambitious international project designed to increase knowledge about the movement of water, heat, and various substances in the sea. The data obtained are expected to be of major importance in predicting future long-term changes in climate.

PRINCIPAL TERMS

- **Coriolis Effect:** the phenomenon, resulting from Earth's rotation, that causes the path of a moving object to curve away from a straight line
- **current:** the horizontal movement of ocean water at a generally uniform depth
- **downwelling:** the sinking of ocean water to a lower depth
- **Ekman layer:** the region of the sea, from the surface to about 100 meters down, in which the wind directly effects water movement
- **ocean circulation:** the worldwide movement of water in the sea
- **thermohaline circulation:** movement of ocean water caused by differences in temperature and salt concentration
- **upwelling:** the rising of ocean water from a depth toward the surface
- **wind-driven circulation:** movement of ocean water caused by frictional interaction with moving air
- **wind stress:** the frictional interaction between moving air and the surface of the ocean

MOVEMENT OF SEAWATER

The World Ocean Circulation Experiment (WOCE), which began in 1990, brought together scientists from around the globe to study the way in which water moves in all parts of the sea. The WOCE was also intended to obtain information on the movement of heat in the ocean, as well as the movement of substances such as salt and oxygen.

The movement of seawater, known as ocean circulation, consists of horizontal and vertical motion. Horizontal movements are known as currents. Vertical motions are known as upwellings and downwellings. Currents vary widely in speed, ranging from a few centimeters per second to as much as 4 meters per second. Surface currents typically move between 5 and 50 centimeters per second, with deeper currents generally moving more slowly. Vertical movement of seawater is much slower, with a typical speed of only a few meters per month.

Ocean circulation is primarily caused by two major factors. Wind-driven circulation is caused by air moving across the surface of the sea. This induces friction, known as wind stress, between the water and the air, thus applying a directional force and setting the water in motion. Thermohaline circulation is caused by differences in temperature and salt concentration. These differences cause variations in the density of seawater, leading to differences in pressure and resulting in motion. Surface currents are mostly caused by wind-driven circulation. Deeper currents and vertical movements are mostly the result of thermohaline circulation.

Several factors are involved in determining the size, shape, and speed of ocean circulation patterns. An important influence is the Coriolis Effect, named for the French scientist Gustave-Gaspard de Coriolis. This effect, caused by the rotation of the planet, causes the actual path of a moving object to be curved away from the straight line path that it would otherwise follow. The Coriolis Effect causes ocean currents to bend to the right in the Northern Hemisphere and to bend to the left in the Southern Hemisphere.

Friction also influences the nature of ocean circulation. Layers of water moving at different speeds produce friction where they meet, forming an intermediate zone of turbulent eddy currents as the two layers interact. This transfers energy from one layer to the other, and causes the faster layer to move more slowly and the slower layer to move more quickly. Friction also occurs between moving water and the continents, and between currents at the bottom of the sea and the ocean floor. This friction, with the attendant resulting turbulence zones, tends to slow the motion of seawater.

Important effects on ocean circulation are seen in the region of the sea known as the Ekman layer, named for the Swedish scientist Vagn Walfrid Ekman (1874–1954). The Ekman layer extends from the surface of the ocean to a depth of about 100 meters. In this layer, the wind has a direct effect on water movement. Wind stress, the Coriolis Effect, and friction between layers of water combine to move the Ekman layer in complex ways.

Water at the surface of the ocean tends to move at an angle of about 45 degrees to the direction of the wind because of the Coriolis Effect. This angle is directed to the right in the Northern Hemisphere and to the left in the Southern Hemisphere. With increasing depth, the water moves more slowly and the angle increases. At a depth where the speed of the water is about 4.3 percent of the surface speed, the water moves in the opposite direction to the wind. The overall effect is that the average movement of the water is at about 90 degrees to the wind.

This movement in the Ekman layer, combined with differences in wind stress, creates areas on the surface of the ocean where water converges or diverges. Where it converges, water sinks in downwellings. Where it diverges, water rises in upwellings. Downwellings and upwellings also occur where the wind blows parallel to the coast of a landmass. They also occur because of differences in temperature and salt concentration, both of which alter the density of the seawater. Cold water and salty water tend to sink, while warm water and less salty water tend to rise.

WORLDWIDE PATTERNS OF OCEAN CIRCULATION

The numerous factors involved in ocean circulation, combined with the irregular shapes of the continents, result in complex patterns of water movement. Although major surface currents have been known since the earliest days of ocean travel, much less is known about deeper currents. Similarly, less is also known about the Southern Hemisphere than the Northern Hemisphere. The WOCE was designed to fill these gaps in scientific knowledge.

Before the WOCE, the basic pattern of surface currents was fairly well understood. In the Northern Hemisphere, strong currents tend to move northward along the eastern coasts of the continents. These include the Gulf Stream-North Atlantic-Norway Current, along North America, and the Kuroshio-North Pacific Current, along Asia. In the Southern Hemisphere, strong currents tend to move northward along the western coasts of the continents. These include the Peru Current, along South America; the Benguela Current, along Africa; and the Western Australia Current, along Australia.

In the regions north and south of the equator, major surface currents move westward. These are the Pacific North Equatorial Current and the Pacific South Equatorial Current, between South America and Asia; the Atlantic North Equatorial Current and the Atlantic South Equatorial Current, between Africa and South America, and the Indian South Equatorial Current, between Australia and Africa.

At the equator, narrow eastward currents are found between the wider westward currents. These are the Pacific Equatorial Countercurrent, between Asia and South America; the Atlantic Equatorial Countercurrent, between South America and Africa; and the Indian Equatorial Countercurrent, between Africa and Asia. Another major surface current is the Antarctic Current, moving eastward around Antarctica.

Although less is known about deep currents, certain broad patterns of movement are now understood from WOCE data. Cold water in the northern part of the Atlantic Ocean sinks in downwellings. This deep, cold water tends to move southward along the eastern coasts of North and South America to join the deep, cold water that sinks in downwellings near Antarctica. This water then tends to flow eastward in a deep current around Antarctica. Some of this water then moves northward along the coasts of Asia and Africa, rising in upwellings as it warms. Overall, the deep ocean currents function something like a continuous conveyor belt system, moving thermal energy in ocean water and distributing it throughout the oceans. Cold water upwelling in the Indian and north Pacific Oceans circulates westward between Australia and Asia, and around the southern tip of Africa to make its way to the North Atlantic. There it sinks again to depths and enters the deep current that takes it back eastward, where it splits and rises again in the Indian and northern Pacific Ocean.

STUDYING OCEAN CIRCULATION

Scientific studies of surface currents began in the eighteenth century in order to aid navigation. Later studies concentrated on the effect of changes in ocean circulation on the weather. The need for a major effort to increase the amount of information known about the movement of seawater became clear in the 1980s. The best models of ocean circulation, based on the available data, failed to describe the conditions actually observed in the sea with complete accuracy.

A major determining factor in the establishment of the WOCE was the development of new techniques for studying ocean circulation. Temperature measurements could be made from a ship without stopping the movement of the vessel using an instrument known as a bathythermograph. Devices designed to drift in the ocean, both on the surface and at specific depths, were developed that could be tracked for months or years. Advanced methods of accurately measuring the concentration of substances present in seawater in very low concentrations were also developed. In addition, computers able to handle the enormous amount of data that the WOCE would generate were made possible through advances in electronic manufacturing methods.

Perhaps the most important new instruments available for the WOCE were satellites capable of obtaining data on ocean circulation. In 1979, the Seasat satellite mission, lasting one hundred days, demonstrated that detection of radar echoes and microwave radiation from the sea could be used to produce detailed information. After years of planning, the WOCE project was ready to begin collecting data in 1990.

Scientists from more than thirty nations participated in the many studies involved in the WOCE. In the United States, the headquarters of the WOCE is located at the Department of Oceanography at Texas Agricultural and Mechanical University. Data collection ended in 1998, but analysis of the information was expected to last until at least 2002, and many scientists expected processing of the data to last until at least 2005.

The first WOCE study began with the launching of the German research ships *Polarstern* and *Meteor* in 1990. These ships collected data in the southern part of the Atlantic Ocean between Antarctica and South Africa. Other early WOCE studies also concentrated on the Southern Hemisphere because this area had been studied in less detail prior to the WOCE than the Northern Hemisphere. Later WOCE studies moved into the Indian Ocean, the North Atlantic Ocean, and the Pacific Ocean.

Satellites used by the WOCE included the ERS series launched by the European Space Agency, the TOPEX/POSEIDON, a joint project of France and the United States, and the Japanese ADEOS. More than one thousand drifting instruments, designed to remain at specific depths far below the surface of the sea, were also used. The movement of these instruments was measured by satellites or by sonic equipment. Tens of thousands of measurements were made at the surface of the ocean as well.

SIGNIFICANCE

The WOCE project is likely to be one of the most important sources of oceanographic data in the early twenty-first century. The official goals of the WOCE included providing a complete description of the general circulation of the ocean, creating a numerical model of ocean circulation for use in advanced computers, accounting for seasonal changes in ocean circulation, obtaining data on the exchange of substances between layers of water in the ocean, providing detailed information on the interaction between the ocean and the atmosphere, and obtaining data on the movement of heat within the ocean.

The most important application of WOCE data is in the study of the effect of ocean circulation on climate changes. This information is expected to aid scientists in predicting the effect on long-term weather patterns of various human activities, such as the increase in carbon dioxide in the atmosphere. Such data will also be useful in predicting natural changes in climate that take place over years or decades.

Several examples of the interaction between ocean conditions and changes in weather are well documented. The Sahel, a region of Africa along the southern fringe of the Sahara Desert, experienced severe droughts in the 1970s and 1980s after having experienced much wetter conditions in the 1950s. These droughts were associated with higher-than-normal surface temperatures in the South Atlantic Ocean, Indian Ocean, and southeast Pacific Ocean, and with lower-than-normal temperatures in the North Atlantic Ocean and most of the Pacific Ocean. Ocean temperature is also an important factor in the formation of tropical cyclones and other powerful storms.

Currents have a powerful effect on weather patterns. The Gulf Stream-North Atlantic-Norway Current brings relatively warm tropical water northward, moderating the climates of eastern North America, Ireland, the British Isles, and the coast of Norway. The Kuroshio-North Pacific Current does the same for Japan and western North America. These warm currents also encourage increased water evaporation, resulting in increased rainfall in these areas. The Peru Current brings cold polar water northward along the western coast of South America, decreasing water evaporation and creating deserts in Peru and Chile. The Benguela Current, running northward along the western coast of Africa, has the same effect in Namibia.

Perhaps the best-known example of the effect of changes in ocean circulation on weather is the El Niño phenomenon. This situation occurs at irregular intervals in the eastern Pacific Ocean. Increased water temperatures, typically 2°C to 8°C higher than normal, are associated with changes in climate. Typical effects seen during an El Niño condition are droughts in Australia, northeastern Brazil, and southern Peru; excessive summer rainfall in Ecuador and northern Peru; severe winter storms in Chile; and warm winter conditions in North America.

The El Niño effect is also associated with large reductions in fish populations along the western coast of South America. During normal conditions, the water near the coast consists of a thin layer of warm, nutrient-poor water above a thick layer of cooler, nutrient-rich water. The top layer is thin enough to allow coastal upwellings to bring nutrients to the surface, supporting marine life. During El Niño conditions, the top layer of warm water is much thicker, preventing nutrients from reaching the surface. Data from the WOCE project are expected to aid in the prediction of climate changes such as El Niño, with the possibility of having a major impact on human activities.

—Rose Secrest

FURTHER READING

Broecker, Wally. *The Great Ocean Conveyor*. Princeton University Press, 2010. Discusses ocean currents, focusing specifically on the great conveyor belt. Written by the great ocean conveyor's discoverer. Explains the conception of this theory and the resulting impact on oceanography. Written in a manner easy to follow with some background in science, yet still relevant to graduate students and scientists.

Di Lorenzo, Emanuele, et al. "North Pacific Gyre Oscillation Links Ocean Climate and Ecosystem Change." *Geophysical Research Letters* 35, no. 108607 (April, 2008). Overview of the discovery of the North Pacific Gyre Oscillation and its consequences for Pacific ecosystems and fisheries industries.

Field, J.G., Gotthilf Hempel, and C.P. Summerhayes. *Oceans 2020: Science, Trends and the Challenge of Sustainability*. Island Press, 2002. A combined work of the Intergovernmental Oceanographic Commission, the International Council of Scientific Unions, the Scientific Committee on Oceanic Research, and the ICSU Scientific Committee on Problems of the Environment in a collection of learned essays that address the state of the oceans and issues for sustainability.

Jacques, Peter. *Globalization and the World Ocean*. Lanham, Md.: Rowman and Littlefield Publishing Group, 2006. Provides a unique analysis of how global marine and atmospheric conditions affect political conditions globally by viewing the world ocean as a single body of water composed of connected regional oceans. Aimed at researchers and students of marine sciences, and environmental and globalization studies.

Jochum, Markus, and Raghi Murtugudde, eds. *Physical Oceanography: Developments Since 1950*. New York: Springer Science + Business Media, 2006. Provide a historic overview of developments in physical oceanography in the last half of the twentieth century with senior scientists in the field authoring individual chapters. Chapter 12, by Karl Wunsch, deals exclusively with the WOCE and its aftermath.

Miller, Robert N. *Numerical Modeling of Ocean Circulation*. New York: Cambridge University Press, 2007. Designed to teach the process of numerical analysis; each chapter includes exercises to practice modeling skills. Discusses models of tropical waters, coastal waters, and shallow waters. Also covers simple and complex numerical modeling. Useful for the advanced undergraduate and graduate student. Requires a strong mathematics background.

Needler, George T. "WOCE: The World Ocean Circulation Experiment." *Oceanus* 35 (Summer, 1992): 74-77. A clear introduction to the goals of WOCE and the methods used to obtain data. Written by the first scientific director of the WOCE, who was instrumental in planning the project.

Pedlosky, Joseph. *Ocean Circulation Theory*. Rev. ed. New York: Springer, 2010. A detailed description of modern models of ocean circulation and the effect it has on climate. An excellent resource for advanced students that demonstrates the importance of the data provided by the WOCE.

Richardson, P. L. "On the History of Meridional Overturning Circulation Schematic Diagrams." *Progress In Oceanography* 76 (2008): 466–486.

Schmittner, Andreas, Chiang, John C.H. And Hemming, Sidney R., eds. *Ocean Circulation. Mechanisms and Impacts.* John Wiley & Sons, 2013. Provides a thorough description of the effects on Earth's climate due to large "turnover" circulation currents, particularly the Atlantic conveyor system.

Vallis, Geoffrey K. *Atmospheric and Oceanic Fluid Dynamics: Fundamentals and Large-Scale Circulation.* New York: Cambridge University Press, 2006. Begins with an overview of the physics of fluid dynamics to provide foundational material on stratification, vorticity, and oceanic and atmospheric models. Part II discusses topics such as turbulence, baroclinic instabilities, and wave-mean flow interactions, while parts III and IV discuss large-scale atmospheric and oceanic circulation, respectively. Best suited for graduate students studying meteorology or oceanography.

Aken, Hendrik Mattheus van. *The Oceanic Thermohaline Circulation: An Introduction.* Springer, 2011. Presents a global hydrographic description of the thermohaline circulation, based on data obtained directly from the WOCE.

Wefer, Gerold, et al. *The South Atlantic: Present and Past Circulation.* Springer Berlin, 2013. An example of modern scientific analysis of data obtained on ocean circulation. Deals with the region of the ocean where the earliest WOCE studies were performed.

Woods, J. D. "The World Ocean Circulation Experiment." *Nature* 314 (April 11, 1985): 501–510. An outstanding introduction to the WOCE project. Discusses the importance of ocean circulation to changes in climate, the failure of models prior to the WOCE to accurately predict ocean conditions, the goals of the WOCE, and the technology that allowed the project to take place.

TIMELINE OF PERIODS AND EVENTS IN CLIMATE HISTORY

START TO QUATERNARY PERIOD

Before 1,000 MYA	Faint young Sun paradox
2,400 MYA	Great Oxygenation Event probably leads to Huronian glaciation perhaps covering the whole globe
650–600 MYA	Later Neoproterozoic Snowball Earth or Marinoan glaciation, precursor to the Cambrian Explosion
517 MYA	End-Botomian mass extinction; like the next two, little understood
502 MYA	Dresbachian extinction event
485.4 MYA	Cambrian–Ordovician extinction event
450–440 MYA	Ordovician–Silurian extinction event, in two bursts, after cooling perhaps caused by tectonic plate movement
450 MYA	Andean-Saharan glaciation
360-260 MYA	Karoo Ice Age
305 MYA	cooler climate causes Carboniferous Rainforest Collapse
251.902 MYA	Permian-Triassic extinction event
199.6 MYA	Triassic–Jurassic extinction event, causes as yet unclear
66 MYA	, perhaps 30,000 years of volcanic activity form the Deccan Traps in India, Or a large meteor impact.
66 MYA	Cretaceous–Paleogene boundary and Cretaceous–Paleogene extinction event, extinction of dinosaurs
55.8 MYA	Paleocene-Eocene Thermal Maximum
53.7 MYA	Eocene Thermal Maximum 2
49 MYA	Azolla event may have ended a long warm period
5.3–2.6 MYA	Pliocene climate became cooler and drier, and seasonal, similar to modern climates.
2.5 MYA	to present Quaternary glaciation, with permanent ice on the polar regions, many named stages in different parts of the world

PLEISTOCENE

All dates are approximate. B-S means this is one of the periods from the Blytt-Sernander sequence, originally based on studies of Danish peat bogs.

120,000–90,000 BP	Abbassia Pluvial wet in North Africa
110,000–10,000 BP	Last glacial period, not to be confused with the Last Glacial Maximum or Late Glacial Maximum below
50,000–30,000 BP	Mousterian Pluvial wet in North Africa

Timeline of Periods and Events in Climate History

26,500–19,000–20,000 BP	Last Glacial Maximum, what is often meant in popular usage by "Last Ice Age"
16,000–13,000 BCE	Oldest Dryas cold, begins slowly and ends sharply (B-S)
12,700 BCE	Antarctic Cold Reversal warmer Antarctic, sea levels rise
12,400 BCE	Bølling oscillation warm and wet in the North Atlantic, begins the Bølling-Allerød period (B-S)
12,400–11,500 BCE	(much discussed) Older Dryas cold, interrupts warm period for some centuries (B-S)
12,000–11,000 BCE	Allerød oscillation warm & moist (B-S)
11,400–9,500 BCE	Huelmo/Mascardi Cold Reversal cold in Southern Hemisphere
11,000–8,000 BCE	Late Glacial Maximum, or Tardiglacial (definitions vary)
10,800–9,500 BCE	Younger Dryas sudden cold and dry period in Northern Hemisphere (B-S)

HOLOCENE

All dates are BCE and approximate. B-S means this is one of the periods from the Blytt-Sernander sequence, originally based on studies of Danish peat bogs.

From 10,000 BCE	Holocene glacial retreat, the present Holocene or Postglacial period begins
9,400 BCE	Pre-Boreal sharp rise in temperature over 50 years (B-S), precedes Boreal
8,500 – 6,900 BCE	Boreal (B-S), rising sea levels, forest replaces tundra in northern Europe
7,500 – 3,900 BCE	Neolithic Subpluvial in North Africa, wet
7,000 – 3,000 BCE	Holocene climatic optimum, or Atlantic in northern Europe (B-S)
6,200 BCE	8.2 kiloyear event cold
5,000–4,100 BCE	Older Peron warm and wet, global sea levels were 2.5 to 4 meters (8 to 13 feet) higher than the twentieth-century average
3,900 BCE	5.9 kiloyear event dry and cold, ends Neolithic Subpluvial in North Africa, expands Sahara Desert
3,000 BCE	- 0 Neopluvial in North America
3,200–2,900 BCE	Piora Oscillation, cold, perhaps not global. Wetter in Europe, drier elsewhere, linked to the domestication of the horse in Central Asia.
2,200 BCE	4.2 kiloyear event dry, lasted most of the 22nd century BCE, linked to the end of the Old Kingdom in Egypt, and the Akkadian Empire in Mesopotamia, various archaeological cultures in Persia and China
1800-1500 BCE	Middle Bronze Age Cold Epoch, a period of unusually cold climate in the North Atlantic region
	Bond Event 2 – possibly triggering the Late Bronze Age collapse
900 – 300 BCE	Iron Age Cold Epoch cold in North Atlantic
250 BCE–400 CE	Roman Warm Period

Common Era/CE

250 BCE–400 CE	Roman Warm Period
	Climate changes of 535-536 (535–536 CE), sudden cooling and failure of harvests, perhaps caused by volcanic dust
900–1300	Medieval warm period, wet in Europe, arid in North America, may have depopulated the Great Plains of North America, associated with the Medieval renaissances in Europe
	Great Famine of 1315–17 in Europe
	Little Ice Age: Various dates between 1250 and 1550 or later are held to mark the start of the Little ice age, ending at equally varied dates around 1850
1460–1550	Spörer Minimum cold
1656–1715	Maunder Minimum low sunspot activity
1790–1830	Dalton Minimum low sunspot activity, cold
	Year Without a Summer (1816), caused by volcanic dust
1850–present	Retreat of glaciers since 1850, instrumental temperature record
	Present and recent past global warming, perhaps to be named the Anthropocene period

GLOSSARY

18O/16O ratio: ratio between two oxygen isotopes that is altered by global average temperatures associated with the advance and retreat of continental glaciers

ablation: the removal of material from a glacier, ice shelf, or other mass of ice through evaporation, melting, or splitting

acid deposition: the depositing of acidic materials on the ground surface through the action of precipitation

acid precipitation: rain or snow that is more acidic than normal, usually because of the presence of sulfuric and nitric acid

acid rain: rain composed of water having a lower-than-normal pH due to having dissolved and reacted with airborne contaminants to produce acidic materials

active sensor: type of remote sensor that emits radiation at a target to study its composition and condition

adiabatic: the effect of changing the temperature of a gas or other fluid solely by changing the pressure exerted on it, without the input or removal of heat energy

advection: horizontal movement of heat carried by the atmosphere, as opposed to the vertical movement associated with convection

aerobic: taking place in or requiring the presence of oxygen

aerosol: fine liquid and solid particles suspended in the atmosphere, typically measuring between 0.01 and 10 microns and either natural or anthropogenic in origin

afforestation: creating forests on lands that was not previously forested

air drainage: the flow of cold, dense air downslope in response to gravity

air parcel: a theoretical house-sized volume of air that remains intact as it moves from place to place

albedo: a measure of the proportion of incoming light or radiation that is reflected from a surface, such as snow, ice, or water; also known as reflectivity

algorithm: a set of instructions used to perform a task

alkaline: having a pH greater than 7 due to a lower concentration of hydrogen (H+) ions than are in neutral water

allogenic sediment: sediment that originates outside the place where it is finally deposited; sand, silt, and clay carried by a stream into a lake are examples

alpine glaciers: large masses of ice found in valleys, on plateaus, and attached to mountains

alternative fuel: clean or renewable fuel that can replace traditional fossil fuels

altimeter: scientific instrument that measures the altitude of an object above a fixed level

anaerobic: taking place in or requiring the absence of oxygen

aneroid barometer: device that uses an aneroid capsule composed of an alloy of beryllium and copper to measure changes in external air pressure

Antarctic Peninsula: a peninsula stretching northward toward South America that contains about 10 percent of the ice of Antarctica

anthropogenic: influenced by humans; the anthropogenic greenhouse effect, also called the enhanced greenhouse effect, refers to the phenomenon as amplified by human activities

anticyclone: a general term for a high-pressure weather system that rotates clockwise in the Northern Hemisphere and counterclockwise in the Southern Hemisphere

anticyclonic: the clockwise rotation of an air mass in the Northern hemisphere, counterclockwise in the Southern hemisphere

aphelion: the time at which the distance between Earth and the sun is smallest; generally occurs on one of the first days of January, two weeks after the December solstice

archaea: a taxonomic group of prokaryotic, single-celled microorganisms similar to bacteria, but evolved differently

archaeological record: the human-made objects and residues of the past that are used by archaeologists to reconstruct history

Arctic oscillation: long-term weather pattern in which the different air pressures in the Arctic and middle latitude regions cause varying weather conditions

astronomical forcing: climatic change triggered by changes in solar luminosity, variation in the Earth's orbit, and bolide impact

Atlantic Meridional Overturning Circulation (AMOC): the official designation of what is commonly called the Gulf Stream.

atmosphere: the thin layer of nitrogen, oxygen, and other gases surrounding Earth, whose density decreases rapidly with height

atmospheric pressure: force exerted on a surface by the weight of air above that surface; measured in force per unit area

aurora: a glowing light display resulting from charged particles from solar wind being pulled into Earth's atmosphere by Earth's magnetic field; most often visible near Earth's North and South Poles

austral: referring to an object or occurrence that is of the Southern Hemisphere

available potential energy: that part of the total potential energy of the atmosphere that is available for conversion to the kinetic energy of atmospheric winds

average: the sum of the numerical values of a particular factor, divided by the number of instances of that factor that have been summed

Baltic Sea: the body of water between Scandinavia and Eastern Europe

bank: an elevated area of land beneath the surface of the ocean

barograph: a graph that records atmospheric pressure in time

barometer: device for measuring atmospheric pressure; some are water-based, some use mercury or an aneroid cell, and some create a line graph of atmospheric pressure

benthic: relating to organisms that inhabit the floors of lakes and seas

benthonic: synonymous with benthic, relating to organisms that inhabit the benthos, or ocean floor region

Bergeron process: precipitation formation in cold clouds whereby ice crystals grow at the expense of supercooled water droplets

bicarbonate: a negatively charged ion, as HCO_3^-, that effectively neutralizes excess hydrogen ions in natural waters, reducing acidity

biochar: another name for charcoal, used specifically for usages of charcoal that involve positive effects such as soil improvement or carbon sequestration

biogenic sediment: sediment that originates from living organisms

biogenic: originating from or formed by living organisms

biome: a geographically defined area of similar plant community structure shaped by climatic conditions

biosphere: the portion of Earth that contains living organisms and ecosystems

blizzard: a winter storm characterized by cold wind having a minimum velocity of 56 kilometers per hour, large amounts of blowing snow, and low levels of visibility

bloom: the sudden or rapid development of a large population of plankton and algae

boreal: referring to an object or occurrence that is of the Northern Hemisphere

brash: splinters that become detached from ice floes and float in the Arctic Ocean

calving: the breaking away of a smaller piece of ice from a larger one

canopy: the upper layers of vegetation or uppermost levels of a forest, where energy, water, and greenhouse gases are actively exchanged between ecosystems and the atmosphere

cap and trade legislation: legislation that places limits on the emission of acid-producing materials, such as sulfur dioxide, while allowing emitters of excess amounts to purchase and utilize the unused allowances of those whose emissions are below the legislated limit

capitalism: an economic system in which the means of production are privately owned and operated with the goal of increasing wealth

carbon cycle: the cyclic mechanism whereby carbon, as carbon dioxide, is removed from the atmosphere, converted to glucose and other compounds in photosynthesis, or to carbonate solids, and subsequently returned to the atmosphere when those materials break down or decompose

carbon dioxide: CO_2, one of many minor gases that are natural components of the atmosphere; the product of the complete oxidation of carbon

carbon sequestration: the process of storing carbon in a stable state to negate carbon's effects on climate

carbon sink: plant growth, mineral formation and synthetic activities that act to remove carbon dioxide and other carbon sources from the environment

carbon-oxygen cycle: the process by which oxygen and carbon are cycled through Earth's environment

catalyst: a substance that increases the rate of a chemical reaction without itself being altered in the process

catalytic surface: an interface or surface that promotes a chemical reaction but is not consumed in that reaction

catastrophist: someone who holds that human society leads to or advances by sudden drastic events rather than by a smooth evolutionary process

Celsius degree: the standard unit of temperature measurement on the Celsius scale, for which the freezing and boiling points of pure water are defined as 0° and 100°, respectively

chemical evolution: the synthesis of amino acids and other complex organic molecules, as the precursors of living systems, by the action of atmospheric lightning and solar ultraviolet radiation on atmospheric gases

chinook/foehn wind: a warm, dry wind on the eastern side of the Rocky Mountains or the Alps

chlorofluorocarbon (CFC): a group of chemical compounds containing carbon, fluorine, and chlorine, used in air conditioners, refrigerators, fire extinguishers, spray cans, and other applications

circulation cell: zones of concentrated air circulation caused by the rotation of the earth and the cyclic distribution of heat through the atmosphere

clastic sediments: sediments composed of durable minerals that resist weathering

clay minerals: any mineral particle less than 2 micrometers in diameter

clay: a mineral group whose particles consist of structures arranged in sandwichlike layers, usually sheets of aluminum hydroxides and silica, along with some potassium, sodium, or calcium ions

climate change: alterations in long-term meteorological averages in a given region or globally

climate controls: the relatively permanent factors that govern the general nature of the climate of a region

climate fluctuations: changes in the statistical distributions used to describe climate states

climate forcing: factors that alter the radiative balance of the atmosphere (the ratio of incoming to outgoing radiation)

climate normals: averages of a climatic variable for a uniform period of thirty years

climate: the sum total of the weather elements that characterize the average condition of the atmosphere over a long period of time for any one region

climatic oscillation: a fluctuation of a climatic variable in which the variable tends to move gradually and smoothly between successive maxima and minima

climatic precession: cycle of variations in the Earth-Sun distance at summer solstice

climatic trend: a climatic change characterized by a smooth monotonic increase or decrease of the average value in the period of record

climatology: the scientific study of climate that depends on the statistical database of weather observed over a period of twenty or more years for a specific location

cloud condensation nuclei: atmospheric particles such as dust that can form the centers of water droplets, increasing cloud cover

cloud seeding: the injection of nucleating particles into clouds to enhance precipitation formation

cold cloud: a visible suspension of tiny ice crystals, supercooled water droplets, or both at temperatures below the normal freezing point of water

cold front: area in which a dominant stream of colder air pushes under a pocket of warmer air, causing the warm air to rise and leading to precipitation at the leading edge of the front

collision-coalescence process: precipitation formation in warm clouds whereby larger droplets grow through the merging of smaller droplets

combustion: reactions by which oxygen and organic materials become carbon dioxide, water and other oxides

commodity: anything that has commensurable value and can be exchanged

compressible fluid: a fluid that can be made to occupy a smaller volume by applying external pressure

condensation: the process by which water, or any other substance, changes from a vapor state to a liquid state, releasing heat into the surrounding air; this process is the opposite of evaporation, which requires the input of heat

confidence: a value assigned to a statistical result that indicates how close the calculated result is expected to be to the true value of the property

constant pressure chart: a chart that shows the altitude of a constant pressure, such as 500 millibars

continental drift: the gradual movement of continental landmasses within Earth's crust, driven by magmatic convection in the underlying mantle layer

continental shelf: the part of the sea floor that is generally gently sloping and extends beneath the ocean from adjacent continents

continental slope: the defining edge of a continental shelf where it drops off sharply toward the ocean's floor

convection: the cyclic movement of matter in fluids such as air or water, by which warmer, less dense matter rises through cooler matter and then descends elsewhere as its heat energy decreases

convective overturn: the renewal of the bottom waters caused by the sinking of surface waters that have become denser, usually because of changes in temperature or salinity

convergence: a tendency of air masses to accumulate in a region where more air is flowing in than is flowing out

conveyor belt current: a large cycle of water movement that carries warm waters from the North Pacific westward across the Indian Ocean, around southern Africa, and into the Atlantic, where it warms the atmosphere, then returns to a deeper ocean level to rise and begin the process again

core ring or core eddy: a mass of water that is spun off of an ocean current by that current's meandering motion

Coriolis force: an apparent force caused by the rotation of the planet, in which objects moving above Earth's surface (such as the wind) deflect to the right in the Northern Hemisphere and to the left in the Southern Hemisphere

corona: the outermost layer of the sun, which extends into space in an irregular pattern surrounding the main body of the star

coronal mass ejection: larger than average burst of solar wind related to deformations and reconfigurations of the sun's magnetic field

cosmic ray: high-energy subatomic particles that are produced by phenomena in space, such as supernovae

cosmogenic isotope: an isotope, possibly radioactive, produced when a cosmic ray strikes the nucleus of an atom

coupled atmosphere-ocean models: computer simulations of alterations in and interactions between Earth's atmosphere and oceans

cryosphere: the portion of Earth's surface that exists below the freezing point of water, including polar ice caps, sea ice, glaciers, snow caps, and permafrost

cumulonimbus cloud: also called "thunderstorm cloud"; a very dense, tall, billowing cloud form that develops an anvil-shaped head due to high-altitude wind shear, and normally accompanied by lightning and heavy precipitation

current: a sustained movement of seawater in the horizontal plane, usually wind-driven

cyclogenesis: the series of atmospheric events that occur during the formation of a cyclone weather system

cyclone: a general term for a low-pressure weather system that rotates counterclockwise in the Northern Hemisphere and clockwise in the Southern Hemisphere

cyclonic: the counterclockwise rotation of an air mass in the Northern hemisphere, clockwise in the Southern hemisphere

dead zones: areas of deepwater oxygen depletion due to surface algal blooms or disruption of thermohaline circulation

decadal: describing a cyclical event that occurs over a period of at least ten years

decay constant: a measure of the radioactivity of an isotope, determined with a Geiger counter

decomposition: process by which organic matter is broken down into its most basic components by microorganisms

deforestation: the loss of forest cover in an area by natural or artificial means

delta: a triangular area with its longest side abutting the sea where a river deposits silt, sand, and clay as it flows into an ocean, lake, or other body of water

dendrochronology: identifying the chronological dates, by year, of past events by coordinating with the growth rings of living trees or trees of known age

dendroclimatology: the study of tree rings as indicators of climatic conditions

density: strictly defined as mass per unit volume, this is a temperature-dependent property as volume changes with temperature, but mass does not

deposition: the process by which loose sediment grains fall out of seawater to accumulate as layers of sediment on the sea floor

desert climate: a hot climate characterized by extremely low and sometimes deficit amounts of precipitation

desertification: the gradual expansion of desert conditions into other climate regions

diffusion: the natural movement of particles through a medium, controlled only by natural vibrations and collisions between particles.

discharge: the volume of water moving through a given flow cross-section in a given unit of time

divergence: a net outflow of air in different directions from a specified region

Dobson spectrophotometer: a ground-based instrument for measuring the total column abundance of ozone at a particular geographic location

doldrums: the equatorial zone where winds are calm and variable and there is heavy thunderstorm rainfall

Dome C: the Antarctic Plateau location of the Concordia Research Station

Doppler radar: a radar method that uses the Doppler shift to measure the speed of targets moving either toward or away from the radar

Doppler shift: a phenomenon in which the wavelength of electromagnetic radiation (or other type of wave) is lengthened by reflection from a surface moving away from the source or shortened by reflection from a surface moving toward the source

dormancy: the portion of the year during which no growth occurs

downburst: a convective windstorm associated with strong thunderstorm systems

downwelling: the bulk movement of water from surface levels to lower depths

drift: a movement similar to a current but more widespread, less distinct, slower, more shallow, and less easily delineated

drought: a long period of no or scarce precipitation

drunken forest: forest that leans at an odd angle as a result of melting permafrost

dry adiabatic lapse rate: the rate at which the temperature of a dry air parcel decreases with altitude as it rises through the atmosphere

dust devil: a rotating column of air rising above a hot ground surface, made visible by the dust it contains; it is much smaller than a tornado, having winds of less than 60 kilometers per hour, and causing little or no damage

dust Veil Index: a numerical index that quantifies the impact of a volcanic eruption's release of dust and aerosols

dynamic mode theory: a theory proposing that enhancement of vertical movement in clouds increases precipitation

East Antarctic ice sheet: ice sheet located east of the Transantarctic Mountains that stores over 60 percent of the world's total freshwater

easterly waves: localized zones of low pressure oriented parallel to the earth's rotational axis and moving from east to west across the ocean; they form an important generative component of tropical weather patterns

eastern trade winds: winds blowing in a northeasterly or southeasterly direction near the equator as a result of the Coriolis effect and the Hadley circulation, so named because they favored sailing ships traveling eastward on trade missions

eccentricity: the departure of an ellipse from circularity; less circularity means greater eccentricity

economic interdependency: a state of affairs in which the economic processes of a group of nations are mutually dependent

ecotone: a transition area between open water and dry land ecosystems

Ekman layer: the region of the sea, from the surface to about 100 meters down, in which the wind directly effects water movement

El Niño-Southern Oscillation (ENSO): a coupled oceanic/atmospheric seesaw that occurs in the equatorial Pacific but often has global climatic consequences

El Niño/La Niña: cyclical increases and decreases, respectively, in Pacific Ocean water temperature that foster shifts in weather patterns worldwide

El Niño: meteorological condition in which the waters of the eastern, tropical Pacific Ocean are warmed by the atmosphere

electrolysis: an industrial process in which electrical current is used to force the reduction of water molecules (H_2O) to hydrogen (H_2) and oxygen molecules (O_2)

electromagnetic waves: the classical form of electromagnetic radiation, produced when electric and magnetic fields come together and interact; can be in the form of radio waves, microwaves, infrared, optical, ultraviolet, x-rays, or gamma rays, depending on their frequency, energy, and wavelength

electromagnetism: relationship between electric energy and magnetic energy responsible for attraction between negatively and positively charged particles

element: one of a number of substances composed entirely of atoms that cannot be broken into smaller particles by chemical means

emission scenario: a set of posited conditions and events, involving climatic conditions and pollutant emissions, used to project future climate change

emissivity: a measure of the ability to radiate absorbed energy

endogenic sediment: sediment produced within the water column of the body in which it is deposited; for example, calcite precipitated in a lake in summer

endosymbiont: an organism living inside a cell or body of another organism

energy from waste: technologies that are designed to produce energy and reduce or eliminate waste at the same time

enhanced greenhouse effect: increased retention of heat in the atmosphere resulting from anthropogenic atmospheric gases

ensemble weather forecasting: repeated use of a single model, run many times using slightly different initial data; the results of the model runs are pooled to create a single "ensemble" weather forecast

environmental lapse rate: the general temperature decrease within the troposphere; the rate is variable but averages approximately 6.5°C per kilometer

epilimnion: a warmer surface layer of water that occurs in a lake during summer stratification; during spring, warmer water rises from great depths, and it heats up through the summer season

equator: an imaginary line, equidistant from the North and South Poles, around the middle of the planet

equator: an imaginary line, equidistant from the North and South Poles, around the middle of the planet where day and night are of equal length

equilibrium: the condition in which no net change occurs within a dynamic system, such as when exactly the same amount of water vapor condenses to liquid as liquid water evaporates to water vapor

equinox: a twice-a-year occurrence during which the tilt of Earth's axis is such that Earth is not tilted toward or away from the sun; the center of the sun is directly aligned with Earth's equator

equivalent potential temperature: the potential temperature an air parcel would have if its water vapor were condensed and the latent heat added to the parcel

533

estuary: an area where the mouth of a river broadens as it approaches the sea, characterized by the mixing of freshwater and saltwater

eukaryote: an advanced cell, containing a nucleus and other membrane-bound organelles

euphotic zone: the upper layer of the oceans, where light-dependent life forms can exist

evaporation: the process by which a liquid changes into a gas

evapotranspiration: the process by which water is transferred from land to air by evaporation from ground-based surfaces and by transpiration from plants and animals.

exosphere: the outermost layer of Earth's atmosphere

extratropical cyclone: a cyclone originating and subsisting outside the tropics

extreme weather events: natural disasters caused by weather, including floods, tornadoes, hurricanes, drought, and prolonged severe hot and cold spells

feedback loops: climatic influences that compound or retard each other, accelerating or decelerating the rate of global warming

feedback: a process in which any change accelerates further changes of the same type (positive feedback) or counteracts itself (negative feedback)

fermentation: the biochemical generation of energy from sugars and organic compounds derived from sugars in the absence of oxygen

firn: the intermediary stage between snow and ice

fjord: a steep-sided narrow inlet eroded into the face of seaside cliff, typical of Scandinavia but found throughout the world

flash floods: rises in water level that occur unusually rapidly, generally because of especially intense rainfall

flood: a rising body of water that overtops its usual confines and inundates land not usually covered by water

fluid: the state of matter characterized by the ability to flow and conform exactly to the shape of its container

fog dissipation: removal of fog by artificial means

food chain: the arrangement of the organisms of an ecological community according to the order of predation in which each consumes the next, usually lower, member as a food source

forecast verification: comparison of predicted weather to observed weather conditions to assess forecasting accuracy and reliability

fossil fuel: combustible materials formed by pressure on plant and animal material over time that are now used as fuels, including coal, oil and natural gas

free oxygen: the element oxygen by itself, not combined chemically with a different element

Freon: trade name for CFCs made by DuPont chemical company

friction: resistance to relative motion caused by contact at a shared surface and influenced by surface irregularities

front: the boundary between two masses of air with different densities and temperatures; usually named for the mass that is advancing (for example, in a cold front, the mass that is colder is moving toward a warmer mass)

fuel alternative: replacement energy source that can be used instead of fuel

fuel: an energy source that is burned to release energy

geographic information system: a network of satellite mapping technologies that can capture detailed images of the land surface

geoid: a dynamic three-dimensional virtual model of Earth in which all points on the 'surface' are at sea level

geomagnetic storm: the effect of variations in solar wind's interactions with Earth's atmosphere; can result in communications disruptions and auroral displays in lower than usual latitudes

geostrophic current: a current resulting from the balance between a pressure gradient force and the Coriolis effect; the current moves horizontally and is perpendicular in direction to both the pressure gradient force and the Coriolis effect

geostrophic wind: a wind resulting from the balance between a pressure gradient force and Coriolis force; the flow produces jet streams and is perpendicular to the pressure gradient force and the Coriolis force

geostrophic: descriptive of wind that occurs when the Coriolis force is in exact balance with the force of a horizontal pressure gradient and therefore blows in a straight line

glacial epoch: an extended period of global temperature reduction and glaciation that generally lasts for millions of years and includes internal glacial and interglacial periods

glacial ice: ice created by the compression of snow, sometimes saturated with meltwater that is refrozen

glacial stage: short-term period of glaciation, generally lasting for less than one million years and alternating with interglacial periods

glaciation: the environmental process by which a terrain becomes covered by a permanent accumulation of ice

glacier: a mass of ice that flows downhill, usually within the confines of a former stream valley

glaze: a coating of ice formed on exposed objects by the freezing of a film of supercooled water deposited by rain, drizzle, or fog

global atmospheric model: computational model of global weather patterns based on a spherical coordinate system representing the entire planet

global dimming: a reduction in the amount of sunlight reaching the surface of the Earth.

global radiative equilibrium: the maintenance of Earth's average temperature through a balance between energy transmitted by the Sun and energy returned to space

global warming potential: a measure of the ability of a greenhouse gas to contribute to global warming, based on its heat-trapping efficiency and its lifetime in the atmosphere

global warming: an overall increase in Earth's average temperature

global: relating to or encompassing the whole Earth

globalization: the worldwide expansion and consequent transformation of socioeconomic interrelationships

Gondwanaland: an ancient supercontinent that geologists theorize broke into at least two large segments; one segment became India and pushed northward to collide with the Eurasian landmass, while the other, Africa, moved westward

gravest empirical mode: concept in which the relationship between oceanic subsurface density and surface elevation is examined

greenhouse effect: a natural process by which water vapor, carbon dioxide, and other gases in the atmosphere absorb heat and reradiate it back to Earth

greenhouse gases (GHGs): atmospheric gases that trap heat within a planetary system rather than allowing it to escape into space; principally carbon dioxide, methane, water and ozone.

greenspace: areas within urban developments characterized by the presence of grass, trees and other plants, maintained using ecologically-friendly or 'green' methods

growing degree-day index: a measurement system that uses thermal principles to estimate the approximate date when crops will be ready for harvest

growing season: the portion of the year during which photosynthesis occurs

guyot: an undersea mountain, or seamount, that has formed by volcanic activity, from which the peak has been eroded through wave action

gyre: the major rotating current system at the surface of an ocean, generally produced by a combination of wind-generated currents and geostrophic currents

Hadley cell: an atmospheric circulation system of air rising near the equator, flowing poleward, descending in the subtropics, and then flowing back toward the equator

Hadley circulation: an atmospheric circulation pattern in which a warm, moist air ascends near the equator, flows poleward, descends as dry air in subtropical regions, and returns toward the equator

hail suppression: a technique aimed at lessening crop damage from hailstorms by converting water droplets to snow to prevent hail formation or, alternatively, by reducing hailstone size

half-life: the time needed for half of a quantity of a radioactive isotope to decay; it is calculated from the decay constant of the specific isotope, not measured directly

halocarbons: the general family of compounds that includes CFCs, HCFCs, HFCs, and other molecules in which carbon atoms are bonded to halogen atoms

halocline: a layer of rapidly changing salinity in north and south polar seas, due to the exclusion of salt from sea ice formation

halon: a compound containing bromine, carbon, chlorine, and fluorine

hazardous: poisonous, corrosive, flammable, explosive, radioactive, or otherwise dangerous to human health

heat budget: the balance between the incoming solar radiation and the outgoing terrestrial reradiation

heat wave: an extended period of abnormally high temperatures

heliosphere: portion of space affected by the presence of the sun or another star

hemispheric model: a numerical model that extends over the whole Northern or Southern Hemisphere, or just one half of the planet

herbicides: substances or preparations for killing plants, such as weeds

heterosphere: a zone of the atmosphere at an altitude of 80 kilometers, including the ionosphere, made up of rarefied layers of oxygen atoms and nitrogen molecules

high-pressure area: region in which the atmospheric pressure is greater than that in the areas around it; represented by H on weather maps

Holocene: the current interglacial, which began 11,700 years ago

homeostasis: the condition in which all components of a multicomponent system are in dynamic balance with each other so that the overall state of the system remains constant

homogeneous climate data: a sequence of values of a climate variable, such as precipitation, which have been observed under the same or similar conditions and with the same or similar measuring equipment; the combination of climate data from two localities that are near each other is often made when considering climate change

homosphere: a major zone of the atmosphere below the heterosphere whose chemical makeup is consistent with the proportions of nitrogen, oxygen, argon, carbon dioxide, and trace gases at sea level; includes the troposphere, stratosphere, and mesosphere

horse latitudes: the belts of latitude approximately 30 degrees north and 30 degrees south of the equator, where the air pressure is high, the winds are very light and the weather is hot and dry

hurricane-force wind: a wind with a speed of 64 knots (118 kilometers per hour) or higher

hurricane: a cyclone that is found in the tropics (between 23.5 degrees north and south of the equator) and that has winds that are equal to or exceed 64 knots, or 74 miles per hour

hydrology: the branch of science dealing with water and its movement in the environment

hydrophyte: a plant that is adapted to aquatic and semi-aquatic environments

hydrosphere: the areas of Earth that are covered by water, including the oceans, seas, lakes, and rivers

hydrostatic balance: the balance between the downward force of gravity and the upward force resulting from the decline in air pressure with altitude

hygroscopic particulates: minute particles that readily take up and retain moisture

Iberian Peninsula: the part of Europe extending from southern France, comprising Spain, Portugal and Malta

ice age: a period during which the average global temperature is reduced, allowing sea ice and glaciers to cover a significant fraction of Earth's surface

ice cap: semipermanent glacial crown of ice atop a mountain or other geologic formation

ice core: a cylinder-shaped piece of ice that is collected by drilling into a glacier; can be used to analyze the history of Antarctica's climate

ice floes: large formations of ice, usually 2.5 to 3.5 meters thick, that float in the waters of the Arctic Ocean

ice sheet: a mass of ice covering a large area of land

ice shelf: a large, flat sheet of freshwater ice formed from a glacier or an ice sheet; floats in a body of water

ice storm: a storm characterized by a fall of freezing rain, with the formation of glaze on objects below

ice stream: a rapidly moving current of freshwater ice flowing from an ice sheet and moving more quickly than the ice that surrounds it; carries ice from the ice sheet

ice-out: synonymous with spring break-up, when winter ice begins to break up as winter ends

iceberg: a large mass of freshwater ice that has broken from an ice shelf or a glacier; floats in a body of water

iceberg: any large block of freshwater ice that has separated from a glacier and floats freely on ocean waters

igloo: a temporary Inuit structure made from blocks of dense snow

incompressible fluid: a fluid whose volume does not change in response to increased pressure

industrialization: the process of transformation from an agrarian society based on animal and human labor to an industrial society based on machines and fossil fuels

inertia: the tendency of physical objects to remain at rest or with unchanging speed and direction unless acted upon by an external force

inertial frame: the frame of reference by which speed and distance measured

infrared radiation: electromagnetic radiation with frequency in the range of 1013 to 1014 Hertz (Hz)

insolation: the amount of solar energy received at a given point on the Earth

Inter-Tropical Convergence Zone (ITCZ): a low-pressure belt, located near the equator, where deep convection and heavy rains occur

interglacial: the period of time between the retreat of existing glaciers and the formation of new glaciers, characterized by the development of warming climate conditions that are unfavorable for glaciation

Inuit: the indigenous peoples of the northern polar regions, whose name means "the People"; often referred to incorrectly as Eskimos, a word from a more southerly native language

inversion: an unusual atmospheric condition in which temperature increases with altitude

ionosphere: portions of the upper atmosphere consisting of part of the mesosphere, thermosphere, and exosphere; characterized by gas ionization through exposure to solar radiation

isobar: on a map, a line connecting two or more points that share the same atmospheric pressure, either at a particular time or, on average, in a particular period

isostacy: equilibrium between the lithosphere of the earth and the liquid layer of rock in the inner layers of the strata

isostatic rebound: a tendency of Earth's continental surfaces to rise after being depressed by continental glaciers, without faulting

isotopes: atoms of an element that have the same number of protons in their nuclei, but different numbers of neutrons; many elemental isotopes are radioactive

Java Trench: one of the deepest areas of the Indian Ocean, located off the southern coast of Java in Indonesia; it is a form of geological canyon created by the upward thrust of mountain ridges from the ocean floor

jökulhlaup: a flood produced by the release of water sequestered by a glacier, most often due to the failure of some type of glacial dam or to subglacial volcanic activity

knot: a unit of nautical distance equivalent to 1.86 kilometers or 1.15 miles

Kyoto Protocol: a 1997 international agreement to limit greenhouse gas emissions

Köppen climate classification system: a system for classifying climate based mainly on average temperature and precipitation

La Niña: meteorological condition in which the waters of the eastern, tropical Pacific Ocean are cooled by a lack of radiation from the atmosphere

lagoon: a body of saltwater separated from the ocean by a bank of sand

lapse rate: the change in a variable with height, often used to discuss changes in temperature with altitude in a context of climate change

late heavy bombardment: a period about 3.9 to 4.0 billion years ago, when Earth was pummeled by debris from space at 1000 times the normal rate, heating the atmosphere and melting the crust

latent heat flux: the flux of thermal energy from land surface to the atmosphere that is associated with evaporation and transpiration of water from ecosystems

liberal institutionalism: a school of thought that focuses on cooperation between countries derived from agreements and organizations

libertarian: based on the principles of minimal government control and maximal personal and corporate freedom

lidar: type of remote sensor that operates similarly to radar but uses lasers instead of radio waves

limestone: a rock containing calcium carbonate that reacts readily with acid rain and tends to neutralize it, being chemically eroded in the process

lithosphere: the portion of Earth consisting of solid rock

local winds: winds that, over a small area, differ from the general pressure pattern owing to local thermal or orographic effects

loess: deposits of very fine grained, wind-blown material often associated with glacial deposits

long-range prediction: a weather forecast for a specific region for a period greater than one week in advance, often supplemented with climatological information

longwave radiation: electromagnetic radiation in the infrared range, having wavelengths longer than those of the visible spectrum, associated with heat and heating

low-pressure area: region where the atmospheric pressure is lower than that in surrounding areas; represented by L on weather maps

lower atmosphere: region of the atmosphere comprising the troposphere and the tropopause, reaching an altitude as high as 19 kilometers, or 12 miles

luminosity of Sun: the total energy output of the Sun every second, measured in watts

löess soils: fine-grained calcareous silt, usually light brown or grey in color, that is easily eroded by wind and water

Madden-Julian oscillation: intraseasonal tropical wave that travels around the globe, causing monsoons and other high-water storms and also suppresses them

magnetosphere: outer layer of the earth's atmosphere constituted by the interaction between the earth's magnetic field and charged particles released by the sun

Malthusian: related to the inevitability of the negative future of technological societies

marine isotope stage: half of a glacial cycle, as identified in the oxygen isotope data from ocean cores; advances are given even numbers, and retreats are given odd numbers

mass balance: the difference between the accumulation of snow and the ablation of ice on a given glacial formation

Maunder Minimum: a period from about 1645 to about 1715 when very few sunspots were observed

mean free path: the average distance traveled by a gas molecule or other free-moving particle between collisions with other molecules or particles.

mean: the sum of the individual values of a series of measurements, divided by the number of measurements

meltwater: liquid water from the thawing of ice and snow

mercury barometer: glass tube of a minimum of 84 centimeters (33 inches), closed at one end, with a mercury-filled pool at the base; the weight of the mercury creates a vacuum at the top of the tube; mercury adjusts its level to the weight of the mercury in the higher column

meridional: parallel with the north-south direction of the meridians of longitude

mesoscale model: a weather forecast for an area of up to several hundred square kilometers in extent on a time scale of between one and twelve hours

mesosphere: the extremely rarefied atmospheric layer at altitudes from 50 to 80 kilometers above the surface, characterized by rapid decreases in temperature

metalliferous: metal-containing, formed from metal salts and metal-bearing minerals

meteorology: the study of changes in temperature, air pressure, moisture, and wind direction in the troposphere; the interdisciplinary scientific study of the atmosphere

microscale: the smallest scale of atmospheric motion, ranging from meters to kilometers

mid-latitude cyclone: a synoptic-scale cyclone found in the mid-latitudes (between 30 and 60 degrees north and south of the equator)

Mid-Ocean Ridge: a ridge of volcanic sea-floor mountains that approximately bisects the Atlantic Ocean, formed by the movement of the east and west Atlantic tectonic plates away from their shared boundary, driven by the upwelling of molten magma

middle atmosphere: region of the atmosphere comprising the mesosphere, mesopause, stratosphere, and stratopause

midlatitudes: the latitudes north and south of the equator between the Tropics of Cancer and Capricorn

Milankovitch cycle: recurring time periods during which the shape of Earth's orbit, the tilt of its axis, and the occurrence of its farthest distance from the Sun all change

mineral: an inorganic substance occurring naturally and having definite physical properties and a characteristic chemical composition that can be expressed by a chemical formula

mineraloid: a solid substance with a constant chemical composition but without a well-ordered crystal structure

MIS 5e: the most recent interglacial before the Holocene, also known as the Eemian, LIG (Last InterGlacial), or Termination II

MIS 11: an interglacial that may be the best analogue for the Holocene; also called the Holsteinian or Termination V

moderation: the effect of buffering against extremes of temperature by the presence of a large body of water

modes: phases of a climatic seesaw; for example, El Niño is the warm mode of the ENSO seesaw, whereas La Niña is the cold mode

monitoring: the systematic observation of an element such as precipitation, sea surface temperatures, or wind speed; such observations are usually made every six hours and sometimes every three hours, hourly, or (conversely) only once daily

monsoon: a seasonal movement of winds into and out of the Indian Ocean region caused by variations of atmospheric pressure over the Indian Ocean and the interior land mass of Asia

Montreal Protocol: a 1987 international agreement to phase out the manufacture and use of ozone-depleting chemicals, especially CFCs

multidecadal oscillation: an alternation spanning several decades between warm and cool periods in oceanic water flows.

multidecadal: occurring over a period of multiple decades

nacreous: having the iridescent appearance of nacre, or 'mother of pearl'

Navier-Stokes equation: an equation describing the flow of air and other fluids

near-infrared radiation: the shortest-wavelength segment of infrared radiation, which has a longer wavelength than visible light; some solar radiation is in the near-infrared frequency range

negative emission: a process that removes carbon dioxide from the atmosphere; technologies include biochar and bioenergy with carbon capture and storage

negative feedback: a cause-and-effect mechanism in which the result tends to decrease the effectiveness of the cause

neoliberalism: a school of economic thought that stresses the importance of free markets and minimal government intervention in economic matters

net radiative heating: the driving force for atmospheric thermodynamics, essentially the difference between heat entering the atmosphere due to solar heating and heat leaving the atmosphere as infrared radiation

neutralization: the adjustment of the concentration of hydrogen ions in solution in order to achieve neutral pH

NGO: acronym for Non-Governmental Organization, generally not-for-profit, charitable and humanitarian in purpose

nipping: process in which ice pushes forcibly against the edge of a ship

nitric acid: an acid formed in rain from nitric oxide gases in the air

nitric oxide gases: gases formed by a combination of nitrogen and oxygen, particularly nitrogen dioxide and nitric oxide

nodule: a lump of mineral rock typically found on the ocean floor

nondeterminism: chaotic, random events that cannot be predicted but that have a significant influence on the development of weather systems

North Atlantic Oscillation (NAO): a seesaw in pressure between the Azores and southwestern Iceland, thought by some scientists to be an expression of the Northern Annular Mode

Northern annular mode (NAM) and Southern Annular Mode (SAM): also called, respectively, the Arctic and Antarctic Oscillations, seesaws in pressure between the latitudes near 45° northern (or southern) latitude and the North (or South) Pole

Northern Ocean: Arctic portion of Earth's oceanic system

Norwegian Sea: the body of water north of the North Sea

nowcasting: a very short-term weather forecast usually for the prediction of rapidly changing, severe weather events within a time of no more than a few hours

NOX: an acronym denoting oxides of nitrogen

nuclear fusion: atomic nuclei join together to form a heavier nucleus; in the sun and other main-sequence stars, hydrogen is fused to form helium

obliquity: the angle of tilt between the earth's rotational axis and an axis perpendicular to the plane of its orbit

observing systems: systems of collectively gathering and analyzing temperature and other atmospheric observations, particularly (in modern times) through the World Meteorological Organization

ocean circulation: the worldwide movement of water in the sea

ocean-atmosphere coupling: the interaction between the sea surface and the lower atmosphere that drives many patterns and changes in Earth's weather systems

off-gassing: the spontaneous emission of entrained or entrapped gases from within natural and artificial sources

opacity: the degree to which a substance or object lets various forms of electromagnetic radiation pass through it.

optics: branch of physics that deals with the properties and characteristics of light

orbital eccentricity: cyclically variant deformities in Earth's orbit

organic matter: carbon-containing compounds produced by life processes

orographic barrier: a mountain or a hilly area that acts as a control on the climate on either side

orography: study of mountains that incorporates assessment of how they influence and are affected by weather and other variables

oscillation: variation of some physical property or condition between two opposing states, much like the rising and falling of a wave between its maximum and minimum heights

overturning circulation: essentially the equivalent of an oceanic convection current, in which warmer, less dense water moves at the ocean's surface in one direction, sinks as it becomes colder, and moves back through the ocean at lower depths.

oxides of nitrogen: several gases that are formed when molecular nitrogen is heated with air during combustion, primarily NO and NO2

oxides of sulfur: gases formed when fuels containing sulfur are burned, primarily SO2

ozone layer: a region in the lower stratosphere, centered about 25 kilometers above the surface of Earth, which contains the highest concentration of ozone found in the atmosphere

ozone: the molecular form of oxygen containing three atoms of oxygen per molecule as O3, as compared to elemental oxygen having the molecular formula O_2

Pacific Decadal Oscillation (PDO): a temperature, pressure, and wind seesaw in the Pacific Ocean

Pacific-North American (PNA) pattern: a seesaw between northern Pacific and North American pressures

pack ice: large, mobile masses of frozen, floating seawater that are not attached to a landform; also known as sea ice

paleobotany: the study and identification of plants preserved in the fossil record

paleoclimate: the climate that existed at a particular period in Earth's distant past

paleoclimatology: study of climate conditions in Earth's ancient and prehistoric past

paleodepth: an estimate of the water depth at which ancient seafloor sediments were originally deposited

paleontology: the study of the flora and fauna that existed in Earth's distant past

Palmer Drought Index: a widely adopted quantitative measure of drought severity that was developed by W. C. Palmer in 1965

parameterization: assigning of measurable factors that simplify the description of complex phenomena that are otherwise difficult to quantify, to enable predictions to be made from the system model

parts per million: number of molecules of a chemical found in one million molecules of the atmosphere

passive sensor: type of remote sensor that detects naturally emitted energy, such as reflected sunlight, from target sources

pedosphere: the soil

perihelion: the time at which the distance between Earth and the sun is largest; generally occurs on one of the first days of July, two weeks after the June solstice

permafrost thawing: defrosting of previously permanently frozen ground, usually in or near the Arctic

permafrost: a subsoil layer in which the entrained water remains in a frozen state regardless of the surface temperatures

pesticides: chemical preparations that kill pests, including unwanted animals, fungi, and plants

pH: a measure of the hydrogen ion concentration, which determines the acidity of a solution; the lower the pH, the greater the concentration of hydrogen ions and the more acidic the solution

photochemical oxidants: pollutants formed in air by primary pollutants undergoing a complex series of reactions driven by light energy

photochemical reaction: a type of chemical reaction that can occur in polluted air driven by the interaction of sunlight with various pollutant gases

photochemical: chemical reactions or processes that are driven by interaction of the reacting materials with light

photodissociation: the condition in which light energy absorbed by a molecule is sufficient to dissociate the bonds between atoms in the molecule, typically caused by light in the ultraviolet range

photolysis: a chemical process in which single atoms are separated fro a molecule by the absorption of light energy

photosynthesis: a metabolic pathway that absorbs inorganic carbon dioxide from the atmosphere and converts it to organic carbon compounds using sunlight as an energy source

phytoplankton: free-floating microscopic aquatic plants that use sunlight to convert carbon dioxide and water into food for themselves and for other organisms in the food chain

pingo: a large, stable ice intrusion of the Arctic tundra terrain, appearing as a large, dome-shaped, earth-covered mound with cracks visible at the top, the core being solid ice

planetary scale: the largest scale of atmospheric motion, covering the entire globe

planetary winds: the large, relatively constant prevailing wind systems that result from Earth's absorption of solar energy and that are affected by Earth's rotation

plankton: microscopic marine plants and animals that live in the surface waters of the oceans; these

floating organisms precipitate the particles that sink to form biogenic marine sediments

plasma: state of matter similar to a gas in which a portion of the molecules have become ionized, giving rise to matter built from free ions and electrons

Pleistocene epoch: an epoch within the geologic timescale, ending at 11,700 years ago

Pleistocene-Quaternary glaciation: current ice age beginning approximately 2.4 million years ago

polar low: a severe, mesocyclonic winter storm that occurs in higher ocean latitudes

polar stratospheric clouds: clouds of ice crystals formed at extremely low temperatures in the polar stratosphere

polar vortex: a closed atmospheric circulation pattern around the South Pole that exists during the winter and early spring; atmospheric mixing between the polar vortex and regions outside the vortex is slow

positive climate feedback loops: self-reinforcing climatic processes, such as increased snow cover increasing planetary albedo, promoting additional cooling and therefore more snow cover

positive feedback: a cause-and-effect mechanism in which the result tends to augment the effectiveness of the cause

potential evapotranspiration: the water needed for growing plants, accounting for water loss by evaporation and transpiration

potential temperature: the temperature an air parcel would have if it were adiabatically expanded or compressed to a standard pressure of 100 kilopascals

prebiotic: relating to the period of time before the appearance of life on Earth

precession: a cyclical change in the orientation of the axis of rotation in a rotating body or system

precipitation: liquid or solid water particles that fall from the atmosphere to the ground

pressure gradient force: a wind-producing force caused by a difference in pressure between two different locations

pressure gradient: a difference in pressure that causes fluids (both liquids and gases) to move from regions of high pressure to regions of low pressure

primary air pollutants: harmful substances that are emitted directly into the atmosphere

primary crystalline rock: the original or first solidified molten rock of Earth

primary production: production of fixed carbon through photosynthesis

primordial isotope: an isotope that has been present on Earth since the planet formed 4.5 billion years ago

primordial solar nebula: an interstellar cloud of gases and dust that condensed by the action of gravitational forces to form the bodies of the solar system about 5 billion years ago

probability: in its simplest definition, the likelihood of a specific event happening, although determining a probability can be a very complex problem as the number of independent variables increases

productivity: the rate at which plankton reproduce in surface waters, which in turn controls the rate of precipitation of calcareous or siliceous shells or tests by these organisms

prokaryote: a primitive cell (bacterium), lacking a nucleus and other membrane-bound organelles

Promethean: related to the inevitability of a positive future for technological societies

proxies: measurable parameters, correlated with climate, that are preserved in the geologic record (for example, oxygen isotope ratios and fossil pollen)

pseudorandom: a series of random numbers or code signals generated by a non-random computational method or algorithm

pulse Doppler: type of radar system that emits waves of electromagnetic energy at an atmospheric target;

543

provides a detailed profile of motion, precipitation, and other conditions and objects

pycnocline: a zone within a body of water, characterized by a rapid rate of change in density

radiation budget: the balance between solar energy entering Earth's atmosphere from space and the heat energy leaving Earth's atmosphere to pass outward into space

radiational cooling: the cooling of Earth's surface and the layer of air immediately above it by a process of radiation and conduction

radiative balance: the balance between incoming and outgoing radiation of a body in space, such as Earth

radiative cooling: the process of emitting energy as longwave radiation to balance the solar energy absorbed as shortwave radiation

rain gauge: an instrument for measuring rainfall, usually consisting of a cylindrical container open to the sky

rain shadow: the region on the lee side of a mountain where precipitation is noticeably less than on the windward side

rawinsonde: a radiosonde tracked by radar in order to collect wind data in addition to temperature, pressure, and humidity

recurrence interval: the average time interval in years between occurrences of a flood of a given magnitude in a measured series of floods

red clays: fine-grained, carbonate-free sediments that accumulate at depths below the CCD in all ocean basins; their red color is caused by the presence of oxidized fine-grained iron particles

reforestation: planting trees to replace forests that have been eliminated either naturally or by human action

regimes: another word for "modes," fitting in with meteorological metaphors such as "fronts"

regional: relating to areas such as the Pacific, the Atlantic, the tropics, and large land areas, such as North America or Australia

relative motion: proportional speed and distance of two or more objects moving within the sane inertial frame of reference

respiration: metabolic reactions and processes to convert organic compounds to energy that release CO_2 as a by-product

reticulate: resembling a more-or-less regular network of lines and cracks

riparian: associated with the ecosystem existing along the shores and banks of natural freshwater bodies such as rivers, streams and lakes

rotation: refers to the spinning of atoms and molecules about a central axis

runoff: that part of precipitation that flows across the land and eventually gathers in surface streams

Sahel: the semiarid southern fringe of the Sahara in West Africa that extends from Mauritania on the Atlantic coast to Chad in the interior

salinity: a measure of the quantity of dissolved solids in ocean water, typically given in parts per thousand by weight

saltation: the process by which grains of sand and other particulate matter are made to move in a leap-frog manner within a moving fluid such as water or air, resulting in the formation of characteristic structures ranging in size from small ripples to giant sand dunes

saltwater wedge: a wedge-shaped intrusion of seawater from the ocean into the bottom of a river; the thin end of the wedge points upstream

saturated: the condition in which air holds the maximum amount of water vapor for the particular conditions of temperature and pressure

savannah: flat grasslands found in the tropics and sub-tropics

scatterometer: active radar that emits high-frequency microwave pulses at a target

sea ice: frozen ocean water

sea-level rise: generally indicates the gradual increase over time of the average sea level due to the influx of meltwater from glaciers and polar icecaps, but also applies locally in events such as storm surge

secondary air pollutants: harmful substances that result from the reaction of primary air pollutants with principal atmospheric components

sedimentary rock: rock formed by the repeated deposition of sediment in a body of water or by the layering of material on land

seesaw: a change in opposite directions, such as high pressure in one region and low pressure in the other

seiche: rocking motion of lake level from one end of the lake to the other following high winds and low barometric pressure; frequently, a seiche will follow a storm event

semiarid climate: a climate of low annual precipitation (250-500 millimeters)

sensible heat flux: the flux of thermal energy that is associated with a rise in temperature

seston: a general term that encompasses all types of suspended lake sediment, including minerals, mineraloids, plankton, and organic detritus

severe thunderstorms: mostly summer convective storms involving microscale rotating winds

shortwave radiation: electromagnetic radiation in the ultraviolet range, having wavelengths shorter than those of the visible spectrum, associated with the transport of energy capable of promoting chemical reactions

soil moisture: water that is held in the soil and that is therefore available to plant roots

solar cycle: an approximately eleven-year-long cycle of varying solar activity; solar cycles are tracked based on the visibility of sunspots

solar flare: a sudden eruption of material from the surface of the Sun, typically associated with sunspot activity, often followed by disturbances in Earth's magnetic field

solar heating: heat acquired by absorption of radiant energy from the Sun

solar irradiance: the amount of light and energy emitted from the Sun at any particular time

solar maximum: the period of time at which the number of sunspots is highest

solar radiation: transfer of energy from the sun to Earth's surface, where it is absorbed and stored

solar ultraviolet radiation: biologically lethal solar radiation in the spectral interval between approximately 0.1 and 0.3 micron (1 micron = 0.0001 centimeter)

solar wind: a stream of charged particles that the sun's atmosphere ejects into space, where it can interact with the magnetic fields of planets

solstice: a twice-a-year occurrence during which the sun appears at its highest point in the sky (once a year as seen from the North Pole and once a year as seen from the South Pole)

Somali Current: a seasonally reversing current that moves between the eastern coasts of Africa and the Arabian Peninsula

Southern Ocean: Antarctic portion of Earth's oceanic system

spectroscopy: study and use of the interactions of light with atoms and molecules, enabling the analysis of molecular structures and properties

speleotherm: stalactites, stalagmites and other features that develop in caves

squall line: any line of vigorous thunderstorms created by a cold downdraft that spreads out ahead of a fast-moving cold front

static mode theory: a theory assuming that natural clouds are deficient in ice nuclei, whereby clouds must be within a particular temperature range and contain a certain amount of supercooled water for cloud seeding to be successful

statistics: the mathematical treatment of a set of collected values in order to determine essential ratios and trends

steam reforming: an industrial process employing steam and pressure to force conversion of methane to methanol and molecular hydrogen

stoma: a pore in the leaf and stem epidermis that is used for gas exchange

stoma: a pore within the surface of leaves, through which various gases such as carbon dioxide and oxygen enter and exit the leaf

storm surge: a rapid rise in lake level associated with low barometric pressure; the water level is frequently "pushed" above a shoreline on one end of the lake and depressed on the opposite end

strait: a narrow waterway connecting two larger bodies of water

stratosphere: part of the atmosphere just above the troposphere that can hold large amounts of aerosols produced by volcanic eruptions for many months

sublimation: the change of physical state of a material from solid directly to gas without passing through an intermediate liquid state

subsidence: in meteorology, the slow descent of air that becomes increasingly dry in the process, usually due to an area of high pressure

subtropical high belt: a high-pressure belt where warm, dry air sinks closer to the surface

sulfur dioxide: a gas whose molecules consist of one sulfur atom and two oxygen atoms, formed by the combustion of sulfur in the presence of oxygen

sulfuric acid: an acid formed as the primary component of acid rain by reaction of sulfur dioxide gas with liquid water in the atmosphere

sunspot cycle: also known as the solar activity cycle, an eleven-year cycle in the number of sunspots and amount of other solar magnetic activity

sunspot minimum/maximum: the time when there is the minimum/maximum number of sunspots during the eleven-year sunspot cycle

sunspot: a cooler area on the sun's surface that appears darker than the surrounding area; a zone of decreased temperature resulting from the complex shape of the sun's magnetic field

supercooled: a liquid cooled below its normal freezing point without crystallizing or becoming solid, typically referring to water

supersaturated: the condition in which the amount of water vapor in air exceeds the maximum amount according to the particular conditions of temperature and pressure

synoptic scale: a scale used to describe high- and low-pressure atmospheric systems that have a horizontal span of 1,000 kilometers (621 miles) or more

synthetic aperture radar: type of radar that emits a high volume of radio waves at a target as it passes overhead, creating a multidimensional image of the target

T Tauri stars: a class of stars that exhibits rapid and erratic changes in brightness

tectonic plate: a segment of Earth's crust that is put into movement by magmatic convection in the underlying mantle layer

teleconnection: a connection between two widely separated regions of the planet that have highly correlated changes in some climatic parameter, usually resulting from a seesaw

temperature inversion: a condition in which a region of warmer occupies a position above its normal location, causing air temperature to increase with increasing elevation from Earth's surface

terrigenous: originating from terrestrial, or land-based, sources

test: an internal skeleton or shell precipitated by a one-celled planktonic plant or animal

thermal inertia: a statement of the rate at which a body of matter approaches thermal equilibrium with its surroundings

thermal infrared radiation: the longest-wavelength segment of infrared radiation, which has a longer wavelength than visible light; the earth emits radiation in the thermal infrared frequency range

thermocline: a well-defined layer of water in a lake separating the warmer and shallower epilimnion from the cooler and deeper hypolimnion

thermohaline circulation: a mode of oceanic circulation that is driven by the sinking of denser water and its replacement at the surface with less dense water

thermohaline cycle: the "great conveyor belt" of ocean currents powered by density gradients created by heat and relative salt content

thermohaline: indicates that both salinity and temperature are determining factors

thermosphere: region of the atmosphere marked by thin gases and ultraviolet radiation; also known as the upper atmosphere

tidal range: the difference in water depth between high and low tides

tipping point: the point at which the transition from one state in a system to another becomes inevitable

topsoil: the uppermost layer of soil on Earth's surface, being the interface between the solid surface and the atmosphere, and the locus of essentially all plant growth

tornado: a violent rotating column of air that forms and extends downward from a cumulonimbus cloud and has the appearance of a funnel, rope, or column that touches the ground

total column abundance of ozone: the total number of molecules of ozone above a 1-centimeter-square area of Earth's surface

Total Ozone Mapping Spectrometer (TOMS): a space-based instrument for measuring the total column abundance of ozone globally

total potential energy: the sum of the internal energy and the gravitational potential energy of an air column

trade winds: winds that blow steadily toward the equator; north of the equator, trade winds blow from the northeast, whereas south of the equator they blow from the southeast

translation: refers to the motion of atoms and molecules moving through three-dimensional space

transpiration: the process by which water in plants is transferred as water vapor into the atmosphere

trench: an extraordinarily deep region of the sea floor where two tectonic plates meet and the one making up the deep sea floor is subducted beneath the less dense, and therefore lighter, continental mass of the adjacent tectonic plate

trilateration: identification of a singular point on a curved surface by the intersection of three lines, analogous to triangulation on a plane surface

tropical climate: a climate generally found in equatorial or tropical regions that exhibit high temperatures and precipitation year-round

tropical cyclone: an area of low pressure that forms over tropical oceans, characterized by extreme amounts of rain, a central area of calm air, and winds that attain speeds of up to 300 kilometers per hour rotating counterclockwise in the Northern Hemisphere and clockwise in the Southern Hemisphere

tropical depression: cyclonic thunderstorms with wind speeds from 36 to 64 kilometers per hour

tropical storm: a thunderstorm with cyclonic winds circulating at speeds of 64 to 118 kilometers per hour

tropical zone: area between the Tropic of Cancer and the Tropic of Capricorn that falls directly under the sun during a part of the year

tropopause: the transition region between the troposphere and the stratosphere

troposphere: an atmospheric region extending from the Earth's surface to about 17 kilometers high over equatorial regions and to about 8 kilometers high over polar regions

trough: a long and relatively narrow area of low barometric pressure

tsunami: the influx of water on a coastline, associated with a sudden, massive displacement of water due to undersea seismic events and landslides

tundra: the treeless region between the northern tree line, characterized by low-growing vegetation and permafrost subsoil

ultraviolet light: electromagnetic radiation having a frequency in the range of 1015 to 1017 Hz

ultraviolet solar radiation: electromagnetic radiation having wavelengths between 4 and 400 nanometers

umiak: a large boat, constructed with animal skins on a wood and bone frame, that the Inuit traditionally used when hunting marine mammals such as seals and whales

unstable air: a condition that occurs when the air above rising air is unusually cool so that the rising air becomes relatively warmer and accelerates upward

upwelling: the process by which colder, deeper ocean water rises to the surface and displaces surface water

urban heat island: a spot on Earth's surface that is significantly warmer than the surrounding area as a result of human alterations to the landscape, primarily cities

urbanization: the replacement of open countryside by housing and other urban developments

varve: an annual layer in a sediment, usually the result of seasonal variation in inputs

vibration: refers to several different spring-like oscillations that bonds within molecules are capable of undergoing, such as stretching and compression, and asymmetric and symmetric scissoring

viscosity: a property of fluids determined by intermolecular attractive forces which in turn determines the resistance of the fluid to motion

volatile outgassing: the release of the gases and liquids, such as argon, water vapor, carbon dioxide, and nitrogen sulfur, trapped within Earth's interior during its formation

volcano: a usually mountainous rift in Earth's crust caused by magma erupting through fissures onto the planet's surface

vortex: the center of a whirling or rotating fluid, typically with low pressure in the center due to loss of the fluid mass in that locus by centrifugal force

Vostok: a Russian Antarctic research station built in 1957 during the First Geophysical Year

Walker circulation: an atmospheric circulation pattern in the Pacific and elsewhere in which hot, moist air rises, travels eastward, cools and dries, descends, and returns westward

warm cloud: a visible suspension of tiny water droplets at temperatures above freezing

warm front: area in which a dominant warm current of air rises over a pocket of cold air to cause condensation and precipitation at the trailing edge of the front

water-based barometer: also known as a storm glass or Goethe barometer, a device with a glass container and a sealed body half full of water; also has a spout that fills with more or less water depending upon atmospheric conditions and their forces

waterspout: a tornado that exists over water; less violent and smaller waterspouts form in fair weather just as dust devils do over dry land

weather: the set of atmospheric conditions obtaining at a given time and place

weathering: the breaking down of rocks by chemical, physical, and biological means

West Antarctic ice sheet: the smallest ice sheet in Antarctica, located west of the Transantarctic Mountains

west coast marine climate: a mild, annually moist climate found along the western coasts of continents

westerlies: belts of wind in midlatitudes that generally blow from west to east

wet adiabatic lapse rate: the rate of temperature decrease with altitude of a rising air parcel in which condensation occurs

wetlands: areas along a coast where the water table is near or above the ground surface for at least part of the year; wetlands are characterized by wet soils, water-tolerant plants, and high biological production

wildfire: spontaneously ignited, naturally occurring fire

wind stress: the frictional interaction between moving air and the surface of the ocean

wind-driven circulation: movement of ocean water caused by frictional interaction with moving air

wind-driven circulation: the surface currents on the ocean that result from winds and geostrophic currents

xylem: the supporting and water-conducting tissue of trees and other vascular plants

GENERAL BIBLIOGRAPHY AND ADDITIONAL READING

"Acid Rain Program." *EPA*. Environmental Protection Agency, 26 Jan. 2015. Web. 4 Mar. 2015.

Ackerman, Steven A., and John A. Knox. *Meteorology: Understanding the Atmosphere*. 3rd ed. Jones and Bartlett Learning, 2012. Provides an overview of the atmosphere and atmospheric phenomena, beginning from the evolution of the early terrestrial atmosphere. Suitable for university-level readers.

Adhiya, Jagat, and Sallie W. Chisholm. "Is Ocean Fertilization a Good Carbon Sequestration Option?" MIT Press, 2001. This white paper presents the case against oceanic fertilization.

Aghedo, A. M., et al. "The Impact of Orbital Sampling, Monthly Averaging, and Vertical Resolution on Climate Chemistry Model Evaluation with Satellite Observations." *Atmospheric Chemistry and Physics* 11, no. 13 (2011): 6493–6514. Describes the combined use of satellite-based sensors and computer climate-change models for a series of assessments performed by the Intergovernmental Panel on Climate Change in 2011.

Aguado, Edward, and James E. Burt. *Understanding Weather and Climate*. 7th ed. Pearson Prentice Hall, 2015. Discusses meteorology and climatology concepts with reference to common, everyday events. Presents conclusions from the IPCC as well as many other scientific studies on climate change. Examines weather events, structure and dynamics of atmosphere, and the past, present, and future climate on earth.

Ahrens, C. Donald, and Robert Henson. *Meteorology Today: An Introduction to Weather, Climate, and the Environment*. Cengage, 2019. One of the most widely used and authoritative introductory textbooks for the study of meteorology and climatology. Explains complex concepts in a clear, precise manner and supports them with numerous images and diagrams. Discusses hurricanes and the mechanisms that generate them extensively.

Ahrens, C. Donald, and Perry J. Samson. *Extreme Weather and Climate*. Brooks Cole, 2010. Provides a nontechnical review of the different types of severe weather, such as hurricanes, tornado-producing storms, and flooding. Discusses the long-term weather patterns that produce such weather.

Ahrens, C. Donald. *Essentials of Meteorology: An Invitation to the Atmosphere*. Brooks/Cole Cengage Learning, 2012. Widely used introductory textbook on atmospheric science; covers a wide range of topics on weather and climate such as tornadoes and thunderstorms, acid deposition and other air pollution topics, humidity and cloud formation, and temperature.

Aken, Hendrik Mattheus van. *The Oceanic Thermohaline Circulation: An Introduction*. Springer, 2011. Presents a global hydrographic description of the thermohaline circulation, based on data obtained directly from the WOCE.

Aldrete, Gregory S. *Floods of the Tiber in Ancient Rome*. Johns Hopkins University Press, 2007.

Alexander, David. *The Sun*. Greenwood Press, 2009. Part of the Greenwood Guides to the Universe series, this easy-to-understand text gives readers the necessary background on the sun before delving into a comprehensive look at recent discoveries and future research goals.

Allen, Gerald R., and D. Ross Robertson. *Fishes of the Tropical Eastern Pacific*. University of Hawaii Press, 1994.

Allen, P. A. *Earth Surface Processes*. Blackwell Science, 2002.

Allen, Robert J., and Charles S. Zender. "Forcing of the Arctic Oscillation by Eurasian Snow Cover." *Journal of Climate* 24, no. 24 (2011): 6528–6539. Describes the factors that contribute to the Arctic Oscillation, including sea level, greenhouse gases, and warm seawater. Argues that snow cover, particularly in northern Europe and Asia, plays an important role in this process.

Alley, Richard B. *Earth: The Operators' Manual*. Norton, 2011. Through storytelling, this book illuminates the history of humankind's use of energy while also discussing the negative effects this has had on Earth. The book also delves into current and future alternative and renewable energy options, such as solar power.

_____. *The Two-Mile Time Machine. Ice Cores, Abrupt Climate Change and Our Future*. Princeton University Press, 2014. Describes the analytical methods and interpretations of physical clues of past climates trapped in ice that is thousands of years old.

Alverson, Keith D., Raymond S. Bradley, and Thomas Pedersen, eds. *Paleoclimate, Global Change, and the Future.* Springer Verlag, 2003. Collection of scholarly papers comparing natural paleoclimate changes with modern anthropogenic trends; discusses ancient civilizations.

American Museum of Natural History. *Ocean.* Dorling Kindersley Limited, 2006. Discusses the geology, circulation, climate, and physical characteristics of the ocean. Covers marine biology and ocean chemistry. Includes a discussion of icebergs and polar ocean circulation. An excellent starting point for anyone learning about oceans and marine ecology. Includes images on each page, an extensive index, a glossary, and references.

American Society of Civil Engineers, Hurricane Katrina External Review Panel. *The New Orleans Hurricane Protection System: What Went Wrong and Why.* ASCE, 2007. *ASCE Library,* doi:10.1061/9780784408933. Accessed 6 Sept. 2017.

Andersen, Stephen O. and Sarma, K. Madhavi *Protecting the Ozone Layer. The United Nations History.* UNEP/Earthscan, 2012. In this book, the authors provide a detailed history and discussion of the effects of chlorofluorocarbons and related compounds on the stratospheric ozone layer, and of the efforts made to curtail the damage.

Anderson, David E., Andrew S. Goudie and Adrian G. Parker. *Global Environments Through the Quaternary. Exploring Environmental Change.* 2nd ed., Oxford University Press, 2013. Provides concise descriptions of several ocean oscillations in relation to their cyclic effects on climatic conditions.

Anderson, John D., Jr. *A History of Aerodynamics and Its Impact on Flying Machines.* Cambridge University Press, 1997.

Anderson, Michael. *Investigating the Global Climate.* The Rosen Publishing Group, 2011. Geared to younger readers, this book nevertheless provides clear, definitive descriptions of the various segments of the global climate system that are a useful starting point for all levels.

Andronache, Constantin, ed. *Mixed-Phase Clouds. Observations and Modeling.* Elsevier, 2018. An advanced text in which the basic principles of cloud formation are discussed and analyzed.

Annin, Peter. *Great Lakes Water Wars.* Island Press, 2018.

"Antarctica Hits Record High Temperature at Balmy 63.5°F." *Reuters,* 1 Mar. 2017, www.reuters.com/article/us-antarctica-temperatures-idUSKBN1684I7. Accessed 1 June 2017.

Appenzeller, C., T. F. Stocker, and M. Anklin. "North Atlantic Oscillation Dynamics Recorded in Greenland Ice Cores." *Science* 282 (1998): 446–449.

Archer, Cristina L., and Ken Caldeira. "Historical Trends in the Jet Streams." *Geophysical Research Letters* 35, no. 24 (2008). Reports an investigation of the change of location of the jet stream in response to global warming.

Archer, David, and David Pierrehumbert, eds. *The Warming Papers. The Scientific Foundation for the Climate Change Forecast.* John Wiley & Sons, 2011. The role of atmospheric carbon dioxide and its role in climate regulation are discussed in a number of contexts in this book.

Archer, David. *Global Warming: Understanding the Forecast.* Wiley, 2012. Devotes a chapter to discussion of methane and its greenhouse effects.

_____. *The Global Carbon Cycle.* Princeton University Press, 2010.

_____. *The Long Thaw: How Humans Are Changing the Next 100,000 Years of Earth's Climate.* Princeton University Press, 2016. Accessible description of the consequences of cryosphere loss. Pitched to a nonscientific audience, resisting alarmist argument, and generally objective.

Arctic Marine Conservation is Not Prepared "Arctic Marine Conservation is Not Prepared for the Coming Melt" *ICES Journal of Marine Science* August 2017. https://www.researchgate.net/publication/306300142_How_predictable_is_the_timing_of_a_summer_ice-free_Arctic_PREDICTING_A_SUMMER_ICE-FREE_ARCTIC. Accessed 2 May 2018.

"Arctic Report Card: Update for 2016." *Arctic Program,* National Oceanic and Atmospheric Administration, http://www.arctic.noaa.gov/Report-Card/Report-Card-2016. Accessed 31 Jan. 2017.

"Arctic Sea Ice Annual Minimum Ties Second Lowest on Record." *NASA,* 15 Sept. 2016, www.nasa.gov/feature/goddard/2016/arctic-sea-ice-annual-minimum-ties-second-lowest-on-record. Accessed 18 Oct. 2016.

Arnold, Dennis G., ed. *The Ethics of Global Climate Change.* Cambridge University Press, 2011. Discusses the ethics of wealthy nations contributing disproportionately to global climate change that affects the poorest nations and the majority of the global population.

Ashton, Mark S., Mary S. Tyrrell, Deborah Spalding, and Bradford Gentry. eds. *Managing Forest Carbon in a Changing Climate*. Springer Science + Business Media, 2012. Provides a close examination of the role of forests and forest management in the regulation of atmospheric carbon dioxide.

"Atmosphere." *National Geographic*. National Geographic Soc., n.d. Web. 27 Apr. 2015.

Austin, B. *Arctic Basin: Results from the Russian Drifting Stations*. Springer, 2010. Discusses the results obtained by manned research stations on the drift ice of the high Arctic. Describes the meteorological, oceanographic, and geophysical observations.

Avidan, Kent and Simon Behrman. *Facilitating the Resettlement and Rights of Climate Refugees. An Argument for Developing Existing Principles and Practices*. Routledge, 2108. This book attempts to construct a contextual framework for the issues of climate refugees as a base for the development of principles and policies for nations that will become climate refugee hosts.

Baer, Hans A. *Global Capitalism and Climate Change. The Need for an Alternative World System*. Alta Mira Press, 2012. Examines the manner in which the current global economic system directs the course of global climate change.

Bahr, Simon. *How Does Climate Change Affect Global Economy?* GRIN Publishing, 2017. An essay examining the natural and anthropogenic causes of climate change and the ramifications for the global economy.

Bai, Xuezhi, et al. "Severe Ice Conditions in the Bohai Sea, China, and Mild Ice Conditions in the Great Lakes During the 2009/10 Winter: Links to El Niño and a Strong Negative Arctic Oscillation." *Journal of Applied Meteorology and Climatology* 50, no. 9 (2011): 1922–1935. The authors conducted a comparative study of the Great Lakes and China's Bohai Sea, focusing on the creation of large amounts of ice caused by negative phases in the Arctic Oscillation.

Bailey, Robert G. *Ecosystem Geography. From Ecoregions to Sites*. 2nd ed., Springer, 2009. Presents a concise description of the various climate regions of Earth, followed by a more granular analysis of the smaller regions comprising the climate zones, and discusses how climate zones and regions could change as the over all climate changes.

Baker, Victor R., editor. *Catastrophic Flooding: The Origin of the Channeled Scabland*. Ross, 1981.

Baker, Victor R., et al. *Flood Geomorphology*. Wiley, 2010.

Ballard, Robert D., and Malcolm McConnell. *Explorations: My Quest for Adventure and Discovery Under the Sea*. New York: Hyperion, 1995. Presents fascinating information about the very deep sea, correcting many previously held notions about it.

Bandy, A. R., and Priestley Conference (7, 1994, Lewisburg, Pa.). "The Chemistry of the Atmosphere: Oxidants and Oxidation in the Earth's Atmosphere ; [Proceedings of the 7th BOC Priestley Conference Organized by the Royal Society of Chemistry ..., Held at Bucknell University, Lewisburg, Pennsylvania, USA on 24–27 June 1994." Royal Society of Chemistry, 1998.

Barker, John Roger, et al. *Advances in Atmospheric Chemistry*. World Scientific, 2017.

Barry, Roger, and Thian Yew Gan. *The Global Cryosphere. Past, Present and Future*. Cambridge University Press, 2011. Individual chapters describe each segment of the terrestrial and marine cryospheres, followed by a discussion of the past conditions and future prospects of that global climatic region.

Barry, Roger G., and Eileen A. Hall-McKim. *Essentials of the Earth's Climate System*. Cambridge University Press, 2014. Describes many aspects of the influence of ocean currents and oscillations as drivers of the state of Earth's climate.

Barry, Roger G., and Peter D. Blanken. *Microclimate and Local Climate*. Cambridge University Press, 2016.

Barry, Roger G., and Richard J. Chorley. *Atmosphere, Weather, and Climate*. 9th ed. Routledge, 2010. Covers the subject of general atmospheric circulation in a thorough but not technically challenging style. Presents a vast number of figures and black-and-white line drawings to illustrate its points. Suitable for advanced high school students.

Bearman, Gerry, et al. *Marine Biogeochemical Cycles*. Elsevier, 2008. Introductory oceanography textbook that covers the processes by which marine sediments are formed, the chemical reactions by which they are modified, and the use of marine sediments in paleoceanographic studies to reconstruct past environments and climates.

Bedient, Philip B., et al. *Hydrology and Floodplain Analysis*. 2019.

Bednaršek, Nina, et al. "Extensive Dissolution of Live Pteropods in the Southern Ocean." *Nature Geoscience*, vol. 5, no. 12, 2012, pp. 881–85.

Behrman, Simon and Avidan Kent, eds. *Climate Refugees. Beyond the Legal Impasse?* Routledge, 2018. An in-depth presentation of the social, economic and legal implications of climate change in regard to climate refugees.

Bell, Gerry. "Impacts of El Niño and La Niña on the Hurricane Season." *Climate.gov*. NOAA, 30 May 2014. Web. 24 Mar. 2015.

Bell, Nigel, and Michael Treshow. *Air Pollution and Plant Life*. Wiley, 2003. Details the impact of a wide variety of pollutants on plant life. Illustrations, figures, tables, bibliography, index, maps.

Benedick, Richard Elliot. *Ozone Diplomacy. New Directions in Safeguarding the Planet*. Harvard University Press, 2009. This enlarged edition provides a deep historical account and analysis of the meshing of science and diplomacy in addressing the problem of depletion of the ozone layer by chlorofluorocarbons.

Bennett, Matthew M., and Neil F. Glasser. *Glacial Geology: Ice Sheets and Landforms*. John Wiley & Sons, 2009. A concise, beginner-friendly overview of the geologic concepts required to understand the icy landforms of Antarctica. Written for undergraduates but accessible to advanced high school students, each chapter includes text boxes, color photographs, illustrations, and suggestions for further reading.

Benson, Keith Rodney, and Philip F. Rhebock. *Oceanographic History: The Pacific and Beyond*. University of Washington Press, 2002.

Benton, Michael J., and Richard J. Twitchett. "How to Kill (Almost) All Life: The End-Permian Extinction Event." *Trends in Ecology and Evolution*, vol. 18, no. 7, 2003, pp. 358–65.

Berner, Elizabeth K., and Robert A. Berner. *The Global Water Cycle: Geochemistry and Environment*. Prentice-Hall, 1986.

Berner, Robert A. *The Phanerozoic Carbon Cycle: CO_2 and O_2*. Oxford University Press, 2004. Discusses climate and atmosphere of the Paleozoic, Mesozoic, and Cenozoic eras. Also covers aspects of weathering and erosion on the carbon cycle. Suited to undergraduates. Contains references for each chapter and indexing.

Berners-Lee, Mike. *How Bad Are Bananas? The Carbon Footprint of Everything*. Greystone, 2011.

Berry, Sandra L., et al. *Green Carbon the Role of Natural Forests in Carbon Storage*. ANU E Press, 2010.

Bertness, Mark D. *Atlantic Shorelines: Natural History and Ecology*. Princeton University Press, 2007.

Berwyn, Bob. "Why Is Antarctica's Sea Ice Growing While the Arctic Melts? Scientists Have an Answer." *InsideClimate News*, 31 May 2016, insideclimatenews.org/news/31052016/why-antarctica-sea-ice-level-growing-while-arctic-glaciers-melts-climate-change-global-warming. Accessed 1 June 2017.

Bielle, David. "Ozone Hole Closing Up, Thanks to Global Action." *Scientific American*. Scientific American, 15 Sept. 2014. Web. 23 Mar. 2015.

Bigg, Grant R. *The Oceans and Climate*. Cambridge Univ. Press, 2006. Written by a noted professor of environmental sciences. Details atmospheric and oceanic circulation patterns, and discusses their influence on meteorological developments. Describes the influence of the atmosphere and the ocean on each other and demonstrates how this interaction influences major ocean-atmosphere oscillations.

Birch, Eugenie Ladner, and Susan M. Wachter. *Rebuilding Urban Places after Disaster: Lessons from Hurricane Katrina*. University of Pennsylvania Press, 2006.

Bischof, Jens. *Ice Drift, Ocean Circulation and Climate Change*. Springer, 2001. Presents and discusses the concept of ice rafting, in which large pieces of Arctic ice floes break away and drift like rafts on the ocean currents, their movement adding to our knowledge of past and present conditions of oceanic circulation.

Black, Ken. *Business Statistics: For Contemporary Decision Making*. 8th ed. Wiley, 2014.

"A Blanket around the Earth." *National Aeronautics and Space Administration*. California Inst. of Technology, n.d. Web. 27 Apr. 2015.

Bluestein, Howard B. *Tornado Alley: Monster Storms of the Great Plains*. New York: Oxford UP, 2006. Print.

Bodri, Louise, and Vladimir Cermak. *Borehole Climatology. A new method on how to reconstruct climate*. Elsevier, 2007.

Bolius, David. *Paleoclimate Reconstructions Based on Ice Cores: Results from the Andes and the Alps*. SVH-Verlag, 2010.

Bolle, Hans-Jurgen, ed. *Mediterranean Climate: Variability and Trends.* Springer, 2003.

Bolsenga, S. J., and C. E. Herdendorf. *Lake Erie and Lake St. Clair Handbook.* Wayne State University Press, 1993.

Bonan, Gordon. *Ecological Climatology: Concepts and Applications.* 3rd ed., Cambridge University Press, 2015. Provides a concise definitive description of Earth's recognized climate zones within the broader context of climatology.

Boon, John D. *Secrets of the Tide: Tide and Tidal Current Analysis and Applications, Storm Surges and Sea Level Trends.* Woodhead Publishing, 2011.

Boslaugh, Sarah. *Statistics in a Nutshell.* 2nd ed. O'Reilly, 2013.

Botts, Lee, et al. *The Great Lakes: An Environmental Atlas and Resource Book.* Govt. of Canada, 2002.

Boucher, Olivier. *Atmospheric Aerosols: Properties and Climate Impacts.* Elsevier, 2015. For advanced readers, this book discusses the role of atmospheric aerosols in the context of cloud physics.

Bourseiller, Philippe, and Jacques Duriex. *Volcanoes.* Harry Abrams, 2008.

Bowen, Mark. *Thin Ice: Unlocking the Secrets of Climate in the World's Highest Mountains.* Macmillan, 2005. This book has been called one of the best yet published on climate change. Part 6 is dedicated to Kilimanjaro. Notes, references, index.

Bradley, Raymond S. *Paleoclimatology: Reconstructing Climates of the Quaternary.* Academic Press, 2015. A definitive introduction to the science of paleoclimatology, and the use of climate proxies.

Bridgman, H.A., John E. Oliver, Michael Glantz, Randall S. Corveny, and Robert Allan. *The Global Climate System: Patterns, Processes and Teleconnections.* Cambridge University Press, 2014. The majority of this book describes the global climate system, while the remainder of the content examines social and economic aspects.

Brimblecombe, Peter, et al., eds. *Acid Rain: Deposition to Recovery.* Dordrecht: Springer, 2010. Print.

Broad, William J., and Dimitri Schidlovsky. *The Universe Below Discovering the Secrets of the Deep Sea.* Paw Prints, 2008. Provides useful information about the very deep sea. Explores the notion that there is little life in the very deep sea. Demonstrates that all sorts of life that have hitherto been undetected inhabit the abyssal depths, sometimes enduring high temperatures from hot flows that would kill most known forms of life.

Broda, E. *The Evolution of the Bioenergetic Processes.* Elsevier, 2014. The author gives a thorough description of the respiration and fermentation processes, and their respective roles in regard to prokaryotes and eukaryotes.

Broecker, W. S. "Does the Trigger for Abrupt Climate Change Reside in the Ocean or the Atmosphere?" *Science* 300 (2003): 1519–1522. Good example of a discussion on some aspects of the natural causes of climate change.

Broecker, Wally. *The Great Ocean Conveyor. Discovering the Trigger for Abrupt Climate Change.* Princeton University Press, 2010. Describes the global oceanic overturning circulation system and how its disruption by climate change may affect the climate of Earth.

Browning, K. A., and R. J. Gurney. *Global Energy and Water Cycles.* Cambridge University Press, 2007. This excellent book features reviews by Europeans and Americans on how weather forecasting models are developed and used.

Bryant, Edward A. *Natural Hazards.* 2nd ed. Cambridge University Press, 2005. Provides a sound scientific treatment for the educated layperson. Readers should have a basic understanding of mathematical principles. Presents many case studies. Contains photographs, tables, figures, and a glossary of terms.

Bullock, A., and A. Acreman. "The Role of Wetlands in the Hydrological Cycle." *Hydrology and Earth System Sciences* 7, no. 3 (2003): 358–389.

Burroughs, William James. *Climate Change in Prehistory: The End of the Reign of Chaos.* Cambridge University Press, 2008. Covers a broad spectrum of data concerning human adaptation to climate change in prehistory, but has chapters specific to the evidence for climate change that is found in ocean sediments.

Byers, Michael. *Who Owns the Arctic? Understanding Sovereignty Disputes in the North.* Vancouver: Douglas & McIntyre Publishers, 2009. Covers many issues of Arctic sovereignty. An especially topical subject since global warming is rapidly freeing up access to Arctic Ocean resources.

Canfield, Donald Eugene. *Oxygen. A Four Billion Year History.* Princeton University Press, 2014. This highly readable book presents a comprehensive overview of atmospheric oxygen and discusses many questions about the origin and significance of atmospheric oxygen.

Carlisle, Rodney P. *Where the Fleet Begins: A History of the David Taylor Research Center*. Naval Historical Center, 1998.

Catling, David C. and Kasting, James F. *Atmospheric Evolution on Inhabited and Lifeless Worlds*. Cambridge University Press, 2017. A complete and comprehensive treatment of the evolution of Earth's atmosphere, with application to the study of other planets.

Çengel Yunus A., et al. *Thermodynamics: An Engineering Approach*. 2019. Sophomore- or junior-level college textbook dealing with heat and energy. Includes a simplified discussion of global warming and the greenhouse effect.

Cessi, Paola. "Voyager: What Would Cause Thermohaline Circulation in the Oceans to Stop?" *Scripps Institution of Oceanography*. Scripps Inst. of Oceanography, UC San Diego, 19 Dec. 2013. Web. 23 Mar. 2015.

Chan, Johnny C. L., and Jeffrey D. Kepert, eds. *Global Perspectives on Tropical Cyclones: From Science to Mitigation*. World Scientific, 2010. Includes chapters on forecasting and modeling tropical cyclones, the effects of climate change on cyclone activity, and approaches to disaster response. Highly technical; best suited to college students and above with some background in meteorology.

Chang, Chih-Pei, and Mong-Ming Lu. "Intraseasonal Predictability of Siberian High and East Asian Winter Monsoon and Its Interdecadal Variability." *Journal of Climate* 25, no. 5 (2012): 1773–1778. Discusses scientific efforts to predict severe weather in East Asia based on the positive and negative phases of Arctic oscillation.

Chang, Julius. *General Circulation Models of the Atmosphere*. Elsevier, 2012. The book covers the fundamentals of general circulation models and their application in atmospheric studies.

Charabe, Yassine, and Salim al-Hatrushi. *Indian Ocean Tropical Cyclones and Climate Change*. Springer, 2010.

Charlier, R. H., and C. W. Finkl. *Ocean Energy: Tide and Tidal Power*. Springer-Verlag, 2009. Discusses fundamentals of oceanic energy harvesting. Describes historical and current technologies, with examples drawn from around the world. Also discusses social, economic, and environmental impacts.

Contains dense, verbose, and technical writing and is therefore best suited for graduate students and researching oceanographers or engineers with some prior knowledge of ocean dynamics.

Chen, Dene-Hern. "Once Written Off for Dead, the Aral Sea Is Now Full of Life." *National Geographic*, National Geographic Society, 16 Mar. 2018, news.nationalgeographic.com/2018/03/north-aral-sea-restoration-fish-kazakhstan/?beta=true.

Chester, Roy. *Marine Geochemistry*. Springer, 2013. Collection of papers discussing the physical and chemical properties of marine sediments and the chemical reactions that occur to those sediments as they undergo diagenesis and are converted into sedimentary rocks.

Christensen, Jen. "The Hidden Health Dangers of Flooding." *CNN*, 31 Aug. 2017, www.cnn.com/2017/08/27/health/health-consequences-flood-waters/index.html. Accessed 22 Sept. 2017.

Christie, Maureen. *The Ozone Layer: A Philosophy of Science Perspective*. Cambridge University Press, 2001. Presents the history of human knowledge about stratospheric ozone in a manner accessible to lay readers. Addresses basic issues of both real-world science and the philosophy of science. Includes figures, references, and index.

Christopherson, Robert W., and Ginger H Birkeland. *Geosystems: An Introduction to Physical Geography*. Pearson, 2018. Presents a discussion of atmospheric and oceanic oscillations as fundamental systems of the global environment. Includes numerous references and links to online resources. Accessible to general readers.

Clarke, Allan J. *An Introduction to the Dynamics of El Niño and the Southern Oscillation*. Burlington, Mass.: Academic Press, 2008. Presents the physics of ENSO, including currents, temperature, winds, and waves. Discusses ENSO forecasting models. Provides good coverage of the influence ENSO has on marine life, from plankton to green turtles. Each chapter has references, and there are a number of appendices and indexing. Best suited for environmental scientists, meteorologists, and similar academics studying ENSO.

Clerbaux, Cathy, Solene Turquety, and Pierre Coheur. "Infrared Remote Sensing of Atmospheric Composition and Air Quality: Toward Operational Applications." *Comptes Rendus Geoscience* 342, nos. 4/5 (2010): 349–356. Discusses the uses of remote sensors for studying pollution's effects on the atmosphere. Reviews satellite-based technologies such as thermal infrared sensors to study the troposphere on a global scale.

Clift, P. D., R. Tada, and H. Zheng, eds. *Monsoon Evolution and Tectonic-Climate Linkage in Asia.* Bath: Geological Soc., 2010. Print.

Clift, Peter D., and R. Alan Plumb. *The Asian Monsoon.* Cambridge University Press, 2008.

Climate Central, Inc. *Global Weirdness: Severe Storms, Deadly Heat Waves, Relentless Drought, Rising Seas and the Weather of the Future.* Vintage Books, 2013.

"Climate Change Indicators: Greenhouse Gases." *EPA*, 6 Oct. 2016, www.epa.gov/climate-indicators/greenhouse-gases. Accessed 2 Feb. 2017.

Dlugokencky, E. J., et al. "Continuing Decline in the Growth Rate of the Atmospheric Methane Burden." *Nature*, 393, 1998, pp. 447–50.

Coe, Angela L., ed. *The Sedimentary Record of Sea-Level Change.* Cambridge University Press, 2003.

Cohen, Stewart J. and Melissa W. Waddell. *Climate Change in the 21st Century.* McGill-Queens University Press, 2009. Discusses many aspects of climate change, with emphasis on the role and function of natural and artificial greenhouse gas sinks.

Coley, David A. *Energy and Climate Change: Creating a Sustainable Future.* John Wiley & Sons, 2008.

Colling, Angela. *Ocean Circulation.* 2nd ed. Butterworth-Heinemann, 2001.

Colman, B. R., and T. D. Potter, eds. *Handbook of Weather, Climate, and Water.* Wiley-Interscience, 2003. Concise and thorough treatment of the hydrologic cycle and its effect on climate. Illustrations, figures, tables, references, index.

Comerio, Mary C. *Disaster Hits Home: New Policy for Urban Housing Recovery.* University of California Press, 1998.

Condie, Kent C. *Earth as an Evolving Planetary System.* 3rd ed. Elsevier, 2016. Provides a complete description of Earth's internal and external structures with regard to the evolution of the planet and its atmosphere.

Congressional Research Service. *Weather Modification: Programs, Problems, Policy and Potential.* University Press of the Pacific, 2004. A thorough, scholarly report, originally prepared in 1978, that reviews the history, technology, activities, and various special aspects of weather modification of interest to the agricultural, scientific, commercial, and governmental fields.

Conkling, Philip W., et al. *The Fate of Greenland: Lessons from Abrupt Climate Change.* MIT Press, 2013.

Conkling, Philip, Richard Alley, Wallace Broecker, and George Danton. *The Fate of Greenland: Lessons from Abrupt Climate Change.* MIT Press, 2013.

Copeland, Sebastian. *The Global Warning.* Earth Aware Editions, 2007. This abundantly illustrated book about Antarctica tends to focus on the potential calamitous effects of global climate change using a clear and simple language.

Cossu, Remo, and Matthew G. Wells. "The Evolution of Submarine Channels under the Influence of Coriolis Forces: Experimental Observations of Flow Structures." *Terra Nova* 25.1 (2013): 65–71. *Academic Search Complete.* Web. 19 Mar. 2015.

Cotton, William R., and Roger A. Pielke. *Human Impacts on Weather and Climate.* 2nd ed. University Press, 2007. A comprehensive overview of weather modification, written as a text for both undergraduate and graduate study in atmospheric and environmental science. Looks at weather modification with regard to both intentional and unintentional effects.

Craig, Richard A. *The Edge of Space: Exploring the Upper Atmosphere.* Doubleday, 1968. Discusses the various methods of measuring atmospheric ozone. Written at a time when understanding of ozone chemistry was far less developed than it is in the twenty-first century, but still useful. Based on an important book in aeronomy, *The Upper Atmosphere: Meteorology and Physics*, which Craig published in 1965, this volume is aimed at a general audience.

Cronin, Thomas M. *Principles of Paleoclimatology.* Columbia University Press, 1999. A standard textbook explaining in detail the methodology of paleoclimatological research. Includes an extensive discussion of the implications of late twentieth century discoveries for global energy policies.

Cropper, Thomas. "Did El Niño Drive the Record Heat of 2015?" *Niskanen Center.* Niskanen Center, 25 Jan. 2016. Web. 1 Feb. 2016.

Crosby, Alfred W. *Children of the Sun: A History of Humanity's Unappeasable Appetite for Energy.* Norton, 2006. More historical than scientific, this book will nonetheless be interesting for science-minded readers who wish to put scientific knowledge of the sun in its broader historical and cultural context, particularly with regard to its energy potential for humanity.

Cumberland, John H., James R. Hibbs and Irving Hoch, eds. *The Economics of Managing Chlorofluorocarbons. Stratospheric Ozone and Climate Issues.* Routledge, 2016. An extensive examination of the history, uses and environmental effects of chlorofluorocarbons.

D'Aleo, Joseph S., and Pamela G. Grube. *The Oryx Guide to El Niño and La Niña.* Greenwood Press, 2002. Examines the causes and effects of the various stages of ENSO. In addition to analysis of the history, economic impacts, and natural forces creating ENSO, reviews attempts to more effectively predict this phenomenon.

Darling, Seth B. and Douglas L. Sisterson. *How to Change Minds About Our Changing Climate.* The Experiment LLC, 2014. Addresses the issue of climate change denial in the face of the demonstrable effects of climate change.

Darrigol, Olivier. *Worlds of Flow: A History of Hydrodynamics from the Bernoullis to Prandtl.* Oxford University Press, 2005.

Dasgupta, KumKum. "Why India Is Failing to Minimise Monsoon Flood Destruction." *Hindustan Times*, 26 July 2017, www.hindustantimes.com/analysis/why-india-is-failing-to-minimise-monsoon-flood-destruction/story-4qC6DuVWacseb-WBwnvLbjI.html. Accessed 22 Sept. 2017.

Dassenakis, Manos, et al. "Remote Sensing in Coastal Water Monitoring: Applications in the Eastern Mediterranean Sea (IUPAC Technical Report)." *Pure and Applied Chemistry 84*, no. 2 (2012): 335–375. Discusses the wide range of applications of satellite-based remote sensing to the study of climate change in coastal areas. Demonstrates how such systems can help scientists study large areas.

Davies, Bethan. "Mapping the World's Glaciers." Antarcticglaciers.org, 25 Nov. 2014. Web. 23 Mar. 2014.

Davis, Richard A., Jr. *Sea-Level Change in the Gulf of Mexico.* Everbest Printing, 2011.

Dawson, Alastair G. *Ice Age Earth: Late Quaternary Geology and Climate.* Routledge, 1992.

De Angelis, Hernan, and Pedro Skvarca. "Glacier Surge After Ice Shelf Collapse." *Science* 299 (2003): 1560–62.

de Boer, Jella Zeilinga, and Donald Theodore Sanders. *Volcanoes in Human History: The Far-Reaching Effects of Major Eruptions.* Princeton University Press, 2002. Provides an overview of volcanism followed by specific volcanoes and eruptions. Includes the interaction of human history and volcanic activity. Covers the Hawaiian Islands, Crete, Vesuvius, Iceland, Tambora, Krakatau, Pelée, Tristan da Cunha, and Mount St. Helens.

De Villiers, Marc. *Windswept: The Story of Wind and Weather.* New York: Walker, 2006. An accessible, scientifically accurate book of popular science that explores the history of human attempts to understand the weather. Contains black-and-white figures and several useful appendices, including two covering tropical cyclone statistics.

Decker, Robert W., and Barbara Decker. *Volcanoes.* 4th ed. Freeman, 2005. An introductory work on the study of volcanoes that contains a concise chapter, with clear illustrations, on the connection between volcanic eruptions and climate.

Dennis, Jerry. *The Living Great Lakes.* Thomas Dunne Books, 2003. Written for the layperson. Covers the geology, natural history, biology, and industry of the Great Lakes. Discusses the structure of the lakes from their formation to the current human impact of resource mining, construction of dams and canals, and introduction of invasive species.

Denny, Mark. *How the Ocean Works. An Introduction to Oceanography.* Princeton University Press, 2012. This book provides an in-depth examination of the effect of the rotational motion of Earth and the Coriolis Effect.

Denny, Mark. *Making the Most of the Anthropocene—Facing the Future.* Johns Hopkins University Press, 2017.

Dessler, Andrew E., and Edward A. Parson. *The Science and Politics of Global Climate Change: A Guide to the Debate.* Cambridge University Press, 2010. Introduction to the issue of climate change, including atmospheric chemistry and other atmospheric properties research. Also discusses the potential future of research in the area of climate change.

Dessler, Andrew. *Introduction to Modern Climate Change* 2nd ed. Cambridge University Press, 2016. Provides a good overview of the science and sociology of climate change.

Di Liberto, Tom. "ENSO + Climate Change = Headache." *Climate.gov.* NOAA, 11 Sept. 2014. Web. 24 Mar. 2015.

Di Lorenzo, Emanuele, et al. "North Pacific Gyre Oscillation Links Ocean Climate and Ecosystem Change." *Geophysical Research Letters* 35, no. 108607 (April, 2008). Overview of the discovery of the North Pacific Gyre Oscillation and its consequences for Pacific ecosystems and fisheries industries.

Diaz, Henry F., and Raymond S. Bradley, eds. *The Hadley Circulation: Present, Past, and Future.* Kluwer Academic, 2004. Essay collection that grew out of a conference examining the Hadley circulation from several different viewpoints. Illustrations, maps, references.

Dobson, G. M. B. "Forty Years' Research on Atmospheric Ozone at Oxford: A History." *Applied Optics* 7 (March, 1968): 387–405. Dobson's personal recollection of the development of the global network for ozone monitoring presents a fascinating account of the scientific work. Notes the unexpected observation in 1956 of low concentrations of ozone until late in the Antarctic spring—the phenomenon, intensified in subsequent years by the presence of radical chlorine from chlorofluorocarbons, now known as the ozone hole.

Dodds, Walter K., and Matt R. Wiles. *Freshwater Ecology: Concepts and Environmental Applications of Limnology.* 2nd ed. Academic Press, 2010. Covers the physical and chemical properties of water, the hydrologic cycle, nutrient cycling in water, as well as biological aspects. Written by two of the leading scientists in freshwater ecology. An excellent resource for college students as each chapter has a short summary of main topics.

Donahue, Michelle Z. "Fast-Growing Moss Is Turning Antarctica Green." *National Geographic*, 19 May 2017, news.nationalgeographic.com/2017/05/antarctica-green-climate-moss-environment/. Accessed 1 June 2017.

Donner, Leo, Wayne Schubert and Richard Somerville, eds. *The Development of Atmospheric General Circulation Models. Complexity, Synthesis and Computation.* Cambridge University Press, 2011. A curated collection of articles describing the historical development of general circulation models and their development since then.

Douglas, R. J. W. *Geology and Economic Minerals of Canada.* Report Number I. Ottawa: Geological Survey of Canada, 1970.

Douglass, Anne, Natalya Kramarova, and Susan Strahan. *Inside the Ozone Hole.* NASA's Goddard Space Flight Center, 2015. PDF file.

Dowdeswell, Julian A. *Atlas of Submarine Glacial Landforms: Modern, Quaternary and Ancient.* Geological Society, 2016.

Dunkel, Ged. *Understanding the Jet Stream. Clash of the Titans.* Authorhouse, 2010. A book that describes the role of the Coriolis Effect in the functioning of the jet stream.

Dunlop, Storm. *Meteorology Manual: The Practical Guide to the Weather.* Haynes, 2014.

_____. *The Weather Identification Handbook.* Lyons, 2003.

Easterbrook, Don. *Evidence Based Climate Science: Data Opposing CO2 Emissions as the Primary Source of Global Warming.* Elsevier, 2011. Presents an evidence-based analysis of the scientific data concerning climate change and global warming. Authored by eight of the world's leading climate scientists who refute the claims embraced by proponents of CO_2 emissions as the cause of global warming. Includes comprehensive citations and references, as well as an extensive bibliography.

Eckert, Michael. *The Dawn of Fluid Dynamics: A Discipline Between Science and Technology.* Wiley-VCH, 2006.

Eddy, J. A. "The Maunder Minimum." *Science* 192 (1976): 1189–1192.

Eddy, John A. *The Sun, Earth, and Near-Earth Space: A Guide to the Sun-Earth System.* National Aeronautics and Space Administration, 2009. Provides a comprehensive but easy-to-understand look at the sun's structure, function, and relationship with Earth.

"The Electromagnetic Spectrum. *National Aeronautics and Space Administration.* NASA, Mar. 2013. Web. 27 Apr. 2015.

Ellis, Erle C., and Peter K. Haff. "Earth Science in the Anthropocene: New Epoch, New Paradigm, New Responsibilities." *EOS, Transactions, American Geophysical Union*, vol. 90, no. 49, 2009, p. 473.

Emanuel, Kerry. *Divine Winds: The History and Science of Hurricanes.* Oxford University Press, 2005. Covers a wide variety of topics on a number of hurricane events and related research. Presents events in chronological order with related topics scattered between. Written in a popular style, but still contains portions with "hard science." Includes multiple useful appendices, a list of sources, further reading, credits, and indexing.

Emelyanov, Emelyan M. *The Barrier Zones in the Ocean.* Berlin: Springer-Verlag, 2005. Suitable as a textbook for advanced oceanography and ocean geochemistry courses, and for research reference. Discusses the properties of forty different ocean barrier zones with regard to salinity, hydrodynamics, temperature, and light, as well as processes that affect sedimentation in the open ocean and in bodies of water such as the Baltic and Mediterranean Seas.

Emerson, Steven, and John Hedges. *Chemical Oceanography and Marine Carbon Cycle.* Cambridge University Press, 2008. Provides a good overview of geochemistry topics in oceanography. Discusses chemical composition, thermodynamics, carbonate chemistry, the carbon cycle, and calcium carbonate sedimentation. Appendices follow the chapters they pertain to. Contains excellent indexing.

Epstein, Alex. *The Moral Case for Fossil Fuels.* Portfolio/Penguin, 2014. An unabashed and unapologetic discussion of the inevitable victory of technological advances in controlling climate change.

Ewing, Mark. "Toyota's Hydrogen Fuel Cell Kenworth Can Revolutionize Heavy Transport." *Forbes*, Forbes Magazine, 9 Aug. 2017, www.forbes.com/sites/markewing/2017/08/09/toyotas-hydrogen-fuel-cell-kenworth-can-revolutionize-heavy-transport/#3957fe276e48.

"Exposure to Air Pollution Increases the Risk of Obesity." *Duke Today.* Office of News and Communications, 19 Feb. 2016. Web. 6 June 2016.

Fabian, Peter and Dameris, Martin. *Ozone in the Atmosphere. Basic Principles, Natural and Human Impacts.* Springer, 2014. This book provides an advanced discussion of the role of ozone in Earth's environment, its interaction with the many other gaseous components if Earth's atmosphere, and how it is affected by many natural and unnatural factors.

Fagan, Brian. *Floods, Famines and Emperors. El Niño and the Fate of Civilizations.* Basic Books, 1999. A readable historical examination of how the El Niño—Southern Oscillation (ENSO) phenomenon have affected civilizations on a global scale throughout the past.

_____. *The Great Warming: Climatic Change and the Rise and Fall of Civilizations.* New York: Bloomsbury Press, 2008.

Fahy, Frank. *Air: The Excellent Canopy.* Chichester, UK: Horwood, 2009. A well-written introduction to the properties of air and the forces that govern atmospheric circulation, designed for nonspecialists. Explains physics concepts using detailed descriptions and analogies rather than equations.

Fears, Darryl. "Without Action on Climate Change, Say Goodbye to Polar Bears." *The Washington Post*, 9 Jan. 2017, www.washingtonpost.com/news/energy-environment/wp/2017/01/09/without-action-on-climate-change-say-goodbye-to-polar-bears/?utm_term=.dbce906341f1. Accessed 8 Nov. 2017.

Fein, Jay S., and Pamela L. Stephens, eds. *Monsoons.* Wiley, 1987.

Fernando, H. J. S. *Handbook of Environmental Fluid Dynamics.* Taylor & Francis, 2013.

Ferreiro, Larrie D. *Ships and Science: The Birth of Naval Architecture in the Scientific Revolution, 1600–1800.* MIT Press, 2007.

Feudale, Laura, and Jagadish Shukla. "Influence of Sea Surface Temperature on the European Heat Wave of 2003 Summer. Part I: An Observational Study." *Climate Dynamics* 36, nos. 9/10 (2011): 1691–1703. A review of observational data on ocean surface temperatures and their contributions to a dangerous heat wave in Europe in 2003.

Field, Christopher B., and Michael R. Rapauch, eds. *The Global Carbon Cycle: Integrating Humans, Climate, and the Natural World.* Island Press, 2004. This compendium of articles from the Scientific Committee on Problems of the Environment (SCOPE) discusses the major atmospheric sources and sinks. The final chapter details possible advantages and limitations of ocean fertilization.

Field, J.G., Gotthilf Hempel, and C.P. Summerhayes. *Oceans 2020: Science, Trends and the Challenge of Sustainability.* Island Press, 2002. A combined work of the Intergovernmental Oceanographic Commission, the International Council of Scientific Unions, the Scientific Committee on Oceanic Research, and the ICSU Scientific Committee on Problems of the Environment in a collection of learned essays that address the state of the oceans and issues for sustainability.

Fisher, Richard V., Grant Heiken, and Jeffrey B. Hulen. *Volcanoes: Crucibles of Change.* Princeton University Press, 1997. Covers a long list of volcanoes, from Mount Vesuvius to Mount St. Helens.

A segment on the effects of volcanic gases helps explain the complex interactions among gaseous, liquid, and solid volcanic matter, the surrounding atmosphere, and solar radiation.

Fitzpatrick, Patrick J. *Contemporary World Issues: Hurricanes.* 2nd ed. ABC-CLIO, 2006. A reference work that provides background material on the issues, people, organizations, statistics, and publications related to hurricanes.

Flannery, Tim. *We Are the Weather Makers* Somerville, MA: Candlewick Press, 2010. Print.

Fleming, Nic. "When a Volcanic Apocalypse Nearly Killed Life on Earth." *BBC Earth*, BBC, 18 Dec. 2015, www.bbc.com/earth/story/20151218-when-a-volcanic-apocalypse-nearly-killed-life-on-earth. Accessed 21 Feb. 2017.

Flemming, Nicholas Coit, Council for British Archaeology, and English Heritage. *Submarine Prehistoric Archaeology of the North Sea: Research Priorities and Collaboration With Industry.* Council for British Archaeology, 2004.

Foukal, P., C. Fröhlich, H. Spruit, and T. M. L. Wigley. "Variations in Solar Luminosity and Their Effect on Earth's Climate." *Nature* 443 (2006): 161–166.

Francis, Peter, and Clive Oppenheimer. *Volcanoes.* 2nd ed. Oxford University Press, 2004. Covers volcanic activity from lava flows to pyroclastic currents. Provides many examples from Hawaii, Italy, Ethiopia, Japan, Mount St. Helens, and Krakatau. Written for the layperson, but some knowledge of geology is helpful.

Frederick, John E. *Principles of Atmospheric Science.* Jones & Bartlett, 2008. Introductory text describing the various fields of atmospheric sciences, including atmospheric chemistry, atmospheric physics, and climatology. Describes techniques and research methods utilized in modern climate and atmospheric research.

Freitag, Bob, et al. *Floodplain Management: A New Approach for a New Era.* Island Press, 2009.

Freytag, Andreas, John J. Kirton, Razeen Sally, and Paolo Savona, eds. *Securing the Global Economy. G8 Global Governance for a Post-Crisis World.* Routledge, 2016. The G* are the eight most affluent nations in the global economy, and this book explores the reasons why the G8 have an ethical responsibility and the ability to address the complex issues of global climate change in order to maintain and develop a stable global economy.

Friedman, Robert Marc. *Appropriating the Weather: Vilhelm Bjerknes and the Construction of a Modern Meteorology.* Cornell University Press, 1989. The author traces the development of meteorological thought that led to a modern understanding of middle-latitude cyclones. The relationship between the development of the commercial applications of meteorology and that of the science itself is explored. The career of the Norwegian meteorologist Vilhelm Bjerknes is central to the discussion.

Fung, Kaiser. *Numbers Rule Your World: The Hidden Influence of Probabilities and Statistics on Everything You Do.* McGraw, 2010.

Gaffney, Vincent, Kenneth Thomson, and Simon Fitch, eds. *Mapping Doggerland: The Mesolithic Landscapes of the Southern North Sea.* Archaeopress, 2007.

Garrison, Tom S. *Essentials of Oceanography.* 8th ed. Brooks/Cole, 2013. Summary of oceanography for introductory college students.

_____. *Oceanography: An Invitation to Marine Science.* Brooks/Cole, Cengage Learning, 2010. Discusses oceanic currents, circulation, and oscillations. Builds a story from the formation of oceans, through waves and tides, to economics and conservation of the ocean, its inhabitants, and its resources. Provides abundant diagrams that aid readers from the layperson to advanced undergraduates.

Gartland, Lisa. *Heat Islands: Understanding and Mitigating Heat in Urban Areas.* Earthscan/James and James, 2008.

Gautier, Catherine. *Oil, Water, and Climate: An Introduction.* Cambridge UP, 2008.

Gettelman, Andrew, and Richard B. Rood. *Demystifying Climate Models. A Users Guide to Earth System Models.* Springer Open, 2016. An excellent introduction to climate modeling, this book first gives a basic introduction to Earth's climate, and then individually describes the many parameters that are used to construct climate models.

Gilding, Paul. *The Great Disruption: How the Climate Crisis Will Transform the Global Economy.* A&C Black, 2012. While not offering more than basic solutions for complex problems related to global climate change, this book nevertheless identifies many of those problems.

Gill, Adrian E., and William L. Donn. *International Geophysics Series Vol. 30. Atmosphere-Ocean Dynamics.*

Elsevier, 2016. This book provides a comprehensive overview of the physical and mathematical principles describing the interaction of the atmosphere and oceans of Earth.

Giupponi, Carlo, and Mordechai Shechter, eds. *Climate Change and the Mediterranean: Socio-economic Perspectives of Impacts, Vulnerability, and Adaptation.* Edward Elgar, 2003.

Glantz, M. H. *Currents of Change: Impacts of El Niño and La Niña on Climate and Society.* 2nd ed. Cambridge University Press, 2001. Print.

Glavin, Terry, and David Suzuki Foundation. "The Last Great Sea a Voyage through the Human and Natural History of the North Pacific Ocean." *The Last Great Sea a Voyage through the Human and Natural History of the North Pacific Ocean,* www.deslibris.ca/ID/416854.

Glennie, Kw. *Petroleum Geology of the North Sea: Basic Concepts and Recent Advances.* 4th ed. John Wiley & Sons, 2009.

Glikson, Andrew Y. *Evolution of the Atmosphere, Fire and the Anthropocene Climate Event Horizon.* Springer, 2014. In this book can be found a detailed discussion of the relationship of oxygen and other atmospheric gases with regard to their natural cycles and the effect of human activities.

Glover, David, William J. Jenkins and Scott C. Doney. *Modeling Methods for Marine Science* Cambridge University Press, 2011. This book provides the basic methodology for the development of general circulation models from first principles, introduced with MATLab.

Goddard Space Flight Center. *Arctic Ozone Watch.* NASA Ozone Watch, 25 Sept. 2013. Web. 23 Mar. 2015.

Goh, Gahyun, and Y. Noh. "Influence of Coriolis Force on the Formation of a Seasonal Thermocline." *Ocean Dynamics* 63.9/10 (2013): 1083–1092. *Energy & Power Source.* Web. 19 Mar. 2015.

Golub, Leon, and Jay M. Pasachoff. *Nearest Star: The Surprising Science of Our Sun.* Cambridge University Press, 2014.

Gombosi, Tamas I. *Physics of the Space Environment.* Cambridge University Press, 2004.

Gomez, Alan. "EPA Proposes New Limits to Ozone Air Pollution." *USA Today.* Gannett, 26 Nov. 2014. Web. 4 Oct. 2014.

Goni, G. J., and Paola Malanotte-Rizzoli. *Interhemispheric Water Exchange in the Atlantic Ocean.* Elsevier B.V., 2003.

Goodkin, N. F., K. A. Hughen, S. C. Doney, and W. B. Curry. "Increased Multidecadal Variability of the North Atlantic Oscillation Since 1781." *Nature Geoscience* 1 (2008): 844–848.

Goosse, Hugues. *Climate System Dynamics and Modeling.* Cambridge University Press, 2015. The individual components and observable mass and energy cycles of Earth are discussed initially, and subsequent chapters analyze and discuss individual feedbacks.

Gore, Al. *An Inconvenient Sequel: Truth to Power.* Rodale, 2017.

Gornitz, Vivien. *Rising Seas. Past, Present, Future.* Columbia University Press, 2013. Presents an overview of sea level rise due to climate change and describes how rising sea levels may affect the stability of the ocean heat conveyor system.

Gosnell, Mariana. *Ice: The Nature, the History, and the Uses of an Astonishing Substance.* Chicago: University of Chicago Press, 2007. Examines in detail ice and its impact on Earth and its inhabitants.

Grady, Wayne, et al. *Great Lakes: The Natural History of a Changing Region.* Greystone Books, 2011. Discusses the natural history, geology, ecology, and conservation of the Great Lakes. A chapter discusses the human impact of invasive species on the ecology of the lakes. Includes further readings, a list of common and scientific names, illustration credits, and indexing.

Gramelsberger, Gabriele, and Feichter, Johann, eds. *Climate Change and Policy. The Calculability of Climate Change and the Challenge of Uncertainty.* Springer, 2011. A curated collection of articles addressing problems of uncertainty about climate modeling and climate change on an individual basis.

Grasshoff, K., et al. *Methods of Seawater Analysis.* Wiley-VCH, 2002. A widely used college text that is completely revised and extended, devoting more attention to the techniques and instrumentation used in determining the chemical makeup of seawater. Illustrations, bibliography, and index.

Grazulis, Thomas P. *The Tornado: Nature's Ultimate Windstorm.* University of Oklahoma Press, 2003.

Gribben, John R., ed. *The Breathing Planet.* Basil Blackwell, 1986. Collection of short articles originally published in the English journal *New Scientist* includes a section on ozone. Chapter titled "Monitoring Halocarbons in the Atmosphere" suggests the difficulties and uncertainties of monitoring substances in the atmosphere.

Gribbin, John. *The Hole in the Sky: Man's Threat to the Ozone Layer*. Rev ed. Bantam Books, 1988. Details concerns about the destruction of the ozone layer. Probably the most balanced of the books published on this topic in the late 1980s.

Grotzinger, John P., and Thomas H. Jordan. *Understanding Earth*. 7th ed. Freeman, 2014.

Grove, Richard, and George Adamson. *El Niño in World History*. Palgrave/Macmillan, 2018. Examines historical climate conditions that can be attributed to the El Niño phenomenon, and the effects on contemporary human societies.

Gruza, George Vadimovich, ed. *Environmental Structure and Function: Climate System, Volumes 1 & 2*. EOLSS Publishers, 2009. A detailed description of atmospheric characteristics and of the various climate zones of Earth.

Guzzi, Rodolfo, ed. *Exploring the Atmosphere by Remote Sensing Techniques*. Springer, 2010. In this collection of lectures, the editor presents various remote sensing practices used in the study of gases, aerosols, and clouds in the atmosphere.

Hadley, G. "Concerning the Cause of the General Trade-Winds." *Philosophical Transactions of the Royal Society of London* 39 (1735): 58–62. George Hadley takes Halley's concept and expands on it.

Haerens, Margaret. *Global Viewpoints: Air Pollution* Farmington Hills: Greenhaven, 2011.

Håkanson, Lars, and M. Jansson. *Principles of Lake Sedimentology* Blackburn Press, 2002. A reference for professionals in the field of lake sedimentology, though parts of it may be accessible to high school students. Focuses on lake sediments in detail. Provides methods of sampling and discusses the influence of lake type and shape on the sediments formed in the lake, the circulation of lake waters, the chemistry of sediments, and the pollution of lakes.

Halley, E. "An Historical Account of the Trade Winds, and Monsoons, Observable in the Seas Between the Tropicks, with an Attempt to Assign the Physical Cause of the Said Winds." *Philosophical Transactions of the Royal Society of London* 16 (1686): 153–168. The original article explaining the Hadley circulation. Illustrations.

Hambrey, Michael, and Jurg Alean. *Glaciers*. Cambridge University Press, 2006.

Hamilton, Kevin, and Wataru Ohfuchi, eds. *High Resolution Numerical Modeling of the Atmosphere and Ocean*. Springer Science+Business Media, 2008. Presents a technical account of the efforts and progress in accurately modeling the climate and weather patterns of the world by computational means. Includes several papers presented by invited speakers to the first international convention focused on this endeavor.

Harmon, Russel S. and Andrew Parker. *Frontiers in Geochemistry*. John Wiley & Sons, 2011. Presents the existence and history of atmospheric oxygen and its formation in a geochemical context, drawing from several academic sources.

Harrison, Roy M. *Principles of Environmental Chemistry*. Royal Society of Chemistry, 2007. Discusses chemistry of the atmosphere, freshwater, oceans, and soils. Also discusses biogeochemical cycling and environmental organic chemistry. Suited for undergraduates and graduate students with a chemistry background.

Harvie, Christopher. *Fool's Gold: The Story of North Sea Oil*. Hamish Hamilton, 1994.

Heikkinen, Niina. "How People Make Summer Hotter." *Scientific American*, 25 Nov. 2014. Web. 24 Mar. 2015.

Heintzenberg, Jost, and Robert J. Charlson. *Clouds in the Perturbed Climate System: Their Relationship to Energy Balance, Atmospheric Dynamics and Precipitation*. MIT Press, 2013. Presents and discusses the complex relationship of clouds in the control of Earth's climate.

Henderson, Bonnie. *Strand: An Odyssey of Pacific Ocean Debris*. Oregon State University Press, 2008.

Henderson-Sellers, Ann, and Kendal McGuffie. "Atmospheric Composition Change: Climate-Chemistry Interactions" *The Future of the World's Climate: A Modeling Perspective*. Elsevier, 2012. 309–66.

Hewitt, C.N. and A.V. Jackson, eds. *Atmospheric Science for Environmental Scientists*. John Wiley & Sons, 2009. An introductory textbook of atmospheric science, discusses many aspects of the structure and evolution of Earth's atmosphere.

Heymann, Matthias, Gramelsberger and Mahoney, Martin, eds. *Cultures of Prediction in Atmospheric and Climate Science. Epistemic and cultural shifts in computer-based modelling and prediction*. Routledge, 2017. Individual chapters in this book present discussions of climate model design and the use of predictions and projections within political and regulatory discussions.

Hidore, John J. *Climatology: An Atmospheric Science.* Prentice Hall, 2010.

Hill, Marquita K. *Understanding Environmental Pollution.* Cambridge University Press, 2010.

Hinch, Stephen W. *Outdoor Navigation with GPS.* Wilderness Press, 2011.

"History of Discovery of the Atmosphere." *UCAR Center for Science Education.* U Corporation for Atmospheric Research, n.d. Web. 27 Apr. 2015.

Hodges, Kip. "Climate and the Evolution of Mountains." *Scientific American* Aug. 2006: 72–79. Print.

Hodgkins, Glenn A., and Ivan C. James. "Historical Ice-Out Dates for Twenty-nine Lakes in New England." U.S. Geological Survey Open-File Report 02–34. U.S. Geological Survey, 2002. Listings of ice-out dates for all years with data and discussion of methods and interpretation. The data listings enable users to perform their own analyses of the data.

Hodgson, Michael. *Basic Essentials: Weather Forecasting.* 3rd ed. Globe Pequot, 2007. Describes how people are trained to make forecasts using human observations of the weather around them. Before attempting to use computer forecasts, people should have a basic understanding of the physical principles operating in the atmosphere.

Hoff, Raymond, et al. "Applications of the Three-Dimensional Air Quality System to Western U.S. Air Quality: IDEA, Smog Blog, Smog Stories, AirQuest, and the Remote Sensing Information Gateway." *Journal of the Air and Waste Management Association* 59, no. 8 (2009): 980–989. Describes an extensive network of remote sensors, based on ground and onboard satellites used to monitor aerosols released into the atmosphere in the United States. The system, which creates three-dimensional images of the atmosphere, is a joint partnership of local, state, and federal governments and private citizens.

Hoffman, Matthew J. *Ozone Depletion and Climate Change: Constructing A Global Response.* State University of New York, 2005. Discusses the challenges of global policies to mitigate climate change. Examines models of climate change due to ozone depletion. Focuses on universal participation and governance of these issues.

Hogan, C. Michael. "Medieval Warm Period." *Encyclopedia of Earth.* Boston U, 1 Jan. 2013. Web. 23 Mar. 2015.

Holland, D. M., et al. "Acceleration of Jakobshavn Isbrae Triggered by Warm Subsurface Ocean Waters." *Nature Geoscience* 1 (2008): 659–664.

Holman, J. Alan. *In Quest of Great Lakes Ice Age Vertebrates.* Michigan State University Press, 2001.

Holman, Jack P. *Heat Transfer.* McGraw-Hill, 2014. College engineering textbook dealing with the various mechanisms of heat transfer.

Holthaus, Eric. "A Texas-Size Flood Threatens the Gulf Coast, and We're So Not Ready." *Grist*, 23 Aug. 2017, grist.org/article/a-texas-size-flood-threatens-the-gulf-coast-and-were-so-not-ready/. Accessed 22 Sept. 2017.

Holton, James R., and Gregory J. Hakim. *An Introduction to Dynamic Meteorology.* 5th ed. Academic, 2013.

Houghton, John T. *The Physics of Atmospheres.* 3rd ed. Cambridge University Press, 2002. Revised textbook with chapters on topics such as remote sensing, numerical modeling, climate change, chaos, and predictability.

Houghton, John. *Global Warming: The Complete Briefing.* 4th ed. Cambridge University Press, 2009. Provides an overview of the evidence for global warming. Several chapters focus on the climate system and seasonal forecasting. Offers accessible coverage of the ENSO system. Includes discussions of how ENSO influences weather and the potential for improved weather modeling based on research of ENSO.

House, Tamzy J. *Weather as a Force Multiplier: Owning the Weather in 2025.* Biblioscholar, 2012. Discusses military weather modification.

Hoyt, Douglas V., and Kenneth H. Schatten. *The Role of the Sun in Climate Change.* Oxford, 1997. Provides an extremely well written and easy-to-follow explanation of how solar variations can affect climate. Extensive bibliography.

Huggel, Christian, Mark Carey, John J. Clague, and Andreas Kaab, eds. *The High-Mountain Cryosphere. Environmental Changes and Human Risks.* Cambridge University Press, 2015. A extensive examination of the global environmental factors that dynamically affect the formation of mountain ice and snow caps and glaciers.

Hughes, Malcolm, and Henry F. Diaz, eds. *The Medieval Warm Period.* Berlin: Springer-Verlag, 1994.

Humphreys, David. *Logjam: Deforestation and the Crisis of Global Government.* Earthscan, 2006. Highly critical of the failure of international organizations to take on the task of preventing deforestation.

Hurrell, J. W., et al., eds. *The North Atlantic Oscillation: Climate Significance and Environmental Impact.* Washington, D.C.: American Geophysical Union, 2003.

Hyman, Andrew. *Principles of Paleoclimatology.* Callisto Reference, 2017.

Ilyina, Tatjana. *The Fate of Persistent Organic Pollutants in the North Sea: Multiple Years Model Simulations of g-HCH, a-HCH and PCB 153.* Springer-Verlag, 2007.

Incropera, Frank P. *Climate Change—A Wicked Problem: Complexity and Uncertainty at the Intersection of Science, Economics, Politics, and Human Behavior.* Cambridge UP, 2016.

Intergovernmental Panel on Climate Change. *Climate Change, 2007—Synthesis Report: Contribution of Working Groups I, II, and III to the Fourth Assessment Report of the Intergovernmental Panel on Climate Change.* Edited by the Core Writing Team, Rajendra K. Pachauri, and Andy Reisinger. Geneva, Switzerland: Author, 2008. Comprehensive overview of global climate change published by a network of the world's leading climate change scientists under the auspices of the World Meteorological Organization and the United Nations Environment Programme.

Irfan, Umair. "The Stunning Price Tags for Hurricanes Harvey and Irma, Explained." *Vox*, 18 Sept. 2017, www.vox.com/explainers/2017/9/18/16314440/disasters-are-getting-more-expensive-harvey-irma-insurance-climate. Accessed 22 Sept. 2017.

Iribarne, Julio V. *Atmospheric Physics.* Springer, 2013. As a short introduction to the field of atmospheric dynamics, this book attempts to provide a comprehensive yet elementary survey of the terrestrial atmosphere and how it functions.

Jacob, Daniel J. *Introduction to Atmospheric Chemistry.* Princeton University Press, 2007. Undergraduate textbook written by a Harvard professor; provides an overview of the new and rapidly growing field of atmospheric chemistry. Illustrations and index.

Jacques, Peter. *Globalization and the World Ocean.* Lanham, Md.: Rowman and Littlefield Publishing Group, 2006. Provides a unique analysis of how global marine and atmospheric conditions affect political conditions globally by viewing the world ocean as a single body of water composed of connected regional oceans. Aimed at researchers and students of marine sciences, and environmental and globalization studies.

Jenkins, Jerry C., et al. *Acid Rain in the Adirondacks: An Environmental History.* Cornell University Press, 2007.

Jochum, Markus, and Raghi Murtugudde, eds. *Physical Oceanography: Developments Since 1950.* New York: Springer Science + Business Media, 2006. Provide a historic overview of developments in physical oceanography in the last half of the twentieth century with senior scientists in the field authoring individual chapters. Chapter 12, by Karl Wunsch, deals exclusively with the WOCE and its aftermath.

Jones, E. Peter, and Leif G. Anderson. "Is the Global Conveyor Belt Threatened by Arctic Ocean Fresh Water Outflow?" Chapter in Dickson, Robert R., Meincke, Jens and Rhines, Peter, eds. *Arctic-Subarctic Ocean Fluxes. Defining the Role of the Northern Seas in Climate.* Springer Science + Business Media B.V., 2008. Provides a detailed analysis of the interaction of Arctic waters with southerly ocean waters and the potential effects of those interactions on the global conveyor system.

Jones, P. D., T. J. Osborn, and K. R. Briffa. "The Evolution of Climate Over the Last Millennium." *Science* 292 (2001): 662–667.

Joughin, Ian, and Richard B. Alley. "Stability of the West Antarctic Ice Sheet in a Warming World." *Nature Geoscience* 4 (2011): 506–513. Analyzes satellite observations and historical data to estimate the magnitude of the contribution of west Antarctic ice sheet melting to past sea level rises, and to model its potential future impact.

Jouzel, Jean, Claude Lorius and Dominique Raynard. *The White Planet. The Evolution and Future of Our Frozen World.* Princeton University Press, 2013. Examines the past, present and future roles of ice on Earth, concluding that climate change and global warming will ultimately have disastrous consequences for human societies on Earth regardless of technological advances.

Jury, Mark R. "An Intercomparison of Observational, Reanalysis, Satellite, and Coupled Model Data on Mean Rainfall in the Caribbean." *Journal of Hydrometeorology* 10, no. 2 (2009): 413–420. Discusses the use of a wide range of technologies to generate data on rainfall trends in the Caribbean in a twenty-one-year period. Different types of data and models were generated based on this data.

Kaper, Hans, and Hans Engler. *Mathematics and Climate*. SIAM, 2013. This book guides the reader through the mathematical basis of climate models, beginning with understanding the use of statistics and continuing through discussions of individual factors affecting the global climate system.

Kaplan, Elliot D., and Christopher Hegarty, eds. *Understanding GPS: Principles and Applications*. 2nd ed. Artech House, 2005.

"Karl Jansky and the Discovery of Cosmic Radio Waves." *National Radio Astronomy Observatory*. Assoc. U, 16 May 2008. Web. 27 Apr. 2015.

Karnosky, David, and Meeting of the International Union of Forestry Research Organizations. *Air Pollution, Global Change and Forests in the New Millennium*. Elsevier, 2006. Overview of the effects of climatic factors that emphasizes the impact of ozone, acid rain, and CO_2 on the growth and health of forests. Illustrations, figures, tables, bibliography, index, maps.

Kaufman, Darrell S., et al. "Continental-Scale Temperature Variability during the Past Two Millennia." *Nature Geoscience* 6 (2013): 339–46. PDF file.

Kaye, Catheryn Berger, and Philippe Cousteau. *Going Blue*. Free Spirit Publishing, 2009. Discusses conservation issues related to oceans, lakes, rivers, and other bodies of water, as well as topics such as pollution, watershed management, coral bleaching, and ocean acidification. Also provides a guideline for action with multiple chapters discussing how students can get involved in water conservation. Lists many resources to help the reader find more information or get started on projects.

Kazlowski, Steve. *The Last Polar Bear: Facing the Truth of a Warming World*. Mountaineers Books, 2008. The author has spent thousands of hours observing and photographing polar bears in the wild.

Kelley, Joseph T., Orrin H. Pilkey, and J. A. G. Cooper. *America's Most Vulnerable Coastal Communities*. Geological Society of America, 2009.

Kennedy, Caitlyn. "Wobbly Polar Vortex Triggers Extreme Cold Air Outbreak." *Climate.gov*. NOAA, 8 Jan. 2014. Web. 23 Mar. 2015.

Kennet, James P. *Marine Geology*. Prentice Hall, 1982.

Kenward, Alyson, Dan Yawitz, Todd Sanford, and Regina Wang. "Heat Islands Cook U.S. Cities Faster Than Ever." *Scientific American*, 22 Aug. 2014. Web. 24 Mar. 2015.

Kerr, J. B., I. A. Asbridge, and W. F J. Evans. "Intercomparison of Total Ozone Measured by the Brewer and Dobson Spectrophotometers at Toronto." *Journal of Geophysical Research* 93 (September, 1988): 11129–11140. Plots of careful ozone measurements and discussion of factors recognized as affecting such measurements over long periods at one site demonstrate, to the seriously suspicious, the variability that makes spotting trends in stratospheric ozone concentration so difficult.

Kininmonth, William. *Climate Change: A Natural Hazard*. Multiscience, 2004. Kininmonth has a career in meteorological and climatological science and policy spanning more than forty-five years. His suspicions that the science and predictions of anthropogenic global warming extend beyond sound theory and evidence were crystallized following the release of the 2001 IPCC assessment report. His book gives information about global and regional monitoring on various scales.

Knight, Jeff R., Chris K. Folland and Adam A. Scaife. "Climate Impacts of the Atlantic Multidecadal Oscillation." *Geophysical Research Letters* 33, no. 17 (September, 2006). Describes some of the climate effects of variations in the AMO.

Knighton, David. *Fluvial Forms and Processes: A New Perspective*. Rev. and updated ed., Hodder Arnold, 1998.

Kokhanovsky, Alexander A., ed. *Light Scattering and Remote Sensing of Atmosphere and Surface*. Vol. 6 in *Light Scattering Reviews*. Springer, 2011. A compilation of papers on remote sensing as it relates to light scattering and the images created by precipitation as influenced by light scattering phenomena.

Kolumban, Hutter, Yongqi Wang, and Irina P. Chubarenko. *Physics of Lakes: Foundation of the Mathematical and Physical Background*. Springer, 2011. Assumes a certain level of mathematical knowledge on the part of the reader, but the mathematical concepts presented are not beyond ready comprehension, beginning with the fundamental equations of lake hydrodynamics and progressing to angular momentum and vorticity.

Komar, Paul D. *Beach Process and Sedimentation.* Prentice Hall, 1998.

Kowahata, Hodaka, and Yoshio Awaya. *Global Climate Change and Response of the Carbon Cycle in the Equatorial Pacific and Indian Oceans and Adjacent Landmasses.* Elsevier, 2006.

Krüger, Tobias. *Discovering the Ice Ages: International Reception and Consequences for a Historical Understanding of Climate.* Brill, 2013.

Kuhlbrodt, T., et al. "On the Driving Processes of the Atlantic Meridional Overturning Circulation." *Reviews in Geophysics* 45 (April 24, 2007).

Kuhn, Gerald G. "The Impact of Volcanic Eruptions on Worldwide Weather." *Twenty-first Century Science and Technology* (Winter, 1997–1998): 48–58. A well-documented article that provides a comprehensive review of the evidence supporting a link between volcanic activity and climatic changes. Gives special attention to a wide range of weather-related abnormalities associated with the aftereffects of specific events.

"La Niña." *National Geographic Education.* National Geographic Society, 1996–2015. Web. 24 Mar. 2015.

Lallanila, Marc. "What Is the Greenhouse Effect? *Live Science,* 12 Apr. 2016, www.livescience.com/37743-greenhouse-effect.html. Accessed 2 Feb. 2017. Le Treut, H., et al. "Historical Overview of Climate Change." In *Climate Change, 2007—The Physical Science Basis: Contribution of Working Group I to the Fourth Assessment Report of the Intergovernmental Panel on Climate Change,* edited by Susan Solomon et al. Cambridge UP, 2007.

Lamb, Hubert Horace. *Climate, History and the Modern World.* Routledge, 2006.

Lambright, W. Henry. *NASA and the Environment. The Case of Ozone Depletion.* NASA SP-2005–4538, 2005. An interesting case study of the problem of ozone depletion, subdivided into eight recognizable stages.

Lang, Kenneth R. *Sun, Earth, and Sky.* Springer, 2006. This beautiful text uses a multitude of illustrations and images to illuminate the workings of the sun and its relationship with Earth. It is easy to understand and perfectly suited for students new to the subject.

Larson, Eric. *Isaac's Storm: A Man, a Time, and the Deadliest Hurricane in History.* Vintage, 2000.

Larssen, Thorjørn, et al. "Acid Rain in China." *Environmental Science & Technology* 40 (2006): 418–25.

Le Sueur, Meridel. *North Star Country.* 2nd ed. University of Minnesota Press, 1998.

Lemke, Peter, and Hans-Werner Jacobi. *Arctic Climate Change: The Acsys Decade and Beyond.* New York: Springer Science+Business Media, 2012. Addresses a number of major topics related to climate change in the Arctic brought on by increasing global temperatures with respect to the role of the Arctic in the global climate system.

Leroux, Marcel. *Dynamic Analysis of Weather and Climate: Atmospheric Circulation, Perturbations, Climatic Evolution.* Springer, 2014.

_____. *Global Warming: Myth or Reality?: The Erring Ways of Climatology.* Praxis Publishing Ltd, 2010. Describes both natural and anthropogenic climate change, with many references. Some discussion is given to the role of solar variations in producing climate change.

Levitan, Ben. *GPS Quick Course: Systems, Technology and Operation.* Althos, 2007.

Li, Tim, and Pang-chi Hsu. *Fundamentals of Tropical Climate Dynamics.* Springer, 2018. Although just one chapter of this book is devoted to the ENSO phenomenon, most of the other chapters include a conversation about the interaction of the ENSO and a number of other phenomena such as the monsoons and other large-scale oscillation phenomena.

"Life and Death after Great Barrier Reef Bleaching." *ARC Centre of Excellence for Coral Reef Studies,* 29 Nov. 2016, www.coralcoe.org.au/media-releases/life-and-death-after-great-barrier-reef-bleaching. Accessed 22 Feb. 2017.

Lighthill, James, and Robert Pearce, eds. *Monsoon Dynamics.* Cambridge University Press, 2009.

Lionello, P., ed. *The Climate of the Mediterranean Region: From the Past to the Future.* Elsevier Science, 2012.

Lloyd, Elisabeth A., and Eric Winsberg, eds. *Climate Modelling. Philosophical and Conceptual Issues.* Springer, 2018. In this book, individual chapters present discussions geared to the interpretation and understanding of climate data from climate models rather than the models themselves.

Lohman, Felix Lüönd and Fabian Mahrt. *An Introduction to Clouds From the Microscale to Climate.* Cambridge University Press, 2016. A thorough introduction to the formation and function of clouds and their role in determining the climate of Earth.

Longshore, David. *Encyclopedia of Hurricanes, Typhoons, and Cyclones.* New York: Facts on File, 2008. A comprehensive reference book containing about four hundred cross-referenced entries covering the science, history, and cultural significance of severe weather phenomena. Contains black-and-white photographs and other images. Suitable for high school readers and older.

Lui, Kevin. "Severe Flooding in South Asia Has Caused More Than 1,200 Deaths This Summer." *Time,* 29 Aug. 2017, time.com/4921340/south-asia-floods-india-mumbai-bangladesh-nepal/. Accessed 22 Sept. 2017.

Lusted, Marcia Amidon. *Extreme Weather Events.* Greenhaven Publishing, 2018.

Lutgens, Frederick K., Edward J. Tarbuck, Dennis G. Tasa, and Redina Herman. *The Atmosphere: An Introduction to Meteorology.* 14th ed. Pearson Education, 2018. An updated version of this textbook, in loose-leaf format, provides students with a general overview of the structure and properties of Earth's atmosphere.

Luthi, Dieter, et al. "High-Resolution Carbon Dioxide Concentration Record 650,000–800,000 Years Before Present." *Nature* 453, no. 7193 (2008): 379–82. Includes a plot of CO_2 concentrations and temperatures over eight glacial cycles. Charts, tables, bibliography.

Macdougall, J. D. *Frozen Earth: The Once and Future Story of Ice Ages.* University of California Press, 2013. A scientific look at the ice ages and their geological impact.

Mackay, Anson, Rick Battarbee, John Birks, and Frank Oldfield, eds. *Global Change in the Holocene.* Routledge, 2014. Chapters by several contributors describe Holocene climate reconstruction from the resources of many different climate proxies.

MacKenzie, Fred T., and Abraham Lerman. *Carbon in the Geobiosphere. Earth's Outer Shell.* Springer, 2006. This book presents descriptions and analyses of the role of carbon dioxide in Earth's history.

Mackwell, Stephen J., et al., eds. *Comparative Climatology of Terrestrial Planets.* University of Arizona Press, 2013.

MacLeish, William H. *The Gulf Stream.* Houghton Mifflin, 1989.

Madigan, Michael T., et al. *Brock Biology of Microorganisms.* 12th ed. Pearson/Benjamin Cummings, 2009. Introductory microbiology textbook. Chapter 17 describes methane production by archaea.

Mahan, A. T. *The Influence of Sea Power upon History, 1660–1783.* 1890. Reprint. Noble Books, 2004.

Malmesheimer, R. W., et al. "Preventing GHG Emissions Through Avoided Land-Use Change." *Journal of Forestry* 106, no. 3 (April/May, 2008). This special issue is devoted entirely to the Society of American Foresters' task force report on forest management solutions for mitigating climate change.

Mander, Jerry, and Edward Goldsmith, eds. *The Case Against the Global Economy and For a Turn Towards Localization.* 2nd ed. Routledge, 2014. The argument is made in this book that the globalized economy is fueling anthropogenic climate change with increasingly catastrophic consequences.

Mann, Michael E. *The Hockey Stick and the Climate Wars. Dispatches from the Front Lines.* Columbia University Press, 2014. This very readable book lays out the science behind global climate change and identifies the conflict between climate change proponents and deniers.

Mann, Michael E., and Lee R. Kump. *Dire Predictions.* D.K., 2008. Print.

Margesin, Rosa, ed. *Permafrost Soils.* Springer-Verlag, 2009. A collection of learned articles describing many aspects of Arctic permafrost soils and the effects of global warming on those soils and their content.

Marris, Emma. *Rambunctious Garden: Saving Nature in a Post-Wild World.* Bloomsbury Publishers, 2011. Describes as an "optimistic book," the author promotes an approach of living with climate change and allowing wild growth to expand into all available niches, rather than attempting to maintain natural wilderness areas in their pristine state by technological intervention to control climate change.

Marshal, John, and R. Alan Plumb. *Atmosphere, Ocean and Climate Dynamics: An Introductory Text.* Elsevier Academic Press, 2008. An excellent introduction to atmospheres and oceans. Discusses the greenhouse effect, atmospheric structure, oceanic and atmospheric circulation, and climate change. Suitable for advanced undergraduates and graduate students with some background in advanced mathematics.

Marshall, Shawn J. *The Cryosphere.* Princeton University Press, 2012. A complete introduction to the physical properties and thermodynamics of ice and snow as the basic matter of the cryosphere and its dynamics.

Marti, Joan, and Gerald Ernst. *Volcanoes and the Environment*. Cambridge University Press, 2005. Presents a discussion of the various ways in which volcanism affects terrestrial, aquatic, and atmospheric environments, from both historical and present-day perspectives. Provides information on volcanoes from the physical mechanisms and behaviors to hazards and effects on humans and nature. Written for students and geologists studying environmental science and geology.

Marzano, Frank S., and Guido Visconti, eds. *Remote Sensing of Atmosphere and Ocean from Space: Models, Instruments, and Techniques*. Springer, 2011. A collection of lectures that focus on space-borne remote-sensing techniques, including microwave, infrared, and passive sensors and weather forecasting practices.

Maslin, Mark. *Global Warming: A Very Short Introduction*. Oxford University Press, 2009. Provides a concise general introduction to the major factors that define the climate of Earth.

Maunder, E. Walter. "The Prolonged Sunspot Minimum, 1645–1715." *Journal of the British Astronomical Society* 32 (1922): 140.

Mbane Biouele, Cesar *Earth's Atmosphere Dynamic Balance Meteorology*. Scientific Research Publishing, 2014. Without becoming overly technical, this book discusses the structure of the atmosphere and many of the physical phenomena that occur there.

McDade, Lucinda. "Plant Communities and Climate in Southern California." *Rancho Santa Ana Botanic Garden*. Rancho Santa Ana Botanic Garden, n.d. Web. 23 Mar. 2015.

McFadden, L., R. Nicholas and E. Penning-Rowsell, eds. *Managing Coastal Vulnerability*. Elsevier Press, 2007. Good reading for those interested in coastal vulnerability areas of the world, including wetlands.

McFadden, Lucy-Ann Adams, Paul Robert Weissman, and T. V. Johnson. *Encyclopedia of the Solar System*. Academic Press, 2007. From the origins of the solar system to modern-day planetary exploration, this comprehensive text covers topics about the sun, solar wind, and the relationship between the sun and Earth.

McGonigal, David. *Antarctica: Secrets of the Southern Continent*. Firefly Books, 2008. Describes the discoveries of the International Polar Year, 2007–2008, in various fields, including geology, geography, and climatology. Discusses potential effects of global warming.

McGranahan, Gordon, and Frank Murray, eds. *Air Pollution and Health in Rapidly Developing Countries* Earthscan, 2003.

McKinney, Frank. *The Northern Adriatic Ecosystem: Deep Time in a Shallow Sea*. Columbia University Press, 2007. Covers the paleogeography of the Adriatic Sea. Discusses the succession of the ecosystem as the sea's geography changed. Discusses oceanography topics such as circulation and sedimentation. Topics are thorough and logically ordered, making this book accessible to undergraduates.

McLeman, Robert A. *Climate and Human Migration. Past Experiences, Future Challenges*. Cambridge University Press, 2014. An advanced text that examines why people migrate, then considers each of the large-scale drivers of forced migration.

Melanotte-Rizzoli, P., ed. *Modern Approaches to Data Assimilation in Ocean Modeling*. Science, 1996. A compilation of articles on oceanographic data assimilation. Contributors focus on developing improved data assimilation techniques for ocean-oriented computer-model development.

Mélieres, Marie-Antoinette, and Chloé Maréchal. *Climate Change: Past, Present, and Future*. John Wiley & Sons, 2015. This book first provides an introduction to the basics of climatology and then investigates the physical evidence of past and present climate change in the context of that basic framework.

Melnikov, I. A. *The Arctic Sea Ice Ecosystem*. Gordon and Breach, 1997. Examines in detail ice and its impact on Earth and its inhabitants.

Mersmann, Katy, and Theo Stein. "Warm Air Helped Make 2017 Ozone Hole Smallest Since 1988." *NASA*, 2 Nov. 2017, www.nasa.gov/feature/goddard/2017/warm-air-helped-make-2017-ozone-hole-smallest-since-1988. Accessed 14 Dec. 2017.

Metcalfe, Sarah, and Dick Derwent. *Atmospheric Pollution and Environmental Change*. Routledge, 2014.

Michaels, Patrick J,. and Robert C. Balling. *Climate of Extremes: Global Warming Science They Don't Want You to Know*. Cato Institute, 2010. Important conservative corrective to growing alarmist projections about crysophere damage. Moderates the predictions and indicates progress in monitoring the cryosphere.

Michaels, Patrick. *Meltdown: The Predictable Distortion of Global Warming by Scientists, Politicians, and the Media.* Cato Institute, 2004. Although Michaels, a climatologist, accepts the reality of global warming, he argues that alarmists have exaggerated its future effects, and he uses the debate over Kilimanjaro's shrinking ice cap to show how science can be distorted for political ends.

Micklin, P. "Desiccation of the Aral Sea: A Water Management Disaster in the Soviet Union." *Science* 241, no. 4844 (September 2, 1988): 1170–1176. Micklin's early work on causes of sea recession and resulting environmental problems; details local water quality improvements and future schemes for preservation of the Aral.

_____. "The Aral Sea Disaster." *Annual Review of Earth and Planetary Science* 35 (2007): 47–72. Excellent collection of data on the water balance, salinities, and hydrology of the Aral Basin; human and ecological consequences of the disaster; improvement efforts by global aid agencies; and engineering mitigation.

Micklin, P., and N. V. Aladin. "Reclaiming the Aral Sea." *Scientific American* 298, no. 4 (April, 2008): 64–71. Outlines the collapse of the Aral Sea due to wasteful irrigation of the desert; the geography of residual lakes; the 2005 dam construction and future of the Amu Dar'ya; and application of lessons learned from the Aral disaster to other regions with similar risks.

Miller, Robert N. *Numerical Modeling of Ocean Circulation.* New York: Cambridge University Press, 2007. Designed to teach the process of numerical analysis; each chapter includes exercises to practice modeling skills. Discusses models of tropical waters, coastal waters, and shallow waters. Also covers simple and complex numerical modeling. Useful for the advanced undergraduate and graduate student. Requires a strong mathematics background.

Miller, Ron. *Chasing the Storm: Tornadoes, Meteorology, and Weather Watching.* Twenty-First Century, 2014.

Milman, Oliver. "Climate Guru James Hansen Warns of Much Worse than Expected Sea Level Rise." *Guardian.* Guardian News and Media, 22 Mar. 2016. Web. 15 Apr. 2016.

Mitra, Abhijit, Kakoli Bannerjee, and Avijit Gangopadhyay. *Introduction to Marine Plankton.* Daya, 2008. Although the authors emphasize plankton in the waters near India, they provide an excellent overall description of plankton, its role in the carbon cycle, fertilization effects, identification, and even culturing. However, the book does not comment directly on oceanic fertilization or acidification.

Mitsch, William J., et al. *Wetlands.* John Wiley & Sons, 2015. A more technical reading for those interested in the science of wetlands.

Mityagina, M. I., O. Y. Larova, and S. S. Karimova. "Multi-sensor Survey of Seasonal Variability in Coastal Eddy and Internal Wave Signatures in the North-Eastern Black Sea." *International Journal of Remote Sensing* 31, nos. 17/18 (2010): 4779–4990. Describes a multisystem approach to the application of remote sensors in the study of coastal areas and waves in the Black Sea. Among the systems employed in this study are synthetic aperture radar, radiometers, infrared sensors, and other active and passive radar systems.

Mlodinow, Leonard. *The Drunkard's Walk: How Randomness Rules Our Lives.* Pantheon, 2008.

Mogil, H. Michael. *Extreme Weather: Understanding the Science of Hurricanes, Tornadoes, Floods, Heat Waves, Snow Storms, Global Warming, and Other Atmospheric Disturbances.* Black Dog & Leventhal Publisher, 2010. A detailed analysis of the science of severe weather and climate change. Provides a review of the long-term weather patterns that produce this weather and suggestions about how society can prepare for future weather disasters.

Möller, Detlev. *Chemistry of the Climate System.* 2nd ed. De Gruyter, 2014. A more advanced book providing detailed information about the chemistry of atmospheric gases and air-borne constituents, using a problem-based approach.

Monmonier, Mark S. *Air Apparent: How Meteorologists Learned to Map, Predict, and Dramatize Weather.* University of Chicago Press, 2016. Using weather maps, this book shows the growth of technology used in weather forecasting.

Monroe, James S., Reed Wicander, and Richard Hazlett. *Physical Geology: Exploring the Earth.* 6th ed. Thomson, 2007.

Montgomery, C. W. *Environmental Geology.* 9th ed. McGraw-Hill, 2010.

"Monthly and Annual U.S. Tornado Summaries." *Storm Prediction Center.* NOAA, 6 Jan 2015. Web. 23 Feb. 2015.

Mooney, Chris. *Storm World: Hurricanes, Politics and the Battle Over Global Warming.* Harcourt, 2007. Provides a clear discussion of the nature of hurricanes and the perceived changes in their nature that are still hotly debated in regard to their relationship with global warming. Very readable, presenting very technical information in a nontechnical manner.

Moreno, José, and Walter C. Oechel, eds. *Anticipated Effects of a Changing Global Environment in Mediterranean-Type Ecosystems.* Springer-Verlag, 1995.

Morozov, Eugene G., Alexander N. Demidov, Roman Y. Tarakanov, and Walter Zenk. *Abyssal Channels in the Atlantic Ocean: Water Structure and Flows.* Springer, 2010.

Mote, Philip W., and Georg Kaser. "The Shrinking Glaciers of Kilimanjaro: Can Global Warming Be Blamed?" *American Scientist* 95 (July/August, 2007): 318–325. This article by a climatologist and glaciologist musters much scientific evidence to show that factors other than global warming are responsible for the retreat of Kibo's ice sheets. Illustrated with photographs, graphs, and a map. Bibliography.

Mozdzynski, George, ed. *Use of High Performance Computing in Meteorology.* World Scientific, 2007. Features discussions on the use of comprehensive observational data in atmospheric and oceanographic computer modeling.

Mukhopadhyay, Ranadhir, Anil K. Ghosh, and Sridhar D. Iyer. *The Indian Ocean Nodule Field: Geology and Resource Potential.* Ed. M. Hale. Elsevier, 2012.

Müller, Rolf, ed. *Stratospheric Ozone Depletion and Climate Change.* RSC, 2012.

Mulvaney, Kieran. *At the Ends of the Earth: A History of the Polar Regions.* Island Press/Shearwater Books, 2001. A history of human interaction with the polar regions and how those regions are affected by the changing climate.

Nadadur, Srikanth S., and John W. Hollingsworth. *Air Pollution and Health Effects.* London: Humana Press, 2015. Print.

NASA Goddard Institute for Space Studies. National Aeronautics and Space Administration, 2012. Web. 21 Aug. 2012. Describes a variety of current research programs in the environmental sciences, physics, and atmospheric chemistry. Also contains descriptions of using atmospheric physics in the study of climate change and global warming.

"NASA, NOAA Analyses Reveal Record-Shattering Global Warm Temperatures in 2015." *Goddard Institute for Space Studies.* NASA, 20 Jan. 2016, www.nasa.gov/press-release/nasa-noaa-analyses-reveal-record-shattering-global-warm-temperatures-in-2015. Accessed 18 Oct. 2016.

National Academy of Sciences. *Methane Generation from Human, Animal, and Agricultural Wastes.* Lulu Com, 2013. Describes the role of disparate sources in methane production.

National Academy of Sciences. *The Future of Atmospheric Chemistry Research. Remembering Yesterday, Understanding Today, Anticipating Tomorrow.* National Academies Press, 2016. This report from the National Academy of Sciences offers a significant account of the state of atmospheric chemistry research.

National Aeronautics and Space Administration (NASA) "NASA Ozone Watch" https://ozonewatch.gsfc.nasa.gov/

National Aeronautics and Space Administration. "Solar Physics." Available at http://solarscience.msfc.nasa.gov/SunspotCycle.shtml. Presents data on the solar cycle, the Maunder Minimum, and other Earth-Sun interactions.

National Research Council, Division of Earth and Life Sciences. *Abrupt Impacts of Climate Change: Anticipating Surprises.* National Academics Press, 2013.

Nebel, Bernard J., and Richard T. Wright. *Environmental Science: Towards a Sustainable Future.* Prentice Hall, 2008. Several chapters describe methane as a GHG and an alternative fuel.

Needler, George T. "WOCE: The World Ocean Circulation Experiment." *Oceanus* 35 (Summer, 1992): 74–77. A clear introduction to the goals of WOCE and the methods used to obtain data. Written by the first scientific director of the WOCE, who was instrumental in planning the project.

Neelin, J. David. *Climate Change and Climate Modeling.* Cambridge University Press, 2011. Discusses how climate data is used to identify climate change and to produce predictive models of future climate states.

Nelleman, Christian, et al., eds. *Blue Carbon. The Role of Healthy Oceans in Binding Carbon.* UNEP/Earthprint, 2009. This "rapid response report" examines the role of effective management of ocean ecosystems as carbon sinks for mitigating climate change.

Neprochnov, Y. P., et al., eds. *Intraplate Deformation in the Central Indian Ocean Basin.* Geological Society of India, 1998.

Nghiem, S. V., et al. "Geophysical Constraints on the Antarctic Sea Ice Cover." *Remote Sensing of Environment*, vol. 181, 2016, pp. 281–92, doi:10.1016/j.rse.2016.04.005. Accessed 2 Feb. 2017.

Nihoul, J. C. J., P. O. Zavialov, and P. Micklin, eds. *Dying and Dead Seas: Climatic Versus Anthropic Causes.* Dordrecht, the Netherlands: Kluwer Academic, 2004. Synthesis of sixteen lectures given at the May, 2003, North Atlantic Treaty Organization Advanced Research Workshop in Belgium on dead or dying internal seas and lakes; assesses the past and present roles of natural and anthropogenic causes.

Nisbet, Robert, John Elder, and Gary Miner. *Handbook of Statistical Analysis and Data Mining Applications.* Elsevier, 2009.

North, Gerald R., and Kwang-Yu Kim. *Energy Balance Climate Models.* Wiley-VCH, 2017. This book provides a mathematics-based presentation of climate phenomena and feedbacks.

Novák, Viliam. *Evapotranspiration in the Soil-Plant-Atmosphere System.* Springer, 2012. Provides detailed information about the movement of water in the evapotranspiration segment of the hydrologic cycle.

Nuccitelli, Dana. *Climatology versus Pseudoscience: Exposing the Failed Predictions of Global Warming Skeptics.* Praeger, 2015.

Nunn, Patrick D. *Climate, Environment and Society in the Pacific During the Last Millenium.* Elsevier, 2007.

"The Ocean Conveyor." *Woods Hole Oceanographic Institution.* Woods Hole Oceanographic Inst., 2014. Web. 23 Mar. 2015.

"The Ocean in a High-CO_2 World." Special Section in *Journal of Geophysical Research* 110, no. C9 (2005). Includes papers on the carbon cycle, as well as articles making arguments for and against oceanic fertilization.

Odenwald, Sten F. *The Twenty-Third Cycle: Learning to Live with a Stormy Star.* Columbia University Press, 2001. Well-referenced layperson's guide to solar activity over a sunspot cycle and the geomagnetic effects of solar storms that occur during high sunspot activity.

Ogawa, Yujiro, Ryo Anma, and Yildirim Dilek. *Accretionary Prisms and Convergent Margin Tectonics in the Northwest Pacific Basin.* Springer Science+Business Media, 2011.

Ogden, Joan. "High Hopes for Hydrogen." *Scientific American* Sept. 2006: 94–99. Print.

O'Hare, Greg, et al. *Weather, Climate and Climate Change Human Perspectives.* Lord, Taylor & Francis, 2014. Discusses the mass and energy effects of air masses and their interaction, before focusing on the roles and description of fronts.

Oke, T.R., G. Christen Mills, and J.A. Voogt. *Urban Climates.* Cambridge University Press, 2017.

Oliver, John E., ed. *The Encyclopedia of Climatology.* Springer, 2004.

O'Loughlin, Karen Fay, and James F Lander. *Caribbean Tsunamis: A 500-Year History from 1498–1998.* Kluwer Academic Publishers, 2010.

Open University. *Ocean Circulation.* 2nd ed. Elsevier Butterworth-Heinemann, 2005. This introductory oceanography textbook covers atmospheric circulation, surface currents, and thermohaline circulation. The theory of fluid flow throughout the ocean is developed with a minimum of mathematics (knowledge of algebra and trigonometry is assumed). Illustrations, figures, tables, maps, references, index.

Oppenheimer, Clive. *Eruptions That Shook the World.* Cambridge University Press, 2011. Describes the mechanics of volcanoes and different types of eruptions. Also describes the effects of eruptions on the atmosphere, humans, and other organisms. Examines specific volcanoes and provides an appendix of major eruptions.

Organization for Economic Cooperation and Development. *Space Technologies and Climate Change: Implications for Water Management, Marine Resources, and Maritime Transport.* Author, 2008. An overview of the various types of technologies used in the study of climate change as it pertains to water resource management. Focuses on space-based systems.

Orsi, Jared. *Hazardous Metropolis: Flooding and Urban Ecology in Los Angeles.* University of California Press, 2004.

Ostmann, Robert. *Acid Rain: A Plague upon the Waters.* Dillon, 1982.

Ott, R. Lyman, and Michael T. Longnecker. *An Introduction to Statistical Methods and Data Analysis.* Brooks, 2010.

Ozsoy, Emin, and Alexander Mikaelyan, eds. *Sensitivity to Change: Black Sea, Baltic Sea, and North Sea.* Kluwer Academic Publishers, 1997.

Paeth, H., et al. "The North Atlantic Oscillation as an Indicator for Greenhouse-Gas Induced Regional Climate Change." *Climate Dynamics* 15, no. 12 (1999): 953–960.

Pap, Judit M., and Peter Fox, eds. *Solar Variability and Its Effects on Climate.* American Geophysical Union, 2004. In addition to Milankovitch cycles, discusses sunspots and changes in solar output in geologic time.

Parson, Edward Anthony. *Protecting the Ozone Layer: Science and Strategy.* Oxford Univ. Press, 2010. Comprehensive technical discussion of efforts to protect the ozone layer undertaken through international cooperation. Chapter 2 is devoted to a review of early stratospheric science. Includes notes, references, and index.

Pearson, M. N. *Trade, Circulation, and Flow in the Indian Ocean World.* 2015.

Pedlosky, Joseph. *Ocean Circulation Theory.* Rev. ed. New York: Springer, 2010. A detailed description of modern models of ocean circulation and the effect it has on climate. An excellent resource for advanced students that demonstrates the importance of the data provided by the WOCE.

Pendergrass, Angeline G., and Dennis L. Hartmann. "Changes in the Distribution of Rain Frequency and Intensity in Response to Global Warming." *Journal of Climate* 27.22 (2014): 8372–83. Print.

Philander, S. George. "Investigating Atmospheric Circulation." *Our Affair with El Niño: How We Transformed an Enchanting Peruvian Current into a Global Climate Hazard.* Princeton University Press, 2005, 177–88.

Pidwirny, Michael. *Understanding Physical Geography.* Our Planet Earth Publishing, 2017. An entry-level textbook of physical geography that presents a complete description of the Coriolis Effect and how it is driven by Earth's rotation.

Pilkey, Orrin, Linda Pilkey-Jarvis and Keith C. Pilkey. *Retreat from a Rising Sea. Hard Choices in an Age of Climate Change.* Columbia University Press, 2016. This book presents a close look at the impending doom of low-lying coastal cities and human societies as sea levels rise.

Pilkington, Ed. "Put Oil Firm Chiefs on Trial, Says Leading Climate Change Scientist." *The Guardian,* Monday, June 23, 2008. Discusses James E. Hansen's return to Congress after he first warned of climate change in 1988. Hansen claims industry groups have been consciously spreading misinformation on climate change that has slowed government reaction to address climate change.

Pinet, Paul R. *Invitation to Oceanography.* Jones & Bartlett, 2009. Discusses geology, biology, chemistry, and physics of the oceans at an introductory college level. Includes a chapter on climate change.

"Polar Bears and Climate Change." *World Wildlife Fund,* 2017, www.worldwildlife.org/pages/polar-bears-and-climate-change. Accessed 8 Nov. 2017.

Polovna, Jeffrey J., Evan A. Howell, and Melanie Abecassis. "Ocean's Least Productive Waters Are Expanding." *Geophysical Research Letters* 35, no. 103618 (February, 2008). Presents the SeaWiFS data, showing that the ocean's oligotrophic regions in the subtropical gyres are expanding at rates exceeding those predicted by computer models.

Polvani, L. M., A. H. Sobel, and D. W. Vaugh, eds. "Stratospheric Polar Vortices." *The Stratosphere: Dynamics, Transport, and Chemistry.* Washington: Amer. Geophysical Union, 2010. Print.

Powell, James Lawrence. *The Inquisition of Climate Science.* Columbia University Press, 2012. An exposé of the most prominent deniers of climate change, focusing on their lack of credentials, the extensive funding received from the industrial complex dependent upon carbon-based fuel consumption, and their inability to provide a valid alternate explanation for global warming.

Prothero, Donald R. *Catastrophes!: Earthquakes, Tsunamis, Tornadoes, and Other Earth-Shattering Disasters.* Johns Hopkins University Press, 2011. Provides a detailed and clear explanation of the many natural and anthropogenic disasters facing our planet. Each chapter is devoted to a different catastrophe, including earthquakes, volcanoes, hurricanes, ice ages, and current climate changes.

Pruppacher, Hans R., and James D. Klett, eds. *Microphysics of Clouds and Precipitation.* 2nd ed. Springer, 2010.

Pugh, David T. *Changing Sea Levels: Effects of Tides, Weather and Climate.* Cambridge University Press, 2008.

Rackley, Stephen A. *Carbon Capture and Storage.* 2nd ed., Elsevier/Butterworth-Heinemann, 2017. An in-depth presentation of the methodologies of artificial carbon sink technologies.

"Radiometric Normalization of Sensor Scan Angle Effects in Optical Remote Sensing Imagery." *International Journal of Remote Sensing* 28, no. 19 (2007): 4453–4469. Describes some of the optical effects, caused by the content of the atmosphere, which can be scanned using satellite-based and airborne remote sensors.

Radovanović, Milan M., Vladan Ducić, and Saumitra Mukherjee. "Climate Changes Instead Of Global Warming." *Thermal Science* 18.3 (2014): 1055–61. Print.

Rahm, Diane. *Climate Change Policy in the United States. The Science, the Politics, and the Prospects for Change.* McFarland, 2009. Provides an overview of the issue of global warming and its anthropogenic causes in relation to climate policy in the United States.

Randall, David A. *Atmosphere, Clouds, and Climate.* Princeton University Press, 2012. An overview of the major atmospheric processes. Covers the function of the atmosphere in the regulation of energy flows and the transport of energy through weather systems, including thunderstorm and monsoons. Examines obstacles in predicting the weather and climate change.

Rao, P. V., ed. *The Indian Ocean: An Annotated Bibliography.* Kalinga, 1998.

Rapp, Donald. *Ice Ages and Interglacials: Measurements, Interpretations, and Models.* Springer, 2009.

Rashid, Haruna, Leonid Polyak and Ellen Mosely-Thompson. *Abrupt Climate Change Mechanisms, Patterns and Impacts.* John Wiley & Sons, 2013.

Raven, Peter Hamilton, et al. *Biology of Plants.* W.H. Freeman and Company Publishers, 2013. Text includes general information on plant growth, tree rings, and the impact of climate change. Illustrations, figures, tables, bibliography, index, maps.

Ravindra, Rasik, et al. "Antarctica." In *Encyclopedia of Snow, Ice, and Glaciers,* edited by Vijay P. Singh, Pratap Singh, and Umesh K. Haritashya. Springer, 2011. A brief but comprehensive overview of Antarctica's ice sheet, glaciers, icebergs, sea ice, and ice cores, complete with color photographs and figures.

Reay, David. *Greenhouse Gas Sinks.* CAB International, 2007. Covers the importance of carbon dioxide, methane and nitrous oxide as greenhouse gases, and describes the function of sinks for each gas.

———. *Nitrogen and Climate Change: An Explosive Story.* Springer, 2015. Discusses the role of nitrogen in climate change and of the oceans as a sink for nitrogen oxides and other greenhouse gases.

Rees, W G. *Physical Principles of Remote Sensing.* Cambridge University Press, 2013. Provides general information about the principles of remote sensing technology and their application to earth sciences (including oceanic studies). The primary types of sensors described are those mounted on satellites.

Repetto, Robert, and Robert Easton. "Climate Change and Damage from Extreme Weather Events." *Environment* 52, no. 2 (2010): 22–33. In this article, the authors describe severe weather events as they relate to global warming and climate change. Repetto and Easton attempt to draw links between global warming and the increase in weather disasters in recent decades.

Rhodes, Frank H.T. *Earth. A Tenant's Manual.* Cornell University Press, 2012. In this book the author presents a scientific view of the past, present and future of Earth's atmosphere and its composition.

Richardson, P. L. "On the History of Meridional Overturning Circulation Schematic Diagrams." *Progress In Oceanography* 76 (2008): 466–486.

Rodó, Xavier and Comín, Francisco A., eds. *Global Climate. Current Research and Uncertainties in the Climate System.* Springer Science + Business Media, 2013. The chapters in this book represent a curated series of articles that focus on individual segments of the global climate system.

Roggema, Rob. *Adaptation to Climate Change: A Spatial Challenge.* Springer, 2009. The chapters of this book present the planned approaches of several European nations for adapting to the challenges that shifting climate zones will present.

Rögner, Matthias. *Biohydrogen.* Walter De Gruyter, 2015.

Rohli, Robert V., and Anthony J. Vega. *Climatology.* Jones & Bartlett, 2008. With excellent images and clear descriptions, this textbook offers an outstanding examination of the various modes of climate variability. The treatment of statistical methods, particularly those involving eigenvalues and principal components, provides a good feel for how they work without requiring the reader to become enmeshed in equations and linear algebra.

Romm, Joseph. *Climate Change. What Everyone Needs to Know.* Oxford University Press, 2016.

Rosenzweig, Cynthia, David Rind, Andrew Lacis, and Danielle Manley, eds. *Lectures in Climate Change, Vol. 1. Our Warming Planet. Topics in Climate Dynamics.* World Scientific, 2018. The chapters in this book are transcripts of individual lectures in climate science, with emphasis on the role of climate feedbacks.

Royal Society. *Ocean Acidification Due to Increasing Atmospheric Carbon Dioxide.* Author, 2005. The Royal Society in the United Kingdom assembled a group of distinguished scientists to summarize the issues of oceanic acidification in this study.

Ruddiman, William F. *Earth's Climate Past and Future.* 3rd ed., W. H. Freeman, 2014. This elementary college textbook has several sections concerning dating methods, their limitations, errors, and resolution. Illustrations, figures, tables, maps, bibliography, index.

_____. *Plows, Plagues, and Petroleum: How Humans Took Control of Climate.* Princeton University Press, 2016. Written for the lay public, this book provides the background and thinking behind the theory that humans have influenced the climate for the last nine thousand years, primarily through agriculture. Illustrations, figures, tables, maps, bibliography, index.

Saha, Kshudiram. *The Earth's Atmosphere: Its Physics and Dynamics.* Springer-Verlag, 2008.

Saha, Pijushkanti. *Modern Climatology.* Allied Publishers Pvt Ltd, 2012. An entry-level textbook that provides an concise, yet thorough, overview of the science of climatology and climate relationships.

Salby, Murray L. *Physics of the Atmosphere and Climate.* Cambridge University Press, 2012. Provides an integrated treatment of the Earth-atmosphere system and its processes. Begins with first principles and continues with a balance of theory and applications as it covers climate, controlling influences, theory, and major applications.

Saltzman, Barry. *Dynamical Paleoclimatology: Generalized Theory of Global Climate Change.* New York: Academic Press, 2002. Treats interactions between abiotic, biotic, and anthropogenic variables; discusses controversies about the magnitude of human climatological impacts.

Samaras, Tim, Stefan Bechtel, and Greg Forbes. *Tornado Hunter: Getting Inside the Most Violent Storms on Earth.* Washington: National Geographic Society, 2009.

Samenow, Jason. "Harvey Is a 1,000-Year Flood Event Unprecedented in Scale." *The Washington Post*, 31 Aug. 2017, www.washingtonpost.com/news/capital-weather-gang/wp/2017/08/31/harvey-is-a-1000-year-flood-event-unprecedented-in-scale/. Accessed 22 Sept. 2017.

Sampath, Nikita. "Mismatched Flood Control System Compounds Water Woes in Southern Bangladesh." *New Security Beat*, Environmental Change and Security Program, Woodrow Wilson International Center for Scholars, 3 Jan. 2017, www.newsecuritybeat.org/2017/01/mismatched-flood-control-system-compounds-water-woes-southern-bangladesh/. Accessed 22 Sept. 2017.

Sánchez-Lavega, Agustín. *An Introduction to Planetary Atmospheres.* CRC Press, 2011. A comprehensive resource discussing many aspects of atmospheric science and evolution.

Santurette, Patrick, and Christo Georgiev. *Weather Analysis and Forecasting.* Academic Press, 2005. Discusses the use of satellite imagery to detect and analyze atmospheric moisture. Identifies factors that indicate severe weather occurrence and discusses the use of water vapor imagery to improve forecasting.

Sarachik, Edward S., and Mark A. Cane. *The El Niño-Southern Oscillation Phenomenon.* Cambridge University Press, 2018. The book offers a thorough description of the history and mechanics of the two components of ENSO, before delving into a technical analysis of the phenomenon.

Sargent, William. *Storm Surge: A Coastal Village Battles the Rising Atlantic.* University Press of New England, 1995.

Sarmiento, Jorge L. and Nicolas Gruber. *Ocean Biogeochemical Dynamics.* Princeton University Press, 2013. Describes the principal mechanisms by which the oceans act as a sink for carbon and other materials.

Satoh, Masaki. *Atmospheric Circulation Dynamics and General Circulation Models.* Springer, 2014.

Savino, John, and Marie Jones. *Supervolcano.* New Page Books, 2007. Discusses the classification of supervolcanoes and their eruptions. Describes the effects of volcanic eruptions on climate and the ecosystem. Includes a glossary, bibliography, further resources, and indexing.

Schimel, David. *Climate and Ecosystems.* Princeton University Press, 2013. One of the "Princeton Primers on Climate" series, this book describes the interaction mechanisms of various ecosystems and climate.

Schmittner, Andreas, John C.H. Chiang and Sidney R. Hemming, eds. *Ocean Circulation. Mechanisms and Impacts.* John Wiley & Sons, 2013. Provides a thorough description of the effects on Earth's climate due to large "turnover" circulation currents, particularly the Atlantic conveyor system.

Schneider, Bonnie. *Extreme Weather: A Guide to Surviving Flash Floods, Tornadoes, Hurricanes, Heat Waves, Snowstorms, Tsunamis and Other Natural Disasters.* Palgrave Macmillan, 2012. Presents vivid explanations of how, when, and why major natural disasters occur. Discusses floods, hurricanes, thunderstorms, mudslides, wildfires, tsunamis, and earthquakes. Presents a guide of how to prepare for and what to do during an extreme weather event, along with background information on weather patterns on natural disasters.

Schneider, Stephen Henry, and Michael D. Mastrandrea. *Encyclopedia of Climate and Weather.* 2nd ed. Oxford University Press, 2011.

Schneider, Tapio, and Adam H. Sobel, eds. *The Global Circulation of the Atmosphere.* Princeton University Press, 2007. Collects papers from the 2004 conference at California Institute of Technology. Covers large-scale atmospheric dynamics, storm-tracking dynamics, and tropical convection zones. A strong understanding of advanced mathematics is required. Best suited for graduate students and professionals.

Schwartz, M. *Encyclopedia of Coastal Science.* Springer, 2005. Contains many articles specific to ocean and beach dynamics. Discusses coastal habitat management topics, hydrology, geology, and topography. Articles may run multiple pages and have diagrams. Each article has bibliographical information and cross referencing.

Seager, Sara. "Atmospheric Circulation." *Exoplanet Atmospheres: Physical Processes.* Princeton University Press, 2010. 211–28.

Segar, Douglas A., and Elaine Stamman, Segar. *Introduction to Ocean Sciences.* Douglas A. Segar, 2012. Comprehensive coverage of all aspects of the oceans, their chemical makeup, and circulation. Readable and well illustrated. Suitable for high school students and above.

Seibold, Eugen, and Wolfgang H. Berger. *The Sea Floor: An Introduction to Marine Geology.* 3rd ed. Springer-Verlag, 2010. Offers an introduction to many topics in marine geology that covers geological structures from the continental shelf to deep-ocean trenches. Discusses processes such as seafloor spreading, the sediment cycle, currents, and pelagic rain.

Seidel, Klaus, and Jaroslav Martinec. *Remote Sensing in Snow Hydrology: Runoff Modelling, Effect of Climate Change.* Springer, 2010. Describes the use of remote sensors to gauge snow- and ice-melting trends and how the resulting melt runoffs affect the nearby and global environments. The book reviews one hundred regions worldwide.

Seinfeld, John H., and Spyros N. Pandis. *Atmospheric Chemistry and Physics. From Air Pollution to Climate Change.* 3rd ed. John Wiley & Sons, 2016. A complete introduction to the structure and chemistry of the atmosphere.

Sene, Kevin. *Flash Floods: Forecasting and Warning.* Springer, 2013.

Serreze, Mark C. *Brave New Arctic. The Untold Story of the Melting North.* Princeton University Press, 2018. Describes the Arctic environment and the climate-related changes to that environment that have occurred since 1980.

Serreze, Mark C., and Roger G. Barry. *The Arctic Climate System.* Cambridge University Press, 2014. Provides a comprehensive, up-to-date assessment of the Arctic climate system for researchers and advanced students.

Service, Robert F. "The Hydrogen Backlash." *Science* 305.5686 (2004): 958–61. Print.

Severin, Tim. *The China Voyage: Across the Pacific by Bamboo Raft.* Addison-Wesley, 1994.

Sharpe, Norean R., Richard De Veaux, and Paul F. Velleman. *Business Statistics: A First Course.* 2nd ed. Pearson, 2013.

Shearer, Christine. *Kivalina: A Climate Change Story.* Haymarket Books, 2011. This book presents a view of the "other side"of climate change, the impacts climate-change denial policies have on real people.

Shennan, Ian, and Julian E. Andrews, eds. *Holocene Land-Ocean Interaction and Environmental Change Around the North Sea.* Special Publication No. 166. Geological Society of London, 2000.

Siebert, Lee, Tom Simkin, and Paul Kimberly. *Volcanoes of the World.* 3rd ed. University of California Press, 2010. Reviewed as "the most comprehensive source on dynamic volcanism," presenting chronological, statistical, environmental, and historical information about volcanism over the past ten thousand years.

Singer, S. Fred, and Dennis T. Avery. *Unstoppable Global Warming: Every Fifteen Hundred Years.* Rowman and Littlefield, 2008. This book is dedicated to those thousands of research scientists who have documented evidence of a fifteen-hundred-year climate cycle over the Earth. Refers throughout to various aspects of monitoring of the global climate.

Singh, V. P. *Hydrology of Disasters.* Springer, 2011.

Sinha, P. C., ed. *Sea Level Rise.* Anmol Publications, 1998.

Skjaerseth, Jon Birger, and Tora Skodvin. *Climate Change and the Oil Industry: Common Problems, Varying Strategies.* Manchester University Press, 2013. While climate change denial still exists, this book describes the internal corporate practices of some major oil companies intended to minimize or eliminate whatever global climate change effects that may result from their activities.

Smith, Jerry E. *Weather Warfare: The Military's Plan to Draft Mother Nature.* Kempton, Ill.: Adventures Unlimited Press, 2006. Covers processes that cause weather events and natural disasters, and our ability to influence these processes. Discusses cloud seeding, electromagnetic wave production, weather modification legislation, contrails, and stratospheric engineering. Written for the general population and lacking in "hard science," but presents a wide range of topics that provoke further research.

Smith, Norman J. *North Sea Oil and Gas, British Industry and the Offshore Supplies Office.* Vol. 7. Edited by John Cubitt. Elsevier, 2011.

_____. *The Sea of Lost Opportunity: North Sea Oil and Gas, British Industry and the Offshore Supplies Office.* Elsevier, 2011.

Smith, Robert Angus. *Air and Rain. The Beginnings of a Chemical Climatology.* Green & Company, 1872.

Soloviev, Alexander, and Roger Lukas. *The Near-Surface Layer of the Ocean: Structure, Dynamics and Applications.* Springer, 2006. Uses the results of major air-sea interaction experiments to present the physics and thermodynamics of this oceanic system, providing a detailed treatment of the surface microlayer, upper-ocean turbulence, thermohaline and coherent structures, and the high-speed wind regime.

Somerville, Richard C. J. *The Forgiving Air: Understanding Environmental Change.* 2nd ed. American Meteorological Society, 2008. Presents a thorough investigation of the relationship between human activities and changes in the atmosphere and global climate. Written by a scientist involved in atmospheric research, but aimed at a nonscientific audience. Chapter 2 offers a discussion of the ozone hole.

Soon, Willie, and Steven H Yaskell. *The Maunder Minimum and the Variable Sun-Earth Connection.* World Scientific, 2004. The bulk of the book is about the Sun's Maunder Minimum, with discussions of climatic changes occurring at that time. Good bibliography.

Speer, James H. *Fundamentals of Tree-Ring Research.* University of Arizona Press, 2010. A thorough presentation of the methodology of tree-rings and their use as environmental and climate proxies.

Spellman, Frank R. *The Science of Air. Concepts and Applications.* 2nd ed. CRC Press, 2009. This book is a comprehensive resource covering the components, dynamics, interactions and uses of air as a resource.

Spencer, Roy W. *Climate Confusion: How Global Warming Hysteria Leads to Bad Science, Pandering Politicians, and Misguided Policies That Hurt the Poor.* Encounter Books, 2008.

Spencer, Roy. *Global Warming.* Roy Spencer, 2012. Web. 23 Aug. 2012. Spencer, a climatologist, used to work for NASA. His website offers alternative ideas to the common view that global warming is caused by human activity.

Sportisse, Bruno. *Fundamentals in Air Pollution: From Processes to Modelling.* Springer, 2009. Discusses issues arising from air pollution such as emissions, the greenhouse effect, acid rain, urban heat islands, and the ozone hole. Topics are well organized and clearly explained, making this text accessible to the layperson, although it was written for the undergraduate.

Spring, Barbara. *The Dynamic Great Lakes.* Independence Books, 2001.

Srivastava, Sanjay, et al. *South Asian Disaster Report 2007.* SAARC Disaster Management Centre, 2008. *SAARC Disaster Management Centre,* saarc-sdmc.nic.in/sdr_p.asp. Accessed 6 Sept. 2017.

Stacey, Frank D., and Paul M. Davis. *Physics of the Earth.* 4th ed. Cambridge University Press, 2008. Discusses Earth's atmosphere in Chapter 2. An appendix provides information on thermodynamics. Well organized, with additional mathematics and physics concepts geared to graduate students. Includes many appendices and student exercises.

Stanley, Steven M., and John A. Luczaj. *Earth System History.* Freeman, 2015. Lays out the history of Earth, including the factors affecting global climates and climate change, over the vastness of geological time. Selected chapters focus on climate change in terms of glaciation and deglaciation.

Steele, John H., Steve A. Thorpe, and Karl K. Turekian, eds. *Marine Chemistry and Geochemistry.* Academic Press, 2010. Pulls articles in from the Encyclopedia of Ocean Sciences to provide a topic-specific compilation of oceanography articles. Discusses the chemistry of seawater, radioactive isotopes in the oceans, pollution, and marine deposits.

Steere, Richard C., ed. *Buoy Technology: An Aspect of Observational Data Acquisition on Oceanography and Meteorology.* University of California Press, 1967. This book, based on presentations at the 1964 Buoy Technology Symposium, features an analysis of engineering developments in buoy technologies and how its evolution generates improved observational data.

Stein, Rüdiger, and Robie W. Macdonald. *The Organic Carbon Cycle in the Arctic Ocean.* Springer Berlin, 2013. Various topics are discussed in relation to the Arctic Ocean carbon dynamics. Covers dissolved organic matter, particulate organic carbon, productivity and growth rates, benthic carbon cycling, and organic carbon burial rates. Summarizes the Arctic carbon cycle and compares it to global cycling.

Steinacker, Reinhold, Dieter Mayer. and Andrea Steiner. "Data Quality Control Based on Self-Consistency." *Monthly Weather Review* 139, no. 12 (2011): 3974–3991. Reviews methods for addressing different types of errors associated with the acquisition of observational data in meteorological studies. Introduces quality control approaches in this field.

Stern, Nicholas. *The Economics of Climate Change: The Stern Review.* New York: Cambridge University Press, 2007. This famous report tackles the charges of those who claim that the economic costs of attempting to prevent climate change are too great.

Stevenson, R. E., and F. H. Talbot, eds. *Oceans.* Time-Life, 1993.

Stewart, Robert H. *Introduction to Physical Oceanography.* A & M University: Robert H. Stewart, 2008. This treatment of oceanography offers extensive discussion of the physics of water movement in the oceans. Intended for upper-level college students.

Stirling, Ian, and Claire L. Parkinson. "Possible Effects of Climate Warming on Selected Populations of Polar Bears (*Ursus Maritimus*) in the Canadian Arctic." *Arctic* 59, no.3 (2006): 261–275. Research paper postulating that climate warming is responsible for Canadian polar bears' declining populations, reduced body condition, and reduced survival among young.

Stirling, Ian. *Polar Bears.* University of Michigan Press, 1999. Written for the general reader by the world's most experienced polar bear researcher. Richly illustrated with photographs by Dan Guravich.

Stocker, Thomas. *Introduction to Climate Modelling.* Springer, 2011. This book presents a discussion of the basic structures and applications of different types of climate models.

Stolarski, Richard S. "The Antarctic Ozone Hole." *Scientific American* 258 (January, 1988): 30–36. Print.

Stowe, Keith. *Exploring Ocean Science.* 2nd ed. John Wiley & Sons, 1996.

Strahler, Alan H., et al. *Physical Geography: Science and Systems of the Human Environment: Canadian Version.* John Wiley & Sons, 2008. A thorough, well-illustrated book containing considerable information about atmospheric processes and issues. Suitable for college students.

Strahler, Alan. *Introducing Physical Geography.* 5th ed., John Wiley & Sons, 2011.

Straka, Jerry M. *Cloud and Precipitation Microphysics: Principles and Parameterizations.* Cambridge University Press, 2009.

Strangeways, Ian. *Precipitation: Theory, Measurement and Distribution.* Cambridge University Press, 2010.

"A Student's Guide to Global Climate Change." *EPA,* 3 Mar. 2016, www3.epa.gov/climatechange/kids/index.html. Accessed 2 Feb. 2017. Walker,

Gabrielle. *An Ocean of Air: Why the Wind Blows and Other Mysteries of the Atmosphere.* Harcourt, 2007.

Sutton, R. T., and D. L. R. Hodson. "Climate Response to Basin-Scale Warming and Cooling of the North Atlantic Ocean." *Journal of Climate* 20 (2007): 891–907. Discusses the use of a climate model to assess the effects of basin-wide warming and cooling of the North Atlantic Ocean.

Sverdrup, Keith A., and E. Virginia Armbrust. *An Introduction to the World's Oceans.* McGraw-Hill, 2009. Includes a clearly written section on the planetary winds and their effects on ocean currents. Maps the general patterns of ocean-current circulation and uses diagrams to explain the Coriolis Effect. Lists suggested readings.

Takahashi, Shin. *The Manga Guide to Statistics.* No Starch, 2009.

Talley, Lynne D., et al. *Descriptive Physical Oceanography: An Introduction.* Academic Press, 2012. Designed to introduce oceanography majors to the field of physical oceanography. Has useful sections on the temperature, salinity, density, light, and sound structure of the ocean. Includes a discussion of the various instruments and methods used for measuring these properties.

Tanvir, Islam, Yongxiang Hu, Alexander Kokhanovsky, and Jun Wang, eds. *Remote Sensing of Aerosols, Clouds and Precipitation* Elsevier, 2017. Describes the methods, theoretical foundations and applications of satellite-based monitoring of aerosols, clouds and precipitation events occurring in Earth's atmosphere.

TAO Project Office of NOAA/Pacific Marine Environmental Laboratory. *Upper Ocean Heat Content and ENSO.* National Oceanic and Atmospheric Administration (NOAA) Web site (http://www.pmel.noaa.gov/tao). Researchers at NOAA and other research institutions work together to track and model current oscillations, including ENSO and NAO. More information about their findings and other ocean-atmosphere interactions are available through links from their Web site.

Tarbuck, Edward J., et al. *Earth: An Introduction to Physical Geology.* Pearson, 2018. Provides a clear picture of earth systems and processes. Suitable for the high school or college reader. Includes an accompanying computer disc in addition to its illustrations and graphics. Bibliography and index.

Taylor, Brian, and James Natland, eds. *Active Margins and Marginal Basins of the Western Pacific.* American Geophysical Union, 1995.

Taylor, F. W. *Elementary Climate Physics.* Oxford University Press, 2014. Readers without significant prior knowledge of climate science should consult this book for help in understanding difficult topics. Illustrations, figures, tables, bibliography, index.

Taylor, Maria. *Global Warming and Climate Change. What Australia Knew and Buried...Then Framed a New Reality for the Public.* Australian National University, 2014. This book describes the manner in which political policies and actions in Australia were used to manipulate public knowledge and awareness of the issue of climate change in Australia and around the world.

Teague, Kevin Anthony, and Nicole Gallicchio. *The Evolution of Meteorology. A Look Into the Past, Present and Future of Weather Forecasting.* Wiley-Blackwell, 2017. A comprehensive presentation of weather forecasting that includes descriptions of barometers and other devices in context.

Tedesco, M., ed. *Remote Sensing of the Cryosphere.* Wiley-Blackwell, 2015. Geared to a technical audience and mathematically knowledgeable readers, this book details the basic principles and methodology of remote sensing applied to the cryosphere.

Thomas, David N. *Frozen Oceans: The Floating World of Pack Ice.* Natural History Museum, 2004. Accessible, illustrated overview of pack ice.

Thomas, David N., and Gerhard Dieckmann, eds. *Sea Ice.* Blackwell, 2010. Examines the role of sea ice in the global ecosystem, particularly in relation to the impact climate change is having on sea ice in the Antarctic and other regions. Fully illustrated with color photographs. A technical collection best suited for college students with some grounding in climate science, oceanography, or geology.

Thurman, Harold V., and Alan P. Trujillo. *Introductory Oceanography.* Prentice Hall, 2008. Provides an introduction to oceanography that is comprehensive but not too technical for the general reader. Very well illustrated and includes some high-quality color maps and diagrams. Each chapter includes questions and exercises and also lists references and suggested readings. Addresses the circulations of the ocean currents according to the major ocean basins, with maps and diagrams for each basin.

Tiner, Ralph W., and CRC Press. *Wetland Indicators: A Guide to Wetland Identification, Delineation, Classification and Mapping.* CRC Press/Taylor & Francis Group, 2017. A good source of information for identification of wetlands.

Tiwary, Abhishek, and Jeremy Colls. *Air Pollution: Measurement, Modelling and Mitigation.* New York: Routledge, 2010. Print.

Trabalka, J.R. And Reichle, D.E., eds. *The Changing Carbon Cycle. A Global Analysis.* Springer Science + Business Media, 2016. The various chapters in this book present scientific discussions of as many aspects of the study of atmospheric carbon dioxide.

Trenberth, Kevin E. "Warmer Oceans, Stronger Hurricanes." *Scientific American* 297, no. 1 (July, 2007): 44–51. Describes how warmer oceans and variations in gyral currents such as the Gulf Stream affect decade-scale circulation patterns and the formation of hurricanes.

Trewby, Mary. *Antarctica: An Encyclopedia from Abbott Ice Shelf to Zooplankton.* Firefly, 2005. General encyclopedia of Antarctica. Explains all the features found on the continent using a clear and simple language.

"Tropics Feel the Heat of Climate Change." *Scientific American.* Scientific American, 31 Jan. 2014. Web. 20 Mar. 2015.

Trujillo, Alan P., and Harold V. Thurman. *Essentials of Oceanography.* 11th ed. Prentice, 2014. Designed to give the student a general overview of the oceans in the first few chapters. Follows up with a well-designed, in-depth study of the ocean—involving chemistry of the ocean, currents, air-sea interactions, the water cycle, and marine biology—and a well-developed section on the practical problems resulting from human interaction with the ocean, such as pollution and economic exploitation.

Turney, Chris. *Ice, Mud, and Blood: Lessons from Climates Past.* Macmillan, 2008. Describes the discoveries derived from ice cores and how these discoveries led to a better understanding of paleoclimates.

U.S. Department of the Interior, Minerals Management Service. *Programmatic Environmental Assessment: Arctic Ocean Outer Continental Shelf Seismic Surveys.* U.S. Department of the Interior Minerals Management Service, Alaska OCS Region, 2006. Provides an overview of seismic surveys and the exploration of the Alaskan continental shelf. Includes alternative scenarios for surveys and their evaluation. Addresses the environmental impact of such surveys.

U.S. Environmental Protection Agency. *The Great Lakes: An Environmental Atlas and Resource Book.* Great Lakes Program Office, 1995.

U.S. Geological Survey. "Lake Ice-Out Data for New England." 2007. Available online at http://me.water.usgs.gov/iceout.html. Contains links to other articles and an interactive map that links to data for each lake. The data are in simple columnar format, making it easy to copy and paste into a spreadsheet.

U.S. National Space Science and Technology Center. Available at http://www.nsstc.org/. The mission of the NSSTC, an arm of NASA, is "to conduct and communicate research and development critical to NASA's mission in support of the national interest, to educate the next generation of scientists and engineers for space-based research, and to use the platform of space to better understand our Earth and space environment and increase our knowledge of materials and processes."

Ulanski, Stan. *The Gulf Stream: Tiny Plankton, Giant Bluefin, and the Amazing Story of the Powerful River in the Atlantic.* University of North Carolina Press, 2008. Discusses the hydrodynamics of the Gulf Stream, followed by the biology, and lastly the impact of this current on humans through history. An interesting compilation of the known information on the Gulf Stream current written for the general public.

University of Colorado at Boulder. "Sea Level Change." Available at http://sealevel.colorado.edu/results.php. Presents tables, maps, time series, and other data on global sea level.

University of East Anglia. "Climatic Research Unit." Available at http://www.cru.uea.ac.uk/. The CRU presents data, information sheets, and the online journal *Climate Monitor.*

University of Illinois. "The Cryosphere Today." Available at http://arctic.atmos.uiuc.edu/cryosphere/. Offers frequently updated data on the current state of Earth's cryosphere.

Usdowski, Eberhard, and Martin Dietzel. *Atlas and Data of Solid-Solution Equilibria of Marine Evaporites.* Springer Verlag, 2013. Offers the reader an illustrated guide to seawater composition, including phase diagrams of seawater processes. Accompanied by a CD-ROM that reinforces the concepts discussed in the chapters.

Vallero, Daniel. *Fundamentals of Air Pollution*. 4th ed. Burlington: Academic Press, 2008. Print.

Vallis, Geoffrey K. *Atmospheric and Oceanic Fluid Dynamics: Fundamentals and Large-scale Circulation*. Cambridge University Press, 2006. Begins with an overview of the physics of fluid dynamics to provide foundational material on stratification, vorticity, and oceanic and atmospheric models. Discusses topics such as turbulence, baroclinic instabilities, wave-mean flow interactions, and large-scale atmospheric and oceanic circulation. Best suited for graduate students studying meteorology or oceanography.

vanLoon, Gary W. and Duffy, Stephen J. *Environmental Chemistry. A Global Perspective*. 4th ed., Oxford University Press, 2017. A well-composed text book that presents a good discussion of chlorofluorocarbons and related compounds.

Vasquez, Tim. *Weather Analysis and Forecasting Handbook*. Weather Graphics Technologies, 2011. Discusses the technology, techniques, and physics principles used in modern weather forecasting. Describes thermal structure and dynamics of weather systems. Also covers model use in weather forecasting and weather system visualization. Easily accessible yet still technical, useful to anyone studying weather forecasting and meteorology.

Ver Berkmoes, Ryan, Thomas Huhti, and Mark Lightbody. *Great Lakes*. Lonely Planet Publishing, 2000.

Visconti, Guido. *Problems, Philosophy and Politics of Climate Science*. Springer, 2018. Presents an introduction to some factors affecting global climate, then addresses the practical and philosophical issues arising in climate observation and modeling.

Voituriez, Bruno. *The Gulf Stream*. UNESCO, 2006. Offers an analytical narrative that examines complex scientific information to describe the causes and dynamics of the Gulf Stream through the history of its discovery and exploration.

Vonnegut, Andrew. *Inside the Global Economy; A Practical Guide*. Rowman & Littlefield, 2018. The first eight chapters of this book identify and describe as many major aspects of the global economy. The next five chapters identify, describe and hypothesize how shifts in underlying aspects such as demographics, technology and global climate might affect the global economy.

Voo, Rob van der. *Paleomagnetism of the Atlantic, Tethys and Iapetus Oceans*. Cambridge University Press, 2005.

Voronin, P., and C. Black. "Earth's Atmosphere as a Result of Coevolution of Geo- and Biospheres." *Russian Journal of Plant Physiology* 54 (2007): 132–136. Covers the evolution of the atmosphere's composition and factors altering the gas composition. Provides background content on photosynthesis and chemolithotrophy. Highly technical. Appropriate for readers with a strong chemistry or geology background.

Wadhams, Peter. *A Farewell to Ice. A Report from the Arctic*. Oxford University Press, 2017. Discusses the role of Arctic and Antarctic ice shields, and how feedback systems function. Special attention is given to the potential for the release of quantities of methane from polar sea floors and Arctic permafrost as Earth's average temperature increases.

Wagner, Günther A. *Age Determination of Young Rocks and Artifacts. Physical and Chemical Clocks in Quaternary Geology and Archaeology*. Springer, 2011. Provides very readable descriptions of the more robust dating methods used by climate scientists and archaeologists, and the manner in which those methods are interpreted.

Walker, Mike. *Quaternary Dating Methods*. John Wiley & Sons, 2013. Based on atomic structure, the chapters discuss the methodology of radiometric dating and the interpretation of annually-banded strata to determine dating chronologies.

Walker, Sally M. *Frozen Secrets: Antarctica Revealed*. Carolrhoda Books, 2010. This highly readable book, written for high school students, focuses on the work of Earth scientists. Explains the techniques used to analyze ice cores, bedrock, water, and even samples of air, and covers what is known about the continent's past and future. Includes a glossary.

Wallace, John M., and Peter V. Hobbs. *Atmospheric Science*. 2nd ed. Academic Press, 2006. Represents a complete study of the atmosphere that covers fundamental physics and chemistry topics as well as specific topics in atmospheric science such as radiative transfer, weather forecasting, and global warming. Contains significant detail and technical writing, but is still accessible to the undergraduate studying meteorology or thermodynamics.

Wang, Bin. *The Asian Monsoon.* Springer, 2007.

Wang, Pao K. *Physics and Dynamics of Clouds and Precipitation.* Cambridge University Press, 2013.

Ward, Peter D. *Under a Green Sky: Global Warming, the Mass Extinctions of the Past, and What They Can Tell Us About Our Future.* Harper Perennial, 2008. A discussion of the Permian extinction, which occurred more than 252 million years ago and killed nearly 97 percent of all living organisms on Earth. Theorizes that a dramatic rise in global temperatures, triggered by an increase in carbon dioxide in the atmosphere, was likely the biggest contributor to this mass extinction.

Ward, Peter Langdon. *What Really Causes Global Warming? Greenhouse Gases or Ozone Depletion?* Morgan James, 2016.

Warner, Jeroen Frank, et al., editors. *Making Space for the River: Governance Experiences with Multifunctional River Flood Management in the U.S. and Europe.* IWA Publishing, 2013.

Warner, Thomas Tomkins *Numerical Weather and Climate Prediction.* Cambridge University Press, 2011. A book describing the mathematical basis of climate prediction and projection.

Warren, E. A., and P. C. Smalley. *North Sea Formation Waters Atlas.* Geological Society, 1994.

Waterlow, Julia. *The Atlantic Ocean.* Raintree Steck-Vaughn, 1997.

Watterson, Ian Godfrey, and Edwin K. Schneider. "The Effect of the Hadley Circulation on the Meridional Propagation of Stationary Waves." *Quarterly Journal of the Royal Meteorological Society* 113, no. 477 (July, 1987): 779–813. Based on Watterson's doctoral dissertation of the same name, this article distills one of the only extended studies of Hadley circulation, presenting its most salient insights and conclusions.

Watts, Alan. *Instant Wind Forecasting.* Adlard Coles Nautical, 2010.

Wefer, Gerold, et al. *The South Atlantic: Present and Past Circulation.* Springer Berlin, 2013. An example of modern scientific analysis of data obtained on ocean circulation. Deals with the region of the ocean where the earliest WOCE studies were performed.

Wells, Neil. *The Atmosphere and Ocean: A Physical Introduction.* 3rd ed. John Wiley & Sons, 2012. The atmosphere and oceans are both fluids circulating on a rotating planet, intimately and profoundly influencing each other. Treats atmospheric and oceanic interactions in a readable, yet thorough, manner. Presents numerous quantitative concepts and equations with graphs or figures that make them understandable to readers with little technical background.

Wennersten, John R., and Denise Robbins. *Rising Tides: Climate Refugees in the Twenty-First Century.* Indiana University Press, 2017. Examines the potential for economic and social disaster arising from large numbers of climate refugees.

Wetzel, R. G., ed. *Limnology.* 3rd ed. Elsevier Science, 2001. A well-written textbook typical of those used by undergraduates and graduates in introductory limnology courses. Covers physical, biological, and chemical aspects of lakes. Assumes a knowledge of high school algebra, chemistry, physics, and biology.

"What Is a Polar Vortex?" *SciJinks.* NASA and NOAA, 18 Mar. 2015. Web. 23 Mar. 2015.

"What Is Ocean Acidification?" *Pacific Marine Environmental Library*, National Oceanic and Atmospheric Administration, www.pmel.noaa.gov/co2/story/What+is+Ocean+Acidification percent3F. Accessed 21 Feb. 2017.

Whish-Wilson, P. "The Aral Sea Environmental Health Crises." *Journal of Rural and Remote Environmental Health* 1, no. 2 (2002): 29–34. Provides general background on the disaster; in-depth health status assessment of the region; list of causes and outcomes of pollution and its effects on humans and the environment; and details of the health community's response.

Wiggert, Jerry D. *Indian Ocean Biogeochemical Processes and Ecological Variability.* American Geophysical Union, 2009.

Williams, Michael. *Deforesting the Earth: From Prehistory to Global Crisis: An Abridgment.* University of Chicago Press, 2006. This massive account by an author already well known for his careful evaluation of forest history is definitive.

Williams, Richard G., and Michael J. Follows. *Ocean Dynamics and the Carbon Cycle: Principles and Mechanisms.* Cambridge University Press, 2011. Provides the fundamentals of oceanography. Discusses both biological and chemical aspects of ocean dynamics. Deals with carbonate chemistry and the carbon cycle in oceans.

Williams, Rob. "Scientists Record New Coldest Temperature on Earth." *The Independent*, 10 Dec. 2013, www.independent.co.uk/news/science/scientists-record-new-coldest-temperature-on-earth-on-the-east-antarctic-plateau-8995135.html. Accessed 1 June 2017.

Winchester, Simon. *Krakatoa: The Day the World Exploded, August 27, 1883*. Harper-Collins Publishers, 2003. An excellent and entertaining description and placement of the violent eruption of Krakatoa within the context and theory of plate tectonics as a planetary process. Well researched. Includes historical observation of atmospheric and climatic events.

Wisconsin State Climatology Office. "Wisconsin Lake Ice Climatologies: Duration of Lake Ice." 2004. Available online at http://www.aos.wisc.edu/~sco/lakes/WI-lake_ice-1.html. Links to data for twenty-seven Wisconsin lakes. The links show duration of ice cover rather than specific dates of freezing and breakup. Lakes Mendota and Monona, with more than 150 years of data, show clearly decreasing trends, as do numerous other lakes.

Wolfson, Richard. *Energy, Environment, and Climate*. W. W. Norton, 2008.

Woods Hole Oceanographic Institution (WHOI) Web site (http://www.whoi.edu/index.html). Offers general information about the ocean, research, education, and resources, including video animations. WHOI researchers work to understand the complexities of the ocean; atmosphere-ocean interaction is just one component of their research.

Woods, J. D. "The World Ocean Circulation Experiment." *Nature* 314 (April 11, 1985): 501–510. An outstanding introduction to the WOCE project. Discusses the importance of ocean circulation to changes in climate, the failure of models prior to the WOCE to accurately predict ocean conditions, the goals of the WOCE, and the technology that allowed the project to take place.

Wooster, Margaret. *Living Waters: Reading the Rivers of the Lower Great Lakes*. Excelsior Editions/State University of New York Press, 2009.

World Meteorological Organization. *Global Atmosphere Watch: Antarctic Ozone Bulletin* 1 (August 28, 2008).

Yuen, Eddie. "The Politics of Failure Have Failed: The Environmental Movement and Catastrophism." Chapter in Lilley, Sasha, McNally, David, Yuen, Eddie and Davis, James *Catastrophism: The Apocalyptic Politics of Collapse and Rebirth*, PM Press, 2012. A philosophical analysis of the use of catastrophic "scare tactics" in an effort to effect political change in regard to climate change.

Zdunkowski, Wilfred, and Andreas Bott. *Thermodynamics of the Atmosphere: A Course in Theoretical Meteorology*. Cambridge University Press, 2004. Written for graduate students and researchers in meteorology and related sciences. Assumes a significant background in mathematics in the discussion of thermodynamic principles as they relate to atmospheric phenomena.

Zhang, Rong, Thomas L. Delworth and Isaac M. Held. "Can the Atlantic Ocean Drive the Observed Multidecadal Variability in Northern Hemisphere Mean Temperature?" *Geophysical Research Letters* 34, no. 2 (January, 2007). Discusses a link between the AMO and Northern Hemisphere temperature variability.

Zhong, B., S. Liang, and B. Holben. "Validating a New Algorithm for Estimating Aerosol Optical Depths Over Land from MODIS Imagery." *International Journal of Remote Sensing* 28, no. 18 (2007): 4207–4214. Describes new algorithms for the estimation of aerosol depth in the atmosphere using remote sensors. Aerosols provide vital clues about elements that lead to climate change.

Ziegler, Karen, Peter Turner, and Stephen R. Daines, eds. *Petroleum Geology of the Southern North Sea: Future Potential*. Geological Society, 1997.

INDEX

2015 Arctic Report Card, 31
ablation, 240, 258, 299-300, 302
absolute temperature scales, 281
absolute zero, 281
acid deposition, 3-7, 62, 75, 434
acidification, 6, 8, 33, 268, 324, 342, 366, 369, 394-395, 478
active sensor, 424, 426, 430-432
Adelie penguins, 402
adiabatic, 63, 65, 69, 93, 193, 195, 228, 388-392
aerobic, 60, 97-98, 130-131
aerodynamics, 225-228
aerosol particles, 11, 104
aerosol veil, 495
afforestation, 134, 230-231, 268
agricultural drought, 196-197
Agulhas Current, 77
air drainage, 69, 72
air parcel, 81, 89-90, 212, 388-392
air separation plants, 53
air-mass classification, 141
albedo feedback, 17-19, 34
albedo flip, 19
aldehydes, 22
Alexander von Humboldt, 168
algorithm, 234, 253, 360-361, 422-423, 428, 498, 500-501
alluvial, 220
alternative fuel, 343-344, 346
altimeter, 118, 333, 428, 432, 438
alveoli, 22
American Geophysical Union, 354, 388
anaerobic, 60, 97-98, 129-130, 266, 322, 470
Andean-Saharan glaciation, 294, 523
Anders Celsius, 168
anemometer, 515
anemometers, 73, 260
aneroid barometer, 118, 120
Antarctic bottom water (AABW), 468
Antarctic ice sheet, 258-261, 297-299, 302-303
Anthropocene, 2, 62, 99, 110-111
anticyclone, 63, 180-184, 217
aphelion, 108, 146, 168, 200, 202
aphotic zone, 371
aquifers, 53, 510

Arabian Sea, 306-307, 349-350
Aral Sea, 25, 27-30, 324
archaea, 343-346
archaeological record, 506, 508-509
Arctic Archipelago, 37, 39
Arctic ice, 17, 31, 33, 40
Arctic oscillation, 77, 79, 328-330
Arctic warming, 19
astronomical forcing, 385
astronomical unit, 201
Atlantic currents, 48, 387
Atlantic heat conveyor, 40-42
Atlantic Indian Basin, 306
Atlantic Multidecadal Oscillation (AMO), 42-44
atmosphere coupling, 207, 369-373
atmospheric boundary layer (ABL), 81
atmospheric chemistry models, 152, 342
atmospheric chemists, 82, 84-86
atmospheric dynamics, 68, 89-91, 108, 169, 174, 329
atmospheric inversions, 14, 92-93
atmospheric opacity, 94-96
atmospheric oxygen, 50, 60, 62, 97-99, 115, 418
atmospheric physics, 92, 99-105, 218, 266, 499
Atmospheric sciences, 105-110, 214, 256, 341, 393
atmospheric thermodynamics, 70, 391
atoll, 26, 382-383
aurora, 54, 73, 97, 102, 114, 200-201, 379
aurora australis, 73, 201
aurora borealis, 73, 201
auroral displays, 73
austral, 348-349
available potential energy, 388, 393
available water, 169, 179, 241
Azolla event, 148, 523

balance of energy, 52
Baltic Sea, 356-357, 359
barograph, 118, 120
barographs, 474
barometer, 73, 106, 118-121, 168, 507
barometric pressure, 106, 118-122, 262, 333, 349, 438
barometry, 118-122
bathymetric contours, 189
Bay of Bengal, 306-309

585

bedrock, 5, 192, 220, 222, 262-263, 303, 313, 320, 409
Benguela Current, 48, 463, 465, 518-519
benthic, 40, 187, 190
benthonic, 187, 190
Bergen school, 66, 499
Bergeron process, 406-407
Bergeron, or air-mass, classification, 141
Bering Strait, 33, 37, 380-382
bicarbonate, 3, 5, 59
biogenic, 319-320, 322, 443-444
biome, 165, 167, 204, 470
black carbon (soot), 19
Black Forest, 477
bleaching event, 368
blizzard, 182, 342, 453-454, 458
bloom, 131, 321-323, 365, 368, 394-395, 427
boreal, 19, 38, 205, 230-231, 348-349
boreal forest, 19, 38, 205, 230-231
boundary layer, 70, 81, 225-226
brash, 35, 38
breaking wave, 9, 461
Brent Spar, 358
bristlecone pine tree chronology, 158
bronchioles, 22
brown bears, 400
buoy systems, 361
buoyancy, 37, 81, 274, 373, 436, 445, 455
burrowing brittle-star, 358-359

calcareous sediments, 444
calving, 240, 299-300, 302
Canada Basin, 39
Canary current, 41, 276
cap and trade legislation, 3
carbon balances, 205
carbon credits, 134
carbon dioxide equivalent, 123-124
carbon emissions, 127, 164
carbon monoxide (CO), 20
carbon sequestration, 126, 128-129, 132, 395
carbon sink, 126, 133-134, 211, 230, 394-395
carbon uptake, 134, 205
carbon-climate feedback, 206
carbon-oxygen cycle, 128-132, 145, 147-148
catalyst, 137, 139, 374, 376
catalytic surface, 403
catastrophic flooding, 221, 224
catastrophist, 135-136, 249

Cenozoic period, 293
centrifugal, 101, 453, 514
centripetal, 101, 513
Channeled Scablands, 220
Chappuis bands, 216
Charles D. Hollister, 188
chemical evolution, 58-59, 61
childhood obesity, 16
Chinook, 242, 245, 514
Christophorus Buys Ballot, 106, 168
circulation cell, 88, 92, 168, 209, 392, 479, 481
cirriform, 172
clathrates, 2, 367
clay, 5, 7, 9, 186, 191, 319-320, 322, 357, 380, 443-444
clean development mechanism (CDM), 134
climate change theories, 145-148
climate lag, 162, 164
climate prediction, 153-156, 251, 330, 412, 448
climate reconstruction, 156-158, 388, 398
climate refugees, 158-160, 180
climate sensitivity, 160-162
climatic oscillation, 162
climatic precession, 236, 396-397
climatic trend, 162, 412
cloud feedback, 173-174
cloud physicists, 407, 505
cloud seeding, 392, 410, 503-506
CO2 concentrations, 20, 156-157, 314, 476
cold climate zone, 165
cold cloud, 406-408, 410
cold fronts, 91, 406, 472
collision efficiency, 408
collision-coalescence process, 407-408
comparative planetology, 74
Concordia, 296-297
conduction, 14, 69, 281, 372, 520
constant pressure chart, 512, 515
continental climate zone, 176
continental drift, 45-46, 148, 380-381
continental ice sheets, 236, 239
continental slope, 45, 380-381
convective instability, 81
convective overturn, 369
conveyor belt current, 76, 78
coral metabolism, 366
coral reefs, 209, 268, 366-368, 477-478
core eddy, 462
core rings, 465

Coriolis acceleration, 512-513
cornucopian, 135-136, 249
cosmic rays, 69-70, 147, 185-186, 460
cosmogenic isotope, 185, 335
Cryogenian period, 294
cummuliform, 172
cumulonimbus cloud, 172, 471-472
cumulus stage, 455
cyanobacteria, 97, 129
cyclogenesis, 181-182

Dalton's law, 282
dating methods, 185-187
dead zones, 365, 368
decadal, 44, 57, 79, 142, 155, 209, 251, 275-277, 346-347, 353-354
decay constant, 185
deep ocean currents, 187-190, 227, 463, 466, 518
deep-sea currents, 188-191
degassing event, 131
deglaciation, 191-193, 317
delta, 29, 307, 320, 380
dendrochronology, 186, 385, 475
dendroclimatology, 145, 334
deoxygenation, 367
depth recorder, 190
desert dust, 9
desertification, 10, 159, 193, 198, 336-337
Devils Hole, 356
dew, 52, 65, 72, 228, 282-283
dew point, 228, 282-283
diagenetic minerals, 323
diatoms, 157, 321-323, 386, 443
dicholorodiphenyl-tricholoethanes (DDT), 28
diffusion, 15, 50, 52, 142, 464
digesters, 345
dirigibles, 66
displaced persons, 159-160
divergence, 181, 184, 444, 482, 512
Dobson spectrophotometer, 219, 374
Dogger Bank, 356-357
doldrums, 48, 286, 369, 371, 382, 481
Dome C, 296-298, 313
Doppler radar, 110, 171, 425, 453, 456-457, 474, 501, 508, 515
Doppler shift, 453
dormancy, 230
downburst, 360, 362, 453
downbursts, 362, 457

downwelling, 275-277, 517-518
Drake Passage, 380-381
drift bottles, 465
drizzle, 229, 408, 453
drought identification, 195-196
drought index, 193, 196
drunken forest, 31, 34
dry climate zone, 165
Dust Bowl, 195
dust devil, 471-473
dust Veil Index, 8, 494
dynamic mode theory, 503, 505

Earth-Sun relations, 200-202
East Greenland Current, 37
easterly waves, 479, 482
eccentricity, 145-147, 236, 292, 294, 312, 385, 396-397
echo, 110, 171, 190, 372-373, 427
ecohydrological regulation, 206
ecosystem dynamics models, 152
ecotone, 510
Edmond Halley, 89, 106, 168
Eemian interglacial period, 157
effective climate sensitivity, 161
Ekman layer, 517-518
Ekman spiral, 76-77
electrolysis, 289-290
electromagnetic waves, 94-96, 281, 428
emissivity, 151
end-Cretaceous event, 387
end-Ordovician extinctions, 387
Endangered Species Act, 400
endosymbiont, 97
Energy balance (EB) models, 152
Enhanced Fujita (EF) Scale, 472
ensemble weather forecasting, 498
environmental lapse rate, 69, 71, 508
epilimnion, 262, 264, 322-323
equation of state, 498
equinox, 64, 147, 200, 202, 385
estuary, 45, 47-48, 371, 394, 451
eukaryote, 97-99
euphotic zone, 394
Eurasian Basin, 39
eustatic sea-level change, 438-440
Evangelista Torricelli, 106, 119, 168
evapotranspiration, 50, 53-54, 193, 196-197, 204, 206

exosphere, 69-70, 102, 115, 253, 425
extratropical cyclones, 57, 213
extreme weather events, 144, 154, 212-214, 475
eye wall, 286

faculae, 326
Faint Early Sun Paradox, 113
feedback loop, 2, 18, 31, 34, 57, 115, 131, 148, 161, 192, 236, 279, 294, 344, 385
fermentation, 97-98, 290, 344
Ferrel cell, 66, 86-88, 90, 106, 168
fertilizer runoff, 6
First Geophysical Year, 297
first law of thermodynamics, 104
fjord, 37, 356, 381
Fjorlie sampler, 451
flash floods, 54, 110, 170, 183, 219, 221, 224, 288-289, 411, 458, 475, 501
floes, 35, 37-38, 40, 436-437
flood-control projects, 357
floodplain, 220, 223-224
fluid dynamic principles, 225, 227
fog dissipation, 503, 505
food chain, 15, 21, 48, 77, 200-202, 374-375, 378, 383-384, 470
forced migration, 159-160
forecast verification, 498
forest decline, 7
forest ecosystems, 419, 477
fossil-fuel combustion, 84, 338, 489
fossilized trees, 158
free electrons, 55, 96, 114
free oxygen, 97-98, 449
freezing rain, 93, 408, 453-454
Freon, 21, 72, 83, 136, 139, 404
friction, 70, 77, 81, 114, 182, 191, 225, 240, 463-464, 513, 515, 517
friction level, 515
frost, 6, 34, 57, 61, 65, 72, 163-164, 230, 402, 495
fuel cells, 249, 291
Fujita (F) scale, 472
funnel, 212, 409, 453, 455, 457, 472-474
fynbos, 167

Galileo, 106, 119, 168
gaseous pollutants, 20, 418
Geochemical Ocean Sections (GEOSECS), 191
geographic information system, 420, 422
geomagnetic storm, 97, 200-201, 379

geomorphic change agents, 29
Georg Wust, 188
George Hadley, 87, 90, 106, 168, 280, 392
geostationary satellites, 109, 170
geostrophic current, 87, 271, 274
geostrophic wind, 63, 65, 67, 512
geosynchronous, 430-432
glacial advance, 26, 313, 334, 396-398
glacial deposits, 237, 295, 396
glacial epoch, 292
glacial retreat, 240, 293, 334
glacial stage, 191-192, 292, 294
glaze, 453-454
global atmospheric model, 498, 501
global cooling, 42, 110, 131, 169, 176, 295, 342, 367, 387, 470, 495
global dimming, 8, 23, 25, 84, 266
global radiative equilibrium, 172
Gondwanaland, 306
governing equations, 499
gravity corer, 191
gravity waves, 73
Great Australian Bight, 307
Great Barrier Reef, 367-369, 382
Great Smoky Mountains, 25
Greenhouse Earth, 148
Greenland ice sheet, 297
Greenpeace, 316, 358
greenspace, 486, 488
ground inversions, 72
growing degree-day index, 69, 75
growing season, 16, 30, 145, 205-207, 230, 476, 503-504
Gulf of Aden, 307
Gulf of Oman, 307
Gustave-Gaspard Coriolis, 88, 106
guyot, 380-381
gyres, 48, 275-277, 306, 308, 363, 382, 444, 463-464, 469

haboob, 514
Hadley cell, 64, 66-67, 86-88, 90, 106, 152, 168, 207, 209, 279-280, 481
Hadley circulation, 209, 242-243, 278-280, 318, 477
hail, 214, 272, 328, 408-409, 454-457, 474, 503-505
hail suppression, 503-505
half-life, 185-186
haloalkanes, 138-140

halocarbons, 136-140, 219, 266
halocline, 369, 371, 468-469
halon, 137, 139, 376
Hartley bands, 216
heat budget, 52, 463-464
heat sink, 104
heat wave, 54, 57, 91, 212-213, 310-311, 330, 362, 395
Heaviside layer, 73
Heinrich Berghaus, 168
Heinrich Wilhelm Dove, 168
hemispheric model, 498, 502
herbicides, 27-28
heterosphere, 52, 55, 69, 72-73, 107
hinterland, 338, 489
holocene, 57, 110, 148, 158, 294, 312-313, 359, 367, 387-388, 398
homosphere, 52, 55, 69, 72, 107
horse latitudes, 273, 369-370
Huggins bands, 216
Humboldt giant squid, 368
Hurricane Agnes, 223, 482
Hurricane Allison, 482
Hurricane Andrew, 287
Hurricane Beulah, 288
Hurricane Celia, 473
Hurricane Harvey, 221, 223, 288
Hurricane Irma, 223
Hurricane Katrina, 223-224, 287, 458, 483
hydric soil, 511
hydrochlorofluorocarbons (HCFCs), 139
hydrodynamics, 225-226, 228, 324, 467
hydrofluorocarbons (HFCs), 138
hydrogen power, 289-291
hydrogen-powered vehicles, 289
hydrologic cycle, 52-54, 142, 206, 263, 272, 284, 324, 406, 409-410
hydrometer, 507
hydrophyte, 510
hydrophytes, 511
hydrosphere, 15, 129, 141-142, 152, 206, 245, 342, 449, 479
hydrostatic balance, 388-389
hygroscopic particulates, 503-504
hypolimnion, 262, 322

Iberian Peninsula, 336-337, 464
ice breakup, 34, 304-305
ice floes, 35, 37-38, 40, 437

ice pellets, 408-409, 504-505
ice shelf collapses, 298-299
ice shelves, 33-34, 179-180, 259, 298-299, 302-303, 402
ice storm, 453-454
ice stream, 259, 300, 302
ice-out, 304-305
ice-out studies, 304
Icehouse Earth, 148
igloo, 35, 37-38
inadvertent weather modification, 503-504
inertia, 142, 177, 228-229
inertial frame, 177
insolation, 419, 480, 506, 508
instrumental climatic observations, 163
interferometer, 216
interferometric radar altimeter, 428
internal combustion engines, 23, 291
internal energy, 52, 391
International Space Station (ISS), 95, 114
Intertropical Convergence Zone (ITCZ), 481
Inuit, 35, 37-38, 399
ionic species, 450
ionization, 52
ionosphere, 55, 69, 73, 95-96, 102, 104, 114-115, 253, 255
isobars, 182-183, 456, 513-514
isostacy, 292-293
isotope chemistry, 296
isotope-ratio records, 186

James Croll, 237, 295, 397
Java Trench, 306-308
jet streams, 48, 63, 73, 90-92, 99, 102, 178, 213, 233, 341
joule, 281, 283
Julian date, 304

kelp, 384
kettles, 320
Kilimanjaro's ice cap, 315-317
killer fogs, 21
killer waves, 362
kinetic energy, 81, 280-281, 388, 391
kinetic properties, 163, 342
knot, 182, 271, 284-287, 453, 512, 514-515
Krakatau, 10, 495-497
krill, 302, 378
Kuroshio-North Pacific Current, 518-519

589

Kwajalein, 382
Kyoto Protocol, 26, 124, 127, 134, 137, 139, 231, 246, 249, 268

lagoon, 45, 47, 380
Lake Agassiz, 1, 42, 340, 387
Lake Vostok, 157
Laki eruption, 10
lamination, 324
landslides, 320-321, 351, 461
lapse rate, 14, 69, 71, 149-150, 160, 388-391, 508
Larsen A, 299
Larsen B, 299, 303
late heavy bombardment, 111
latent energy, 52
leaf litter, 476
lichenometry, 186
lidar, 424-427
limestone, 3, 5, 112, 115, 262, 319-320, 368, 441
lithosphere, 141, 206, 245, 292, 342, 479
little egret, 337
local sea-level change, 438-440
local winds, 512, 514
Lomonosov Ridge, 39
long-range prediction, 498, 502
low-pressure air masses, 92
luminosity of Sun, 325

Madden-Julian oscillation, 328-329
Makganyene glaciation, 294
malaria, 478
Malaysian-Australian monsoon, 349
Malthusian, 135
Mandab Straits, 307
mangrove swamps, 47-48
Mariana Trench, 383-384
marine isotope stage, 236-237, 292, 295, 312, 396-397
marine isotope stage (MIS), 237, 295, 397
maritime climate, 331-332
mathematical models, 109, 170, 255, 361
Maunder Minimum, 201, 252, 325-327, 448, 460
Mauritius, 307
mean free path, 50, 52
mean global air temperature, 210
mean sea level, 333, 438, 442
mean surface air temperature, 144
Medieval Grand Maximum, 327
Medieval Warm Period (MWP), 333-334
medium-range forecast, 501

megacities, 337-338, 487-490
megalopolis, 337, 488
meridional, 40-41, 44, 280, 339-341, 363-364, 469, 521
meridional overturning circulation (MOC), 339-340
mesopause, 13, 52, 55, 70, 107, 424-425, 484
mesoscale, 57, 341, 498, 501
mesoscale model, 498
metalliferous, 443-444
methanogenic archaea, 344
methanogens, 344
microscale, 174, 212-213, 341, 404
microscale rotating winds, 212
Mid-Atlantic Ridge, 46, 189
Mid-Holocene Maximum, 57
mid-latitude cyclone, 182-183
Mid-Ocean Ridge, 443-444
Middle America Trench, 381
midlatitudes, 87, 90, 107, 232, 363, 465, 494
Milankovitch cycles, 111, 115, 294-295, 311, 313, 385, 388, 398
Milankovitch hypothesis, 294
Milky Way galaxy, 95, 201
mineraloid, 320-321, 323
MIS 5e, 312
moderation, 331
modes, 311, 346-348
moist maritime (M), 141
Montreal Protocol, 15, 21-22, 26, 85, 107, 124, 127, 137-140
Moosehead Lake, 304-305
Mountain of Greatness, 315
multidecadal, 42-45, 277, 334-335, 354
multidecadal oscillation, 42-44, 277

nacreous, 403-404
named storm, 482
Nanook of the North, 38
Nansen bottles, 450-451
Nansen Cordillera, 39
NAO Index, 353-354
National Data Buoy Center, 445
National Research Council, 1, 150, 162, 165
National Snow and Ice Data Center, 34
Navier-Stokes equations, 152, 226
neap tide, 364
near-polar orbit, 430
net carbon loss, 205

net radiative heating, 69, 71
neutralization, 4-5
Newtonian fluids, 225
Nilometers, 168
Ninety East Ridge, 308
nipping, 301
Niskin bottle, 451
nitric acid aerosol, 3
nitric oxide (NO), 21
nitric oxide gases, 4
nodule, 310, 323, 449
nondeterminism, 498
North American monsoon, 349
North Atlantic Drift, 41, 48, 77, 463-465
North Atlantic drift (NAD), 41
North Atlantic Oscillation (NAO), 346, 353-354
Northern annular mode (NAM), 346
Northern Ocean, 424
nowcasting, 498, 501
NOX, 202, 403
nuclear fusion, 200
numerical weather prediction, 342, 498-502

obliquity, 145-147, 292, 294
observational data, 360-363
occluded front, 183, 232
ocean biomes, 470
ocean dynamics, 92, 178, 234, 363-364, 452
ocean structure, 369-373
ocean tides, 47, 439
oceanic oscillations, 76-80
oceanic rivers, 48
off-gassing, 12
offshore oil well, 358
one-dimensional models, 151
opacity, 94-96
opaline frustules, 322
orbital eccentricity, 146, 236, 292, 312, 396
organic matter, 5-6, 40, 97, 109, 129, 266, 321-324, 358, 386
orographic barrier, 506, 508
orography, 244, 348, 350
overturning circulation, 41-42, 165, 277, 339-341, 364, 470, 521
oxygen isotope ratios, 154, 237, 295, 334
ozone depletion, 56, 138, 141, 269, 342, 354, 374-379, 404, 406, 506
ozone holes, 107, 374-379
ozone shield, 15, 85

Pacific Decadal Oscillation (PDO), 79, 346
Pacific Rim, 383-384
Pacific-North American (PNA) pattern, 346
pack ice, 301, 303, 436-437
paleobotany, 156
paleoclimate, 17, 144, 156-157, 161, 174, 385-388
paleoclimate change, 144, 385-387
paleoclimatologists, 85, 421-422
paleoflood hydrology, 222
paleofloods, 222
paleontology, 145, 156
Paleoproterozoic era, 294
Paleozoic era, 106, 262, 294
Palmer Drought Index, 193, 196
Panel on Climate Change (IPCC), 42, 111, 124, 143, 161, 335, 446
Pangaea, 45-46, 381
Panthalassa, 45-46, 381
parameterization, 151-153, 234-235, 411
particulates, 14, 22, 84, 425-426, 429, 503-504
passive sensors, 426, 429-431
paternoster lakes, 320
perihelion, 108, 146-147, 168, 200, 202
permanent ice, 142, 165, 292, 311, 523
Permian extinction, 366-367, 423
Permian-Triassic boundary, 98, 366
Persian Gulf, 24, 306-309, 370
persistent haze, 25
Peru Current, 382, 518-519
Peru-Chile Trench, 381
pesticides, 27-28, 30, 462
pH, 3-7, 12, 14, 16, 409, 444
photic zone, 371
photochemical oxidants, 12-13
photochemical reaction, 12, 14, 22, 62
photochemical smog, 14, 22
photodissociation, 69, 72, 417-418
photolysis, 417-418
photometric techniques, 16
photosynthetic bacteria, 291
photovoltaics, 202
physics of weather, 388-393
pilot balloon, 515
Pinatubo, 107, 110, 171, 433, 494, 496
pingo, 35, 39, 269
planetary scale, 341
planetary winds, 463-464, 466
plasma, 96, 101, 200-201
Pleistocene climate, 395-398

591

Pleistocene epoch, 27, 29, 262, 315, 357, 396, 398
Pleistocene-Quaternary glaciation, 292
Pliocene epoch, 28, 356, 396
polar bear, 33, 37, 366-367, 398-402, 437
polar cell, 64, 66, 86-88, 90, 279
polar jet stream, 67, 90
polar low, 360-362
polar stratospheric clouds (PSCs), 403-404
Polarstern, 32, 304, 519
prebiotic, 58-60
precession, 145-146, 236, 292, 294, 312, 385, 396-397
pressure gradient force, 63, 65, 87, 271-272, 512-513
primary air pollutants, 83
primary crystalline rock, 449-451
primary production, 365
primordial isotope, 185
primordial solar nebula, 58, 60-61
prokaryote, 97, 99
Promethean, 135
pulse Doppler, 424-425
pycnocline, 369, 371

Quaternary period, 27, 193, 396, 523

radiation budget, 102, 108, 282-284, 434
radiational cooling, 69
radiative balance, 149, 172, 278
radiative cooling, 71-72, 228, 483-484
radio waves, 73, 94-96, 102, 107, 114, 253, 281, 424-427, 456
radiometer, 100, 427-432, 446
radiosondes, 74, 456, 500
rain forests, 230, 243, 266, 417-420, 477-478
rain gauge, 109, 170-171, 407, 409-410
rawinsonde, 453, 456, 512, 515
recurrence interval, 219, 221-222
Red Sea, 307, 309, 370-371
reforestation, 134, 230-231, 268
regimes, 205, 346
regression, 43, 438, 441
relative humidity, 101, 283-284
relative motion, 81, 177
Reynolds number, 226
rice cultivation, 26, 268
ridges, 183, 190, 306-308, 382, 393, 436, 443
rifting, 295
ringed seal, 399-400

riparian, 219, 510-511
Ross Ice Shelves, 303
Rossby waves, 67
Royal Society of London, 280
rudists, 368

Saffir-Simpson scale, 286
Sahel, 11, 193, 195, 243, 509, 519
salinometer, 451
saltation, 187-188, 190
saltwater wedge, 369, 371
sandbars, 47
Sargasso Sea, 49, 273, 334, 466
satellite data, 34, 217, 267, 351, 430, 432-433, 437
saturated, 131, 134, 238, 282-283, 350, 408
savannah, 477
Savonics rotor current meter, 190
sawtimber, 231
scattering, 9, 11, 16, 100, 216-217, 372, 425-426, 429, 493
scattering layers, 372
scatterometer, 424, 427, 432
sea sediments, 144, 443-444
sea surface temperatures (SST), 445-446
sea-level rise, 57, 159-160, 268, 510-511
seasonal changes, 54, 57, 269, 401, 419, 446-447, 470, 519
seawater composition, 448-452
seaweed, 49, 384
Sebago Lake, 304
Secchi disk, 373
second law of motion, 498
sediment cores, 145, 157, 186, 237, 396-397
sediment drilling, 145
sedimentary rock, 145-146, 262, 319, 386, 420-421, 445
seesaw, 79, 207-209, 346-347
seismic measurements, 358
semiarid climate, 243, 507, 509
seston, 320
severe storms, 20, 330, 389, 428, 452-458, 472, 485, 496
Seychelles, 307
shear zone, 66
shortwave radiation, 16, 56, 99, 104, 106, 205, 228, 483
siliceous sediments, 444
Silver Pit, 356

sinkhole, 320, 402
sirocco, 514
sixth extinction, 387
skin cancer, 21, 202, 378
smog, 8, 10, 13-14, 17, 21-22, 26, 84, 93-94, 229, 266, 378, 429, 487, 489
snowball Earth theory, 294
soil chemistry, 7
soil moisture, 193, 195-197, 351
solar activity, 52, 70, 73, 200-201, 325-326, 335, 434, 459-460
solar cells, 202, 516
solar cycle, 151, 200-201, 252, 326, 459-460
solar flare, 201, 459
solar irradiance, 150, 459
solar luminosity, 327, 385-386
solar maximum, 201, 459
solar power, 201-203, 290
solar ultraviolet radiation, 58-59, 62
solar wind, 60, 70, 73, 102, 200-201, 203
solstice, 64, 147, 200, 202, 396
Somali Current, 306, 308-309
soot, 2, 11, 19-20, 22, 84, 107, 149
sounders, 74
South Indian Basin, 306
Southern Ocean, 300-303, 339, 369, 395, 424, 428-429, 444
Southern Oscillation Index (SOI), 207
space weather, 201-202
spatial resolution, 430-431
spectrophotometers, 217, 219
speleotherm, 334
spring break-up, 304
spring tide, 333, 364
squall line, 232, 455, 472
stalagmite, 185, 334
static mode theory, 503, 505
static properties, 163
statistical means, 141
stomatal index, 158
storm frequency, 213
strait, 33, 37, 272, 307, 356-357, 380-382
Strait of Dover, 356-357
Straits of Hormuz, 307
stratiform, 172
stratospheric dust, 10
streamline, 225-226
sublimation, 196, 240, 299, 315-316
suborbital rockets, 74

subsidence, 14, 87, 93-94, 193, 220, 320, 392, 439-440
subsidence inversion, 14, 93-94, 392
subsidence-type inversions, 14
subtropical jet stream, 65, 67, 90-91
sulfur emissions, 4, 6
sulfuric acid aerosols, 4, 110, 171, 367, 493, 496
sulfurous smog, 14
summer stratification, 262
sun-synchronous orbit, 430
sunspot cycle, 201, 325, 375, 459-460
sunspot minimum/maximum, 325
sunspots, 200-201, 325-327, 388, 434, 459-460
supercomputers, 151-152, 500-501
supercooled water droplets, 406-407, 409
supersaturated, 282-283
superstorms, 2
surface ocean currents, 77, 364, 462-467
synoptic scale, 182, 341
synthetic aperture radar, 424, 427, 430, 432

T Tauri stars, 58
tabular icebergs, 299
taiga, 509
Tambora, 8, 10, 495, 497
tarns, 239, 320
Tauranga, 116-117
tectonic plates, 39, 45-46, 320, 380, 383, 443
tectonics, 57, 148, 306, 384, 386, 438-439, 441-442, 497
teleconnection, 245, 346-347
telegraph, 507
temperate forest, 230-231
temperate zone, 165, 230, 309, 478
temperature gradients, 65, 67, 213
temperature inversion, 14, 21, 69, 71-72, 81, 86, 93-94, 392, 484, 495
terrestrial ecosystems, 133-134, 207, 372, 419, 482
terrigenous, 443-444
Tertiary period, 28
Tethys Sea, 45-46
theodolite, 515
thermal equilibrium, 101, 104, 228
thermal properties, 141, 163
thermocline, 178, 262, 264, 322, 369-370, 468-470
thermohaline circulation (THC), 364, 468-470
thermometers, 73, 144, 170, 326

Three-Dimensional General Circulation Models, 152
Thunderstorm Project, 456
Tibetan Plateau, 67, 245, 349-350
tidal range, 45, 47, 333, 380, 382, 438
tidal surge, 461
tipping point, 1, 3, 19, 31, 34
topsoil, 197, 401
Tornado Alley, 456, 458, 473
Total Ozone Mapping Spectrometer (TOMS), 375
total potential energy, 388, 391-392
toxaphene, 28
transgression, 438, 441
translation, 280, 448
trench, 306-308, 356-357, 374, 380-381, 383-384
Tropic of Capricorn, 309, 479
tropical climate, 116, 175, 331, 477-478, 507
tropical cyclone, 57, 183-184, 268, 284, 310, 453, 456, 479, 482, 519
tropical depression, 284, 286, 479, 482
tropical ecology, 310
tropical moist climates, 165
tropical rain forests, 230, 243, 419-420, 478
Tropical Rainfall Monitoring Mission (TRMM), 109
tropical storms, 47, 99, 213, 221, 311, 329, 433, 448, 461, 473, 479, 482
tropical zone, 352, 479-481
trough, 182-183, 238-239, 241, 349-350, 393, 482
tsunami, 224, 289, 333, 360-362, 383, 411, 438, 458, 461, 465, 475
tundra, 35, 38-39, 165, 205, 244, 344, 386, 401-402, 509
turbidity currents, 188
Tuvalu, 159, 333, 442
typhoons, 181, 183-184, 213, 284-288, 433, 448, 456-457, 482

ultraviolet (UV) radiation, 202
ultraviolet light, 15, 26, 50, 69, 74, 95, 114-115, 140, 215, 404
ultraviolet solar radiation, 56, 375
umiak, 35, 37
Umkehr, 217
unstable air, 455, 472
unstable atmosphere, 390
urban heat island (UHI), 486-490

urban smog, 10
urbanization, 231, 338, 342, 486-489

Valdaian, 237, 295
vapor pressure, 282-283
varve, 185-186, 239, 264, 323-324
Vega, 294, 348
velocity meter, 222
vertical mixing, 14, 52, 93, 339
vibration, 280
Vikings, 508
viscosity, 225, 227
volatile chlorine compounds, 137, 139
volatile outgassing, 58-59, 61
volcanic ash, 70, 100, 104, 145, 321, 386, 432, 492-493, 495
Vostok, 157, 260, 297

W. C. Palmer, 193, 196
Walker cell, 67
Walker circulation, 63, 207-209
Walker cycle, 91
walruses, 32-33, 400, 437
warm cloud, 407
water need, 196-197
water shortages, 211
waterspout, 472
wave equation, 95
weather balloon, 95, 102, 360, 425, 484, 500
weather front, 73, 91, 108, 169, 454, 481
weather modification, 391-392, 503-506
weather phenomenon, 213, 452
weather satellites, 74, 430-434, 501
weathering, 9, 62, 131, 236, 295, 319, 321, 398, 449-451
West African monsoon, 349
West Antarctic ice sheet, 258-260, 298-299, 303
west coast marine climate, 175, 244, 507
West Wind Drift, 382
westerlies, 48, 66, 88, 90, 363-364, 464
Western Australia Current, 518
wet adiabatic lapse rate, 388, 390-391
whales, 35, 37, 49, 384, 400
whirlwinds, 472-473, 493
wildfires, 155, 213, 289, 336-337
William Ferrel, 66, 88, 90, 106, 168, 392
wind power, 516
wind stress, 517-518
wind-driven circulation, 87, 271-272, 517

windblown dust, 5, 386
windmills, 516
winter tornadoes, 473
World Climate Research Programme, 362
World Meteorological Organization (WMO), 250
World Ocean Circulation Experiment, 445, 516-521
Wright brothers, 226

xylem, 475-476

Younger Dryas, 1, 42, 57, 364, 386, 470

Zambezi, 307
zero-dimensional energy-balance models, 151
zooplankton, 144, 298, 321, 378, 383, 393-395